Improve Your Grade!

Access included with any new book.

Visit Hazard City! REGISTER NOW!

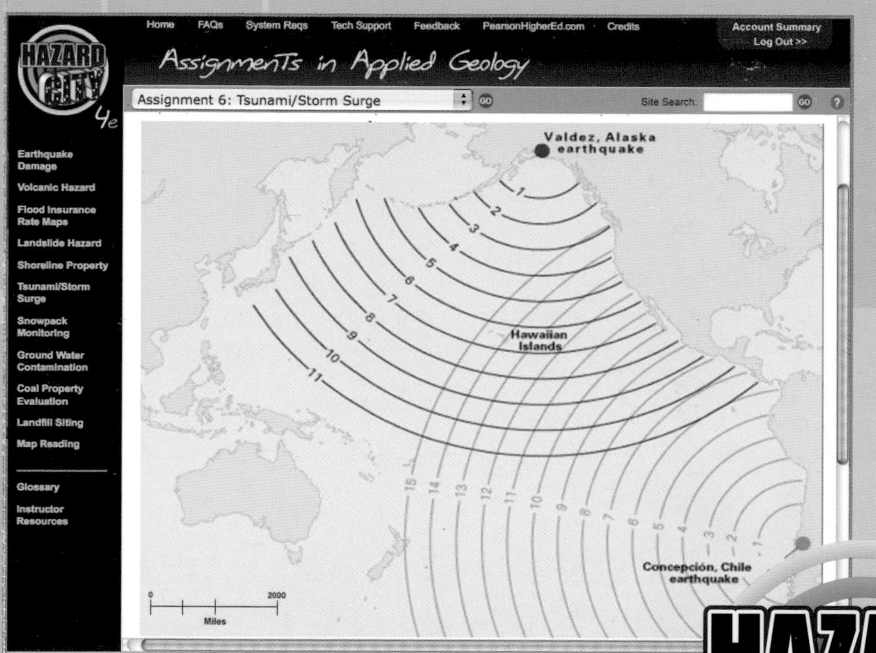

Registration will let you:

- Access all Hazard City assignments.
- Step into the role of a practicing geologist.
- Analyze potential disasters in the fictional town of Hazard City.
- Learn at your own pace using the 11 assignments.

HAZARD CITY

www.hazardcity.com

TO REGISTER

1. Go to www.hazardcity.com
2. Click "Register."
3. Follow the on-screen instructions to create your login name and password.

Your Access Code is:

TO LOG IN

1. Go to www.hazardcity.com
2. Enter your login name and password.
3. Click "Log In."

Hint:
Remember to bookmark the site after you log in.

Technical Support:
http://247pearsoned.custhelp.com

Earth's Processes as Hazards, Disasters, and Catastrophes

THIRD EDITION

Natural Hazards

Edward A. Keller
University of California, Santa Barbara

Duane E. DeVecchio
University of California, Santa Barbara

with assistance from
Robert H. Blodgett

Prentice Hall
Boston Columbus Indianapolis New York San Francisco Upper Saddle River
Amsterdam Cape Town Dubai London Madrid Milan Munich Paris Montréal Toronto
Delhi Mexico City São Paulo Sydney Hong Kong Seoul Singapore Taipei Tokyo

Acquisitions Editor, Geology: *Andrew Dunaway*
Marketing Manager: *Maureen McLaughlin*
Project Editor: *Crissy Dudonis*
Assistant Editor: *Sean Hale*
Editorial Assistant: *Michelle White*
Marketing Assistant: *Nicola Houston*
VP/Executive Director, Development: *Carol Trueheart*
Development Editor: *Dan Schiller*
Managing Editor, Geosciences and Chemistry: *Gina M. Cheselka*
Project Manager: *Edward Thomas*
Full Service/Composition: *Laserwords.*
Production Editor, Full Service: *Patty Donovan*
Art Director: *Derek Bacchus*
Cover and Interior Design: *Randall Goodall/Seventeenth Street Studios*
Senior Technical Art Specialist: *Connie Long*
Art Studio: *Precision Graphics*
Photo Research Managers: *Elaine Soares and Billy Ray*
Photo Researcher: *Caroline Commins*
Senior Manufacturing and Operations Manager: *Nick Sklitsis*
Operations Specialist: *Maura Zaldivar*
Senior Media Producer: *Angela Bernhardt*
Senior Media Supervisor: *Liz Winer*
Associate Media Project Manager: *David Chavez*

Cover and Title Page Photo: Lightning flashes in a cloud of volcanic ash during the eruption of Iceland's Eyjafjallajokull volcano on April 18th, 2010 (© Artic-Images/Corbis)

Pearson Prentice Hall
Pearson Education, Inc.
Upper Saddle River, New Jersey 07458

Printed in the United States of America
10 9 8 7 6 5 4 3 2 1

Library of Congress Cataloging-in-Publication Data
Keller, Edward A.
Natural hazards : earth's processes as hazards, disasters, and catastrophes / Edward A. Keller, Duane E. DeVecchio ; with assistance from Robert H. Blodgett.
p. cm.
ISBN-13: 978-0-321-66264-4
ISBN-10: 0-321-66264-4
1. Natural disasters. I. DeVecchio, Duane E. (Duane Edward), 1970- II. Blodgett, Robert H., 1951- III. Title.
GB5014.K45 2012
551—dc22
2010047451

Prentice Hall
is an imprint of

www.pearsonhighered.com

ISBN-10: 0-321-66264-4 / ISBN-13: 978-0-321-66264-4

To Valery Rivera-Keller, who recognized early on the importance of writing about hazards and encouraged me to write the first edition.

E.A.K.

For the victims of the 2010 earthquakes in Haiti and Chile, and the floods in Pakistan: may the lessons learned from your tragic losses illuminate the path of future scientific awareness and help improve the global building standard, such that no people should face the hardship that you have endured.

D.E.D.

About our Sustainability Initiatives

This book is carefully crafted to minimize environmental impact. The materials used to manufacture this book originated from sources committed to responsible forestry practices. The paper is FSC ® certified. The binding, cover, and paper come from facilities that minimize waste, energy consumption, and the use of harmful chemicals.

Pearson closes the loop by recycling every out-of-date text returned to our warehouse. We pulp the books, and the pulp is used to produce items such as paper coffee cups and shopping bags. In addition, Pearson aims to become the first climate neutral educational publishing company.

The future holds great promise for reducing our impact on Earth's environment, and Pearson is proud to be leading the way. We strive to publish the best books with the most up-to-date and accurate content, and to do so in ways that minimize our impact on Earth.

Prentice Hall
is an imprint of

About the Authors

EDWARD A. KELLER

Ed Keller is a professor, researcher, writer, and most important mentor and teacher to undergraduate and graduate students. Ed's students are currently working on earthquake hazards, how waves of sediment move through a river system following a disturbance, and geologic controls on habitat to endangered southern steelhead trout.

Ed was born and raised in California. He received bachelor's degrees in geology and mathematics from California State University at Fresno and a master's degree in geology from the University of California at Davis. It was while pursuing his Ph.D. in geology from Purdue University in 1973 that Ed wrote the first edition of *Environmental Geology*, a text that became the foundation of an environmental geology curriculum in many colleges and universities. He joined the faculty of the University of California at Santa Barbara in 1976 and has been there ever since, serving multiple times as chair of both the Environmental Studies and Hydrologic Science programs. In that time he has been an author on more than 100 articles, including seminal works on fluvial processes and tectonic geomorphology. Ed's academic honors include the Don J. Easterbrook Distinguished Scientist Award, Geological Society of America (2004); the Quatercentenary Fellowship from Cambridge University, England (2000); two Outstanding Alumnus Awards from Purdue University (1994, 1996); a Distinguished Alumnus Award from California State University at Fresno (1998); and the Outstanding Outreach Award from the Southern California Earthquake Center (1999).

Ed and his wife Valery, who brings clarity to his writing, love walks on the beach at sunset and when the night herons guard moonlight sand at Arroyo Burro Beach in Santa Barbara.

DUANE E. DEVECCHIO

Duane DeVecchio is currently a researcher and adjunct professor at the University of California at Santa Barbara, where he earned his Ph.D. in geology. Since starting his graduate education, Duane has devoted a significant amount of time to becoming an effective communicator of science. He is a passionate teacher and feels strongly that students need to develop critical thinking skills, which will enable them to evaluate for themselves data presented in graphs and tables from various sources. This is particularly important today, when the Internet and cable television offer accessibility to vast amounts of information, yet the validity of this information is often questionable or misleading. Integrating data from current and relevant research into core curriculum to illustrate the methodology and rigor of scientific investigations is essential to his teaching strategy.

Duane has a broad field-based background in the geological sciences, and likes to share stories about his many months living alone in mobile trailers in the mountains and deserts so he could study rocks and landforms for his research. For his master's degree and post-master's research he conducted structural and stratigraphic analysis, as well as numerical dating of volcanic and volcaniclastic rocks in southeast Idaho and the central Mojave Desert of California, which record the Miocene depositional and extensional histories of these regions. His Ph.D. research was aimed at resolving fault slip rates and quantifying the earthquake hazard presented by several active fault-related folds growing beneath urbanized Southern California. Duane's current research interests focus on the timing and rates of change of Earth's surface due to depositional and erosional processes that result from climate change and tectonics. When Duane is not teaching or conducting research, he enjoys whitewater rafting, rock climbing, snowboarding, and camping with his partner Christy.

Brief Contents

Contents

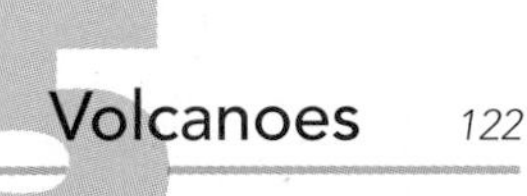

7 Mass Wasting 206

8 Subsidence and Soils 242

14 Impacts and Extinctions 472

Preface

Natural Hazards: Earth Processes as Hazards, Disasters, and Catastrophes, Third Edition is an introductory-level survey intended for university and college courses that are concerned with earth processes that have direct, and often sudden and violent, impacts on human society. The text integrates principles of geology, hydrology, meteorology, climatology, oceanography, soil science, ecology, and solar system astronomy. The book is designed for a course in natural hazards for nonscience majors, and a primary goal of the text is to assist instructors in guiding students who may have little background in science to understand physical earth processes as natural hazards and their consequences to society.

In revising the third edition of this book we take advantage of the greatly expanding amount of information regarding natural hazards, disasters, and catastrophes. Since the second edition was published, many natural disasters and catastrophes have occurred. In 2010 alone, a drought, heat wave, and air pollution from wildfires killed several thousand people in Moscow; a 7.0 earthquake in Haiti killed more than 200,000 people and forced several million more into temporary camps; and flooding in Pakistan killed more than 1500 people and displaced as many as 20 million. Two of these events had a common denominator — the earthquake and flood were catastrophes largely because of human processes interacting with natural processes. In other words they were largely, in terms of lives lost and property damaged, disasters caused by humans. With proper preparation many fewer lives would have been lost, and damages could have been greatly reduced.

On a global scale, climate change is causing glaciers, ice caps, and permafrost to melt; the atmosphere and oceans to warm; and sea levels to rise more rapidly than originally forecast. These changes are caused in part by human activities, primarily the burning of fossil fuels, which releases vast quantities of carbon dioxide and other gases into the atmosphere each day. The interaction between humans and earth processes has never been clearer, nor has the need for understanding these processes as hazards for our economy and society been greater. This edition of *Natural Hazards* seeks to explain the earth processes that drive hazardous events in an understandable way, illustrate how these processes interact with our civilization, and describe how we can better adjust to their effects.

A central thesis to this text is that we must first understand that earth processes are not, in and of themselves, hazards. Earthquakes, floods, volcanic eruptions, wildfires, and other processes have occurred for millennia, indifferent to the presence of people. Natural processes become hazards when they impact humanity. Ironically, it is human behavior that often causes the interactions with these processes to become disasters or, worse, catastrophes. Most important is the unprecedented increase in human population in the past 50 years linked to poor land-use decisions.

In addition to satisfying a natural curiosity about hazardous events, there are additional benefits to studying natural hazards. An informed citizenry is one of our best guarantees of a prosperous future. Armed with insights into linkages between people and the geologic environment, we will ask better questions and make better choices. On a local level we will be better prepared to make decisions concerning where we live and how best to invest our time and resources. On a national and global level we will be better able to advise our leaders on important issues related to natural hazards that impact our lives.

Distinguishing Features of the Third Edition

We have incorporated into this edition of *Natural Hazards* a number of features designed to support the student and instructor.

A BALANCED APPROACH

Although the interest of many readers will naturally focus on natural hazards that threaten their community, state, or province, the globalization of our economy, information access, and human effects on our planet require a broader, more balanced approach to the study of natural hazards. A major earthquake in Taiwan affects trade in the ports of Seattle and Vancouver; the economy of Silicon Valley in California affects the price of computer memory in Valdosta, Georgia, and Halifax, Nova Scotia. Because of these relationships we provide examples of hazards from throughout the United States as well as throughout the world.

This book discusses each hazardous process as both a natural occurrence and a human hazard. For example, the discussion of earthquakes balances the description of their characteristics, causes, global distribution, estimated frequency, and effects with a description of engineering and nonstructural approaches to reduce their effects on humans, including actions that communities and individuals can take.

FIVE FUNDAMENTAL CONCEPTS

The five concepts introduced in Chapter 1 and revisited in every chapter are designed to provide a memorable, transportable, and conceptual framework of understanding that students can carry with them throughout their lives to make informed choices about their interaction with and effect upon earth processes.

1. **Hazards are predictable from scientific evaluation.** Most hazardous events and processes can be monitored and mapped and their future activity predicted based upon the frequency of past events, patterns in their occurrence, and types of precursor events.
2. **Risk analysis is an important component in our understanding of the effects of hazardous processes.** Hazardous processes are amenable to risk analysis based

upon the probability of an event occurring and the consequences resulting from that event.

3. **Linkages exist between different natural hazards as well as between hazards and the physical environment.**
   Hazardous processes are linked in many ways. For example, earthquakes can produce landslides and giant sea waves called tsunamis, and hurricanes often cause flooding and coastal erosion.

4. **Hazardous events that previously produced disasters are now producing catastrophes.**
   The magnitude, or size, of a hazardous event as well as its frequency, or how often it occurs, may be influenced by human activity. As a result of increasing human population and poor land-use practices, events that used to cause disasters are now often causing catastrophes.

5. **Consequences of hazards can be minimized.**
   Minimizing the potential adverse consequences an d effects of natural hazards requires an integrated approach that includes scientific understanding, land-use planning and regulation, engineering, and proactive disaster preparedness.

SURVIVOR STORIES AND PROFESSIONAL PROFILES

Many of the chapters contain personal stories of someone who has experienced the effects of a hazardous event, such as a flood, earthquake, or wildfire, and a profile of a scientist or other professional who has worked with a particular hazard. Most of us in our lifetimes will experience (directly or indirectly) a flood, wildfire, volcanic eruption, tsunami, or major hurricane, and we are naturally curious as to what we will see, hear, and feel. For example, a scientific description of wildfire does not convey the fear and anxiety when a fire is rushing toward your home (see the Survivor Story about wildfire, experienced by one of your authors, Ed Keller, in Chapter 13). Likewise, the stream gauge records for the Rio Grande River do not give us the sense of excitement and fear felt by Jason Lange and his friends when their spring-break canoe trip in West Texas nearly turned to tragedy (see Survivor Story in Chapter 6). To fully appreciate natural hazards we need both scientific knowledge and human experience. As you read the survivor stories, ask yourself what you would do in a similar situation, especially once you more fully understand the hazard. Knowledge from reading this book could save your life someday, as it did for Tilly Smith and her family on the beach in Phuket, Thailand, during the Indian Ocean tsunami (see Case History in Chapter 4).

People study and work with natural hazards for many reasons—scientific curiosity, monetary reward, excitement, or the desire to help others deal with events that threaten our lives and property. As you read the professional profiles, think about these people's motivation, the type of work that they do, and how that work contributes to increasing human knowledge or to saving people's lives and property. For example, for many years Bob Rasely worked as a geologist with the U.S. Natural Resources Conservation Service studying what happens to hillslopes after wildfires in Utah (see Professional Profile in Chapter 7). For Bob, geology had long been both a vocation and avocation. He maintains an intellectual curiosity as to how the Earth works and a practical goal of predicting the likelihood and location of debris flows following a fire. Working with other state and federal scientists, Bob develops plans for hillslope recovery and helps communities downslope from burned land establish warning systems to protect lives and property.

Nearly all the survivor stories and professional profiles are based on interviews conducted exclusively for *Natural Hazards* by Kathleen Wong, a science writer from Oakland, California, or Chris Wilson, a former journalism student at the University of Virginia.

MAJOR NEW MATERIAL IN THE THIRD EDITION

The third edition benefited greatly from feedback from instructors using the previous edition, and many of the changes reflect their thoughtful reviews. New material for the third edition includes the following:

- NEW! A chapter on plate tectonics that reflects its overarching importance to Earth science.
- NEW! Revisiting the Fundamental Concepts remind students of the unifying theme of the "Five Fundamental Concepts" introduced in the first chapter, identifying how each particular hazard relates to the five fundamental concepts.
- NEW! A Closer Look uses real-life events and data to enhance the understanding and comprehension of not only the hazard, but also the mitigation that coincides with the hazard.
- NEW! Did You Learn questions at the end of each chapter give students the ability to track their comprehension of the learning goals stated at the beginning of each chapter.
- Coverage of the most recent disasters on Earth is provided, such as the earthquakes in Haiti, China, Italy, and Chile; 2010 Icelandic volcanic eruptions; 2010 debris flows in California; 2010 severe winter weather; and more.
- Revised design throughout the book includes dramatic changes to the chapter openers, which feature all new photos and content.
- Updated art program showcases new figures, illustrations, and photos throughout. Each image has been reviewed for accuracy and relevance, focusing on its educational impact.

FEATURES OF THE TEXT

This text is sensitive to the study needs of students. Each chapter is clearly structured to help students understand the material and effectively review the major concepts. Each chapter is organized with the following study aids:

- Learning Objectives on the first page of each chapter state clearly what students should be able to do upon completing the study of the topic.
- Selected features, including Case Studies, Closer Look boxes, Survivor Stories, and Professional Profiles, are added where appropriate to help students relate topics in the text to the world around them.
- A Summary, followed by a discussion of the five fundamental concepts introduced in Chapter 1 to the specific hazard at hand, reinforces the major points of each chapter and is designed to help students refocus on the important concepts.

- Detailed references are supplied for each chapter to provide instructors and students with additional sources of information and to give credit to the scholars who did the research reported in the chapter.
- Key terms are highlighted in **bold** where they are first discussed in detail in the text, and these terms are listed at the end of each chapter. This feature will help students identify important concepts and terminology necessary to better understand the chapter. Other useful technical terms that may be new to students are indicated in *italics*. Instructors may inform their students that some of the italicized terms are actually key terms for hazards in their local area. The authors encourage students to refer to the glossary at the end of the book for definitions of both the **key terms** and *useful terms*.
- Review questions at the end of each chapter will help students focus on the important subject matter.
- Critical thinking questions at the end of each chapter are designed to stimulate students to think about some of the important issues in each chapter and apply the information to both their lives and society.
- The appendices in *Natural Hazards*, Third Edition provide additional information useful in helping students understand some of the more applied aspects of geology as related to natural hazards. This information may be most useful with laboratory and field exercises.

For the Instructor

Instructor Resource Center
Everything instructors need, where they want it. The Pearson Prentice Hall Instructor Resource Center is accessed online and helps instructors be more effective by saving them time and effort. All digital resources can be found in one, well-organized, easy-to-access place. The materials in the Instructor Resource Center include

- All textbook images as JPEGs and in a PowerPoint® presentation
- Pre-authored Lecture Outline PowerPoint presentations, which outline the concepts of each chapter with embedded art and can be customized to fit instructors' lecture requirements
- The Instructor Resource Manual and Test Bank as MS Word files provide outlines and objectives, classroom discussion topics, and answers to end-of-chapter questions. The test bank includes hundreds of multiple-choice, true/false, and short-answer test questions on the science of natural hazards and their effects on the world and humankind.
- The TestGen software for both PC and Mac

For the Student

Hazard City: Assignments in Applied Geology, Fourth Edition Online access to *Hazard City* is included in every text, providing instructors with meaningful, easy-to-assign, and easy-to-grade assignments for students. Based on the fictional town of Hazard City, the assignments put students in the role of a practicing geologist—gathering and analyzing real data, evaluating risk, and making assessments and recommendations.

- **Substantial, critical thinking assignments**—Each activity takes 30 to 90 minutes. The activities require students to gather and analyze real data, participate in real issues, encounter uncertainty, and make decisions.
- **NEW Assessment Questions**—Assignable, assessable multiple-choice questions have been added to each *Hazard City* assignment. The questions engage students in higher order thinking to apply the concepts and processes learned in the assignments to the broader world in different contexts.
- **Student flexibility for submitting answers**—Easy to assign and assess via the website and use of gradetracker. Using worksheets that are downloadable from the *Hazard City* website, students can print out, complete, and submit their answers to the instructor. Students are also able to answer assignable multiple-choice questions directly from the *Hazard City* website that are computer graded, and feed to a course gradebook.
- **Thoroughly class tested**—All activities have been refined through testing in both the traditional and online classrooms.
- The activities found in *Hazard City* are
  - **Earthquake Damage Assessment:** Students research the effects of earthquakes on buildings, explore Hazard City, and determine the number of people needing emergency housing given an earthquake of specific intensity.
  - **Volcanic Hazard Assessment:** Researching volcanic hazards, collecting field information, and decision making are all used to determine the potential impact of a volcanic eruption on different parts of Hazard City.
  - **Flood Insurance Rate Maps:** Flood insurance premiums are estimated using a flood insurance rate map, insurance tables, and site characteristics.
  - **Landslide Hazard Assessment:** Students research the factors that determine landslide hazard at five construction sites and make recommendations for development.
  - **Shoreline Property Assessment:** Students visit four related waterfront building sites—some developed and some not—and analyze the risk each faces due to shoreline erosion processes.
  - **Tsunami/Storm Surge:** Students learn about the causes and effects of tsunamis and storm surges, then develop a hazard assessment response in case of inundation by water.
  - **Snowpack Monitoring:** Students utilize climatic data to estimate variables that are key to flood control and water supply management.
  - **Ground Water Contamination:** Students use field and laboratory data to prepare a contour map of the water table, determine the direction of ground water flow, and map a contaminated area.
  - **Coal Property Evaluation:** The potential value of a mineral property is estimated by learning about mining and property evaluation and then applying that knowledge in a resource calculation.

- **Landfill Siting:** Students use maps and geological data to determine if any of five proposed sites meets the requirements of the state administrative code for landfill siting.
- **Map Reading:** Builds map-reading skills and gives students the confidence they need to solve map-based problems in later assignments.

Acknowledgments

Many individuals, companies, and agencies have helped make this book a success. In particular, we are indebted to the U.S. Geological Survey and the National Oceanic and Atmospheric Administration for their excellent natural hazard programs and publications.

To authors of papers cited in this book, we offer our thanks and appreciation for their contributions. Without their work, this book could not have been written. We must also thank the thoughtful scholars who dedicated valuable time reviewing chapters of this book. The following reviewer's comments helped us significantly improve the third edition:

Stephen P. Altaner, *University of Illinois*
Hassan Boroon, *California State University at Los Angeles*
Tim Dolney, *Penn State University–Altoona College*
Brian A. Hampton, *Michigan State University*
Brenda Hanke, *Northern Kentucky University*
Richard W. Hurst, *California State University at Los Angeles*
Robin Whatley, *Columbia College Chicago*

Thanks also to the following reviewers who helped us put together the first and second editions:

Bob Abrams, *San Francisco State University*
David Best, *Northern Arizona University*
Beth Nichols Boyd, *Yavapai College*
Katherine V. Cashman, *University of Oregon*
Jennifer Cole, *Northeastern University*
Bob Davies, *San Francisco State University*
John Hermance, *Brown University*
Jean M. Johnson, *Shorter College*
Jamie Kellogg, *University of California at Santa Barbara*
Stephen Kesler, *University of Michigan at Ann Arbor*
Guy King, *California State University at Chico*
Timothy Kusky, *Saint Louis University*
Gabi Lakse, *University of California at San Diego*
Thorne Lay, *University of California at Santa Cruz*
William P. Leeman, *Rice University*
Alan Lester, *University of Colorado at Boulder*
Lawrence McGlinn, *Rensselaer University*
Rob Mellors, *San Diego State University*
Richard Minnich, *University of California at Riverside*
Stephen A. Nelson, *Tulane University*
Mark Ouimette, *Hardin Simmons University*
Gigi Richard, *Mesa State College*
Jennifer Rivers, *Northeastern University*
Tim Sickbert, *Illinois State University*
Don Steeples, *University of Kansas*
Paul Todhunter, *University of North Dakota*
Robert Varga, *College of Wooster*
Tatiana Vislova, *University of St. Thomas*
John Wyckoff, *University of Colorado at Denver*

Special thanks to Kathleen Wong and Chris Wilson for locating and interviewing many of the natural hazard professionals and survivors. All of us are especially grateful to these people who told us their stories. Thanks also to Jim Kennett for reviewing the case study on the Younger Dryas Boundary event; to Joel Michaelsen for reviewing the chapters on meteorology, hurricanes, and extratropical cyclones; and to Rob Thieler for his review of the case study on Folly Island. Ed would like to thank Tanya Atwater, William Wise, and Frank Spera for their assistance in preparing the content on plate tectonics, rocks and minerals, and impacts, respectively. Thanks to Kenji Satake, David Cramer, Gordon Wells, Theresa Carpenter, and David Rogers for providing photographs and graphics for use in the third edition.

Special thanks are extended to Professor Bob Blodget who as coauthor worked very hard on the first two editions. Much of the third edition reflects his excellent contributions.

We are particularly indebted to our publisher and editors at Pearson Prentice Hall. First and foremost we are grateful to our editors Andrew Dunaway and Crissy Dudonis, whose contributions and encouragement made this book possible. Our appreciation is extended to our production project managers Gina Cheselka and Ed Thomas, and production editor Patty Donovan. We would also like to thank Linda Benson, copy editor; Connie Long, art specialist; Caroline Commins, photo acquisitions; Sean Hale, assistant editor; and Michelle White, editorial assistant, for their hard work. Art was professionally rendered by Precision Graphics and Spatial Graphics.

Last, but certainly not least, Ed would like to thank his wife Valery for her encouragement and support.

Duane would like to pay special thanks to his family, his partner Christy and her family, and to all of the students who continue to ask good questions reinforcing and reminding him of the importance of science today.

Edward A. Keller

Duane E. DeVecchio

Santa Barbara, California

1 Introduction to Natural Hazards

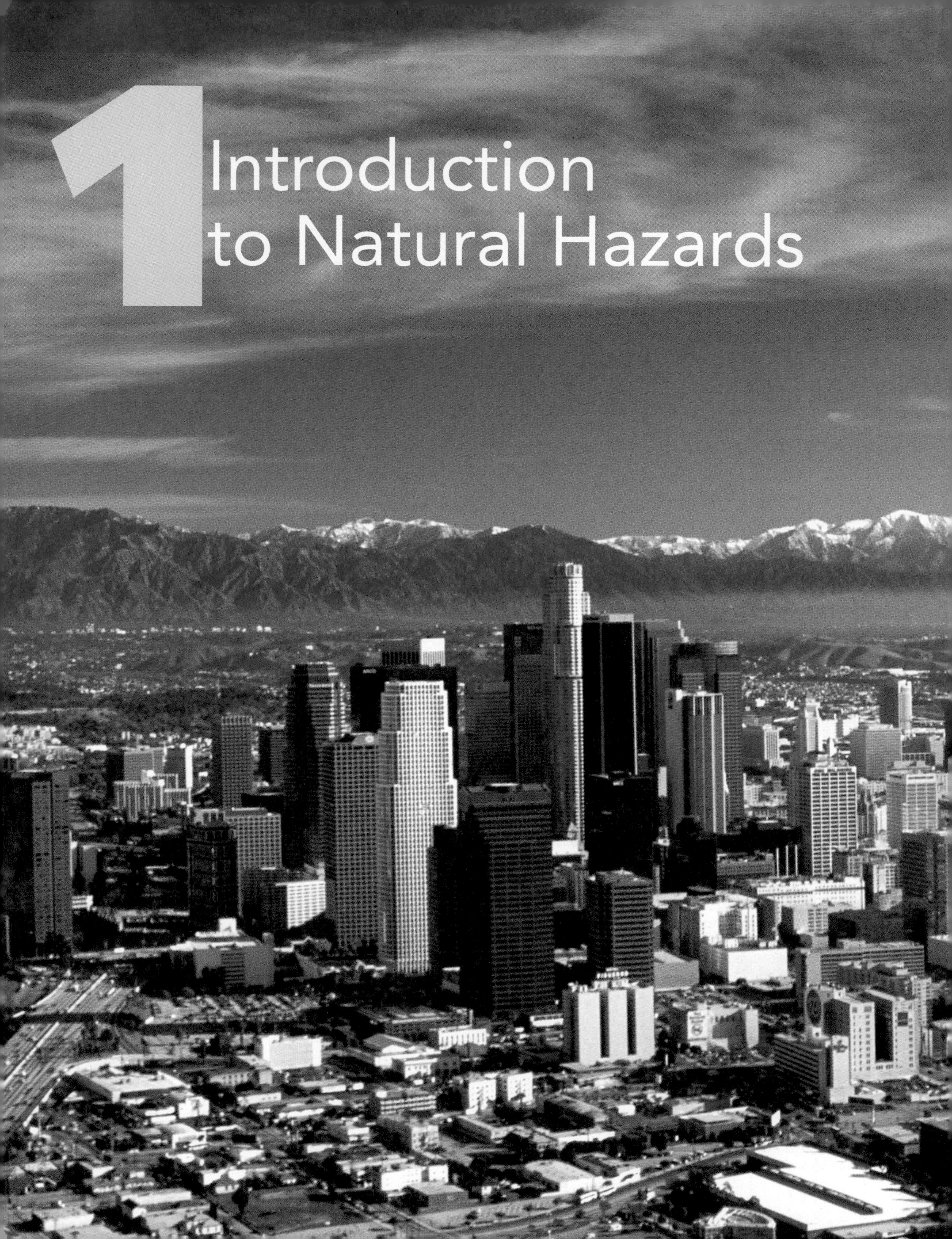

Earthquake in Haiti, 2010: Was This a Human-Caused Catastrophe?

One of the primary principles in this book is that as a result of increasing human population and poor land-use choices, what were formerly disasters many are now catastrophes. Haiti has been recognized for more than a decade as an environmental catastrophe waiting to happen. The population of Haiti has increased dramatically in recent decades, and about 90 percent of the country has been deforested (Figure 1.1). As a result, it is extremely vulnerable to hurricanes and other high-intensity storms during which runoff flooding and landslides are very frequent on land that has been deforested. Haiti is also vulnerable to large earthquakes with three major earthquakes occurring there since 1750.

Haiti is the poorest country in the Northern Hemisphere with an annual income per person of only a few percentage points of that of the United States. Four hurricanes struck in 2008, occurring one after another with a space of about one month. Lands denuded of trees responded quickly to torrential downpours and hill slopes went crashing into homes. Much of Haiti is mountainous and landslides are a constant problem. By September, when hurricane Ike, the last of the storms, hit, Haiti was devastated. Nearly 800 people died, and much of the country's harvest of food crops had been destroyed. The situation was grim.

On January 12, 2010 the Haiti earthquake occurred and about 240,000 people were killed

On January 12, 2010, an earthquake occurred that killed about 240,000 people. The storms of 2008 pale in the number of deaths compared to the earthquake in 2010. This was an appalling loss of life for an earthquake of magnitude 7.0 (Figure 1.2). For example, the magnitude 7.1 earthquake that struck the San Francisco area in 1989 caused 63 deaths with property damage of about $6 billion. The Loma Prieta earthquake that struck San Francisco injured about 4000 people. In contrast the Haiti event injured approximately 300,000 people. So why did two earthquakes of about the same magnitude cause such vastly different casualties?[1,2]

When the damage was surveyed after the earthquake, some buildings were found that did not collapse. Those buildings had evidently been designed and built with earthquakes in mind and function

LEARNING OBJECTIVES

Natural processes such as volcanic eruptions, earthquakes, floods, and hurricanes become hazards when they threaten human life and property. As population continues to grow, hazards, disasters, and catastrophes become more common. An understanding of natural processes as hazards requires some basic knowledge of earth science. Your goals in reading this chapter should be to

- know the difference between a disaster and a catastrophe.
- know the components and processes of the geologic cycle.
- understand the scientific method.
- understand the basics of risk assessment.
- recognize that natural hazards that cause disasters are generally high-energy events caused by natural earth processes.
- understand the concept that the magnitude of a hazardous event is inversely related to its frequency.
- understand how natural hazards may be linked to one another and to the physical environment.
- recognize that increasing human population and poor land-use practices compound the effects of natural hazards and can turn disasters into catastrophes.

◀ **Los Angeles California and the San Gabriel Mountains** Like Port au Prince, the capital city of Haiti destroyed by an earthquake on January 12, 2010, Los Angeles is located on the boundary of two tectonic plates that grind past one another producing large earthquakes and mountains. Faults in the southern California area are capable of producing earthquakes equivalent in size to the one that devastated Port Au Prince and killed more than 200,000 people. When the "big one" does hit southern California, how will Los Angeles fair? *(Richard Wong/Alamy)*

▲ **FIGURE 1.1 DEFORESTATION IN HAITI** Satellite image of the boarder between Haiti (left) and the Dominican Republic (right) taken in 2002. It is estimated that approximately 98% of the original Haitian forest cover has been removed. As you will see throughout this text, denuded landscapes, such as this, are extremely susceptible to the affects of many natural hazards. *(NASA)*

adequately. This suggests that if more of the buildings in Haiti had been constructed properly, the damage would have been much less. The conclusion offered by the study of the buildings and damage suggests that the massive loss of human life in the Haiti earthquake can be attributed to the failure to attain earthquake-resistant design and construction of buildings. Some of the problems observed varied from construction of heavy unsupported block walls to a lack of rebar (reinforcing field and concrete) columns that were constructed with unreinforced concrete and thus provided insufficient support. In addition there was concrete that was poorly compacted and utilized the wrong materials such as sands from the marine environment that would corrode more easily than the more desirable river sands. The end result was that the buildings in the capital of Haiti and other areas affected by the earthquake could not withstand the shaking from a magnitude 7 event.[3]

Thus the answer to our question of why were there so many deaths from the earthquake is easy to find. There is a very heavy human footprint in the catastrophic loss of life. Had the buildings been constructed properly, the number of lives lost would have been much less. If the population of the country had not grown so fast, in particularly with respect to the number of young people (thousands of schools collapsed during the earthquake), then loss of life also would have been less.

As Haiti rebuilds with international support, ways must be found to build the next generation of structures to better withstand a seismic shaking. This does not necessarily have to raise the cost of construction beyond what can be paid, but the consequences of another damaging earthquake are far too likely not to take appropriate steps and be proactive in earthquake hazard preparation.[3]

1.1 Why Studying Natural Hazards is Important

Since 1995, the world has experienced a devastating tsunami in the Indian Ocean that was produced by one of the three largest earthquakes in recorded history; catastrophic flooding in Venezuela, Bangladesh, and central Europe; the strongest El Niño on record; and deadly earthquakes in Haiti, Japan, India, Iran, and Turkey. Since 1995, North America has experienced deadly hurricanes in the Gulf Coast, Guatemala, and Honduras; record-setting wildfires in Arizona, Colorado, Utah, and California; the worst outbreak of tornadoes in Oklahoma history; a record-matching series of four

◀ **FIGURE 1.2 COLLAPSED BUILDINGS IN HAITI** A front end loader is seen ready for clean up work in front of the destroyed National Palace in Port-au-Prince, Haiti. Poor construction practices, of even important government buildings, were largely to blame for the great loss of life that resulted from the relatively modest earthquake that occurred in 2010. *(St. Felix Evans/Reuters/Newscom)*

hurricanes within 6 weeks in Florida and the Carolinas; a paralyzing ice storm in New England and Quebec; record-setting hail in Nebraska; and a rapid warming of the climate, especially in Alaska and northern Canada. These events are the result of enormous forces that are at work both inside and on the surface of our planet. In this book, we will explain these forces, how they interact with our civilization, and how we can better adjust to their effects. Although we will describe most of these forces as *natural hazards*, we can at the same time be in awe of and fascinated by their effects.

PROCESSES: INTERNAL AND EXTERNAL

In our discussion of these natural hazards, we will use the term *process* to mean the physical, chemical, and biological ways by which events, such as volcanic eruptions, earthquakes, landslides, and floods affect Earth's surface. Some of these processes, such as volcanic eruptions and earthquakes, are the result of internal forces deep within Earth. Most of these internal processes are explained by the theory of plate tectonics, one of the basic unifying theories in science. In fact, *tectonic plates*, large surface blocks of the solid Earth, are mapped by identifying zones of earthquakes and active volcanism.

Other processes associated with natural hazards result from external forces that are at or very near Earth's surface. For example, energy from the Sun warms Earth's atmosphere and surface, producing winds and evaporating water. Wind circulation and water evaporation are responsible for forming Earth's climatic zones and for driving the movement of water in the hydrologic cycle. These forces are in turn directly related to hazardous processes, such as violent storms and flooding, as well as coastal erosion. Still other external processes, such as landsliding, are the result of gravity acting on hillslopes and mountains. Gravity is the force of physical attraction to a mass—in this case the attraction of materials on Earth's surface to the entire mass of Earth. Because of gravitational attraction, objects on hillslopes or along the bed of a river naturally tend to move downslope. As a result, water that falls as precipitation on mountain slopes moves down hill by gravity on its way back to the ocean.

Thus, the processes we consider to be hazards result from natural forces such as the internal heating of Earth or from external energy from the sun. The energy released by natural processes varies greatly. For example, the average tornado expends about 1000 times as much energy as a lightning bolt, whereas the volcanic eruption of Mount St. Helens in May 1980 expended approximately a million times as much energy as a lightning bolt. The amount of solar energy Earth receives each day is about a trillion times that of a lightning bolt. However, it is important to keep in mind that a lightning bolt may strike a tree, igniting a tremendous release of energy in a forest fire, whereas solar energy is spread around the entire globe.

Events such as earthquakes, volcanoes, floods, and fires are natural processes that have been occurring on Earth's surface since long before it was populated by humans. These natural processes become hazardous when human beings live or work in their path. We often use the terms *hazard, disaster*, or *catastrophe* to describe our interaction with these natural processes.

HAZARD, DISASTER, OR CATASTROPHE

A **natural hazard** is a natural process and event that is a potential threat to human life and property. The process and events themselves are not a hazard but become so because of human use of the land. A **disaster** is a hazardous event that occurs over a limited time span in a defined area. Criteria for a natural disaster are (1) ten or more people killed, (2) 100 or more people affected, (3) a state of emergency is declared, and (4) international assistance is requested. If any one of these applies, an event is considered a *natural disaster*.[4,5] A **catastrophe** is a massive disaster that requires significant expenditure of money and a long time (often years) for recovery to take place. Hurricane Katrina, which flooded the city of New Orleans and damaged much of the coastline of Mississippi in 2005, was the most damaging, most costly catastrophe in the history of the United States. Recovery from this enormous catastrophe is taking years.

Natural hazards affect the lives of millions of people around the world. All areas of the United States are at risk from more than one hazardous Earth process (Figure 1.3). Not shown in Figure 1.3 are the areas prone to blizzards and ice storms; the coastlines that have experienced tsunamis during the past century; the areas regularly affected by wildfires; the regions that have experienced drought, subsidence, or coastal erosion; or the craters made by the impacts of asteroids and comets. No area is considered hazard free.

During the past few decades, natural disasters such as earthquakes, floods, and hurricanes have killed several million people on this planet; the average annual loss of life has been about 80,000. Financial loss from natural disasters now exceeds $50 billion per year; that figure does not include social losses such as loss of employment, mental anguish, and reduced productivity. Two individual disasters, a hurricane accompanied by flooding in Bangladesh in 1970 and an earthquake in China in 1976, each claimed more than 300,000 lives. The Indian Ocean tsunami in 2004 resulted in at least 230,000 deaths, and another hurricane that struck Bangladesh in 1991 claimed 145,000 lives (Figure 1.4). Hurricane Katrina in 2005 destroyed much of the coastal development in Mississippi and Louisiana and caused the flooding of New Orleans. Katrina killed more than 1600 people and inflicted at least $100 billion in damages, making the storm the largest financial catastrophe from a natural hazard in United States history. An earthquake in Pakistan in 2005 claimed more than 80,000 lives, destroyed many thousands of buildings, and caused

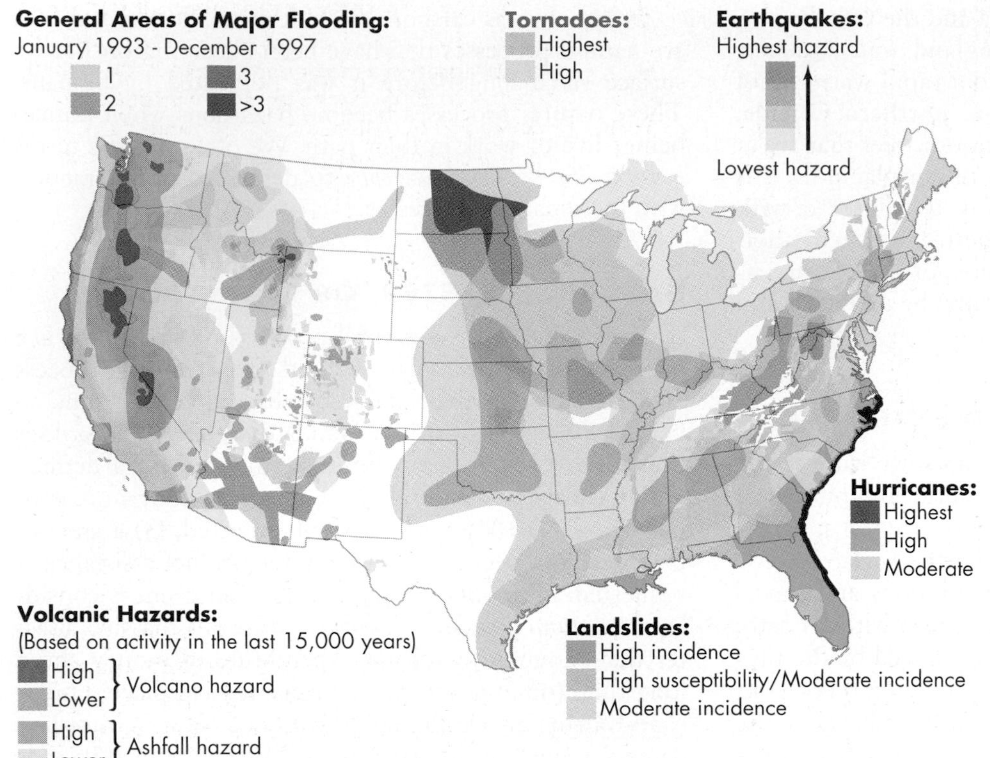

FIGURE 1.3 MAJOR HAZARDS IN THE UNITED STATES Areas of the United States at risk for earthquakes, volcanoes, landslides, flooding, hurricanes, and tornadoes. Almost every part of the country is at risk for one of the six hazards considered here. A similar map or set of maps is available for Canada. *(U.S. Geological Survey)*

extreme property damage (Figure 1.5). These catastrophes along with the Haiti earthquake discussed in the opening of this chapter were caused by natural processes and forces that have always existed. Hurricanes are caused by atmospheric disturbance, and Earth's internal heat drives the movement of tectonic plates, which causes earthquakes and sometimes tsunamis. The impact of these events has been affected by human population density and land-use patterns.

During the 1990s, a record number of great catastrophes occurred worldwide. Although there are several hundred disasters from natural hazardous events each year, only a few are classified as a great catastrophe—one that results in deaths or losses so great that outside assistance is required.[6] During the past half century, there has been a dramatic increase in natural disasters, as illustrated in Figure 1.6. Of these events, flooding is the major killer of people (more than 9000 deaths/yr worldwide), followed by earthquakes, volcanic eruptions, and windstorms. Since 1974, about 2.5 million people have died as a result of natural hazards. The vast majority of deaths were in developing countries. Asia suffers the greatest losses, with about

FIGURE 1.4 KILLER HURRICANE Flooded fields and villages adjacent a river in Bangladesh the day after the 1991 hurricane struck the country killing approximately 145,000 people. *(Defence Visual Information Center)*

◀ FIGURE 1.5 **DEVASTATING EARTHQUAKE** People searching for victims in a 10-story building that collapsed in a major earthquake in Islamabad, Pakistan, in 2005. By the onset of winter, more than 87,000 people had died and more than 3 million were homeless. *(Warrick Page/Corbis)*

three-fourths of the total deaths and nearly one-half of the economic losses. It could have been even worse were it not for improvements in warning systems, disaster preparedness, and sanitation following disasters.[6,7] Nevertheless, economic losses have increased at a faster rate than have the number of deaths.

In an effort to address the increasing number of casualties and increased property damage from natural hazards, the United Nations designated the 1990s as the International Decade for Natural Hazards Reduction. The objectives of the UN program were to minimize loss of life and property damage resulting from natural hazards. Reaching this objective requires measures to mitigate both specific physical hazards and the biological hazards that often accompany them.[8]

The term **mitigation**, which means to reduce the effects of something, is often used by scientists, planners, and policy makers in describing disaster preparedness efforts. For example, after earthquakes and floods, water supplies may be contaminated by bacteria, increasing the

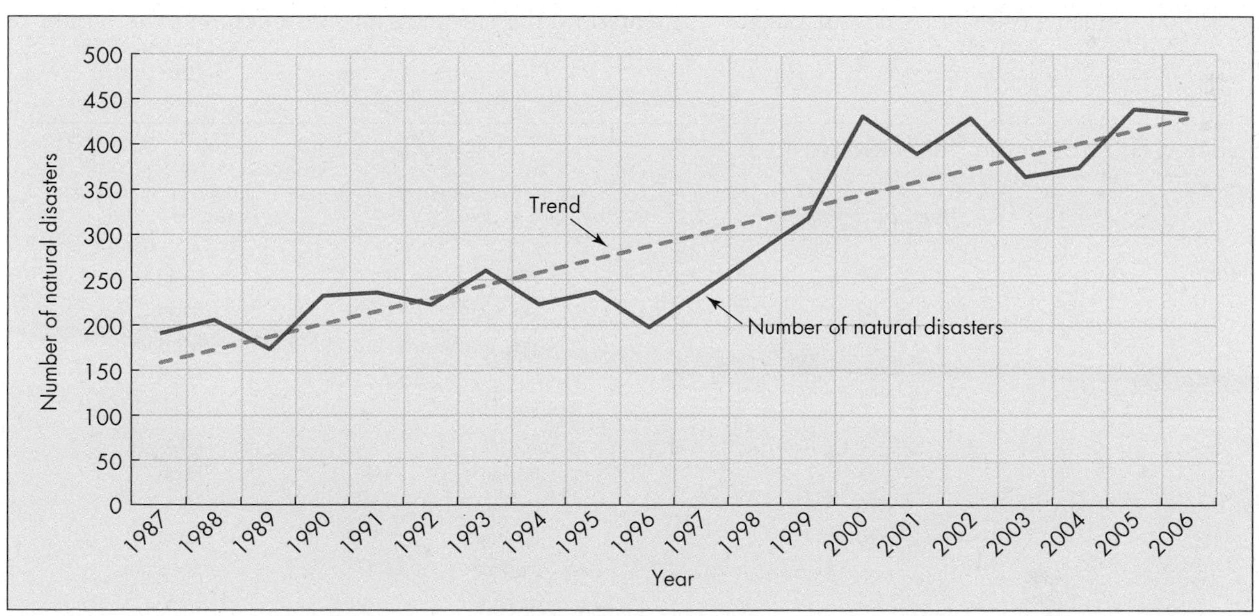

▲ FIGURE 1.6 **INCREASING NUMBERS OF WORLDWIDE NATURAL DISASTERS** The operating definition or criteria for a disaster are 10 or more people killed, 100 or more people affected, state of emergency declared, and international assistance requested. If any one of these applies, an event is considered a natural disaster by the Center for Research on the Epidemiology of Disasters (CRED). *(Modified by CRED 2007. Hoyois, P., Below, R., Scheuren, J-M., and Guha-Sapir, D. Annual Disaster Statistical Review: Numbers and Trends, 2006. University of Lonvain, Brussels, Belgium.)*

spread of diseases. To mitigate the effects of this contamination, a relief agency or government may deploy portable water treatment plants, disinfect water wells, and distribute bottled water.

DEATH AND DAMAGE CAUSED BY NATURAL HAZARDS

When we compare the effects of various natural hazards, we find that those that cause the greatest loss of human life are not necessarily the same as those that cause the most extensive property damage. The largest number of deaths each year is associated with tornadoes and windstorms, although heat waves, lightning, floods, and hurricanes also take a heavy toll (Table 1.1). Loss of life from earthquakes can vary considerably from one year to the next, because a single great quake can cause tremendous human loss. For example, in 1994, the large, but not great, Northridge earthquake in the Los Angeles area killed some 60 people and inflicted at least $20 billion in property damage. The next great earthquake in a densely populated part of California could inflict $100 billion in damages while killing several thousand people.[8]

The property damage caused by each type of hazard is considerable. Floods, landslides, expansive soils, and frost each causes in excess of $1.5 billion in damage each year in the United States. Surprisingly, expansive soils, that is, those soils that shrink and swell, are one of the most costly hazards, causing more than $15 billion in damages annually.

Natural disasters cost the United States between $10 billion and $50 billion annually; the average cost of a single major disaster is as much as $500 million. A catastrophe, similar to Hurricane Katrina, can cost many hundreds of billions of dollars. Because the population is steadily increasing in high-risk areas, such as along coastlines, we can expect this number to increase significantly.[9]

An important aspect of all natural hazards is their potential to produce a catastrophe, which has been defined as any situation in which the damages to people, property, or society in general are sufficient that recovery and/or rehabilitation is a long, involved process. Natural hazards vary greatly in their potential to cause a catastrophe (see Table 1.1). Floods, hurricanes, tornadoes, earthquakes, volcanic eruptions, large wildfires, and heat waves are the hazards most likely to create catastrophes. Landslides, because they generally affect a smaller area, have only a moderate catastrophe potential. Drought also has a moderate catastrophe potential because although it may cover a wide area, there is usually plenty of warning time before its worst effects are felt. Hazards with a low catastrophe potential include coastal erosion, frost, lightning, and expansive soils.[10]

The effects of natural hazards change with time because of changes in land-use patterns. Urban growth can influence people to develop on marginal lands, such as steep hillsides and floodplains. This trend is especially a problem in areas surrounding major cities in developing nations. In addition to increasing population density, urbanization can also change the physical properties of earth materials by influencing drainage, altering the steepness of hillslopes, and removing vegetation. Changes in agricultural, forestry, and mining practices can affect rates of erosion and sedimentation, the steepness of hillslopes, and the nature of

TABLE 1.1

Effects of Selected Hazards in the United States

Hazard	Deaths per Year	Occurrence Influenced by Human Use	Catastrophe Hazard Potential
Flood	86	Yes	High
Earthquake[a]	50+?	Yes	High
Landslide	25	Yes	Medium
Volcano[a]	<1	No	High
Coastal erosion	0	Yes	Low
Expansive soils	0	No	Low
Hurricane	55	Perhaps	High
Tornado and windstorm	218	Perhaps	High
Lightning	120	Perhaps	Low
Drought	0	Perhaps	Medium
Frost and freeze	0	Yes	Low
Heat wave	10s to 100s	Yes	High
Wildfire[b]	<10	Yes	High
Extraterrestrial impact	0	No	Very high

[a] Estimate based on recent or predicted loss over 150-year period. Actual loss of life and/or property could be much greater.

[b] Deaths mostly firefighters.

Source: Modified after White, G. F., and Haas, J. E. 1975. Assessment of research on natural hazards. *Cambridge, MA: MIT Press.*

vegetative cover. Overall, damage from most natural hazards in the United States is increasing, but the number of deaths from many hazards is decreasing because of better prediction, forecasting, and warning.

1.2 Role of History in Understanding Hazards

A fundamental principal for understanding natural hazards is that they are repetitive events, and therefore the study of their history provides much-needed information for any hazard-reduction plan. You will recall that poor building practices in Haiti led to the great loss of life. Whether we are studying earthquakes, floods, landslides, or volcanic eruptions, knowledge of historic events and recent geologic history of an area is vital to our understanding and assessment of the hazard. For example, if we wish to evaluate the flooding history of a particular river, one of the first tasks is to identify floods that have occurred in the historic and recent prehistoric past. Useful information can be obtained by studying aerial photographs and maps as far back as the record allows. In our reconstruction of previous events, we can look for evidence of past floods in stream deposits. Commonly these deposits contain organic material such as wood or shells that may be radiocarbon dated to provide a chronology of ancient flood events. This chronology can then be linked with the historic record of high flows to provide an overall perspective of how often the river floods and how extensive the floods may be. Similarly, if we are studying landslides in a particular area, an investigation of both historic and prehistoric landslides is necessary to better predict future landslides. Geologists have the tools and training to "read the landscape" and evaluate prehistoric evidence for natural hazards. Linking the prehistoric and historic records extends our perspective of time when we study repetitive natural events.

In summary, before we can truly appreciate the nature and extent of a natural hazard, we must study in detail its historic occurrence as well as any geologic features that it may produce or affect. These geologic features may be landforms, such as channels, hills, or beaches; structures, such as geologic faults, cracks, or folded rock; or earth materials, such as lava flows, meteorites, or soil. Any prediction of the future occurrence and effects of a hazard will be more accurate if we can combine information about historic and prehistoric behavior with a knowledge of present conditions and recent past events, including land-use changes.

To fully understand the natural processes we call hazards, some background knowledge of the geologic cycle—processes that produce and modify earth materials such as rocks, minerals, and water—is necessary. In the next few sections, we will discuss the geologic cycle and then introduce five concepts that are fundamental to understanding natural processes as hazards.

1.3 Geologic Cycle

Geologic conditions and materials largely govern the type, location, and intensity of natural processes. For example, earthquakes and volcanoes do not occur at random across Earth's surface; rather, most of them mark the boundaries of tectonic plates. The location of landslides, too, is governed by geologic conditions. Slopes composed of a weak rock, such as shale, are much more likely to slip than those made of a strong rock, such as granite. Hurricanes, although not themselves governed specifically by geology, will have differing effects depending on the geology of the area they strike. An understanding of the components and dynamics of the geologic cycle will explain these relationships.

Throughout much of the 4.6 billion years of Earth's history, the materials on or near Earth's surface have been created, maintained, and destroyed by numerous physical, chemical, and biological processes. Continuously operating processes produce the earth materials, land, water, and atmosphere, necessary for our survival. Collectively, these processes are referred to as the **geologic cycle,** which is really a group of subcycles that includes

- the tectonic cycle
- the rock cycle
- the hydrologic cycle
- the biogeochemical cycles

THE TECTONIC CYCLE

The term *tectonic* refers to the large-scale geologic processes that deform Earth's crust and produce landforms such as ocean basins, continents, and mountains. Tectonic processes are driven by forces deep within Earth. To describe these processes, we must use information about the composition and layering of Earth's interior and about the large blocks of the solid Earth that we call *tectonic plates.* The **tectonic cycle** involves the creation, movement, and destruction of tectonic plates. It is responsible for the production and distribution of rock and mineral resources invaluable to modern civilization, as well as hazards such as volcanoes and earthquakes. The tectonic cycle and its linkages to hazards is the subject of Chapter 2.

THE ROCK CYCLE

Rocks are aggregates of one or more *minerals.* A mineral is a naturally occurring, crystalline substance with defined properties (see Appendix A for a discussion of minerals). The **rock cycle** is the largest of the geologic subcycles, and it is linked to all the other subcycles. It depends on the tectonic cycle for heat and energy, the biogeochemical cycle for materials, and the hydrologic cycle for water. Water is

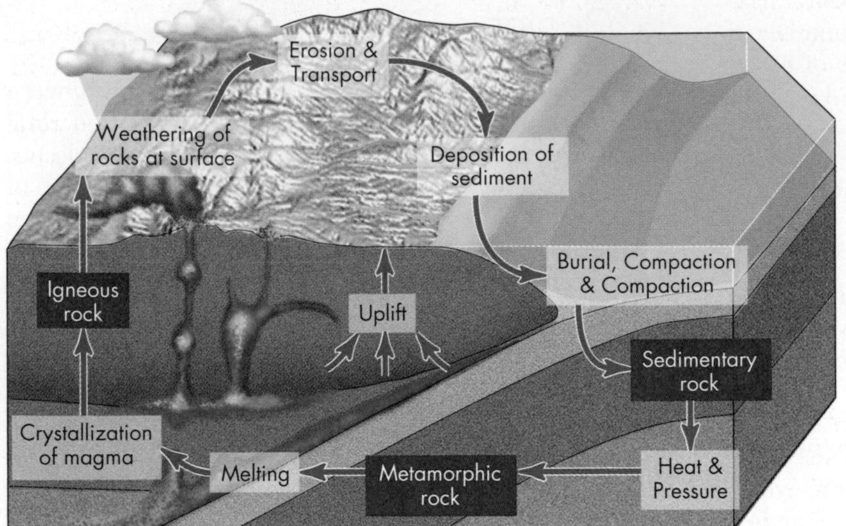

◀ **FIGURE 1.7 THE ROCK CYCLE** Idealized rock cycle showing the three types of rocks and important processes that form them.

then used in the processes of weathering, erosion, transportation, deposition, and lithification of sediment.

Although rocks vary greatly in their composition and properties, they can be classified into three general types, or families, according to how they were formed in the rock cycle (Figure 1.7); rocks and their properties are discussed in Appendix B. We may consider the rock cycle as a worldwide rock-recycling process driven by Earth's internal heat, which melts the rocks subducted in the tectonic cycle (see Chapter 2). *Crystallization* of molten rock produces *igneous rocks* beneath and on Earth's surface. Rocks at or near the surface break down chemically and physically by *weathering* to form particles known as *sediment.* These particles vary in size from fine clay to very large pieces of boulder-size gravel. Sediment formed by weathering is then transported by wind, water, ice, and gravity to depositional basins such as the ocean. When wind or water currents slow down, ice melts, or material moving by gravity reaches a flat surface, the sediment settles and accumulates by a process known as *deposition.* The accumulated layers of sediment eventually undergo *lithification* (conversion to solid rock), forming *sedimentary rocks.* Lithification takes place by cementation and compaction as sediment is buried beneath other sediment. With deep burial, sedimentary rocks may be metamorphosed (altered in form) by heat, pressure, or chemically active fluids to produce *metamorphic rocks.* Metamorphic rocks may be buried to depths where pressure and temperature conditions cause them to melt, beginning the entire rock cycle again. As with any other Earth cycle, there are many exceptions or variations from the idealized sequence. For example, an igneous or metamorphic rock may be altered into a new metamorphic rock without undergoing weathering or erosion (Figure 1.7), or sedimentary and metamorphic rocks may be uplifted and weathered before they can continue on to the next stage in the cycle. Finally, there are other sources of sediment that have a biological or chemical origin and types of metamorphism that do not involve deep burial. Overall, the type of rock formed in the rock cycle depends on the rock's environment.

THE HYDROLOGIC CYCLE

The movement of water from the oceans to the atmosphere and back again is called the **hydrologic cycle** (Figure 1.8). Driven by solar energy, the cycle operates by way of

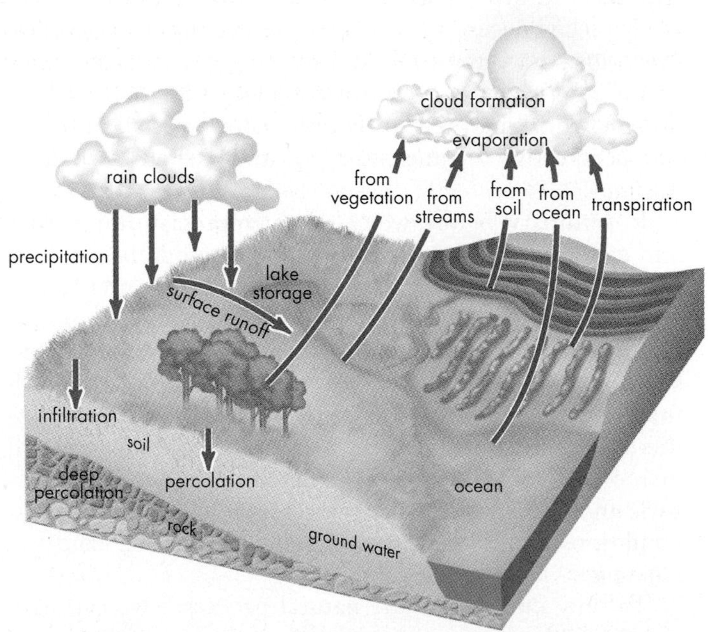

▲ **FIGURE 1.8 HYDROLOGIC CYCLE** Idealized diagram showing the hydrologic cycle's important processes and transfer of water. *(Modified after Council on Environmental Quality and Department of State. 1980.* The Global 2000 Report to the President. Vol. 2)

evaporation, precipitation, surface runoff, and subsurface flow, and water is stored in different compartments along the way. These compartments include the oceans, atmosphere, rivers and streams, groundwater, lakes, and ice caps and glaciers (Table 1.2). The *residence time*, or estimated average amount of time that a drop of water spends in any one compartment, ranges from tens of thousands of years or more in glaciers to 9 days in the atmosphere.

As you can see from studying Table 1.2, only a very small amount of the total water in the cycle is active near Earth's surface at any one time. Although the combined percentage of water in the atmosphere, rivers, and shallow subsurface environment is only about 0.3 percent of the total, this water is tremendously important for life on Earth and for the rock and biogeochemical cycles. This surface or near-surface water helps move and sort chemical elements in solution, sculpt the landscape, weather rocks, transport and deposit sediments, and provide our water resources.

BIOGEOCHEMICAL CYCLES

A **biogeochemical cycle** is the transfer or cycling of a chemical element or elements through the atmosphere (the layer of gases surrounding Earth), lithosphere (Earth's rocky outer layer), hydrosphere (oceans, lakes, rivers, and groundwater), and biosphere (the part of the Earth where life exists). It follows from this definition that biogeochemical cycles are intimately related to the tectonic, rock, and hydrologic cycles. The tectonic cycle provides water from volcanic processes as well as the heat and energy required to form and change the earth materials transferred in biogeochemical cycles. The rock and hydrologic cycles are involved in many processes that transfer and store chemical elements in water, soil, and rock.

Biogeochemical cycles can most easily be described as the transfer of chemical elements through a series of storage compartments or reservoirs (e.g., air, soil, groundwater, vegetation). For example, carbon (in the form of CO_2) is exhaled by animals, enters the atmosphere, and is then taken up by plants. When a biogeochemical cycle is well understood, the rate of transfer, or *flux*, among all the compartments is known. However, determining these rates on a global basis is a very difficult task. The amounts of such important elements as carbon, nitrogen, and phosphorus in each compartment, and their rates of transfer among compartments, are known only approximately.

1.4 Fundamental Concepts for Understanding Natural Processes as Hazards

The five concepts described below are basic to an understanding of natural hazards. These fundamental concepts serve as a conceptual framework for our discussion of each natural hazard in subsequent chapters of this book.

1. **Hazards are predictable from scientific evaluation.** Natural hazards, such as earthquakes, volcanic eruptions, landslides, and floods, are natural processes that can be identified and studied using the scientific method. Most hazardous events and processes can be monitored and mapped, and their future activity predicted, on the basis of frequency of past events, patterns in their occurrence, and types of precursor events.
2. **Risk analysis is an important component in our understanding of the effects of hazardous processes.** Hazardous processes are amenable to *risk analysis*, which estimates the probability that an event will occur and the consequences resulting from that event. For example, if we were to estimate that in any given year, Los Angeles has a 5 percent chance of a moderate earthquake, and if we know the consequence of that earthquake in terms of loss of life and damage, then we can calculate the potential risk to society.

TABLE 1.2

The World's Water Supply (Selected Examples)

Location	Surface Area (km^2)	Water Volume (km^2)	Percentage of Total Water	Estimated Average Residence Time
Oceans	361,000,000	1,230,000,000	97.2	Thousands of years
Atmosphere	510,000,000	12,700	0.001	9 days
Rivers and streams	—	1200	0.0001	2 weeks
Groundwater; shallow	130,000,000	4,000,000	0.31	Hundreds to many thousands of years to depth of 0.8 km
Lakes (freshwater)	855,000	123,000	0.009	Tens of years
Ice caps and glaciers	28,200,000	28,600,000	2.15	Up to tens of thousands of years and longer

Source: Data from U.S. Geological Survey.

3. **Linkages exist between different natural hazards as well as between hazards and the physical environment.**
Hazardous processes are linked in many ways. For example, earthquakes can produce landslides and giant sea waves called tsunamis, and hurricanes often cause flooding and coastal erosion.

4. **Hazardous events that previously produced disasters are now producing catastrophes.**
The magnitude, or size, of a hazardous event as well as its frequency, or how often it occurs, may be influenced by human activity. As a result of increasing human population and poor land-use practices, events that used to cause disasters are now often causing catastrophes.

5. **Consequences of hazards can be minimized.**
Minimizing the potential adverse consequences and effects of natural hazards requires an integrated approach that includes scientific understanding, land-use planning and regulation, engineering, and proactive disaster preparedness.

CONCEPT 1

Hazards Are Predictable from Scientific Evaluation

SCIENCE AND NATURAL HAZARDS

Science is a body of knowledge that has resulted from investigations and experiments, the results of which are subject to verification. The method of science, often referred to as the scientific method, has a series of steps. The first step is the formulation of a question. With respect to a hazardous event, a geologist may ask: Why did a landslide occur that destroyed three homes? In order to explore and answer this question, the geologist will spend time examining the slope that failed. She may notice that a great deal of water is emerging from the base of the slope and landslide. If the geologist also knows that a waterline is buried in the slope, she may refine the question to ask specifically: Did the water in the slope cause the landslide? This question is the basis for a hypothesis that may be stated as follows: The landslide occurred because a buried water main within the slope broke, causing a large amount of water to enter the slope, reducing the strength of the slope materials, and causing the landslide. Often a series of questions or multiple hypotheses are tested.

The hypothesis is thus a possible answer to our question and is an idea that may be tested. In our example, we may test the hypothesis that a broken water main caused a landslide by excavating the slope to determine the source of the water. In science, we test hypotheses in an attempt to disprove them. That is, if we found that there were no leaking water pipes in the slope on which the landslide occurred, we would reject the hypothesis and develop and test another hypothesis. Use of the scientific method has improved our understanding of many natural earth processes, including flooding, volcanic eruptions, earthquakes, hurricanes, and coastal erosion. In our scientific study of natural processes, we have identified where most of these processes occur, their range of magnitude, and how frequently they occur. We have also mapped the nature and extent of hazards. Coupled with knowledge of the frequency of past events, we can use such maps to predict when and where processes such as floods, landslides, and earthquakes will occur in the future. We have also evaluated patterns and types of precursor events. For example, prior to large earthquakes there may be foreshocks and prior to volcanic eruptions patterns of gas emissions may signal an imminent eruption.

HAZARDS ARE NATURAL PROCESSES

Since the dawn of human existence, we have been obligated to adjust to processes that make our lives more difficult. We humans apparently are a product of the Pleistocene ice ages, which started more than 1.8 million years ago (Table 1.3). The Pleistocene and following Holocene Epoch experienced rapid climatic changes—from relatively cold, harsh glacial conditions as recently as a few thousand years ago to the relatively warm interglacial conditions we enjoy today. Learning to adjust to harsh and changing climatic conditions has been necessary for our survival from the very beginning.

Events we call natural hazards are natural Earth processes. They become hazardous when people live or work near these processes, and when land-use changes such as urbanization or deforestation amplify their effect. To reduce damage and loss of life, it is imperative to identify potentially hazardous processes and make this information available to planners and decision makers. However, because the hazards that we face are natural and not the result of human activities, we encounter a philosophical barrier whenever we try to minimize their adverse effects. For example, when we realize that flooding is a natural part of river dynamics, we must ask ourselves if it is wiser to attempt to control floods or to simply make sure that people and property are out of harm's way when they occur.

Although it is possible to control some natural hazards to a certain degree, many are completely beyond our control. For example, although we may have some success in temporarily preventing damage from forest fires by using controlled burns and advanced firefighting techniques, we will never be able to prevent earthquakes. In fact, we may actually worsen the effects of natural processes simply by labeling them as hazardous. Rivers will always flood. Because we choose to live and work on floodplains, we have labeled floods as hazardous processes. This label has led to attempts to control them. Unfortunately, as we will discuss later, some flood-control measures actually intensify the effects of flooding, thereby increasing the very hazard we are trying to prevent (see Chapter 6). The best

TABLE 1.3

Geologic Time with Important Events

Era	Period		Epoch	Million Years before Present	Events: Life	Events: Earth	Million Years before Present	True Scale (Million Years before Present)
Cenozoic	Neogene	Quaternary	Holocene	0.0114	• Extinction event • Modern humans	Ice age; Formation of Transverse Ranges, CA		Cenozoic
			Pleistocene	1.8	• Early humans		1.8	
		Tertiary	Pliocene	5.3		Formation of Andes Mountains		
			Miocene	23		Collision of India with Asia forming Himalayan Mountains and Tibetan Plateau		
	Paleogene		Oligocene	34	• Grasses • Whales			
			Eocene	56	• Extinction event			
			Paleocene	65	• Mammals expand	Rocky Mountains form	65	
Mesozoic	Cretaceous			146	• Dinosaur extinction[1], extinction event • Flowering plants	Emplacement of Sierra Nevada granites (Yosemite National Park)		Mesozoic
	Jurassic			200	• Birds	• Supercontinent Pangaea begins to break up		
	Triassic			251	• Extinction event • Mammals • Dinosaurs		251	
Paleozoic	Permian			299	• Extinction event • Reptiles	• Ice age		Paleozoic
	Carboniferous			359	• Coal swamps • Extinction event	Appalachian Mountains form		
	Devonian			416	• Trees			
	Silurian			444	• Land plants • Extinction event			
	Ordovician			488	• Fish			
	Cambrian			542	• Explosion of organisms with shells		542	
Precambrian time					• Multicelled organisms	• Ice age		Precambrian
				2500	• Free oxygen in atmosphere and ozone layer in stratosphere	• Ice age		
				3500	• Primitive life (first fossils)			
				4000		• Oldest rocks		
				4600		• Age of Earth	4600	4600

[1]Many scientists believe that not all dinosaurs became extinct but that some dinosaurs evolved to birds.

approach to hazard reduction is to identify hazardous processes and delineate the geographic areas where they occur. Every effort should be made to avoid putting people and property in harm's way, especially for those hazards, such as earthquakes, that we cannot control.

FORECAST, PREDICTION, AND WARNING OF HAZARDOUS EVENTS

Predicting Changes in the Earth System The idea that "the present is the key to the past," called **uniformitarianism**, was popularized in 1785 by James Hutton (referred to by some scholars as the father of geology) and is heralded today as a fundamental concept of Earth sciences. As the name suggests, uniformitarianism holds that processes we observe today also operated in the past (flow of water in rivers, formation and movement of glaciers, landslides, waves on beaches, uplift of the land from earthquakes, and so on). Uniformitarianism does not demand or even suggest that the magnitude (amount of energy expended) and frequency (how often a particular process occurs) of natural processes remain constant with time. We can infer that for as long as Earth has had an atmosphere, oceans, and continents similar to those of today, the present processes were operating.

In making inferences about geologic events, we must consider the effects of human activity on the Earth system and what effect these changes to the system as a whole may have on natural Earth processes. For example, small streams, with drainage areas of a few to several 10s of km^2, flood regardless of human activities, but human activities, such as paving the ground in cities, increase runoff and the magnitude and frequency of flooding. That is, after the paving, floods of a particular size are more frequent, and a particular rainstorm can produce a larger flood than before the paving. Therefore, to predict the long-range effects of flooding, we must be able to determine how future human activities will change the size and frequency of floods. In this case, *the present is the key to the future.* For example, when environmental geologists examine recent landslide deposits (Figure 1.9) in an area designated to become a housing development, they must use uniformitarianism to infer where there will be future landslides, as well as to predict what effects urbanization will have on the magnitude and frequency of future landslides. We will now consider linkages between processes in what is called environmental unity or "you can't do just one thing."

The principle of **environmental unity** states that one action causes others in a chain of actions and events. For example, if we remove the native vegetation from a steep slope, or build a large structure on top of the slope, a landslide may occur. The slide mass may move into a canyon and dam a stream, which will cause the stream flow to back up forming a lake. The dam of landslide material could be overtopped and quickly eroded, causing a flood. Alternatively the dam may become saturated and collapse, producing a muddy flow of water and debris downstream as a destructive debris flow. Either event could destroy homes and perhaps kills people down the valley. Thus, modifying and destabilizing the slope set off a chain or series of events that changes the landscape environment where people live. Thus, when we consider the potential effects of a particular event, we need to think about what other events may happen and take appropriate precautions.

A **prediction** of a hazardous event such as an earthquake involves specifying the date, time, and size of the event. This is different from predicting where or how often a particular event such as a flood will occur. A **forecast**, on the other hand, has ranges of certainty. The weather forecast for tomorrow may state there is a

◀ **FIGURE 1.9 URBAN DEVELOPMENT** The presence of a landslide on this slope suggests that the slope is not stable and further movement may occur in the future. This is a "red flag" for future development in the area. *(Edward A. Keller)*

40 percent chance of showers. Learning how to predict hazardous events so we can minimize human loss and property damage is an important endeavor. For some natural hazards we have enough information to predict or forecast events accurately. When there is insufficient information available, the best we can do is locate areas where hazardous events have occurred and infer where and when similar future events might take place. If we know both the probability and the possible consequences of an event occurring at a particular location, we can assess the risk that the event poses to people and property, even if we cannot accurately predict when it will next occur.

The effects of a hazardous event can be reduced if we can forecast or predict it, and if we can issue a warning. Attempting to do this involves most or all of the following elements:

- Identifying the location where a hazardous event is likely to occur.
- Determining the probability that an event of a given magnitude will occur.
- Observing any precursor events.
- Forecasting or predicting the event.
- Warning the public.

Location For the most part, we know where a particular kind of event is likely to occur, and we can accurately map where they have occurred (see Appendix C for a discussion of maps and images useful in hazard analysis). On a global scale, the major zones for earthquakes and volcanic eruptions have been delineated by mapping (1) where earthquakes have occurred, (2) the extent of recently formed volcanic rocks, and (3) the locations of recently active volcanoes. On a regional scale, we can use past eruptions to identify areas that are likely to be threatened in future eruptions. This risk has been delineated for several Cascade volcanoes, such as Mt. Rainier, as well as for several volcanoes in Japan, Italy, Colombia, and elsewhere. On a local scale, detailed mapping of soils, rocks, groundwater conditions, and surface drainage may identify slopes that are likely to fail or where expansive soils exist. In most cases, we can predict where flooding is likely to occur from the location of the floodplain and from mapping the extent of recent floods.

Probability of Occurrence Determining the probability of a particular event in a particular location within a particular time span is an essential part of a hazard prediction (see Appendix D for a discussion of how geologists determine the all-important chronology necessary to evaluate the time element of hazard evaluation and probability of occurrence). For many rivers, we have sufficiently long records of flow to develop probability models that can reasonably predict the average number of floods of a given magnitude that will occur in a decade. Likewise, droughts may be assigned a probability on the basis of past rainfall in the region. However, these probabilities are similar to the chances of throwing a particular number on a die or drawing an inside straight in poker—the element of chance is always present. Although the 10-year flood may occur on average only once every 10 years, it is possible to have several floods of this magnitude in any one year, just as it is possible to throw two straight sixes with a die.

Precursor Events Many hazardous events are preceded by *precursor events*. For example, the surface of the ground may creep (that is, move slowly for a long period of time) prior to an actual landslide. Often, the rate of creep increases up to the final failure and landslide. Volcanoes sometimes swell or bulge before an eruption, and often earthquake activity increases significantly in the area. Foreshocks or unusual uplift of the land may precede earthquakes.

Identification of precursor events helps scientists predict when and where a major event is likely to happen. Thus, measurements of landslide creep or swelling of a volcano may lead authorities to issue a warning and evacuate people from a hazardous area.

Forecast When a forecast of an event is issued, the certainty of the event is given, usually as the percent chance of something happening. When we hear a forecast of a hazardous event, it means we should be prepared for the event.

Prediction It is sometimes possible to accurately predict when certain natural events will occur. Flooding of the Mississippi River, which occurs in the spring in response to snowmelt or large regional storm systems, is fairly common. Hydrologists with the National Weather Service usually can predict when the river will reach a particular flood stage or water level. When a hurricane is spotted far out to sea and tracked toward the shore, we can predict when and where it will likely strike land. Large ocean waves, called tsunamis, that are generated by undersea earthquakes, landslides, and other disturbances may also be predicted once they form. In the past several decades, the prediction of the arrival time for tsunamis in the Pacific Ocean has been fairly successful. Even a short-notice prediction of a hazardous event, such as a tornado, motivates us to act to reduce potential consequences before the event happens.

Warning After a hazardous event has been predicted or a forecast has been made, the public must be warned. The flow of information leading to the **warning** of a possible disaster—such as a large earthquake or flood—should move along a predefined path (Figure 1.10). People do not always welcome such warnings, especially when the predicted event does not come to pass. In 1982, after geologists advised that a volcanic eruption near Mammoth Lakes, California, was quite likely, the advisory caused loss of tourist business and apprehension for the residents. The eruption did not occur and the advisory was eventually lifted. In July 1986, a series of earthquakes occurred over a

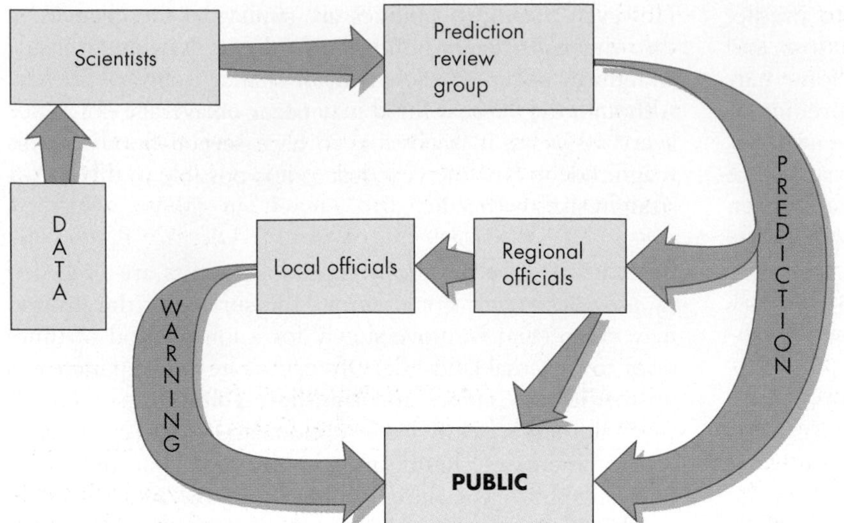

◀ **FIGURE 1.10 HAZARD PREDICTION OR WARNING** Possible flow path for issuance of a prediction or warning for a natural disaster.

4-day period near Bishop, California. The quakes began with minor magnitude 3 tremors and culminated in a damaging magnitude 6.1 shock. Investigators concluded that there was a high probability of a larger quake in the vicinity in the near future and issued a warning. Local business owners, who feared the loss of summer tourism, thought that the warning was irresponsible; in fact, the predicted quake never materialized.

Incidents of this kind have led some people to conclude that scientific predictions are worthless and that advisory warnings should not be issued. Part of the problem is poor communication between the investigating scientists and the news media. Newspaper, television, Internet, and radio reports may fail to explain the evidence or the probabilistic nature of natural hazard prediction, leading the public to expect completely certain statements about what will happen. Although scientists are not yet able to predict volcanic eruptions and earthquakes accurately, it would seem that they have a responsibility to publicize their informed judgments. An informed public is better able to act responsibly than an uninformed public, even if the subject makes people uncomfortable. Ship captains, who depend on weather advisories and warnings of changing conditions, do not suggest that they would be better off not knowing about an impending storm, even though the storm might veer and miss the ship. Just as weather warnings have proved useful for planning, official warnings of hazards such as earthquakes, landslides, and floods are also useful to people deciding where to live, work, and travel.

Consider once more the prediction of a volcanic eruption in the Mammoth Lakes area of California. The location and depth of earthquakes suggested to scientists that molten rock was moving toward the surface. In light of the high probability that the volcano would erupt and the possible loss of life if it did, it would have been irresponsible for scientists not to issue an advisory. Although the predicted eruption did not occur, the warning led the community to develop evacuation routes and consider disaster preparedness. This planning may prove useful, for it is likely that a volcanic eruption will occur in the Mammoth Lakes area in the future. The most recent event occurred only 600 years ago! As a result of the prediction, the community is better informed than it was before and thus better able to deal with an eruption when it does occur.

CONCEPT 2

Risk Analysis is an Important Component of Our Understanding of the Effects of Hazardous Processes

Before rational people can discuss and consider adjustments to hazards, they must have a good idea of the risk that they face in various circumstances. The field of risk assessment is rapidly growing, and its application in the analysis of natural hazards probably should be expanded.

The **risk** of a particular event is defined as the product of the probability of that event occurring times the consequences should it occur.[11] Consequences (damages to people, property, economic activity, public service, and so on) may be expressed on a variety of scales. If, for example, we are considering the risk of earthquake damage to a nuclear reactor, we may evaluate the consequences in terms of radiation released, which then can be related to damages to people and other living things. In any such assessment, it is important to calculate the risks for various possible events—in this example, for earthquakes of various magnitudes. A large event has a lower probability than does a small one, but its consequences are likely to be greater. Determining *acceptable risk* is more complicated, for the risk that society or individuals are willing to take depends upon the situation. Driving an automobile is fairly risky, but most of us accept that risk as part of living in a modern world. On the other hand, for many people the acceptable risk from a nuclear power plant is very low because they consider almost any risk of radiation poisoning unacceptable. Nuclear power plants are controversial because many people perceive them as high-risk facilities. Even though the probability of an accident owing to a geologic hazard such as an earthquake may be quite low for most power plants, the consequences could be devastating, resulting in a relatively high risk.

Institutions, such as the government and banks, approach the topic of acceptable risk from an economic point of view rather than a personal perception of the risk. For example, a bank will consider how much risk it can tolerate with respect to flooding. The federal government may require that any property that receives a loan from it not have a flood hazard that exceeds 1 percent per year, that is, protection up to and including the 100-year flood.

On an individual level, it is important to recognize that you have some degree of choice regarding the level of risk you are willing to live with. For the most part, you can choose where you are willing to live. If you move to the North Carolina coast, you should realize you are putting yourself in the path of potentially deadly hurricanes. If you choose to live in Los Angeles, it is highly likely that in your lifetime you will experience an earthquake. So why do people live in hazardous areas? Perhaps you were offered an excellent job in North Carolina, or the allure of warm weather, mountains, and the ocean drew you to Los Angeles. Whatever the case, we as individuals must learn to weigh the pros and cons of living in a given area and decide whether or not it is worth the risk. This assessment should consider factors such as the level of potential devastation caused by an event, the frequency of the event, and the extent of the geographic area at risk, and it should compare these factors to the potential benefits of living in the high-risk area. In this way, we determine our own acceptable risk, which may vary from person to person.

A frequent problem of risk analysis is lack of reliable data for analyzing either the probability or the consequences of an event. It can be difficult to assign probabilities to geologic events such as earthquakes and volcanic eruptions, because the known record of past events is often inadequate.[11] Similarly, it may be difficult to determine the consequences of an event or series of events. For example, if we are concerned with the consequences of releasing radiation into the environment, we need a lot of information about the local biology, geology, hydrology, and meteorology, all of which may be complex and difficult to analyze. Despite these limitations, risk analysis is a step in the right direction. As we learn more about determining the probability and consequences of a hazardous event, we should be able to provide the more reliable analyses necessary for decision making.

CONCEPT 3

Linkages Exist Between Natural Hazards and Between Hazards and the Physical Environment

Linkages between natural processes that are hazardous to people generally fall into two categories. First, many of the hazards themselves are linked. For examples, hurricanes are often associated with flooding, and intense precipitation associated with hurricanes can cause erosion along the coast and landslides on inland slopes. Volcanic eruptions on land are linked to mudflows and floods, and eruptions in the ocean are linked to tsunamis. Second, natural hazards are linked to earth materials. Exposures of shale, a type of sedimentary rock composed of loosely cemented and compacted clay particles, are prone to landslides. On the other hand granite, a type of igneous rock that is generally strong and durable, is prone to sliding along large fractures within the rock.

CONCEPT 4

Hazardous Events That Previously Produced Disasters Are Now Producing Catastrophes

Early in the history of our species, our struggle with earth processes was probably a day-to-day experience. However, our numbers were neither great nor concentrated, so losses from hazardous earth processes were not very significant. As people learned to produce and maintain a larger and, in most years, more abundant food supply, the population increased and became concentrated near sources of food. The concentration of population and resources also increased the effect of periodic earthquakes, floods, and other potentially hazardous natural processes. This trend has continued, so that many people today live in areas that are likely to be damaged by hazardous earth processes or that are susceptible to the effect of such processes in adjacent areas. Because an increase in population puts a greater number of people at risk and also forces more people to settle in hazardous areas, the need increases for planning to minimize losses from natural disasters.

EXAMPLES OF DISASTERS IN DENSELY POPULATED AREAS

Mexico City is the center of the world's second most populous urban area. Approximately 23 million people are concentrated in an area of about 2300 km^2 (890 $mi.^2$). Families in this area average five members, and an estimated one-third of the families live in a single room. The city is built on ancient lakebeds that amplify the shaking in an earthquake, and parts of the city have been sinking several centimeters per year from the pumping of groundwater. The subsidence has not been uniform, so the buildings tilt and are even more vulnerable to earthquake shaking.[12] In September 1985, Mexico endured a magnitude 8.0 earthquake that killed about 10,000 people in Mexico City.

Another example comes from the Izmit, Turkey, earthquakes in 1999. These quakes killed more than 17,000 people because they took place near a heavily populated area where many buildings were poorly constructed.

One reason that the Mexico City and Izmit earthquakes caused such great loss of life was that so many people were living in the affected areas. If either of these quakes had occurred in less densely populated areas, fewer deaths would have resulted.

HUMAN POPULATION GROWTH

The world's population has more than tripled in the past 70 years. Between 1830 and 1930, the world's population doubled from 1 to 2 billion people. By 1970, it had nearly doubled again, and by the year 2000 there were about 6 billion people on Earth (see Case Study 1.1). This rapid increase in population is sometimes called the *population bomb*, because the exponential growth of the human population results in an explosive increase in numbers (Figure 1.11). *Exponential growth* means that the population grows each year not by the addition of a constant number of people; rather, it grows by the addition of a constant percentage of the current population. However, it is the exponential growth itself that is hazardous to our survival. With it comes more exposure to hazardous natural processes, increased pollution, reduced availability of food and clean drinking water, and a greater need for waste disposal. The question is: Will our planet even be able to support that many people?

There is no easy answer to the population problem, but the role of education is paramount. As people (particularly women, worldwide) become more educated, the population growth rate tends to decrease. As the rate of literacy increases, population growth is reduced. Given the variety of cultures, values, and norms in the world today, it appears that our greatest hope for population control is, in fact, through education.[13]

MAGNITUDE AND FREQUENCY OF HAZARDOUS EVENTS

The *impact* of a hazardous event is in part a function of the amount of energy released, that is, its **magnitude**, and the interval between occurrences, that is, its **frequency**. Its impact is also influenced by many other factors, including climate, geology, vegetation, population, and land use. In general, the frequency of an event is inversely related to the magnitude. Small earthquakes, for example, are more common than large ones (Case Study 1.2). A large event, such as a massive forest fire, will do far more damage than a small, contained burn. However, such events are much less frequent than smaller burns. Therefore, although planners need to be prepared for large, devastating events, the majority of fires suppressed will be smaller ones.

Land use may directly affect the magnitude and frequency of events. Take the Mississippi River, for example. For years we have tried to reduce the threat of floods by building levees along the river. However, the effect of levees has been to constrict the width of the river, and the reduced width creates a bottleneck that raises the height of floodwaters upstream. In effect, our efforts to reduce floods may actually be *causing* larger, more frequent floods upstream.

Four of the deadliest catastrophes resulting from natural hazards in recent years were Hurricane Mitch and the flooding of the Yangtze River in China, both in 1998; the Indonesian tsunami in 2004 that killed about 230,000 people; and Hurricane Katrina in 2005 that killed at least 1600 Americans. Hurricane Mitch, which devastated Central America, caused approximately 11,000 deaths, whereas the floods in the Yangtze River resulted in nearly 4000 deaths. Land-use changes made the damage from these events particularly severe. For example, Honduras has lost nearly one-half of its forests, and an 11,000 km^2 (4200 mi.^{2}) fire occurred in the region prior to the hurricane. As a result of deforestation and the fire, hillsides washed away and with

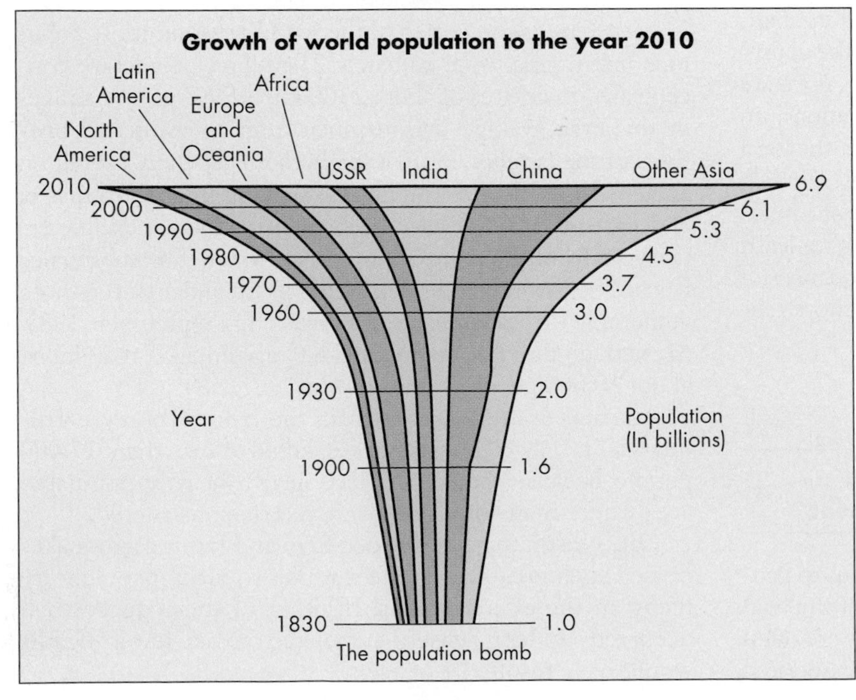

FIGURE 1.11 THE POPULATION BOMB The world population in 2007 was about 6.7 billion. *(Modified after U.S. Department of State)*

Human Population through History

The increase in the number of people on our planet can be related to various stages of human development (Table 1.4). When we were hunter-gatherers, our numbers were small and growth rates were low. With the development of agriculture, population growth rates increased by several hundred times as the result of a stable food supply. During the early industrial period (A.D. 1600 to 1800), growth rates increased again by about 10 times. Since the Industrial Revolution, with modern sanitation and medicine, the growth rates increased another 10 times. Human population reached 6 billion in 2000, and by 2013 it will be 7 billion. That is 1 billion new people in only 13 years. By comparison, total human population reached 1 billion only about A.D. 1800, after more than 40,000 years of human history!

Population Growth and the Future

Because Earth's population is increasing exponentially, many scientists are concerned that in the twenty-first century, it will be impossible to supply resources and a high-quality environment for the billions of people added to the world population. Increasing population at local, regional, and global levels compounds nearly all hazards, including floods, landslides, volcanic eruptions, and earthquakes.

In the future, we may be able to mass-produce enough food from nearly landless agriculture or to use artificial growing situations. However, enough food does not solve the problems of the space available to people and maintaining or improving their quality of life. Some studies suggest that the present population is already above a comfortable carrying capacity for the planet. *Carrying capacity* is the maximum number of people Earth can hold without causing environmental degradation that reduces the ability of the planet to support the population.[13]

Most of the population growth in the twenty-first century will be in developing nations. India will likely have the greatest population of all countries by 2050, which will be about 18 percent of the total world population; China will have about 15 percent. By 2050, these two countries will have more than a third of the total world population![14]

The news regarding human population growth is not all bad—for the first time in the past 50 years the rate of increase in human population is decreasing. Growth may have peaked at 85 million people per year in the late 1980s, and by 2001, the increase was reduced to 74 million new people per year. This reduction is a milestone in human population growth and is encouraging.[15] From an optimistic point of view, it is possible that our global population of 6 billion persons in 2000 may not double again. Although population growth is difficult to project because of variables such as agriculture, sanitation, medicine, culture, and education, by the year 2050, human population is forecast to be between 7.3 and 10.7 billion. Population reduction is most likely related to education of women, the decision to marry later in life, and the availability of modern birth-control methods. However, until the growth rate is zero, population will continue to grow. If the rate of growth is reduced to 0.7 percent per year, that is, one-half the current rate of 1.4 percent, human population will still double in 100 years.

TABLE 1.4

How We Became 6 Billion+

40,000–9000 B.C.: HUNTERS AND GATHERERS

Population density about 1 person per 100 km^2 of habitable areas[a]; total population probably less than a few million; average annual growth rate less than 0.0001 percent (doubling time about 700,000 years)

9000 B.C.–A.D. 1600: PREINDUSTRIAL AGRICULTURAL

Population density about 1 person per 3 km^2 of habitable areas (about 300 times that of the hunter and gatherer period); total population about 500 million; average annual growth rate about 0.03 percent (doubling time about 2300 years)

A.D. 1600–1800: EARLY INDUSTRIAL

Population density about 7 persons per 1 km^2 of habitable areas; total population by 1800 about 1 billion; annual growth rate about 0.1 percent (doubling time about 700 years)

A.D. 1800–2000: MODERN

Population density about 40 persons per 1 km^2; total population in 2000 about 6.1 billion; annual growth rate in 2000 about 1.4 percent (doubling time about 50 years)

[a] *Habitable area is assumed to be about 150 million square kilometers (58 million square miles).*
Source: Modified after Botkin, D. B. and Keller, E. A. 2000. Environmental science, *3rd ed. New York: John Wiley and Sons.*

CASE STUDY 1.2

The Magnitude-Frequency Concept

The *magnitude-frequency concept* asserts that there is generally an inverse relationship between the magnitude of an event and its frequency. In other words, the larger the flood, the less frequently such a flood occurs. The concept also includes the idea that much of the work of forming the Earth's surface is done by events of moderate magnitude and frequency, rather than by common processes with low magnitude and high frequency or extreme events of high magnitude and low frequency.

As an analogy to the magnitude-frequency concept, consider the work of logging a forest done by resident termites, human loggers, and elephants. The termites are numerous and work quite steadily, but they are so small that they can never do enough work to destroy all the trees. The people are fewer and work less often, but being stronger than termites they can accomplish more work in a given time. Unlike the termites, the people can eventually fell most of the trees (Figure 1.A). The elephants are stronger still and can knock down many trees in a short time, but there are only a few of them and they rarely visit the forest. In the long run, the elephants do less work than the people and bring about less change.

In our analogy it is humans who, with a moderate expenditure of energy and time, do the most work and change the forest most drastically. Similarly, natural events with a moderate energy expenditure and moderate frequency are often the most important shapers of the landscape. For example, most of the sediment carried by rivers in regions with a subhumid climate, like most of the eastern United States, is transported by flows of moderate magnitude and frequency. However, there are many exceptions. In arid regions, for example, much of the sediment in normally dry channels may be transported by rare high-magnitude flows produced by intense, but infrequent rainstorms. Along the barrier-island coasts of the eastern United States, high-magnitude storms often cut inlets that cause major changes in the pattern and flow of sediment.

▲ FIGURE 1.A **HUMAN SCALE OF CHANGE** Human beings, with our high technology, are able to down even the largest trees in our old-growth forests. The lumberjack shown here is working in a national forest in the Pacific Northwest. *(William Campbell/Corbis)*

them went farms, homes, roads, and bridges. In central China, the story is much the same; in recent years, the Yangtze River basin has lost about 85 percent of its forest as a result of timber harvesting and conversion of land to agriculture. These land-use changes have probably made flooding on the Yangtze River much more common than it was previously.[16]

The Indonesian tsunami in 2004 killed so many people in part because of the tremendous recent increase in human population of coastal areas and development of coastal tourism. For the first time, thousands of tourists were among the casualties (see Chapter 4).

Damages from Hurricane Katrina in 2005 were extensive in Mississippi and Louisiana where coastal development has increased in recent decades. In these areas, wetlands that could buffer storm waves have been damaged or removed by human activity and development (see Chapters 8 and 10).

These and other catastrophes caused by natural processes may be an early warning of things to come. It is apparent that human activities are likely increasing the effects of natural disasters. In recognition of this, China has banned timber harvesting in the upper Yangtze River basin, prohibited unwise use of the Yangtze floodplain, and has allocated several billion dollars for reforestation. The lesson being learned is that if we wish to minimize damages from natural hazards, we need to rehabilitate land with the goal of achieving sustainable development. This approach will ensure that future generations have equal access to the resources that our planet offers.[16] In many parts of the world, this goal will be difficult to achieve, given the pressure of human population growth.

CONCEPT 5

Consequences of Hazards Can Be Minimized

The ways in which we deal with hazards are too often primarily *reactive*—following a disaster we engage in search and rescue, firefighting, and providing emergency food, water, and shelter. There is no denying that these activities reduce loss of life and property and need to be continued. However, a move to a higher level of hazard reduction will require increased efforts to *anticipate* disasters and their effects. Land-use planning that limits construction in hazardous locations, hazard-resistant construction, and hazard modification or control (such as flood control channels) are some of the adjustments that anticipate future disastrous events and may reduce our vulnerability to them.[8]

REACTIVE RESPONSE: IMPACT OF AND RECOVERY FROM DISASTERS

The effect of a disaster upon a population may be either direct or indirect. *Direct effects* include people killed, injured, dislocated, or otherwise damaged by a particular event. *Indirect effects* are generally responses to the disaster. They include emotional distress, donation of money or goods, and the paying of taxes levied to finance the recovery. Direct effects are felt by fewer individuals, whereas indirect effects affect many more people.[17,18]

The stages of recovery following a disaster are emergency work, restoration of services and communication lines, and reconstruction (Figure 1.12). We can see these stages in recovery activities following the 1994 Northridge earthquake in the Los Angeles area. Restoration began almost immediately with the repair of roads and utilities using funds from federal programs, insurance companies, and other sources that arrived in the first few weeks and months after the earthquake. The damaged areas then moved quickly from the restoration phase to the reconstruction I period, which lasted until the year 2000.

As we now move through the reconstruction II period, it is important to remember lessons from two past disasters: the 1964 Anchorage, Alaska, earthquake and the flash flood that devastated Rapid City, South Dakota, in 1972. Restoration following the Anchorage earthquake began almost immediately in response to a tremendous influx of dollars from federal programs, insurance companies, and other sources approximately one month after the earthquake. Reconstruction moved rapidly as everyone tried to obtain as much of the available funds as possible. In Rapid City, the restoration did not peak until approximately 10 weeks after the flood, and the community took time to carefully think through the best alternatives. As a result, Rapid City today uses land on the floodplain in an entirely different way, and the flood hazard is much reduced. Conversely, in Anchorage the rapid restoration and reconstruction were accompanied by little land-use planning.

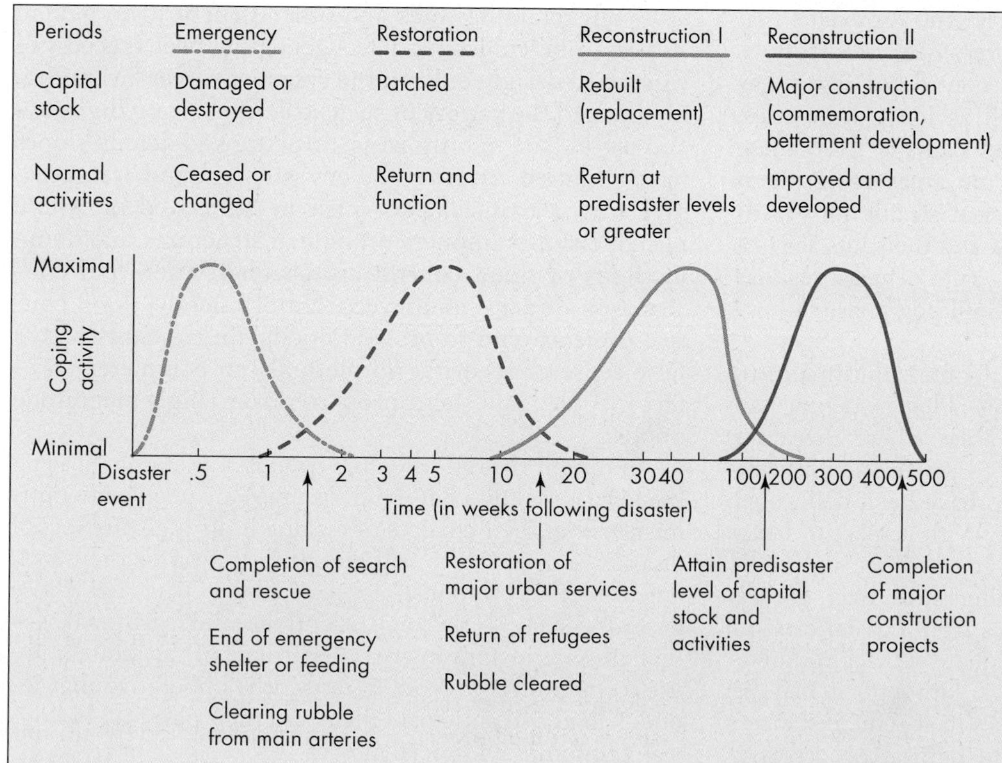

FIGURE 1.12 RECOVERY FROM DISASTER Generalized model of recovery following a disaster. The first 2 weeks after a disaster are spent in a state of emergency, in which normal activities cease or are changed. During the following 19 weeks, in the restoration phase, normal activities return and function, although perhaps not at predisaster levels. Finally, during reconstruction II, major construction and development are under way, and normal activities have returned. *(From Kates, R. W., and Pijawka, D. 1977. From rubble to monument: The pace of reconstruction. In* Disaster and reconstruction, *ed. J. E. Haas, R. W. Kates, and M. J. Bowden, pp. 1–23. Cambridge, MA: MIT Press)*

Apartments and other buildings were hurriedly constructed across areas that had suffered ground rupture; the ground was prepared by simply filling in the cracks and regrading the land surface. By ignoring the potential benefits of careful land-use planning, Anchorage has made itself vulnerable to the same type of earthquake that struck in 1964. In Rapid City, the floodplain is now a greenbelt with golf courses and other such facilities—a change that has reduced the flood hazard.[10,17,18]

In the Northridge case, the effects of the earthquake on highway overpasses and bridges, buildings, and other structures have been carefully evaluated. The goal has been to determine how more vigorous engineering standards for construction of new structures or strengthening of older structures can be implemented during the reconstruction II period (Figure 1.12). Future moderate to large earthquakes are certain to occur again in the Los Angeles area. Therefore, we need to continue efforts to reduce earthquake hazards.

ANTICIPATORY RESPONSE: AVOIDING AND ADJUSTING TO HAZARDS

The options we choose, individually or as a society, for avoiding or minimizing the effects of disasters depend in part on our perception of hazards. A good deal of work has been done in recent years to try to understand how people perceive various natural hazards. This understanding is important because the success of hazard reduction programs depends on the attitudes of the people likely to be affected by the hazard. Although there may be adequate perception of a hazard at the institutional level, this perception may not filter down to the general population. This lack of awareness is particularly true for events that occur infrequently; people are more aware of situations such as brush or forest fires that may occur every few years. Standard procedures, as well as local ordinances, may already be in place to control damage from these events. For example, homes in some areas of southern California are roofed with shingles that do not burn readily, they may have sprinkler systems, and their lots are frequently cleared of brush. Such safety measures are commonly noticeable during the rebuilding phase following a fire.

One of the most environmentally sound adjustments to hazards involves **land-use planning**. That is, people can avoid building on floodplains, in areas where there are active landslides, or in places where coastal erosion is likely to occur. In many cities, floodplains have been delineated and zoned for a particular land use. With respect to landslides, legal requirements for soil engineering and engineering geology studies at building sites may greatly reduce potential damages. Damages from coastal erosion can be minimized by requiring adequate setback of buildings from the shoreline or sea cliff. Although it may be possible to control physical processes in specific instances, land-use planning to accommodate natural processes is often preferable to a technological fix that may or may not work.

Insurance is another option that people may exercise in dealing with natural hazards. Flood insurance is common in many areas, and earthquake insurance is also available. Just because insurance is available, however, doesn't mean it is practical (or ethical) to build in an earthquake-prone area without careful engineering, knowing that you will most likely have to use your insurance at some point. In fact, because of large losses following the 1994 Northridge earthquake, several insurance companies announced they would no longer offer the insurance.

Evacuation is an important option or adjustment to the hurricane hazard in the states along the Gulf of Mexico and along the eastern coast of the United States. Often there is sufficient time for people to evacuate provided they heed the predictions and warnings. However, if people do not react quickly and the affected area is a large urban region, then evacuation routes may be blocked by residents leaving in a last-minute panic.

Disaster preparedness is an option that individuals, families, cities, states, or even entire nations can implement. Of particular importance is the training of individuals and institutions to handle large numbers of injured people or people attempting to evacuate an area after a warning is issued.

Attempts at *artificial control of natural processes* such as landslides, floods, and lava flows have had mixed success. Seawalls constructed to control coastal erosion may protect property to some extent, but over a period of decades they tend to narrow or even eliminate the beach. Even the best-designed artificial structures cannot be expected to adequately defend against an extreme event, although retaining walls and other structures to defend slopes from landslides have generally been successful when well designed. Even the casual observer has probably noticed the variety of such structures along highways and urban land in hilly areas. Structures to defend slopes have a limited effect on the environment and are necessary where artificial cuts must be excavated or where unstable slopes impinge on human structures. Common methods of flood control are channelization and construction of dams and levees. Unfortunately, flood control projects tend to provide floodplain residents with a false sense of security; no method can completely protect people and their property from high-magnitude floods.

An option that all too often is chosen is simply bearing the loss caused by a natural disaster. Many people are optimistic about their chances of making it through any sort of disaster and therefore will take little action in their own defense. This response is particularly true for those hazards—such as volcanic eruptions and earthquakes—that are rare in a given area. Regardless of the strategy we use to minimize or avoid hazards, it is imperative that we understand and anticipate them and their physical, biological, economic, and social effects.

1.5 Many Hazards Provide a Natural Service Function

It is ironic that the same natural events that take human life and destroy property also provide us with important benefits, sometimes referred to as *natural service functions*. For example, periodic flooding of the Mississippi River supplies nutrients to the floodplain and creates the fertile soils used for farming. Flooding, which causes erosion on mountain slopes, also delivers sediment to beaches and flushes pollutants from estuaries in the coastal environment. Landslides may bring benefits to people when their debris forms dams, creating lakes in mountainous areas. Although some landslide dams will collapse and cause hazardous downstream flooding, if a dam remains stable it can provide valuable water storage and an important aesthetic resource. Volcanic eruptions have the potential to produce real catastrophes; however, they also provide us with numerous benefits. They often create new land, as in the case of the Hawaiian Islands, which are completely volcanic in origin (Figure 1.13). Nutrient-rich volcanic ash may settle on existing soils and quickly become incorporated, creating soil suitable for both crops and wild plants. Earthquakes can also provide us with valuable services. When rocks are pulverized during an earthquake, they may form an impervious clay zone known as *fault gouge* along a geological fault. In many places, fault gouge has formed natural subsurface barriers to groundwater flow; these barriers then pool water or create springs that are important sources of water. In addition, most of the hydrocarbon traps that contain the worlds petroleum resources are produced by deformation along faults that create large zones where oil and gas accumulate, without which the industrial revolution would never have happened (Figure 1.14). Finally, earthquakes are important in mountain building and thus are directly responsible for many of the scenic landscapes of the western United States.

(a)

(b)

▲ **FIGURE 1.13 NEW LAND FROM VOLCANIC ERUPTION** New land being added to the island of Hawai'i. (a) The cloud of steam and acidic gases in the central part of the photograph is where hot lava is entering the sea. *(Edward A. Keller)* (b) Close-up of an advancing lava front near the gas cloud. *(Edward A. Keller)*

◀ **FIGURE 1.14 HYDROCARBON TRAP** Faulting in the Santa Barbara Channel has produced a linear fault trap, along which oil rigs at the surface are situated above. *(Nik Wheeler/Corbis)*

1.6 Global Climate Change and Hazards

Global and regional climatic change, often associated with global warming, is likely to affect the incidence of hazardous natural events such as storms, landslides, drought, and fires. How might climatic change affect the magnitude and frequency of these events? With global warming, sea level is rising as the heating of seawater expands the volume of the ocean, and glacial ice is melting. As a result, coastal erosion will increase. Climate change may shift food production areas as some receive more precipitation and others receive less. Deserts and semiarid areas will likely expand, and warmer northern latitudes could become more productive. Such changes could lead to global population shifts, which might bring about wars or major social and political upheavals.

Global warming will feed more energy from warmer ocean water into the atmosphere; this energy is likely to increase the severity and frequency of hazardous weather such as thunderstorms and hurricanes. In fact, this trend may already be underway—a recent analysis of global extreme-weather events by a United Nations panel indicates that since the 1950s there has been an increase in heavy precipitation events in mid-latitude regions and, since the 1970s, a likely increase in the intensity of hurricanes.[19]

Summary

Natural hazards are responsible for causing significant death and damage worldwide each year. Processes that cause hazardous events include those that are internal to the Earth, such as volcanic eruptions and earthquakes that result from Earth's internal heat, and those that are external to the Earth, such as hurricanes and global warming, which are driven by energy from the sun.

Natural processes may become hazards, disasters, or catastrophes when they interact with human beings. Central to an understanding of natural hazards is awareness that hazardous events result from natural processes that have been in operation for millions and possibly billions of years before humans experienced them. These processes become hazards when they threaten human life or property and should be recognized and avoided.

Hazards involve repetitive events. Thus a study of the history of these events provides much-needed information for hazard reduction. A better understanding and more accurate prediction of natural processes come by integrating historic and prehistoric information, present conditions, and recent past events, including land-use changes.

Geologic conditions and materials largely govern the type, location, and intensity of natural processes. The geologic cycle creates, maintains, and destroys earth materials by physical, chemical, and biological processes. Subcycles of the geologic cycle are the tectonic cycle, rock cycle, hydrologic cycle, and various biogeochemical cycles. The tectonic cycle describes large-scale geologic processes that deform Earth's crust, producing landforms such as ocean basins, continents, and mountains. The rock cycle may be considered a worldwide earth-material recycling process driven by Earth's internal heat, which melts the rocks subducted in the tectonic cycle. Driven by solar energy, the hydrologic cycle operates by way of evaporation, precipitation, surface runoff, and subsurface flow. Biogeochemical cycles can most easily be described as the transfer of chemical elements through a series of storage compartments or reservoirs, such as air or vegetation.

Five fundamental concepts establish a philosophical framework for studying natural hazards:

1. Hazards are predictable from scientific evaluation.
2. Risk analysis is an important component in our understanding of the effects of hazardous processes.
3. Linkages exist between different natural hazards as well as between hazards and the physical environment.
4. Hazardous events that previously produced disasters are now producing catastrophes.
5. Consequences of hazards can be minimized.

Key Terms

biogeochemical cycle (p. 9)
catastrophe (p. 3)
disaster (p. 3)
environmental unity (p. 12)
forecast (p. 12)
frequency (p. 16)
geologic cycle (p. 7)
hydrologic cycle (p. 8)
land-use planning (p. 20)
magnitude (p. 16)
mitigation (p. 5)
natural hazard (p. 3)
prediction (p. 12)
risk (p. 14)
rock cycle (p. 7)
tectonic cycle (p. 7)
uniformitarianism (p. 12)
warning (p. 13)

Review Questions

1. What forces drive internal and external earth processes?
2. What is the distinction between a natural hazard, disaster, and catastrophe?
3. Which natural hazards are likely to be more deadly, more likely to cause property damage, and more likely to become catastrophes?
4. Explain why the effects of natural hazards are not constant over time.
5. Why is history so important in understanding natural hazards?
6. What kinds of information must be assembled to make hazard predictions?
7. Describe the components and interactions involved in the geologic cycle.
8. What are the five fundamental concepts for understanding natural processes as hazards?
9. Explain the scientific method as it is applied to natural hazards.
10. Explain why calling something a "natural" hazard may act as a philosophical barrier to dealing with it.
11. What are the elements involved in making a hazard forecast and warning?
12. Explain why two 10-year floods might occur in the same year.
13. What is a precursor event? Give some examples.
14. Explain the magnitude-frequency concept.
15. How do risk and acceptable risk differ?
16. Explain how population growth increases the number of disasters and catastrophes.
17. Describe the differences between direct and indirect effects of disasters.
18. What are the stages of disaster recovery? How do they differ?
19. Describe four common adjustments to natural hazards.
20. What are natural service functions?

Critical Thinking Questions

1. How would you use the scientific method to test the hypothesis that sand on the beach comes from the nearby mountains?
2. It has been argued that we must control human population because otherwise we won't be able to feed everyone. Even if we could feed 10 to 15 billion people, would we still want a smaller population? Why or why not?
3. Considering that events we call natural hazards are natural processes that have been occurring on Earth for millions of years, how do you think we should go about trying to prevent loss of life from these events? Think about the choices society has, from attempting to control and prevent hazards to attempting to keep people out of harm's way.
4. Global warming is a major concern today. Discuss how global warming might influence the magnitude or frequency of hazardous events, disasters or catastrophes caused by natural hazards.

2 Internal Structure of Earth and Plate Tectonics

Written with the assistance of Tanya Atwater

Case History

Two Cities on a Plate Boundary

California straddles the boundary between two tectonic plates, which are discussed in detail in this chapter. That boundary between the North American and Pacific plates is the notorious San Andreas fault (Figure 2.1). A *fault* is a fracture along which one plate has moved relative to the other, and the San Andreas fault is a huge zone of fracturing hundreds of kilometers long. Two major cities, Los Angeles to the south and San Francisco to the north, are located on opposite sides of this fault. San Francisco was nearly destroyed by a major earthquake in 1906, which led to the identification of the fault. Many of the moderate to large earthquakes in the Los Angeles area are on faults related to the San Andreas fault system. Most of the beautiful mountain topography in coastal California near both Los Angeles and San Francisco is a direct result of processes related to movement on the San Andreas fault. However, this beautiful topography comes at a high cost to society. Since 1906, earthquakes on the San Andreas fault system or on nearby faults, undoubtedly influenced by the plate boundary, have cost hundreds of lives and many billions of dollars in property damage. Construction of buildings, bridges, and other structures in California is more expensive than elsewhere because they must be designed to withstand ground shaking caused by earthquakes. Older structures have to be *retrofitted,* or have changes made to their structure, to withstand the shaking, and many people purchase earthquake insurance in an attempt to protect themselves from the "big one."

In about 20 million years the cities will be side by side

Los Angeles is on the Pacific plate and is slowly moving toward San Francisco, which is on the North American plate. In about 20 million years, the cities will be side by side. If people are present, they might be arguing over which is a sub urb of the other. Of course, there will still be a plate boundary between the Pacific and North American plates 20 million years from now, because large plates have long geologic lives, on the order of 100 million years. However, the boundary may not be the San Andreas fault. The plate boundary will probably have moved eastward, and the topography of what is now California may be somewhat different. In fact, some recent earthquake activity in California, such as the large 1992 Landers earthquake, east of the San Andreas fault, may be the beginning of a shift in the plate boundary.

LEARNING OBJECTIVES

The surface of Earth would be much different—relatively smooth, with monotonous topography—if not for the active tectonic processes within Earth that produce earthquakes, volcanoes, mountain chains, continents, and ocean basins.[1] In this chapter, we focus directly on the interior of Earth. Your goals in reading this chapter should be to

- understand the basic internal structure and processes of Earth
- know the basic ideas behind and evidence for the theory of plate tectonics
- understand the mechanisms of plate tectonics
- understand the relationship of plate tectonics to natural hazards

◀ **European Alps** Satellite photo of the Bernese Alps, which is only a small part of the European Alps. The European Alps are as wide as 200 km (130 mi.), are 725 km (450 mi.) long, and reach an elevation in excess of 4000 m (15,000 ft.). The mountain range formed when the African tectonic plate collided with the Eurasian plate between 30 and 5 million years ago. Since that time, the Alps have been deeply eroded by rivers and glaciers. *(NASA)*

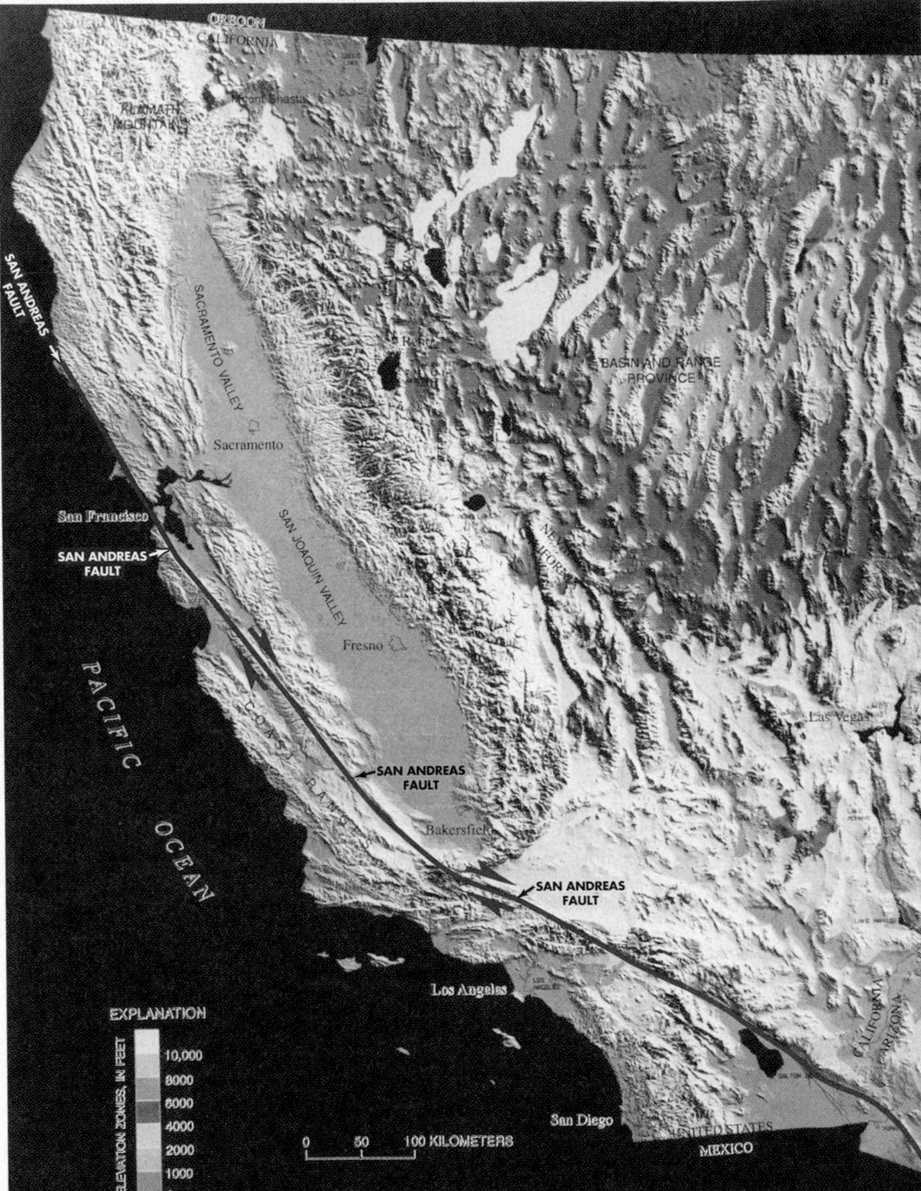

◀ **FIGURE 2.1 SAN ANDREAS FAULT** False color topographic map of the Western United States showing the location of the San Andreas fault in California. Red arrows show relative plate motions on either side of USGS the fault. *(R. E. Wallace/National Earthquake Information Center. U.S.G.S.)*

2.1 Internal Structure of Earth

You may be familiar with the situation comedy *Third Rock from the Sun,* a phrase that refers to our planet Earth. Far from being a barren rock, Earth is a complex dynamic planet that in some ways resembles a chocolate-covered cherry. That is, Earth has a rigid outer shell, a solid center, and a thick layer of liquid that moves around as a result of dynamic internal processes. The internal processes are incredibly important in affecting the surface of Earth. They are responsible for the largest landforms on the surface: continents and ocean basins. The configuration of the continents and ocean basins in part controls the oceans' currents and the distribution of heat carried by seawater in a global system that affects climate, weather, and the distribution of plant and animal life on Earth. Finally, Earth's internal processes are also responsible for regional landforms including mountain chains, chains of active volcanoes, and large areas of elevated topography, such as the Tibetan Plateau and the Rocky Mountains. The high topography that includes mountains and plateaus significantly affects both global circulation patterns of air in the lower atmosphere and climate, thereby directly influencing all life on Earth. Thus, our understanding of the internal processes of Earth is of much more than simply academic interest. These processes are at the heart of producing the multitude of environments shared by all living things on Earth.

The Earth Is Layered and Dynamic Earth (Figure 2.2a) has a radius of about 6,300 km (4,000 mi.) (Figure 2.2b). Information regarding the internal layers of the Earth is shown in Figure 2.2b. We can consider the internal structure of Earth in two fundamental ways:

(a)

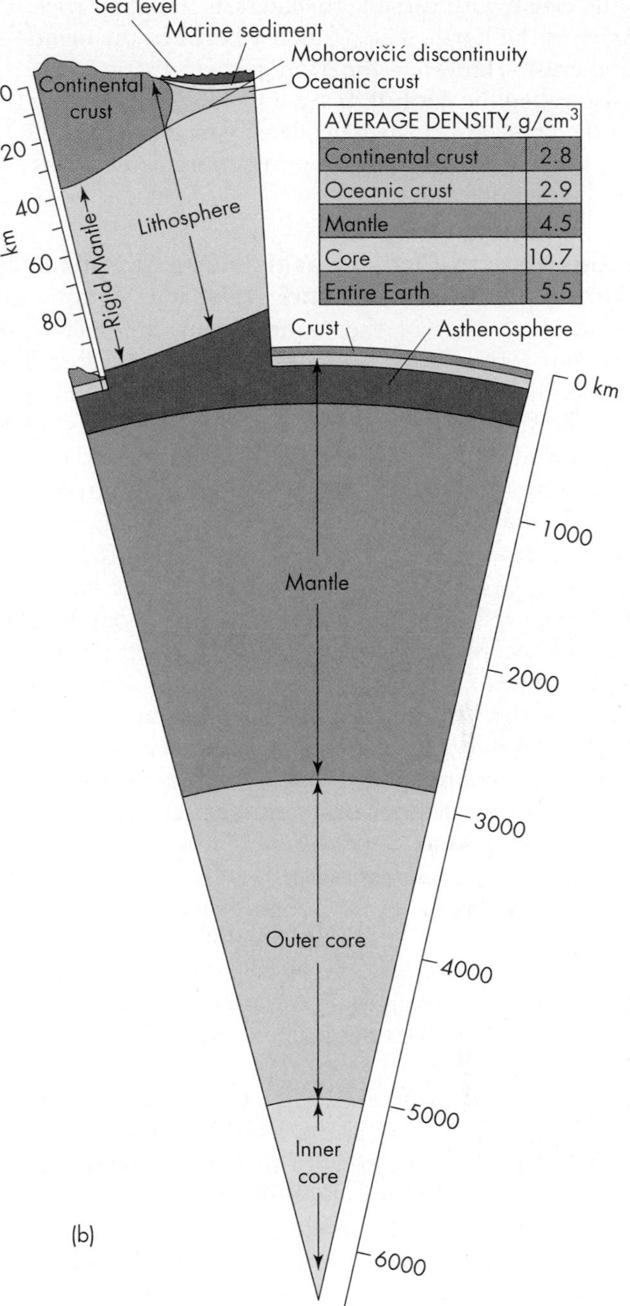

(b)

▲ **FIGURE 2.2 EARTH AND ITS INTERIOR** (a) False color topographic image of the Earth. The different colors reflect the elevation of Earth's surface with red being the highest and blue being the lowest. *(National Geophysical Data Center, NOAA)* (b) Idealized diagram showing the internal structure of Earth and its layers extending from the center to the surface. Notice that the lithosphere includes the crust and part of the mantle, and the asthenosphere is located entirely within the mantle. Properties of the various layers have been estimated on the basis of (1) interpretation of geophysical data (primarily seismic waves from earthquakes); (2) examination of rocks thought to have risen from below by tectonic processes; and (3) meteorites, thought to be pieces of an old Earthlike planet. *(From Levin, H. L. 1986.* Contemporary physical geology, *2nd ed. Philadelphia: Saunders)*

- by composition and density (heavy or light)
- by physical properties (e.g., solid or liquid, weak or strong)

Our discussion will explore the two ways of looking at the interior of our planet. Some of the components of the basic structure of Earth[1] are

- A solid **inner core** with a thickness of more than 1,300 km (808 mi) that is roughly the size of the moon but with a temperature about as high as the temperature of the surface of the sun.[2] The inner core is believed to be primarily metallic, composed mostly of iron (about 90 percent by weight), with minor amounts of elements such as sulfur, oxygen, and nickel.
- A liquid **outer core** with a thickness of just over 2000 km (1243 mi.) with a composition similar to that of the inner core. The average density of the inner and outer core is approximately 10.7 grams per cubic centimeter (0.39 pounds per cubic inch). The maximum near the center of Earth is about 13 g/cm^3 (0.47 lb./in.^{3}). By comparison, the density of water is 1 g/cm^3 (0.04 lb./in.^{3}) and the average density of Earth is approximately 5.5 g/cm^3 (0.2 lb./in.^{3}).
- The **mantle,** nearly 3000 km (1864 mi.) thick, surrounds the outer core and is composed mostly of solid iron- and magnesium-rich silicate rocks. The average density of the mantle is approximately 4.5 g/cm^3 (0.16 lb./in.^{3}), which is less than one-half that of the underlying core.

- The **crust,** with variable thickness, is the outer rock layer of the Earth. The boundary between the mantle and crust is known as the Mohorovičić discontinuity (also called the **Moho**). It separates the lighter rocks of the crust with an average density of approximately 2.8 g/cm^3 (0.10 lb./in.^{3}) from the more dense rocks of the mantle below.

Continents and Ocean Basins Have Significantly Different Properties and History Within the uppermost portion of the mantle, near the surface of Earth, our terminology becomes more complicated. For example, the cool, strong outermost layer of Earth is also called the **lithosphere** (*lithos* means "rock"). It is much stronger and more rigid than the material underlying it, the **asthenosphere** (*asthenos* means "without strength"), which is a hot and slowly flowing layer of relatively weak rock. The lithosphere averages about 100 km (62 mi.) in thickness, ranging from a few kilometers (1 to 2 mi.) thick beneath the crests of mid-ocean ridges to about 120 km (75 mi.) beneath ocean basins and 20 to 400 km (13 to 250 mi.) beneath the continents. The crust is embedded in the top of the lithosphere. Crustal rocks are less dense than the mantle rocks below, and oceanic crust is slightly denser than continental crust. Oceanic crust is also thinner: The ocean floor has a uniform crustal thickness of about 6 to 7 km (3.7 to 4.4 mi.), whereas the crustal thickness of continents averages about 35 km (22 mi.) and may be up to 70 km (44 mi.) thick beneath mountainous regions. Thus, the average crustal thickness is less than 1 percent of the total radius of Earth and can be compared to the thin skin of a tangerine. Yet it is this layer that is of particular interest to us because we live at the surface of the continental crust.

In addition to differences in density and thickness, continental and oceanic crusts have very different geologic histories. Oceanic crust of the present ocean basins is less than approximately 200 million years old, whereas continental crust may be several billion years old. Three thousand kilometers (1865 mi.) below us, at the core-mantle boundary, processes may be occurring that significantly affect our planet at the surface. It has been speculated that gigantic cycles of convection occur within Earth's mantle, rising from as deep as the core-mantle boundary up to the surface and then falling back again. The concept of **convection** is illustrated by heating a pan of hot water on a stove (Figure 2.3). Heating the water at the bottom of the pan causes the water to become less dense and more unstable, so it rises to the top. The rising water displaces denser, cooler water, which moves laterally and sinks to the bottom of the pan. It is suggested that Earth's layers contain convection cells and operate in a similar fashion.

A complete cycle in the mantle may take as long as 500 million years.[1] Mantle convection is fueled by Earth's internal heat, which results from the original heat of formation of the planet, heat generated by crystallization of the core, and heat supplied by radioactive decay of elements (such as uranium) scattered throughout the mantle. Let us now examine some of the observations and evidence that reveal the internal structure of Earth.

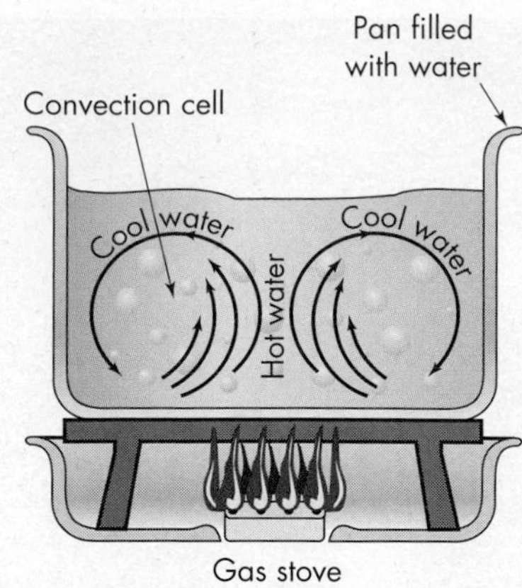

▲ FIGURE 2.3 **CONVECTION** Idealized diagram showing the concept of convection. As the pan of water is heated, the less dense hot water rises from the bottom to displace the denser cooler water at the top, which then sinks down to the bottom. This process of mass transport is called convection, and each circle of rising and falling water is a convection *cell.*

2.2 How We Know About the Internal Structure of Earth

What We Have Learned about Earth from Earthquakes Our knowledge concerning the structure of Earth's interior arises primarily from our study of seismology. **Seismology** is the study of earthquakes and the passage of seismic waves through Earth.[3] When a large earthquake occurs, seismic energy is released and seismic waves move both through Earth and along its surface. The properties of these waves are discussed in detail in Chapter 3 with earthquake hazards.

Some waves move through solid and liquid materials, whereas others move through solid, but not liquid materials. The rates at which seismic waves propagate (move) are on the order of a few kilometers per second (1 or 2 miles per second). Their actual velocity varies with the properties of the materials through which the waves are propagating (moving). When the seismic waves encounter a boundary, such as the mantle-core boundary, some of them are *reflected* back. Others cross the boundary and are *refracted* (change the direction of propagation). Still others fail to propagate through the liquid outer core (Figure 2.4). Thousands of seismographs (instruments that record seismic waves) are stationed around the world. When an earthquake occurs, the reflected and refracted waves are recorded when they emerge at the surface. Study of these waves has been a powerful tool for deducing the

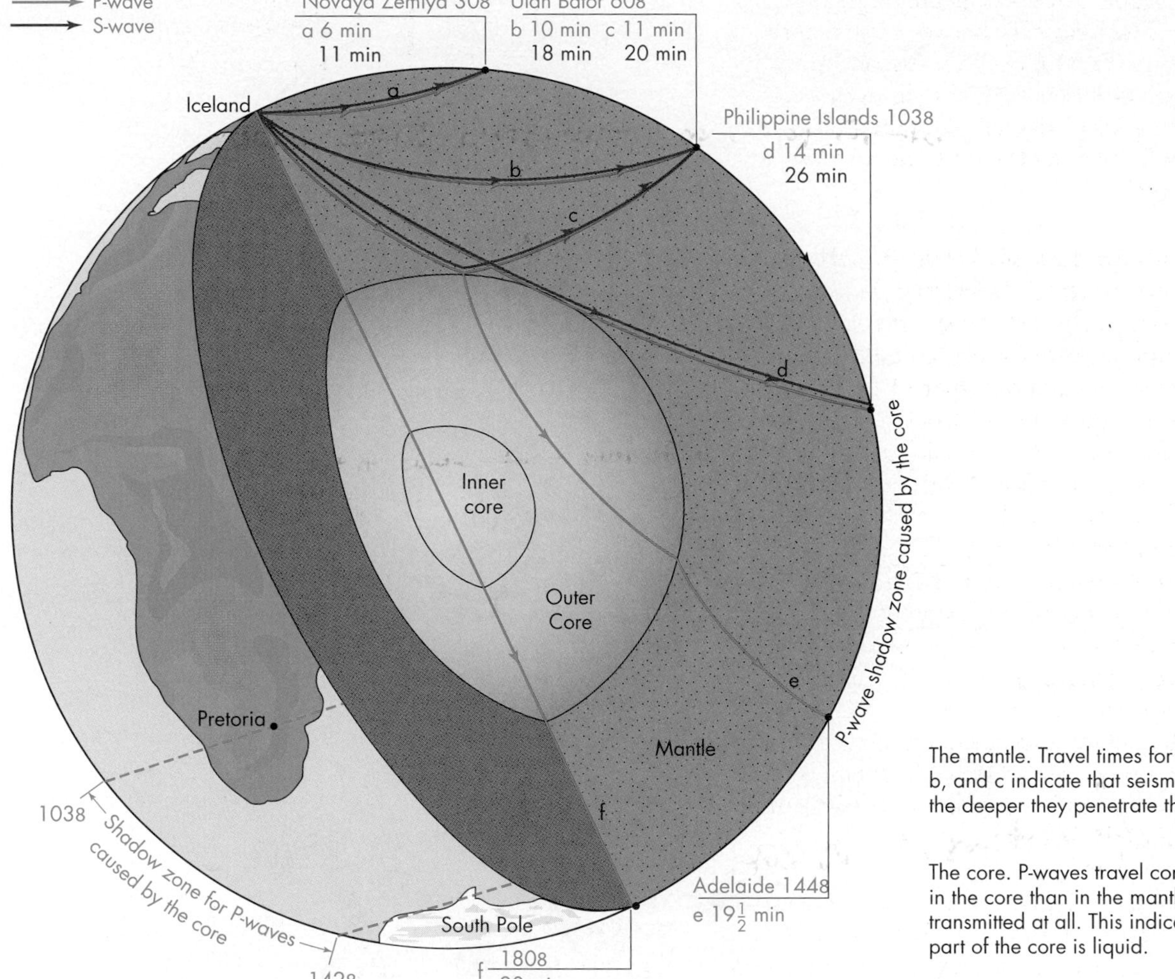

▲ **FIGURE 2.4 SEISMIC STRUCTURE OF EARTH** By studying the travel times of different seismic waves, which propagate at different rates and through different mediums, the physical properties and composition of Earth can be inferred.

layering of the interior of Earth and the properties of the materials found there (Figure 2.4).

In summary, the boundaries that delineate the internal structure of Earth are determined by studying seismic waves generated by earthquakes and recorded on seismographs around Earth. As seismology has become more sophisticated, we have learned more and more about the internal structure of Earth and are finding that the structure can be quite variable and complex. For example, we have been able to recognize

- where magma, which is molten rock material beneath Earth's surface, is generated in the asthenosphere
- the existence of slabs of lithosphere that have apparently sunk deep into the mantle
- the extreme variability of lithospheric thickness, reflecting its age and history

2.3 Plate Tectonics

The term *tectonics* refers to the large-scale geologic processes that deform Earth's lithosphere, producing landforms such as ocean basins, continents, and mountains. Tectonic processes are driven by forces within the Earth. These processes are part of the tectonic system, an important subsystem of the Earth system.

MOVEMENT OF THE LITHOSPHERIC PLATES

What Is Plate Tectonics? The lithosphere is broken into large pieces called *lithospheric plates* that move relative to one another (Figure 2.5a).[4] Processes associated with the creation, movement, and destruction of these plates are collectively known as **plate tectonics.**

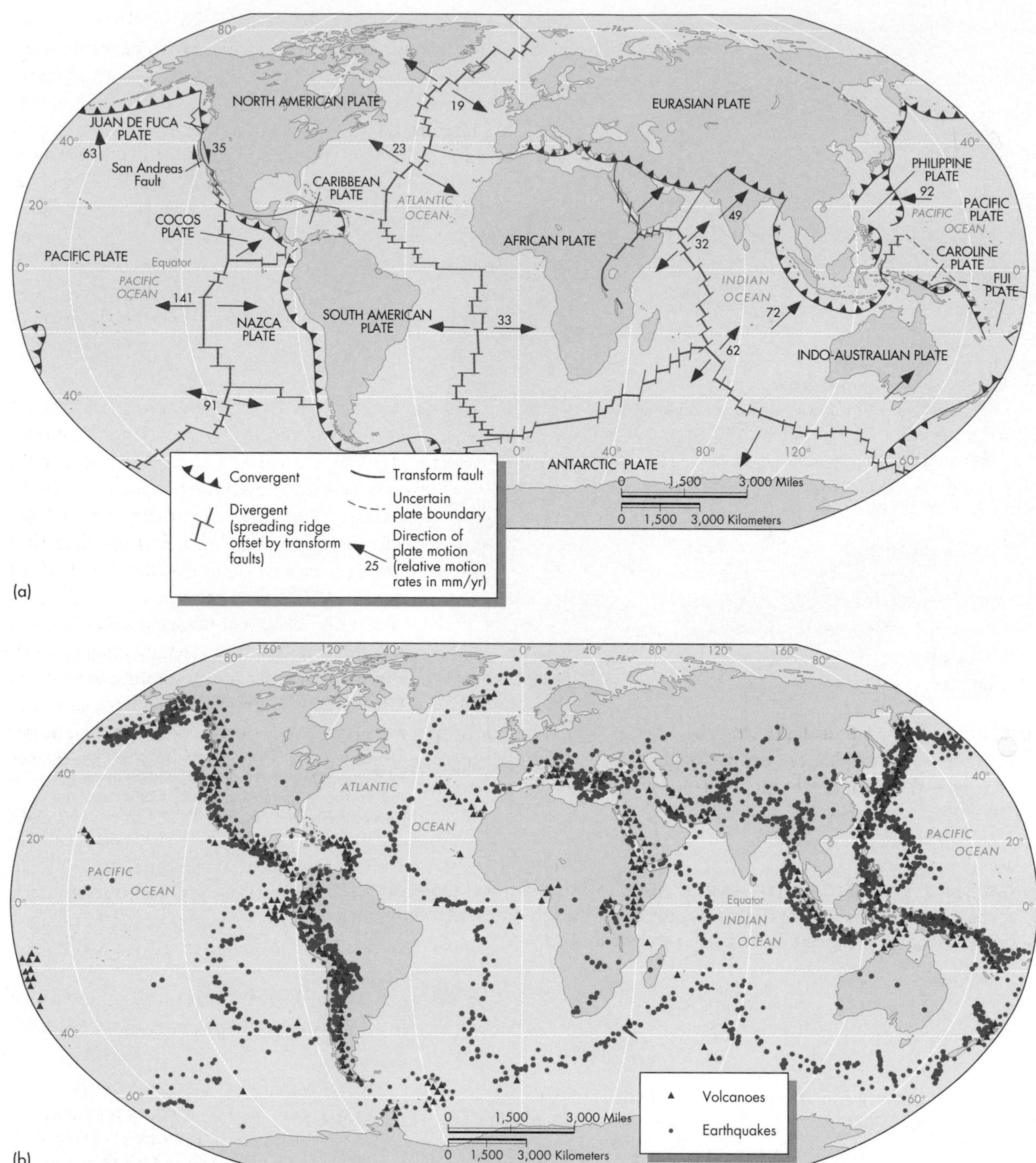

▲ **FIGURE 2.5 EARTH'S PLATES** (a) Map showing the major tectonic plates, plate boundaries, and direction of relative plate movement. *(Modified from Christopherson, R. W. 1994.* Geosystems, *2nd ed. Englewood Cliffs, NJ: Macmillan)* (b) Volcanoes and earthquakes: Map showing location of volcanoes and earthquakes. Notice the correspondence between this map and the plate boundaries. *(Modified after Hamblin, W. K. 1992. Earth's dynamic systems, 6th ed. New York: Macmillan)*

Locations of Earthquakes and Volcanoes Define Plate Boundaries A lithospheric plate may include both a continent and part of an ocean basin or an ocean region alone. Some plates are very large and some are relatively small, although they are significant on a regional scale. For example, the Juan de Fuca plate off the Pacific Northwest coast of the United States (Figure 2.5a), which is relatively small, is responsible for many of the earthquakes in northern California and all of the volcanoes in Oregon and Washington (Figure 2.5b). The boundaries between lithospheric plates are geologically active areas. Most earthquakes and many volcanoes are associated with these boundaries. In fact, plate

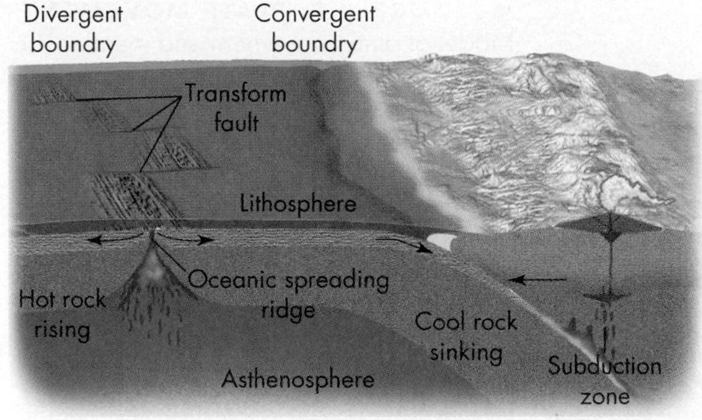

▲ **FIGURE 2.6 MODEL OF PLATE TECTONICS** Diagram of the model of plate tectonics. New oceanic lithosphere is being produced at the spreading ridge (divergent plate boundary). Elsewhere, oceanic lithosphere returns to the interior of Earth at a convergent plate boundary (subduction zone). *(Modified from Lutgens, F., and Tarbuck, E. 1992. Essentials of geology. New York: Macmillan)*

boundaries are defined by the areas in which concentrated seismic activity occurs (Figure 2.5b). Over geologic time, plates are formed and destroyed, cycling materials from the interior of Earth to the surface and back again at these boundaries (Figure 2.6). The continuous recycling of tectonic processes is collectively called the *tectonic cycle*.

Seafloor Spreading Is the Mechanism for Plate Tectonics As the lithospheric plates move over the asthenosphere, they carry the continents embedded within them.[5] The idea that continents move is not new; it was first suggested by German scientist Alfred Wegener in 1915. The evidence he presented for **continental drift** was based on the congruity of the shape of continents, particularly those across the Atlantic Ocean, and on the similarity in fossils found in South America and Africa. Wegener's hypothesis was not taken seriously because there was no known mechanism that could explain the movement of continents around Earth. The explanation came in the late 1960s, when **seafloor spreading** was discovered. In seafloor regions called **mid-oceanic ridges,** or **spreading centers,** new crust is continuously added to the edges of lithospheric plates (Figure 2.6, left). As oceanic lithosphere is added along some plate edges (spreading centers), it is destroyed along other plate edges, for example, at **subduction zones** (areas where one plate sinks beneath another and is destroyed) (Figure 2.6, right). Thus continents do not move *through* oceanic crust; rather they are *carried along with it* by the movement of the plates. Also, because the rate of production of new lithosphere at spreading centers is balanced by consumption at subduction zones, the size of Earth remains constant, neither growing nor shrinking.

Sinking Plates Generate Earthquakes The concept of a lithospheric plate sinking into the upper mantle is shown in diagrammatic form in Figure 2.6. When the wet, cold oceanic crust comes into contact with the hot asthenosphere, magma is generated (see Chapter 5). The magma rises back to the surface, producing volcanoes, such as those that ring the Pacific Ocean basin, over subduction zones. The path of the descending plate (or *slab*, as it sometimes is called) into the upper mantle is clearly marked by earthquakes. As the oceanic plate subducts, earthquakes are produced both between it and the overriding plate and within the interior of the subducting plate. The earthquakes occur because the sinking lithospheric plate is relatively cooler and stronger than the surrounding asthenosphere; this difference causes rocks to break and seismic energy to be released.[6]

The paths of descending plates at subduction zones may vary from a shallow dip to nearly vertical, as traced by the earthquakes in the slabs. These dipping planes of earthquakes are called **Wadati-Benioff zones** (Figure 2.7). The very existence of Wadati-Benioff zones is strong evidence that subduction of rigid "breakable" lithosphere is occurring.[6]

Plate Tectonics Is a Unifying Theory The theory of plate tectonics is to geology what Darwin's origin of species

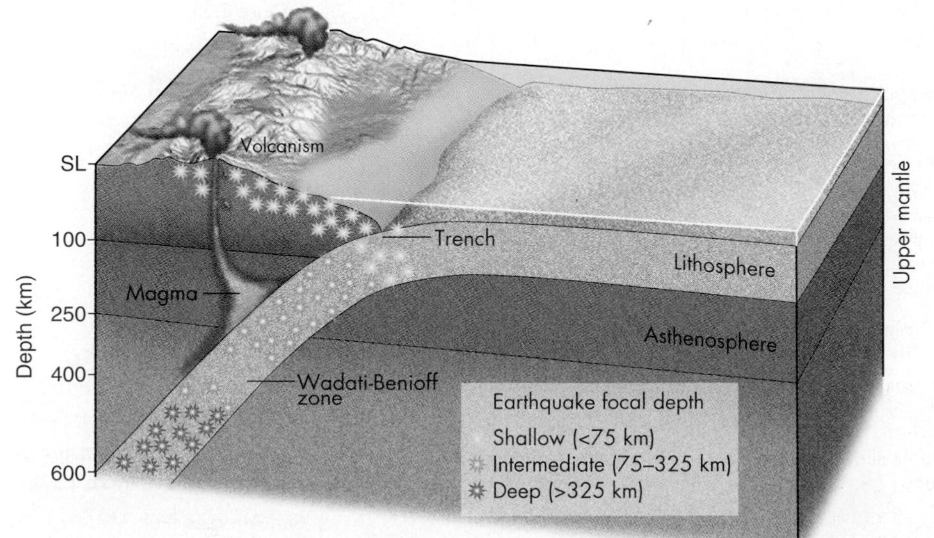

◀ **FIGURE 2.7 SUBDUCTION ZONE** Idealized diagram of a subduction zone showing the Wadati-Benioff zone, which is an array of earthquake foci from shallow to deep that delineates the subduction zone and the descending lithospheric plate.

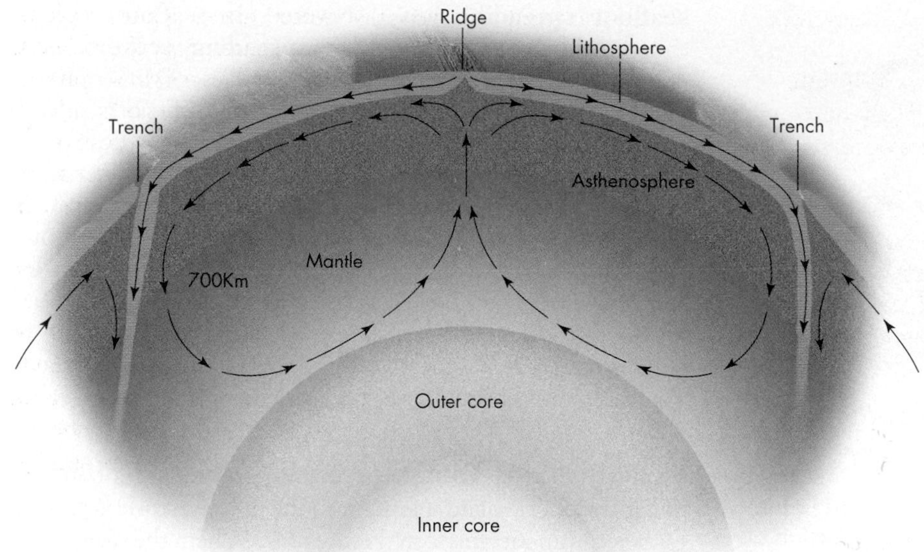

FIGURE 2.8 PLATE MOVEMENT Model of plate movement and mantle. The outer layer (or lithosphere) is approximately 100 km (approximately 62 mi.) thick and is stronger and more rigid than the deeper asthenosphere, which is a hot and slowly flowing layer of relatively low-strength rock. The oceanic ridge is a spreading center where plates pull apart, drawing hot, buoyant material into the gap. After these plates cool and become dense, they descend at oceanic trenches (subduction zones), completing the convection system. This process of spreading produces ocean basins, and mountain ranges often form where plates converge at subduction zones. A schematic diagram of Earth's layers is shown in Figure 2.2b. *(Grand, S. P. 1994. Mantle shear structure beneath the Americas and surrounding oceans.* Journal of Geophysical Research *99:11591–621. Modified after Hamblin, W. K. 1992.* Earth's dynamic systems, *6th ed. New York: Macmillan)*

is to biology: a unifying concept that explains an enormous variety of phenomena. Biologists now have an understanding of evolutionary change. In geology, we are still seeking the exact mechanism that drives plate tectonics, but we think it is most likely convection within Earth's mantle. As rocks are heated deep in Earth, they become less dense and rise. Hot materials, including magma, leak out, and are added to the surfaces of plates at spreading centers. As the rocks move laterally, they cool, eventually becoming dense enough to sink back into the mantle at subduction zones. This circulation is known as convection, which was introduced in Section 2.1. Figure 2.8 illustrates the cycles of convection that may drive plate tectonics.

TYPES OF PLATE BOUNDARIES

There are three basic types of plate boundaries: divergent, convergent, and transform (Figure 2.6 and Table 2.1). These boundaries are not narrow cracks as shown on maps and diagrams,[6] but are zones that range from a few to hundreds of

TABLE 2.1

Types of Plate Boundaries: Dynamics, Results, and Examples

Plate Boundary	Plates Involved	Dynamics	Results	Example
Divergent	Usually oceanic	Spreading. The two plates move away from each other and molten rock rises up to fill the gap.	Mid-ocean ridge forms and new material is added to each plate.	African and North American plate boundary (Figure 2.5a) Mid-Atlantic Ridge
Convergent	Ocean-continent	Oceanic plate sinks beneath continental plate.	Mountain ranges and a subduction zone are formed with a deep trench. Earthquakes and volcanic activity are found here.	Nazca and South American plate boundary (Figure 2.5a) Andes Mountains Peru-Chile Trench
Convergent	Ocean-ocean	Older, denser, oceanic plate sinks beneath the younger, less dense oceanic plate.	A subduction zone is formed with a deep trench. Earthquakes and volcanic activity are found here.	Fiji plate (Figure 2.5a) Fiji Islands
Convergent	Continent-continent	Neither plate is dense enough to sink into the asthenosphere; compression results.	A large, high mountain chain is formed, and earthquakes are common.	Indo-Australian and Eurasian plate boundary (on land) (Figure 2.5a) Himalaya Mountains
Transform	Ocean-ocean or continent-continent	The plates slide past one another.	Earthquakes are common and may result in some topography.	North American and Pacific plate boundary (Figure 2.11) San Andreas fault

kilometers across. Plate boundary zones are narrower in ocean crust and broader in continental crust.

Divergent boundaries occur where new lithosphere is being produced and neighboring parts of plates are moving away from each other. Typically this process occurs at mid-ocean ridges, and the process is called seafloor spreading (see Figure 2.6). Mid-ocean ridges form when hot material from the mantle rises up to form a broad ridge typically with a central rift valley. It is called a *rift valley*, or *rift*, because the plates moving apart are pulling the crust apart and splitting, or rifting, it. Molten volcanic rock that is erupted along this rift valley cools and forms new plate material. The system of mid-oceanic ridges along divergent plate boundaries forms linear submarine mountain chains that are found in virtually every ocean basin on Earth.

Convergent boundaries occur where plates collide. If one of the converging plates is oceanic and the other is continental, an oceanic-continental plate collision results. The higher-density oceanic plate descends, or subducts, into the mantle beneath the leading edge of the continental plate, producing a subduction zone (Figure 2.9a). The convergence or collision of a continent with an ocean plate can result in compression. *Compression* is a type of stress, or force per unit area. When an oceanic-continental plate collision occurs, compression is exerted on the lithosphere, resulting in shortening of the surface of Earth, like pushing a table cloth to produce folds. Shortening can cause folding, as in the table cloth example, and faulting, or displacement of rocks along fractures to thicken the lithosphere. This process of deformation produces major mountain chains and volcanoes such as

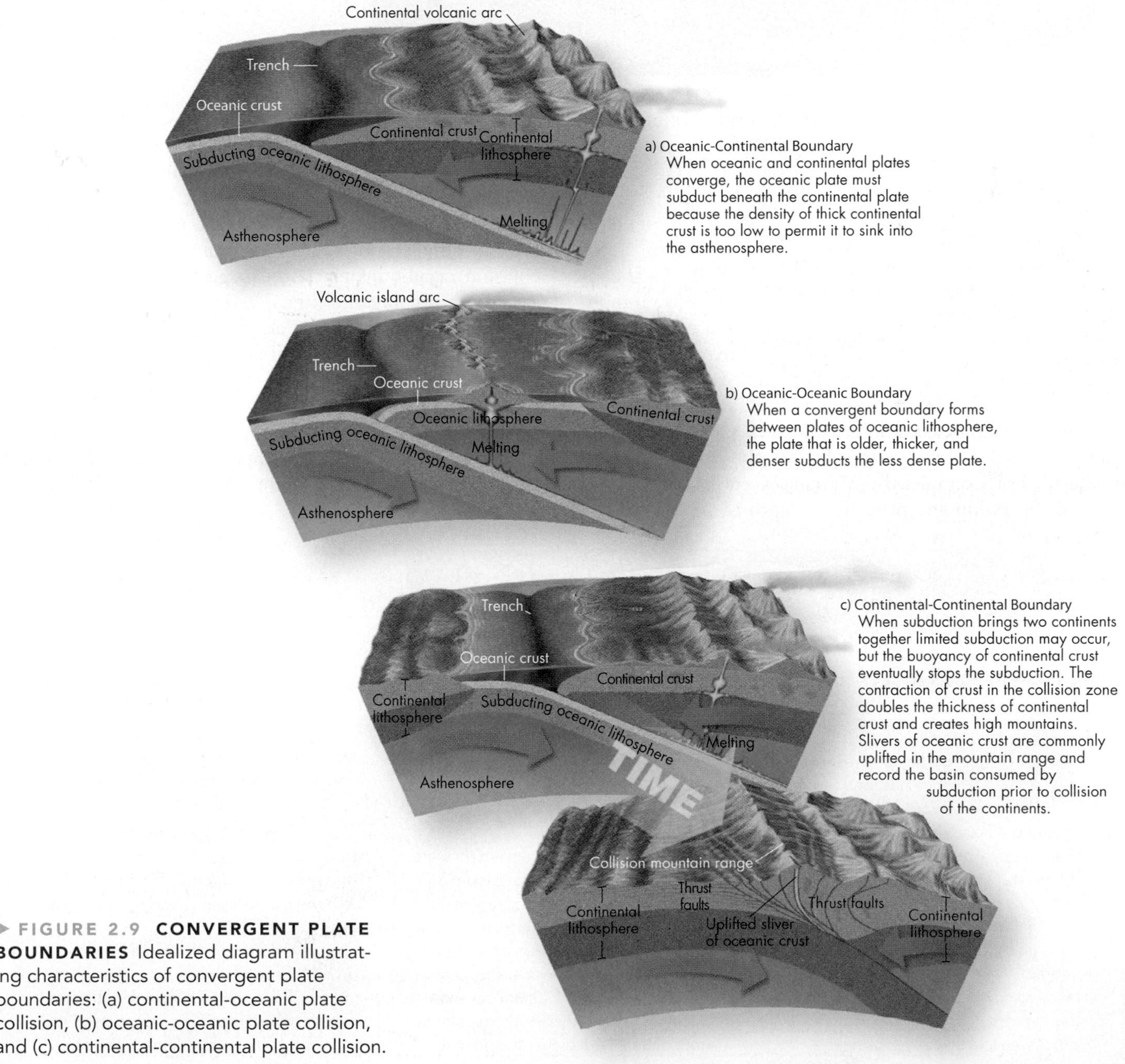

▶ FIGURE 2.9 **CONVERGENT PLATE BOUNDARIES** Idealized diagram illustrating characteristics of convergent plate boundaries: (a) continental-oceanic plate collision, (b) oceanic-oceanic plate collision, and (c) continental-continental plate collision.

◀ **FIGURE 2.10 MOUNTAINS IN ITALY** Mountain peaks (the Dolomites) in southern Italy are part of the Alpine mountain system formed from the collision between Africa and Europe. *(Edward A. Keller)*

the Andes in South America and the Cascade Mountains in the Pacific Northwest of the United States (see A Closer Look 2.1). If two oceanic lithospheric plates collide (oceanic-to-oceanic plate collision), one plate subducts beneath the other, and a subduction zone and arc-shaped chain of volcanoes known as an *island arc* are formed (Figure 2.9b) as, for example, the Aleutian Islands of the North Pacific. A **submarine trench,** a relatively narrow, usually several thousand km long and several km deep depression on the ocean floor, is often formed as the result of the convergence of two colliding plates with subduction of one. The trench is located seaward of a subduction zone associated with an oceanic-continental plate or oceanic-oceanic plate collision. Submarine trenches are sites of some of the deepest oceanic waters on Earth. For example, the Marianas trench at the center edge of the Philippine plate has a depth of 11 km (7 mi.), which is deep enough for Mount Everest to fit into. Other major trenches include the Aleutian trench south of Alaska and the Peru-Chile trench west of South America. If the leading edges of both plates contain relatively light, buoyant continental crust, subduction into the mantle of one of the plates is difficult. In this case, a continent-to-continent plate collision occurs, in which the edges of the plates collide, causing shortening and lithospheric thickening due to folding and faulting (Figure 2.9c). Where the two plates join is known as a *suture zone.* Continent-to-continent collision has produced some of the highest mountain systems on Earth, such as the Alpine and Himalayan mountain belts (Figure 2.10). Many older mountain belts were formed in a similar way; for example, the Appalachians formed during an ancient continent-to-continent plate collision 250 to 350 million years ago.

A CLOSER LOOK 2.1

The Wonder of Mountains

Mountains' awesome presence has long fascinated people. We are now discovering a fascinating story concerning their origin. The story removes some of the mystery as to how mountains form, but it has not removed the wonder. The new realization that mountains are systems (see Chapter 1) resulting from the interaction among tectonic activity (that leads to crustal thickening), the climate of the mountain, and Earth surface processes (particularly erosion) has greatly expanded our knowledge of how mountains develop.[7,8] Specifically, we have learned the following:

- Tectonic processes at convergent plate boundaries lead to crustal thickening and initial development of mountains. The mean (or average) elevation that a mountain range attains is a function of the uplift rate, which varies from less than 1 mm to about 10 mm per year (0.04 to 0.4 in. per year). The greater the rate of uplift, the higher the point to which the mean elevation of a mountain range is likely to rise during its evolution.
- As a mountain range develops and gains in elevation, it begins to modify the local and regional climate by blocking storm paths and producing a "rain shadow" in which the mountain slopes on the rain-shadow side receive much less rainfall than does the other side of the mountain. As a result, rates of runoff and erosion on the side of the rain shadow are less than for the other

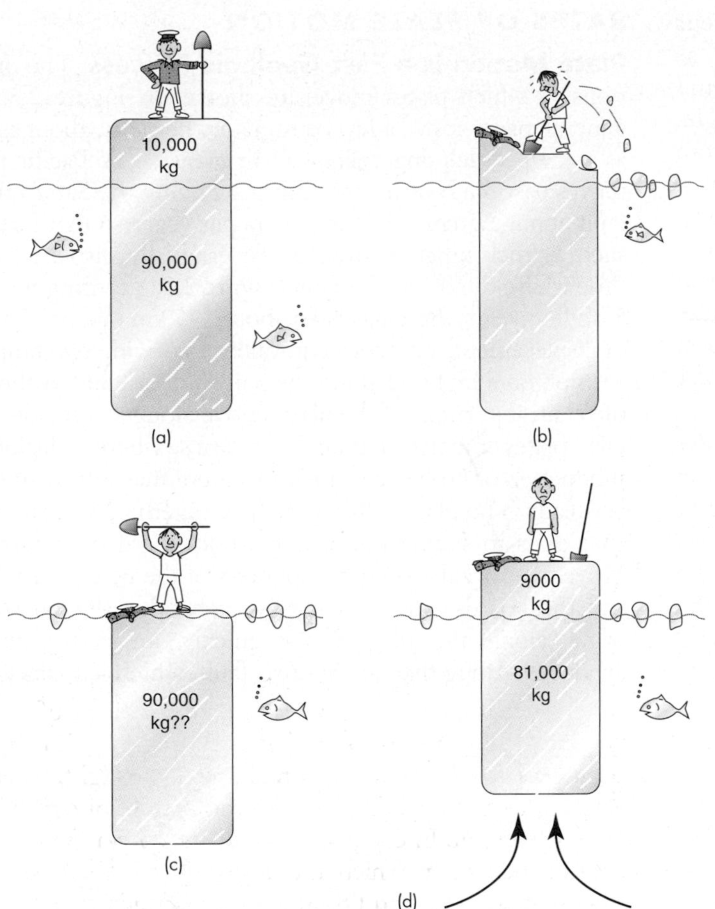

◀ **FIGURE 2.A ISOSTASY** Idealized diagram showing the principle of isostatic uplift. Admiral Frost is left adrift on an iceberg and is uncomfortable being so far above the surface of the water (a). He decides to remove the 10,000 kg (22,046 lb.) of ice that are above the waterline on the iceberg on which he is standing (b). Were it not for isostatic (buoyant) uplift, Admiral Frost would reach his goal (c). However, in a world with isostasy, uplift results from removal of the ice, and there is always one-tenth of the iceberg above the water (d). What would have happened if Admiral Frost had elected to remove 10,000 kg of ice from only one-half of the area of ice exposed above the sea? Answer: The maximum elevation of the iceberg above the water would have actually increased. Similarly, as mountains erode, isostatic adjustments also occur, and the maximum elevation of mountain peaks may actually increase as a result of the erosion alone! *(From Keller, E. A., and Pinter, N. 1996. Active tectonics. Upper Saddle River, NJ: Prentice Hall)*

side. Nevertheless, the rate of erosion increases as the elevation of the mountain range increases, and eventually the rate of erosion matches the rate of uplift. When the two match, the mountain reaches its maximum mean elevation, which is a dynamic balance between the uplift and erosion. At this point, no amount of additional uplift will increase the mean elevation of the mountains above the dynamic maximum. However, if the uplift rate increases, then a higher equilibrium mean elevation of the range may be reached. Furthermore, when the uplift ceases or there is a reduction in the rate of uplift, the mean elevation of the mountain range will decrease.[7] Strangely, the elevations of individual peaks may still increase!

- Despite erosion, the elevation of a mountain peak in a range may actually increase. This statement seems counterintuitive until we examine in detail some of the physical processes resulting from erosion. The uplift that results from the erosion is known as isostatic uplift. **Isostasy** is the principle whereby thicker, more buoyant crust stands topographically higher than crust that is thinner and denser. The principle governing how erosion can result in uplift is illustrated in Figure 2.A. The fictitious Admiral Frost has been marooned on an iceberg and is uncomfortable being far above the surface of the water. He attempts to remove the ice that is above the waterline. Were it not for isostatic (buoyant) uplift, he would have reached his goal to be close to the water line. Unfortunately for Admiral Frost, this is not the way the world works; continuous isostatic uplift of the block as ice is removed always keeps one-tenth of the iceberg above the water. So, after removing the ice above the waterline, he still stands almost as much above the water line as before.[7]
- Mountains, of course, are not icebergs, but the rocks of which they are composed are less dense than the rocks of the mantle beneath. Thus, they tend to "float" on top of the denser mantle. Also, in mountains, erosion is not uniform but is generally confined to valley walls and bottoms. Thus, as erosion continues and the mass of the mountain range is reduced, isostatic compensation occurs, and the entire mountain range rises in response. As a result of the erosion, the maximum elevation of mountain peaks actually may increase, whereas the mean elevation of the entire mountain block decreases. As a general rule, as the equivalent of 1 km (0.6 mi.) of erosion across the entire mountain block occurs, the mean elevation of mountains will rise approximately five-sixths of a kilometer (one-half mile).

In summary, research concerning the origin of mountains suggests that they result in part from tectonic processes that cause the uplift, but they also are intimately related to climatic and erosional processes that contribute to the mountain-building process. Erosion occurs during and after tectonic uplift, and isostatic compensation to that erosion occurs for millions of years. This is one reason it is difficult to remove mountain systems from the landscape. For example, mountain systems such as the Appalachian Mountains in the southeastern United States were originally produced by tectonic uplift several hundred million years ago when Europe collided with North America. There has been sufficient erosion of the original Appalachian Mountains to have removed them as topographic features many times over were it not for continued isostatic uplift in response to the erosion.

Transform boundaries, or transform faults, occur where the edges of two plates slide past each other, as shown in Figure 2.6. If you examine Figures 2.5a and 2.6, you will see that a spreading zone is not a single, continuous rift but a series of rifts that are offset from one another along connecting transform faults. Although the most common locations for transform plate boundaries are within oceanic crust, some occur within continents. A well-known continental transform boundary is the San Andreas fault in California, where the rim of the Pacific plate is sliding horizontally past the rim of the North American plate (see Figures 2.1 and 2.11).

Locations where three plates border one another are known as **triple junctions.** Figure 2.11 shows several such junctions: Two examples are the meeting point of the Juan de Fuca, North American, and Pacific plates on the West Coast of North America (this is known as the Mendocino triple junction) and the junction of the spreading ridges associated with the Pacific, Cocos, and Nazca plates west of South America.

RATES OF PLATE MOTION

Plate Motion Is a Fast Geologic Process The directions in which plates move are shown on Figure 2.5a. In general, plates move a few centimeters per year, about as fast as some people's fingernails or hair grows. The Pacific plate moves past the North American plate along the San Andreas fault about 3.5 cm per year (1.4 in. per year), so that features such as rock units or streams are gradually displaced over time where they cross the fault (Figure 2.12). During the past 5 million years, there has been about 175 km (about 110 mi.) of displacement, a distance equivalent to driving two hours at 55 mph on a highway along the San Andreas fault. Although the central portions of the plates move along at a steady slow rate, plates interact at their boundaries, where collision or subduction or both occur, and movement may not be smooth or steady. The plates often get stuck together. Movement is analogous to sliding one rough wood board over another. Movement occurs when the splinters of the boards break off and the boards move quickly by each other. When rough edges along the plate move quickly, an earthquake is produced. Along the San Andreas fault, which is a transform

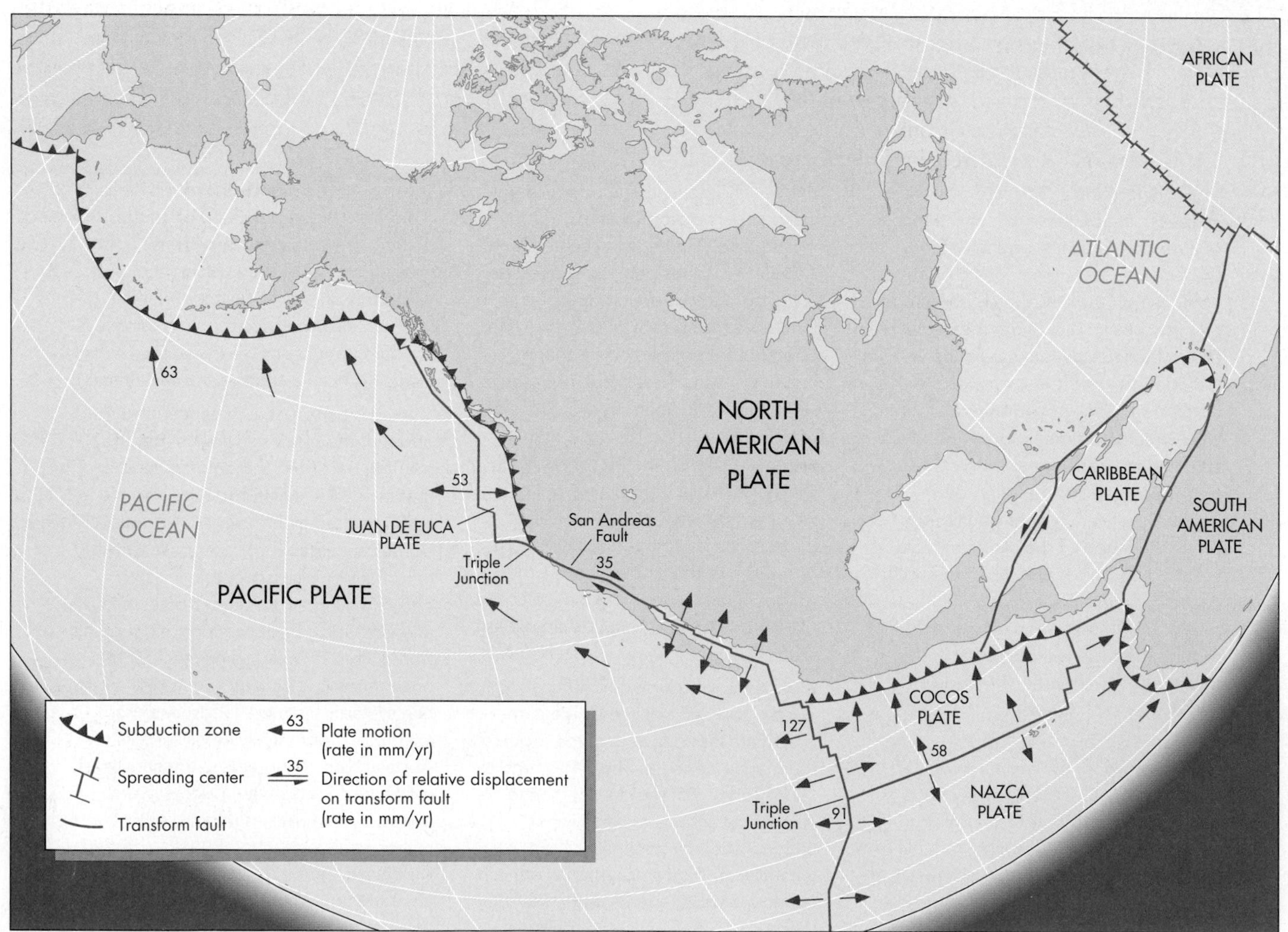

▲ **FIGURE 2.11 NORTH AMERICAN PLATE BOUNDARY** Detail of boundary between the North American and Pacific plates. *(Courtesy of Tanya Atwater)*

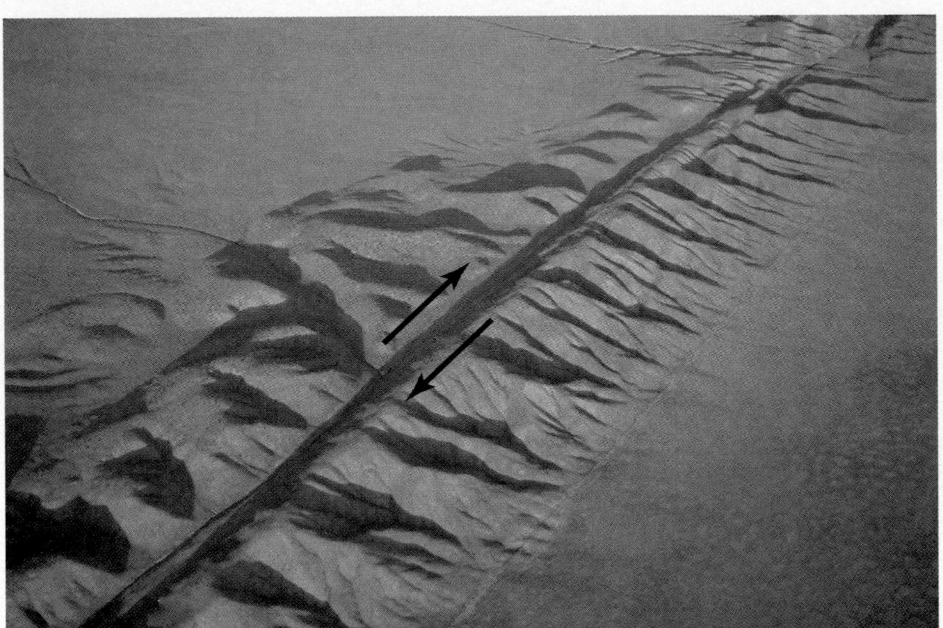

◀ **FIGURE 2.12 THE SAN ANDREAS FAULT** The fault is visible from the lower left to upper right diagonally across the photograph, as if a gigantic plow had been dragged across the landscape. *(James Balog)*

plate boundary, the displacement is horizontal and can amount to several meters during a great earthquake. During an earthquake in 1857 on the San Andreas fault, a horse corral across the fault was reportedly changed from a circle to an "S" shape. Fortunately, such an event generally occurs at any given location only once every 100 years or so. Over long time periods, rapid displacement from periodic earthquakes and more continuous slow "creeping" displacements add together to produce the rate of several centimeters of movement per year along the San Andreas fault.

2.4 A Detailed Look at Seafloor Spreading

When Alfred Wegener proposed the idea of continental drift in 1915, he had no solid evidence of a mechanism that could move continents. The global extent of mid-oceanic ridges was discovered in the 1950s, and in 1962 geologist Harry H. Hess published a paper suggesting that continental drift was the result of the process of seafloor spreading along those ridges. The fundamentals of seafloor spreading are shown in Figure 2.6. New oceanic lithosphere is produced at the spreading ridge (divergent plate boundary). The lithospheric plate then moves laterally, carrying along the embedded continents in the tops of moving plates. These ideas produced a new major paradigm that greatly changed our ideas about how Earth works.[3,6,9]

The validity of seafloor spreading was established from three sources: (1) identification and mapping of oceanic ridges, (2) dating of volcanic rocks on the floor of the ocean, and (3) understanding and mapping of the paleomagnetic history of ocean basins.

PALEOMAGNETISM

We introduce and discuss Earth's magnetic field and paleomagnetic history in some detail in order to understand how seafloor spreading and plate tectonics were discovered. Earth has had a magnetic field for at least the past 3 billion years[2] (Figure 2.13). The field can be represented by a

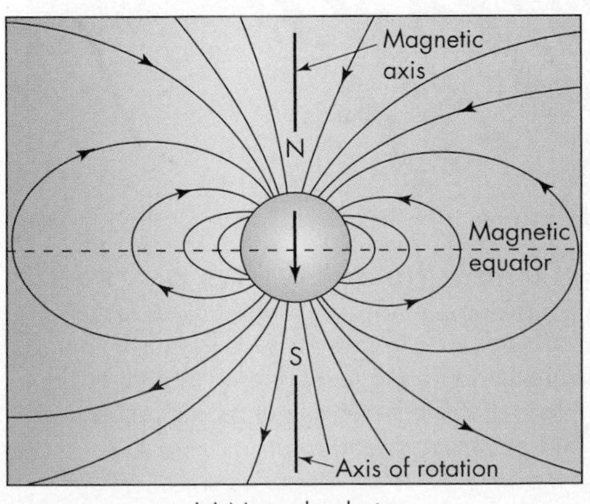

(a) Normal polarity

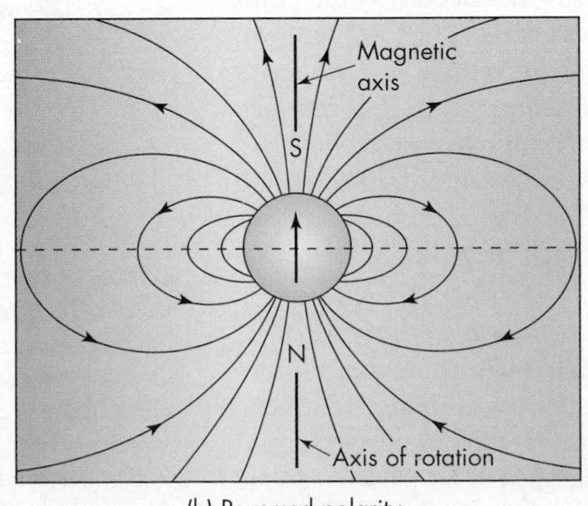

(b) Reversed polarity

◀ **FIGURE 2.13 MAGNETIC REVERSAL** Idealized diagram showing the magnetic field of Earth under (a) normal polarity and (b) reversed polarity. *(From Kennett, J. 1982.* Marine geology. *Englewood Cliffs, NJ: Prentice Hall)*

dipole magnetic field with lines of magnetic force extending from the South Pole to the North Pole. A dipole magnetic field is one that has equal and opposite charges at either end. Convection occurs in the iron-rich, fluid, hot outer core of Earth because of compositional changes and heat at the inner-outer core boundary. As more buoyant material in the outer core rises, it starts the convection (see Figure 2.3). The convection in the outer core, along with the rotation of Earth that causes rotation of the outer core, initiates a flow of electric current in the core. This flow of current within the core produces and sustains Earth's magnetic field.[2,3]

Earth's magnetic field is sufficient to permanently magnetize some surface rocks. For example, volcanic rock that erupts and cools at mid-oceanic ridges becomes magnetized at the time it passes through a critical temperature. At that critical temperature, known as the *Curie point*, iron-bearing minerals (such as magnetite) in the volcanic rock orient themselves parallel to the magnetic field. This is a permanent magnetization known as *thermoremnant magnetization*.[3] The term **paleomagnetism** refers to the study of the magnetism of rocks at the time their magnetic signature formed. It is used to determine the magnetic history of Earth.

The magnetic field, based on the size and conductivity of Earth's core, must be continuously generated or it would decay away in about 20,000 years. It would decay because the temperature of the core is too high to sustain permanent magnetization.[2]

Earth's Magnetic Field Periodically Reverses Before the discovery of plate tectonics, geologists working on land had already discovered that some volcanic rocks were magnetized in a direction opposite to the present-day field, suggesting that the polarity of Earth's magnetic field was reversed at the time the volcanoes erupted and the rocks cooled (Figure 2.13b). The rocks were examined for whether their magnetic field was normal, as it is today, or reversed relative to that of today, for certain time intervals of the Earth's history. A chronology for the last few million years was constructed on the basis of the dating of the "reversed" rocks. You can verify the current magnetic field of the Earth by using a compass; at this point in Earth's history, the needle points to the north magnetic pole. During a period of reversed polarity, the needle would point south! The cause of **magnetic reversals** is not well known, but it is related to changes in the convective movement of the liquid material in the outer core and processes occurring in the inner core. Reversals in Earth's magnetic field are random, occurring on average every few hundred thousand years. The change in polarity of Earth's magnetic field takes a few thousand years to occur, which in geologic terms is a very short time.

What Produces Magnetic Stripes? To further explore the Earth's magnetic field, geologists towed magnetometers, instruments that measure magnetic properties of rocks, from ships and completed magnetic surveys. The paleomagnetic record of the ocean floor is easy to read because of the fortuitous occurrence of the volcanic rock basalt (see Chapter 5) that is produced at spreading centers and forms the floors of the ocean basins of Earth. The rock is fine-grained and contains sufficient iron-bearing minerals to produce a good magnetic record. The marine geologists' discoveries were not expected. The rocks on the floor of the ocean were found to have irregularities in the magnetic field. These irregular magnetic patterns were called anomalies or perturbations of Earth's magnetic field caused by local fields of magnetized rocks on the seafloor. The anomalies can be represented as stripes on maps. When mapped, the stripes form quasi-linear patterns parallel to oceanic ridges. The marine geologists found that their sequences of stripe width patterns matched the sequences established by land geologists for polarity reversals in land volcanic rocks. Magnetic survey data for an area southwest of Iceland are shown on Figure 2.14. The black stripes represent normally magnetized rocks and the intervening white stripes represent reversed magnetized rocks.[10] Notice that the stripes are not

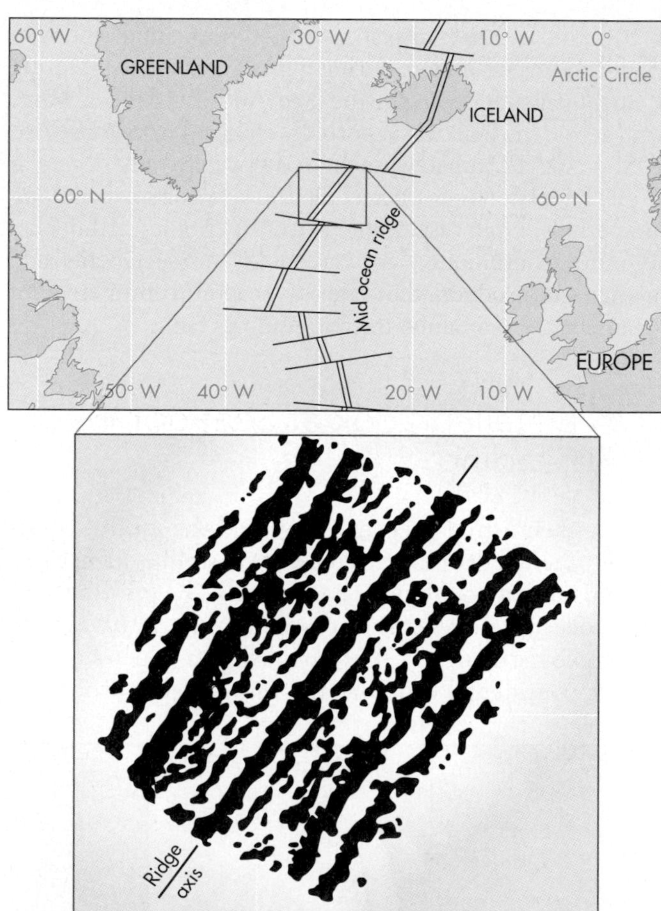

▲ **FIGURE 2.14 MAGNETIC ANOMALIES ON THE SEAFLOOR** Map showing a magnetic survey southwest of Iceland along the Mid-Atlantic Ridge. Positive magnetic anomalies are black (normal) and negative magnetic anomalies are white (reversed). Note that the pattern is symmetrical on the two sides of the mid-oceanic ridge. *(From Heirtzler, J. R., Le Pichon, X., and Baron, J. G. 1966. Magnetic anomalies over the Reykjanes Ridge.* Deep-Sea Research *13:427–43)*

evenly spaced but have patterns that are symmetrical on opposite sides of the Mid-Atlantic Ridge (Figure 2.14).

Why is the Seafloor No Older than 200 Million Years? The discovery of patterns of magnetic stripes at various locations in ocean basins allowed geologists to infer numerical dates for the volcanic rocks. Merging the magnetic anomalies with the numerical ages of the rocks produced the record of seafloor spreading. The spreading of the ocean floor, beginning at a mid-oceanic ridge, could explain the magnetic stripe patterns.[11] Figure 2.15 is an idealized diagram showing how seafloor spreading may produce the patterns of magnetic anomalies (stripes). The pattern shown is for the past several million years, which includes several periods of normal and reversed magnetization of the volcanic rocks. Black stripes represent normally magnetized rocks, and brown stripes are rocks with a reversed magnetic signature. Notice that the most recent magnetic reversal occurred approximately 0.7 million years ago. The basic idea illustrated by Figure 2.15 is that rising magma at the oceanic ridge is extruded, or pushed out onto the surface, through volcanic activity, and the cooling rocks become normally magnetized. When the field is reversed, the cooling rocks preserve a reverse magnetic signature, and a brown stripe (Figure 2.15) is preserved. Notice that the patterns of magnetic anomalies in rocks on both sides of the ridge are mirror images of one another. The only way such a pattern might result is through the process of seafloor spreading. Thus, the pattern of magnetic reversals found on rocks of the ocean floor is strong evidence that the process of spreading is happening. Mapping of magnetic anomalies, when combined with age-dating of the magnetic reversals in land rocks creates a database that suggests exciting inferences; Figure 2.16 shows the age of the ocean floor as determined from this database. The pattern, showing that the youngest volcanic rocks are found along active mid-oceanic ridges, is consistent with the theory of seafloor spreading. As distance from these ridges increases, the age of the ocean floor also increases, to a maximum of about 200 million years, during the early Jurassic period (see Table 1.3). Thus, it appears that the present ocean floors of the world are no older than 200 million years. In contrast, rocks on continents are often much older than Jurassic, going back about 4 billion years, almost 20 times older than the ocean floors! We conclude that the thick continental crust, by virtue of its buoyancy, is more stable at Earth's surface than are rocks of the crust of the ocean basins. Continents form by the processes of accretion of sediments, addition of volcanic materials, and collisions of tectonic plates carrying continental landmasses. We will continue this discussion when we consider the movement of continents during the past 200 million years. However, it is important to recognize that it is the pattern of magnetic stripes that allows us to reconstruct

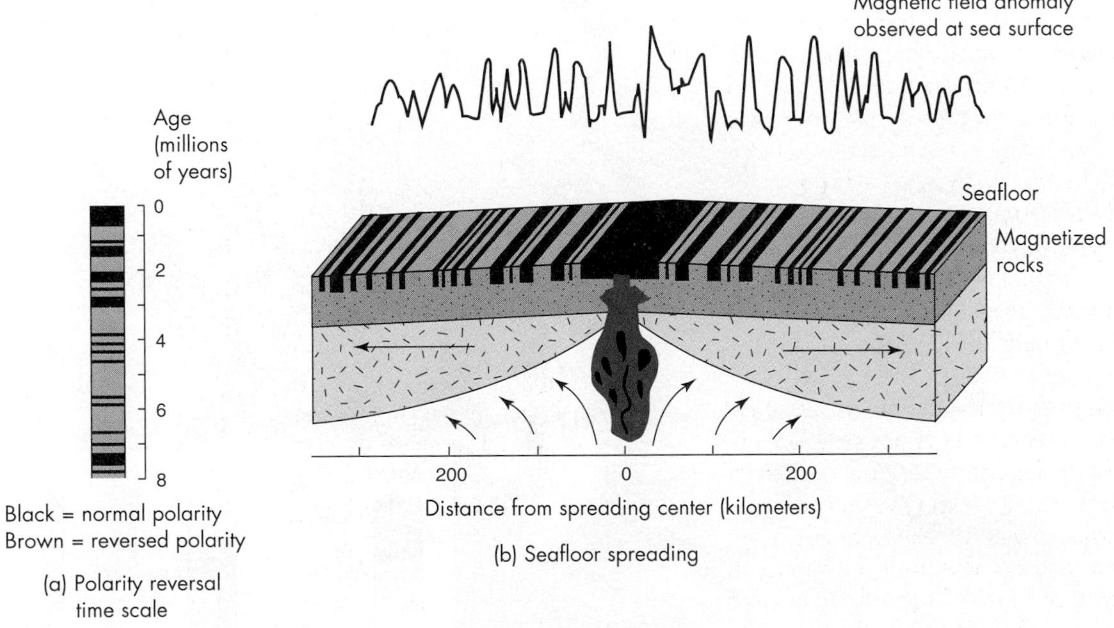

▲ **FIGURE 2.15 MAGNETIC REVERSALS AND SEAFLOOR SPREADING** Idealized diagram showing an oceanic ridge and the rising of magma, in response to seafloor spreading. As the volcanic rocks cool, they become magnetized. The black stripes represent normal magnetization; the brown stripes are reversed magnetization. The record shown here was formed over a period of several million years. Magnetic anomalies (stripes) are a mirror image of each other on opposite sides of the mid-oceanic ridge. Thus, the symmetrical bands of the normally and reversely magnetized rocks are produced by the combined effects of the reversals and seafloor spreading. *(Courtesy of Tanya Atwater)*

how the plates and the continents embedded in them have moved throughout history.

HOT SPOTS

What Are Hot Spots? There are a number of places on Earth called **hot spots,** characterized by volcanic centers resulting from hot materials produced deep in the mantle, perhaps near the core-mantle boundary. The partly molten materials are hot and buoyant enough to move up through mantle and overlying moving tectonic plates.[3,6] An example of a continental hot spot is the volcanic region of Yellowstone National Park. Hot spots are also found in both the Atlantic and Pacific Oceans. If the hot spot is anchored in the slow-moving deep mantle, then, as the plate moves over a hot spot, a chain of volcanoes is produced. Perhaps the best example of this type of hot spot is the line of volcanoes forming the Hawaiian-Emperor Chain in the Pacific Ocean (Figure 2.17a). Along this chain, volcanic eruptions range in age from present-day activity on the big island of Hawai'i (in the southeast) to more than 78 million years ago near the northern end of the Emperor Chain. With the exception of the Hawaiian Islands and some coral atolls (ringlike coral islands such as Midway Island), the chain consists of submarine volcanoes known as *seamounts.* Seamounts are islands that were eroded by waves and submarine landslides and subsequently sank beneath the ocean surface. As seamounts move farther off the hot spot, the volcanic rocks the islands are composed of cool and the oceanic crust they are on becomes denser and sinks.

Seamounts constitute impressive submarine volcanic mountains. In the Hawaiian Chain, the youngest volcano is Mount Loihi, which is still a submarine volcano, presumably directly over a hot spot, as idealized on Figure 2.17b. The ages of the Hawaiian Islands increase to the northwest, with the oldest being Kauai, about 6 million years old. Notice in Figure 2.17a that the line of seamounts makes a sharp bend at the junction of the Hawaiian and Emperor Chains. The age of the volcanic rocks at the bend is about 43 million years, and the bend is interpreted to represent a time when plate motions changed.[12] If we assume that the hot spots are fixed deep in the mantle, then the chains of volcanic islands and submarine volcanoes along the floor of the Pacific Ocean that get older farther away from the hot spot provide additional evidence to support the movement of the Pacific plate. In other words, the ages of the volcanic islands and submarine volcanoes could systematically change as they do only if the plate is moving over the hot spot.

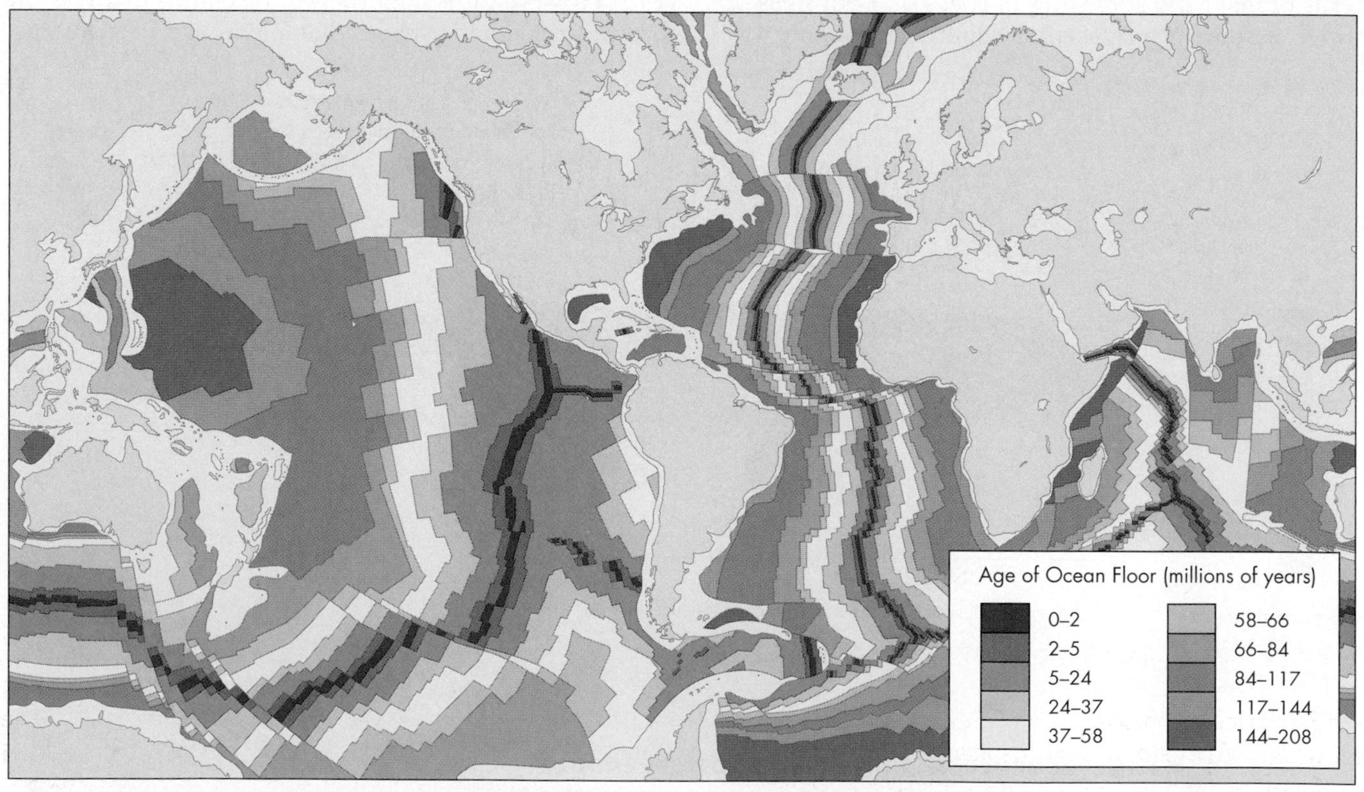

▲ FIGURE 2.16 **AGE OF THE OCEAN FLOOR** Age of the seafloor is determined from magnetic anomalies and other methods. The youngest ocean floor (red) is located along oceanic ridge systems, and older rocks are generally farther away from the ridges. The oldest ocean floor rocks are approximately 180 million years old. *(From Scotese, C. R., Gahagan, L. M., and Larson, R. L. 1988. Plate tectonic reconstruction of the Cretaceous and Cenozoic ocean basins.* Tectonophysics *155:27–48)*

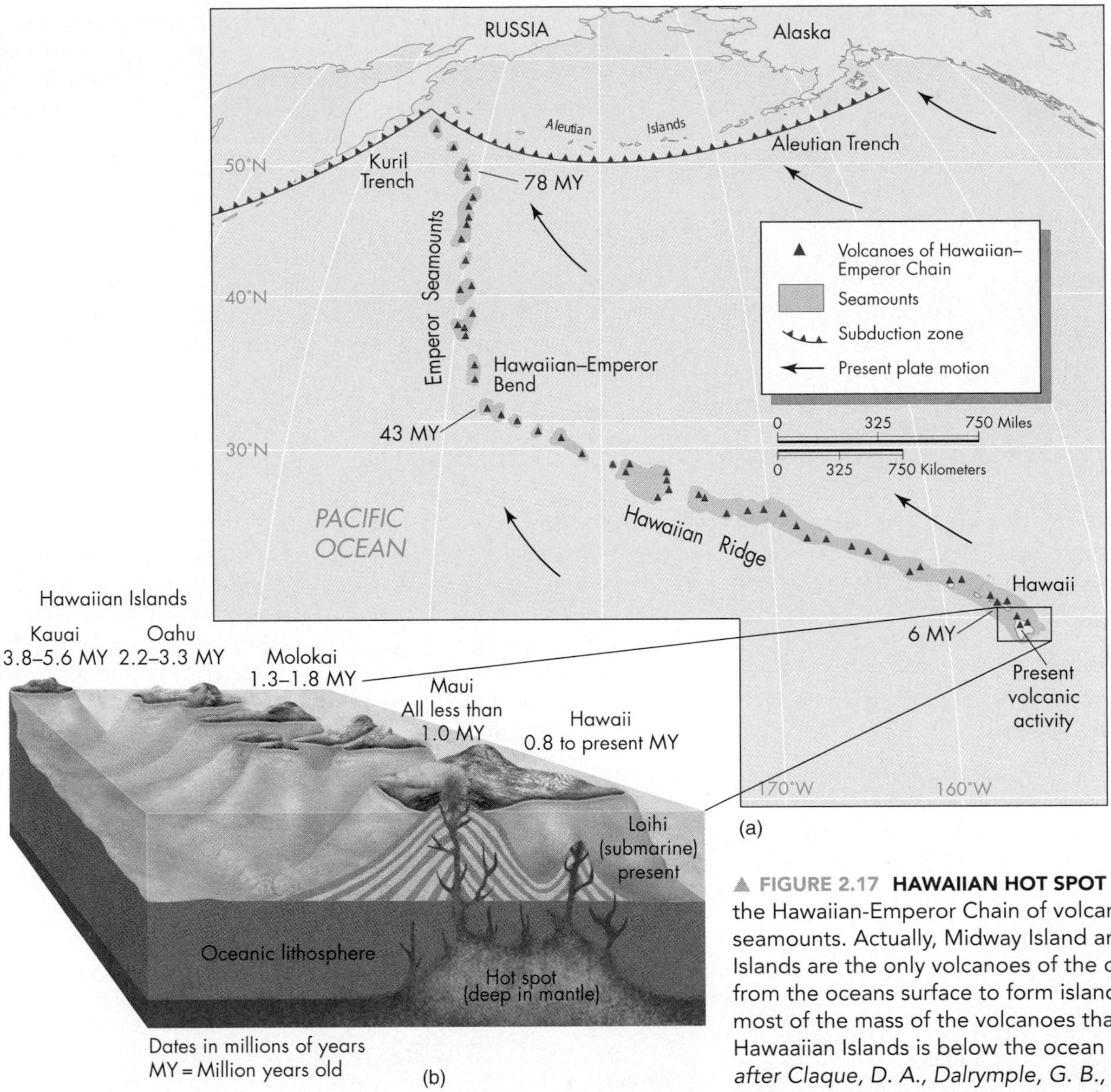

▲ **FIGURE 2.17 HAWAIIAN HOT SPOT** (a) Map showing the Hawaiian-Emperor Chain of volcanic islands and seamounts. Actually, Midway Island and the Hawaiian Islands are the only volcanoes of the chain that protrude from the oceans surface to form islands. Notice that most of the mass of the volcanoes that form the Hawaaiian Islands is below the ocean surface. *(Modified after Claque, D. A., Dalrymple, G. B., and Moberly, R. 1975. Petrography and K-Ar ages of dredged volcanic rocks from the western Hawaiian Ridge and southern Emperor Seamount chain. Geological Society of America Bulletin 86:991–98)* (b) Sketch map showing the Hawaiian Islands, which range in age from present volcanic activity to about 6 million years old on the island of Kauai. *(From Thurman, Oceanography, 5th ed. Columbus, OH: Merrill, plate 2)*

2.5 Pangaea and Present Continents

Plate Tectonics Shapes Continents and Dictates the Location of Mountain Ranges Movement of the lithospheric plates is responsible for the present shape and location of the continents. There is good evidence that the most recent global episode of continental drift, driven by seafloor spreading, started about 180 million years ago, with the breakup of a supercontinent called Pangaea (this name, meaning "all lands," was first proposed by Wegener). Pangaea (pronounced pan-jee-ah) was enormous, extending from pole to pole and over halfway around Earth near the equator (Figure 2.18). Pangaea had two parts (Laurasia to the north and Gondwana to the south) and was constructed during earlier continental collisions. Figure 2.18a shows Pangaea as it was nearly 180 million years ago. Seafloor spreading over the past 200 million years separated Eurasia and North America from the southern landmass, Eurasia from North America, and the southern continents (South America, Africa, India, Antarctica, and Australia) from one another (Figure 2.18b–d). The Tethys Sea, between Africa and Europe-Asia (Figure 2.18a–c), closed as part of the activity that

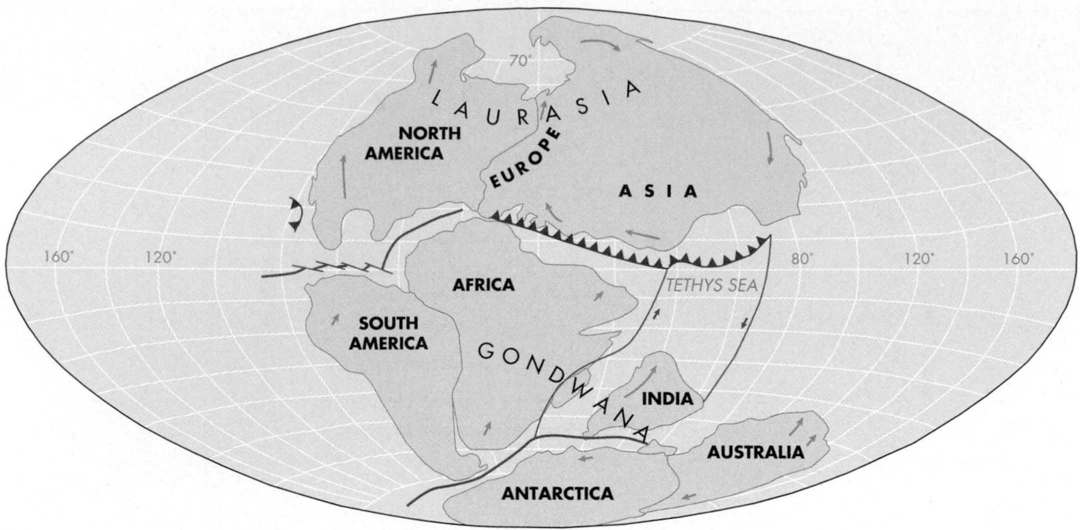

(a) 180 million years ago

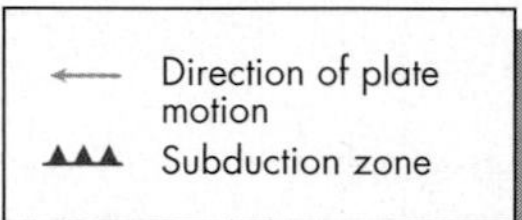

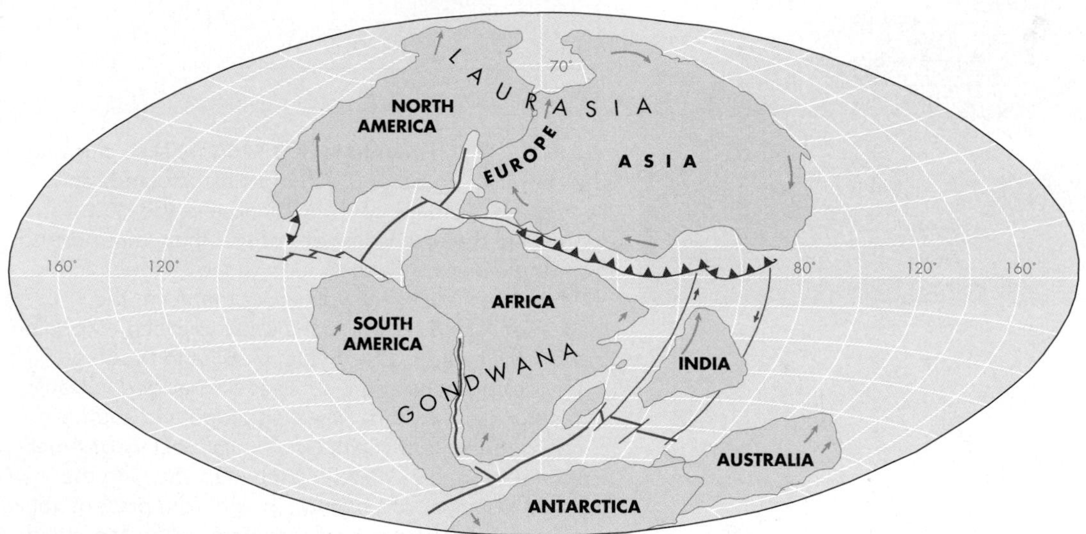

(b) 135 million years ago

▲ FIGURE 2.18 **TWO HUNDRED MILLION YEARS OF PLATE TECTONICS** (a) The proposed positions of the continents at 180 million years ago, (b) 135 million years ago, (c) 65 million years ago, and (d) at present. Arrows show directions of plate motion. See text for further explanation of the closing of the Tethys Sea, the collision of India with China, and the formation of mountain ranges. *(From Dietz, R. S., and Holden, J. C. 1970. Reconstruction of Pangaea: Breakup and dispersion of continents, Permian to present.* Journal of Geophysical Research *75(26):4939–56. Copyright by the American Geophysical Union. Modifications and block diagrams from Christopherson, R. W. 1994.* Geosystems, *2nd ed. Englewood Cliffs, NJ: Macmillan)*

NORTH AMERICA
EUROPE
ASIA
AFRICA
SOUTH AMERICA
INDIA
AUSTRALIA
ANTARCTICA
160°
120°
80°
120°
160°

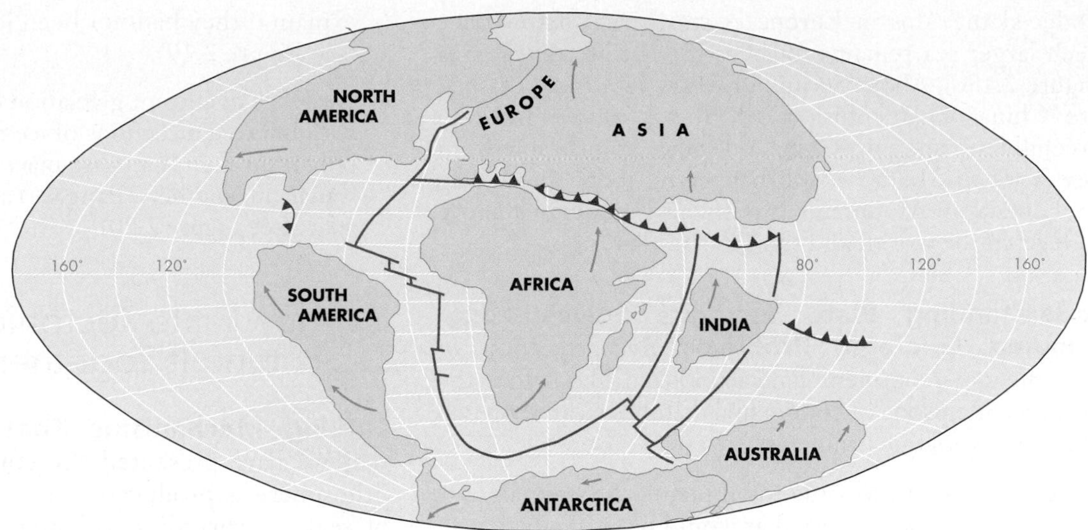

(c) 65 million years ago

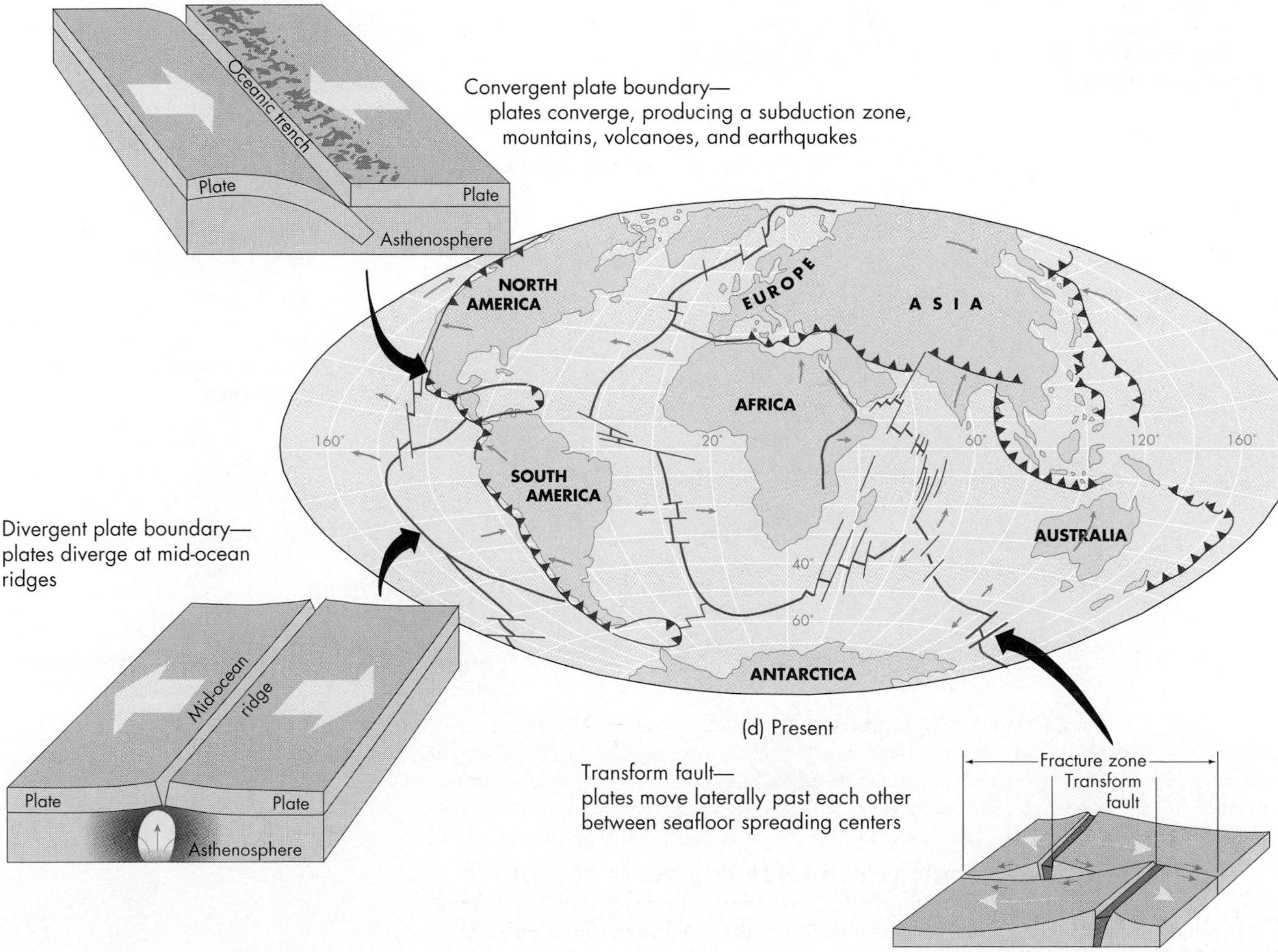

Oceanic trench
Plate
Plate
Asthenosphere
Convergent plate boundary—
plates converge, producing a subduction zone, mountains, volcanoes, and earthquakes
NORTH AMERICA
EUROPE
ASIA
AFRICA
SOUTH AMERICA
AUSTRALIA
ANTARCTICA
160°
20°
60°
120°
160°
40°
60°
Divergent plate boundary—
plates diverge at mid-ocean ridges
Mid-ocean ridge
Plate
Plate
Asthenosphere
Transform fault—
plates move laterally past each other between seafloor spreading centers
Fracture zone
Transform fault

(d) Present

▲ **FIGURE 2.18 (CONTINUED)**

produced the Alps in Europe. A small part of this once much larger sea remains today as the Mediterranean Sea (Figure 2.18d). About 50 million years ago, India crashed into China. That collision, which has caused India to forcefully intrude into China a distance comparable from New York to Miami, is still happening today, producing the Himalayan Mountains (the highest mountains above sea level in the world) and the Tibetan Plateau.

Understanding Plate Tectonics Solves Long-Standing Geologic Problems. Reconstruction of what the supercontinent Pangaea looked like before the most recent episode of continental drift has cleared up two interesting geologic problems:

- Occurrence of the same fossil plants and animals on different continents that would be difficult to explain if they had not been joined in the past (see Figure 2.19).
- Evidence of ancient glaciation on several continents, with inferred directions of ice flow that makes sense only if the continents are placed back within Gondwanaland (southern Pangaea) as it was before splitting apart (see Figure 2.20).

2.6 How Plate Tectonics Works: Putting It Together

Driving Mechanisms That Move Plates Now that we have presented the concept that new oceanic lithosphere is produced at mid-oceanic ridges because of seafloor spreading and that old, cooler plates sink

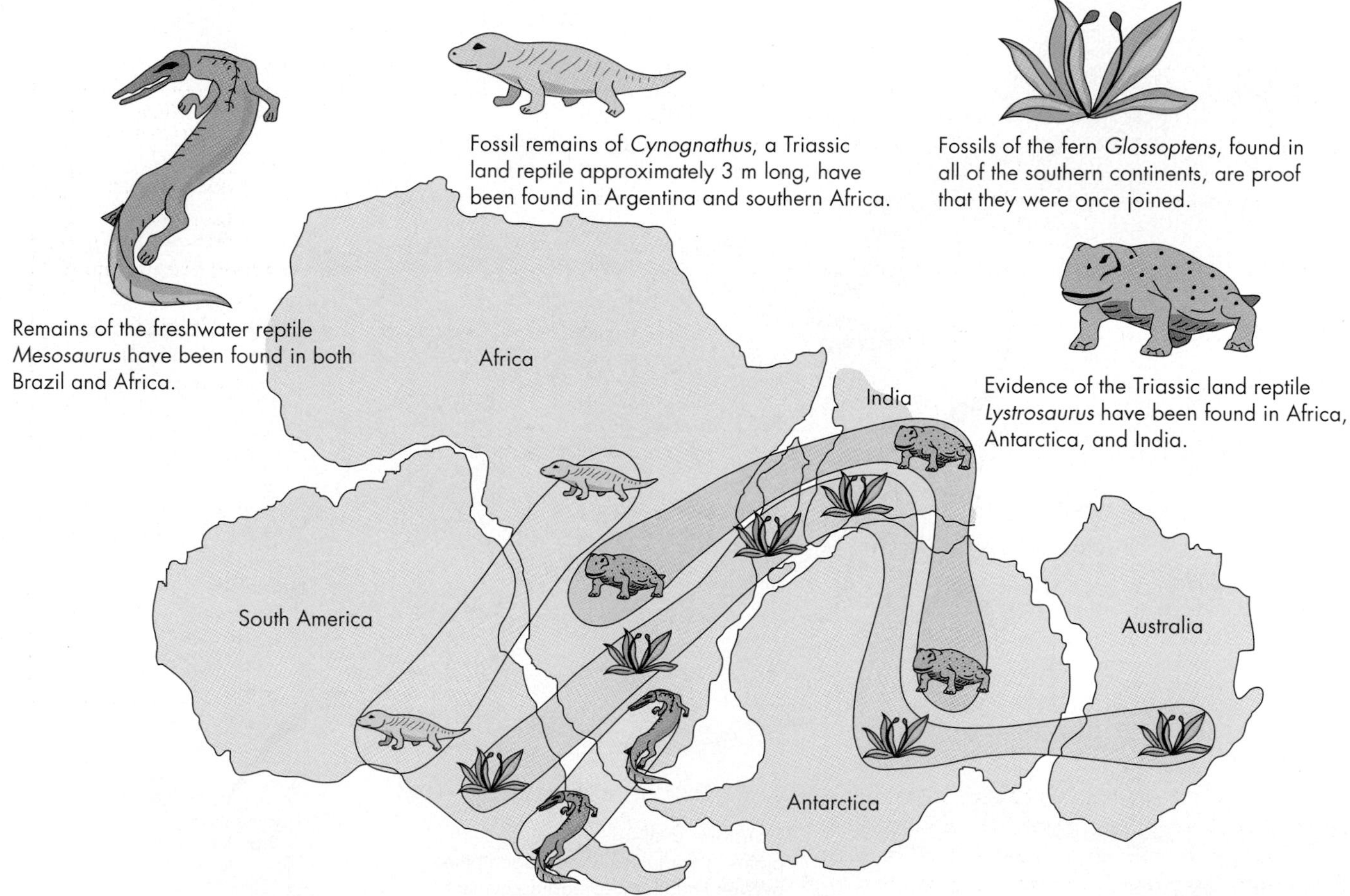

▲ FIGURE 2.19 **PALEONTOLOGICAL EVIDENCE FOR PLATE TECTONICS**
This map shows some of the paleontological (fossil) evidence that supports continental drift. It is believed that these animals and plants could not have been found on all of these continents were they not once much closer together than they are today. Major ocean basins would have been physical barriers to their distribution. *(From Hamblin, W. K. 1992.* Earth's dynamic systems, *6th ed. New York: Macmillan)*

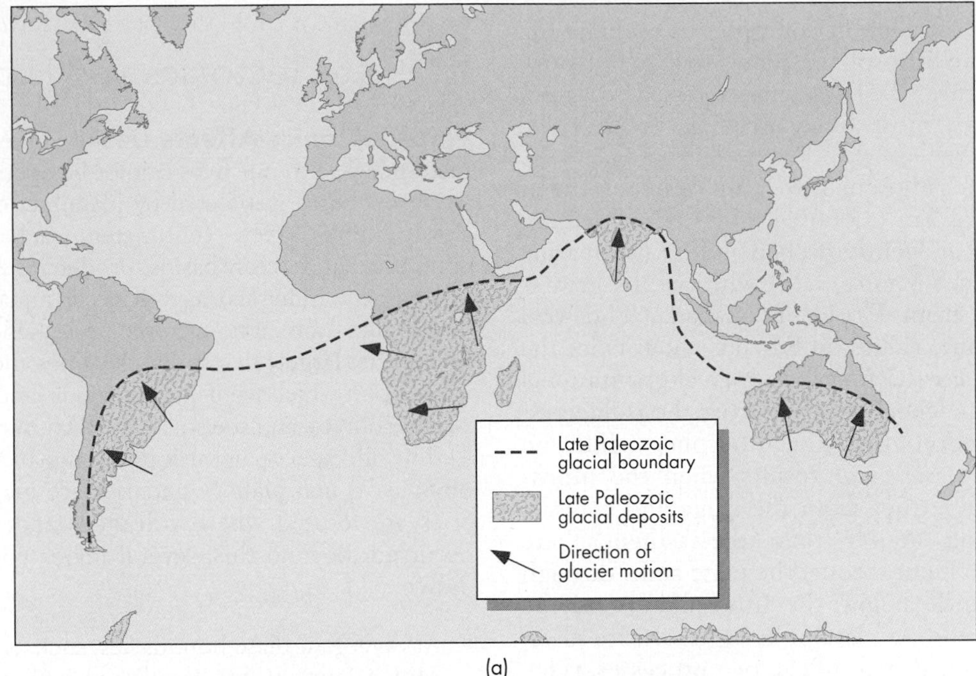

(a)

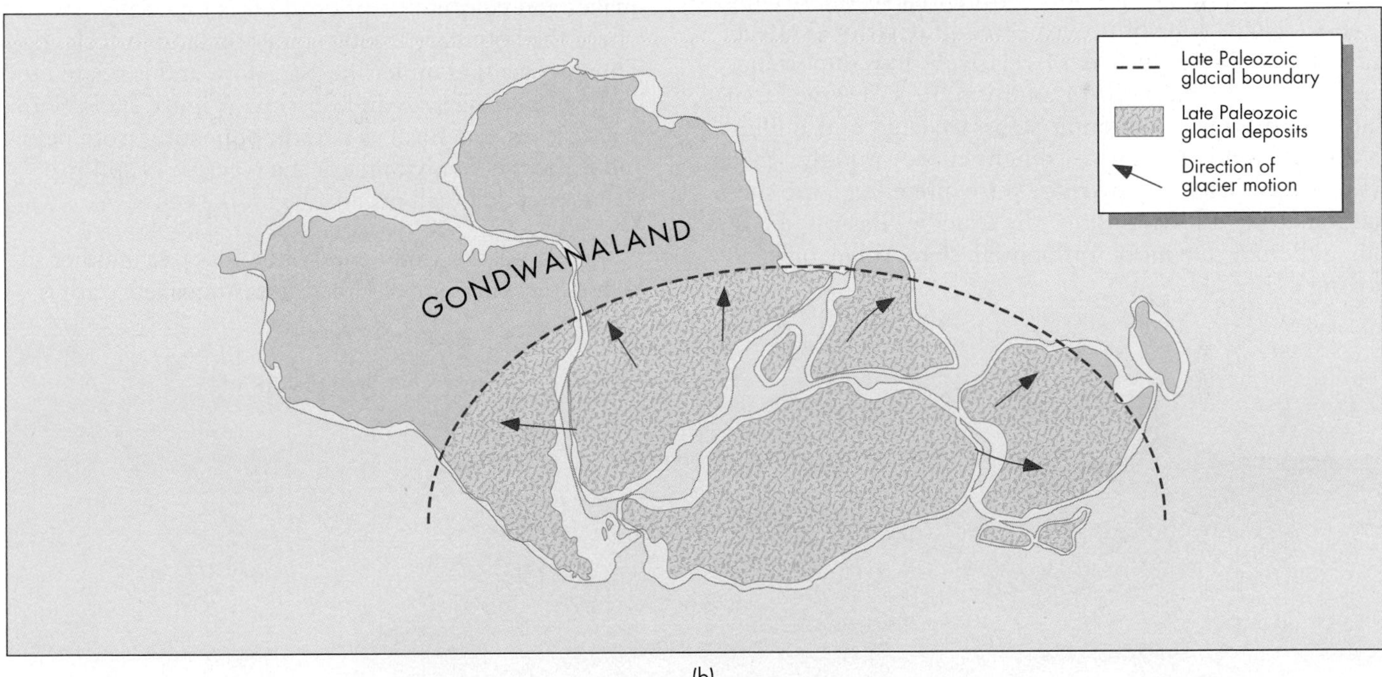

(b)

▲ **FIGURE 2.20 GLACIAL EVIDENCE FOR PLATE TECTONICS** (a) Map showing the distribution of evidence for late Paleozoic glaciations. The arrows indicate the direction of ice movement. Notice that the arrows are all pointing away from ocean sources. Also these areas are close to the tropics today, where glaciation would have been unlikely in the past. These Paleozoic glacial deposits were formed when Pangaea was a supercontinent, before fragmentation by continental drift. (b) The continents are restored (it is thought that continents drifted north away from the South Pole). Notice that the arrows now point outward as if moving away from a central area where glacial ice was accumulating. Thus, restoring the position of the continents produces a pattern of glacial deposits that makes much more sense. *(Modified after Hamblin, W. K. 1992.* Earth's dynamic systems, *6th ed. New York: Macmillan)*

into the mantle at subduction zones, let us evaluate the forces that cause the lithospheric plates to actually move and subduct. Figure 2.21 is an idealized diagram illustrating the two most likely driving forces, ridge push and slab pull.

The mid-oceanic ridges or spreading centers stand at elevations of 1 to 3 km (3,000 to 9,000 ft.) above the ocean floor as linear, gently arched uplifts (submarine mountain ranges; see Figure 2.22) with widths greater than the distance from Florida to Canada. The total length of mid-oceanic ridges on Earth is about twice the circumference of Earth. *Ridge push* is a gravitational push, like a gigantic landslide, away from the ridge crest toward the subduction zone (the lithosphere slides on the asthenosphere). *Slab pull* results when the lithospheric plate moves farther from the ridge and cools, it gradually becoming denser than the asthenosphere beneath it. At a subduction zone, the plate sinks through lighter, hotter mantle below the lithosphere, and the weight of this descending plate pulls on the entire plate, resulting in slab pull. Which of the two processes, ridge push or slab pull, is the more influential of the driving forces? Calculations of the expected gravitational effects suggest that ridge push is of relatively low importance compared with slab pull. In addition, it is observed that plates with large subducting slabs attached and pulling on them tend to move much more rapidly than those driven primarily by ridge push alone (e.g., the subduction zones surrounding the Pacific Basin). Thus, slab pull may be more influential than ridge push in moving plates.

2.7 Plate Tectonics and Hazards

Plate Tectonics Affects Us All The importance of the tectonic cycle to our lives cannot be overstated. Everything living on Earth is affected by plate tectonics. As the plates slowly move a few centimeters each year, so do the continents and ocean basins, producing zones of resources (oil, gas, and minerals), as well as earthquakes and volcanoes (see Figure 2.5b). The tectonic processes occurring at plate boundaries largely determine the types and properties of the rocks upon which we depend for our land, our mineral and rock resources, and the soils on which our food is grown.

The linkage of hazardous events to plate tectonics is obvious. When plate boundaries are mapped, the boundaries are located where volcanoes and earthquakes have occurred. Beyond this, several major conclusions may be drawn:

- At divergent plate boundaries, such as along the Mid-Atlantic Ridge, the dominant hazards are earthquakes and volcanic eruption. This is especially true where the boundary briefly comes on land in Iceland. There, volcanoes underline glacial ice and is related to both volcanic hazards and flooding (Figure 2.23). Subglacial lakes may form as a result of heating from below and if they suddenly burst or if a volcanic eruption occurs, flooding will occur.
- Along boundaries where one plate slides past another such as the San Andreas fault, the earthquake hazard is

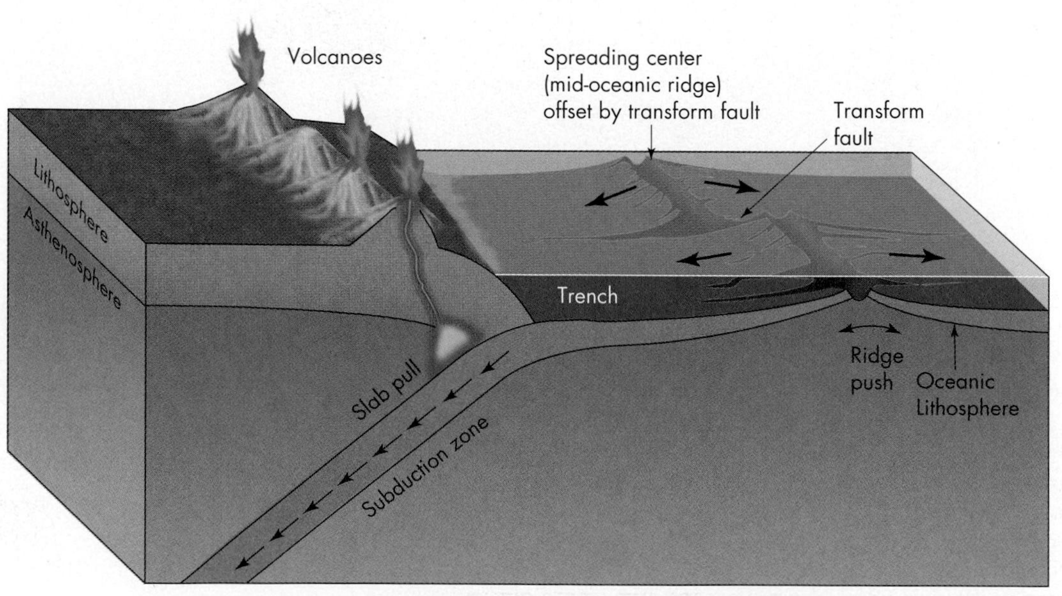

▲ FIGURE 2.21 **PUSH AND PULL IN MOVING PLATES** Idealized diagram showing concepts of ridge push and slab pull that facilitate the movement of lithospheric plates from spreading ridges to subduction zones. Both are gravity driven. The heavy lithosphere falls down the mid-oceanic ridge slope and subducts down through the lighter, hotter mantle. *(Modified after Cox, A., and Hart, R. B. 1986.* Plate tectonics. *Boston: Blackwell Scientific Publications)*

▲ **FIGURE 2.22 MID-ATLANTIC RIDGE** Image of the Atlantic Ocean basin showing details of the seafloor. Notice that the width of the Mid-Atlantic Ridge is about one-half the width of the ocean basin. *(Heinrich C. Berann/National Geographic Image collection)*

▲ FIGURE 2.23 **FIRE AND ICE** The Vatnajökull ice cap covers approximately 8% of the volcanic island of Iceland. During volcanic eruptions, melting of the ice cap can produce enormous floods like the one that occurred in 2010 when the Icelandic volcano Eyjafjallajökull erupted. *(Wild Wonders of Europe/O Haarberg/Nature Picture Library)*

appreciable. There are a number of such boundaries around the world in places such as Southern California, Haiti, and New Zealand (see Figure 2.5a). Associated with these plate boundaries are often hilly or mountainous areas where rainfall may be higher and thus hazards such as landslides and flooding are also present.

- Convergent plate boundaries where one plate dives beneath another or two collide are areas particularly prone to natural hazards. Where subduction zone volcanism occurs, particularly explosive volcanoes are typical. Thus we see the string of volcanoes in Alaska, the Pacific Northwest, New Zealand, Japan, and the western coast of South America. Where two continental land masses collide and neither can be subducted, the highest topography in the world has been produced (Figure 2.24). Along with the earthquake hazard there is increasing precipitation and an abundance of high, steep slopes prone to landslides and flooding. Where India has collided with Asia for the past 50 million years, a high plateau (Tibetan Plateau) has formed and this high topography affects the occurrence of the summer monsoon as well as the regional climate. Thus plate tectonics can affect the occurrence of severe storms and related flooding and erosion. To a lesser extent, the same forces have produced the high plateau area of the Rocky Mountains that affects the regional climate and the monsoonal storms of the southwestern United States.

In summary, we see that plate tectonics is linked to many geologic processes operating at the surface of the earth that are linked to natural hazards.

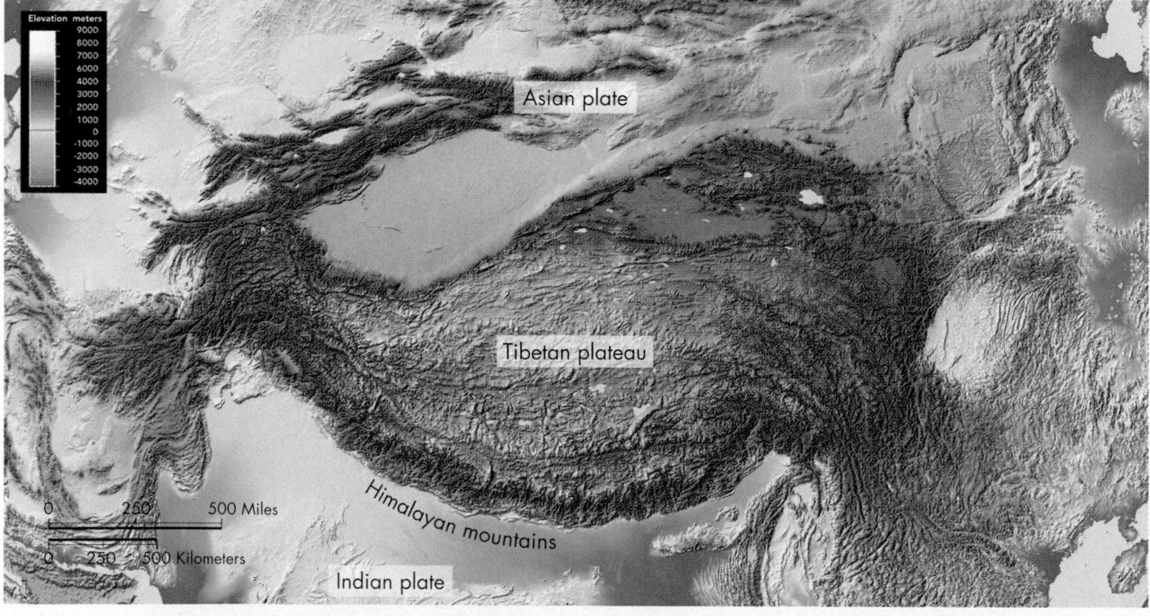

▲ FIGURE 2.24 **CONTINENTAL COLLISION** False color topographic image showing continental collision between the Indian and Asian plates. Collision of the two plates has uplifted the Tibetan plateau to average of 5000 m (16,400 ft.) and numerous peaks in the Himalayan Mountains to heights greater than 8000 m (26,000 ft.).

Summary

Our knowledge concerning the structure of Earth's interior is based on the study of seismology. Thus we are able to define the major layers of Earth, including the inner core, outer core, mantle, and crust. The uppermost layer of Earth is known as the lithosphere, which is relatively strong and rigid compared with the soft asthenosphere found below it. The lithosphere is broken into large pieces called plates that move relative to one another. As these plates move, they carry along the continents embedded within them. This process of plate tectonics produces large landforms, including continents, ocean basins, mountain ranges, and large plateaus. Oceanic basins are formed by the process of seafloor spreading and are destroyed by the process of subduction, both of which result from convection within the mantle.

The three types of plate boundaries are divergent (mid-oceanic ridges, spreading centers), convergent (subduction zones and continental collisions), and transform faults. At some locations, three plates meet in areas known as triple junctions. Rates of plate movement are generally a few centimeters per year.

Evidence supporting seafloor spreading includes paleomagnetic data, the configurations of hot spots and chains of volcanoes, and reconstructions of past continental positions.

The driving forces in plate tectonics are ridge push and slab pull. At present, we believe the process of slab pull is more significant than ridge push for moving tectonic plates from spreading centers to subduction zones.

Plate tectonics is extremely important in determining the occurrence and frequency of volcanic eruptions, earthquakes, and other natural hazards.

Key Terms

asthenosphere (p. 28)
continental drift (p. 31)
convection (p. 28)
convergent boundary (p. 33)
crust (p. 28)
divergent boundary (p. 33)
hot spot (p. 40)
inner core (p. 27)
isostasy (p. 35)
lithosphere (p. 28)
magnetic reversal (p. 38)
mantle (p. 27)
mid-oceanic ridge (p. 31)
Moho (p. 28)
outer core (p. 27)
paleomagnetism (p. 38)
plate tectonics (p. 29)
seafloor spreading (p. 31)
seismology (p. 28)
spreading center (p. 31)
subduction zone (p. 31)
submarine trench (p. 34)
transform boundary (p. 36)
triple junction (p. 36)
Wadati-Benioff zone (p. 31)

Review Questions

1. What are the major differences between the inner and outer cores of Earth?
2. How are the major properties of the lithosphere different from those of the asthenosphere?
3. What are the three major types of plate boundaries?
4. What is the major process that is thought to produce Earth's magnetic field?
5. Why has the study of paleomagnetism and magnetic reversals been important in understanding plate tectonics?
6. What are hot spots?
7. What is the difference between ridge push and slab pull in the explanation of plate motion?

Critical Thinking Question

1. Assume that the supercontinent Pangaea (see Figure 2.17) never broke up. Now deduce how Earth processes, landforms, and environments might be different from how they are today with the continents spread all over the globe. *Hint:* Think about what the breakup of the continents did in terms of building mountain ranges and producing ocean basins that affect climate and so forth.

3 Earthquakes

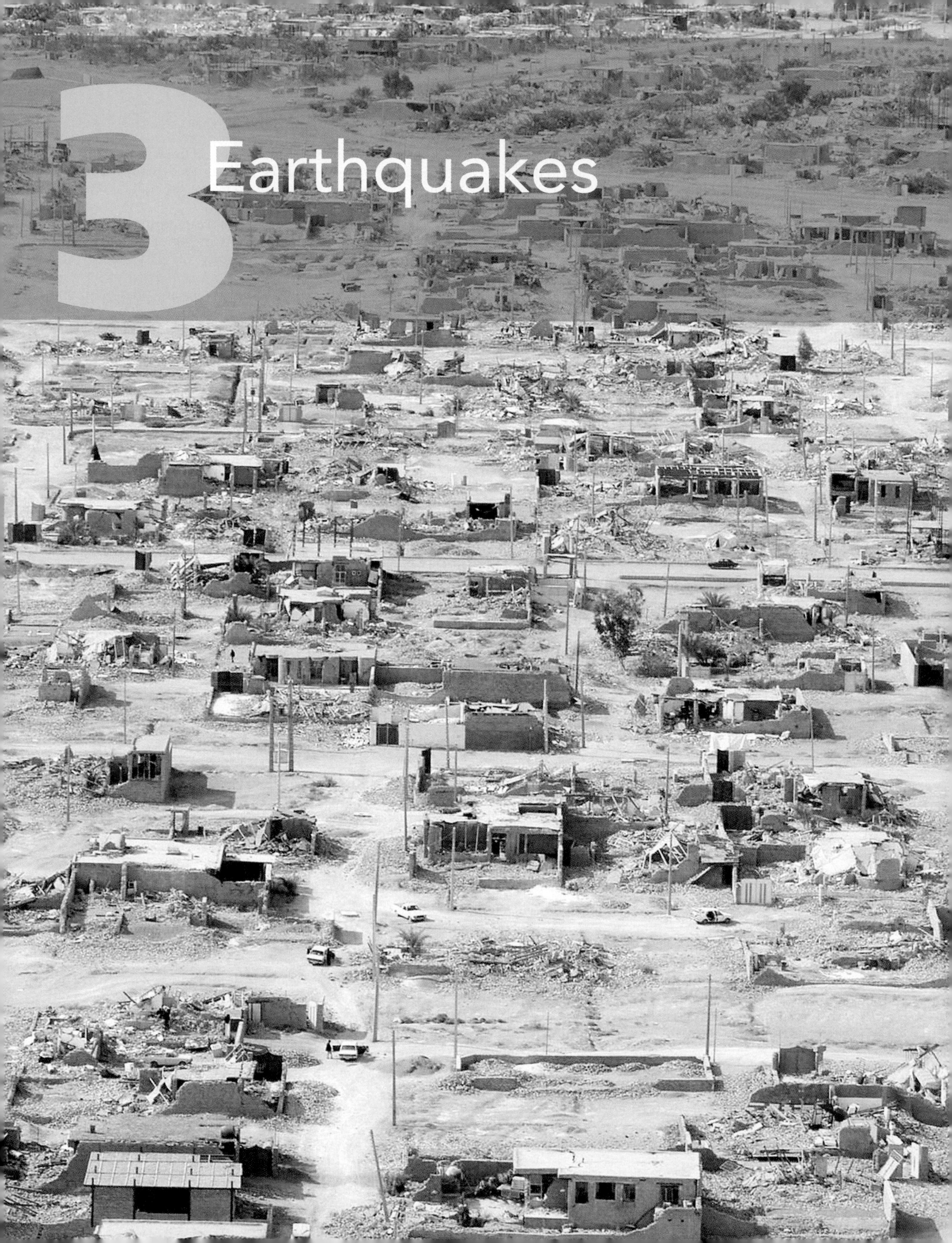

Earthquake Catastrophes: Lessons Learned

Catastrophic **earthquakes** are devastating events that can destroy large cities and take thousands of lives in a matter of seconds. Here we use the term *catastrophic* as introduced earlier to designate an earthquake that kills a large number of people or causes extensive damage (see Chapter 1). A sixteenth-century earthquake in China reportedly claimed 850,000 lives. More recently, a 1923 earthquake near Tokyo killed 143,000 people, and a 1976 earthquake in China killed several hundred thousand. On October 8, 2005, a catastrophic earthquake struck northern Pakistan. Although the epicenter was in Pakistan, extensive damage also occurred in Kashmir and India (Figure 3.1a). More than 80,000 people were killed and over 30,000 buildings collapsed. Entire villages were destroyed, some buried by landslides triggered by the violent shaking. A catastrophic continental earthquake occurred in the Sichuan province of China in 2008, killing about 87,500 people. Some villages were completely buried by landslides, and more than 5 million buildings collapsed (Figure 3.1b).[1–3] On January 12, 2010, an earthquake struck Haiti and killed about 240,000 people.[4] The recent earthquakes that took thousands of lives have a common dominator: They all caused tremendous loss of life because the buildings people were in collapsed. Homes, schools, hospitals, and industrial buildings were subject to collapse in strong shaking because they were not constructed with earthquakes in mind or the required construction codes were ignored to presumably save money. In short, most deaths were human induced. We have experience with strong shaking in California (Northridge in 1994), and more recently in Chile (February 27, 2010), where an earthquake about 500 times as strong as the Haiti earthquake a month earlier killed not hundreds of thousands but hundreds. A earthquake several times smaller than Northridge struck the town of L'Aqila in Italy in 2009, and many of the buildings collapsed, killing about 300 people. A similar earthquake in California probably would have caused no or very few deaths. The Italian event was a catastrophe because the buildings were

The earthquakes in recent years that took thousands of lives have a common dominator: they all caused tremendous loss of life because the buildings people were in collapsed or in short most deaths were human induced.

LEARNING OBJECTIVES

Earthquakes are serious natural hazards that affect people across the globe, sometimes at long distances from where the quakes occur. They are especially dangerous because seismologists, the scientists who study earthquakes, cannot predict them in time for evacuations or other precautions. Your goals in reading this chapter should be to

- understand how scientists measure and compare earthquakes.
- be familiar with earthquake processes such as faulting, tectonic creep, and the formation of seismic waves.
- know which global regions are most at risk for earthquakes and why they are at risk.
- know and understand the effects of earthquakes such as shaking, ground rupture, and liquefaction.
- identify how earthquakes are linked to other natural hazards such as landslides, fires, and tsunamis.
- know the important natural service functions of earthquakes.
- know how human beings interact with and affect the earthquake hazard.
- understand how we can minimize seismic risk, and recognize adjustments we can make to protect ourselves.

◀ **Poorly constructed buildings collapse** These buildings in Bam in southwestern Iran were constructed of masonry and unable to withstand the violent shaking of a **M** 6.6 earthquake in 2003. Collapse of masonry buildings was also widespread in the 2005 **M** 7.6 Bhuj quake in India. *(Getty Images)*

(a)

(b)

(c)

(d)

▲ **FIGURE 3.1 EARTHQUAKE DAMAGE** (a) Survivors rescue an injured neighbor from collapsed buidlings in Balakot, Pakistan, in 2005. Death toll from the quake exceeded 80,000 people and millions were left without food and shelter. *(B.K. Bangash/AP Images)*; (b) collapsed buildings from 2008 earthquake in China that killed about 87,500 people; *(Greg Baker/AP Images)* (c) earthquake damage in Italy that killed several hundred people; *(Ho/Reuters/Newscom)* (d) the brown area in the center of this image is the landslide that buried more than 500 homes in the Las Colinas neighborhood of Santa Tecla, a western suburb of San Salvador, the capital of El Salvador. The landslide, which killed more than 500 people, was triggered by a January 2001 earthquake. *(U.S. Geological Survey)*

not constructed to sustain moderate seismic shaking (Figure 3.1c).

Building regulations and zoning played an important role in determining the extent of damage from these earthquakes. Other factors, such as how deep the earthquake rupture was and the nature of the soil and rock, were also important in explaining the pattern of damage. However, the dominant factor in the number of deaths was the degree to which buildings could withstand the shaking. Criminal complaints have been lodged against some construction companies when experts conclude that poor building materials were used or building codes were ignored. The pattern of damage following some earthquakes supports these complaints—many older homes were shaken, but not damaged, whereas newer homes were reduced to rubble. Finally, in El Salvador (catastrophic earthquake in 2001), as in many other earthquakes, it was not just how structures were built, but where they were built that affected the number of casualties. For example, the Las Colinas neighborhood in Santa Tecla suffered the greatest loss of life from quake-related landslides mainly from one giant slide (Figure 3.1d). The extensive damage appears to have resulted from building homes below a slope of loose earth material and from poor land-use practices. Balsamo Ridge, the hill above the community, is composed of unstable volcanic ash and other granular materials that are susceptible to landslides. Local authorities and environmentalists also claim that deforestation and greed were directly responsible for intensifying the disaster. Residents of Las Colinas had pleaded with the government to block construction of mansions on the hillside above them. The residents correctly argued that fewer plants covering the bare ground would leave them vulnerable to landslides. Their pleas were ignored, the mansions went up, and during the earthquake, the slopes came down.

Although we cannot control the geologic environment or the depth of an earthquake, there is much we can do to avoid excessive damage and loss of life. The deaths in Haiti, Italy, El Salvador, and China are especially tragic because much of the devastation could have been avoided if building regulations had been adhered to, and protecting lives had been more important than building cheaper structures or making more money. Sound planning techniques and earthquake engineering of buildings can prevent unnecessary damage and loss of life. For examples, the 1994 Northridge earthquake shook the Los Angeles Basin, killing 60 people while causing at least $20 billion in property damages, and the 1989, the Loma Prieta earthquake (San Francisco region) killed about 63 people. As tragic as these were, a similar earthquake in Pakistan, India, or Mexico could kill at least several tens of thousands of people. The difference is that homes in California are constructed to better withstand earthquake shaking.[3] Therefore, the lesson is that where buildings are designed to withstand shaking and are built properly, as many are in Chile and California, deaths from collapsed buildings are many times lower than in areas where construction is substandard.[5]

3.1 Introduction to Earthquakes

Worldwide, people feel an estimated 1 million earthquakes a year. However, few of these quakes are noticed very far from their source, and even fewer are considered major earthquakes. Table 3.1 lists significant earthquakes in the United States in the past 200 years. Even if you have never experienced an earthquake, you may wonder what actually happens when one occurs. To understand the effects of earthquakes, we need to learn how they are measured and how one earthquake can be compared with another. Earthquakes are compared by the amount of energy released, their **magnitude**, and by the effects of ground motion on people and structures, their **intensity**.

TABLE 3.1

Significant Earthquakes in the United States

Year	Locality	Damage (millions of dollars)	Number of Deaths
1811–1812	New Madrid, Missouri	Unknown	Unknown
1886	Charleston, South Carolina	23	60
1906	San Francisco, California (includes deaths from fire)	524	>3000
1925	Santa Barbara, California	8	13
1933	Long Beach, California	40	115
1940	Imperial Valley, California	6	9
1952	Kern County, California	60	12
1959	Hebgen Lake, Montana (damage to timber and roads)	11	28
1964	Prince William Sound, Alaska (includes tsunami deaths and damage near Anchorage and on U.S. West Coast)	500	125
1965	Seattle-Tacoma, Washington	13	7
1971	Sylmar (San Fernando), California	553	65
1983	Coalinga, California	31	0
1983	Borah Peak, central Idaho	15	2
1987	Whittier Narrows, California	358	8
1989	Loma Prieta (San Francisco area), California	6000	63
1992	Landers, California	271	3
1994	Northridge, California	20,000	60
2001	Nisqually (Puget Sound), Washington	2300	1
2002	Denali Fault, south-central Alaska	(sparsely populated area)	0
2006	Kiholo Bay, Hawai'i, Hawai'i	200	0
2010	Southernmost California and Northern Mexico	~100	several in Mexico

Source: U.S. Geological Survey Earthquake Hazards Program (http://earthquakes.usgs.gov).

EARTHQUAKE MAGNITUDE

A CNN or AP news report on an earthquake generally gives information about where the earthquake started. This location, known as the **epicenter**, is the place on the surface of Earth above where the ruptured rocks broke to produce the earthquake (Figure 3.2). The point of initial breaking or rupturing within the Earth is known as the **focus**, or hypocenter, of the earthquake and is directly below the epicenter.

News reports also indicate the size of the quake with a decimal number (e.g., 6.8) that refers to the moment magnitude of the earthquake. **Moment magnitude** is determined from an estimate of the area that ruptured along a fault plane during the quake, the amount of movement or slippage along the fault, and the rigidity of the rocks near the focus of the quake.

Prior to the use of moment magnitude, seismologists described earthquake energy with a scale developed for local use in Southern California by the famous seismologist Charles Richter. Although some news reports still refer to the "Richter scale," it is no longer in common use by seismologists. Both the older Richter scale and the newer moment magnitude scale are logarithmic. This means that the increase from one integer to the next is not linear. For example, the ground displacement during a magnitude 6 earthquake is approximately 10 times as great as displacement during a magnitude 5 earthquake.

Except for very large earthquakes, the magnitude on the Richter scale is approximately equal to moment magnitude. Because of this correlation, we will refer to the size of an earthquake simply as its magnitude, **M**, without designating Richter or moment magnitude.

It is interesting to look at how often earthquakes of various magnitudes occur. Earthquakes are given descriptive adjectives based on their magnitude (Table 3.2). For example, most damaging earthquakes are described as major (**M** 7–7.9) or strong (**M** 6–6.9). *Major earthquakes* are capable of causing widespread and serious damage. *Strong earthquakes* can also cause considerable damage depending on their location and the nature of earth materials. Fortunately, the most powerful *great earthquakes* (**M** 8 or higher) are uncommon; the worldwide average is one per year (Table 3.2). In contrast, more than 3,500 minor earthquakes with a magnitude of less than 3 occur each day. Most of these quakes are too small or too remote to be felt by people.

It is also interesting to see how earthquake magnitude is related to the amount of ground motion, or shaking, that occurs during a quake. We will discuss this relationship in general terms here and in more detail later. Ground motion, in either a vertical or horizontal direction, is recorded by an instrument known as a **seismograph**.

The amount of ground motion that takes place from an earthquake is related to its magnitude, depth, and the geologic environment in which it occurs. As earthquake

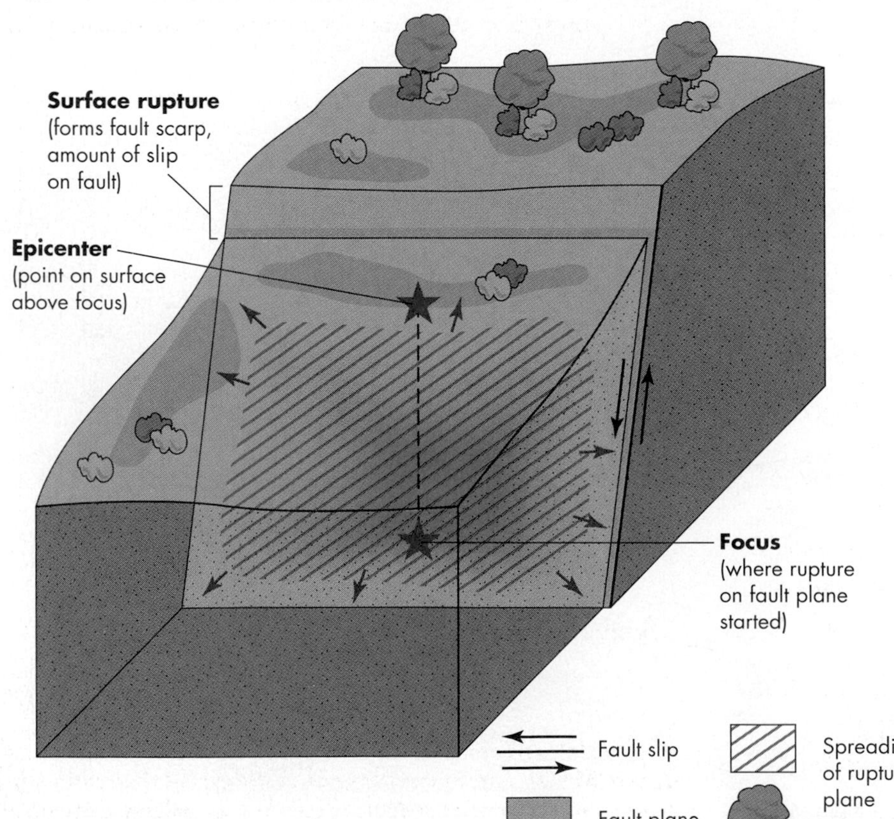

FIGURE 3.2 BASIC EARTHQUAKE FEATURES Block diagram showing fault plane (light tan surface); amount of displacement, rupture area (gray, closely spaced diagonal lines); focus (lower red star); and epicenter (upper red star). Rupture starts at the focus and propagates (red arrows) up, down, and laterally. During a major or great earthquake, slip may be 2 to 20 m (7 to 70 ft.) along a fault length of 100 or more kilometers (60 or more miles). Rupture area may be 1000 km^2 (400 mi.^{2}) or more.

TABLE 3.2

Frequency of Earthquakes by Descriptor Classification

Descriptor	Average Annual Magnitude	Number of Events
Great	8 and higher	1
Major	7–7.9	17
Strong	6–6.9	134
Moderate	5–5.9	1319
Light	4–4.9	13,000 (estimated)
Minor	3–3.9	130,000 (estimated)
Very Minor	2–2.9	1,300,000 (estimated) (approx.150 per hour)

Source: U.S. Geological Survey. 2010. Earthquake facts and statistics. http://earthquake.usgs.gov/earthquakes/eqarchives/year/eqstats.php. Accessed 11/26/2010.

magnitudes increase, the amount of ground motion changes less drastically than the amount of energy released (Table 3.3). As this table illustrates, the difference between a **M** 6 and a **M** 7 earthquake is considerable. Although the amount of ground motion, or shaking, from a **M** 7 earthquake is 10 times as great as that from a **M** 6 earthquake, the amount of energy released is 32 times as much! The energy released from a **M** 6 earthquake is about equivalent to a 30-kiloton nuclear explosion, compared to about a 1-megaton explosion for a **M** 7 earthquake. Another way of looking at this is that it would take 32 **M** 6 earthquakes to release as much energy as one **M** 7 quake.

EARTHQUAKE INTENSITY

The moment magnitude scale provides a quantitative way of comparing earthquakes. In contrast, earthquake intensity is often indicated with the qualitative **Modified Mercalli scale**. The 12 categories on this scale are assigned Roman numerals (Table 3.4). Each category contains a description of how people perceived the shaking from a quake, and the extent of damage to buildings and other human-made structures. For example, the 1971 Sylmar earthquake in the San Fernando Valley, California, had a single magnitude (6.7), but its Mercalli intensity varied from I to XI depending on how close you were to the epicenter and on the local geologic conditions (Figure 3.3).

TABLE 3.3

Change in Ground Motion and Energy Released from an Incremental Change in Earthquake Magnitude

Units of Magnitude Change	Ground Motion Change[a]	Change in Amount of Energy Released
1	10 times	About 32 times
0.5	3.2 times	About 5.5 times
0.3	2 times	About 3 times
0.1	1.3 times	About 1.4 times

[a] As reflected in the maximum amplitude of seismic waves on a standard seismograph. Amplitude is the distance that a seismic wave is displaced from a baseline (zero line) that is established when no seismic waves are detected.

Source: U.S. Geological Survey 2010. Earthquake facts and statistics http://earthquake.usgs.gov/earthquakes/eqarchives/year/eqstats.php. Accessed 11/26/2010.

Earthquake intensities are commonly shown on maps. Conventional Modified Mercalli Intensity maps, such as Figure 3.3, take days or even weeks to complete. They are based on questionnaires sent to residents near the epicenter, newspaper articles, and reports from damage assessment teams.

Recently the U.S. Geological Survey (USGS) began experimenting with the use of the Internet to collect intensity information. Its Web site (http://earthquake.usgs.gov/eqcenter/dyfi.php) solicits e-mail reports from people in the United States who have recently felt an earthquake. These reports are used to prepare an online *Community Internet Intensity Map* that is updated every few minutes after a large earthquake.

One of the major challenges during a damaging earthquake is to quickly determine where the damage is most severe. This information is now available in parts of California, the Pacific Northwest, and Utah, where there are dense networks of high-quality seismograph stations. These networks transmit direct measurements of ground motion as soon as the shaking stops. This information is known as *instrumental intensity*, and it is used to immediately produce a *shake map* that shows both intensity of

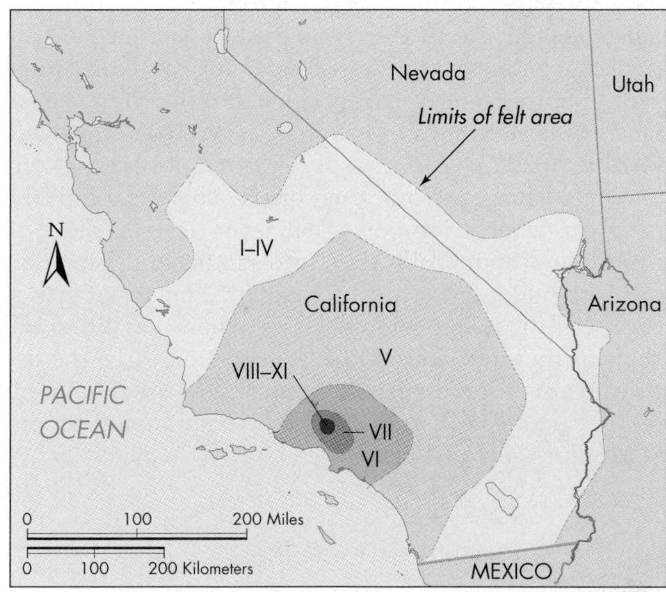

▲ FIGURE 3.3 **INTENSITY OF SHAKING** Modified Mercalli Intensity map for the 1971 Sylmar, California, earthquake (**M** 6.7), determined after the earthquake. The red and orange areas experienced the most intense shaking. See Table 3.4 for an explanation of the Roman numerals. *(U.S. Geological Survey. 1974.* Earthquake Information Bulletin *6[5])*

TABLE 3.4

Abbreviated Modified Mercalli Intensity Scale

Intensity	Effects
I	Felt by few people under especially favorable conditions.
II	Felt by only a few persons at rest, especially on upper floors of buildings.
III	Felt quite noticeably indoors, especially on upper floors of buildings. Many people do not recognize it as an earthquake. Standing vehicles may rock slightly. Vibration feels like the passing of a truck.
IV	During the day felt indoors by many, outdoors by few. At night some awakened. Dishes, windows, doors disturbed; walls make cracking sound. Sensation like heavy truck striking building; standing vehicles rock noticeably.
V	Felt by nearly everyone; many awakened. Some dishes and windows broken. Unstable objects overturned.
VI	Felt by all; many frightened. Some heavy furniture moved; a few instances of fallen plaster or damaged chimneys. Damage slight.
VII	Damage negligible in buildings of good design and construction; slight to moderate in well-built ordinary structures; considerable in poorly built or badly designed structures; some chimneys broken. Noticed by vehicle drivers.
VIII	Damage slight in specially designed structures; considerable damage in ordinary substantial buildings with partial collapse; damage great in poorly built structures; fall of chimneys, factory stacks, columns, monuments, and walls. Heavy furniture overturned. Disturbs vehicle drivers.
IX	Damage considerable in specially designed structures; well-designed frame structures thrown out of plumb. Damage great in substantial buildings, with partial collapse. Buildings shifted off foundations. Underground pipes broken.
X	Some well-built wooden structures are destroyed; most masonry and frame structures with foundations destroyed; train rails bent.
XI	Few, if any, masonry structures remain standing. Bridges destroyed. Underground pipelines taken out of service. Train rails bent greatly.
XII	Damage total. Waves seen on ground surfaces. Lines of sight and level are distorted. Objects thrown into the air.

Source: Modified after U.S. Geological Survey Earthquake Hazards Program http://earthquake.usgs.gov/learning/topics/mercalli.php. Accessed 6/12/07

shaking and potential damage. Figure 3.4 shows shake maps for the 1994 **M** 6.7 Northridge, California, earthquake and the 2001 **M** 6.8 Nisqually earthquake centered in Puget Sound, Washington. Notice that the magnitudes of the two earthquakes are similar, but the intensity of the shaking was greater for the Northridge event. The technology to produce and distribute shake maps in the minutes following an earthquake was made available in 2002.[6] The cost of seismographs is small relative to damages from earthquake shaking, and the arrival of emergency personnel is critical in the first minutes and hours following an earthquake if people in collapsed buildings are to be rescued. The shake map is also useful in helping locate areas where gas lines and other utilities are likely to be damaged. Clearly, the use of this technology, especially in our urban areas vulnerable to earthquakes, is a desirable component of our preparedness for earthquakes.

3.2 Earthquake Processes

As discussed in Chapter 2, the Earth is a dynamic, evolving system in which plate tectonic processes form ocean basins, continents, and mountain ranges. These processes, including both earthquakes and volcanoes, are most active along the boundaries of lithospheric plates (Figure 3.5). Specifically, earthquakes occur along planes of weakness in Earth's crust, usually in the presence of strong tectonic forces.

PROCESS OF FAULTING

The process of fault rupture, or *faulting*, can be compared to sliding one rough board past another. Friction along the boundary between the boards may temporarily slow their motion, but rough edges break off and motion occurs at various places along the plane. Similarly, one lithospheric plate moving past another is slowed by friction along their boundary. This "braking action" stresses the rocks along the boundary. As a result, these rocks undergo strain or deformation. When stress on rocks exceeds their ability to withstand it, referred to as their *strength*, the rocks rupture along a fault. A **fault** is a fracture or fracture system where rocks have been displaced; that is, Earth's crust on one side of the fracture or fracture system has moved relative to the other side. The long-term rate of movement is known as the *slip rate* and is often recorded as millimeters per year or meters per 1000 years. When a rupture begins, it starts at the focus and then propagates up, down, and laterally along the fault plane during the earthquake. The sudden rupture of the rocks produces shock waves, called *seismic waves*, which can shake the ground. In other words, an earthquake releases the pent-up energy of the strained rocks as a series of vibrational pulses or waves of energy. Faults are

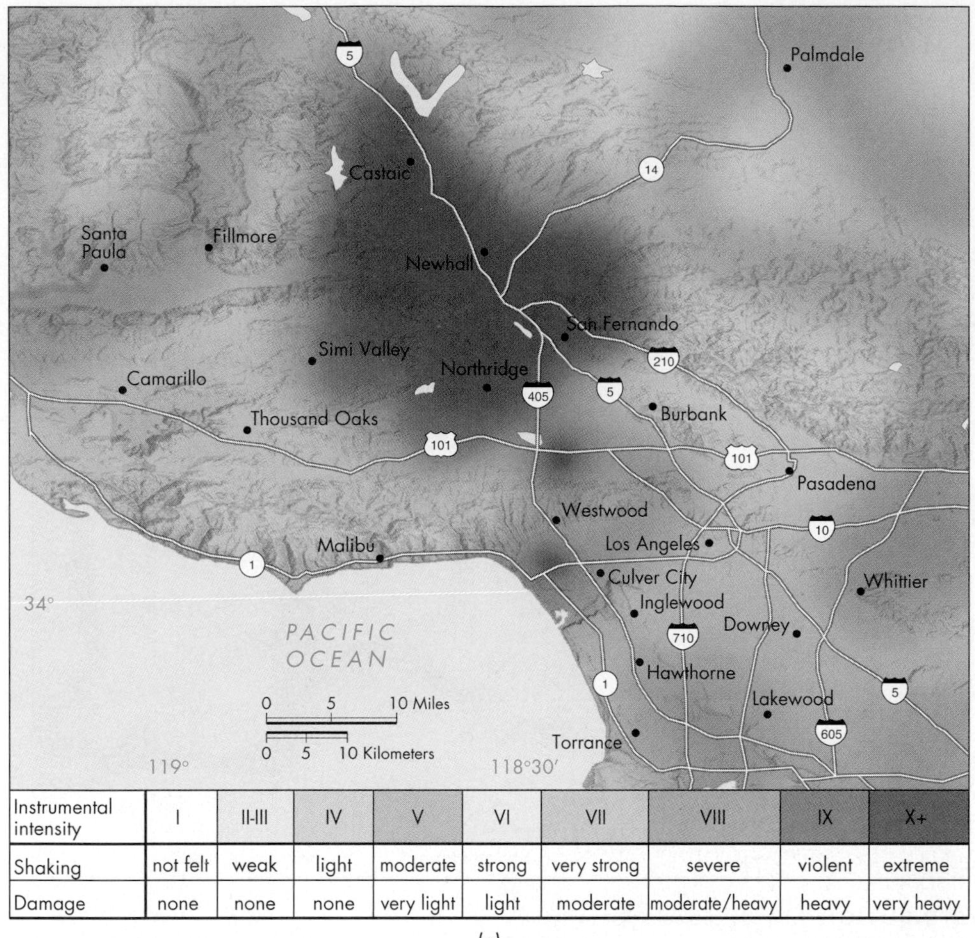

Instrumental intensity	I	II-III	IV	V	VI	VII	VIII	IX	X+
Shaking	not felt	weak	light	moderate	strong	very strong	severe	violent	extreme
Damage	none	none	none	very light	light	moderate	moderate/heavy	heavy	very heavy

(a)

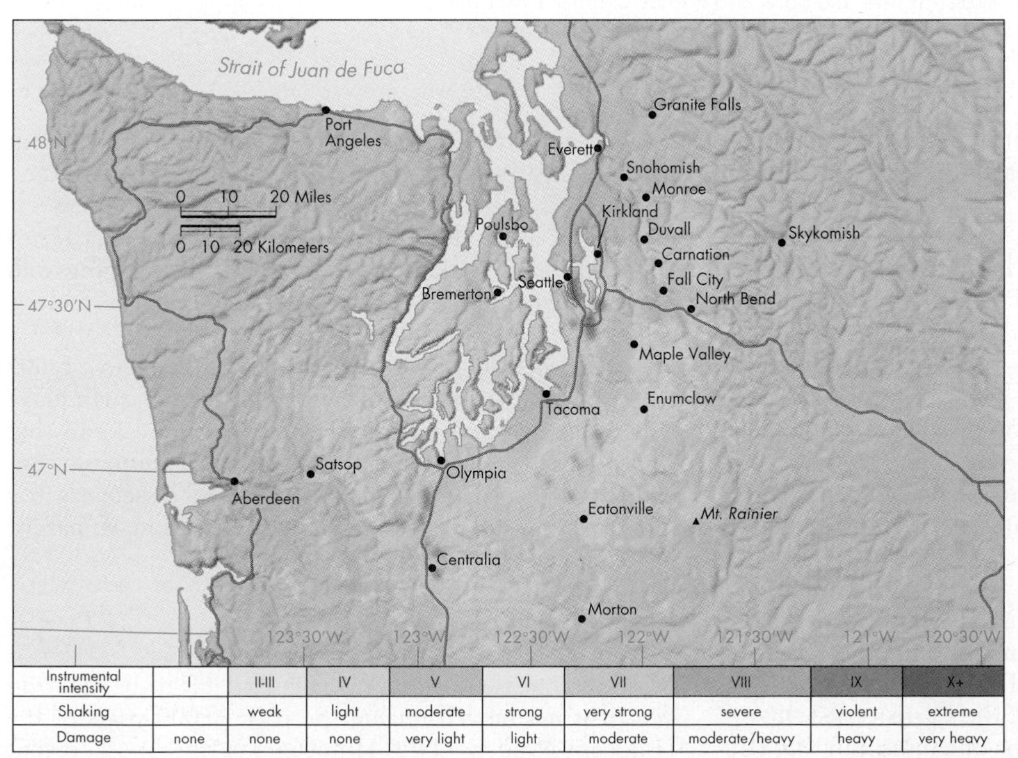

Instrumental intensity	I	II-III	IV	V	VI	VII	VIII	IX	X+
Shaking		weak	light	moderate	strong	very strong	severe	violent	extreme
Damage	none	none	none	very light	light	moderate	moderate/heavy	heavy	very heavy

(b)

◀ **FIGURE 3.4 SHAKE MAPS: REAL-TIME INSTRUMENTAL INTENSITY OF SHAKING** (a) Instrumental Intensity map for the 1994 Northridge, California, earthquake (**M** 6.7) prepared as the earthquake occurred. *(U.S. Geological Survey and courtesy of David Wald)*; (b) The 2001 Nisqually, Washington, earthquake (**M** 6.8). *(Pacific Northwest Seismograph Network, University of Washington)*

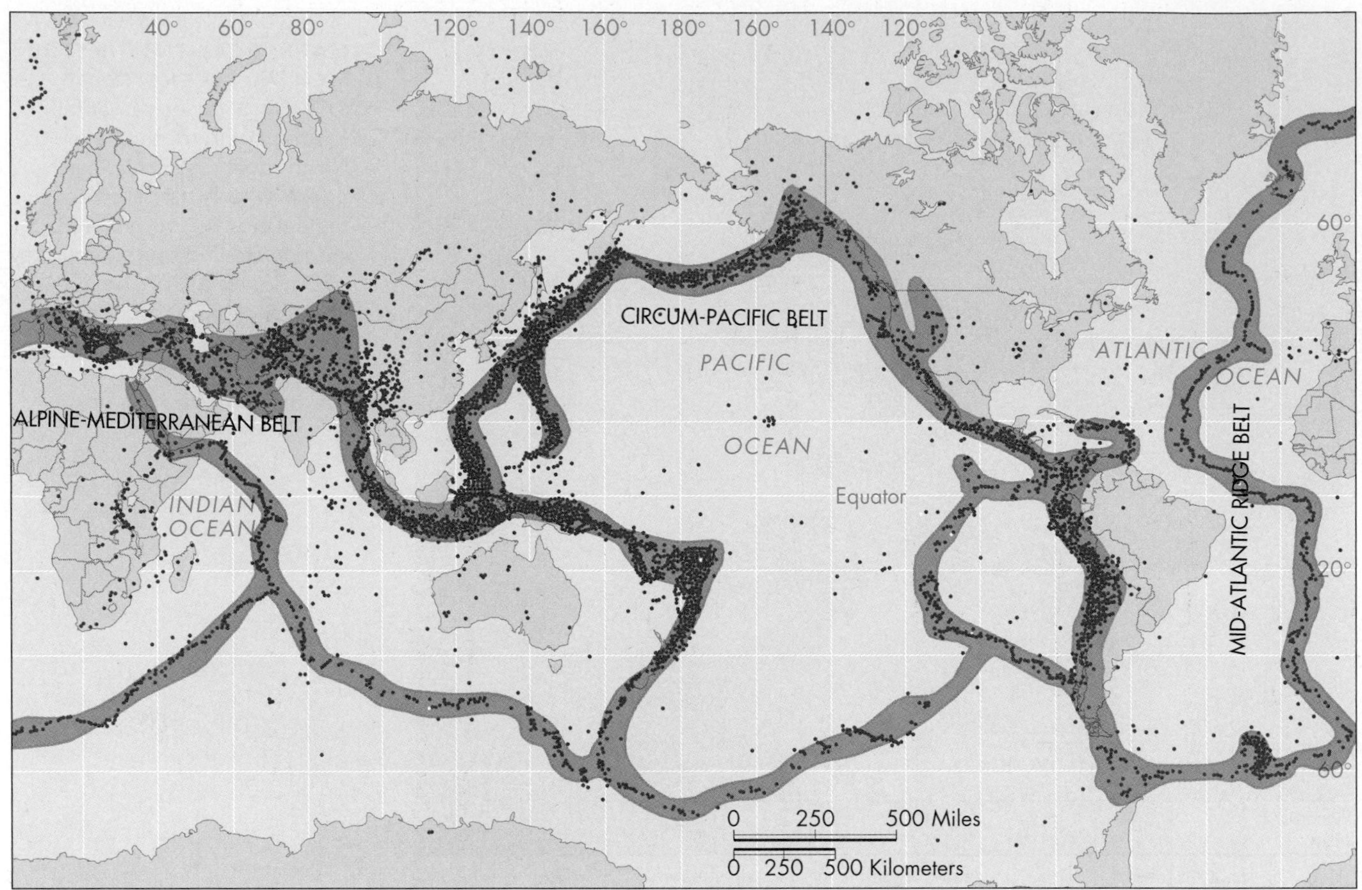

▲ **FIGURE 3.5 EARTHQUAKE DISTRIBUTION** Map of global seismicity (1963–1988, M 5+) showing epicenters of plate boundary (heavy concentration of dots) and intraplate (isolated dots) earthquakes. For the locations and names of Earth's tectonic plates, refer to Figure 2.5a. *(Courtesy of USGS National Earthquake Information Center)*

therefore *seismic sources*, and identifying them is the first step in evaluating the risk of an earthquake, or the seismic risk, in a given area.

Fault Types There are two basic types of geologic faults, strike-slip and dip-slip, which are distinguished by the direction in which the earth materials move. In a *strike-slip fault*, a block of the crust moves mainly in a horizontal direction, and in a *dip-slip fault*, a block of the crust moves mainly in a vertical direction (Figure 3.6). A dip-slip fault is classified as a *reverse fault* or a *normal fault* depending on which way the earth material on one side of the fault has moved relative to the material on the other side of the fault (Figure 3.6bc).

Geologists use centuries-old mining terminology to distinguish reverse and normal faults. Many early underground mines were dug at an incline into the Earth to mine mineralized fault zones. The mine, like the fault plane, was between two blocks of the Earth's crust. Imagine that the fault plane is the floor of a mine and that you are walking down the fault into the Earth (Figure 3.6). Miners called the block of the crust below their feet the *footwall* and the block hanging above their heads the *hanging-wall*. If motion on a geologic fault is such that the hanging-wall moves up relative to the footwall, then the fault is called a *reverse fault* if the fault plane is inclined at an angle steeper than 45 degrees or a *thrust fault* if the angle is 45 degrees or less. If the hanging-wall moves down relative to the footwall, then the fault is called a *normal fault*.

Until recently, it was thought that most active faults could be mapped because their most recent earthquake would cause surface rupture. However, we now know that some active faults, referred to as *blind faults*, do not extend to the surface (Figure 3.7). This discovery has made it more difficult to evaluate the earthquake hazard in some areas.

FAULT ACTIVITY

Most geologists consider a particular fault to be an *active fault* if it has moved during the past 10,000 years of the Holocene Epoch. The Holocene is the most recent epoch of the Quaternary Period of Earth history. It was preceded by the Pleistocene Epoch of the Quaternary Period. Much of our present-day landscape developed

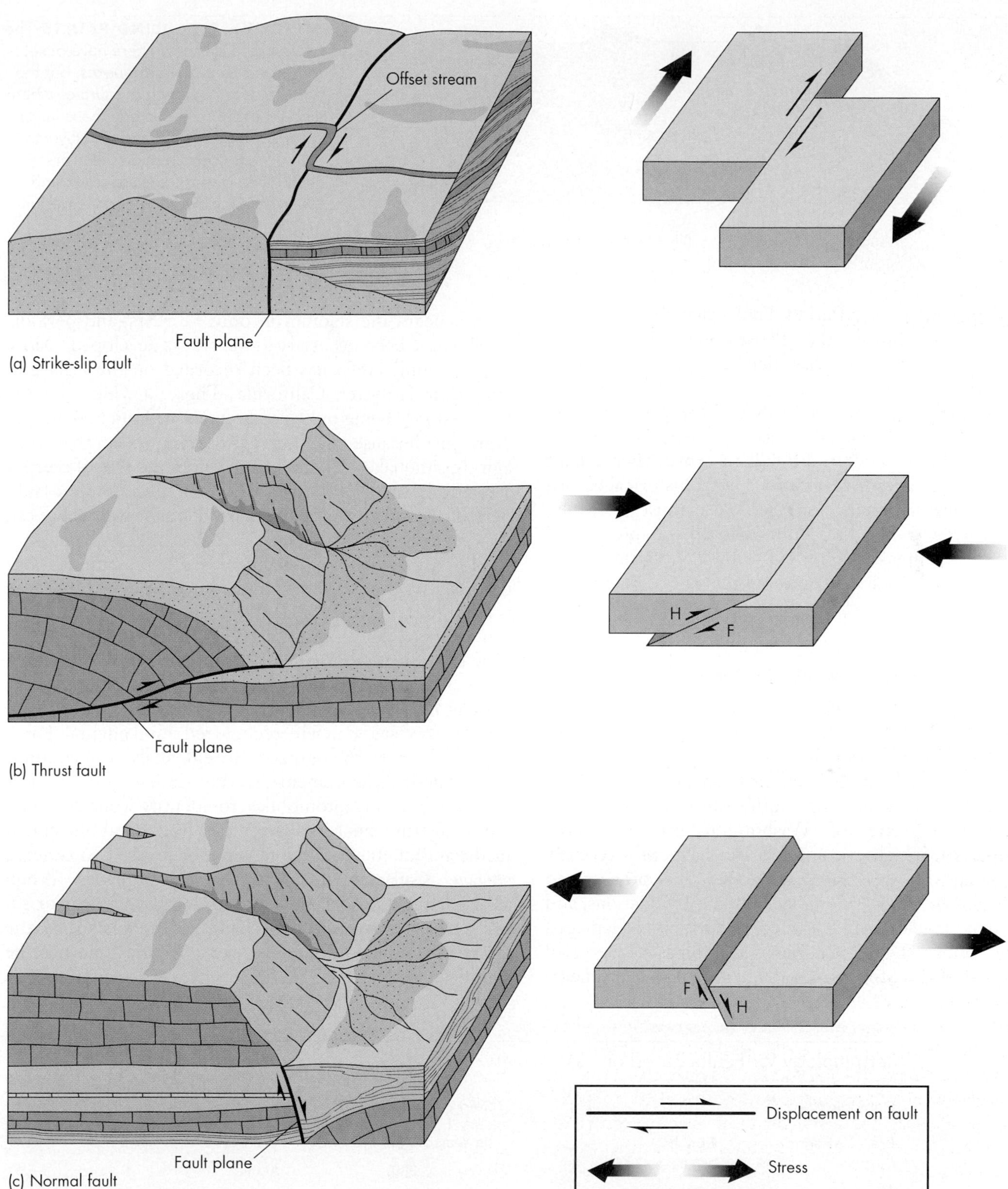

▲ **FIGURE 3.6 TYPES OF GEOLOGIC FAULTS** Three common types of faults and their effects on the landscape. Green and tan diagrams on the left show the landscape after movement along the fault (thick black line). (a) A strike-slip fault with horizontal displacement along the fault plane; (b) a thrust fault in which the hanging-wall (H) above the fault has moved up and over the footwall (F) below the fault; and (c) a normal fault in which the hanging-wall (H) on the right side of the fault has dropped down. Gray diagrams in the right column show the directions of stress (thick arrows) and displacement (thin half arrows) along each type of the fault.

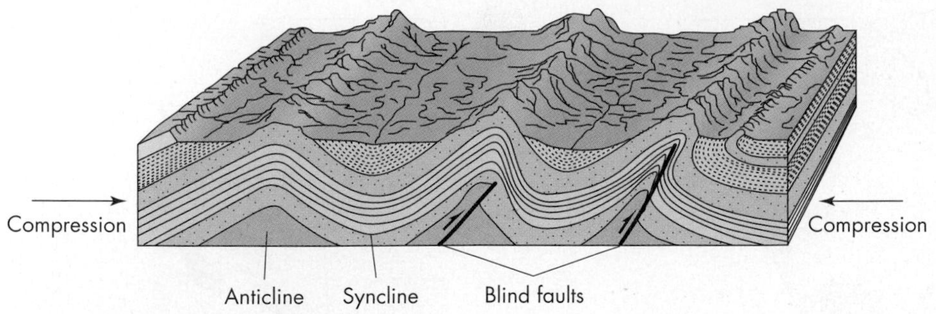

◀ FIGURE 3.7 **BLIND FAULTS** The two blind faults shown here have offset rock layers causing earthquakes, but the rupture does not reach the surface where it could easily be detected. These faults are associated with folded rock layers produced by compression of Earth's crust. The upward folds are called anticlines and downward folds are called synclines. *(Modified after Lutgens, F., and Tarbuck, E. 1992.* Essentials of geology, *4th ed. New York: Macmillan)*

during the Quaternary Period. Faults that show evidence of movement during the Pleistocene, but not the Holocene, Epoch are classified as *potentially active* (Table 3.5).

Faults that have not moved during the past 2 million years are generally classified as *inactive*. However, we emphasize that it is often difficult to prove when a fault was last active, especially if there is not a historical record of earthquakes. In many cases, geologists must determine the *paleoseismicity* of the fault, that is, the prehistoric record of earthquakes. They do so by identifying faulted earth materials and determining the age of the most recent displacement.

TECTONIC CREEP AND SLOW EARTHQUAKES

Some active faults exhibit **tectonic creep**, that is, gradual movement along a fault, which is not accompanied by felt earthquakes. For example, tectonic creep along the Cascadia subduction zone beneath southwestern British Columbia and the state of Washington produces **slow earthquakes**, which are not felt and have only recently been detected.[7] Also called *fault creep*, this process can slowly damage roads, sidewalks, building foundations, and other structures (Figure 3.8). Tectonic creep has damaged culverts under the University of California at Berkeley football stadium. Movement of 2.2 cm (1.3 in.) was measured beneath the stadium in only 11 years, and periodic repairs have been necessary as the cracks developed.[8] More rapid tectonic creep has been recorded on the Calaveras fault near Hollister, California. There, a winery on the fault is slowly being pulled apart at about 1 cm (0.4 in.) per year.[9] Just because a fault creeps does not mean that damaging earthquakes will not occur. Often the rate of creep is a relatively small portion of the total slip rate on a fault; periodic sudden displacements producing earthquakes can also be expected.

Slow earthquakes are similar to other earthquakes in that they are produced by fault rupture. The big difference is that the rupture, rather than being nearly instantaneous, can last from days to months. The moment magnitude of slow earthquakes can be in the range of 6 to 7 because a large area of rupture is often involved, although the amount of slip is generally small (a centimeter or so). Slow earthquakes are a newly recognized fundamental Earth process. They are recognized through analysis of continuous geodetic measurement or GPS, similar to the devices that are used in automobiles to identify your location. These instruments can differentiate horizontal movement in the millimeter range and have been used to observe surface displacements from slow earthquakes. When slow earthquakes occur frequently, say every year or so, their total contribution to changing the surface of the earth to produce mountains over geologic time may be significant.[10]

TABLE 3.5

Terminology for Faults Based on Last Activity

Geologic Age				Start of Time Interval	Fault Activity
Era	Period	Epoch		(in Years before Present)	
Cenozoic			Historic	200	Active
		Holocene	Prehistoric	10,000[a]	
	Quaternary	Pleistocene		1,650,000[a]	Potentially Active
	Tertiary			65,000,000	
Pre-Cenozoic time				4,600,000,000 (Age of the Earth)	Inactive

[a]Dates used for regulatory purposes. Actual dates for these geologic time intervals have changed (see Chapter 1).

Source: After California State Mining and Geology Board Classification, 1973.

(a)

(b)

▲ FIGURE 3.8 **TECTONIC CREEP** (a) Slow, continual movement along the San Andreas fault has split this concrete drainage ditch at Almaden Vineyards south of Hollister, California. For scale, there is a person's shoe and blue pant leg in the upper-left corner of the image. *(James A. Sugar/NGS Image Collection)*; (b) creep along the Hayward fault (a branch of the San Andreas fault) is slowly deforming the football stadium at the University of California, Berkeley, from goalpost to goalpost. *(Richard Allen)*

SEISMIC WAVES

When a fault ruptures, rocks break apart suddenly and violently, releasing energy in the form of seismic waves. These waves radiate outward in all directions from the focus, like ripples on a pond after a pebble has hit the water. It is the passage of these waves through the ground that we perceive as an earthquake, although the term *earthquake* also refers to the fault rupture that gives rise to the waves.

Some of the seismic waves generated by fault rupture travel within the body of the Earth and others travel along the surface. There are two types of body waves, P waves and S waves.

P waves, also called compressional or primary waves, are the faster of the two (Figure 3.9a). They can travel through a solid, liquid, and gas. P waves travel much more quickly through solids than through liquids. The average velocity for P waves through Earth's crust is 6 km (3.7 mi.) per second in contrast to 1.5 km (0.9 mi.) per second through water. Interestingly, when P waves reach Earth's surface and are transmitted into the air, people and other animals may be able to hear a fraction of them.[11] However, the sound that some people hear when an earthquake is approaching is the loud noise of objects vibrating, not the actual P wave.

S waves, also called shear or secondary waves, can travel only through solid materials (Figure 3.9b). They travel more slowly than P waves and have an average velocity through Earth's crust of 3 km (1.9 mi.) per second. S waves produce an up-and-down motion (sideways shear) at right angles to the direction that the wave is moving. This movement is similar to the whipping back and forth of a large jump rope being held by two people on a playground. When liquids are subjected to sideway shear, they are unable to spring back, explaining why S waves can't move through liquids.

When P and S waves reach the land surface, complex **surface waves** form and move along Earth's surface. These waves travel more slowly than either P or S waves, and they cause much of the damage near the epicenter. Because surface waves have a complex horizontal and vertical ground movement or rolling motion, they may crack walls and foundations of buildings, bridges, and roads (Figure 3.9c). People caught near the epicenter of a strong earthquake have reported seeing these waves rippling across the land surface. One type of surface wave, called a *Love wave*, causes horizontal shaking that is especially damaging to foundations.

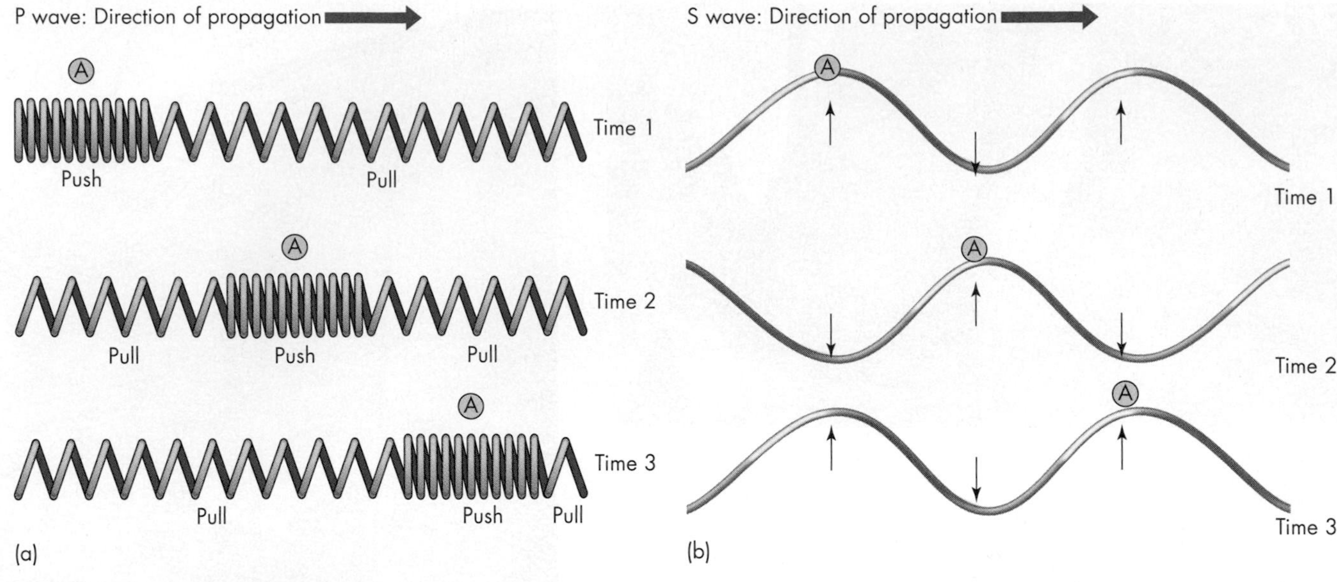

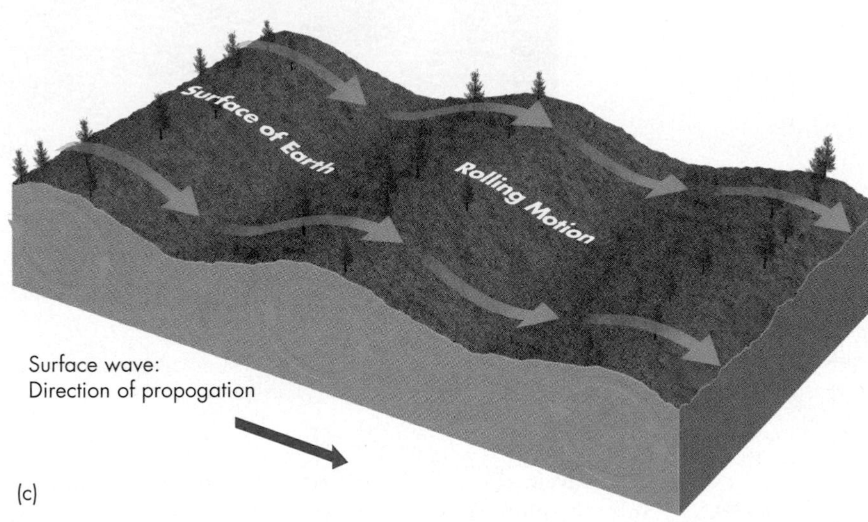

▲ **FIGURE 3.9 SEISMIC WAVES** Idealized diagram showing the behavior of the three major types of earthquake waves: P, S, and surface. (a) A P wave is a compressional wave like a sound wave. In this example, the coil compresses to form wave A as it passes from left to right from time 1 at the top of the illustration to time 3 on the bottom. (b) An S wave is a shear wave in which atoms in the solid earth vibrate at right angles to the direction that the wave is moving. In this example, vibrations move the coil up and down as wave A moves from left to right from time 1 at the top of the illustration to time 3 on the bottom. (c) A surface wave produces a rolling motion along the land surface as atoms vibrate in the solid earth at right angles to the direction that the wave is moving. In the type of surface wave shown here, the atoms vibrate upward and downward as the wave passes through from left to right.

3.3 Earthquake Shaking

Three important factors determine the shaking you will experience during an earthquake: (1) earthquake magnitude, (2) your location in relation to the epicenter and direction of rupture, and (3) local soil and rock conditions. In general, strong shaking may be expected from earthquakes of moderate magnitude (**M** 5 to 5.9) or larger. It is the strong motion from earthquakes that cracks the ground and makes the Earth "rock and roll" to damage buildings and other structures.

DISTANCE TO THE EPICENTER

To determine the distance to the epicenter, it must first be located using information about the P and S waves detected by seismographs (Figure 3.10a and b). The written or digital record of these waves is called a *seismogram*. On a seismogram, the P and S waves appear as an oscillating line that looks similar to an electrocardiograph (EKG) of a person's heartbeat (Figure 3.10c). Because P waves travel faster than S waves, they will always appear first on a seismogram. Seismologists use the difference between the time that the first P and S waves arrive (S–P) from a quake to determine how far away the epicenter is from the seismograph. For example, in Figure 3.10c, the S waves arrived 50 seconds after the P waves. This 50-second delay would occur if both types of waves traveled from an epicenter about 420 km (261 mi) away. Seismographs scattered across the globe will thus record different arrival times for P and S waves from an earthquake. The seismograph stations that are farthest from the epicenter will observe the greatest difference between P and S wave arrival times. This relationship was observed for the 1994 Northridge, California, earthquake (Figure 3.10d).

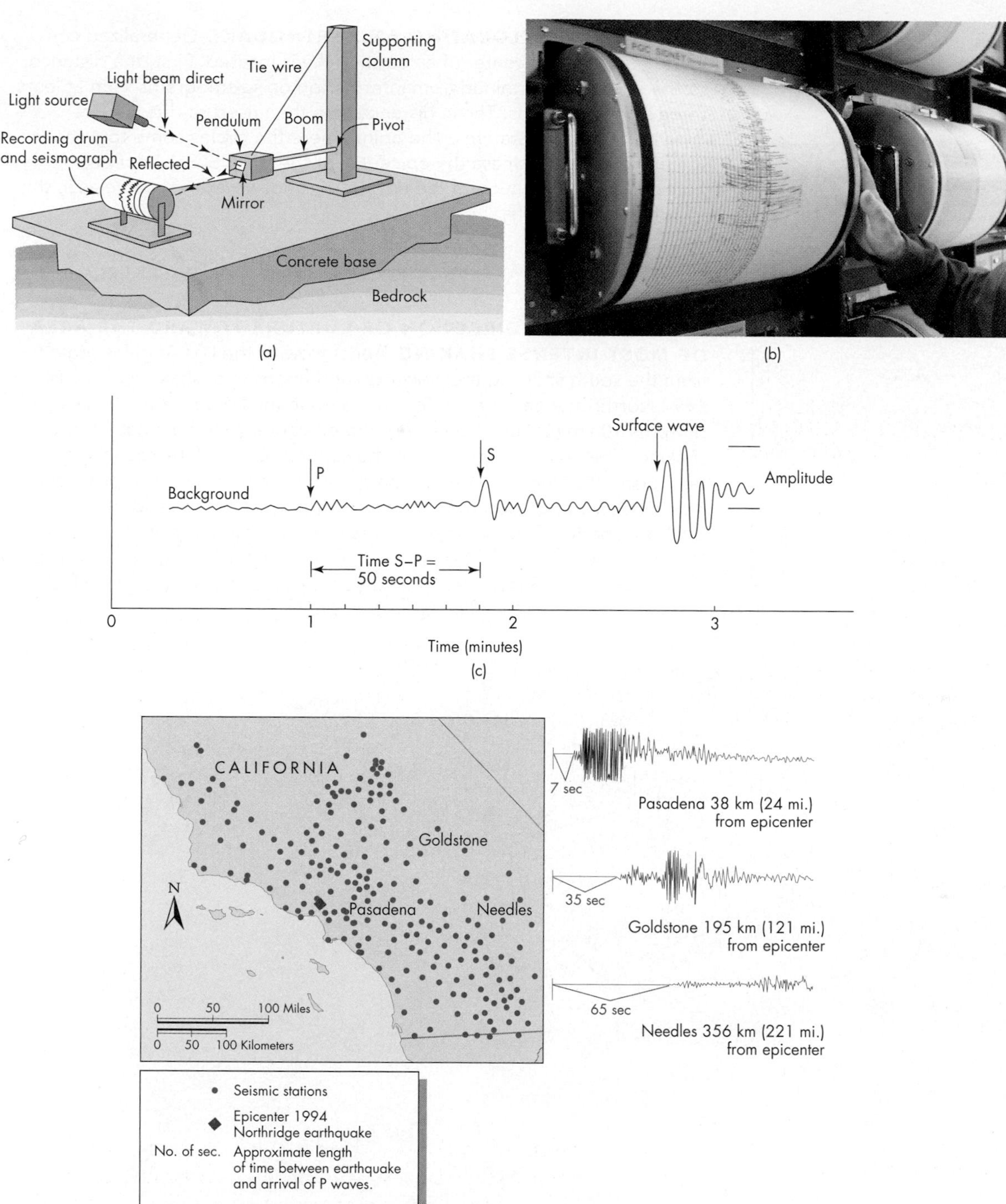

▲ **FIGURE 3.10 SEISMOGRAPH** (a) Simple seismograph showing how it works. (b) Modern seismograph at the Pacific Geoscience Centre near Victoria, British Columbia, showing a recording drum on which a seismogram is made. The earthquake is the February 28, 2001, Nisqually, Washington, earthquake (**M** 6.8). (*Ian McKain/AP Images*) (c) Idealized seismogram for an earthquake. The P, S, and surface waves can be identified by changes in the amplitude of the waves on the graph. The S–P time of 50 seconds tells us that the earthquake epicenter was about 420 km (261 mi.) from this seismograph. (d) Differences in arrival time and amount of shaking at three seismic stations located from 38 to 356 km (24 to 221 mi.) from the 1994 Northridge, California, earthquake. Notice that as you get farther away from the epicenter, the seismic waves take longer to reach the seismograph and the amplitude of the waves on the seismograph decreases. In general, the greater the amplitude of the waves on the seismogram, the stronger the shaking of the ground. *(Modified from Southern California Earthquake Center)*

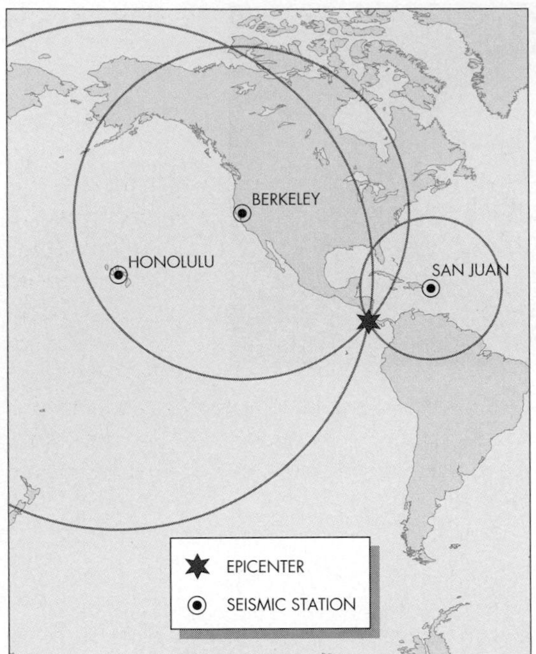

◀ **FIGURE 3.11 LOCATING AN EARTHQUAKE** Generalized concept of how the epicenter of an earthquake is located. First, the distance to the event is determined from information on seismograms from at least three seismic stations. Those distances are then used to draw circles around the seismic stations. The point where the circles come together is the epicenter, in this case the epicenter of the 2001 El Salvador earthquake. Accurate location of the epicenter is not always as simple as in this hypothetical example.

▼ **FIGURE 3.12 DIRECTION OF RUPTURE TOWARD THE AREA OF MOST INTENSE SHAKING** Aerial view of the Los Angeles region from the south showing the epicenter and intensity of shaking of the **M** 6.7 1994 Northridge earthquake. The red area in the center of the bull's-eye patterns 10 km (6 mi.) northwest of the epicenter had the most intense shaking. The bottom of the figure shows the focus and the section of the fault plane that ruptured during the quake. Colors on the fault plane show how much slippage occurred. Greatest slippage is in the reddish-purple bull's-eye pattern northwest of the focus. Rupture started at the focus and followed the direction of the white arrow to the northwest. *(U.S. Geological Survey. 1996.* USGS response to an urban earthquake, Northridge '94. *U.S. Geological Survey Open File Report 96–263)*

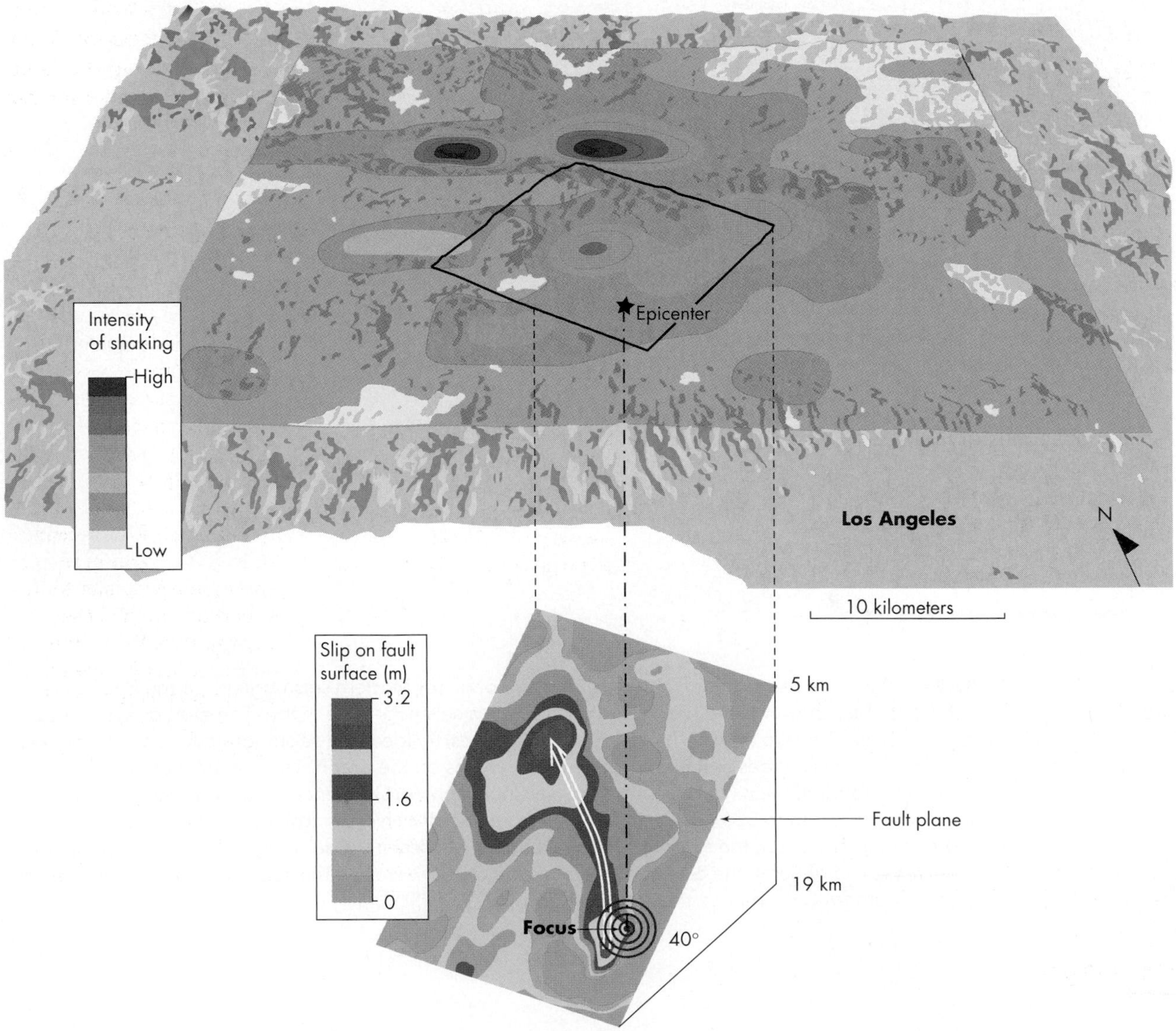

The difference between P and S arrival times (S–P) at seismographs in different locations can be used to locate the epicenter of an earthquake. To accomplish this, a distance to the epicenter is calculated for each of three seismographs, and the respective values are used for the radius of a circle drawn around each seismic station. These circles will intersect in one location—the epicenter. The process of locating a feature using distances from three points is called *triangulation*. Thus the epicenter of an earthquake in Central America can be located by triangulation from seismic stations in Honolulu, Hawai'i; Berkeley, California; and San Juan, Puerto Rico (Figure 3.11).

DEPTH OF FOCUS

In addition to distance, the depth of an earthquake influences the amount of shaking. Recall that the place within the Earth where the earthquake starts is the focus. In general, the deeper the focus of the earthquake, the less shaking that will occur at the surface. For relatively deep earthquakes. the seismic waves lose much of their energy before they reach the surface. This loss of energy, referred to as *attenuation*, occurred in the 2001 Nisqually, Washington, earthquake (**M** 6.8). The Nisqually quake occurred along the Cascadia subduction zone at a depth of 52 km (32 mi.). In comparison, there was less attenuation and more shaking in the 1994 Northridge quake (**M** 6.7), which had a shallower focus depth of 19 km (12 mi.) (Figure 3.12).

DIRECTION OF RUPTURE

A third factor that influences the amount of shaking is the direction that the rupture moves along the fault during the earthquake. Although rupture may proceed in many directions from the focus, the path of greatest rupture can focus earthquake energy. This behavior, known as **directivity**, contributes to the amplification of seismic waves and thus to increased shaking. For example, in the 1964 Northridge quake, the path of greatest rupture on the fault was to the northwest (Figure 3.12). This trend caused the most intense shaking to occur northwest of the epicenter rather than directly over the focus.

SUPERSHEAR

Supershear occurs when the propagation of rupture is faster than the velocity of shear-waves or surface waves produced by the rupture (Figure 3.13). The effect is analogous to supersonic aircraft that break the sound barrier, producing a sonic boom. Supershear can produce shock waves that produce strong ground motion along the fault. The increased shear and shock may significantly increase

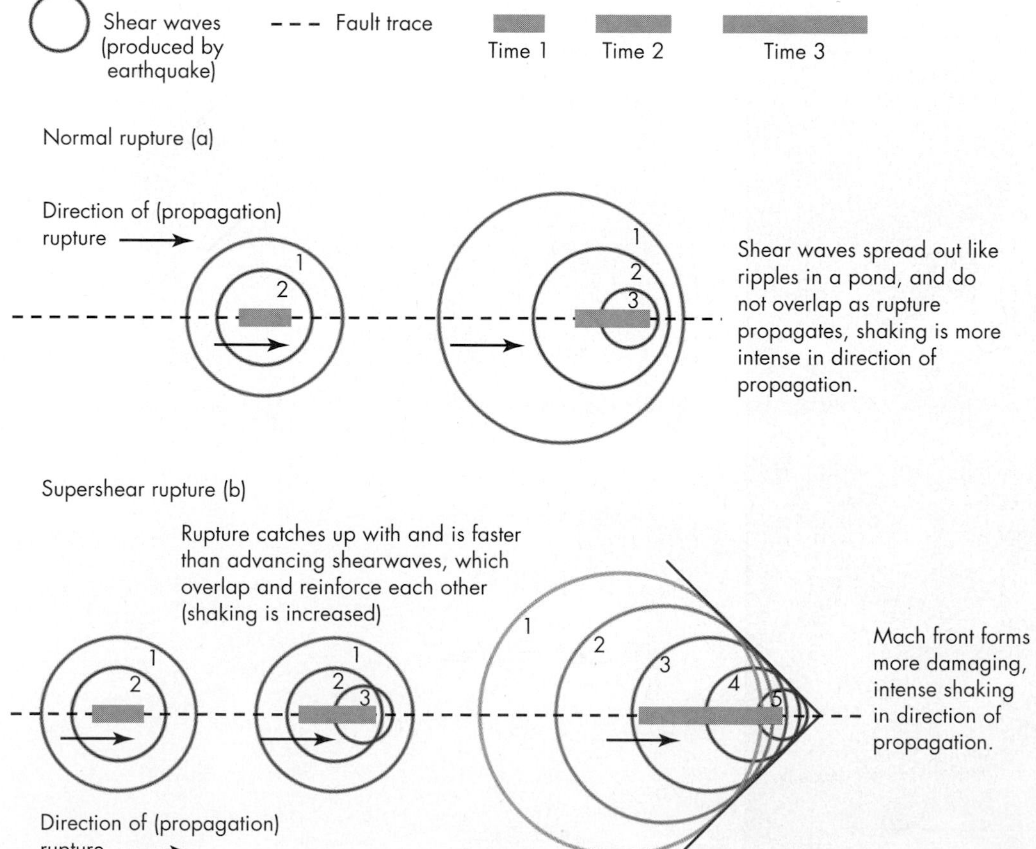

◀ **FIGURE 3.13** **SUPERSHEAR** Idealized diagram of the supershear process that occurs when the rupture propagation is faster than the speed of the shear waves produced by the rupture. A Mach front is produced where shaking is more intense and earthquake damage may be greater. The earthquake Mach front is analogous to the ratio of the speed of a jet plane to the speed of sound in the atmosphere. When the sound barrier is broken, a sonic boom is produced. *(Modified after New Scientist, 2009.)*

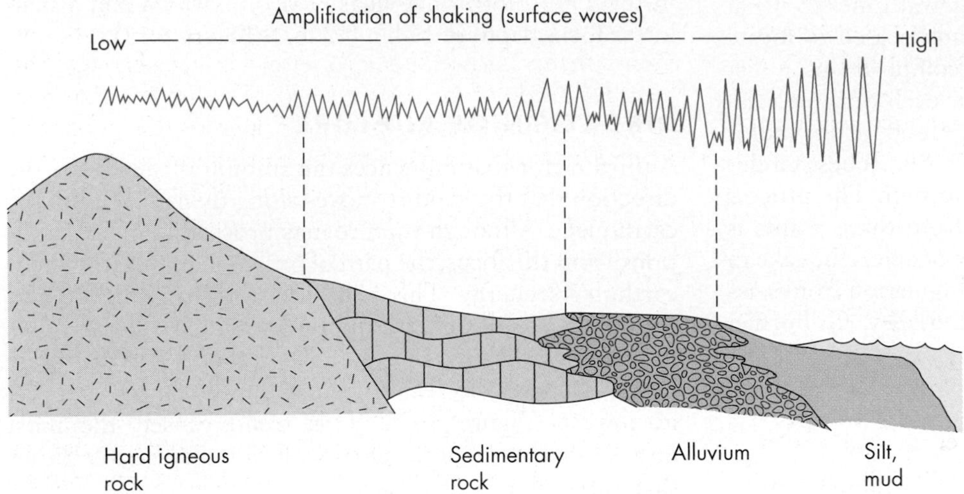

◀ **FIGURE 3.14 MATERIAL AMPLIFICATION OF SHAKING** Generalized relationship between near-surface earth material and amplification of shaking during an earthquake. The material amplification is highest in water-saturated sediment.

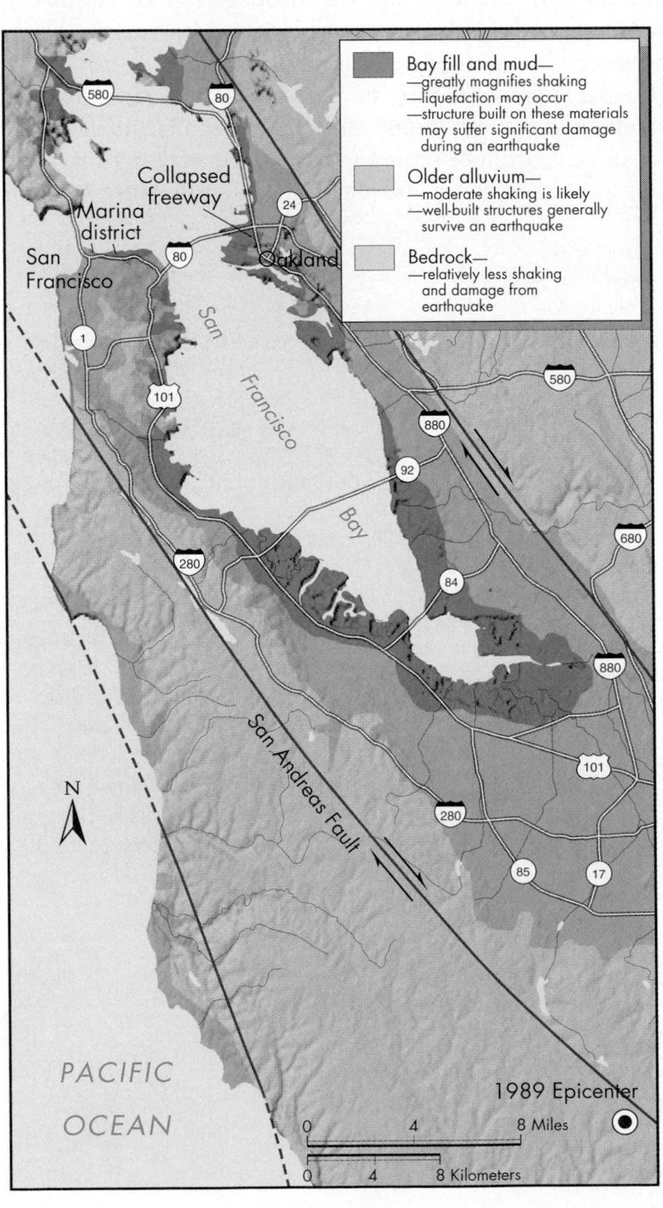

◀ **FIGURE 3.15 LOMA PRIETA EARTHQUAKE** Map of earth materials in the San Francisco Bay region, showing the San Andreas fault zone and the epicenter of the 1989 **M** 6.9 earthquake (southeast part of map). Fault zones are shown as solid and dashed red lines with the direction of relative movement indicated by arrows. The most severe shaking was on the muddy bay shore (bright orange) where there are natural deposits of mud and the bay has been filled in to create new land. Shaking caused the collapse of the Nimitz Freeway in Oakland and damage to buildings in the Marina district of San Francisco (northwest part of map). *(Modified after T. Hall from U.S. Geological Survey)*

the damage from a large earthquake. Supershear has been observed or suspected during several past earthquakes, including the 1906 San Francisco earthquake and the 2002 Denali earthquake in Alaska. Apparently, supershear is most likely to occur with strike-slip earthquakes that rupture a long straight fault segment of several tens to a hundred or more kilometers in length.[12]

LOCAL GEOLOGIC CONDITIONS

The nature of the local earth materials and geologic structure strongly influences the amount of ground motion. Earth materials of various types behave differently in an earthquake. This difference is related to their degree of consolidation. Seismic waves move faster through consolidated bedrock than they do through unconsolidated sediment or soil. They slow even further if the unconsolidated material has a high water content. For example, seismic waves typically slow down as they move from bedrock to stream deposits of sand and gravel, called *alluvium*, and then slow again as they move through coastal deposits of mud (Figure 3.14). As P and S waves slow down, the energy that was once directed forward is transferred to the vertical motion of the surface waves. This effect, known as **material amplification**, strongly influences the amount of ground motion experienced in an earthquake.

For example, in the 1989 Loma Prieta earthquake (**M** 6.9) in Northern California, the most severe ground shaking was experienced along the shore of San Francisco Bay (Figure 3.15). Intense ground motion collapsed the upper deck of the Nimitz Freeway in Oakland, killing 41 people and caused extensive damage in the Marina district of San Francisco (Figure 3.16, Figure 3.17). The portion of the freeway that collapsed was built on bay fill, mud that had been dumped into San Francisco Bay to create new land. Likewise, the Marina district was created after the 1906 San Francisco earthquake by filling in the shoreline with debris from damaged buildings and with mud pumped from the bottom of the bay.[13] The unconsolidated nature of the bay fill and mud, combined with

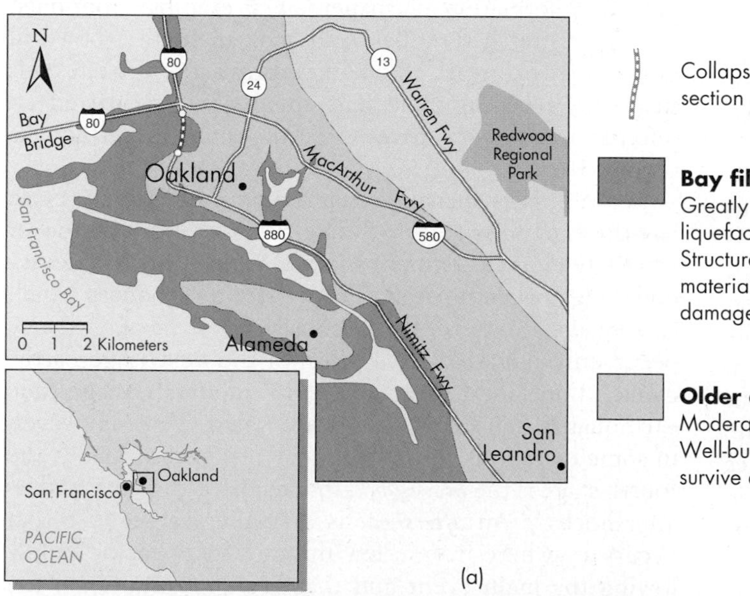

◀ **FIGURE 3.16 COLLAPSE OF A FREEWAY** (a) Generalized geologic map of part of the east shore of the San Francisco Bay showing bay fill and mud deposits (orange), and older alluvium (tan). The section of collapsed Nimitz Freeway (I-880) is above the word "Oakland" in the northwest part of the map. *(Modified after Hough, S. E., et al., 1990,* The role of sediment-induced amplification in the collapse of the Nimitz freeway during the October 17, 1989 Loma Prieta earthquake, Nature *344: 853–55. Copyright © Macmillan Magazines Ltd., 1990. Used by permission of the author)* (b) Upper deck of the Nimitz Freeway collapsed onto lower deck in the 1989 Loma Prieta earthquake. The road level for the lower deck was approximately at the level of the break in the concrete column. *(Dennis Laduzinsky)*

◀ **FIGURE 3.17 EARTHQUAKE DAMAGE** Damage to buildings in the Marina district of San Francisco resulting from the 1989 **M** 6.9 Loma Prieta earthquake. The poorly supported first floor has shifted from intense shaking because of material amplification. *(John K. Nakata/US Geological Survey, Denver)*

its high water content, caused the increased shaking. Amazingly, because of material amplification, the greatest shaking occurred 100 km (more than 60 mi.) north of the epicenter.

The deaths of more than 8000 people in the 1985 Michoacán earthquake (**M** 8.0) are another tragic example of this effect. This quake further demonstrated that buildings constructed on materials that are likely to accentuate and increase seismic shaking are extremely vulnerable to earthquakes, even if the event is centered several hundred kilometers away. Much of Mexico City is built on mud deposits from ancient Lake Texcoco (Figure 3.18a). When seismic waves struck the unconsolidated mud, the amplitude of surface shaking appears to have increased by a factor of 4 or 5. More than 500 buildings fell down from the intense, regular shaking.[14] In many buildings, the amplified shaking collapsed upper floors onto lower ones like a stack of pancakes[15] (Figure 3.18b). Local geologic structures can also influence the amount of shaking. For example, troughlike geologic structures, such as synclines and fault-bounded sedimentary basins, can focus seismic waves the way a magnifying lens focuses sunlight. This causes severe shaking in some areas and less intense shaking in others.

3.4 The Earthquake Cycle

Observations of the 1906 San Francisco earthquake (**M** 7.8) led to a hypothesis known as the **earthquake cycle**. The earthquake cycle proposes that there is a drop in elastic strain after an earthquake and a reaccumulation of strain before the next event.

Strain is deformation resulting from stress. *Elastic strain* may be thought of as deformation that is not permanent, provided that the stress is eventually released. When the stress is released, the elastically deformed material returns to its original shape. If the stress is not released and continues to increase, the deformed material eventually ruptures, making the deformation permanent. For example, continued stress on a stretched rubber band or a bent archery bow will cause them to break. When they break, the broken ends snap back, releasing their pent-up energy. A similar effect, referred to as *elastic rebound*, occurs after an earthquake (Figure 3.19).

Seismologists speculate that a typical earthquake cycle has three or four stages.[16] The first is a long period of inactivity along a segment of a geologic fault. In the second stage, accumulated elastic strain produces small earthquakes. A third stage, consisting of *foreshocks*, may occur only hours or days prior to the next large earthquake. Foreshocks are small- to moderate-magnitude earthquakes that occur before the main event. However, in some cases, this third stage may not occur. Finally, the fourth stage is the *mainshock*, the major earthquake, and its aftershocks.[16] An *aftershock* is a smaller earthquake that occurs anywhere from a few minutes to a year or so following the main event and that has an epicenter in the vicinity of the mainshock epicenter. Although this cycle is hypothetical and periods between major earthquakes are variable, these stages have been identified in many large earthquakes.

3.5 Geographic Regions at Risk from Earthquakes

Earthquakes are not randomly distributed. Most occur in well-defined zones along the boundaries of Earth's tectonic plates (see Figure 3.5). In the United States, the areas with the highest earthquake risk include the Pacific coastal areas of California, Oregon, Washington, Alaska, and Hawai'i; an area along the California-Nevada border; and the territories of Puerto Rico and the Virgin

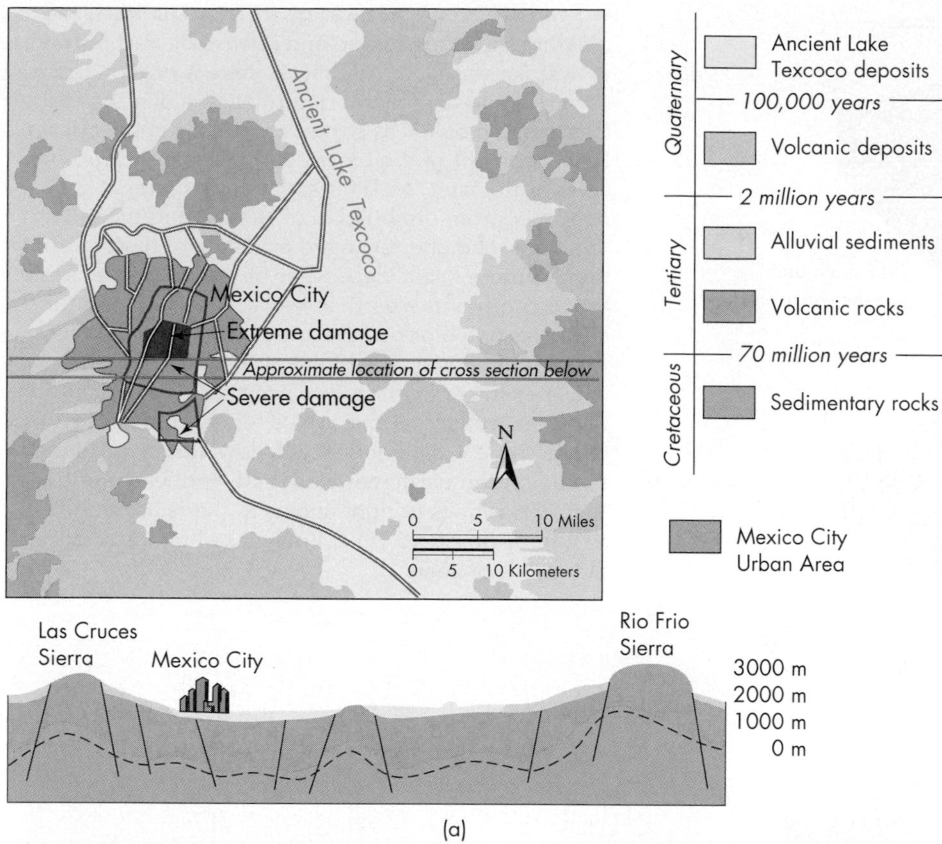

(a)

(b)

◀ **FIGURE 3.18 EARTHQUAKE DAMAGE TO MEXICO CITY** (a) Generalized geologic map and cross section of Mexico City and surrounding area. The deposits of ancient Lake Texcoco shown in yellow are where the greatest damage occurred in the 1985 **M** 8.0 Michoacán earthquake. Extreme damage from the quake occurred in the solid red area, and severe damage took place in the area outlined by the red line. Cross section extends to the west beyond the map. (b) The Benito Juárez Hospital in Mexico City that collapsed during the quake. *(T.C.Hanks & Darrell Herd/U.S. Geological Survey, Denver)*

Islands (Figure 3.20). For Canada, the risk is greatest in British Columbia and the Yukon Territory (Figure 3.21). This distribution is not surprising, since, with the exception of Hawai'i, these areas are on or very close to plate boundaries. What may be more surprising are the high-risk zones in South Carolina, the central Mississippi River Valley, and the St. Lawrence River Valley. These areas are not close to present-day plate boundaries, but they have experienced large *intraplate earthquakes*.

PLATE BOUNDARY EARTHQUAKES

Earthquakes occur along all three types of plate boundaries—convergent, divergent, and transform. The world's great earthquakes with magnitude 9+ are usually associated with subduction zones. In the past 100 years, earthquakes greater than **M** 9 have been associated with subduction along what is called the mega thrust that helps define the zone (see Figure 3.8a). These events are called **megathrust earthquakes**. The great Chile earthquake of

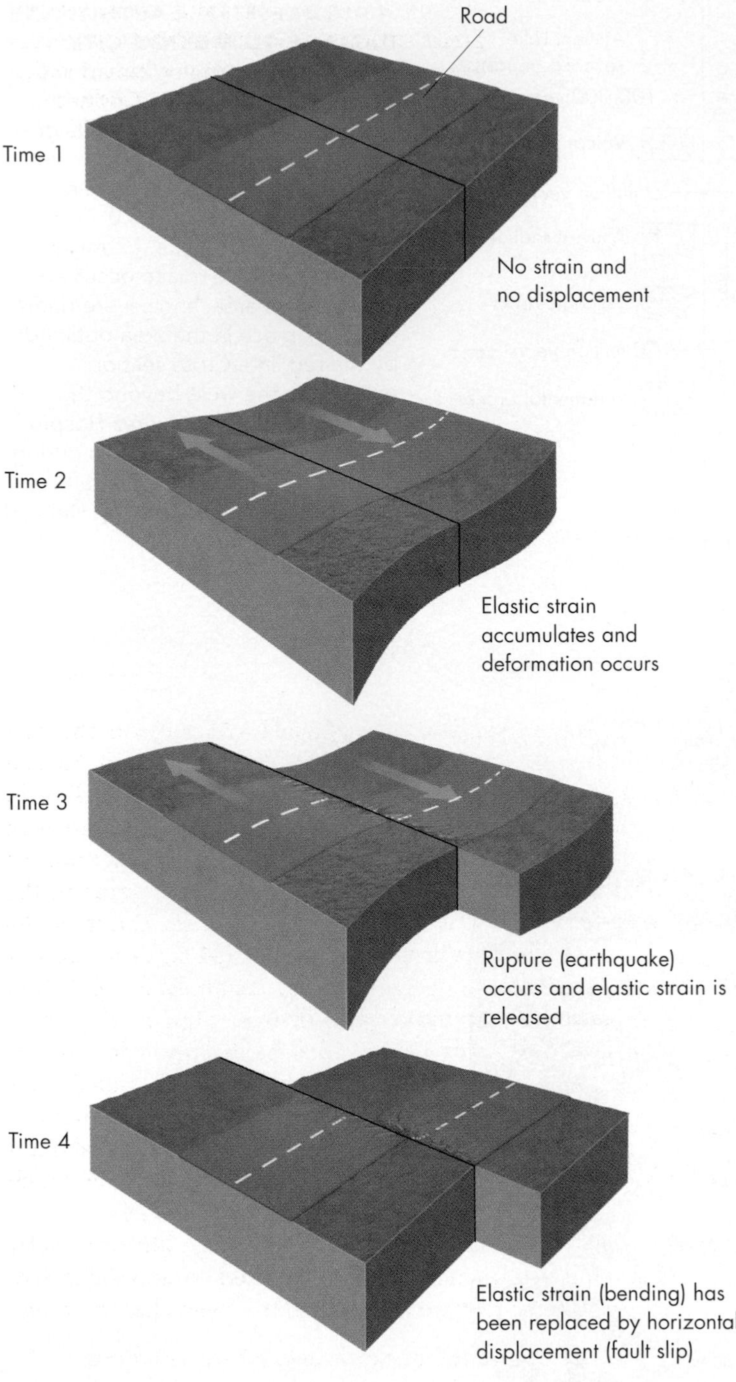

◀ FIGURE 3.19 **ELASTIC REBOUND** Idealized diagram showing buildup and release of elastic strain on a strike-slip fault. At Time 1, there is no significant accumulation of strain or deformation along the fault. Elastic strain builds as stress is gradually applied from the movement of the blocks of the crust on opposite sides of the fault. At Time 2, the earth material has deformed from the buildup of strain along the fault. The bent road and fence are evidence of the deformation. Rupture takes place when strain buildup exceeds the strength of the earth material. At Time 3, rupture occurs as one block of the crust moves and displaces the earth material. The offset of the road and fence indicates the amount of displacement. Ground motion takes place from seismic waves released in the earthquake. Time 4 is immediately following the rupture as the deformed earth material, road, and fence rebound to new positions at right angles to the fault. Progression from Time 1 to Time 4 may take hundreds to thousands of years.

1960 was a megathrust event of **M** 9.5 that killed about 1650 people. The earthquake occurred before subduction and the theory of plate tectonics theory was proposed. The 2010 **M** 8.8 Chile earthquake, while smaller than the 1960 one was most likely on the megathrust fault system that defines the down-going slab in the subduction zone offshore. According to the U S Geological Survey, the 2010 earthquake was generated at the gently sloping fault that separates the down-going Nazca plate to the east from the overriding South American plate to the west (see Figure 2.4). The regional rate of convergence is about 7 meters per century (70 mm per year, or 70 meters per 1000 years). The mostly offshore rupture exceeded 500 km (310 mi.) parallel to the coast (along the subduction zone). The rupture began deep (35 km or 21 mi.) beneath the coast and spread to the west, north, and south. As the rupture propagated, the fault slip generated intense earthquake shaking. The earthquake killed about 500 people. Fault slip warped the ocean floor, generating a damaging local tsunami in Chile coastal areas.[17]

In the western United States, earthquakes are common along the transform San Andreas fault zone and the

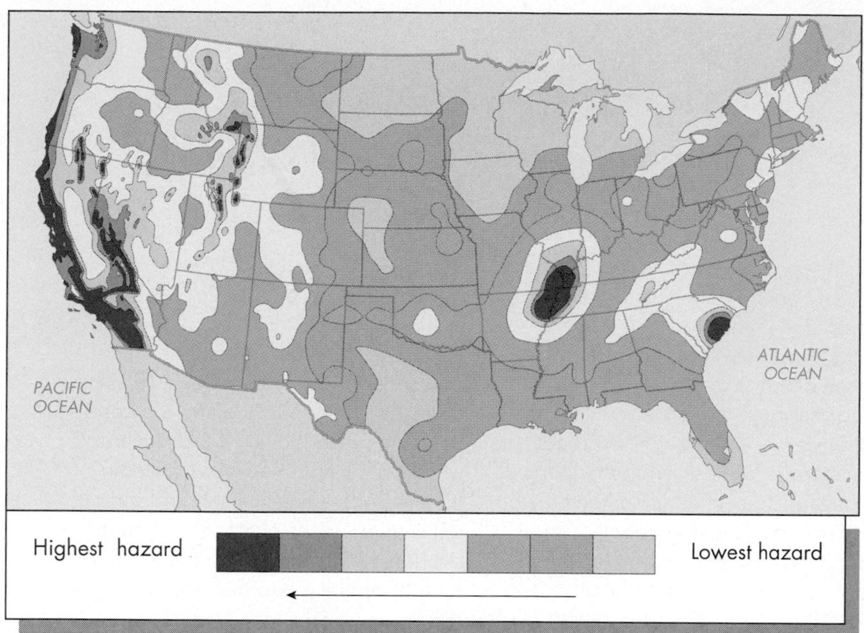

◀ **FIGURE 3.20 SEISMIC HAZARD IN CONTIGUOUS UNITED STATES** A probabilistic approach to the seismic hazard in the contiguous United States. Colors indicate level of hazard. *(From U.S. Geological Survey Fact Sheet FS-131-02, 2002)*

convergent Cascadia and Aleutian subduction zones (see Table 3.1). Whereas California and Alaska are famous for their earthquakes, Nevada, Utah, Idaho, Montana, Wyoming, Oregon, and Washington also experience quakes.

California straddles two lithospheric plates, the Pacific plate west of the San Andreas fault zone and the North American plate to the east. The motion of these plates in essentially opposite directions along the San Andreas and related faults results in frequent damaging earthquakes.

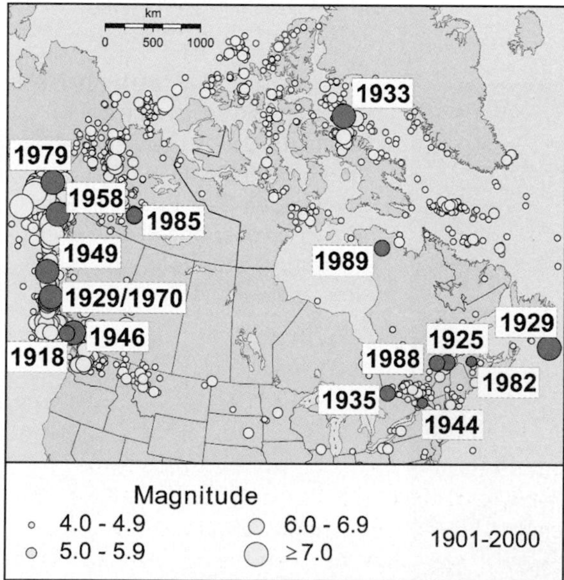

▲ **FIGURE 3.21 HISTORIC EARTHQUAKES IN CANADA** Map of significant twentieth-century earthquakes in Canada. *(Courtesy of Earth Sciences Information Centre, Canada's national earth sciences library)*

The 1989 Loma Prieta earthquake (**M** 6.9) on the San Andreas fault system south of San Francisco caused 63 deaths, 3757 injuries, and an estimated $5.6 billion in property damage. Both deaths and injuries would have been significantly greater if many people had not stayed home to avoid the crowds and congestion of the third game of the World Series in Oakland. Neither the Loma Prieta earthquake nor the Northridge earthquake (**M** 6.7) was considered a great earthquake. A great earthquake (**M** 8 and higher) occurring today in a densely populated part of Southern California could inflict $100 billion in damage and kill several thousand people.[18] Thus, neither quake was the anticipated "big one" in California. Both plate boundary earthquakes were initially puzzling because they did not produce a surface rupture. They demonstrated how much we still need to learn about earthquake processes.

This story about the Loma Prieta quake illustrates how observing a sequence of events related to an earthquake does not mean that we understand the causes behind them:

> Shortly before the quake struck, a 2-year-old boy was out playing in his yard and discovered how to turn on the lawn sprinklers. His mother was not amused and took him to his room for disobeying her instructions. Shortly after she had returned to the yard the Earth began to tremble violently. She heard her child yell and ran back into the house to find him terrified. He then remained very quiet until his father came home from work. On seeing his father, the child's first words were an emotional "Daddy, don't turn on the sprinkler!"[19]

The more we know about the probable location, magnitude, and effects of an earthquake, the better we can estimate the damage that is likely to occur and make the necessary plans for minimizing loss of life and property.

SURVIVOR STORY 

Magnitude 8.8 Earthquake and Tsunami on the Coast of Chile

Dr. Barry Keller (Figure 3.Aa) is a consulting geologist, hydrologist, and geophysists who earned his PhD at the University of California at Santa Barbara where I had the opportunity to meet him. He contributes to professional projects encompassing everything from industrial to governmental applications. He has academic and professional experience of more than 25 years in geological reconnaissance, earthquake seismology, geophysical exploration, soil and groundwater investigation and remediation, storm water management, dredge material study design, and hazardous waste planning. Dr. Keller has a home in Santa Barbara and another in Pichilemu, a beach town in central Chile known as a surfer's paradise. It is a rare opportunity for a professional who knows the science of earthquakes to write from personal experience. The following is Dr. Keller's story.

Strong earthquakes are a recurring hazard in some parts of the world, including the rim of the Pacific Ocean. Surviving a strong earthquake entails two basic phases. The first is the shaking itself and the second is the following days when basic infrastructure such as roads, water, electricity, and telephones may be out of service.

First Phase. The most important thing that can be done to prepare to survive strong shaking is to build strong buildings, as well as well thought out roads, water storage facilities, and other parts of the infrastructure. In addition, the time of day that an earthquake happens and where people happen to be is an important factor in the number of fatalities and injuries.

Second Phase. Storage of emergency supplies such as food, water, candles, flashlights with batteries, and generators with fuel are very helpful in the following days. To this might be added cash, since our ubiquitous sources of money, ATMs, do not work when the power is out.

My wife, Rosemarie, and I survived the M 8.8 earthquake and accompanying tsunami on the coast of Central Chile in February, 2010. The west coast of South America is above the subduction zone in which various oceanic plates are consumed beneath the South American continental plate, so very large earthquakes are a recurring fact of life. This particular part of the subduction zone had not ruptured in more than 100 years, and so was identified as a "seismic gap," liable to have a "BIG ONE."

As a geophysicist who studies such risks for environmental reviews, I was well aware of the seismic gap and the potential for large earthquakes and tsunamis when we built our house in Pichilemu (Figure 3Ab), a beach resort and surfing destination on the coast at 34.5° south latitude, which is similar to the north latitude of central California. Accordingly, the house was designed for seismic safety, built of poured concrete and concrete block with lots of rebar. It is on a hill at about 120 m elevation, well above the elevations typically reached by tsunamis (Figure 3.Ab)

In the dark of that very early Saturday morning, I was already awake, having gone to the bathroom and checked my watch at 3:22 A.M. The quake occurred at 3:34 A.M. The motion may have built up for a few seconds, then was VERY strong for about a minute—I was quickly thinking "This is for real!" I tried to check my watch again (as my seismology professor, Charlie Richter, taught us!), but the lights went out. Pretty strong motion was felt for about 2–3 more minutes. When the

FIGURE 3.A SURVIVORS OF THE 2010 CHILEAN EARTHQUAKE Barry Keller and his wife Rosie in front of the Golden Gate Bridge in San Francisco near one of Dr. Keller's field study sites. *(Barry & Rosemarie Keller)*

◀ **FIGURE 3.B THE KELLER'S HOME IN PICHILEMU CHILE** The home was constructed with poured concrete and rebar to strengthen the structure in case of an earthquake and it is situated at 120 m elevation; well above the height of an earthquake generated tsunami. *(Barry & Rosemarie Keller)*

quake "went off," I started to regain my belief that my life expectancy was more than a few seconds, and that my roof was NOT about to collapse; it was quite interesting to me as a geologist who studies such risks. However, it was quite terrifying to most other people, including my wife.

The feeling of the strong motion was similar to flying in a plane through severe turbulence, but with more high frequency jolts. I sat on the corner of the bed, holding on. The good news is that the house came through with no structural damage. However, items on shelves were strewn all over and there was lots of broken glass on the floor, so it is good advice to put on shoes before walking around in the dark in such a situation.

Within minutes, lots of people came up the hill to the roundabout at the bottom of our driveway. Many more were on the next hill over, the official evacuation spot. The lessons of the Sumatra tsunami were taken to heart by the Chilenos, and most people rapidly headed for high ground. It was a calm night with a full moon, which made evacuation easier, and people stayed until dawn. People continued to camp on both hills for several days.

That night, information was minimal, with only one local radio station broadcasting, but without outside news, and an AM station from Argentina on the air, again with minimal news. It was Sunday before we knew the extent of the damage in the rest of the country.

The tsunami struck the coast within 20 minutes of the earthquake. Damage at Pichilemu included complete destruction of a surf school (*Escuela de Surf*), restaurant, and cabañas at Punta de Lobos, a famous surf spot. In the lower downtown part of Pichilemu, there was widespread flooding, but the tsunami did not hit with enough force to destroy buildings. Only one life was lost, a child camped on the beach, so the evacuation was very effective, especially considering that this was the final weekend of summer vacation and there were lots of people at the coast.

Farther south along the coast there was much more destruction from the tsunami, including severe damage to a Chilean Navy base at Talcahuano harbor, near Concepción.

We had no electricity for two days (until we left for Santiago—it stayed out three more). My house has a booster pump for water, so without power we had no water pressure—although we had plenty of water in a storage tank (I have a boat bilge pump to get the water out, but it is very slow) and I had lots of bottles of water stashed, because the power goes out often. We had cell phone the first day, then (according to the local scuttlebutt) the gasoline for the emergency generator ran out, so the phones went out.

Aftershocks were generally pretty sharp, with almost continuous motion for many hours. Daily aftershocks continued for weeks, including an **M** 6.9 very local one that cracked some plaster at our house. The aftershocks were nerve-wracking, especially for the non-geologists, and caused a great deal of mental stress for the local populace, if not much physical damage.

Most modern structures fared well in the quake, minimizing loss of life. Many older adobe structures were damaged, including historic churches, government buildings, a museum, and the funeral parlor shown here however, at 3:30 A.M. on a Saturday morning, most such structures were not occupied, again minimizing loss of life. The timing was also fortuitous for major roads, which suffered cracking and fallen bridges, but, with little traffic at that hour, no loss of life.

—EDWARD KELLER

INTRAPLATE EARTHQUAKES

Although much less common than plate boundary earthquakes, intraplate earthquakes can be large and extremely damaging. Because they occur less often, there is generally a lack of preparedness, and buildings may not be able to withstand strong shaking. At least two earthquakes with **M** 7.5+ occurred in the winter of 1811–1812 in the central Mississippi Valley. The quakes nearly destroyed the town of New Madrid, Missouri, and killed an unknown number of people. They were felt in nearly every city of eastern North America from New Orleans to Quebec City in Canada, an area of more than a million square miles.[20] Seismic waves from these quakes rang church bells in Boston, more than 1600 km (1000 mi.) away (Figure 3.22a)! Associated ground motion produced intense surface deformation over a wide area from Memphis, Tennessee, 230 km (140 mi.) north, to the confluence of the Mississippi and Ohio Rivers. As a result, forests were flattened, fractures in the ground opened so wide that people had to cut down trees to cross them, and land sank several meters, causing local flooding. Journal and newspaper reports also indicate that local uplift of the land surface actually caused the Mississippi River to reverse its flow for a short time.[21]

These earthquakes occurred along the New Madrid seismic zone, a seismically active portion of a geologic structure known as the Mississippi Embayment (Figure 3.22b). The embayment is a downwarped area of the Earth's crust where the lithosphere is relatively thin. At least two hypotheses exist to explain this thinning, one that it took place near the end of the Proterozoic eon, around 600 million years ago, when a divergent plate boundary developed in the southeastern United States, and the other that it is related to the passage of the North American plate over a mantle hot spot during the Cretaceous period, around 95 million years ago. In either case, the observations of landforms and rates of uplift in the Mississippi Valley indicate that the seismic activity is recent, perhaps less than 10,000 years ago.

The *recurrence interval*, or time between events, for large earthquakes in the Mississippi Embayment is estimated to be several hundred years.[21,22] With material amplification, the New Madrid seismic zone appears to be capable of producing intensities commonly associated with great earthquakes in California. Thus, the interior of the North American plate is far from "stable." In recognition of this, the Federal Emergency Management Agency (FEMA) and affected states and municipalities have adopted a new building code designed to mitigate major earthquake hazards.

Another large, damaging intraplate earthquake (**M**~7.3) occurred on the night of August 31, 1886, near Charleston, South Carolina (Figure 3.22a). This earthquake killed about 60 people and damaged or destroyed most buildings in Charleston. More than 102 buildings were completely destroyed and nearly 14,000 chimneys fell, many because of poor construction following a great fire in 1838. The quake was felt from Canada to Cuba and as far west as Arkansas.[20] Effects of the earthquake were reported at distances exceeding 1000 km (620 mi.) from the epicenter.

Intraplate earthquakes in the rest of the United States are generally more damaging and felt over a much larger area than similar magnitude earthquakes in California. Because the rocks in the eastern United States are generally stronger and less fractured, they can more efficiently transmit earthquake waves.

3.6 Effects of Earthquakes and Linkages with Other Natural Hazards

Shaking is not the only cause of death and damage during earthquakes. Many earthquakes cause other hazards and are thus an excellent example of how natural hazards are often linked. The primary effects of an earthquake are those caused directly by fault movement, such as ground shaking, with its effects on people and structures, and surface rupture. Secondary effects are those that subsequently result from the faulting and shaking. These include liquefaction of the ground, regional changes in land elevation, landslides, fire, tsunamis, and disease. We will discuss the effects of tsunamis, large ocean waves that are often generated by earthquakes, in Chapter 4.

SHAKING AND GROUND RUPTURE

The immediate effects of a catastrophic earthquake can include violent ground shaking accompanied by widespread surface rupture and displacement of Earth's surface. Although most surface cracks produced by earthquakes are the result of liquefaction and landslides (discussed later), a major surface rupture may occur along the fault that was the source for the earthquake. This rupture commonly creates a low cliff, called a *fault scarp*, which may extend for kilometers along the fault (Figure 3.23). Contrary to many Hollywood movies, the Earth rarely cracks open and then closes during an earthquake.

The 1906 San Francisco earthquake (**M** 7.8) produced 6.5 m (slightly over 21 ft.) of horizontal displacement along the San Andreas fault north of San Francisco. Studies after the quake indicate that portions of the Bay area reached a maximum Modified Mercalli Intensity of XI.[11] At this intensity, surface accelerations can snap and uproot large trees and knock people to the ground. This level of shaking may damage or collapse large buildings, bridges, dams, tunnels, pipelines, and other rigid structures.[23]

The great 1964 Prince William Sound, Alaska, earthquake (**M** 9.2) caused extensive damage to transportation systems, railroads, airports, and buildings. Both the 1989 Loma Prieta (**M** 6.9) and 1994 Northridge earthquakes (**M** 6.7) were smaller than the great Prince William Sound earthquake, yet their damage was far more costly. In

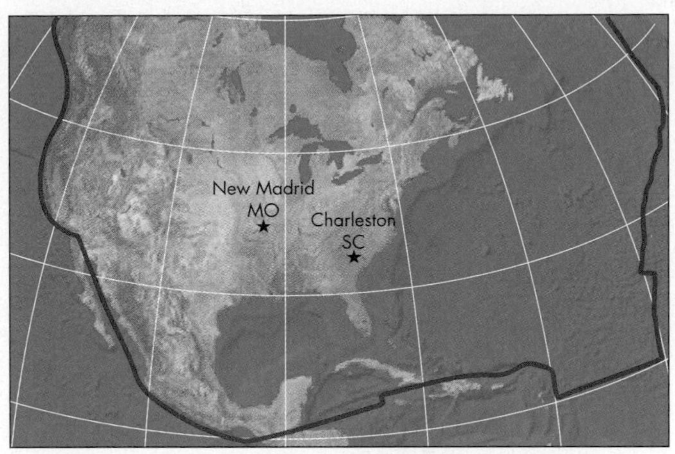

(a)

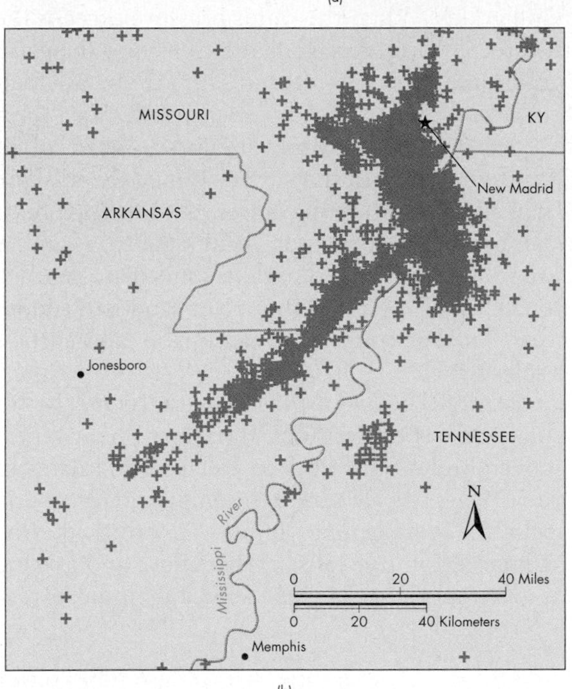

(b)

◀ FIGURE 3.22 **NEW MADRID SEISMIC ZONE** (a) The city of New Madrid Missouri and Charleston South Carolina were the sites of several large earthquakes even though the cities are located thousands of kilometers from the nearest North American plate boundary (red lines). (b) This zone is the most earthquake-prone region in the United States east of the Rocky Mountains. Epicenters of recorded earthquakes since 1974 are shown as crosses. New Madrid is located in the northeastern part of the map. *(U.S. Geological Survey)*

◀ FIGURE 3.23 **FAULT SCARP** This small cliff, called a fault scarp, was produced by the rupturing of Earth's crust during the 1992 **M** 7.3 Landers earthquake in California. The rupture could be traced for a distance of 70 km (43 mi.) in the Mojave Desert. *(Edward A. Keller)*

inflation-adjusted dollars, the Loma Prieta quake was 5 times as costly as the Alaska quake and the Northridge quake was 20 times as costly. With losses of at least $20 billion, the Northridge earthquake was one of the most expensive hazardous events in U.S. history.[14] *The Northridge earthquake caused so much damage because there was so much there to be damaged.* Most of the damage was in the San Fernando Valley northwest of Los Angeles, a highly urbanized area with a high population density that experienced intense shaking.

Violent ground shaking can be especially damaging to buildings if the horizontal motion is particularly strong or if the frequency of the shaking matches the natural vibrational frequency of the building. Shaking is commonly measured as *ground acceleration* and is compared to the overall acceleration of gravity. The matching of vibrational frequencies is called **resonance** and can affect buildings a significant distance from the epicenter. One- and two-story buildings typically have high vibrational frequencies in contrast to tall buildings that have low vibrational frequencies. In general, high shaking frequencies damage low buildings and low shaking frequencies damage tall buildings.

Ground shaking can also be a significant hazard to dams and levees. Partial failure of the Van Norman Dam in the 1971 **M** 6.7 Sylmar earthquake came perilously close to flooding more than 80,000 people downstream of the dam.[14] Parts of the dam slid into the reservoir, which fortunately at the time was only about half full of water. People living downstream were evacuated, and the water level was lowered in the reservoir as a precaution. Likewise, in Northern California, ground shaking from an earthquake in the San Francisco Bay area could produce intense shaking in the Sacramento River Delta. This shaking could cause the failure of levees along the river, resulting in widespread flooding and potential disruption of the water supply to Southern California.

LIQUEFACTION

During earthquakes, intense shaking can cause a near-surface layer of water-saturated sand to change rapidly from a solid to a liquid. This effect, called **liquefaction**, is most common in **M** 5.5 and greater earthquakes when there is strong shaking in Holocene sediment, that is, deposits less than about 10,000 years old.[23] Layers of solid sand can liquefy into a soupy mixture of sand and water as seismic waves pass through the earth material. Vibrations from the passing seismic waves increase water pressure in the space between sand grains. When the water pressure exceeds the downward-directed pressure from the overlying material, the sand grains and other objects buried in the sand effectively "float," like pieces of vegetables suspended in a thick stew. This creates a sand slurry that flows upward, often squirting through cracks to the surface. Once the shaking slows and the water pressure decreases, the liquefied sediment re-compacts and becomes solid again. A similar change from solid to liquid and back to solid again can also occur in "quick sand" on a beach or river sand bar. Liquefaction in earthquakes commonly causes the land surface to shift or subside.

The liquefaction of sand layers around 10 m (30 ft.) below the surface has caused four-story apartment buildings to tip over, highway bridges to collapse, and dams to fail (Figure 3.24). It has also brought empty underground tanks, pipelines, and bridge pilings floating to the surface.[24] Liquefaction can cause slope failure and ground subsidence over large areas. One amazing indicator of

◀ FIGURE 3.24 **BUILDING COLLAPSE FROM LIQUEFACTION** These apartment buildings collapsed as a result of liquefaction of sandy sediment beneath Niigata, Japan, in 1964. The **M** 7.5 earthquake that caused the liquefaction killed 26 people. *(NOAA National Geophysical Data Center)*

liquefaction are "sand blows," small mounds of sand formed at the surface where water from liquefied layers squirts out of the ground.

REGIONAL CHANGES IN LAND ELEVATION

Vertical deformation of the land surface is a hazard linked to some large earthquakes. This deformation includes both regional uplift and subsidence of Earth's surface. These changes in elevation can cause substantial damage in coastal areas and along streams and can raise or lower the groundwater table.

The great 1964 Prince William Sound earthquake (**M** 9.2) provided a dramatic example of regional deformation in Alaska. This quake caused vertical deformation over an area of more than 250,000 km^2 (97,000 $mi.^2$), a land surface slightly larger than the state of Vermont.[25] Within this area, uplift of as much as 10 m (30 ft.) exposed and killed coastal marine life, lifted docks out of the water, and shifted the shoreline away from fish canneries and neighboring homes. In other areas, subsidence of as much as 2.4 m (8 ft.) resulted in the partial flooding of several communities. Both the uplift and subsidence produced changes in the groundwater table.

LANDSLIDES

Two of the most closely linked natural hazards are earthquakes and landslides. Earthquakes are one of the most common triggers for landslides in hilly and mountainous areas. Landslides triggered by earthquakes can be extremely destructive and cause great loss of life. For example, a giant landslide triggered by the 1970 Peru earthquake buried the cities of Yungay and Ranrahirca. An estimated 20,000 of the 70,000 people who died in the quake were killed by the landslide. Both the 1964 Alaska earthquake and the 1989 Loma Prieta earthquake caused extensive landslide damage to buildings, roads, and other structures. The 1994 Northridge earthquake and associated aftershocks triggered thousands of landslides.

A giant landslide from the side of a mountain was triggered by the 2002 Alaska earthquake (Figure 3.25). Thousands of other landslides were also triggered by the earthquake, most of which were on the steep slopes of the Alaska Range.

The 2008 **M** 7.9 earthquake in China caused many landslides. Large landslides (Figure 3.26) buried villages and blocked rivers, forming "earthquake lakes" that produced a serious flood hazard, should the landslide dam fail by overtopping or erosion. Channels were excavated to slowly drain lakes and reduce the flood hazard.[26]

As mentioned in the opening case history, a large landslide associated with the January 13, 2001, El Salvador earthquake (**M** 7.7) buried the community of Las Colinas, killing hundreds of people. Tragically, the landslide could probably have been avoided if the slope that failed hadn't previously been cleared of vegetation for the construction of luxury homes (see Figure 3.1d).

FIRES

Fire is another major hazard linked to earthquakes. Shaking of the ground and surface displacements can break electrical power and natural gas lines, thus starting fires. The threat from fire is even greater because firefighting equipment may be damaged; streets, roads, and bridges may be blocked; and essential water mains broken. In

◀ **FIGURE 3.25 EARTHQUAKE-TRIGGERED LANDSLIDES** This giant landslide was one of thousands triggered by the 2002 **M** 7.9 Alaska earthquake. Landslide deposits cover part of the Black Rapids Glacier. *(U.S. Geological Survey/U.S. Department of the Interior)*

◀ FIGURE 3.26 **2008 M 7.9 EARTHQUAKE IN CHINA** This earthquake produced a huge landslide that blocked a river, producing an "earthquake lake." Channels in the landslide dam were excavated with explosives to lower the lake level and reduce a potential flood hazard by overtopping or erosion. *(Color China Photo/AP Images)*

individual homes and other buildings, appliances such as gas water heaters may be knocked over, causing gas leaks that are ignited. Earthquakes in both Japan and the United States have been accompanied by devastating fires (Figure 3.27). The San Francisco earthquake of 1906 has repeatedly been referred to as the "San Francisco Fire" because 80 percent of the damage was caused by firestorms that ravaged the city for several days. The 1989 Loma Prieta earthquake also caused large fires in the Marina district of San Francisco.

Most fires after an earthquake are not wildfires because they start in urban areas where gas and electrical lines are located. However, in a warm, dry area, such as Southern California, it is possible for an urban fire to spread to wild lands, causing a wildfire.

DISEASE

Outbreaks of diseases are sometimes associated with large earthquakes. They may be caused by a loss of sanitation and housing, contaminated water supplies, disruption of public health services, and the disturbance of the natural environment. Earthquakes also rupture sewer and water lines, causing water to become polluted by disease-causing organisms.

An interesting example of how an earthquake-related disturbance can result in a disease outbreak occurred in Southern California. Desert soils in the southwestern United States and northwestern Mexico contain spores of a fungus that causes a respiratory illness known as *valley fever*. Landslides from the 1994 Northridge earthquake

◀ FIGURE 3.27 **EARTHQUAKE AND FIRE** Fires associated with the 1995 **M** 6.9 Kobe, Japan, earthquake caused extensive damage to the city. *(Corbis)*

raised large volumes of desert dust containing these fungal spores. Winds carried the dust and spores to urban areas such as Simi Valley, where an outbreak of valley fever occurred. More than 200 cases of the disease were diagnosed within 8 weeks after the earthquake, 16 times the normal number of cases. Fifty of these people were hospitalized and three died.[27]

3.7 Natural Service Functions of Earthquakes

When one is confronted with the death and destruction caused by large earthquakes, it is difficult to imagine that there could be any benefits from such disasters. However, earthquakes, like many other natural hazards, do provide natural service functions. They contribute to the development of groundwater and energy resources, the formation and exposure of valuable mineral resources, the development of landforms, and to local reductions in the danger of future large quakes.

GROUNDWATER AND ENERGY RESOURCES

Geologic faults produced by earthquakes strongly influence the underground flow of water, oil, and natural gas (see Figure 1.14). Fault zones with coarsely broken earth materials can act as preferential paths for fluid movement. These faults may serve as conduits for the downward flow of surface water or as zones through which groundwater comes back to the surface as springs. Barton Springs in Austin, Texas, is one of many major springs that discharge from fault zones.

In other settings, faulting creates natural underground dams to slow or redirect the flow of water, oil, or natural gas. These dams develop where faulting pulverizes rock to form an impervious clay barrier and where faulting moves impervious earth materials alongside earth materials containing water, oil, or natural gas. Such subsurface dams are responsible for oases in arid areas of Southern California and for many underground accumulations of oil and gas in Texas, Oklahoma, Alberta, and elsewhere.

MINERAL RESOURCES

Faulting related to earthquakes may be responsible for the accumulation or exposure of economically valuable minerals. Mineral deposits commonly develop along cracks in Earth associated with faulting. These mineral-filled cracks, called *veins*, can be the source of precious metals such as gold, silver, and platinum. Veins associated with large fault zones may produce enough minerals to be economically viable for extraction.

LANDFORM DEVELOPMENT

Earthquakes can form scenic landforms over long intervals of geologic time. Uplifts of earth materials along geologic faults can produce hills, rugged mountain ranges, and coastal terraces. A spectacular example of the role that earthquakes have played in creating scenic landforms is found in Grand Teton National Park in northwestern Wyoming. In the past several million years, earthquakes along faults both inside and outside the park have allowed the Teton Range to rise and Jackson Hole to subside, creating a scenic mountain front.

The fault zones themselves may become part of landforms if they contain broken rock that is more easily eroded than earth materials in surrounding areas. Easily eroded fault zones commonly become valleys cut by streams and rivers. For example, the course of the Hudson River in southeastern New York State appears to be controlled by geologic faulting in the underlying bedrock and the Rio Grande Rift Valley in New Mexico is the result of normal faulting.

FUTURE EARTHQUAKE HAZARD REDUCTION

Frequent small earthquakes may help prevent larger ones in the same area. If we assume that small earthquakes release pent-up energy, then the reduction in elastic strain may lower the chance of a devastating earthquake along a particular fault. In fact, scientists try to identify areas along active faults that have not experienced earthquakes in a long time. As discussed later, these areas, called "seismic gaps," can have the greatest potential for producing large earthquakes in the future.[16]

3.8 Human Interaction with Earthquakes

As a natural hazard, an earthquake cannot be prevented or controlled; we can only devise ways to minimize the death and destruction it causes. That does not mean, however, that we cannot cause earthquakes ourselves. In fact, several human activities are known to increase or cause earthquake activity. Damage from these earthquakes is regrettable, but the lessons learned may help control or stop large catastrophic earthquakes in the future.

EARTHQUAKES CAUSED BY HUMAN ACTIVITY

Three ways that the actions of people have caused earthquakes are as follows:[28]

- Loading Earth's crust, as in building a dam and reservoir
- Injecting liquid waste deep into the ground through disposal wells
- Creating underground nuclear explosions

Water Reservoirs Construction of a large, deep reservoir behind a dam on a river may create, or induce, earthquakes. The huge weight of the water in a new reservoir can create or extend fractures in adjacent rock and increase water pressure in the surrounding groundwater. These effects may produce new faults or lubricate and activate existing faults. In the United States, several hundred local tremors occurred in the decade following the completion of Hoover Dam and the filling of Lake Mead in Arizona and Nevada. Most of these quakes were very small, but one was **M** 5 and two were about **M** 4.[28] In India and China, reservoir filling has triggered earthquakes with a magnitude of more than 6.0. Two of these quakes killed hundreds of people and severely damaged the reservoir dams.[23]

Deep Waste Disposal In the early 1960s, an unplanned experiment by the U.S. Army provided the first direct evidence that injecting fluids into the Earth can cause earthquakes. From April 1962 to November 1965, the Denver, Colorado, area experienced several hundred earthquakes, far more than normally occur over this length of time. The largest quake, a **M** 4.3, caused sufficient shaking to knock bottles off shelves in stores. Geologist David Evans traced the source of the earthquakes to the Rocky Mountain Arsenal, a chemical warfare plant on the northeast side of Denver. Liquid waste from the plant was being injected under pressure down a deep disposal well 3600 m (11,800 ft.) into Earth. The injected liquid increased underground fluid pressure and caused slippage of numerous fractures in metamorphic rock. Evans demonstrated a high correlation between the rate of waste injection and the location and timing of the earthquakes. When the injection of waste stopped, so did the earthquakes (Figure 3.28).[29] Deep well injection has also triggered earthquakes in Ohio and Texas.[30,31]

Recognizing that the injection of fluids into the Earth could trigger an earthquake was an important development because it drew attention to the relationship

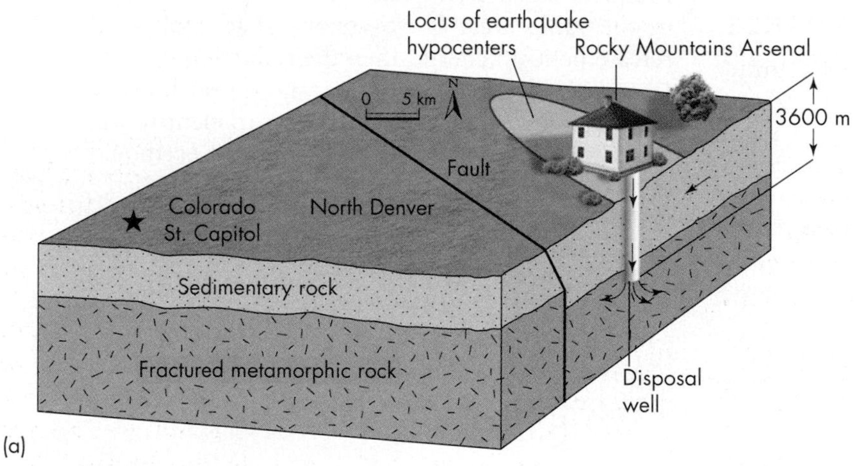

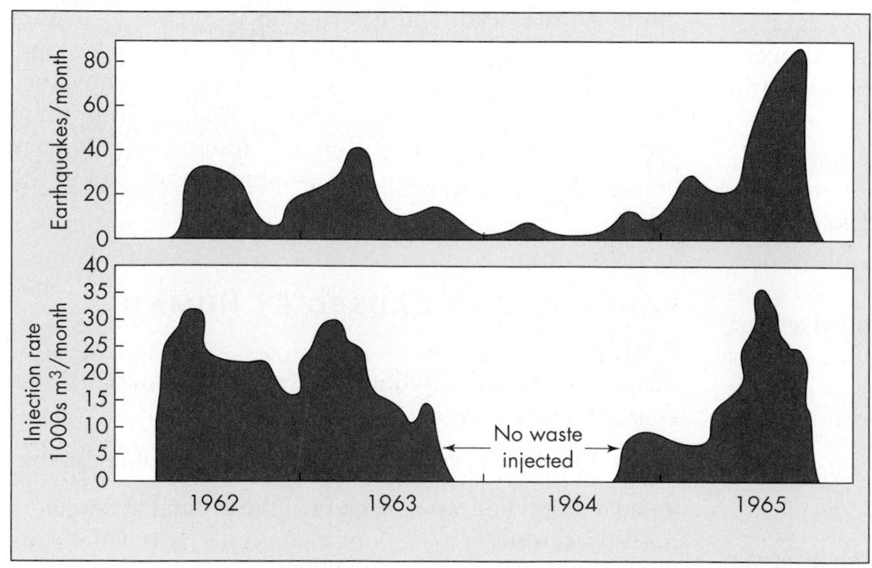

◀ **FIGURE 3.28 HUMAN-CAUSED EARTHQUAKES** (a) A red line encompasses an area most of the earthquakes occurred and a natural fault is present about 10 km southwest of the disposal well. (b) Injection of waste stopper in 1966 and the well was permanently sealed in 1986. Earthquake activity has settled back to the natural frequency. An earthquake of magnitude 4.2 occurred in 1982, suggesting the faulting remains active. Data from www.geosurvey.state.co.us. Rocky Mountain Aresenal—The most famous example of induced earthquakes access 12/5/10. *(After Evans, D. M. 1966. Man-made earthquakes in Denver. Geotimes 10: 11–18. Reprinted by permission)*

between fluid pressure and earthquakes. Subsequent studies of subduction zones and active fold belts suggest that high fluid pressures are present in many areas where earthquakes occur. One hypothesis is that fluid pressure rises in an area until rocks break, triggering an earthquake and discharging fluid upward. After the event, fluid pressure begins to build up again, beginning another cycle.

Nuclear Explosions Underground tests of nuclear weapons in Nevada have triggered numerous earthquakes with magnitudes as large as 6.3.[28] An analysis of the aftershocks suggests that these explosions may have released some natural strain within Earth. This has led to discussions by scientists as to whether nuclear explosions might be used to prevent large earthquakes by releasing strain before it reaches a critical point. Even if such a technique were feasible and legal liability and environmental issues were resolved, the explosive release of strain on one fault might trigger unanticipated earthquakes on other faults.

3.9 Minimizing the Earthquake Hazard

One important reason that earthquakes cause great damage and loss of life is that they often strike without warning. Because of this, a great deal of research is being devoted to anticipating earthquakes. The best we can do at present is to forecast the likelihood that an earthquake will occur in a particular area or on a particular segment of a fault. These forecasts use probabilistic methods to determine the risk. A forecast will say that an earthquake of a given magnitude or intensity has a high probability of occurring in a given area or fault segment within a specified number of years. These forecasts assist planners who are considering seismic safety measures or people who are deciding where to live. However, long-term forecasts do not help residents of a seismically active area anticipate and prepare for a specific earthquake. An actual prediction specifying the time and place of an earthquake would be much more useful, but the ability to make such predictions has eluded us. Predicting earthquakes depends to a large extent on observation of natural phenomena or changes in the Earth that precede an event.

THE NATIONAL EARTHQUAKE HAZARD REDUCTION PROGRAM

In the United States, the USGS as well as university and other scientists are developing a National Earthquake Hazard Reduction Program. The major goals of the program are as follows:[32]

- Develop *an understanding of the earthquake source.* This goal requires obtaining information about the physical properties and mechanical behavior of faults and developing quantitative models of the physics of the earthquake process.
- Determine *earthquake potential.* This goal requires detailed study of seismically active regions to determine their paleoseismicity, identify active faults, and determine rates of deformation (see A Closer Look 3.1). This information is used to calculate probabilistic forecasts and to develop methods for intermediate- and short-term predictions of earthquakes.
- Predict *effects of earthquakes.* This goal requires obtaining information needed to predict ground rupture and shaking and an earthquake's effects on buildings and other structures. This information is used to evaluate losses associated with the earthquake hazard (see Case Study 3.2).
- Apply *research results.* The program educates individuals, communities, states, and the nation about earthquake hazards. This goal requires better planning for earthquakes and ways to reduce loss of life and property.

ESTIMATION OF SEISMIC RISK

Several types of hazard maps are used to communicate earthquake risk. The simplest maps portray relative hazard by showing the locations of epicenters of historic earthquakes of various magnitudes. More informative maps communicate the probability of a particular event or the amount of shaking likely to occur. For example, a seismic-shaking hazard map of California (Figure 3.29) shows areas in purple through red that have a chance of experiencing a damaging earthquake in a 30-year period. Within the colored area, the red and orange regions have the greatest seismic hazard because they are likely to experience the most intense shaking in the next 30 years. Regional earthquake-hazard maps are also helpful in establishing property zoning restrictions and determining insurance rates. Figure 3.30 shows the probability of a **M** 6.7 or larger earthquake occurring on faults in the San Francisco Bay region. There is a 62 percent probability of at least one **M** 6 to 7 or greater earthquake occurring before 2032.[39] The probability is based on past earthquake activity, fault displacement, and slip rate.

In addition to mapping, the state of California is now classifying faults as a method for assessing seismic risk. Geologic faults are classified on the maximum moment magnitude the fault can produce and on the slip rate of the fault. Slip rate is determined by the amount of movement along a fault averaged over thousands of years and numerous earthquakes. It is not actual slippage that takes place along a fault each year. Although classifying faults provides more information than simply determining if a fault has been active in the past 10,000 years, the long-term slip

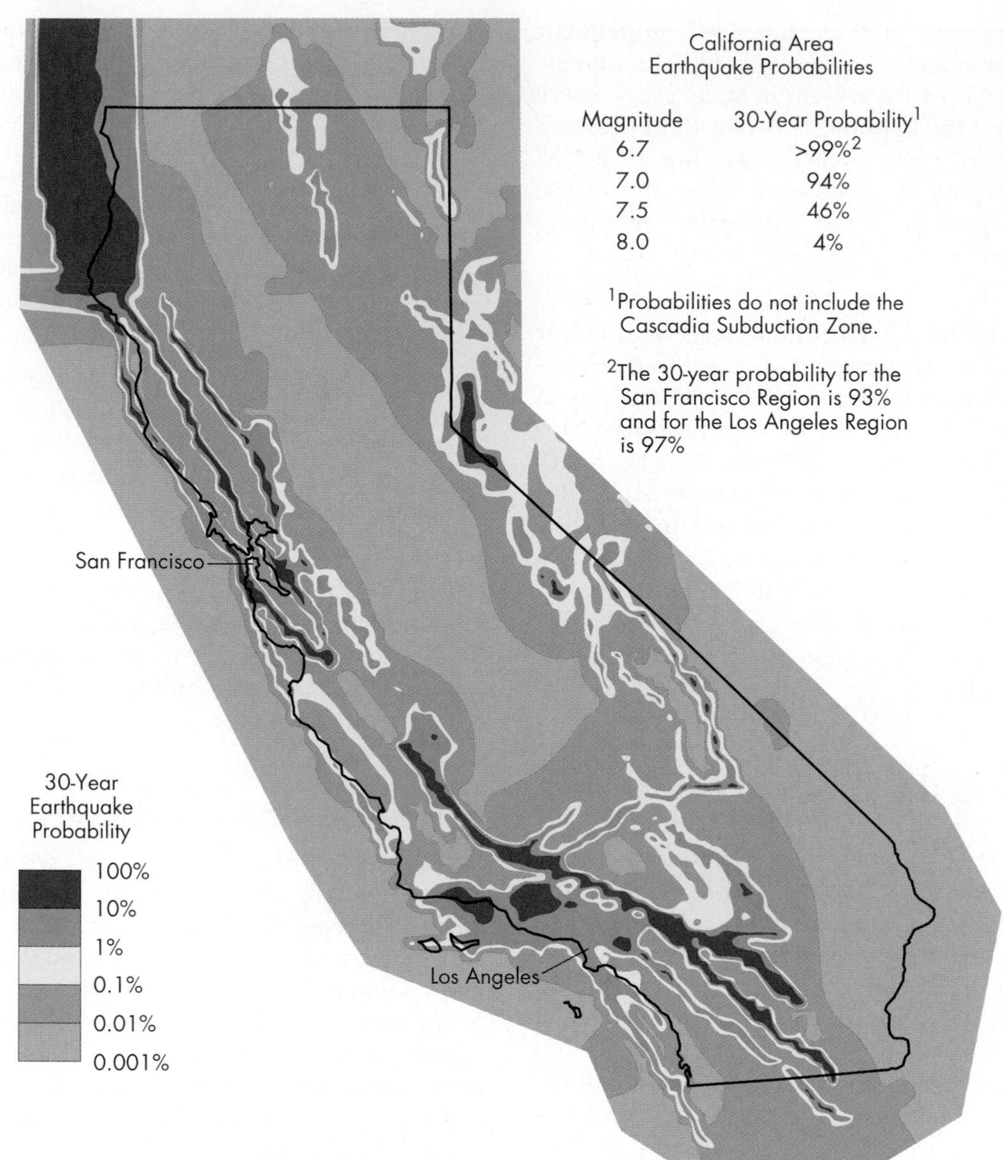

▲ FIGURE 3.29 **PROBABILITIES OF EARTHQUAKES IN CALIFORNIA OVER NEXT 30 YEARS** (*Field, E. H., Milner, K. R., and the 2007 Working Group on California Earthquake Probabilities. 2008.* Forecasting California's earthquakes—What can we expect in the next 30 years? *U. S. Geological Survey Fact Sheet. 2008-3027*

rates of most major faults in North America are unknown or determined only by information collected at one or two sites.[40]

SHORT-TERM PREDICTION

The short-term prediction of earthquakes is an active area of research. Predictions must rely on *precursors*, events or changes that occur before the mainshock of the earthquake. Unlike a forecast, a true earthquake prediction specifies a relatively short time period (days to weeks) in which the event is likely, an estimated magnitude for the quake, a relatively limited geographic area, and a probability of its occurring. The basic procedure for predicting earthquakes was once thought to be as easy as "one-two-three."[41] First, deploy instruments to detect potential precursors of a future earthquake; second, detect and recognize the precursors that tell you when and how big an earthquake will be; and third, after review of your data, publicly predict the earthquake. Unfortunately, earthquake prediction is much more complex than first thought.[41]

Japanese seismologists made the first attempts at earthquake prediction using the frequency of small *microearthquakes* (**M** less than 2), repetitive surveys to detect tilting of the land surface, and measurements of changes in the local magnetic field of Earth. They found that earthquakes in the areas they studied were nearly always accompanied by swarms of microearthquakes several months before the major shocks. Furthermore, they found that ground tilt correlated strongly with earthquake activity.[28]

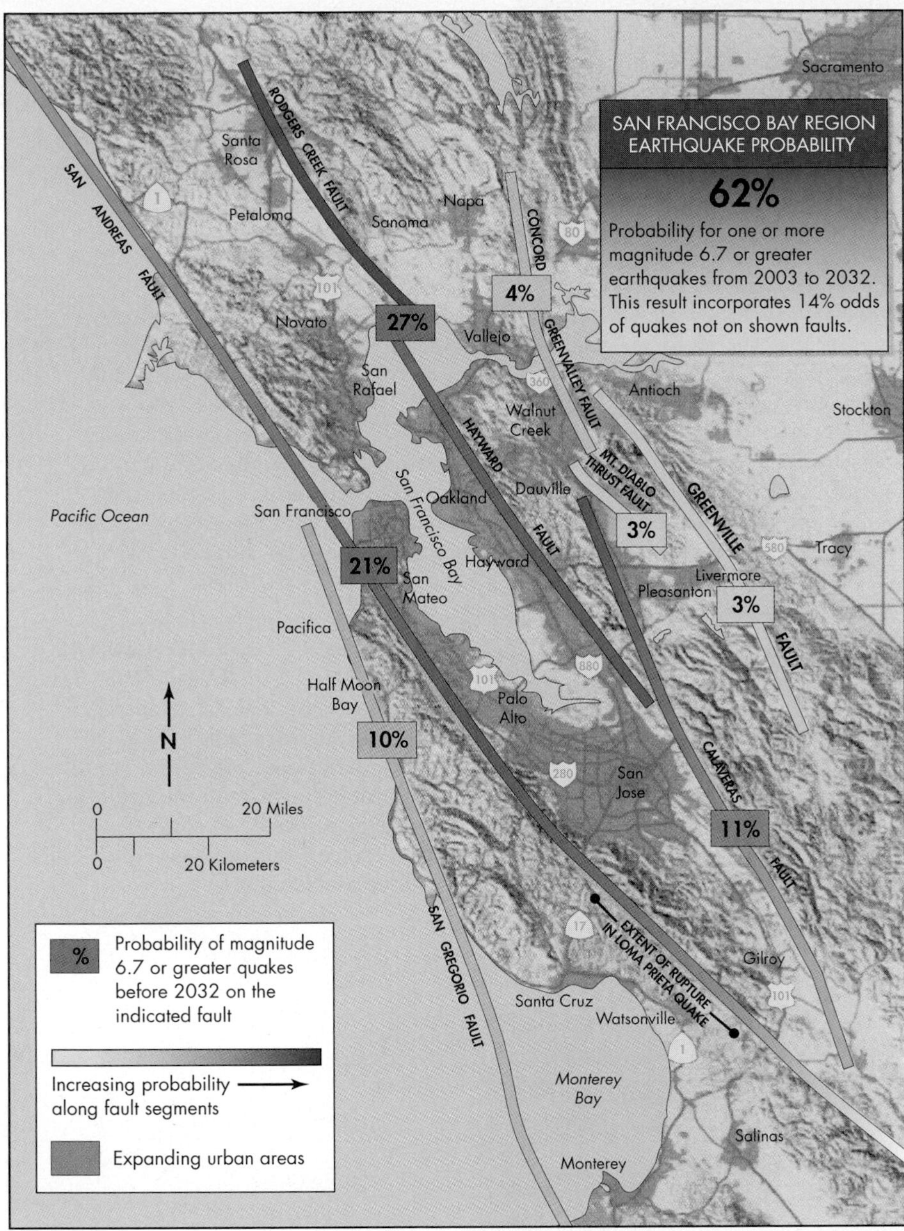

FIGURE 3.30 SHAKING-HAZARD MAP Probability of at least one **M** 6.7 or greater earthquake occurring before 2032 on any fault in the San Francisco Bay region. Small colored boxes show the probabilities of the quake occurring on each of the major faults. *(From U.S. Geological Survey Fact Sheet 039-03)*

Chinese scientists predicted a major earthquake (**M** 7.0) in 1975, saving thousands of lives. Although this prediction appears to have been successful by chance, because Chinese scientists had issued many unsuccessful predictions and also subsequently failed to predict major quakes, the result was beneficial.[40] The predicted earthquake destroyed or damaged about 90 percent of the buildings in Haicheng, China. Most of the town's 9000 people were saved because of a massive evacuation from potentially unsafe housing. The short-term prediction was based primarily on a series of progressively larger foreshocks that began 4 days prior to the main event.

Unfortunately, foreshocks do not precede most large earthquakes and there is no method for distinguishing a foreshock from any other earthquake. In 1976, one of the deadliest earthquakes in recorded history (**M** 7.5) struck near the mining town of Tangshan, China, killing more than 240,000 people. There were no foreshocks.

In addition to foreshocks, many phenomena have been proposed as precursors for earthquakes. They range from lunar tides to unusual animal behavior. Reports of animal behavior have included unusual barking of dogs, chickens that refuse to lay eggs, horses or cattle that run in circles, rats perched on power lines, and snakes crawling out in the winter and freezing. To date, no scientific studies show a correlation between unusual animal behavior or lunar tides and earthquakes. Earthquake prediction is still a complex problem. Even if reliable precursors can be identified, dependable short-range predictions are many years away. Such predictions, if they are possible, will

A CLOSER LOOK 3.2

Paleoseismic Earthquake Hazard Evaluation

The primary purpose of earthquake hazard assessment is to provide a safeguard against loss of life and major structural failures, rather than to limit damage or to maintain functional aspects of society, including roads and utilities.[33] The basic philosophy behind the seismic hazard's evaluation is to have what is known as a **design basis ground motion**, defined as the ground motion that has a 10 percent chance of being exceeded in a 50-year period (equivalent to 1 chance in 475 of being exceeded each year). This is determined in California by a series of maps that show strong motion PGA (peak ground acceleration) for earthquake waves of 0.3 second period, for one- and two-story buildings and 1.0 second period for buildings taller than two stories.[34] The U. S. Geological Survey has programs to predict peak acceleration and velocity of shaking through generation of an earthquake-planning scenario for a particular fault. For example, Figure 3.B shows a scenario for a **M** 7.2 event on the San Andreas fault at San Francisco, California.[35]

Assessment of earthquake hazard at a particular site includes identification of the **tectonic framework** (geometry and spatial pattern of faults or seismic sources), in order to predict earthquake slip rate and ground motion (Figure 3.C). The process of predicting slip rate and ground motion from a given earthquake may be illustrated by considering a hypothetical example. Figure 3.Da

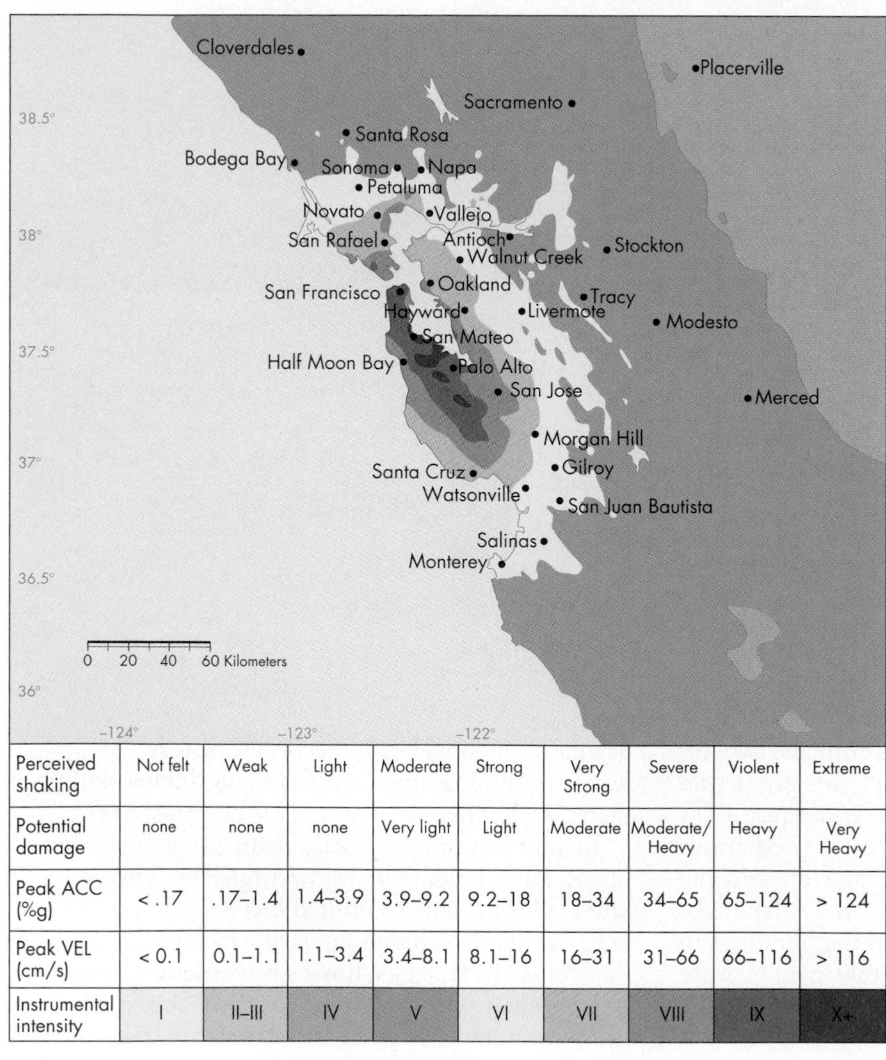

Perceived shaking	Not felt	Weak	Light	Moderate	Strong	Very Strong	Severe	Violent	Extreme
Potential damage	none	none	none	Very light	Light	Moderate	Moderate/ Heavy	Heavy	Very Heavy
Peak ACC (%g)	< .17	.17–1.4	1.4–3.9	3.9–9.2	9.2–18	18–34	34–65	65–124	> 124
Peak VEL (cm/s)	< 0.1	0.1–1.1	1.1–3.4	3.4–8.1	8.1–16	16–31	31–66	66–116	> 116
Instrumental intensity	I	II–III	IV	V	VI	VII	VIII	IX	X+

◄ **FIGURE 3.B EARTHQUAKE PLANNING SCENARIO** This scenario is for a **M** 7.2 event on the San Andreas fault in San Francisco, California. Instrumental Intensity refers to the Mercalli Scale (see Table 3.4). *(http://earthquake.usgs.gov/research/hazmaps/scenario/)*

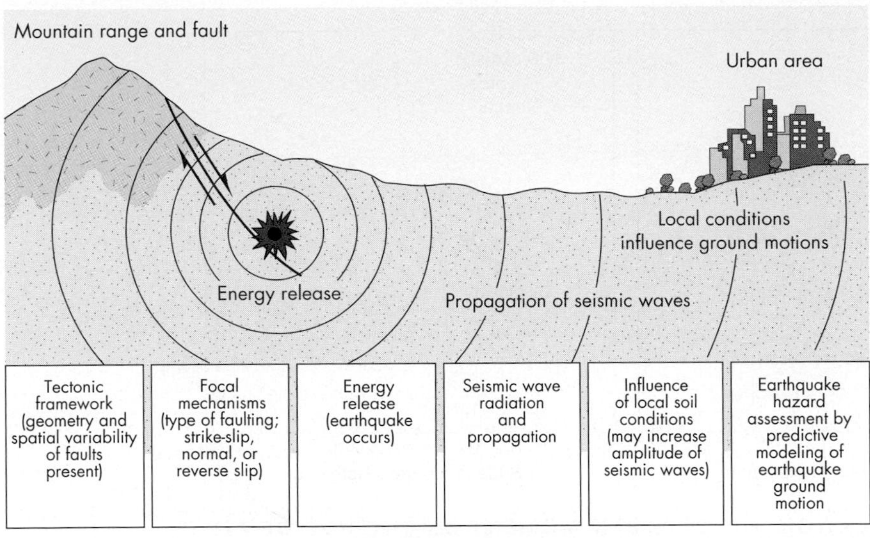

◀ **FIGURE 3.C ASSESSMENT OF EARTHQUAKE HAZARD BY MODELING OF GROUND MOTION** (After Vogel, 1988. In A. Vogel and K. Brandes (eds.), *Earthquake prognostics.* Brannschweig/Weisbaden: Friedr. Vieweg & Sohn, 1–13.)

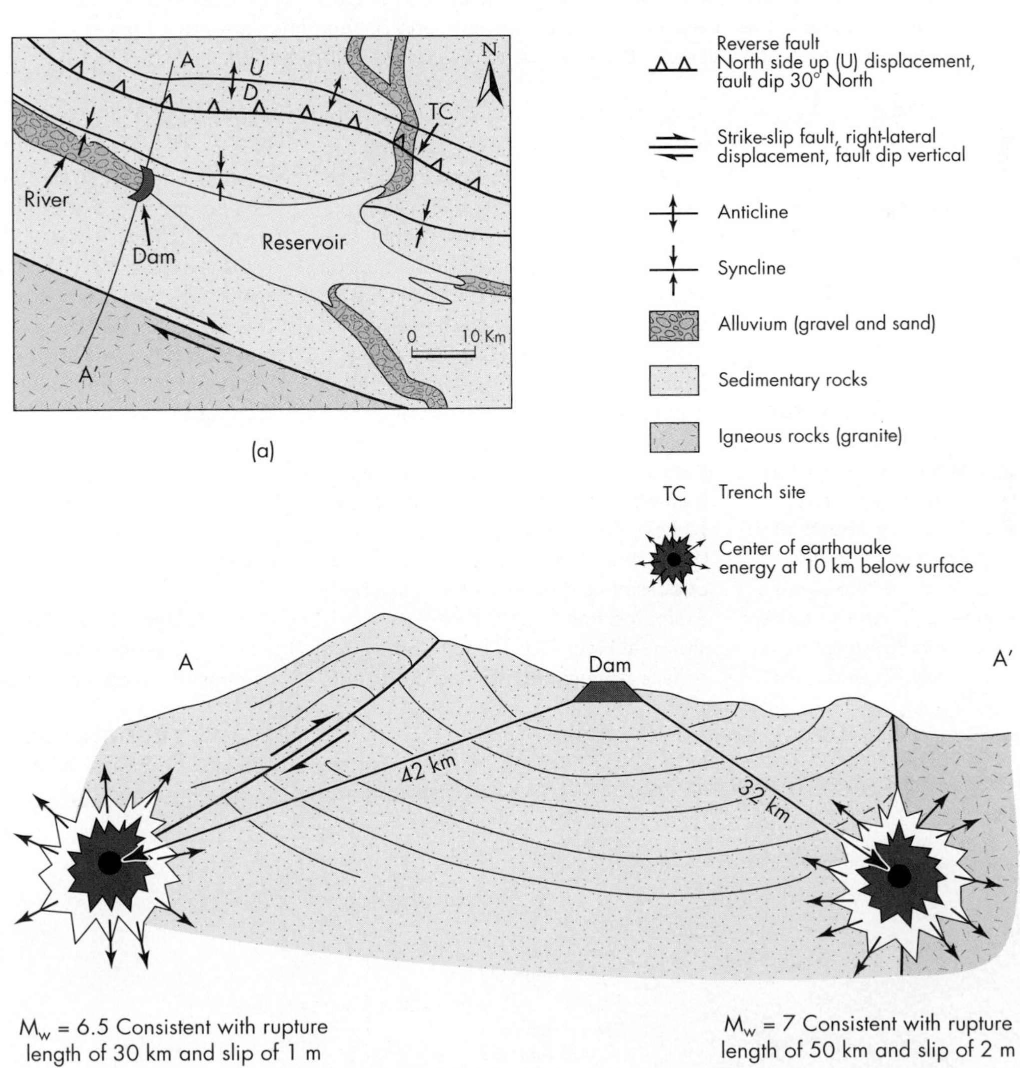

▲ **FIGURE 3.D TECTONIC FRAMEWORK FOR A HYPOTHETICAL DAM SITE** (a) Geologic map. (b) Cross section A-A′, showing seismic sources and distances of possible ruptures to the dam.

(continued)

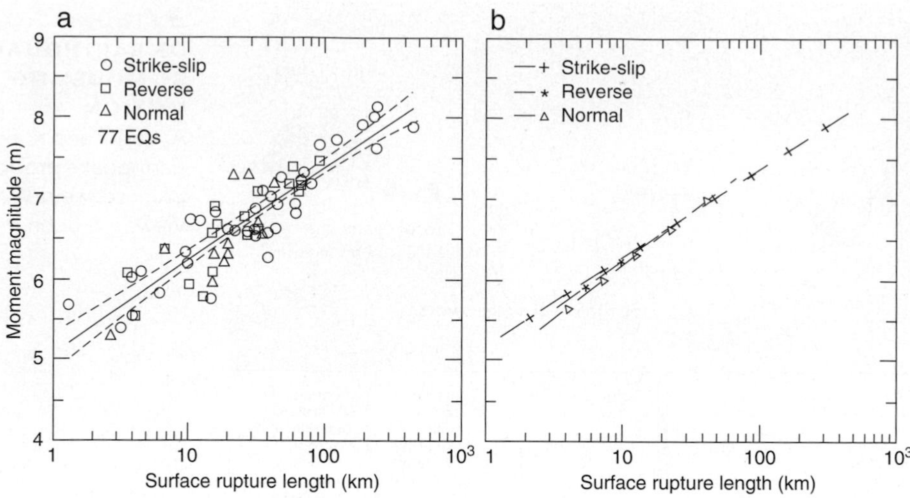

▲ **FIGURE 3.E RELATIONSHIP BETWEEN MOMENT MAGNITUDE OF AN EARTHQUAKE AND SURFACE RUPTURE LENGTH** Data for 77 events are shown in (a). Solid line is the "best-fit" or regression line, and dashed lines are error bars at significant difference between the lines. *(After Wells and Coppersmith, 1994 [39].)*

shows an example of a dam and reservoir site. The objective is to predict strong ground motion at the dam from several seismic sources (faults) in the area. The tectonic framework shown consists of a north-dipping reverse fault and an associated fold (an anticline) located to the north of the dam, as well as a right-lateral strike-slip fault located to the south of the dam. Figure 3.Db shows a cross section through the dam, illustrating the geologic environment, including several different earth materials, folds, and faults. Assuming that earthquakes would occur at depths of approximately 10 km (6 mi.), the distances from the dam to the two seismic sources (the reverse fault and the strike-slip fault) are 42 km and 32 km (27 mi and 20 mi.), respectively. Thus, for this area, two focal mechanisms are possible: reverse faulting and strike-slip faulting.

Another step in the process is to estimate the largest earthquakes likely to occur on these faults. Assume that fieldwork in the area revealed ground rupture and other evidence of faulting in the past, suggesting that, on the strike-slip fault, approximately 50 km (32 km) of fault length might rupture in a single event, with right-lateral strike-slip motion of 2 m. The fieldwork also revealed that the largest rupture likely on the reverse fault would be 30 km of fault length, with vertical displacement of about 1.5 m. Given this information, the magnitudes of possible earthquake events can be estimated from graphs such as those shown in Figure 3.E.[35] Fifty kilometers of surface rupture are associated with an earthquake of approximately **M** 7. Similarly, for the reverse fault with surface rupture length of 30 km, the magnitude of a possible earthquake is estimated to be M_W 6.5. Notice that, in Figure 3.E, the regression line that predicts the moment magnitude is for strike-slip, normal, and reverse faults. Statistical analyses have suggested that the relation between moment magnitude and length of surface rupture is not sensitive to the style of faulting.[36]

Slip rate (the vertical component) and average return period may be estimated from geologic data collected in a trench excavated across a fault. Figure 3.F shows the stratigraphy (trench log) for the site shown in Figure 3.Da. The slip rate is about 1 mm/yr (1 m per 1,000 years), and the average return period is about 1300 years.

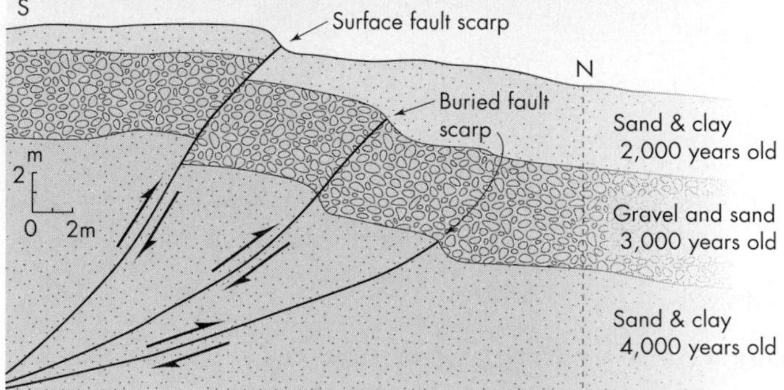

◀ **FIGURE 3.F TRENCH LOG** (stratigraphy) for the trench site on Figure 3.Da.

The Denali Fault Earthquake: Estimating Potential Ground Rupture Pays Off

On November 3, 2002, a magnitude 7.9 earthquake occurred in early afternoon on the Denali fault in a remote section of south-central Alaska. Historical movement along the fault was primarily horizontal and the 2002 earthquake was no exception. Along the fault zone, the quake produced a surface rupture of about 340 km (210 mi.) in length and a horizontal displacement of up to 8 m (25 ft.) (Figure 3.G). Although the quake caused thousands of landslides, intense shaking, and many areas of liquefaction, there was little structural damage and no deaths—primarily because few people live in the affected area.[37]

The Denali fault earthquake demonstrated the value of seismic hazard assessments for geologic faults. Geologists studied the fault in the early 1970s as part of planning for the Trans-Alaskan pipeline. Where the pipeline crosses the Denali fault, the investigators determined that the fault zone is several hundred meters wide and could experience a 6-m (20-ft) horizontal displacement in a **M** 8 earthquake. These estimates were used in the engineering design of the pipeline across the fault zone. In anticipation of a future earthquake, the pipeline was elevated on long horizontal steel beams with Teflon shoes that allowed it to slide horizontally up to 6 m (20 ft.). These slider beams were installed along a zigzag path for the pipeline for added flexibility and horizontal movement during an earthquake (Figure 3.G). As anticipated, the 2002 earthquake occurred within the mapped fault zone and caused the pipeline to shift horizontally about 4.3 m (14 ft.). Because of advanced engineering design, the pipeline suffered little damage and there was no oil spill from the quake.

Although the $3 million cost for the design and construction of the pipeline across the fault may have seemed excessive in 1970, in retrospect it was money well spent. Today the pipeline delivers approximately 17 percent of the domestic oil supply for the United States and transports close to $25 million worth of oil each day. A pipeline break caused by an earthquake would have a significant and costly impact on our oil supply. Current estimates are that the repair of the pipeline and cleanup of the oil spill alone would cost several million dollars.[38]

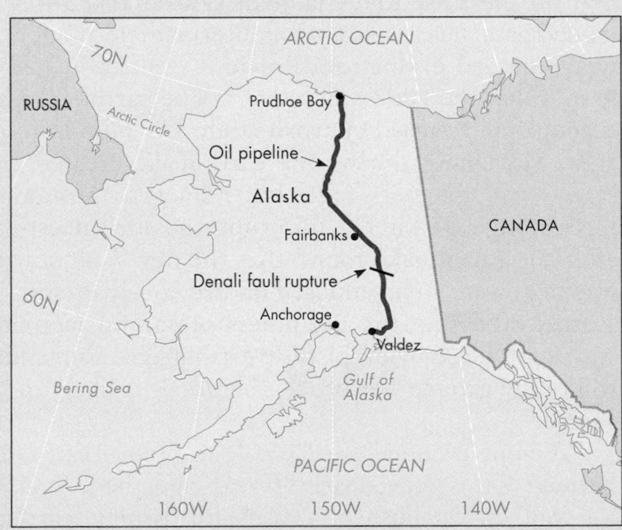

▲ FIGURE 3.G **TRANS-ALASKA OIL PIPELINE SURVIVES A M 7.9 EARTHQUAKE** In its construction, the Alaskan oil pipeline was designed to withstand several meters of horizontal displacement where it crossed the Denali fault south of Fairbanks. That design was put to the test in 2002 when it shifted horizontally 4.3 m (14 ft.) during the **M** 7.9 Denali fault earthquake. Built-in bends of the pipeline on slider beams with teflon shoes accommodated the earthquake rupture as designed. This was strong confirmation of the importance of estimating potential ground rupture and taking action to mitigate it. *(U.S. Geological Survey/U.S. Department of the Interior)*

most likely be based upon precursory phenomena such as the following:

- Patterns and frequency of earthquakes, such as foreshocks discussed earlier with the Haicheng prediction
- Deformation of the ground surface
- Seismic gaps along faults
- Geophysical and geochemical changes in Earth

Ground Deformation Changes in the land elevation, referred to as *uplift* and *subsidence*, which occur rapidly or are unusual may help predict earthquakes. For example, more than 120 km (75 mi.) of the western Japanese coast experienced several centimeters of slow uplift in the decade prior to the 1964 Niigata earthquake (**M** 7.5).[42] A similar uplift occurred over a 5-year period prior to the 1983 Sea of Japan earthquake (**M** 7.7).[42,43]

Uplifts of 1 to 2 m (3 to 7 ft.) also preceded large Japanese earthquakes in 1793, 1802, 1872, and 1927. These uplifts were recognized by sudden withdrawals of the sea from the land. On the morning of the 1802 earthquake, the sea suddenly withdrew about 300 m (1000 ft.) in response to an uplift of about 1 m (3 ft.). Four hours later, the earthquake struck, destroying many houses and uplifting the land another meter.[42]

Seismic Gaps Areas along an active fault zone that are likely to produce large earthquakes but have not produced one recently are known as *seismic gaps*. These areas are believed to be temporarily inactive while they store elastic strain for future large earthquakes.[16] The seismic-gap concept was the basis for what may have been the first scientific, long-term earthquake forecast. In 1883, the multitalented USGS geologist G. K. Gilbert forecast a catastrophic earthquake for the Salt Lake City segment of the Wasatch fault based on evidence of its inactivity.[40] Since 1965, the seismic-gap concept has been used successfully in intermediate-range forecasts for at least 10 large, plate-boundary earthquakes. One of these quakes took place in Alaska, three in Mexico, one in South America, and three in Japan. In the contiguous United States, seismic gaps along the San Andreas fault in California include one near Fort Tejon that last ruptured in 1857 and one along the Coachella Valley, a segment that has not produced a great earthquake for several hundred years. Both gaps are likely to produce a great earthquake in the next few decades.[16,44]

Geophysical and Geochemical Phenomena Local changes in Earth's gravity, magnetic field, and ability to conduct electrical currents have been associated with earthquakes. Changes have also been observed in groundwater levels, temperature, and chemistry.[45] Many of these changes may occur before an earthquake as rocks expand and fracture, and the fractures fill with water. For example, changes in the ability of earth materials to conduct an electrical current, referred to as *electrical resistivity*, have been reported before earthquakes in the United States, Eastern Europe, and China.[33] Also, significant increases in the radon content of well water were reported in the month or so prior to the 1995 Kobe, Japan, earthquake (**M** 6.9).[45]

THE FUTURE OF EARTHQUAKE PREDICTION

We are still a long way from a working, practical methodology to reliably predict earthquakes. However, a good deal of information is currently being gathered concerning possible precursor phenomena associated with earthquakes. To date the most useful precursor phenomena have been patterns of earthquakes, particularly foreshocks, and recognition of seismic gaps. Optimistic scientists around the world today believe that we will eventually be able to make consistent long-range forecasts (decades to centuries), intermediate-range forecasts (months to decades), and short-range predictions (hours to weeks) for the general locations and magnitudes of large, damaging earthquakes.

Although progress on short-range earthquake prediction has not matched expectations, intermediate- to long-range forecasting, including hazard evaluation and probabilistic analysis of areas along active faults, has progressed faster than expected (see Professional Profile 3.3).[46] For example, the 1983 Borah Peak earthquake (**M** 6.9) on the Lost River fault in central Idaho has been lauded as a success story for intermediate-range earthquake hazard evaluation. Previous evaluation had suggested that the fault was active.[47] The earthquake killed 2 people and caused approximately $15 million in damages. Movement during the earthquake created a fault scarp several meters high and numerous ground fractures along a 36 km (22 mi.) rupture zone. Investigation after the earthquake found that the new fault scarp and fractures were superimposed on previously existing fault scarps, validating the usefulness of careful mapping of scarps produced by prehistoric earthquakes. Remember—where the ground has broken before, it is likely to break again!

We hope eventually to be able to predict earthquakes. Because of this hope, nearly 30 years ago, the U.S. Geological Survey established a plan for issuing earthquake predictions and more urgent earthquake warnings (Figure 3.31). This plan requires independent scientific review before information is transmitted to government officials and the public.

Government officials and news media that give publicity and thus credence to predictions that have not been scientifically reviewed do a great disservice to the community. This precept was demonstrated in 1990 when a pseudoscientific prediction for an earthquake in New Madrid, Missouri, was acted on by some government and business leaders and publicized by local, state, and national

PROFESSIONAL PROFILE 3.4

Andrea Donnellan, Earthquake Forecaster

Andrea Donnellan (Figure 3.H) talks about earthquakes the same way meteorologists talk about the weather: as dynamic, interconnected systems with their own peculiar set of rules. So it is only appropriate that she refers to her work as earthquake "forecasting".

Specifically, Donnellan and her colleagues have installed hundreds of high-precision global positioning system (GPS) receivers across Southern California, a network known as the Southern California Integrated GPS Network (SCIGN), in order to study movements in the tectonic plates. The receivers, which Donnellan says are more highly sophisticated than commercial GPS devices such as those used in cell phones, can measure millimeter-scale slips for faults and give scientists valuable data for understanding how and why earthquakes take place. Donnellan's work goes far beyond studying just the earthquakes themselves. In fact, most of the work occurs in between all the moving and shaking.

"I look at the quiet part of the earthquake cycle," Donnellan said. "Each earthquake changes where the next earthquake will be."

Like meteorologists, Donnellan and her colleagues have reproduced earthquake systems using supercomputers in order to understand how the dynamic systems change over time.

"My focus is on modeling earthquake systems," she said. "We want to treat it like weather, where the system is always changing."

But Donnellan's research consists of far more than just sitting in front of a computer all day. In fact, her studies have taken her to places across the globe, from Antarctica to Mongolia to Bolivia, to name a few. Her current research has taken her across much of Southern California, where she and her team are studying the San Andreas fault.

Back in the lab, Donnellan is also involved in an ambitious project called QuakeSim, in which scientists are developing sophisticated, state-of-the-art computer models of earthquake systems.

Donnellan said the results of the project will eventually be accessible to anyone, and that schools are already using some of the software for educational purposes.

In terms of the practical implications for Donnellan's work, her research is already giving scientists a much clearer idea of where earthquakes may next strike, allowing for localities to prepare much further in advance. Although, like the weather, earthquakes will probably never be totally predictable, a close study of the ever-shifting activity beneath the ground by Donnellan and her colleagues is making the picture a lot clearer.

—CHRIS WILSON

FIGURE 3.H **GEOPHYSICIST DEVELOPS EARTHQUAKE FORECASTING METHOD** Dr. Andrea Donnellan, deputy manager of NASA's Jet Propulsion Laboratory Earth and Space Sciences Division, leads a team of scientists working on earthquake forecasting. She was co-winner of Women in Aerospace's 2003 Outstanding Achievement Award and twice a finalist in the NASA astronaut selection process. *(NASA)*

media—even after the prediction was rejected by the National Earthquake Prediction Evaluation Council.[48] Schools and businesses closed, public events were cancelled, people evacuated their homes, and more than 30 television and radio vans converged on the predicted epicenter.[49] Widely publicizing earthquake predictions that have not been independently vetted by seismologists is the geologic equivalent of yelling "fire" in a crowded movie theater.

EARTHQUAKE WARNING SYSTEMS

Technically it is feasible to develop an earthquake warning system that would provide up to about 1 minute of warning to the Los Angeles area prior to the arrival of damaging earthquake waves from an event several hundred kilometers away. Such a system would be based on the principle that radio waves travel much faster than seismic waves. For nearly 20 years, the Japanese have had such a system that provides earthquake warnings for their high-speed trains; derailment of one of their "bullet" trains by an earthquake could kill hundreds of people.

A proposed system for California involves a sophisticated network of seismometers and transmitters along the San Andreas fault. This system would first sense motion associated with a large earthquake and then send a warning

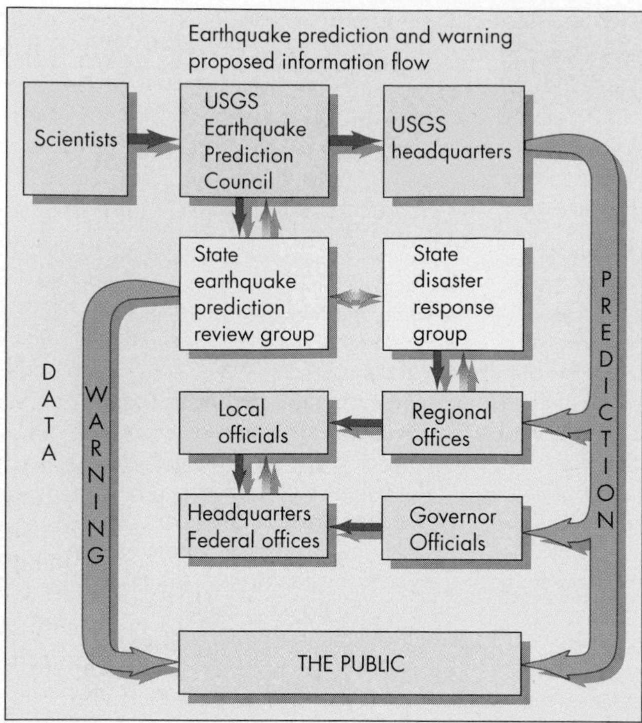

FIGURE 3.31 ISSUING A PREDICTION OR WARNING Flowchart for the United States plan for issuance of an earthquake prediction and warning. The plan calls for a National Earthquake Prediction Evaluation Council of seismologists from government agencies, universities, and the private sector to evaluate predictions and warnings before they are passed on to state and local officials, and the public. *(From McKelvey, V. E. 1976.* Earthquake prediction—opportunity to avert disaster. *U.S. Geological Survey Circular 729)*

to Los Angeles, which would then relay the warning to critical facilities, schools, and the general population (Figure 3.32). The warning time would range from as little as 15 seconds to as long as 1 minute depending upon the location of the earthquake epicenter. This interval could be enough time for many people to shut down machinery and computers and take cover.[50] Remember that an earthquake warning system is not a prediction tool, because it warns that an earthquake has already occurred.

Because the warning time is so short, some people believe that the damage to scientific credibility caused by false alarms would be far greater than the benefits of a brief warning. In the Japanese system, it is estimated that around 5 percent of the warnings turn out to be false alarms. Others have expressed concern for liability issues resulting from false alarms, warning system failures, and damage and suffering resulting from actions taken as the result of false early warning.

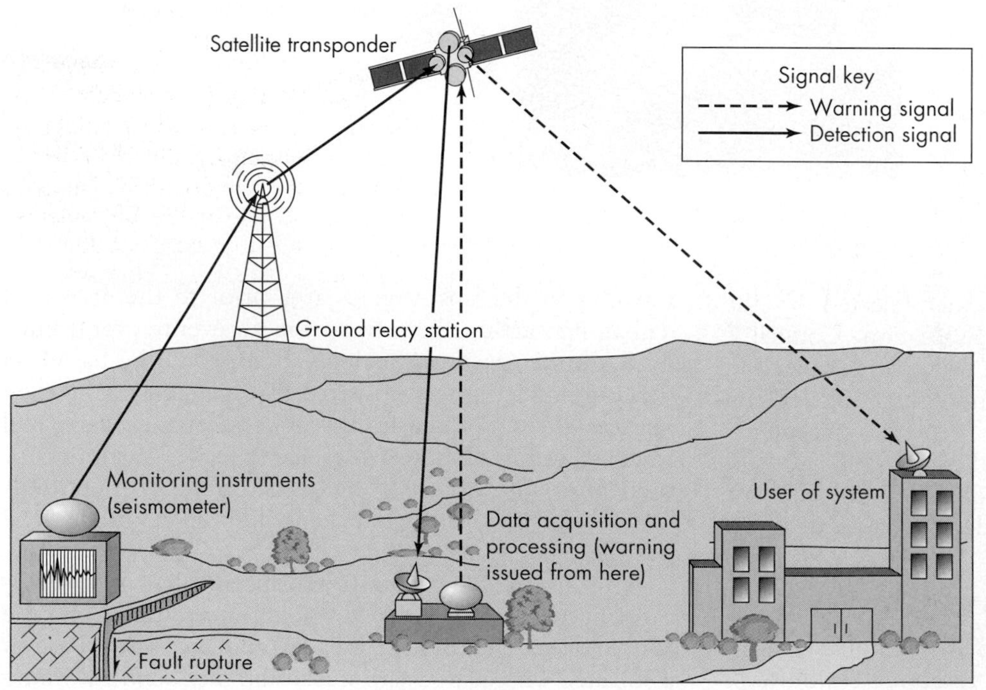

FIGURE 3.32 EARTHQUAKE WARNING Idealized diagram showing an earthquake warning system using radio waves, which travel faster than seismic waves. Once an earthquake is detected, a signal is sent ahead of the seismic shaking to warn people and facilities. The warning time, generally measured in seconds, depends on how far away an earthquake occurs. It could be long enough to shut down critical facilities and for people to take cover. *(After Holden, R., Lee, R., and Reichle, M. 1989. Technical and economic feasibility of an earthquake warning system in California, California Division of Mines and Geology Bulletin 101)*

3.10 Perception of and Adjustment to the Earthquake Hazard

PERCEPTION OF THE EARTHQUAKE HAZARD

The experience that the Earth is not so firm in places is disconcerting to people who have felt even a moderate earthquake. The large number of people, especially children, who suffered mental distress following the San Fernando and Northridge earthquakes in California attests to the emotional and psychological effects of earthquakes. These events caused a number of families to move away from Los Angeles.

One community's experience with a large earthquake does not always translate into increased preparedness for a community in another area. For example, in the 1994 Northridge earthquake (**M** 6.7), intense shaking caused part of the local seismograph network to malfunction. This malfunction delayed emergency response efforts because the exact location of the epicenter was not immediately available. The intense shaking also caused many unreinforced freeway bridges and buildings to collapse. A year later, the residents of Kobe, Japan, experienced nearly identical problems in a **M** 6.9 earthquake. Delays caused by communication problems and infrastructure damage kept the government from mounting a quick and effective response to the disaster. Even though Japan is one of the most earthquake-prepared countries in the world, much of the emergency relief in the Kobe quake did not arrive until about 10 hours after the earthquake!

Another example of the problems our modern society experiences with earthquakes comes from two large (**M** 7.6 and **M** 7.2) quakes in Turkey in 1999. Turkey was second only to China in the number of lethal earthquakes in the twentieth century. The first of the 1999 earthquakes occurred on August 17 and leveled thousands of concrete buildings. One-quarter of a million people were left homeless and approximately 17,118 people died. Although Turkey has a relatively high standard for new construction to withstand earthquakes, many modern buildings collapsed from the intense seismic shaking and older buildings were left standing (Figure 3.33). As with the experience of the residents of Ahmedabad, India, in the 2001 Bhuj earthquake (**M** 7.7), there is suspicion in Turkey that poor construction practices contributed to the collapse of the newer buildings. There are allegations that some of the Turkish contractors bulldozed collapsed buildings soon after the earthquake to remove evidence of shoddy construction. If that allegation is true, these contractors also tied up bulldozers that could have been used to help rescue people trapped in other collapsed buildings.

The lessons learned from Northridge, Kobe, and Turkey were bitter ones. Our modern society is vulnerable to catastrophic loss from large earthquakes. Older, unreinforced concrete buildings or buildings not designed to withstand strong ground motion are most vulnerable. In Kobe, reinforced concrete buildings constructed with improved seismic building codes experienced little damage compared to those constructed earlier. Minimizing the hazard requires new thinking about it.

COMMUNITY ADJUSTMENTS TO THE EARTHQUAKE HAZARD

Since it is clearly impossible to avoid all human habitation in earthquake-prone areas, certain steps must be taken by countries, states, communities, and individuals to adjust to the earthquake hazard. These steps include

◄ **FIGURE 3.33 COLLAPSE OF BUILDINGS IN TURKEY** Damage to the town of Golcuk in western Turkey from the **M** 7.6 Izmit earthquake in August 1999. The very old mosque on the left remains standing, whereas many modern buildings collapsed. *(Enric Marti/ AP Images)*

careful location of critical facilities, structural protection, education, and increased insurance and relief measures. Furthermore, individuals can also take steps to protect themselves. The extent to which these adjustments occur depends in part on people's perception of the hazard.

Location of Critical Facilities Buildings and other structures, often referred to as *facilities*, that are critical to the community must be located as safely as possible. These structures include hospitals, schools, utility plants, communication systems, and police and fire stations. Selecting safe locations first requires site-specific investigation of earthquake hazards such as the liquefaction, ground motion, and landslide potential. This involves *microzonation*, the detailed location and land-use planning of areas having various earthquake hazards. Microzonation is necessary because the ground's response to seismic shaking can vary greatly within a small area. In urban areas, where individual property values may exceed millions of U.S. dollars, detailed maps of ground response are required to accomplish microzonation. These maps help engineers and architects design buildings and other structures that can better withstand seismic shaking. Clearly, microzonation requires a significant investment of time and money; however, it provides the basic information needed to adequately predict the ground motion at a specific site. Although we may never be able to prevent injury and death from earthquakes, we can reduce the number of casualties through the safe location of critical facilities.

▲ FIGURE 3.34 **SEISMIC RETROFIT OF UNIVERSITY BUILDING** The cross braces and horizontal bracing on the building exterior were added to University Hall on the University of California, Berkeley, campus to help the structure withstand shaking in the next major earthquake. *(Robert H. Blodgett)*

Structural Protection The often repeated statement "earthquakes don't kill people, buildings kill people" succinctly expresses the importance of building design and construction in reducing the hazard of earthquakes. In regions where there is a significant earthquake hazard, buildings, bridges, pipelines, and other structures must be constructed to withstand at least moderate shaking. To accomplish this, communities must adopt and enforce building codes with earthquake sections similar to those in the International Building Code and International Residential Code. In establishing seismic provisions in building codes, engineers try to balance reducing risk to human life with the high construction costs of earthquake-resistant design.[40]

Of equal importance to seismic provisions in building codes is the inspection and strengthening of existing structures to withstand the ground motion of large earthquakes. Making engineering changes to existing structures, referred to as *retrofitting*, is a costly but extremely important activity (Figure 3.34). For example, a recent study showed that more than half of the hospitals in Los Angeles County could collapse in a large earthquake.[40] The cost of retrofitting these buildings is likely to be more than $8 billion. Retrofitting is especially critical in developing countries and in U.S. communities such as Memphis, Tennessee, that have only recently included earthquake-resistant design in their building codes.

Earthquake-resistant design and retrofitting of buildings, which first began in the 1920s, has generally been successful in the United States. The effect of these improvements in construction can be appreciated by comparing two events of approximately equal magnitude, the 1988 Armenia (**M** 6.8) and the 1994 Northridge (**M** 6.7) earthquakes. The loss of life and destruction in Armenia was staggering (at least 25,000 people were killed, compared with 60 in California) and some towns near the epicenter were almost totally destroyed. Most buildings in Armenia were constructed of unreinforced concrete and instantly crumbled into rubble, crushing or trapping their occupants. This is not to say that the Northridge quake was not a catastrophe. It certainly was; the Northridge earthquake left 25,000 people homeless, caused several freeway overpasses to collapse, injured at least 8000 people, and inflicted many billions of dollars in damages to buildings and other structures. However, since most buildings in the Los Angeles basin are constructed with wood frames or reinforced concrete, thousands of deaths in Northridge were avoided.

Education As with most other natural hazards, education is an important component of preparedness at the community level. This educational effort could include the public distribution of pamphlets and videos; workshops and training sessions for engineers, architects, geologists, and community planners; and the availability of information on the Internet. In areas of high seismic or tsunami risk, education also takes place through earthquake and tsunami drills in schools and earthquake disaster exercises for government officials and emergency responders.

Increased Insurance and Relief Measures Insurance and relief measures are vital to help a community, state, or country recover from an earthquake. Losses from a major earthquake can be huge. This fact was graphically demonstrated in the 1906 San Francisco earthquake (**M** 7.8) when only 6 of the 65 insurance companies were able to pay their liabilities in full.[11] Although risk assessment and the insurance industry have greatly improved in the past century, the potential losses from a large earthquake in a densely urbanized area such as Los Angeles, the San Francisco Bay area, Seattle, New York City, or Boston are enormous. The $15.3 billion in insurance claims paid in the 1994 Northridge quake (**M** 6.7) does not come close to the $200 billion estimate of losses in the 1995 Kobe quake (**M** 6.9).[14] The state of California has partially addressed this problem with state-subsidized earthquake insurance through the nonprofit California Earthquake Authority. Even with state subsidies, barely 25 percent of California residents have earthquake insurance, in part because of the high deductible.[14] Unlike flood insurance and crop insurance, there is no federally subsidized earthquake insurance. Federal officials and the insurance industry have yet to agree on a national strategy to deal with catastrophic earthquake loss.

Development of methods to estimate the potential impact of earthquakes is important to insurance companies, government agencies, and homeowners.[51] The estimation of loss from earthquakes and other hazards can now be accomplished with "Hazards U.S." (HAZUS), a FEMA computer program. This free software can be downloaded from the FEMA Web site for use on a personal computer.

PERSONAL ADJUSTMENTS: BEFORE, DURING, AND AFTER AN EARTHQUAKE

Individual actions before, during, and after a major earthquake can reduce casualties and property damage. Close to 150 million people in 39 U.S. states live in seismically active areas. In damage alone, billions of dollars could be saved if our buildings and contents were better secured to withstand shaking from earthquakes.

If you live in a seismically active area, a home safety check could increase the probability that you and your property survive a large earthquake. This inspection should include checking chimneys and foundations for reinforcement and the security of large objects, such as water heaters, that might fall over in an earthquake.[52] Probably the most important thing is to plan exactly what you will do should a large earthquake occur. This plan might include teaching your family to "drop, cover, and hold on," which means to "drop" to the ground; take "cover" under a sturdy desk, table, or bed away from falling objects; and "hold on" to something until the shaking stops. Contrary to some suggestions, standing in most doorways is not a good option.[11] Preparations should also include maintaining an adequate stock of bottled water, food, medical supplies, batteries, and spare cash.[52]

During the quake, remember what you have learned in this chapter. P waves will arrive first, causing initial vibrations followed by heavy shaking from the S and surface waves. If there has been at least 20 seconds between the P and S waves, then the quake is from a distant source. The noise of the ground vibrating will be deep and loud and may begin suddenly like a loud clap of thunder if you are close to the epicenter.[44] Books, dishes, glass, furniture, and other objects may come crashing down. Car alarms will go off and there may be explosive flashes from electrical transformers and falling power lines. The length of shaking will vary with the magnitude. For example, shaking during the **M** 6.7 Northridge quake lasted around 15 seconds, whereas shaking during the **M** 7.8 San Francisco quake lasted nearly 2 minutes.

After the shaking stops, you may feel dizzy and sick.[53] Take several deep breaths, look around, and then leave the building, carefully watching for fallen and falling objects. Do not leave the building until the shaking stops. More injuries occur when people attempt to move to a different location inside a building or try to leave. Turn off the main gas line and do not light matches or lighters. Move to an open area, away from fallen power lines and buildings or trees that might fall during aftershocks. The number and intensity of aftershocks are also related to the magnitude of the quake. For example, in the **M** 6.9 Loma Prieta quake, there were 2 **M** 5, 4 **M** 4, and 65 **M** 3 aftershocks.[54] The largest aftershock is typically at least 1 magnitude value lower than the mainshock. Overall, the most hazardous period for aftershocks is in the minutes, hours, and day following the mainshock. Both the number and intensity of aftershocks will generally decrease in the days and weeks following the mainshock.

The good news is that in most developed countries, large earthquakes are survivable. In areas of greatest hazard, such as California, buildings are generally constructed to withstand earthquake shaking and wood-frame houses seldom collapse. In the United States, earthquake-proof construction is progressively less common as one moves from west to east. Many brick houses, buildings, and other structures in New England, New York City, and Washington, D.C., will not withstand shaking in a large earthquake.[19] Therefore, it is important to be well informed and prepared for earthquakes.

REVISITING THE FUNDAMENTAL CONCEPTS

Earthquakes

The chapter on earthquakes is the first of several in this book that directly discuss a specific natural hazard. Following the summary in those chapters will be a short discussion of how that chapter relates to the fundamental concepts introduced in Chapter 1:

1. **Hazards Are Predictable from a Scientific Evaluation**
2. **Risk Assessment Is an Important Component of Our Understanding of the Effects of Hazardous Processes**
3. **Linkages Exist between Natural Hazards**
4. **Hazardous Events That Previously Produced Disasters Are Now Producing Catastrophes**
5. **Consequences of Hazards May Be Minimized**

1) Earthquake prediction has come a long way in recent decades. We are able to make long-term predictions as to which faults will most likely have earthquakes and what the return period of those earthquakes will probably be. We have not come to the point at which we can make accurate short-term predictions of, say, a few weeks to a few months, within an open framework of a few days and specify the magnitude of the event, where it will occur, and when. However, in terms of very short predictions, there have been some successes based upon precursory events that occur hours to days prior to a large earthquake.

2) Risk assessment is a mature area of research for earthquakes. We have a good sense of the probability of an earthquake of a particular magnitude happening over a relatively long period of time (30 years or so), and we are capable of measuring the consequences of such an earthquake. These are the two components that go into risk assessment, and it is an area in which a lot of progress has been and continues to be made.

3) A number of well-known linkages exist between earthquakes and other natural hazards and events. For example, we know that earthquakes cause a great number of landslides, and that submarine earthquakes that displace marine waters may produce a tsunami. Health risks also increase following large earthquakes, as a result of the spread of disease from pathogens in polluted water.

4) The impact of earthquakes on society is growing rapidly because human population in areas with an earthquake hazard is increasing exponentially. Also, in many places, we are not giving enough attention to land-use decisions that include constructing our buildings to better withstand seismic shaking. As a result, the deaths and damage from a moderate earthquake vary considerably, depending upon where it occurs in the world. For example, a magnitude 6.5 earthquake in California might cause a few deaths and billions of dollars of property damage. A similar-sized earthquake in parts of the Middle East, India, China, and other areas may kill several tens of thousands of people. Therefore, earthquakes that used to produce disasters are now producing catastrophes.

5) The earthquake hazard can be minimized through a variety of measures that include building better structures to withstand earthquakes; educating people as to the consequences of earthquakes; and developing better science to predict earthquakes, particularly in the relative short term. That is an objective that has not been reached. We have a good idea where earthquakes occur, and many earthquake faults have been mapped. However, we also know that there are many hidden faults that can produce large, damaging earthquakes. Therefore, the most prudent aspect of minimizing consequences with respect to human life is, in fact, through simply using the good engineering design and construction criteria (now available) and making sure that contractors, when they build buildings, actually follow those instructions.

QUESTIONS TO THINK ABOUT

1. Do the five fundamental principles adequately cover what we need to know about the earthquake hazard? What could be added?
2. Where does paleoseismic evaluation (geologic study of past earthquakes) fit in the five principles?
3. Specifically, how could a catastrophic earthquake destabilize a country? What are your assumptions?

Summary

Large earthquakes release a tremendous amount of energy. Seismologists measure this energy on a magnitude (**M**) scale. On this scale, an increase from one whole number to the next represents a 10-fold increase in the amount of shaking and a 32-fold increase in the amount of energy released. The USGS National Earthquake Information Center and seismic observatories rapidly calculate a preliminary magnitude for a large earthquake and later revise the number after further analysis.

After an earthquake, scientists determine the intensity of its effects on people and structures. Earthquake intensity varies with the severity of shaking and is affected by proximity to the epicenter, the local geological environment, and the engineering of structures. Intensity is described on the qualitative Modified Mercalli scale using reports of people's experience and property damage, or as instrumental intensity using a dense network of seismographs. Information on intensity helps focus emergency efforts on areas that have experienced the most intense shaking. Other measurements, such as the amount of ground acceleration, are needed to design structures that can withstand shaking in future quakes.

Earthquakes create new fractures or occur on existing fracture systems. A fault is formed where a quake has displaced the Earth along a fracture. Displacement can be mainly horizontal, as along a strike-slip fault, or mainly vertical, as on a dip-slip fault. On thrust faults, a common type of low-angle, dip-slip fault, displacement is both upward and horizontal. Faults may reach Earth's surface, creating a fault scarp or may remain buried as a blind fault.

A fault is usually considered active if it has moved during the past 10,000 years and potentially active if it has moved during the past 2 million years. Some faults exhibit tectonic creep, a slow displacement not accompanied by felt earthquakes.

Before an earthquake, strain builds up in the rocks on either side of a fault as the sides pull in different directions. When the stress exceeds the strength of the rocks, they rupture, and waves of energy, called seismic waves, radiate outward in all directions from the ruptured surface of the fault.

Seismic waves are vibrations that compress (P) or shear (S) the body of the Earth or travel across the ground as surface waves. Although P waves travel the fastest, the S and surface waves cause most of the shaking and damage. The severity of shaking of the ground and buildings is affected by the type and thickness of earth material present; the direction in which the fault ruptured; the depth of the earthquake focus; and for buildings, their engineering design.

Buildings highly subject to damage are those that (1) are constructed on unconsolidated sediment, artificially filled land, or water-saturated sediment, all of which tend to amplify shaking; (2) are not designed to withstand significant horizontal acceleration of the ground; or (3) have natural vibrational frequencies that match the frequencies of the seismic waves.

Seismologists have proposed a four-stage earthquake cycle for large earthquakes. The first stage is a period of seismic inactivity, during which elastic strain builds up in the rocks along a fault. This stage is followed by a stage of increased seismicity as the strain locally exceeds the strength of the rocks, initiating local faulting and small earthquakes. The third stage, which does not always occur, consists of foreshocks. Finally, the fourth stage is the *mainshock*, which occurs when the fault segment ruptures, producing the elastic rebound that generates seismic waves.

Most earthquakes occur on faults near tectonic plate boundaries, such as the San Andreas fault in California, the Cascadia subduction zone in the Pacific Northwest, and the Aleutian subduction zone in Alaska. Intraplate earthquakes are also common in Hawai'i, the western United States, the southern Appalachians and South Carolina, and in the northeastern United States. Some of the largest historic earthquakes in North America occurred within the plate in the central Mississippi Valley in the early 1800s.

The primary effect of an earthquake is violent ground motion accompanied by fracturing, which may shear or collapse large buildings, bridges, dams, tunnels, pipelines, levees, and other structures. Other effects include liquefaction, regional subsidence, uplift of the land, landslides, fires, tsunamis, and disease. Earthquakes also provide natural service functions such as enhancing groundwater and energy resources, exposing or contributing to the formation of valuable minerals deposits and, in some cases, potentially reducing the chance of future large quakes.

Human activity has locally increased earthquake activity by fracturing rock and increasing water pressure underground below large water reservoirs, by raising fluid pressure in faults and fractures through the deep-well disposal of liquid waste, and by setting off underground nuclear explosions. The accidental damage caused by the first two activities is regrettable, but understanding how we have caused earthquakes may eventually help us control or stop large natural earthquakes.

Reduction of earthquake hazards requires detailed mapping of geologic faults, the cutting of trenches to determine earthquake frequency, and detailed mapping and analysis of earth materials sensitive to shaking. It also requires new methods for predicting, controlling, and adjusting to earthquakes. Adjustments include improving structural design to better withstand shaking, retrofitting existing structures, microzonation of areas of seismic risk, and updating and enforcing building codes.

Being able to predict the location, date, time, and magnitude of earthquakes has been a long-term goal of seismologists. Accomplishing this goal is many years away. To date, scientists have been able to make long- and intermediate-term forecasts for earthquakes using probabilistic methods, but not consistent, accurate short-term predictions. A potential problem of predicting earthquakes is that their pattern of occurrence is often variable, with clusters of events separated by longer periods of time with reduced earthquake activity.

Warning systems and earthquake prevention are not yet reliable alternatives to earthquake preparedness. A satellite and sensor-based tsunami warning system being deployed in the world's oceans will improve tsunami preparedness in some coastal areas. More communities must develop emergency plans to respond to a predicted or unexpected catastrophic earthquake, especially in areas of seismic risk outside of California. Such plans should include earthquake education, disaster response, drills, and improved insurance coverage. At a personal level, individuals who live in or visit areas of seismic risk can learn how to "duck, cover, and hold on" before the next large earthquake. Preparation both reduces the earthquake hazard and eases recovery.

Key Terms

design basis ground motion (p. 84)
directivity (p. 65)
earthquake (p. 51)
earthquake cycle (p. 68)
epicenter (p. 54)
fault (p. 56)
focus (p. 54)
intensity (p. 53)
liquefaction (p. 76)
magnitude (p. 53)
material amplification (p. 67)
megathrust earthquake (p. 69)
Modified Mercalli scale (p. 55)
moment magnitude (p. 54)
P wave (p. 61)
S wave (p. 61)
resonance (p. 76)
seismograph (p. 54)
slow earthquake (p. 60)
surface wave (p. 61)
supershear (p. 65)
tectonic creep (p. 60)
tectonic framework (p. 84)

Review Questions

1. What is the difference between the epicenter and focus of an earthquake?
2. What does moment magnitude measure? How is it related to the amount of shaking and the amount of energy released by an earthquake?
3. What is instrumental intensity? How is it related to a shake map?
4. Explain how faulting occurs.
5. How are active and potentially active faults defined?
6. What is paleoseismicity?
7. What is the difference in the rates of travel of P, S, and surface waves? How is this difference important in locating earthquakes?
8. How do seismologists locate earthquakes?
9. How does the depth of an earthquake's focus relate to shaking and damage?
10. What types of earth materials amplify seismic waves? How is material amplification related to earthquake damage?
11. Explain the earthquake cycle and elastic rebound.
12. What are foreshocks and aftershocks?
13. Where are earthquakes most likely to occur in the world? In North America?
14. List the major effects of earthquakes.
15. How do plate boundary and intraplate earthquakes differ?
16. Why aren't the largest earthquakes always the most damaging and most deadly?
17. What are the effects of earthquake-induced liquefaction?
18. Why do disease outbreaks sometimes follow major earthquakes and other large natural disasters?
19. How can earthquakes be beneficial?
20. How can humans cause earthquakes?
21. What kinds of information are useful in assessing seismic risk?
22. What kinds of phenomena may be precursors for earthquakes?
23. What is the difference between an earthquake prediction and a forecast?
24. What kinds of adjustments can a community make to the earthquake hazard?
25. What is retrofitting?

Critical Thinking Questions

1. You live in an area that has a significant earthquake hazard. Public officials, the news media, and citizens are debating whether an earthquake warning system should be developed. Some people are worried that false alarms will cause a lot of problems, and others point out that the response time may be very long. What are your views on this? Should public funds be used to finance an earthquake warning system, assuming such a system is feasible? What are potential implications if a warning system is not developed and a large earthquake results in damage that could have been partially avoided with a warning system in place?
2. You are considering buying a home on the California coast. You know that earthquakes are common in the area. What questions would you want to ask before purchasing the home? For example, consider the effects of earthquakes, the relationship of shaking to earth material, and the age of the structure. What might you do to protect yourself (both financially and physically) if you did decide to buy the house?
3. You are working for the Peace Corps in a developing country where most of the homes are built out of unreinforced bricks. There has not been a large damaging earthquake in the area for several hundred years, but earlier there were several earthquakes that killed thousands of people. How would you present the earthquake hazard to the people living where you are working? What steps might be taken to reduce the hazard?

Assignments in Applied Geology: Earthquake Damage Assessment

THE ISSUE:

In 1897, the Twin Fork fault produced an earthquake that caused ground motion in Hazard City equivalent to IX on the Modified Mercalli Intensity scale. A recent major earthquake in California has caused the citizens of Hazard City to recall the earthquake history of their town, and they worry about the impact that another IX earthquake would have on their homes.

YOUR TASK:

You have been hired to investigate this situation. To complete your study, you must clearly understand what level IX on the Modified Mercalli Intensity scale means in terms of building impact and human actions; go on a tour of Hazard City to learn where the residential neighborhoods are located, what types of buildings are found there, and the number of residents in each neighborhood; and consult reference materials to learn how the residential buildings found in the neighborhoods of Hazard City will respond if they are shaken by a ground motion equivalent to a Mercalli Intensity of IX. It is up to you to determine the methods of your investigation.

AVERAGE COMPLETION TIME:

1 hour

4 Tsunamis

Indonesian Tsunami

Prior to Sunday, December 26, 2004, few people knew what the Japanese word *tsunami* meant. That changed in the span of only a few hours, as close to 230,000 people were killed, many hundreds of thousands injured, and millions were displaced in more than a dozen countries surrounding the Indian Ocean. With no warning system in place, residents of coastal area after coastal area around the Indian Ocean were struck by a series of tsunami waves without notice.

The source of this tsunami was the largest earthquake on Earth in the past 4 decades that struck on the morning of December 26, just off the west coast of the Indonesian island of Sumatra (Figure 4.1). Now assigned a magnitude of at least 9.1, this earthquake caused the most damaging tsunami in recorded history.[1] Great earthquakes of this magnitude occur along tectonic plate boundaries where one plate is diving down below another in a process known as subduction. In this case, the giant Indian and Australian plates are being subducted to the northeast beneath the Burma microplate. In this earthquake, the Burma microplate moved about 20 m (65 ft.) to the west-southwest along a gently inclined subduction zone (Figure 4.2).[1] Because this was a large amount of displacement along the thrust faults in the subduction zone, geologists classify this type of earthquake as a "megathrust event." Seismic waves generated by the rupture caused several minutes of shaking on nearby islands. The total length of the rupture along the subduction zone was more than 1500 km (930 mi.), about the length of the state of California.[1] Not only did the seafloor shift about 20 m (65 ft.) horizontally, but it also rose several meters vertically.[1] Displacement of the sea bottom disturbed and displaced the overlying waters of the Indian Ocean, and a series of tsunami waves were generated. The effect is like that of throwing a large boulder in your bathtub and watching the rings or waves of water spread out. In this case, though, the boulder came up from the bottom, but the result was the same, and waves radiated outward, moving at high speed across the Indian Ocean (Figure 4.3).

Now assigned a magnitude of at least 9.1, this earthquake caused the most damaging tsunami in recorded history

Unlike the Pacific Ocean, there was no tsunami warning system for the Indian Ocean, and people were mostly caught by surprise. Scientists on duty at the Pacific Tsunami Warning Center in Hawai'i recognized that the earthquake could potentially produce a tsunami, but the geophysical techniques to calculate the size of an earthquake of such magnitude were not immediately available to them.[2] Once they determined that a tsunami

◀ **2004 Indonesian Tsunami** Flattend houses in the tsunami-struck city of Banda Aceh, Indonesia. *(Kim Kyung/Newscom)*

LEARNING OBJECTIVES

In this chapter, we focus on one of Earth's most destructive natural hazards—the tsunami. Sometimes incorrectly called tidal waves, these ocean waves are both fascinating in their behavior and awesome in their power. Tsunamis are common in some coastal regions and rare in others. Although they have long been known to cause disasters and catastrophes, the hazard posed by tsunamis has generally been underestimated. For years, scientists attempted to get public officials to expand tsunami warning systems to ocean basins outside of the Pacific Ocean. It took the catastrophic deaths and devastation of the Indonesian tsunami of 2004 for many governments and communities to take the tsunami hazard more seriously. However, as often occurs after disasters and catastrophes, translating the increased hazard awareness into improved warning, preparedness, and mitigation is proceeding at an excruciatingly slow pace. This chapter will examine the natural tsunami process and assess the hazard that these waves pose to people. Your goals in reading the chapter should be to

- understand the process of tsunami formation and development.
- understand the effects of tsunamis and the hazards they pose to coastal regions.
- know what geographic regions are at risk for tsunamis.
- recognize the linkages between tsunamis and other natural hazards.
- know what nations, communities, and individuals can do to minimize the tsunami hazard.

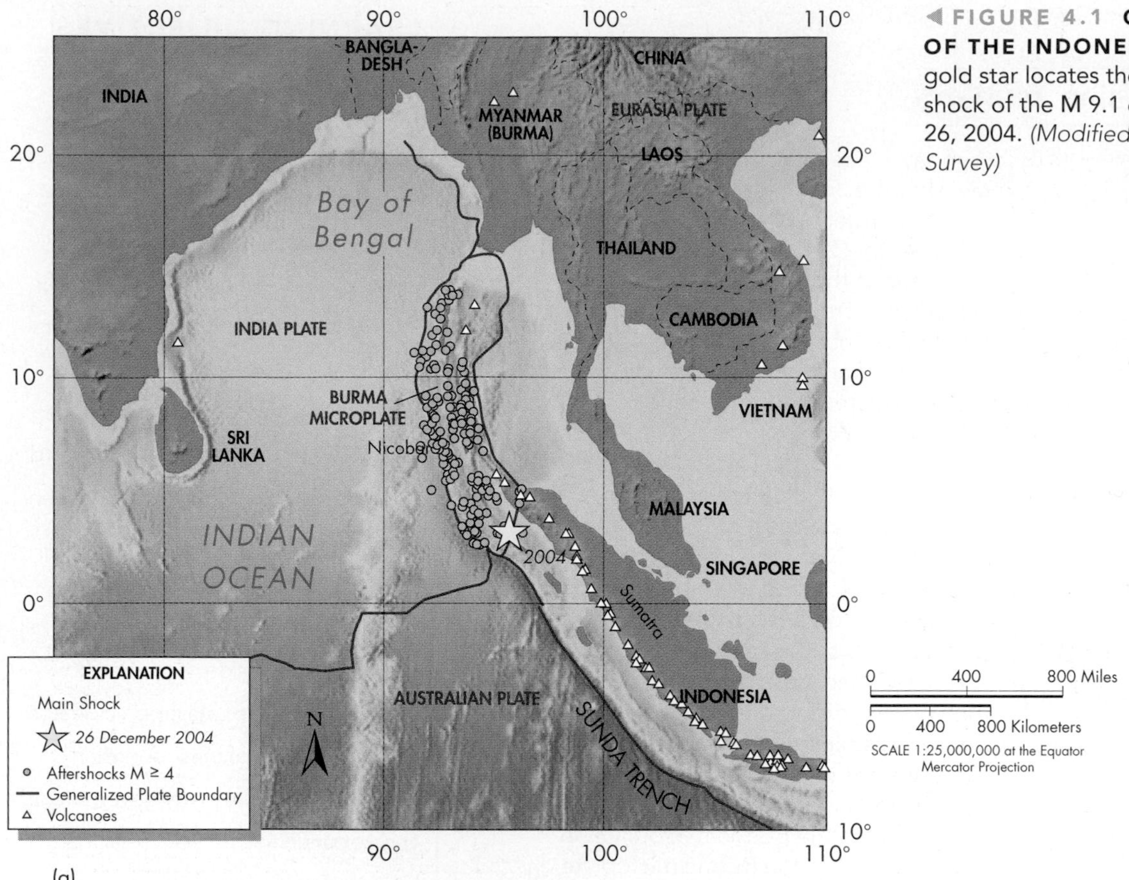

FIGURE 4.1 GEOLOGIC SETTING OF THE INDONESIAN TSUNAMI The gold star locates the epicenter of the mainshock of the M 9.1 earthquake on December 26, 2004. *(Modified after U.S. Geological Survey)*

was likely, the scientists managed to contact other scientists in Indonesia and have the U.S. State Department relay its concerns to some nations surrounding the Indian Ocean. Unfortunately, the warnings did not reach authorities in time to take action, and, even if they had, there was no system in place for directly notifying coastal residents. Had it been possible for warnings to be received, tens of thousands of lives could have been saved because the tsunami waves took several hours to reach some of the coastlines where people died (Figure 4.3). Even a warning of a half hour or so would have

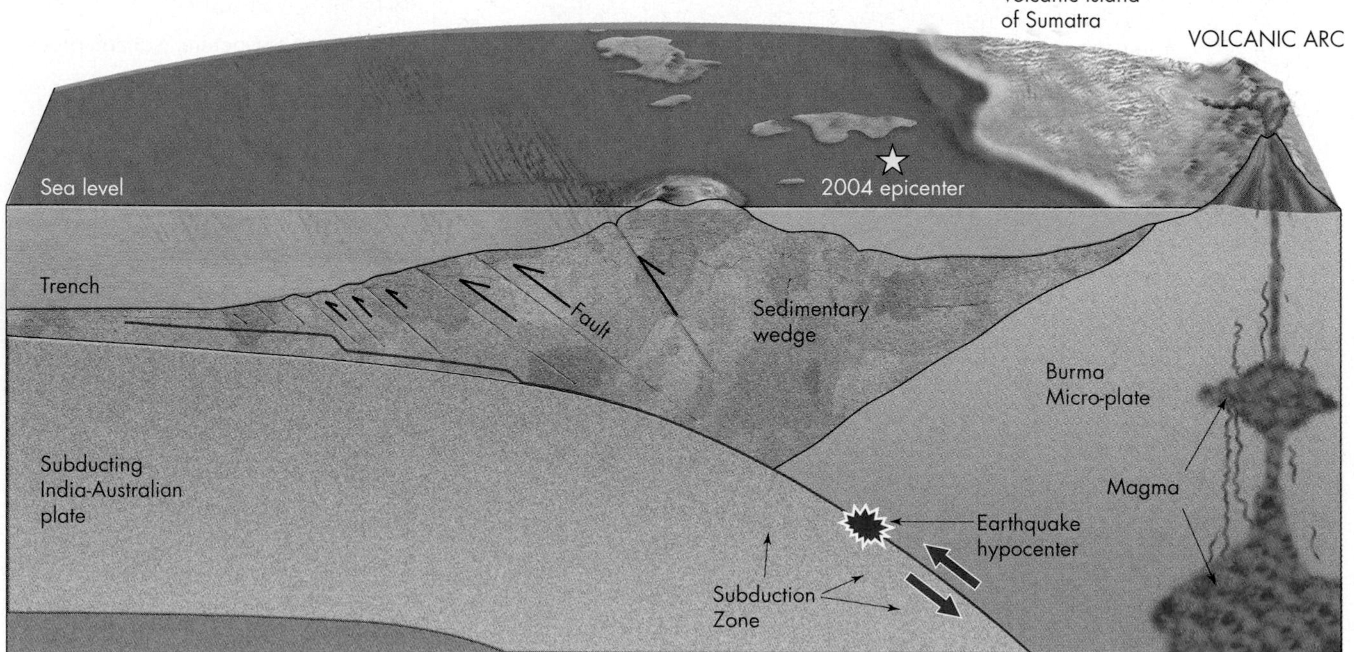

FIGURE 4.2 EARTHQUAKE IN SUBDUCTION ZONE Idealized cross section of where a subduction zone earthquake might occur. *(Modified from United Kingdom Hydrographic Office).*

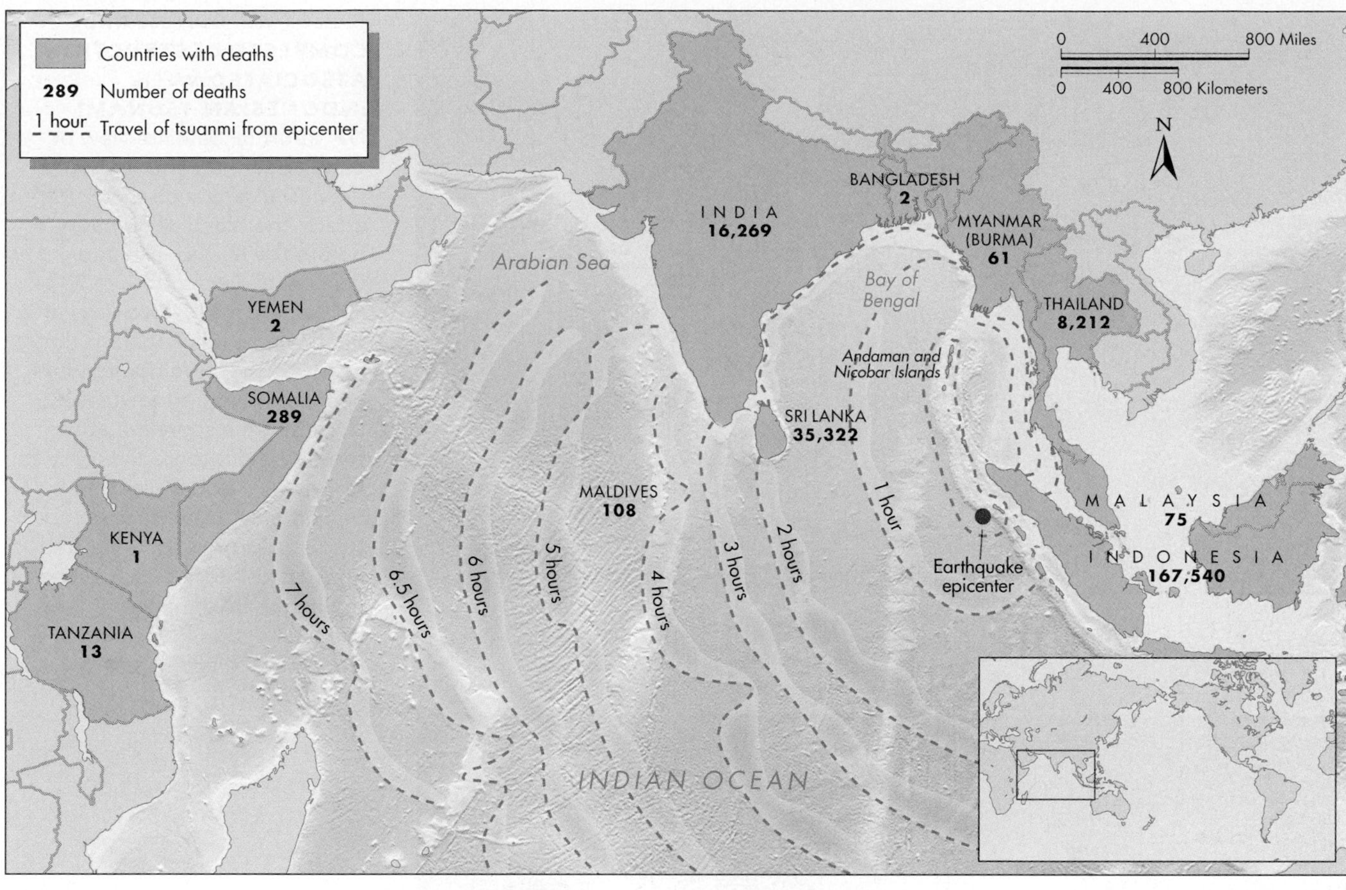

▲ FIGURE 4.3 **2004 TSUNAMI KILLED PEOPLE ON BOTH SIDES OF INDIAN OCEAN** Map showing the path of the tsunami produced by a **M** 9.1 earthquake off the east coast of Sumatra, Indonesia, on December 26, 2004. Shown with blue dashed lines is the location of the lead tsunami wave each hour after the earthquake. Note that the tsunami waves took approximately 7 hours to reach Somalia on Africa's east coast. Most deaths were in Indonesia, where the first local tsunami wave arrived an hour or less after the earthquake. The number of deaths shown in each country is the total number of people dead or still missing. (Causalities summarized in Telford, J., and Cosgrave, J. 2006. *Joint evaluation of the international response to the Indian Ocean tsunami: Synthesis report.* London: Tsunami Evaluation Coalition; T*sunami travel time data from NOAA)*

been sufficient to move many people from low-lying coastal areas.[3] The first sensor on the seafloor to detect a tsunami in the Indian Ocean was installed in 2006, and others will follow. Sirens are now in place in some coastal areas around the Indian Ocean to warn people if a tsunami is coming.

More than three-quarters of the deaths were in Indonesia, which experienced both the intense shaking from the earthquake and the inundation by tsunami waves within less than an hour. In contrast, the first tsunami wave took between 1 1/2 and 2 hours to reach Sri Lanka and India where many thousands died, and more than 7 hours to reach Somalia on the east coast of Africa (Figure 4.3). The first of many tsunami waves reached other locations earlier or later, depending on their distance from the rupture in the subduction zone.

At the northern tip of the island of Sumatra, the Indonesian provincial capital of Banda Aceh was nearly destroyed (Figure 4.4). Destruction was caused by shaking from the earthquake, the force of the tsunami waves, and flooding as the land subsided following the earthquake. Coastal subsidence of a tectonic plate is common in megathrust events along a subduction zone (see Chapter 8, Figure 8.12).

Tourist areas in the region were hard hit, especially in Thailand, where several thousand tourists from Europe, the United States, and elsewhere were killed. Most visitors, and many first-generation residents, were unfamiliar with tsunamis. They did not know how to recognize that a tsunami was about to take place or what to do if one occurred. This was true for many people in Phuket, Thailand (Figure 4.5). Some people seemed to be mesmerized by the approaching waves, while others ran in panic.

On some coastlines, everything wasn't quite as bleak. In Thailand, a 10-year-old British girl sounded the warning in time for 100 people to evacuate a resort beach. Only a few weeks before going to Thailand, she had received a lesson in school about plate tectonics, earthquakes, and tsunamis. As part of that lesson, she learned that, sometimes, the sea recedes prior to the arrival of a tsunami wave. That is precisely what she observed, and her screaming to get off the beach eventually convinced her mother and others to take action. Thanks to her persistence, the beach was

(a)

(b)

◀ **FIGURE 4.4 NEARLY COMPLETE DESTRUCTION ASSOCIATED WITH INDONESIAN TSUNAMI OF 2004** (a) Satellite view of Banda Aceh, the Indonesian provincial capital on the northern end of the island of Sumatra, before the earthquake and tsunami of December 26, 2004. *(Digital Globe)* (b) Two days after the tsunami, it is apparent that nearly all the development has been damaged or destroyed. Note that the shoreline at the top of the photograph has been extensively eroded, leaving behind only a few small islands. Large parts of the city flooded because of subsidence caused by the earthquake that combined with inundation from the tsunami. *(Digital Globe)*

successfully evacuated. Her mother later stated she did not even know what a tsunami was, but that her daughter's school lesson had saved their lives.

On another beach in Sri Lanka, a scientist on vacation witnessed a small wave rise up and inundate the hotel swimming pool. In the next 20 minutes, the sea level dropped by around 7 m (23 ft.). The scientist recognized these events as a sign that a big wave was coming and sounded the alarm. A hotel employee then used a megaphone to warn people to get off the beach. Many people had gone down to the beach out of curiosity to see the exposed seafloor. Shortly thereafter, the sea rose 7 m (23 ft.) above its normal level. Fortunately, most people were either near hotel stairs or had escaped to higher floors. None of the staff or guests drowned, but several people on the ground floor were swept out and survived only by clinging to palm trees in the hotel garden.[4]

In the Andaman and Nicobar Islands in the northeastern Indian Ocean, as well as in parts of Indonesia, some native people have a collective memory of tsunamis. When the earthquake occurred, they applied that knowledge and moved to high ground, saving entire small tribes on some islands.

In Khao Lak, Thailand, it was elephants, not people, who sounded the warning and saved lives in the 2004 tsunami.[5] The elephants, which normally give tourists rides, started trumpeting about the time the **M** 9.1 earthquake occurred in the subduction zone more than 600 km (370 mi.) to the

▲ FIGURE 4.5 **TOURISTS RUNNING FOR THEIR LIVES** Man in foreground is looking back at a brown wave of the 2004 Indonesian tsunami rushing toward him. The wave is higher than the building behind him at this Phuket, Thailand, resort. Many people living there, as well as the tourists, did not initially think the wave would inundate the area where they were, and, when the waves arrived, they thought they would be able to outrun the rising water. In some cases, people did escape, but all too often they drowned. *(John Russell/ZUMA Press)*

west. The elephants then became agitated again about 1 hour later. Those elephants that were not taking tourists for rides broke loose from their strong chains and headed inland. Other elephants carrying tourists on their backs ignored their handlers and climbed a hill behind the beach resort, saving the tourists from the fate that fell upon about 4000 people who were killed by the tsunami at the resort. When handlers recognized the advancing tsunami, they got other elephants to lift tourists onto their backs with their trunks and proceed inland. The elephants did so even though they were accustomed only to people mounting them from a wooden platform. Tsunami waves then surged about 1 km (0.5 mi.) inland. The elephants stopped just beyond where the waves ended their destructive path.

The question is: Did the elephants know something that people did not? Animals have sensory ability that differs from humans. It's possible the elephants heard the earthquake, because earthquakes produce sound waves with low tones referred to as *infrasonic sound*. Some people can sense infrasonic sound, but they don't generally perceive it as a hazard. The elephants may also have sensed the motion as the land vibrated from the earthquake. In either case, the elephants fled inland away from the advancing tsunami. Although the link between the elephants' sensory ability and their behavior is still speculative, the actions of the elephants nevertheless saved at least a dozen lives.[5]

Numerous reports received after the 2004 Indonesian tsunami indicate the vital role of education in tsunami preparedness. Thousands of lives would have been saved had more people recognized from Earth's behavior that a tsunami was likely. Those people who felt the large earthquake would have known that a tsunami might be coming and could have evacuated to higher ground. Even thousands of miles away where the seismic waves were not felt, there were still signals of what was about to happen. As the British schoolgirl knew, if the water suddenly recedes from the shore and exposes the sea bottom, you can expect it to come back as a tsunami wave. This is the signal to move to higher ground. People should also be informed that tsunamis are seldom one wave but are, in fact, a series of waves, with later ones many times larger and more damaging than earlier ones. The education of people close to where a tsunami may originate is particularly important, as waves may arrive in 10 to 15 minutes following an earthquake. Geologists have warned that it is likely that another large tsunami will be generated by earthquakes offshore of Indonesia in the next few decades.[6,7]

Since 2004, an ocean bottom tsunami warning system consisting of sensors that recognize tsunami waves is in place in the Indian Ocean. In addition, much more has been done to educate people about the tsunami hazard. A tsunami in 2009 generated by a **M** 8.1 earthquake struck American Samoa and other islands in the region (Figure 4.6). The number of lives lost was about 200, and this was fewer than might have occurred without the program of tsunami awareness and education. People got the word out about the tsunami hazard and what to look for, and that resulted in evacuations that saved lives. However, with even more education and improved warning linked to evacuation processes saved, more lives could have been saved.[8]

▲ FIGURE 4.6 **2009 TSUNAMI DAMAGE** An officer with the American Samoa Police Department helps out with the clean up at Asili Village, located on the western side of American Samoa Thursday Oct. 1, 2009. An 8.3 earthquake on Tuesday Sept. 29, 2009 triggered a 3 m (10 ft.) tsunami that destroyed many homes and businesses on American Samoa and killed over 139 people. *(Eugene Tanner/AP Images)*

4.1 Introduction

Tsunamis (the Japanese word for "large harbor waves") are produced by the sudden vertical displacement of ocean water.[3] These waves are a serious natural hazard that can cause a catastrophe thousands of kilometers from where they originate. They may be triggered by several types of events, including a large earthquake that causes a rapid uplift or subsidence of the seafloor; an underwater landslide that may be triggered by an earthquake; the collapse of part of a volcano that slides into the sea; a submarine volcanic explosion; and an impact in the ocean of an extraterrestrial object, such as an asteroid or comet. Asteroid impact can produce a **mega-tsunami**, a wave that is about 100 times higher than the largest tsunami produced by an earthquake, and one that could put hundreds of millions of people at risk.[3] Fortunately, the frequency of large asteroid impacts is low. Of the previously mentioned potential causes, tsunamis produced by earthquakes are by far the most common.

Damaging tsunamis in historic time have been relatively frequent and mostly in the Pacific Basin. Recent examples include the following:[9, 10]

- The 1755 (~**M** 9) Lisbon, Portugal, earthquake produced a tsunami that, along with the earthquake and resulting fire, killed an estimated 20,000 people. Tsunami waves crossed the Atlantic Ocean and amplified to heights of 7 m (23 ft.) or more in the West Indies.[9]
- The 1883 violent explosion of the Krakatoa volcano in the Sundra Strait between Java and Sumatra caused the top of the volcano to collapse into the ocean. This sudden collapse produced a giant tsunami more than 35 m (116 ft.) high that destroyed 165 villages and killed more than 36,000 people.[10]
- The 1946 (**M** 8.1) Aleutians (Alaska) earthquake produced a tsunami in the Hawaiian Islands that killed about 160 people.
- The 1960 (**M** 9.5) Chile earthquake triggered a tsunami that killed 61 people in Hawai'i after traveling for 15 hours across the Pacific Ocean.
- The 1964 (**M** 9.2) Alaska earthquake generated a deadly tsunami that killed about 130 people in Alaska and California.
- The 1993 (**M** 7.8) earthquake in the Sea of Japan caused a tsunami that killed 120 people on Okushiri Island, Japan.
- The 1998 (**M** 7.1) Papua New Guinea earthquake triggered a submarine landslide that produced a tsunami that killed more than 2100 people.
- The 2004 (**M** 9.1) Sumatran earthquake generated a tsunami that killed about 230,000 people.
- The 2009 (**M** 8.1) Samoan earthquake generated a tsunami that killed about 200 people.
- The 2010 (**M** 8.8) Chile earthquake generated a tsunami that killed about 700 people in coastal towns.

HOW DO EARTHQUAKES CAUSE A TSUNAMI?

An earthquake can cause a tsunami by movement of the seafloor and by triggering a landslide. Seafloor movement is probably the more common of these two mechanisms. This movement occurs when the seafloor sits on a block of the Earth's crust that shifts up or down during a quake. In general, it takes a **M** 7.5 or greater earthquake to create enough displacement of the seafloor to generate a damaging tsunami. The upward or downward movement of the seafloor displaces the entire mass of water, from the sea bottom to the ocean surface. This starts a four-stage process that eventually leads to landfall of tsunami waves on the shore (Figure 4.7):

1. For example, if an earthquake rupture uplifts the seafloor, the water surface above the uplift initially forms an elongate dome parallel to the geologic fault. That dome collapses and generates a tsunami wave. Oscillations of the water surface and, in some cases, aftershocks along the fault produce additional waves. These waves radiate outward like the pattern made by a pebble thrown into a pond of water.
2. In the deep ocean, the tsunami waves move very rapidly and are spaced long distances apart. Their velocity is equal to the square root of the product of the acceleration of gravity and the water depth. The acceleration of gravity is approximately 10 m/sec^2 and the average water depth in the deep ocean is 4000 m (13,000 ft.). Thus, when we do the math, we see that if we take the product of 10 m/sec^2 and 4000 m and then take the square root of that number, we arrive at a velocity of 200 m/sec. Converting 200 m/sec to km/hour, we find that tsunamis travel at 720 km (450 mi.) per hour, close to the average airspeed for the most recent Boeing 737 jet airliner! In the deep ocean, the spacing between the crests of tsunami waves may be more than 100 km (60 mi.), and the height of the waves is generally less than 1 m (3 ft.). Sailors rarely notice a passing tsunami in the deep ocean.
3. As the tsunami nears land, the water depth decreases, so the velocity of the tsunami also decreases. Near land, the forward speed of a tsunami may be about 45 km (28 mi.) per hour—too fast to outrun, but not nearly as

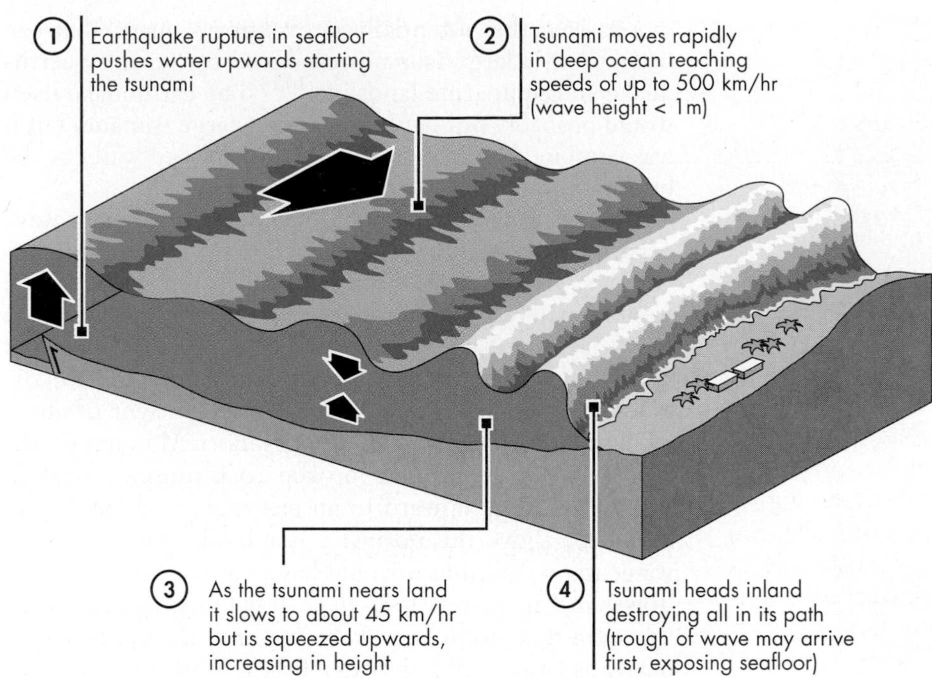

◀ **FIGURE 4.7 TSUNAMI DAMAGE** Idealized diagram showing the process of how a tsunami is produced by an earthquake. *(Modified after the United Kingdom Hydrographic Office)*

fast as in the open ocean. This decrease in velocity also decreases the spacing between wave crests, that is, the wavelength. As the water slows down and piles up, the height of the waves increases.

4. When the first tsunami wave reaches the shore and moves inland, it may be several meters to several tens of meters high and destroys everything in its path. Sometimes, the trough of the wave may arrive first, and this exposes the seafloor. When tsunamis strike the coast, they generally do not come in as a giant breaking wave as you might see on the surfing beaches of the north shore of Oahu or the California coast. Instead, when the wave arrives, it is more like a very strong and fast-rising increase in sea level. On the rare occasions when tsunamis do break, they may appear as a vertical wall of turbulent water. The movement of a tsunami inland is called the **runup** of the wave. Runup refers to the furthest horizontal and vertical distance that the largest wave of a tsunami moves inland. Once a wave has moved to its farthest extent inland, most of the water then returns back to the open ocean in a strong and often turbulent flow (see chapter opener photograph). A tsunami can also generate other types of waves, known as *edge waves*, that travel back and forth parallel to the shore. The interaction between edge waves and additional incoming tsunami waves can be complex. As result of this interaction, wave amplification may occur, causing the second or third tsunami wave to be even larger than the first (see Survivor Story 4.1, p.108). Most commonly a series of tsunami waves will strike a particular coast over a period of several hours.[11]

When an earthquake ruptures and uplifts the seafloor close to land, both distant and local tsunamis may be produced (Figure 4.8). The initial fault displacement and uplift lift the water above mean sea level. This uplift creates potential energy that drives the horizontal movement of the waves. The uplifted dome of water collapses downward and expands radially splitting into two waves. One wave, known as a **distant tsunami**, travels out across the deep ocean at high speed. Distant tsunamis can travel thousands of kilometers across the ocean to strike remote shorelines with little loss of energy. The second wave, known as the **local tsunami**, heads in the opposite direction toward the nearby land. A local tsunami can arrive quickly following an earthquake, giving coastal residents and visitors little warning time. When the initial tsunami wave is split, each (distant and local) tsunami has a wave height about one-half of that of the original dome of water.[11]

An example of a local tsunami comes from the island of Okushiri in western Japan (Figure 4.9). On July 12, 1993, a **M** 7.8 earthquake in the Sea of Japan produced a local tsunami that extensively damaged Anoae, a small town on the southern tip of the island. The town was struck several times by 15 to 30 m (50 to 100 ft.) waves coming around both sides of the island. There was virtually no warning because the earthquake epicenter was very close to the island (Figure 4.9). The huge waves arrived only 2 to 5 minutes after the earthquake, killing 120 people and causing $600 million in property damage.[12]

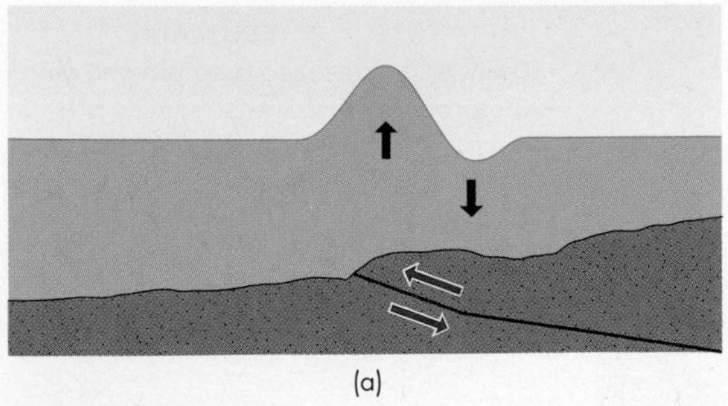

(a)

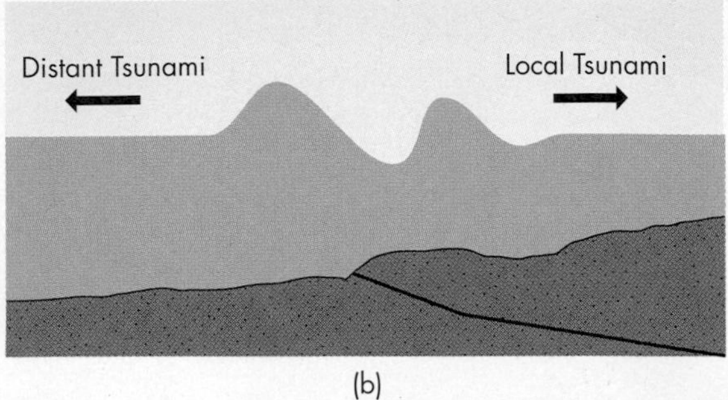

(b)

▲ FIGURE 4.8 **DISTANT AND LOCAL TSUNAMIS** (a) Fault displacement lifts the water above mean sea level, which creates potential energy that drives the horizontal propagation of the waves. In these diagrams, wave height is greatly exaggerated compared to the depth of the water. Actual height of the waves is at most a meter or so, but it is spread out to tens or several hundreds of kilometers in length. (b) The initial wave is split into a tsunami that travels out across the deep ocean (distant tsunami) and another tsunami that travels toward nearby land (local tsunami). The wave heights are also split, and each (distant and local) tsunami has a height about one-half the original wave in (a). *(Modified after U.S. Geological Survey; http://walrus.wr.usgs.gov/tsunami/basics.html)*

HOW DO LANDSLIDES CAUSE A TSUNAMI?

Although most large tsunamis are produced by fault rupture along subduction zones at tectonic plate boundaries, landslides have also generated huge tsunamis. These landslides can take place underwater, where they are referred to as *submarine landslides*, or they can be large rock avalanches that fall from mountains into the sea. In most cases, the landslides are triggered by an earthquake. For example, in 1998, a **M** 7.1 earthquake occurred off the north shore of the island of New Guinea. The earthquake was felt at Sissano Lagoon, and shortly thereafter a tsunami arrived with little warning. In less than an hour, coastal villages were swept away, leaving 12,000 people homeless and more than 2000 dead. The tsunami waves, which reached heights of 15 m (50 ft.), appear to have resulted primarily from a submarine landslide triggered by the earthquake.[13] The epicenter of the earthquake was only about 50 km (30 mi.) offshore so there was little, if any, warning time for coastal residents. This event emphasized the potentially devastating damage that can occur from a large tsunami produced by a nearby earthquake and submarine landslide.[10,13] The earthquake itself would probably not have generated a large tsunami, but it was combined with a landslide that displaced water at the bottom of the sea.

Perhaps the most famous landslide-induced giant wave occurred at Lituya Bay, Alaska, in 1958. The landslide was set in motion by a **M** 7.7 earthquake on a nearby fault. Approximately 30.5 million cubic meters of rock fell from a cliff into the bay, instantly displacing a huge volume of seawater.[9] That volume of earth material would fill the National Mall in Washington, D.C., to a height of about 25 m (80 ft.) (the height of the Hirshhorn Museum on the Mall). The huge mass of broken rock caused waters in the bay to surge upward to an elevation of about 524 m (1720 ft.) above the normal water level.[9] This splash of water was so tall that it would have washed over the Sears Tower in Chicago with 82 m (270 ft.) to spare! Although the wave that swept out of the bay into the ocean wasn't nearly as large, it did pick up a fishing boat with two people onboard. The boat's captain estimated that the tsunami wave was 30 m (100 ft.) high when it lifted the boat over the sand spit at the mouth of the bay. Two other fishing boats in the bay disappeared after being caught in the wave.[14] A panoramic view of the bay taken a few days after the tsunami shows that the erosive force of the wave stripped the shoreline bare of trees (Figure 4.10). In the area beyond the direct splash effect of the fallen rock, the high waterline above the barren shoreline indicates that the local tsunami had a runup height of 20 to 70 m (65 to 230 ft.).[9]

4.2 Regions at Risk

Although all ocean and some lake shorelines are at risk for tsunamis, some coasts are more at risk than others. The heightened risk comes from the geographic location of a coast in relation to potential tsunami sources, such as earthquakes, landslides, and volcanoes. Coasts in proximity to a major subduction zone or directly across the ocean basin from a major subduction zone are at greatest risk. For the purpose of this discussion, a major subduction zone is considered one that is capable of periodically generating a great earthquake of **M** 9 or greater. Ruptures produced by these great earthquakes may extend for 1000 km (625 mi.) or more along the subduction zone and produce significant uplift of the seafloor. The greatest tsunami hazard, with return periods of several hundred years, is adjacent to those major subduction zones with a convergence of a few centimeters per year. These include the Cascadia subduction zone in the Pacific Northwest, the Chilean trench along southwestern South America, and the subduction zones off the coast of Japan.[15]

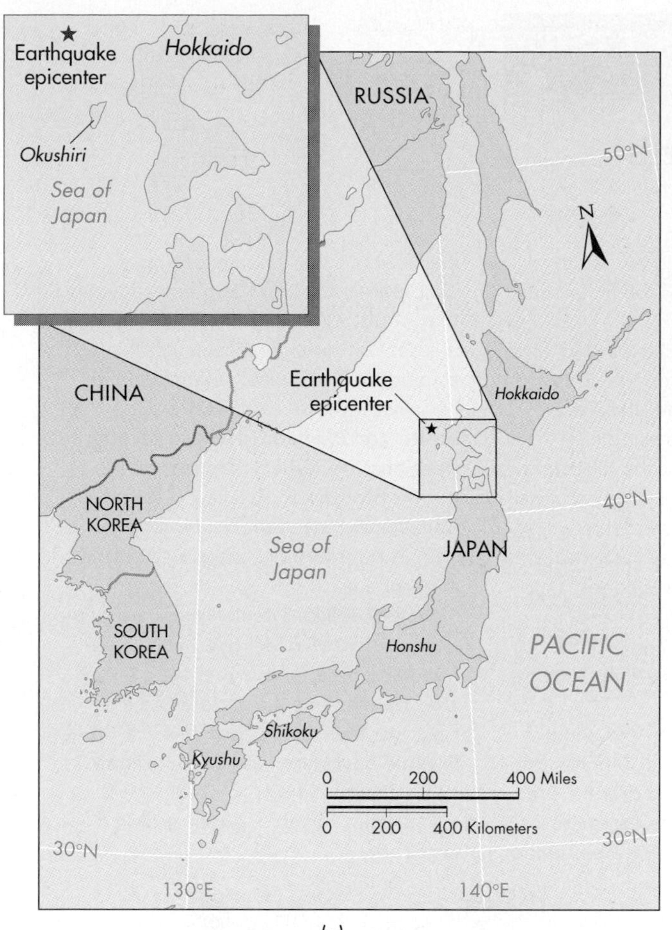

▲ **FIGURE 4.9 EARTHQUAKE-PRODUCED LOCAL TSUNAMI** (a) Location of the 1993 Hokkaido-Nanse-Oki earthquake (**M** 7.8). *(From Hokkaido Tsunami Survey Group. 1993. Tsunami devastates Japanese coastal region.* EOS, Transactions of the American Geophysical Union *74(37):417, 432)* (b) Aerial view of the town of Anoae at the southern tip of Okushiri Island, Japan, following inundation by a local tsunami from the 1993 earthquake. Note the extensive damage to the shoreline area and fires burning from leaking propane and kerosene used for heating. *(Sankei Shimbun/Corbis)*

◀ **FIGURE 4.10 LANDSLIDE-PRODUCED GIANT WAVE** Lituya Bay, Alaska, following local tsunami. Exposed shoreline of the bay was produced as trees were removed by the erosive force of the wave. *(DJ Miller/US Geological Survey)*

SURVIVOR STORY 4.1

Tsunami in the Lowest Country on Earth

Boxing Day, December 26, 2004, was a hot sunny day in the Maldives, a beautiful group of atolls in the Indian Ocean west of India (Figure 4.A). The Republic of Maldives is the lowest country on Earth; its highest point is only 2.4 m (8 ft.) above sea level. Dave Lowe was working in his office at a resort on South Ari Atoll when he heard a strange bump on the door, and, outside, people were screaming "the children." He opened the door and to his horror, the ocean was level with the island and a boiling, frothing wall of water was bearing straight down on him.

"There was a strange mist that looked like a fog, so I stopped breathing and tried to decide where to run."

The resort had no two-story buildings, and he was just a meter (3 ft.) above sea level! He ran to the reception area where there were pillars to hold onto. Others were there screaming as the first waves began to hit.

"Three glass windows exploded, and, within seconds, the water was up to my waist. I couldn't tell if the island was sinking or if the sea was rising."

Dave made his way to the reception counter, grabbing two children who were being swept out to sea, and threw them up onto the counter. Wave after wave hit, and water passed over his head.

"I knew I was going to die and blacked out from fear. I came conscious when a receptionist was yelling 'What is happening?"'

As quickly as the water came up, it was gone, leaving fish flopping on the lobby floor and seaweed draped everywhere. The resort guests regrouped. Then Dave saw a second wave coming. It was worse than the first. People desperately tried to hang onto anything they could. Wave after wave came for the next 6 hours, as the sea gradually calmed down. Guests suffered in the hot sun, and the resort staff built a shelter for 15 children whose parents were missing. When darkness came, people stood watch, looking for more waves under a full moon. No one slept, and, in the morning, there was debris everywhere.

"It looked like *Titanic, Lost, Lord of the Flies*, and *Survivor* all rolled into one." Dave and other employees were evacuated by seaplane 2 days later. As they flew over the islands, they saw the unbelievable devastation, and several staff broke down crying.

They had survived, but they didn't know how.

▲ FIGURE 4.A **RESORT EMPLOYEES CLIMB DOWN FROM TSUNAMI REFUGE** The roof of this South Ari Atoll resort in the Maldive Islands was the highest place around to take refuge from the multiple tsunami waves that flooded the island. On islands closer to the earthquake epicenter, this building would have been destroyed by the waves. *(SIPA Press)*

Tsunamis can range in height from a few centimeters to 30 m (100 ft.) or more. Those that have a runup height of at least 5 m (16 ft.) are considered significant tsunamis and are commonly produced by high-magnitude earthquakes and associated submarine slides. A recent assessment of global tsunami hazard ranks the hazard from relatively low to greatest based upon the return period of a significant tsunami (at least 5-m [16-ft.] runup) (Figure 4.11). Keep in mind that this map is highly generalized because runup from tsunami varies considerably with the shape of the seafloor immediately offshore and with the type of topography and vegetation landward of the beach. Both the offshore and shoreline topography can greatly accentuate tsunami wave height. Also, not enough is known about global seismicity to accurately predict return periods of earthquakes with **M** 9 or above—those quakes that are most likely to produce a significant tsunami. Examination of the map reveals that most zones of greatest hazard surround the Pacific Ocean. This is no surprise, given that most of the major subduction zones are found on the margins of the Pacific. Other regions judged to have the greatest hazard are parts of the Mediterranean, as well as the northeastern side of the Indian Ocean. The good news about the risk from significant tsunamis is that almost all of them will cause substantial damage only within a couple of hundred kilometers from their source. However, the very largest subduction-zone earthquakes may cause substantial damage several thousands of kilometers from the source. That certainly was the case with the 2004 Indonesian tsunami, which raced across the Indian Ocean causing death and destruction as far away as the east coast of Africa. However, most of the death and destruction were much closer to the earthquake epicenter (see Figure 4.3).

An interesting example of damage from a distant earthquake comes from three centuries ago. In the year 1700, a tsunami generated by an estimated **M** 9 earthquake on the Cascadia subduction zone reached the shores of Japan approximately 12 hours later (Figure 4.12). You might wonder how we are able to determine the magnitude of the earthquake and determine that it created a tsunami in Japan. This detective story started with geologic investigations in North America that found logs and soil that were buried below a tsunami deposit sometime after 1660. The logs and other plant material in the soil were radiocarbon dated to determine the time that they were last alive, and the deposits showed evidence of subsidence at the time of the earthquake. Based upon the geologic evidence, the length of rupture was estimated to be about 1000 km (620 mi.).[16–18] The evidence from Japan consists of historical descriptions of a tsunami with a 1 to 5 m (3 to 16 ft.) runup height that took place in 1700. Knowing the time of arrival of the tsunami in Japan, scientists have inferred that the tsunami originated at the Cascadia subduction zone about 9:00 P.M. PST on January 26, 1700.[16]

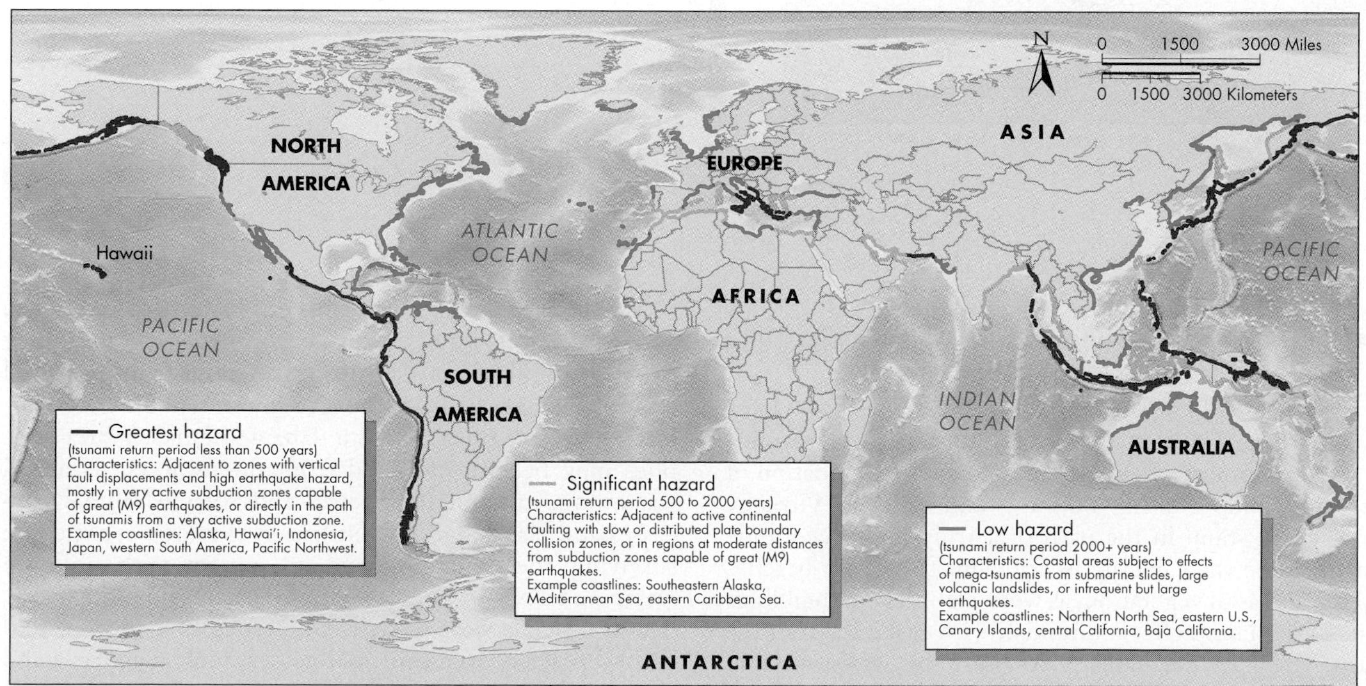

▲ FIGURE 4.11 **GLOBAL TSUNAMI HAZARD** Map of the relative hazard of coastlines to experience a tsunami that is at least 5 m (16 ft.) high. *(Modified from Risk Management Solutions. 2006.* 2004 Indian Ocean Tsunami report. *Newark, CA: Risk Management Solutions, Inc.)*

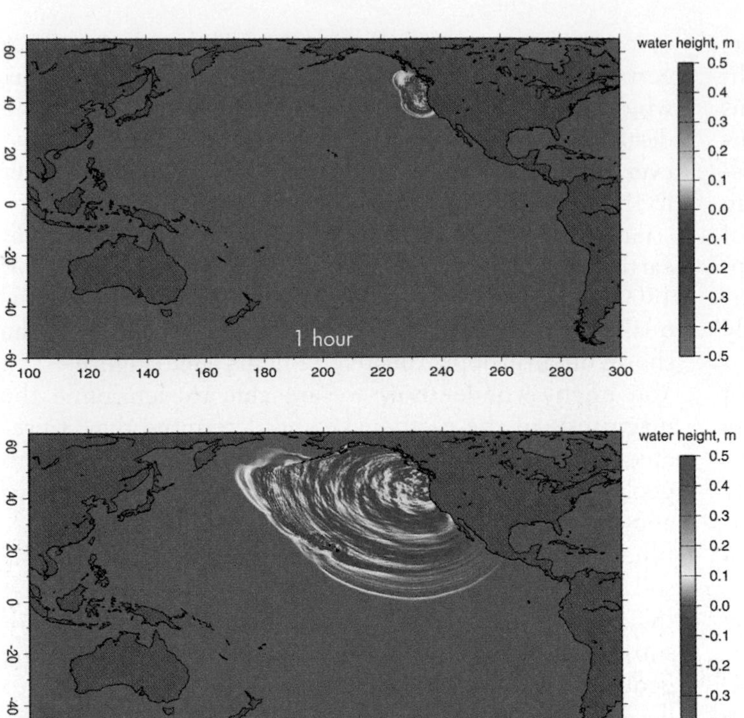

◀ **FIGURE 4.12** **TSUNAMI OF 1700** This output from a computer model of the January 1700 tsunami is based on arrival times at several sites in Japan. The maps show wave height and location 1 hour, 6 hours, and 12 hours after an ~**M** 9 earthquake occurred on the Cascadia subduction zone. Such an earthquake would produce a fault displacement of around 19 m (60 ft.) over a distance of about 1100 km (680 mi.) from British Columbia south to Northern California. *(Courtesy of Kenji Satake)*

4.3 Effects of Tsunamis and Linkages with other Natural Hazards

The effects of tsunamis are both primary and secondary. Primary effects are related to the inundation of the water and resulting flooding and erosion. Virtually nothing can stand in the path of a truly high-magnitude tsunami. The wave energy is sufficient to tear up beaches and most coastal vegetation, as well as homes and buildings in its path. These effects diminish with distance from the coast. Much of the damage to both the landscape and human structures results from the tremendous amount of debris carried by the water as it moves inland and then back again to the ocean. What is left behind is often bare, eroded ground and areas covered with all sorts of human and natural debris (Figure 4.13).

Secondary effects of tsunamis occur in the hours, days, and weeks following the event. Immediately following the tsunami, fires may start in urban areas from ruptured natural gas lines or from the ignition of flammable chemicals that were released from damaged tanks. Water supplies may become polluted from floodwaters, damaged wastewater treatment systems, and rotting animal carcasses and plants. Disease outbreaks may occur, as people surviving the tsunami come in contact with polluted water and soil. In the 2004 Indonesian tsunami, public health officials were initially concerned that there would be outbreaks of waterborne illnesses, such as malaria and cholera. Fortunately, this did not become a serious problem because of the quick action of relief agencies and the destruction of mosquito breeding grounds by saltwater flooding. What did occur is a pneumonia-like disease known as "tsunami lung," a condition that developed in

(a)

(b)

▲ FIGURE 4.13 **DAMAGE AND DEBRIS FROM 2004 INDONESIAN TSUNAMI** (a) In some areas, the tsunami removed all but the most sturdy buildings, such as this mosque in Aceh Province, Indonesia. *(Spencer Platt/Getty Images)* (b) In other areas, such as this part of the Indonesian resort town of Pangandaran, the tsunami piled up huge amounts of human and natural debris. This made it difficult for these soldiers to locate victims. *(Dimas Ardian/Getty Images)*

people who inhaled bacteria in muddy saltwater. Because the disease was not initially recognized and there were few antibiotics or medical personnel available for treatment, many of the patients developed advance lung infections that resulted in paralysis. Fortunately, treatment with antibiotics eventually brought the infections under control.[19]

There are several linkages between tsunamis and other natural hazards. Tsunamis are obviously closely linked to submarine and coastal earthquakes and landslides, as well as island volcanic explosions and oceanic impacts of asteroids and comets. For earthquake-generated tsunamis, coastal communities near the epicenter experience casualties and property damage from both the ground shaking of the earthquake and inundation by the tsunami. Powerful tsunami waves interact with coastal processes to change the coastline through erosion and deposition of sediment. There was dramatic evidence of this interaction following the 2004 tsunami in the area near Banda Aceh, Indonesia (see Figure 4.4). A combination of erosion caused by the tsunami and subsidence caused by the earthquake has altered the coastal area to the point that it scarcely resembles what it was prior to the event.

4.4 Natural Service Functions of Tsunamis

Large tsunamis are capable of causing catastrophic death and destruction. It is hard to imagine much in the way of benefits from such disasters. The movement of vast amounts of seawater on land undoubtedly brings many chemicals from the ocean to the land, and that may have long-term effects on ecosystems that might otherwise be deprived of nutrients. Tsunamis also bring ashore a large volume of sediment that, over a long period, contributes to the general development of the landscape. In the coming years, studies of landscape and ecosystem changes following the 2004 Indonesian tsunami may reveal other natural service functions of tsunamis.

4.5 Human Interaction with Tsunamis

We cannot prevent or control tsunamis, and few, if any, tsunamis have been influenced by human activities. The magnitude and frequency of these events are not in any way tied to human activity. As with hurricanes and rising sea level, people who move to coasts that have an elevated risk for tsunamis are increasing the likelihood that they will be affected by this natural process.

Damage from tsunamis is regrettable, but there are some lessons from past tsunamis that can be applied to reduce the damage from future events. For example, we can plant buffer zones of trees along a coast that will absorb some of the impact of incoming tsunami waves. We might also consider the strength of our buildings to withstand the onslaught of moderate tsunamis and move other structures inland where they are less likely to be damaged.

4.6 Minimizing the Tsunami Hazard

A number of strategies are available for minimizing the tsunami hazard and these include

- detection and warning
- structural control

- construction of tsunami runup maps
- land use
- probability analysis
- education
- tsunami-ready status

DETECTION AND WARNING

Since nearly all large tsunamis are associated with giant earthquakes, our first warning comes from an earthquake in an offshore area that is large enough to produce a tsunami. We have the capability to detect distant tsunamis in the open ocean and accurately estimate their arrival time to within a few minutes. This information has been used to create a successful tsunami warning system in the Pacific Ocean.

A tsunami warning system has three components: a network of seismographs to accurately locate and determine the depth and magnitude of submarine and coastal earthquakes, automated tidal gauges to measure unusual rises and falls of sea level, and a network of sensors connected to floating buoys (Figure 4.14). Surface buoys with a bottom sensor, known as a *tsunameter*, detect small changes in the pressure exerted by the increased volume of water as a tsunami passes overhead (Figure 4.14a). This information is relayed by satellite to a warning center and is combined with tidal gauge information to predict tsunami arrival times. For example, an underwater earthquake in Hawai'i could produce a tsunami that would radiate outward across the entire Pacific Ocean and arrive at different times in California, Alaska, Japan, and Papua New Guinea (Figure 4.14b). Following the Indonesian tsunami, similar systems are being created in the Indian and Atlantic Oceans, including warning sensors for Puerto Rico and the east coast of the United States and Canada.

For a local tsunami that strikes land very close to the source of the earthquake, there may be little warning time. People close to the source, though, will probably feel the earthquake and can immediately move inland to higher ground. Certainly, if one observes the water receding, it is a sign to run inland. Some coastal communities in Hawai'i, Alaska, the Pacific Northwest, British Columbia, Japan, and elsewhere also have warning sirens to alert people that a tsunami may soon arrive.

STRUCTURAL CONTROL

Tsunamis that are even a meter or 2 high have such power that houses and small buildings are unable to withstand their impact.[9] However, building designs for larger structures, such as high-rise hotels and critical facilities, can be engineered in such a way as to greatly reduce or minimize the destructive effects of a tsunami. For example, the city of Honolulu, Hawai'i, has special requirements for construction of buildings in areas with a tsunami hazard. However, for most hazard areas, the current building codes and guidelines do not adequately address the effect of a tsunami on buildings and other structures.[20]

TSUNAMI RUNUP MAPS

Following a tsunami, it is a fairly straightforward procedure to produce a **tsunami runup map** that shows the level to which the water traveled inland. Such a map was created for the island of Oahu, Hawai'i, following the 1946 tsunami that originated from an earthquake in Alaska's Aleutian Islands (Figure 4.15). That map illustrates the tremendous variability of runup height from a typical tsunami. The runup from the 1946 tsunami varied from about 0.6 m (2 ft.) in Honolulu to 11 m (36 ft.) at Kena Point and Makapuu Point on opposite ends of the island.

Before a tsunami strikes, a community can produce a hazard map that shows the area that is likely to be inundated by a given height. Huntington Beach, California, one of a number of American and Canadian communities that has a significant tsunami hazard, has prepared such a map (Figure 4.16). Although the probability of a large tsunami in Southern California is relatively low, the consequences of such an event could be catastrophic. For example, the entire city of Huntington Beach has an elevation of less than 33 m (100 ft.), and three-quarters of the city is less than 8 m (25 ft.) above sea level. Huntington Beach itself faces to the southwest and, as such, is vulnerable to a tsunami from either the south or west. Such an event, for example, might originate in Alaska, Japan, or the South Pacific. The beach attracts up to 100,000 people or more each day during the summer, and the city is giving serious consideration to the possibility of a tsunami. Many other cities and areas have produced tsunami runup maps, and this trend will undoubtedly continue in the future.

LAND USE

In the aftermath of the 2004 Indonesian tsunami, scientists discovered that tropical ecology played a role in determining tsunami damage. Where the largest waves struck the shore, there was inevitably massive destruction (Figure 4.17). In other areas, the damage was not uniform. Some coastal villages were destroyed, whereas others were less damaged. Those villages that were spared destruction were partly protected from the energy of the tsunami by either a coastal mangrove forest or several rows of plantation trees that reduced the velocity of incoming water.[21] Coast land-use and land-cover studies have documented the advantage of locating villages

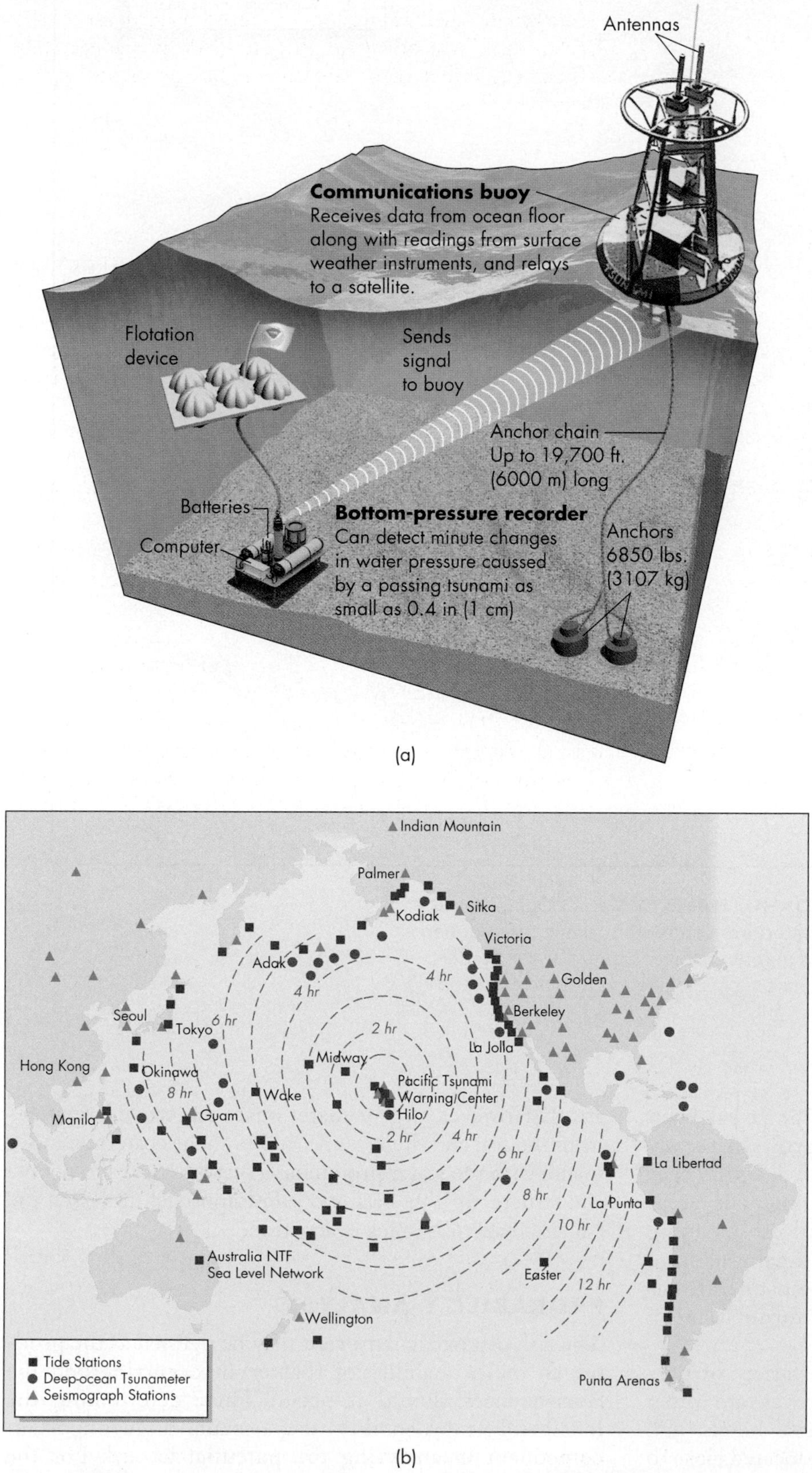

◀ **FIGURE 4.14 TSUNAMI WARNING SYSTEM** (a) A surface buoy and bottom sensor to detect a tsunami. (b) Travel time (each band is 1 hour) for a tsunami generated in Hawai'i. The wave arrives in Los Angeles in about 5 hours. It takes about 12 hours for the waves to reach South America. Locations of 30 DART tsunameter buoys are shown with red dots. At least 14 additional DART tsunameter buoys are planned for the Pacific and Atlantic Oceans. *(NOAA)*

▲ **FIGURE 4.15 TSUNAMI RUNUP ON OAHU, HAWAI'I** Map of Oahu, Hawai'i, showing runup of the 1946 tsunami that originated from an earthquake in the Aleutian Islands, Alaska. Maximum runups are indicated in feet. *(Modified after Walker, D. 1994.* Tsunami facts. *SOEST Technical Report 94-03. School of Ocean and Earth Science and Technology)*

inland behind a coastal forest (Figure 4.18). It has been suggested that places where the forest grows naturally may be somewhat protected from wave attack by the near shore and land topography. Thus development in areas with a tsunami hazard might look to land-use planning that includes identifying where serious tsunami damage has occurred in the past and avoiding the most hazardous places. That is using the principle of uniformitarianism discussed in Chapter 1.

With rapid coastal development for a variety of purposes, including tourism, coastal mangroves are often removed and replaced by homes, high-rise hotels, and other buildings. These structures are often located close to the beach and, thus, are vulnerable to tsunami attack. Although it may not be practical to move tourist areas inland, planting or retaining native vegetation could provide a partial buffer from a small-to-moderate tsunami attack. However, large tsunami waves can erode large trees from the soil and their trunks and branches can then move inland with the water producing a serious hazard to people and structures. The best and safest approach to lessen the tsunami hazard is quick evacuation.

PROBABILITY ANALYSIS

The risk of a particular event may be defined as the product of the probability of that event occurring and the consequences should it occur. Thus, determining the likelihood or probability of a tsunami is an important component in analyzing the potential hazard. For the most part, the hazard analysis for tsunamis has relied upon evidence from past tsunamis to determine the hazard rather than to attempt to calculate the probability of a future tsunami. The more deterministic approach is

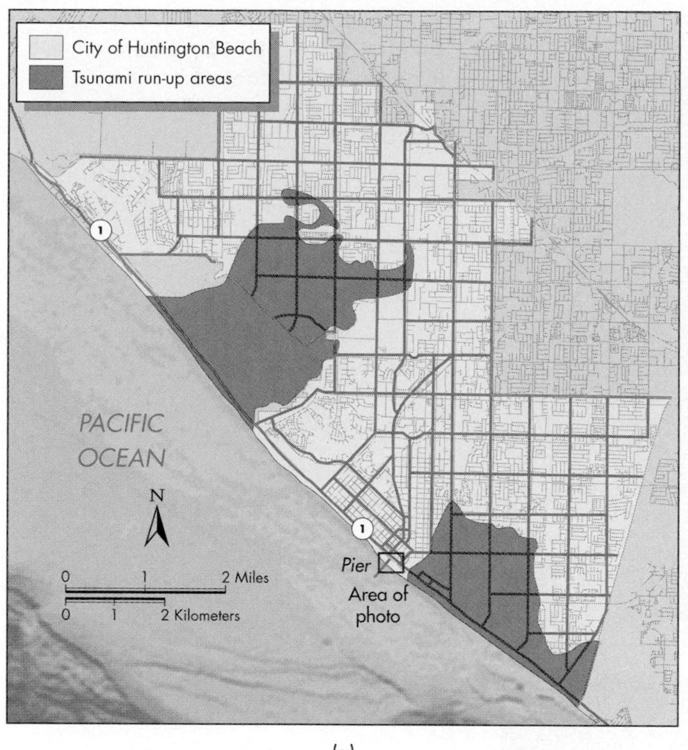

(a)

(b)

▲ **FIGURE 4.16 HUNTINGTON BEACH, CALIFORNIA, TSUNAMI RUNUP MAP** (a) Map showing areas likely to be inundated by tsunami runup in red. These areas are more susceptible to run-up due to their lower elevations with respect to yellow regions *(Courtesy City of Huntington Beach)* (b) The popular beach with pier and area to the southeast. *(Kenneth & Gabrielle Adelman/California Coastal Records Project)*

simply to derive a tsunami inundation map, such as that shown in Figure 4.16 and use that to develop evacuation maps. In contrast, analysis of the probability of a tsunami hazard provides information not only on the likelihood of the event, but also other aspects as well, such as location of the event, the extent of the runup, and the possible severity of damage. The approach taken in developing a probabilistic analysis of the tsunami hazard is to

- Identify and specify the potential earthquake sources and their associated uncertainties.
- Specify relationships that will either attenuate or reduce tsunami waves as they travel from the source area.
- Apply probabilistic analysis to the tsunami hazard similar to what is currently being done for earthquake hazard analysis. However, with tsunamis, the analysis must consider the fact that tsunamis can originate from multiple sources, including ones that are far away from the location where the hazard is present.

This probabilistic approach to tsunami hazard assessment is still being developed. One difficulty is that tsunamis at a particular location are generally rare events. If there is not sufficient activity in the past to develop the probability of future events, then the analysis depends upon a statistical technique known as Monte Carlo simulation. This technique has been used to simulate the behavior of earthquakes and tsunamis. The overall objective is to determine tsunami return periods and probabilities for both distant and local sources. To apply this technique, the Monte Carlo simulation selects a random sample of earthquakes of various magnitudes and determines the tsunamis that would be propagated from these quakes. Using these tsunamis, a mathematical model is then constructed to estimate the tsunami amplitude or runup along a particular coast. This must be done for each of the potential seismic sources for that particular coastline.[22]

EDUCATION

Education concerning the tsunami hazard is critical to minimizing risk (see Professional Profile 4.2). This was shown dramatically during the 2004 Indonesian tsunami when numerous lives were saved because people recognized that the receding seawater was a warning sign of an impending tsunami. Likewise, it is important to educate coastal residents and visitors as to the difference between a **tsunami watch**, which is a notification that an earthquake that can cause a tsunami has occurred, and an actual **tsunami warning** that a tsunami has been detected and is spreading across the ocean toward their area. For a distant

(a)

(b)

◀ FIGURE 4.17 **TSUNAMI DAMAGE TO TREES** (a) Before and (b) after 2004 Indonesian tsunami removed many trees. *(CRISP/Centre for Remote Imaging, Sensing & Processing)*

tsunami, there will likely be several hours before the waves arrive after a warning has been issued. In a local tsunami, there may be little warning time, and, so, more attention must be given to nature's warning systems, such as earthquake shaking or the water receding from a shoreline. People must also be taught that tsunamis come in a series of waves, and that the second and third waves may be larger than the first one. Finally, people must be told that the water returning to the ocean once a wave has runup is just as dangerous as the incoming water.

TSUNAMI-READY STATUS

In order for a community to be prepared, or **tsunami ready**, it must

- Establish an emergency operation center with 24-hour capability.
- Have ways to receive tsunami warnings from the National Weather Service, Canadian Meteorological Centre, Coast Guard, or other agencies.
- Have ways to alert the public.
- Develop a tsunami preparedness plan with emergency drills.
- Promote a community awareness program to educate people concerning a tsunami hazard.

These guidelines for tsunami-readiness status are being applied to a number of communities in California and other areas. For example, the University of California, Santa

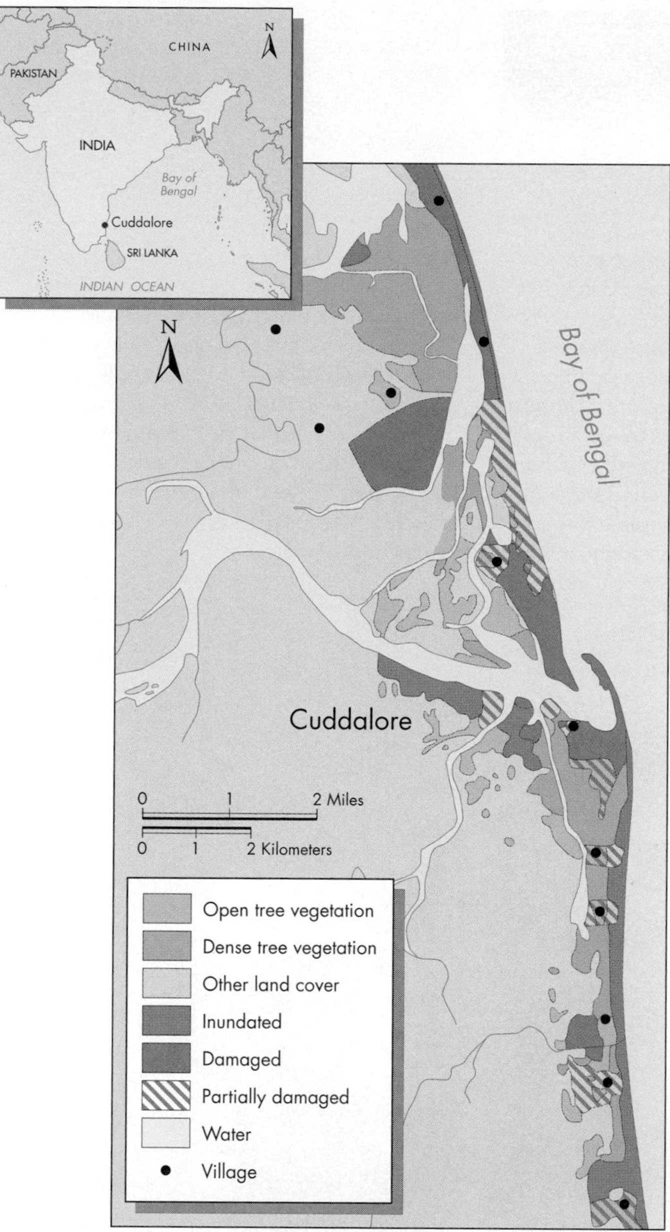

▲ **FIGURE 4.18 TREES PROVIDE SOME PROTECTION FROM TSUNAMI DAMAGE** Tree cover pre-tsunami with post-tsunami damages to the land and villages in the Cuddalore District of India. *(Modified after Danielsen, F., Serensen, M. K., Olwig, M. F., Selvam, V., Parish, F., Burgess, N. D., Hiraishi, T., Karunagaran,V. M., Rasmussen, M. S., Hansein, L. B., Quarto, A., and Suryadiputra. N. 2005. The Asian tsunami: A protective role for coastal vegetation.* Science *310:643)*

Barbara, and the City of Huntington Beach, California (Figure 4.16) have been certified as tsunami ready.

The educational component is of particular importance. Most people don't even know if a tsunami watch or warning is issued. For example, in 2005, there was an earthquake far away from the city of Santa Barbara in the Pacific. As it turned out, a tsunami did not occur, but a tsunami watch was placed up and down the California coastline. Nothing was said about the size of the possible tsunami, and, so, some people, on hearing the notice, drove to the top of a nearby mountain pass thousands of feet above sea level. Those locations were great for views, but people certainly did not need to drive that far or that high to evacuate the potential danger zone. There was no plan in place to directly observe the tsunami, and the news media reported that some people perched on a sea cliff at night, whereas others climbed palm trees to see if the wave was coming. This experience in Santa Barbara, and other West Coast communities, suggests that even many "tsunami-ready" communities are not adequately prepared for the tsunami that will eventually strike their coasts.

4.7 Perception and Personal Adjustment to Tsunami Hazard

The preceding discussion suggests that many people do not know the signs of an approaching tsunami or what to do if a watch or warning is issued. From a personal perspective, if a warning is issued, you can take the following actions:

- Although all earthquakes do not cause tsunamis, they might. If you feel a strong earthquake and are at the beach, leave the beach and low-lying coastal area immediately.
- If the trough of a tsunami wave arrives first, the ocean will recede. This is one of nature's warning signs that a wave is on the way and that you should run from the beach for higher ground.
- Although a tsunami may be relatively small at one location, it may be much larger nearby. Therefore do not assume that all locations are safe because of an absence of dangerous waves elsewhere.
- Tsunamis generally consist of a series of waves, and there can be up to an hour between waves. Therefore, stay out of dangerous areas until a notice that all is clear is given by the proper authorities.
- As coastal communities gain tsunami-readiness status, they will have warning sirens; if you hear a siren, move away from the beach to higher ground (at least 20 m [60 ft.] above sea level) and listen for emergency information.
- If you are aware that a tsunami watch or warning has been issued, do not go down to the beach to watch the tsunami. If you can actually see the wave, you are probably too close to escape. These waves look small out at sea because of the vast distance to the horizon. Remember that these waves move fast and can be deadly. A 2-m (6-ft.) tall person is very small compared to a 15-m (45-ft.) tsunami wave (Figure 4.19).

PROFESSIONAL PROFILE 4.2

Jose Borrero—Tsunami Scientist

Tsunami scientist Jose Borrero has witnessed firsthand the combined powers of land and sea unleashed (Figure 4.B). As a researcher with the University of Southern California Tsunami Research Center, he has traveled to areas hit by some of the most massive natural disasters in recent decades.

When a 10-meter (30-ft.) tsunami killed 2100 people in New Guinea in 1998, Borrero and his colleagues were there just a week later to study the extent of the wave damage. "It looked like a hurricane came through. We saw dead bodies on the beach, ghost towns that looked completely bombed out—it was shocking," Borrero says.

The researchers traveled up and down the coast looking for flattened trees, measuring waterlines on house walls, and asking local people what they remembered about the timing and size of the waves. Back home, they used their wave damage data, seismometer measurements, and computer models to reconstruct the events leading up to the tsunami. Their work revealed that an underwater landslide triggered by a relatively small earthquake was what had created such a large and powerful tsunami.

After the 2004 Indonesian earthquake and tsunami, Borrero and a team of National Geographic filmmakers were among the first outsiders to reach one of the hardest hit areas—the city of Banda Aceh, Indonesia. What Borrero saw horrified him: "It was the worst of the worst of what I saw in New Guinea, except that, instead of being confined to just one town, it went on for 200 miles." In some areas, 10-m (30-foot) waves had pushed boats onto balconies and stripped the trees off the side of a mountain 24 m (80 ft.) up.

Back home, Borrero and colleagues used the data to develop a report about the risk Sumatra faces from future earthquakes and tsunamis along the next segment of the earthquake fault. "If you know what's possible, you can make a plan. We can give that information directly to cities and towns so they can make evacuation routes and public awareness programs," he says.

Tsunami education, Borrero says, is critical for coastal areas located directly on top of earthquake faults that produce tsunamis. "Tsunami warning systems only work for areas that are more than 2 hours away from the area of ground shaking. The only way to alert people is through education. If you feel an earthquake, and you're near the coast, don't sit and wait—head for high ground."

Growing up surfing in California, Borrero always wanted a career in which he could work directly with the ocean. "There's this primal fear people have of giant walls of water. Many people have told me they have this recurring nightmare where they're trapped in a tsunami and can't escape. I've never had dreams like that; maybe understanding it keeps me from worrying too much."

—KATHLEEN WONG

▲ FIGURE 4.B **DR. JOSE BORRERO** A research professor at the University of Southern California Department of Civil Engineering Tsunami Research Group, Dr. Borrero was part of an international damage assessment team that studied the effects of the 2004 Indonesian tsunami. Dr. Borrero is shown here with boats that were destroyed by the tsunami in Banda Aceh, Indonesia *(Dr. Jose Borrero/USC Tsunami Research Group)*

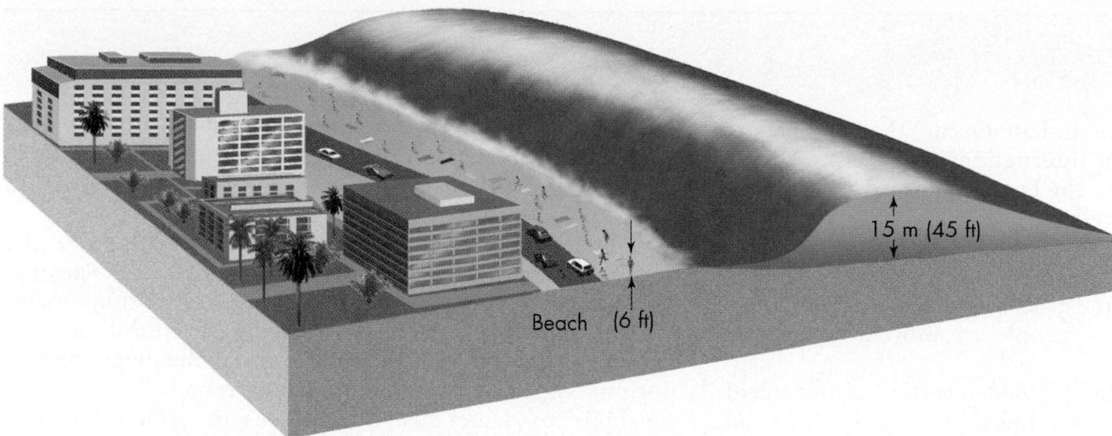

FIGURE 4.19 A PERSON VERSUS A TSUNAMI Tsunami height can be huge compared to the waves we normally see on a beach. (See also Figure 4.5).

REVISITING THE FUNDAMENTAL CONCEPTS

Tsunami

1. **Hazards Are Predictable from a Scientific Evaluation**
2. **Risk Assessment Is an Important Component of Our Understanding of the Effects of Hazardous Processes**
3. **Linkages Exist between Natural Hazards**
4. **Hazardous Events That Previously Produced Disasters Are Now Producing Catastrophes**
5. **Consequences of Hazards May Be Minimized**

1. Although we cannot yet predict the earthquakes or landslides that may produce tsunamis, once a tsunami has been generated, there are ocean-bottom detection devices in the Pacific and other oceans of the world that can be used to predict when waves are likely to reach a particular coastline thousands of miles away from the earthquake that generated a tsunami. That system is still being developed in many places. It works well in the Pacific but less so in some of the other oceans of the world.

2. The risk from a tsunami, as with other hazards, may be defined as the probability of an event occurring times the consequences. Our understanding of earthquakes has progressed to the point that we can assign probability of occurrence of great earthquakes likely to produce a tsunami. We also have the capability of determining what might be inundated and what might be the consequences in terms of property damage and loss of life. Such risk assessment has been done for some locations where there is a tsunami hazard. As that work progresses, a better understanding of the risk associated with tsunamis will unfold.

3. There are significant linkages between tsunamis and other natural hazards. For example, a tsunami may be linked to severe coastal erosion processes and also to secondary effects, such as fires and disease, which may result from pollution of water and damage to coastal facilities from fires generated from spilled flammable materials.

4. The human population in coastal areas likely to experience tsunamis is growing rapidly in many regions. In some cases, it is the building of large cities in coastal areas, and, in others, it is the growth of the tourist industry that brings people to islands and other places that may have a tsunami hazard. Even though the incidence of tsunamis has probably not increased, their impact on people has. As a result, what used to be disasters may now be catastrophes. In addition, coastal areas may be denuded of natural vegetation and wetlands that formerly produced a barrier between the coast and inland development. As a result, the choices we make in land use will affect the consequences of tsunami waves.

5. The consequences of tsunamis can be minimized through better education that informs people of the natural signals that a tsunami may arrive, even if no earthquake is felt. Cities also can become tsunami ready through education and the use of sirens and other warning devices on radio and television and other media, such as cell phones, to alert people that waves are arriving and to not visit or evacuate low-lying coastal areas. In some areas, buffers of coastal vegetation may be planted or preserved to minimize tsunami damage. Past tsunamis have taught us that where trees are planted or exist naturally in the coastal environment between the coast (beach) and more inland areas, vegetation reduces wave heights of tsunami waves moving through coastal areas, thus possibly reducing the hazard inland. The best way to reduce the hazard is prompt evacuation.

QUESTIONS TO THINK ABOUT

1. How might we better educate people about the tsunami hazard?
2. Why might it be difficult for people who live far from a serious earthquake hazard to appreciate and prepare for tsunami damage?
3. What would be the potential economic gains and losses of large-scale planting of trees to provide a buffer from tsunamis?

Summary

The 2004 tsunami in the Indian Ocean that claimed at least 230,000 lives was an international wake-up call that we are not yet prepared for the tsunami hazard. Of primary importance is to ensure we have a tsunami warning system available in the world's major ocean basins. This warning system must be designed to reach both coastal residents and visitors. The system must be coupled with an effective education program, so that people are more aware of the hazard.

A tsunami is produced by the sudden vertical displacement of ocean water. There are several possible processes that can produce tsunamis, including underwater landslides, submarine volcanic explosions, and impact of extraterrestrial objects. However, the major source of large damaging tsunamis over the past few millennia has been giant earthquakes associated with the major subduction zones on Earth. These tsunamis have formed where geologic faulting ruptures the seafloor and displaces the overlying water. When this happens, both a distant and local tsunami may be produced. Distant tsunamis can travel thousands of kilometers across the ocean to strike a remote shoreline. On the other hand, a local tsunami heads toward a nearby coast and can strike with little advance warning.

Effects of a tsunami are both primary and secondary. The primary effects are related to the powerful water from the tsunami runup that results in flooding and erosion. Virtually nothing can stand in the path of a large-magnitude tsunami. In 2004 in Indonesia, huge concrete barriers were moved by the force of the waves. Secondary effects of a tsunami include a potential for water pollution, fires in urban areas, and disease to people surviving the event.

Tsunamis are linked to other natural hazards. For example, they are obviously tightly linked to the earthquakes that cause them, and, thus, their effects are often combined with the ground shaking, fires, and subsidence associated with the quakes. Tsunami waves also interact with coastal processes to change the coastline through erosion and deposition of sediment. Following an earthquake and tsunami, a coastal area may scarcely resemble what it was prior to the event.

A number of strategies are available to minimize the tsunami hazard. These include detection and warning, structural control, construction of tsunami runup maps, land-use practices, probability analysis, education, and achieving tsunami-ready status. Of these, the detection and warning system is of paramount importance. We have the capability to detect distant tsunamis in open ocean and accurately estimate their arrival time to within a few minutes. It is more difficult to provide adequate warning for local tsunamis, as tsunami wave formation quickly follows an earthquake or earthquake-produced landslide. In this case, warning sirens can alert people that a tsunami may soon arrive.

Without adequate education, watches and warnings are often ineffective because many people do not know how to recognize a tsunami or take appropriate action to save themselves and others. Through education, people can learn the natural warning signs of an approaching tsunami. This may include the earthquake shaking itself and withdrawal of seawater prior to arrival of the large wave. People must understand that tsunamis come in a series of waves, and that the second or third wave may be the largest. In addition, the water returning to the ocean following tsunami inundation can cause as much damage as the runup of the incoming waver.

Along coasts with great or significant tsunami hazard, most communities have not adequately prepared for this underestimated natural hazard Adequate preparation includes improved perception of the hazard, development of ways to alert the public, preparation and implementation of a tsunami preparedness plan, and promotion of community awareness and education concerning the hazard.

Key Terms

distant tsunami (p. 105)
local tsunami (p. 105)
mega-tsunami (p. 104)
runup (p. 105)
tsunami (p. 104)
tsunami ready (p. 116)
tsunami runup map (p. 112)
tsunami warning (p. 115)
tsunami watch (p. 115)

Review Questions

1. What is a tsunami?
2. How do natural processes cause a tsunami? What is the primary process?
3. What is the difference between a distant and local tsunami?
4. What are the major effects of a tsunami?
5. Explain the relationship between plate tectonics and tsunamis.
6. How are tsunamis detected in the open ocean?
7. What is the difference between a tsunami watch and tsunami warning?
8. What are the primary and secondary effects of tsunamis?
9. Describe the methods used to minimize the tsunami hazard.
10. What is meant by tsunami-ready status?

Critical Thinking Questions

1. You are placed in charge of developing an education program with the objective of raising a community's understanding of tsunamis. What sort of program would you develop and what would it be based upon?
2. What do you think the role of the media should be in helping to make people more aware of the tsunami hazard? How should scientists be involved in increasing the perception of this hazard?
3. You live on a coastal area that is subject to large, but infrequent, tsunamis. You are working with the planning department of the community to develop tsunami-ready status. What issues do you think are most important in obtaining this status and how could you convince the community that it is necessary or in their best interest to develop tsunami-ready status?

Assignments in Applied Geology: Tsunami/Storm Surge Assessment

THE ISSUE

Tsunamis and the storm surge associated with hurricanes are two of the most devastating natural hazards a coastal population can experience. Low lying, often unprotected populated areas along shorelines can easily fall victim to the powerful force of moving water. Until recently, these hazards have passed quietly under the radar of a fair majority of the public. However, the accompanying maps show just how frequently these destructive hazards can influence the citizens of the United States.

YOUR TASK

Due to recent events such as the December 26th, 2004 tsunami and destruction related to Hurricane Katrina, the village council of Ocean Village, a small town approximately 50 miles east of Hazard City where the Clearwater River enters the ocean, have become concerned about the vulnerability of their fair village. As a qualified tsunami and hurricane risk assessment geologist, they have hired you as a consultant, as they work to develop a hazard assessment response plan to the village's possible inundation by water.

5 Volcanoes

Mt. Unzen, 1991

Japan has 19 active volcanoes distributed throughout most of the country's landmass. Just over 200 years ago, Mt. Unzen in southwestern Japan erupted and killed an estimated 15,000 people when a volcanic avalanche triggered a tsunami. The volcano then lay dormant until June 3, 1991, when another violent eruption resulted in the evacuation of thousands of people. By the end of 1993, Mt. Unzen had produced about 0.2 km^3 (0.05 $mi.^3$) of lava. It had also produced more than 8000 **pyroclastic flows,** fast-moving steams of extremely hot gas, ash, and large rock fragments—more than any other volcano in recent time, making it a natural laboratory for studying such flows (Figure 5.1).[1,2] Several of these flows were especially devastating because they entered urban areas along a specially designed channel constructed to contain volcanic mudflows. Far larger than any mudflows, the pyroclastic flows overran the channel, incinerating buildings and burying many homes in volcanic debris (Figure 5.2).

"It had also produced more than 8000 pyroclastic flows"

The Mt. Unzen story exemplifies several of the fundamental principles discussed in Chapter 1:

- The 1991 eruption occurred at a dormant volcano that had been the source of previous historic eruptions (Concept 1).
- Hazardous processes of pyroclastic flows, debris avalanches, and tsunami were linked (Concept 3).
- Evacuation of thousands of people reduced potential loss of life (Concept 5).

LEARNING OBJECTIVES

There are about 1500 active volcanoes on Earth, almost 400 of which have erupted in the past century. While you are reading this paragraph, at least 20 volcanoes are erupting on our planet. Volcanoes occur on all seven continents as well as in the middle of the ocean. When human beings live in the path of an active volcano, the results can be devastating. Your goals in reading this chapter should be to

- know the different types of volcanoes and their associated features.
- understand the relationship of volcanoes to plate tectonics.
- know what geographic regions are at risk from volcanoes.
- know the effects of volcanoes and how they are linked to other natural hazards.
- recognize the potential benefits of volcanic eruptions.
- understand how we can minimize the volcanic hazard.
- know what adjustments we can make to avoid death and damage from volcanoes.

◀ **MT. Unzen** The large mountain in the top center is Mt. Unzen, one of Japan's 19 active volcanoes. Dormant for more than 2 centuries until 1991, the volcano threatens parts of Fukae Town and Shimabara City in the foreground. In the 1991 eruption, searing hot pyroclastic flows killed 43 and burned a school and more than 600 homes. This 1993 aerial photograph shows a channel constructed to divert volcanic mud flows to the sea. *(Michael S. Yamashita/Corbis)*

▲ FIGURE 5.1 **PYROCLASTIC FLOW FROM MT. UNZEN** The gray cloud on the forested slope of Mt. Unzen at the top of the photo is moving rapidly downslope toward a village in the June 1991 eruption. The cloud is a pyroclastic flow of extremely hot gases, volcanic ash, and large rocks. Both the firefighters in the truck and the individual are running for their lives. Fortunately, the flow stopped before reaching both the firefighters and the village. *(AP Images)*

▲ **FIGURE 5.2 DAMAGE FROM PYROCLASTIC FLOWS** Three large pyroclastic flows came down the Mitzunashi River valley into urban areas on the slopes of Mt. Unzen in 1991. Here a flow overtopped a channel constructed to contain volcanic mudflows and destroyed many homes and buildings on both sides of the channel. *(Micheal S. Yamashita)*

5.1 Introduction to Volcanoes

Volcanic activity, or volcanism, is directly related to plate tectonics, with most volcanoes being located near plate boundaries (see Section 2.3).[3] This relationship exists because spreading or sinking lithospheric plates at plate boundaries interact with other earth materials to produce molten rock, called **magma** within the Earth, or **lava** when it erupts onto the Earth's surface. Volcanoes are constructed around an opening, called a *vent*, through which lava and other volcanic materials are extruded onto the surface.

Approximately two-thirds of all active volcanoes on Earth are located along the "Ring of Fire" that surrounds the Pacific Ocean (Figure 5.3). This ring has formed above the subduction zones where the Pacific, Nazca, Cocos, Philippine, and Juan de Fuca plates are sinking below adjacent plates. Volcanoes are also found where tectonic plates spread apart at mid-ocean ridges, such as the Mid-Atlantic Ridge in the center of the Atlantic Ocean,

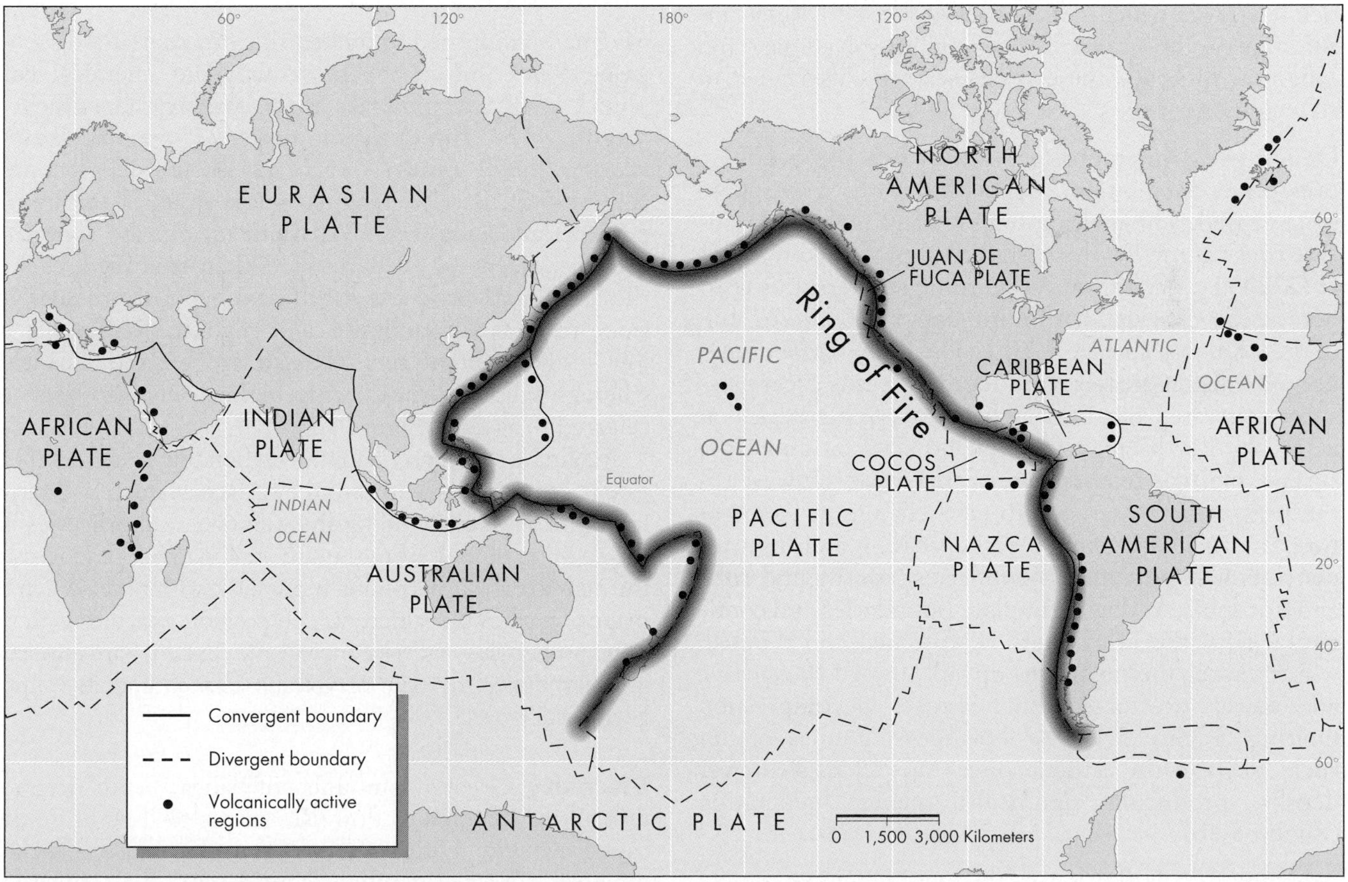

▲ **FIGURE 5.3 THE "RING OF FIRE"** The thick orange line outlines the Ring of Fire surrounding the Pacific plate. Although no fire is present along the ring of fire, the ring of fire contains most of the world's explosive volcanoes. *(Modified from Costa, J. E., and Baker, V. R. 1981. Surficial geology. Originally published by John Wiley and Sons, New York and republished by Tech Books, Fairfax, VA)*

and along continental rift zones, such as the Rio Grande rift, a down-faulted valley that extends from Colorado through New Mexico into Chihuahua, Mexico.

The sinking and spreading of tectonic plates produce magma that varies widely in composition from place to place. Understanding these variations helps explain the types, shapes, and eruption styles of volcanoes.

HOW MAGMA FORMS

It is a common misconception that the plates move around on an ocean of molten rock. This is not ture. With the exception of the outer core deep within Earth's interior, the planet is composed of solid rock (see Section 2.2). So then, where does the magma come from that forms volcanoes at the surface?

Most magmas come from the asthenosphere, the weak layer that underlies the lithospheric plates (see Section 2.1). The reason the asthenosphere is weak and able to flow, allowing rigid plates to move around the globe, is that although it is not actually molten, it is close to its melting temperature. This means that the asthenosphere is more prone to melting than the overlying lithosphere or lower mantle.[4] Figure 5.4 illustrates the three principle ways in which silicate rocks can melt: (1) decompression, (2) addition of volatiles (e.g., dissolved gas), and (3) addition of heat. These processes are closely tied to plate tectonic processes.

1. **Decompression melting** occurs when the overlying pressure exerted on hot rock within the asthenosphere is decreased. The melting temperature for mantle silicate rocks at the Earth's surface is approximately 1200° C (2200° F).[4] Within most tectonic settings, this temperature is exceeded within the upper 100 km (62 mi.) of the Earth's surface.[4] However, like temperature, pressure also increases with depth. It is this great pressure, generated by the weight of overlying rock that keeps the Earth's mantle in a solid state far above the surface melting temperature of the rock. However, if the pressure is released without a change in temperature, the rocks will melt. Decompression melting happens at divergent plate boundaries, continental rifts, and hot spots. At divergent plate boundaries and continental rifts, the lithosphere is stretched and thinned, which causes the mantle to upwell toward the surface and lower pressures, which causes decompression melting (Figure 5.4a). At hot spots, plumes of hot rocks deep in the asthenosphere upwell to shallower depths, producing significant magma generation (Figure 5.4b).

2. **Addition of volatiles** lowers the melting temperature of rocks by helping to break chemical bonds within silicate minerals. Consequently, rocks that are close to their melting temperature will melt in the presence of added volatiles. **Volatiles** are chemical compounds, such as water (H_2O) or carbon dioxide (CO_2), that evaporate easily and exist in a gaseous state at the Earth's surface. Such volatiles are often incorporated into minerals formed on the sea floor or within the oceanic lithosphere and are released when oceanic plates are subducted. The released volatiles rise upward from the subducting slab and interact with the formerly dry asthenosphere and induce melting and volcanism at the Earth's surface (Figure 5.4c). Subduction zone volcanism results in most of the on-land volcanoes around the world and is exemplified by the "Ring of Fire," which surrounds the Pacific plate (Figure 5.3).

3. *Addition of heat* to rocks will induce melting if the temperature of the rocks exceeds the melting temperature of silicate rocks at that depth. As magmas rise from the asthenosphere, they transfer heat to the overlying rocks (Figure 5.4d). This heat may induce lithospheric melting, which can be assimilated into the rising magma, changing its composition. This process is likely widespread, but the degree to which assimilation occurs is variable, depending on the rate at which the magma rises toward the surface.

MAGMA PROPERTIES

Magma is composed of melted silicate rocks and dissolved gases. Silica (SiO_2), the most abundant chemical compound in silicate minerals, is the primary constituent of magma. The three major types of magma—*basaltic*, *andesitic*, and *rhyolitic*—vary in their silica content (Figure 5.5). Basaltic magma is low in silica (around 50 percent), and andesitic and rhyolitic magma are high in silica (60 percent to 70 percent). When solidified, each of these magmas produces a different type of igneous volcanic rock: basalt, andesite, and rhyolite. Rocks that are low in silica content, such as basalt, are referred to as *mafic*, whereas volcanic rocks with higher silica content are referred to as *felsic* (Figure 5.5).

Magma is less dense than the surrounding rock, and therefore it rises buoyantly toward the surface where it will form a lava flow or volcano. Along the way, chemical processes and the addition of melt from the surrounding rock tend to cause the melt to evolve, becoming more felsic with time. When the magma reaches the surface its composition, together with two other properties—viscosity and volatile content—are important in determining the type of volcano formed and its eruptive behavior (Figure 5.5).

Viscosity Greater amounts of silica make it more difficult for magma to flow. Resistance to flow in fluids is called *viscosity*. Fluids that do not flow easily, such as molasses or refrigerated honey, are like silica-rich magma—they have a high viscosity. Other fluids, such as water, warm honey, and low-silica magma, have a low viscosity. As the example of honey suggests, viscosity is typically influenced by temperature as well as composition (Figure 5.5). For example, a low silica content basaltic lava

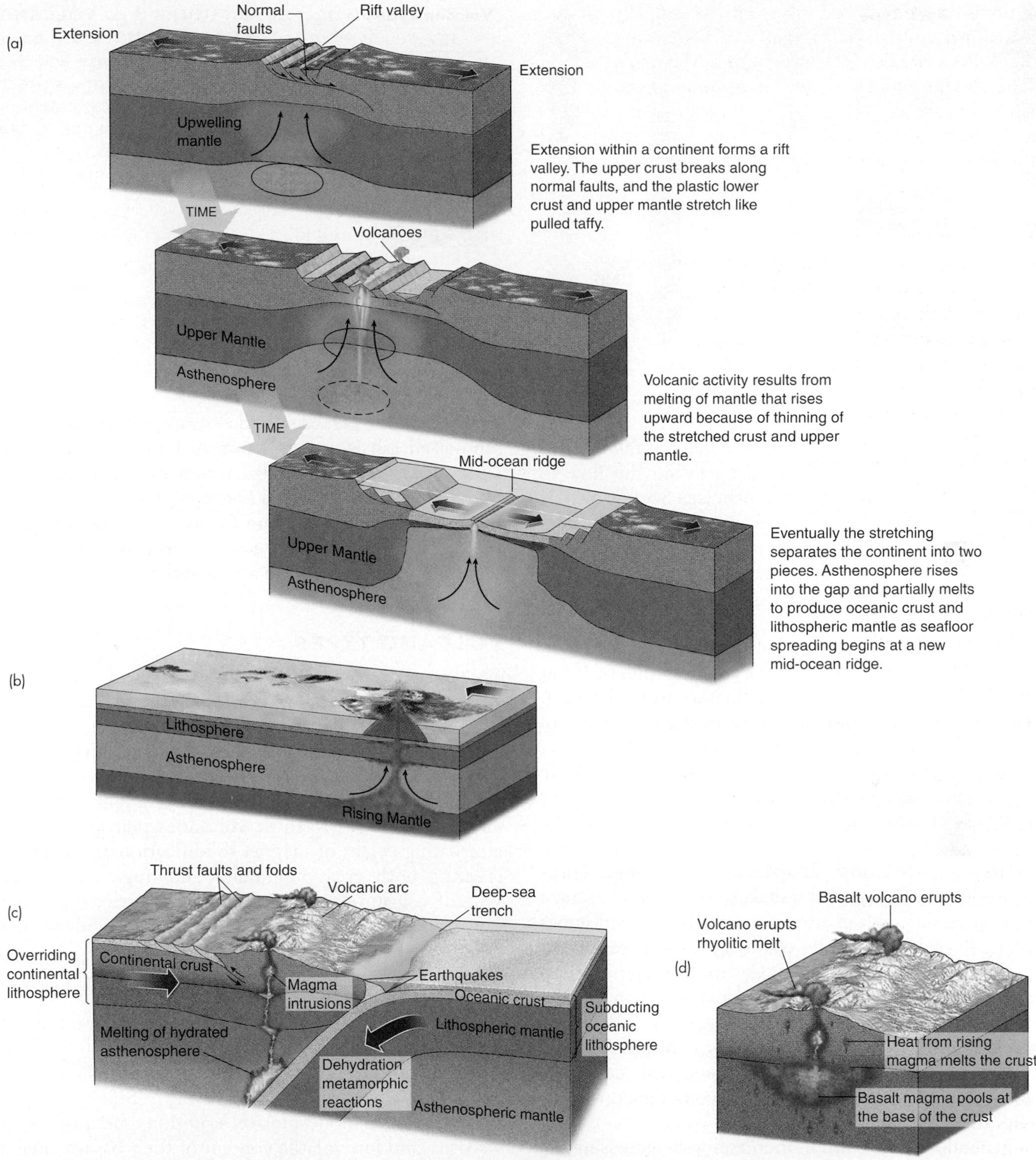

▲ **FIGURE 5.4 MAGMA FORMATION** Magma is generated primarily within the asthenosphere. (a) Decompression melting occurs when the overburden pressure on asthenospheric rocks is decreased by thinning of the overlying lithospheric plate, causing upwelling of superheated rocks. (b) Decompression melting also occurs when superheated rocks upwell from deep within the asthenosphere as part of a hot spot. (c) Melting at subduction zone due to addition of volatiles to the asthenosphere released from the down-going oceanic plate. (d) Melting of the lithosphere due to the addition of heat carried by rising magma.

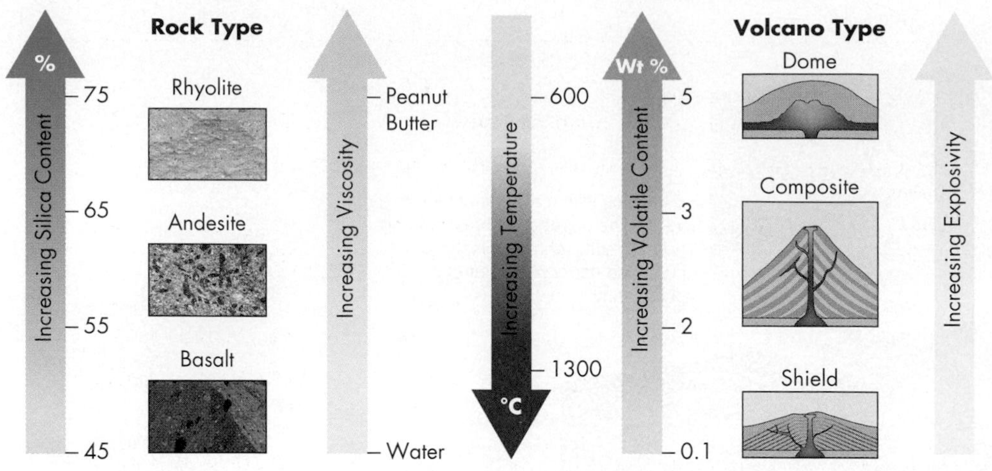

◀ **FIGURE 5.5 VOLCANO CHARACTERISTICS** The eruptive character, shape, and type of volcano that forms is directly related to the silica content of the magma, which defines the rock type and affects the viscosity and volatile content of the melt.

flow may have a low viscosity when it erupts onto the surface; however, as it flows down the flank of the volcano and cools, its viscosity will increase causing the flow to move more slowly and change in form (see Section 5.3).

The variability in magma viscosity strongly influences both the mobility of the magma in the subsurface and the velocity and form it takes if it reaches the surface as a lava flow. Rhyolitic lava flows have high viscosity, move slowly, may be a hundred feet thick, are generally restricted to the vent region, and form steep-sided domes. In contrast, basaltic lava flows can move rapidly, are often thin (< 3m or 10 ft.), and may travel 10s of kilometers from the vent. These differences in flow characteristics affect the shape of the volcano formed by different lava types. More importantly, the viscosity is typically correlated to the volatile content of the magma, which is responsible for whether or not volcanic eruptions are peaceful or violent (Figure 5.5).

Volatile Content and Eruptive Behavior Uncorking a bottle of champagne or a shaken carbonated beverage is a good analogy often used to explain why volcanoes erupt explosively. When the champagne is depressurized by uncorking, the dissolved CO_2 molecules (volatiles) come out of solution from the liquid to form rapidly expanding bubbles, causing the liquid to vigorously squirt from the bottle. Similarly, a high concentration of dissolved volatiles within the magma will cause an explosive eruption when the melt is decompressed when it reaches the Earth's surface.

In general, volatile content increases with increasing silica content (Figure 5.5). Andesitic-to-rhyolitic magma has more dissolved gas (2–5 weight percent) than basaltic magma (<1 wt. %) Thus, volcanoes with andesitic-to-rhyolitic magma are more prone to explosive eruptions than volcanoes erupting basaltic lava (Figure 5.5). In the former case, rapid bubble formation and degassing of the more viscous magma during an eruption actually breaks apart the molten material into small ash fragments, or into frothy foam that later solidifies as pumice. These are examples of *pyroclastic debris*, or *tephra*, a term that includes all types of volcanic materials that are explosively ejected from a volcano. Debris particles range from fine dust to sand-sized ash less than 2 mm (0.1 in.) to small gravel-sized *lapilli* 2 to 64 mm (0.1 to 2.5 in.), to large angular *blocks* and smooth-surfaced *bombs* greater than 64 mm (2.5 in.). Accumulation of tephra forms a **pyroclastic deposit** (Greek *pyro*, "fire," and *klastos*, "broken") that can be cemented or fused to form a *pyroclastic rock*.

VOLCANO TYPES

Volcanoes are landforms that vary greatly in size and shape (Table 5.1). Some volcanoes build up over time with the eruption of low-viscosity basaltic magma that accumulates like a pile of wax at the bottom of a dripping candle. Other volcanoes are built up of cinder-like magma fragments that form as lava shoots like a fountain into the air. Finally, there are those volcanoes that grow through alternating cycles of magma solidification and explosion, creating both great mountains and huge craters. Each type of volcano has a characteristic eruptive style that is due primarily to the viscosity and water content of the magma.

Shield Volcanoes The largest volcanoes in the world are **shield volcanoes** (Table 5.1). They are common in the Hawaiian Islands, Iceland, and some islands in the Indian Ocean. In profile, a shield volcano appears as a broad arc like a warrior's shield (Figure 5.6). Shield volcanoes generally have nonexplosive eruptions because of low viscosity and low volatile content of their basaltic magma. When a shield volcano erupts, lava tends to flow down the sides of the volcano rather than exploding violently. Accumulations of thousands of thin basaltic lava flows construct the broad gentle slopes of the volcano (Figure 5.6). Although shield volcanoes are much wider than they are tall, they are still among the tallest mountains on Earth when measured from their bases, which are often located on the ocean floor. By contrast, a volcano constructed of viscous lava, such as Mount Rainier in Washington State, has a height similar to its width (Figure 5.7).

TABLE 5.1

Types of Volcanoes

Volcano Type	Shape	Eruption Type	Examples
Shield volcano	Gentle arch, or shield shape, with shallow slopes; built up of many lava flows	Lava flows, tephra ejections	Mauna Loa, Hawai'i Figure 5.6
Composite volcano, or stratovolcano	Cone-shaped with steep sides; built of lava flows and pyroclastic deposits	Combination of lava flows and explosive activity	Mt. Unzen, Japan p. XX photograph Mt. Fuji, Japan Figure 5.9 Mount St. Helens, Washington Figure 5.30
Volcanic dome	Dome shaped	Highly explosive	Mt. Lassen, California Figure 5.7
Cinder cone	Cone shaped with steep sides and summit crater	Tephra (mostly ash) ejection	SP Crater, Arizona Figure 5.11 Parícutin, Mexico Figure 5.12 Mt. Helgafell, Iceland

▲ FIGURE 5.6 **SHIELD VOLCANO** Panoramic view of Mauna Loa, the largest shield volcano on the Big Island of Hawai'i and in the world. Notice the gently sloping profile of the volcanoe; similar to that of a warrior's shield lying on the ground. Dark patches are less vegetated regions where more recent low-viscosity basaltic lavas have flowed down the flank of the mountain. Note reddish basaltic cinder cones in the foreground. *(Duane DeVecchio)*

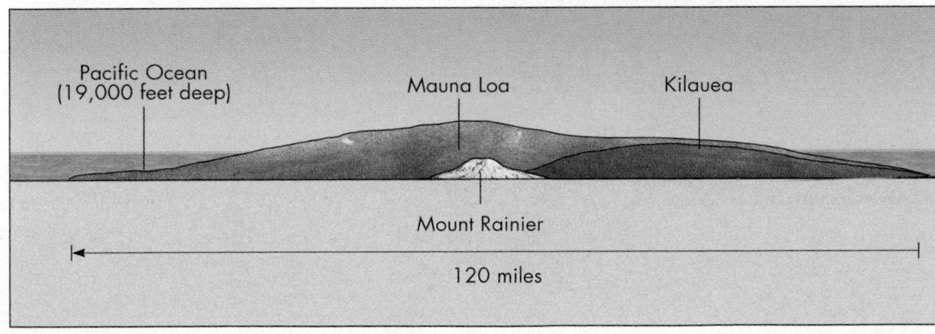

◀ FIGURE 5.7 **VOLCANO DIMENSIONS** Comparison between the shape and size of a shield volcano (Mauna Loa and Kilauea) and Mt. Rainier, one of the largest composite volcanoes in the Cascade Range.

(a)

(b)

▲ FIGURE 5.8 **LAVA TUBES** Magma can move many kilometers below ground in tunnels called lava tubes that form below cooled crusts of basaltic lava. Both active and drained tubes can be hazards when roofs collapse. (a) Holes, called skylights, develop in the roof of some active tubes such as this one on the Kilauea volcano in Hawai'i. *(Jeffrey B. Judd/U.S. Geological Survey, Denver)* (b) Older drained tubes, such as the Thurston (Nahuku) lava tube shown here near the summit of Kilauea, are attractions for tourists. *(Volcano Hazards Team)*

In addition to flowing down the sides of a volcano, lava can move away from its source in other ways. Magma can move for many kilometers underground in *lava tubes* that are often very close to the surface (Figure 5.8). The rock walls of these tubes insulate the magma, keeping it hot and fluid. After the lava cools and crystallizes to rock, the lava tubes may be left behind as long, sinuous cavern systems. They form natural pipes for movement of groundwater and may cause engineering problems when they are encountered during construction projects.

Composite Volcanoes Known for their beautiful cone shapes (Figure 5.9), Mount Fuju in Japan and Mount St. Helens and Mt. Rainier in the United States are examples of **composite volcanoes**. The magma of composite volcanoes is generally more viscous and has a higher volatile content than that of shield volcanoes (see Figure 5.5), resulting in a mixture of explosive activity and lava flows. As a result, these volcanoes, also called *stratovolcanoes*, are composed of layers of both pyroclastic deposits and lava flows. Composite volcanoes erupt a variety of lava types from basalts to lavas that are intermediate between andesite and rhyolite.

Don't let the beauty of these mountains fool you because they tend to be explosive and are common along the rim of the Pacific ocean. Composite volcanoes are responsible for most of the death and destruction caused by volcanoes throughout history. As the 1980 eruption of

◄ FIGURE 5.9 **COMPOSITE VOLCANO** Mt. Shasta, in Northern California, is the most voluminous volcano in the Cascade Range (85 cubic miles) and reaches an elevation 4, 322 m (14,179 ft.). *(Duane E. DeVecchio)*

◀ **FIGURE 5.10 ERUPTION OF LASSEN PEAK** Lassen Peak, east of Redding, California, is the southernmost volcano in the Cascades. This eruption took place in June 1914. *(BF Loomis/National Park Service)*

Mount St. Helens demonstrated, they can produce gigantic horizontal (lateral) blasts, similar in form to the blast of a shotgun. We should consider such volcanoes armed and dangerous.

Volcanic Domes Characterized by highly viscous magma, **volcanic domes** like Lassen Peak and Mono Craters in California sometimes exhibit highly explosive eruptions. Lassen Peak's last series of eruptions, from 1914 to 1917, included a tremendous horizontal (lateral) blast that affected a large area (Figure 5.10). Following a large eruption, such as the one at Mount St. Helens, smaller domes consisting of degassed felsic lava often form within the volcanic crater created during the eruption.

Cinder Cones Built up by the accumulation of tephra near a volcanic vent, **cinder cones** are relatively small volcanoes composed of nut- to fist-sized pieces of vesicular red or black lava.[5] Tephra from extinct cinder cones is the "lava rock" used widely in commercial landscaping. Also called *scoria cones*, cinder cones are common on the flanks of larger volcanoes, along normal faults, and along cracks or fissures (Figure 5.11).

The Parícutin cinder cone in the Itzicuaro Valley of central Mexico, about 320 km (200 mi.) west of Mexico City, offered a rare opportunity to observe the birth and rapid growth of a volcano at a location where none had existed before (Figure 5.12). On February 20, 1943, following several weeks of carthquakes and sounds like thunder coming from beneath the surface of Earth, an astounding event occurred. As Dionisio Pulido was preparing his cornfield for planting, he noticed that a hole he had been trying to fill for years had opened in the ground at the base of a knoll. As Señor Pulido watched, the surrounding ground swelled, rising more than 2 meters (more than 6 ft.), while sulfurous gases and ash began billowing from the hole. By that night, the hole was ejecting glowing-red rock fragments high into the air. By the next day, the cinder cone had grown to 10 m (33 ft.) high as rocks and ash continued to be blown into the sky by the eruption. After only 5 days, the cinder cone had grown to more than 100 m (330 ft.) high. In June 1944, a fissure opened in the base of the now 400-m (1300-ft.) high cone. The fissure produced a basaltic lava flow that overran the nearby village of San Juan Parangaricutiro, leaving little but the church steeple exposed. No one was killed in these eruptions, and within a decade, Parícutin cinder cone became a dormant volcano. Nevertheless, during the 9 years of eruption, more than a billion m^3 (1.3 billion $yd.^3$) of ash and 700 million m^3 (900 billion $yd.^3$) of lava were erupted from Señor Pulido's cornfield. Crops failed, especially when buried by ash faster than they could grow, and livestock became sick and died. Although several villages were relocated to other areas, some residents have moved back to the vicinity of Parícutin. Locating property boundaries has been difficult because markers are generally covered by ash and lava, and this has resulted in land ownership disputes.[5]

VOLCANIC FEATURES

Volcanoes are complex systems, and a "volcano" is really much more than a mountain that expels lava and pyroclastic debris from its top. Volcanoes or volcanic areas commonly include craters, calderas, volcanic vents, geysers, and hot springs. A volcano or volcanic area may exhibit one, some, or all of these features.

◀ **FIGURE 5.11 CINDER CONE** SP Crater, a cinder cone with a small crater at the top in the San Francisco Peaks volcanic field near Gray Mountain, Arizona. The lava flow in the upper part of the photograph originated from the base of the cinder cone and extends about 6 km (4 mi.). *(Micheal Collier)*

Craters, Calderas, and Vents Depressions commonly found at the tops of volcanoes are **craters**, which form explosively during the eruption, removing the upper portion of the volcanic cone. They are usually a few kilometers (around a mile) in diameter. Gigantic depressions formed during explosive ejection of magma and subsequent collapse of the upper cone are known as **calderas**. The violent eruptions that form calderas are discussed below. Calderas may be 20 or more kilometers (12 or more miles) in diameter and contain volcanic vents as well as other volcanic features such as gas vents and hot springs. A **volcanic vent** is any opening through which lava and pyroclastic debris are erupted. Vents may be roughly circular or they may be elongated cracks called *fissures*. Some extensive fissure eruptions have produced huge accumulations of nearly horizontal lava flows called *flood basalts*. The best-known accumulation of flood basalts in the United States is the Columbia Plateau in Washington, Oregon, and the Snake River Plain in Idaho (Figure 5.13). These flood basalts cover about 130,000 km^2 (50,000 mi.^{2}), an area slightly larger than the state of Mississippi.

Hot Springs and Geysers In some volcanic areas, hydrologic features such as hot springs and geysers are common. When groundwater comes into contact with hot rock, it becomes heated and can discharge at the surface as a *hot spring*, or *thermal spring*. Less commonly, groundwater boils in an underground chamber to produce periodic, steam-driven releases of steam and hot water at the surface, a

◀ **FIGURE 5.12 RAPID GROWTH OF A VOLCANO** In this 1943 photograph, Parícutin cinder cone in central Mexico is erupting a cloud of volcanic ash and gases. A basaltic lava flow has nearly buried the village of San Juan Parangaricutiro, leaving only the church steeple exposed. *(Tad Nichols)*

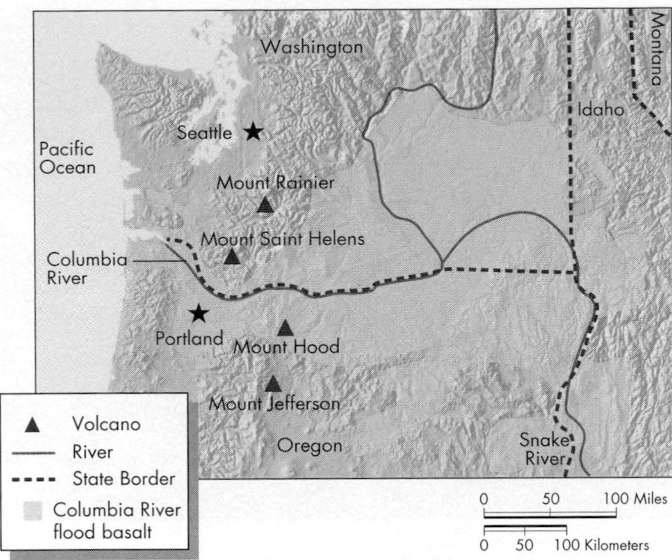

▲ FIGURE 5.13 **COLUMBIA PLATEAU FLOOD BASALTS** From approximately 17 to 15.5 million years ago, thousands of basalt flows poured out onto the surface of the Pacific Northwest, blanketing the states of Washington, Oregon, and Idaho in volcanic rock that locally reaches a thickness of 3,500 m (11,000 ft).

phenomenon called a *geyser*. World-famous geyser fields are found in Iceland, New Zealand, and Yellowstone National Park in Wyoming (Figure 5.14). People are rarely injured in geyser eruptions because their locations are known and periodic eruptions are expected. However, in Yellowstone and probably other geyser fields, larger steam-driven explosions fracture bedrock, produce craters, and could cause personal injury.[6] Fortunately such explosions are relatively infrequent and most locations where they might occur are remote.[5]

Caldera Eruptions Calderas are produced by very rare, but extremely violent, eruptions (Figure 5.15). Although none have occurred anywhere in the United States in the past few hundred thousand years, at least 10 such eruptions have occurred in the past million years, three of which were in North America. Although significant caldera eruptions have occurred in historic time, such as the 1815 Tambora and 1883 Krakatau explosions in Indonesia, the most recent huge caldera eruption occurred 27,000 years ago in New Zealand. Today this caldera is the site of Lake Taupo, a famous trout fishery on the North Island. The caldera eruption 27,000 years ago ejected between 1000 and 2000 km^3 (240 and 480 mi^3) of volcanic ash. Such huge eruptions have been termed *supervolcanos* by the media, but this term is not formally accepted by volcanologists.

A large caldera-forming eruption may explosively extrude more than 100 km3 (240 $mi.^3$) of pyroclastic debris consisting mostly of ash. A quantity this large would cover the island of Manhattan to a height of about 1.6 km (1 mi.)—four times the height of the Empire State Building. This volume is approximately 1000 times the amount of ash ejected by the 1980 eruption of Mount St. Helens! Such an eruption could produce a caldera more than 10 km (6 mi.) in diameter with a crater larger than Crater Lake in Oregon (Figure 5.16). Ash deposits from such an eruption could be 100 m (300 ft.) thick near the crater's rim and a meter (3 ft.) or so thick 100 km (60 mi.) away from the source.[7] The most recent caldera-forming eruptions in North America occurred about 640,000 years

(a)

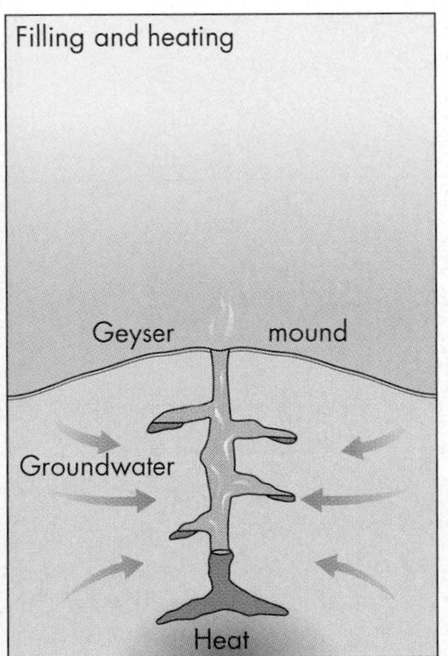

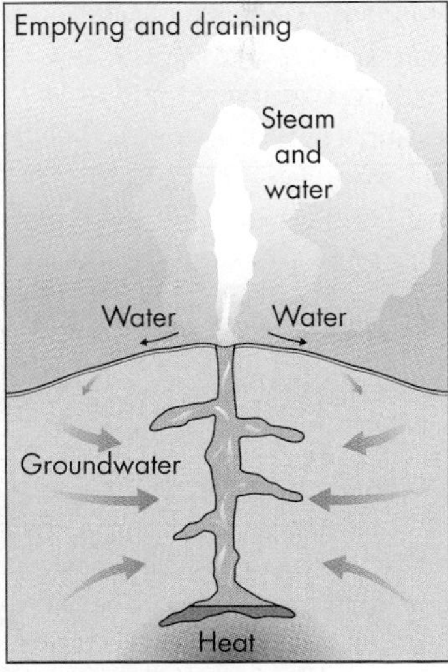

(b)

Groundwater fills subsurface geyser's irregular tubes and water is heated until it flashes to steam and erupts.

Eruption partly empties subsurface tubes and recycled groundwater starts the filling process again. Water running off precipitates a white silica-rich mineral called geyserite that forms the geyser mound.

▲ FIGURE 5.14 **HOW A GEYSER WORKS** Eruption of Old Faithful Geyser, Yellowstone National Park, Wyoming. Although the name implies predictability, eruption intervals vary from day to day and year to year and often change after earthquakes. *(James Randklev/Getty Images)*

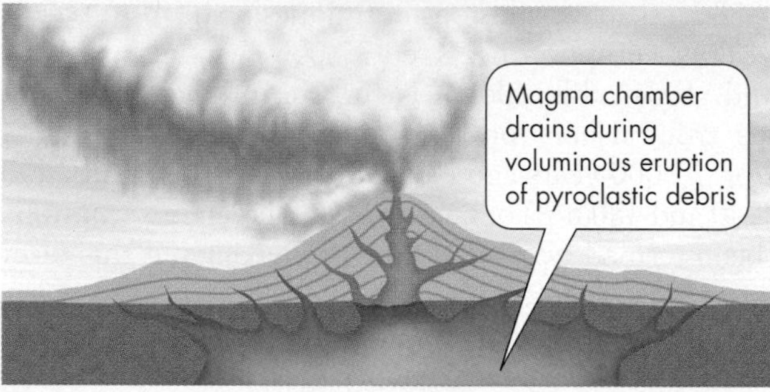

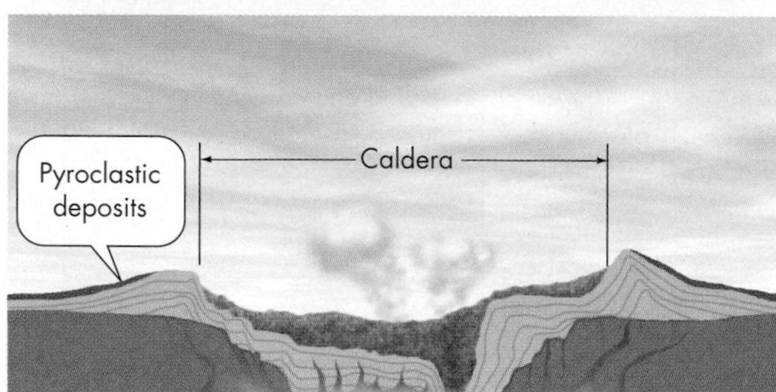

▲ FIGURE 5.15 **LARGE CALDERAS FORM BY EXPLOSION AND COLLAPSE** Large calderas, such as the one illustrated here, generally form by explosion and subsequent collapse of the magma chamber below a composite volcano. The resulting depression may be 20 km (12 mi.) or more in diameter. This process may occur once, like Crater Lake in Oregon, or multiple times, like the formation of resurgent calderas in Yellowstone National Park, Wyoming. *(Modified after Smith, G. A., and Pun, A. 2006. How does Earth work? Physical geology and the process of science. Upper Saddle River, NJ: Pearson Prentice Hall)*

ago at Yellowstone National Park in Wyoming and 700,000 years ago in Long Valley, California. The latter event produced the Long Valley caldera and covered a large area with ash (Figure 5.17). Although the most recent volcanic eruptions at Long Valley were about 400 years ago, there still are

▲ FIGURE 5.16 **CRATER LAKE CALDERA** Aerial view of Crater Lake, Oregon, which was produced by an explosive eruption of Mount Mazama around 7700 years ago. Crater Lake is 8 by 10 km (5 by 6 mi.) across and almost 600 m (2000 ft.) deep, making it the second deepest lake in North America. Originally a composite volcano around 1200 m (4000 ft.) higher than it is today, the explosion and subsequent collapse of the caldera took place in an eruption that was 42 times greater than the 1980 eruption of Mount St. Helens. Each year close to 500,000 people visit Crater Lake National Park. *(US Geological Survey/ US Department of the Interior)*

potential hazards from a future volcanic eruption (Figure 5.18). In the early 1980s, measurable uplift of the land accompanied by swarms of earthquakes up to **M** 6 suggested that magma was moving upward at Long Valley. This prompted the U.S. Geological Survey (USGS) to issue a potential volcanic hazard warning that was subsequently lifted. Although activity at Long Valley has subsided, the future remains uncertain.

The main events in a caldera-producing eruption can occur quickly—in a few days to a few weeks—but intermittent, lesser-magnitude volcanic activity can linger on for a million years. Caldera-producing eruptions at Yellowstone have left us hot springs and geysers, including Old Faithful, whereas the Long Valley eruption has left us a potential volcanic hazard. In fact, both sites are still capable of producing volcanic activity because magma is present at shallow depths beneath the caldera floors. Both are considered *resurgent calderas* because their floors have slowly domed upward since the explosive eruptions that formed them.

VOLCANO ORIGINS

We established earlier that the causes of volcanic activity are directly related to plate tectonics. More specifically, the tectonic setting determines the type of volcano that will be present (Figure 5.19).

1. **Mid-ocean ridges and continental rifts**
   Volcanism at mid-ocean ridges produces basaltic magma derived directly from the asthenosphere. Approximately three-quarters of all lava erupted on Earth is extruded

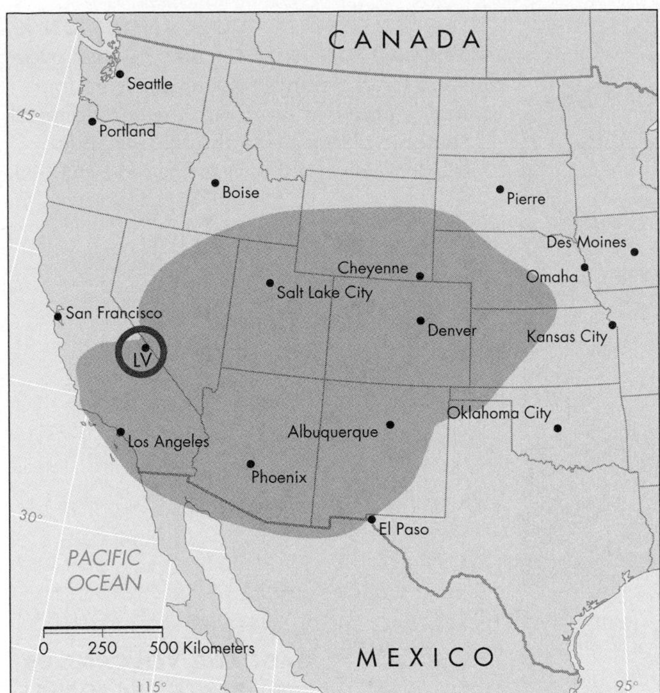

▲ **FIGURE 5.17 WIDESPREAD ASH FALL HAZARD** Area in orange was covered by ash from the Long Valley caldera eruption approximately 700,000 years ago. Red circle around Long Valley (LV) has a radius of 120 km (75 mi.) and encloses the area subject to at least 1 m (3 ft.) of downwind ash accumulation if a similar eruption were to occur again. *(From Miller, C. D., Mullineaux, D. R., Crandell, D. R., and Bailey, R. A. 1982. Potential hazards from future volcanic eruptions in the Long Valley—Mono Lake area, east-central California and southwest Nevada—A preliminary assessment. U.S. Geological Survey Circular 877)*

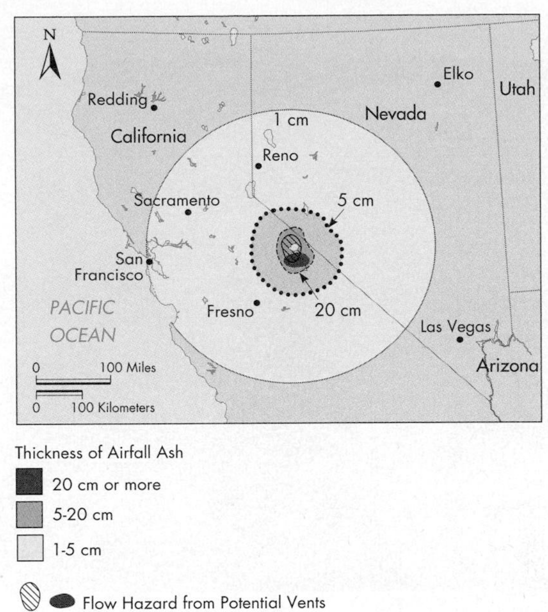

▲ **FIGURE 5.18 CALDERA HAZARDS** Potential hazards from a volcanic eruption at the Long Valley caldera near Mammoth Lakes, California. Red area and diagonal lines show the hazard from flows of all types (pyroclastic, debris, mud, lava) out to a distance of approximately 20 km (12 mi.) Lines surrounding the larger orange and yellow hazard areas represent potential ash thicknesses: 20 cm (8 in.) at the 35 km (22 mi.) dashed line; 5 cm (2 in.) at the 85 km (53 mi.) dotted line; and 1 cm (0.4 in.) at the 300 km (190 mi.) solid line. These hazard estimates assume an explosive eruption of approximately $1\ km^3$ $(0.24\ mi.^3)$ of tephra from the vicinity of recently active vents. *(From Miller, C. D., Mullineaux, D. R., Crandell, D. R., and Bailey, R. A. 1982. Potential hazards from future volcanic eruptions in the Long Valley—Mono Lake area, and east-central California and southwest area, and east-central Nevada—A preliminary assessment. U.S. Geological Survey Circular 877)*

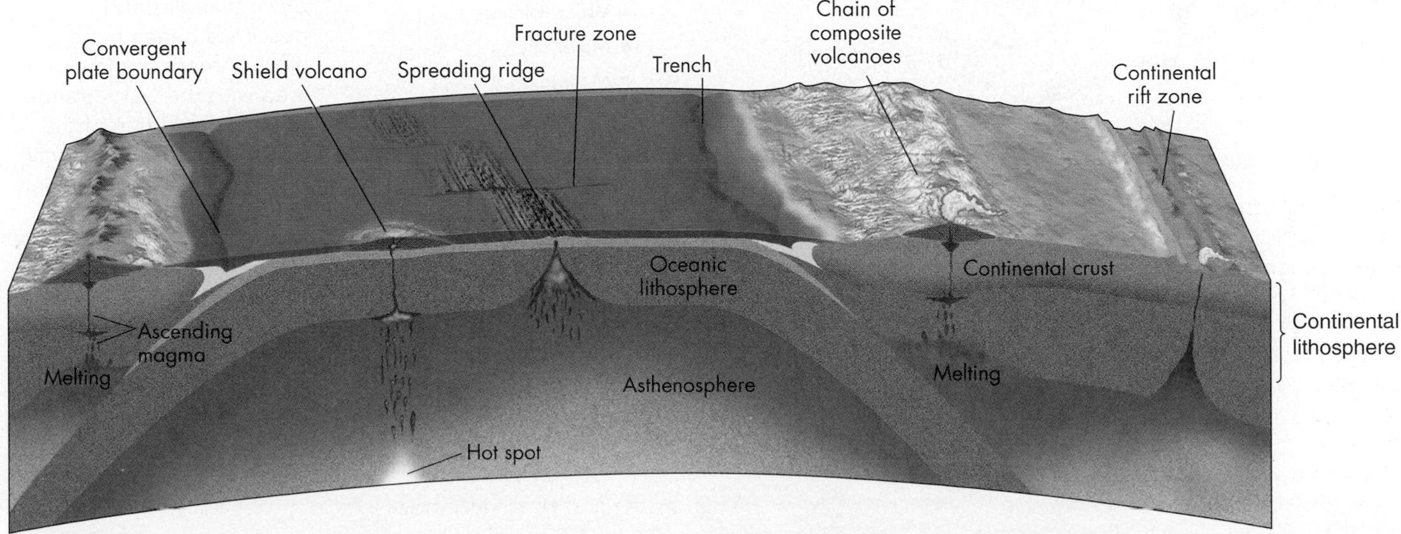

▲ **FIGURE 5.19 VOLCANIC ACTIVITY AND PLATE TECTONICS** With the exception of hot spots, melting of the asthenosphere to produce volcanoes at Earth's surface occurs in response to the relative the movement of lithospheric plates. Due to different melting processes along different types of plate boundaries, the type and eruptive behavior of volcanoes will also be different.

◀ **FIGURE 5.20 VOLCANOES ON A SPREADING RIDGE** Icelandic shield volcano (background) and large fissure (normal fault) associated with the spreading of tectonic plates along the Mid-Atlantic Ridge. *(University of Washington Libraries)*

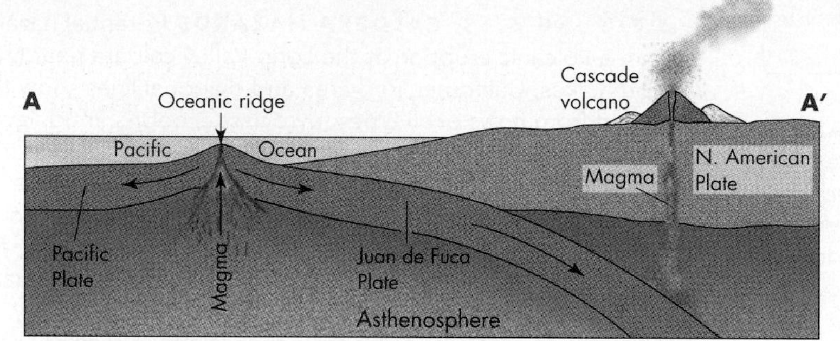

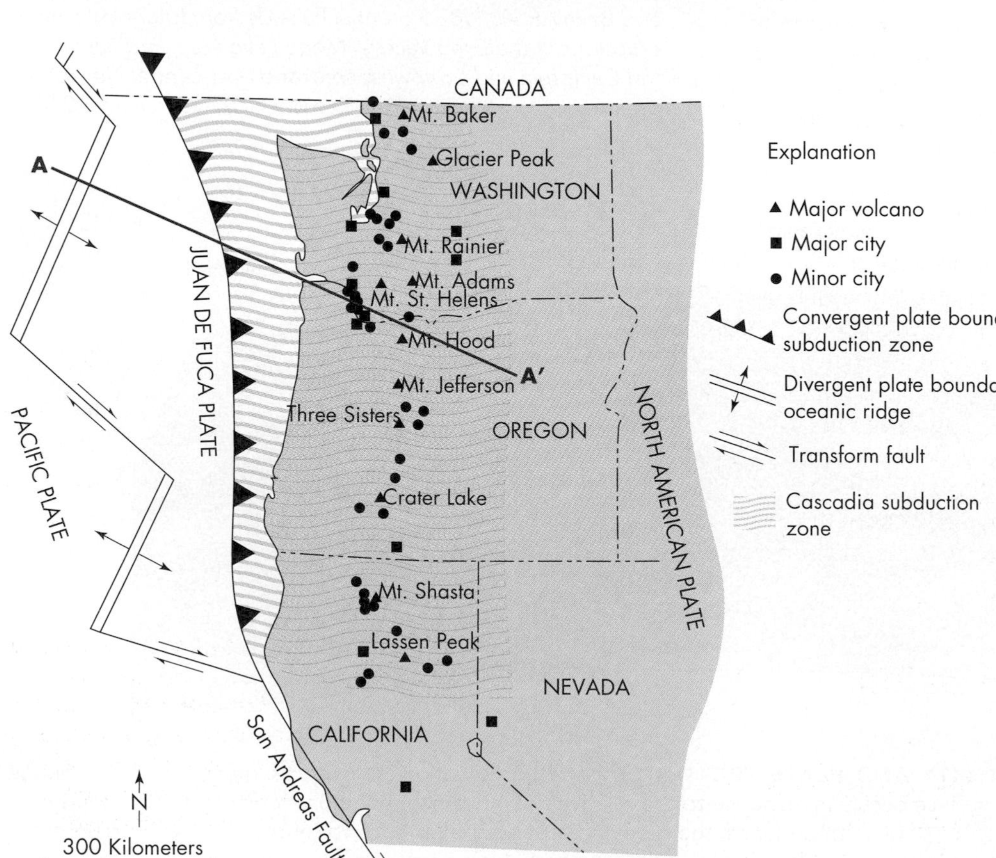

◀ **FIGURE 5.21 CASCADE VOLCANOES AND PLATE TECTONICS** Cross section (top) and map (bottom) of the plate tectonic setting of the Cascade Range showing major volcanoes and cities in their vicinity. The Juan de Fuca plate, off the coast of northern California, Oregon, Washington, and Vancouver Island in British Columbia, is being subducted below the North American plate (in green) along the Cascadia subduction zone. This subduction zone starts offshore along the line with the black triangles and descends beneath the North American plate in the area with the wavy orange lines. Dehydration of the down-going Juan de Fuca plate produces partial melt in the overlying asthenosphere causing the Cascade volcanoes on the North American plate. *(Modified after Crandell, D. R., and Waldron, H. H. 1969. Disaster preparedness. Office of Emergency Preparedness)*

from undersea mid-ocean ridges.[8] This magma mixes very little with other materials except basaltic oceanic crust. Therefore, the resultant lavas are composed almost entirely of low-viscosity *basalt.* Where these spreading ridges occur on land, such as in Iceland, shield volcanoes are formed and large rifts occur (Figure 5.20).

2. **Subduction zones**
   Composite volcanoes are associated with subduction zones and thus are the most common type found around the Pacific Rim. For example, volcanoes in the Cascade Range of Washington, Oregon, and California are derived from the Cascadia subduction zone (Figure 5.21). These volcanoes are commonly andesitic in composition and have silica content intermediate between basaltic oceanic crust and the more silica-rich continental crust (see Figure 5.5). The higher silica content and higher volatile content make these volcanoes explosively unpredictable. More than 80 percent of the volcanic eruptions in historic times have come from volcanoes above subduction zones.[9]

3. **Hot spots beneath the oceans**
   Shield volcanoes are formed above hot spots in the oceanic lithosphere. For example, Hawaiian volcanoes, located well within the Pacific plate, have been built up from the sea floor through submarine eruptions of basaltic lava similar to those at mid-ocean ridges. This magma appears to be derived from a hot spot that has been fairly stationary for many millions of years. The Pacific plate is moving roughly northwest over this hot spot. With time, this movement has produced a chain of volcanic islands running northwest to southeast (see Chapter 2). The island of Hawai'i is presently near the hot spot and is experiencing active volcanism and growth. Islands to the northwest, such as Molokai and Oahu, have moved off the hot spot, since their volcanoes are no longer active.

4. **Hot spots beneath continents**
   Caldera-forming eruptions occur in this tectonic setting. They may be extremely explosive and violent, and they are associated with rhyolitic magma. Rhyolite has a high silica content produced when rising magma from the asthenosphere melts and mixes with felsic continental crust. As mentioned earlier, the most recent caldera-forming eruption in North America occurred about 640,000 years ago at Yellowstone National Park. This explosive eruption of rhyolitic magma has been linked to a still-active hot spot under the North American plate.

5.2 Geographic Regions at Risk from Volcanoes

Like earthquakes, volcanoes are intimately related to plate tectonics and most occur along the "Ring of Fire" surrounding the Pacific Ocean basin (see Figure 5.3). Areas outside the Ring of Fire can also experience volcanoes, because of hot spot activity (Hawai'i) or because they are located on a portion of the mid-ocean ridge (Iceland). East Africa has volcanism related to the rifting, or pulling away from one another, of three tectonic plates. Also, large, infrequent eruptions at Long Valley and those centered in Yellowstone National Park in North America indicate that these areas may be at risk in the future. The highest risk of local volcanic activity in the contiguous United States and Canada is in the mountainous regions of the Pacific Coast and at Yellowstone (Figure 5.22). Other isolated areas of the Southwest are at risk, whereas the eastern two-thirds of the United States and Canada is risk free for local volcanic activity. However, the effects of a large caldera explosion in the western United States or Canada would likely be felt far from the source, in the form of ash fall and ash clouds in the atmosphere.

5.3 Effects of Volcanoes

Worldwide, 50 to 60 volcanoes erupt each year.[1] With the exception of Kilauea volcano on the big island of Hawai'i, which has been erupting continuously since 1983, eruptions in the United States occur two or three times per year, mostly in Alaska. Eruptions are

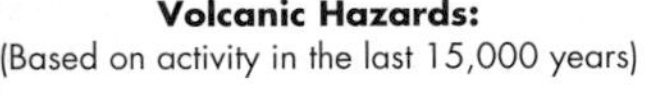

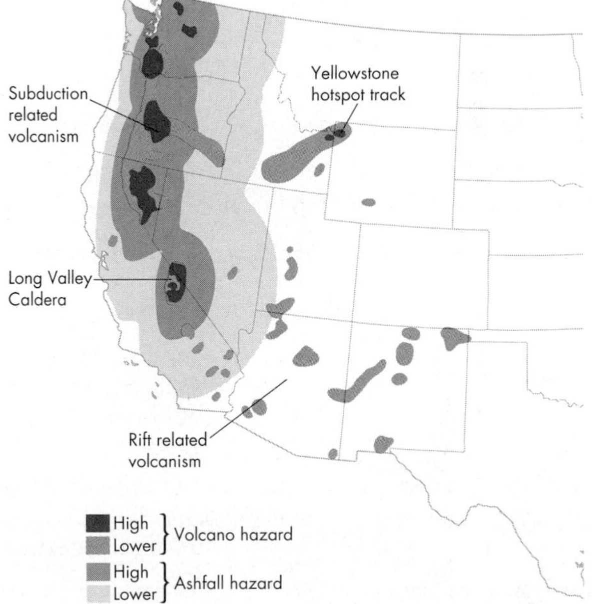

▲ **FIGURE 5.22 UNITED STATES VOLCANIC HAZARD** Volcanic hazard for the contiguous United States based on activity during the past 15,000 years. Red colors show high and lower risk of local volcanic activity, whereas gray colors show regions at risk of receiving 5 cm or more of ash fall from large explosive eruptions. The hazard belt in Washington State continues north into southern British Columbia. *(U.S. Geological Survey)*

often in sparsely populated areas of the world, causing little, if any, loss of life or economic damage. However, when an eruption takes place near a densely populated area, the effects can be catastrophic (Table 5.2).[10] Approximately 500 million people on Earth live close to volcanoes, and as the human population grows, more and more people are living on the flanks, or sides, of active or potentially active volcanoes. In the past 100 years, nearly 100,000 people have been killed by volcanic eruptions—approximately 28,500 lives were lost in the 1980s alone.[1,10] Densely populated countries with many active volcanoes, such as Japan, Mexico (especially near Mexico City), the Philippines, and Indonesia, are particularly vulnerable.[2] Several active or potentially active volcanoes in the western United States are near cities with populations of more than 500,000 people (Figure 5.23).

Volcanic hazards include the *primary effects* of volcanic activity that are the direct results of an eruption, and *secondary effects*, which may be caused by the primary effects. Lava flows, pyroclastic activity such as ash fall, pyroclastic flows and lateral blasts, and release of volcanic gases are primary effects. Secondary effects include debris flows, mudflows, landslides or debris avalanches, floods, fires, and tsunamis (discussed in Chapter 4) and at the planetary level, large eruptions can cause global cooling of the atmosphere for a year or so.[7]

LAVA FLOWS

A **lava flow** is one of the most familiar products of volcanic activity. Lava flows result when magma reaches the surface and overflows the central crater or erupts from a volcanic vent along the flank of the volcano. The three major types of lava take their names from the volcanic rocks they form: basaltic—by far the most abundant of the three—andesitic, and rhyolitic.

Lava flows can be quite fluid and move rapidly or be relatively viscous and move slowly. Basaltic lavas, which have lower viscosity and higher eruptive temperatures, are the fastest and can move 15–35 km (10-20 mi.) per hour near the vent. Called *pahoehoe* (pronounced pa-hoy-hoy), these lavas have a smooth, sometimes ropey surface

TABLE 5.2

Selected Historic Volcanic Events

Volcano or City	Year	Effect
Vesuvius, Italy	A.D. 79	Destroyed Pompeii and killed 16,000 people. City was buried by volcanic activity and rediscovered in 1595.
Skaptar Jokull, Iceland	1783	Killed 10,000 people (many died from famine) and most of the island's livestock. Also killed some crops as far away as Scotland.
Tambora, Indonesia	1815	Global cooling; killed 10,000 people and 80,000 starved; produced "year without a summer."
Krakatau, Indonesia	1883	Tremendous explosion; more than 36,000 deaths from tsunami.
Mount Pelée, Martinique	1902	Ash flow killed 30,000 people in a matter of minutes.
La Soufrière, St. Vincent	1902	Killed 2000 people and caused the extinction of the Carib Indians.
Mount Lamington, Papua New Guinea	1951	Killed 6000 people.
Villarica, Chile	1963–64	Forced 30,000 people to evacuate their homes.
Mount Helgafell, Heimaey Island, Iceland	1973	Forced 5200 people to evacuate their homes.
Mount St. Helens, Washington, United States	1980	Debris avalanche, lateral blast, and mudflows; killed 57 people, destroyed more than 100 homes.
Nevado del Ruiz, Colombia	1985	Eruption generated mudflows that killed at least 22,000 people.
Mount Unzen, Japan	1991	Ash flows and other activity killed 41 people and burned more than 125 homes. More than 10,000 people evacuated.
Mount Pinatubo, Philippines	1991	Tremendous explosions, ash flows, and mudflows combined with a typhoon killed more than 740 people; several thousand people evacuated.
Montserrat, Caribbean	1995	Explosive eruptions, pyroclastic flows; south side of island evacuated, including capital city of Plymouth; several hundred homes destroyed.
Chaitén, Chile	2008	Explosive eruptions, pyroclastic flows: 5000 people evacuated and disrupted aviation in South America for weeks.
Eyjafjallajokull, Iceland	2010	Large ash emission: disrupted air travel in the United Kingdom and Northern Europe for several weeks.

Source: Data partially derived from Ollier, C. 1969. Volcanoes. Cambridge, MA: MIT Press.

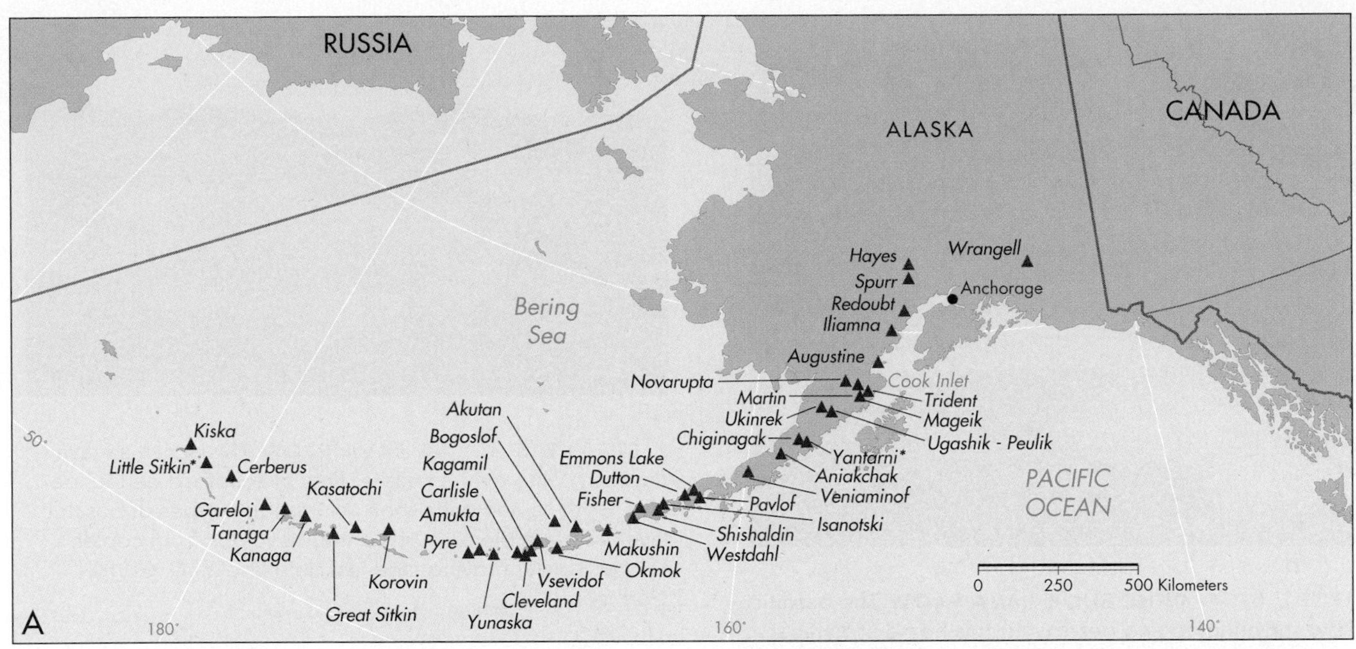

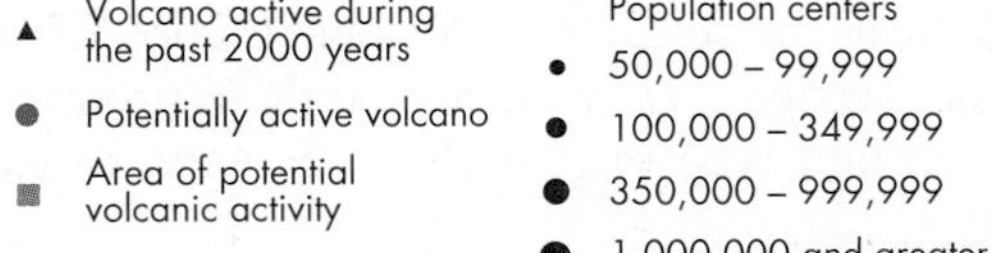

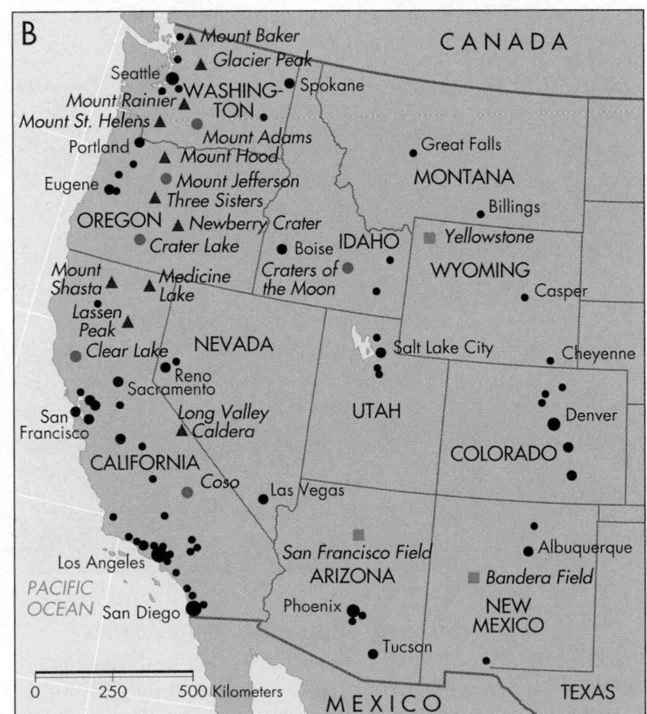

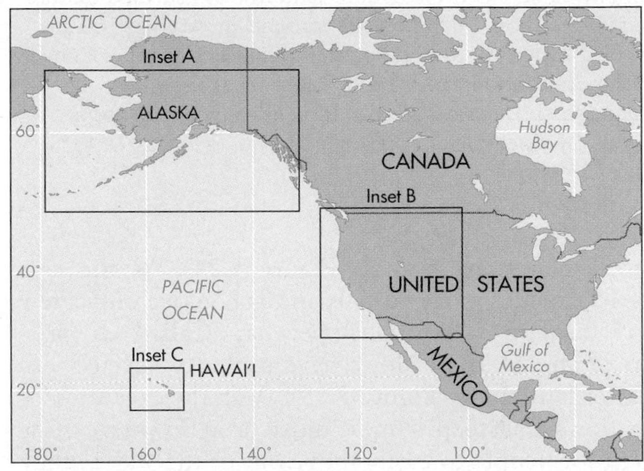

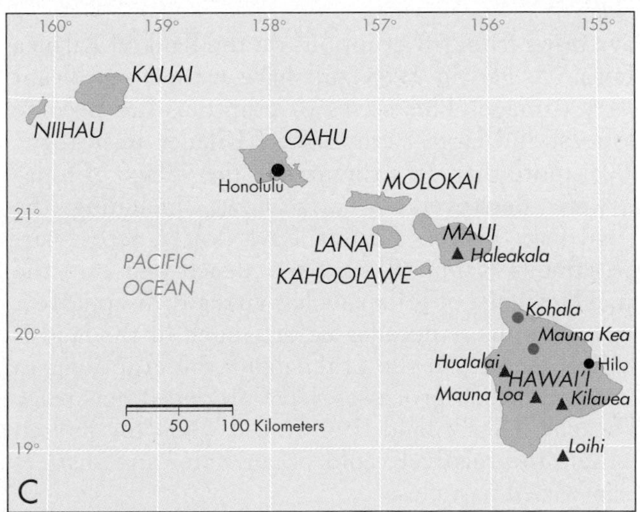

▲ **FIGURE 5.23 LOCATIONS OF VOLCANOES IN THE UNITED STATES** Index maps show locations of active and potentially active volcanoes and nearby population centers (not all labeled) of the United States. There are at least 11 active and potentially active volcanoes in British Columbia and the Yukon Territory. *(From Wright, T. L., and Pierson, T. C. 1992. U.S. Geological Survey Circular 1073)*

▲ FIGURE 5.24 **PAHOEHOE LAVA FLOW** The basaltic lava flow surrounding and destroying this home at Kalapana, Hawai'i, in 1990 has the characteristic smooth surface texture of most pahoehoe. Such flows destroyed more than 100 structures, including the National Park Service Visitors Center. Some pahoehoe flows are referred to as ropy lava because the surface texture can look like a piece of rope lying side by side as if it were coiled. *(Paul Richards/UPI/ Bettmann/Corbis)*

▲ FIGURE 5.25 **AA LAVA FLOW** This active aa lava flow is moving over an earlier flow of solidified pahoehoe lava. Both aa and pahoehoe are common types of basaltic lava. Aa has a blocky surface that develops from cooler, thicker, slower moving, and less fluid lava. *(J.D. Griggs/ Corbis)*

texture when they harden (Figure 5.24). As the lava cools, it becomes more viscous and may move only a few meters (around 6 to 10 ft.) per day. Called *aa* (pronounced ah-ah), these flows have a blocky surface texture after hardening (Figure 5.25). With the exception of some flows on steep slopes, most lava flows are slow enough for people to easily move out of the way as they approach.[11]

Lava flows from rift eruptions on the flank of Kilauea in Hawai'i began in 1983 and have caused significant property damage. This series of eruptions has become the longest and largest eruption of Kilauea in history.[3] By 2005, more than 50 structures in the village of Kalapana were destroyed by lava flows, including the National Park Visitors Center. Lava flowed across part of the famous Kaimu Black Sand Beach and into the ocean. The village of Kalapana has virtually disappeared, and it will be many decades before much of the land is productive again. On the other hand, the eruptions, in concert with beach processes, have produced new black sand beaches. Black sand is produced when the molten lava enters the relatively cold ocean water and shatters into sand-sized particles.

PYROCLASTIC ACTIVITY

Pyroclastic activity refers to explosive volcanism in which tephra is physically blown from a volcanic vent into the atmosphere. Several types of pyroclastic activity can occur. In *volcanic ash eruptions,* a tremendous quantity of fine-grained rock and volcanic glass shatters, and gas is blown high into the air by volcanic explosions (Figure 5.26). These particles are carried downwind and settle to produce an **ash fall**. In a second type of pyroclastic activity, an explosion

▲ FIGURE 5.26 **MOUNT VESUVIUS THREATENS NAPLES** Shown here is the last eruption of Mount Vesuvius, one of three active volcanoes close to Naples, Italy. This February 1944 eruption has been followed by an unusually long period of quiet. Even more dangerous is the Campi Flegrei volcano to the west of Naples, which has had the largest volcanic eruption in Mediterranean history. Today more than 3 million people live in volcanic areas surrounding the Bay of Naples and 1 million people live on the slopes of Mount Vesuvius. *(Getty Images)*

destroys part of the volcano as gas, ash, and rock fragments are blown horizontally from the side of the mountain. Known as **lateral blasts**, these eruptions can eject debris at tremendous speeds and be extremely destructive.

Pyroclastic Flows Some of the most lethal aspects of volcanic eruptions are **pyroclastic flows**. They are avalanches of hot pyroclastic materials—ash, rock, volcanic glass fragments, and gas—that are blown out of a vent and move rapidly down the sides of the volcano (see Figure 5.1). Pyroclastic flows are also known as ash flows, hot avalanches, or *nuée ardentes* (French for "glowing clouds"). Pyroclastic flows may be as hot as hundreds of degrees Celsius and move as fast as 160 km (100 mi.) per hour down the sides of a volcano, incinerating everything in their path.[5] Direct measurement of the temperatures of pyroclastic flows on Montserrat and at Mt. Unzen ranged from 99° to 600°C (210° to 1100°F).[12] Overall, these flows have killed more people than any other volcanic hazard in the past 2000 years (Figure 5.27).[12]

Pyroclastic flows can be catastrophic if a populated area is in their path. A tragic example occurred in 1902 on the Caribbean island of Martinique. On the morning of May 8, a flow of hot, glowing ash, steam, and other gases roared down Mount Peleé and through the town of St. Pierre, killing an estimated 30,000 people. A jailed prisoner was one of only two survivors, and he was severely burned and horribly scarred. Reportedly, he spent the rest of his life touring circus sideshows as the "Prisoner of St. Pierre." The second survivor was a shoemaker who ran inside a building. Although he was burned, he escaped being suffocated by the ash that killed nearly everyone else. Flows like these have occurred on volcanoes of the Pacific Northwest, Alaska, and Japan in the past and can be expected in the future.

▲ FIGURE 5.27 **PLASTER CASTS OF VOLCANO VICTIMS** In A.D. 79, pyroclastic flows from Mt. Vesuvius destroyed the Roman towns of Pompeii and Herculaneum, killing more than 16,000 people. Archeologists have excavated the remains of some 2000 people asphyxiated in the eruption. Plaster casts of molds of the victims reveal adults, children, and dogs in their death positions. Pompeii was buried by up to 3 m (10 ft.) of ash and pumice, and Herculaneum was excavated from beneath 20 m (65 ft.) of volcanic debris. *(Bruno Morandi/AGE Fotostock)*

Ash Fall Volcanic ash eruptions can cover hundreds or even thousands of square kilometers with a carpet of volcanic ash. Ash eruptions create several hazards:

- Vegetation, including crops and trees, may be destroyed.
- Surface water may be contaminated by sediment. The very fine particles clog the gills of fish and kill other aquatic life. Chemical coatings on the ash can cause a temporary increase in the acidity of the water, which may last for several hours after an eruption ceases.
- Structural damage to buildings may occur as ash piles up on roofs (Figure 5.28). As little as 1 centimeter (0.4 in.) of ash can place an extra 2.5 tons of weight on an average house with a 140 m^3 (1500 $ft.^3$) roof.

◀ FIGURE 5.28 **VOLCANIC TEPHRA ON BUILDINGS** Fall out of ash and cinders may completely bury houses and increase the load on walls and roofs causing structural collapse. Shown here are numerous homes, some of which are completely buried, following the 1973 Heimaey eruption in Iceland. *(Kai Honkanen/Photo Alto/Alamy)*

- Health hazards, such as irritation of the respiratory system and eyes, are caused by contact with volcanic ash and associated caustic fumes.[10]
- Engines of jet aircraft may "flame out" as melted silica-rich ash forms a thin coating of volcanic glass in the engines. For example, in 1989, a KLM 747 jet on the way to Japan flew though a cloud of volcanic ash from Redoubt Volcano in Alaska. Power to all four of its jet engines was lost and the plane began a silent, 4270 m (14,000 ft.) before the engines could be restarted.[13] Since this event, Federal Aviation Administration (FAA) guidelines limit air traffic near volcanic ash clouds. In April 2010, an ash plume rising 9.5 km (6 mi.) into the atmosphere above the erupting Eyjafjallajokull volcano on the island of Iceland caused the shutdown of airspace above Northern Europe, the United Kingdom, Ireland, Sweeden and Norway. This represented the largest aerial closure in Europe since World War II and caused the cancellation of approximately 95,000 flights, stranding hundreds of thousands of travelers and costing the airline industry more than a billion dollars during a one-week period (Figure 5.29).[14]

POISONOUS GASES

A number of gases, including water vapor, *carbon dioxide* (CO_2), *carbon monoxide* (CO), *sulfur dioxide* (SO_2) and *hydrogen sulfide* (H_2S) are emitted during volcanic activity. Water and carbon dioxide make up more than 90 percent of all emitted gases. Toxic concentrations of hazardous volcanic gases rarely reach populated areas. A notable tragic exception occurred at Lake Nyos, a deep crater lake on a dormant volcano in the Cameroon highlands of West Africa. Over a period of time, carbon dioxide escaping from the crater floor accumulated in the lake sediment and the bottom waters of the lake. On an August night in 1986, with little warning other than a loud rumbling, Lake Nyos released a misty cloud of dense, mainly carbon dioxide gas. Nearly odorless, the gas cloud flowed from the volcano into valleys below, displacing the air. As the cloud lost its water droplets, it became invisible, spreading silently through five villages. The cloud traveled 23 km (more than 14 mi.) from Lake Nyos, suffocating 1742 people, an estimated 3000 cattle, and numerous other animals (Figure 5.30a and Figure 5.30b).[15]

Since 1986, Lake Nyos has continued to accumulate carbon dioxide in the bottom of the lake, and another release could occur at any time.[15] Although the lake area was to remain closed to all but scientists studying the hazard, thousands of people are returning to farm the land. Scientists have installed an alarm system at the lake that will sound if carbon dioxide levels become high. They have also installed a pipe from the lake bottom to a degassing fountain on the surface of the lake that allows the carbon dioxide gas to escape slowly into the atmosphere (Figure 5.30c). The hazard is being slowly reduced as the single degassing fountain is now releasing a little more carbon dioxide gas than is naturally seeping into the lake. Additional pipes with degassing fountains will be necessary to adequately reduce the hazard.

Sulfur dioxide, a gas that smells like gunpowder, is often released from volcanoes. It can be a direct hazard to people, and evacuations or area closures are ordered if large amounts are released. For example, in 2007 part of Crater Rim Drive at the Kilauea volcano in Hawai'i was closed because of high levels of the gas. Sulfur dioxide can react in the atmosphere to produce *acid rain* downwind of an eruption. Toxic concentrations of some chemicals emitted as

(a)

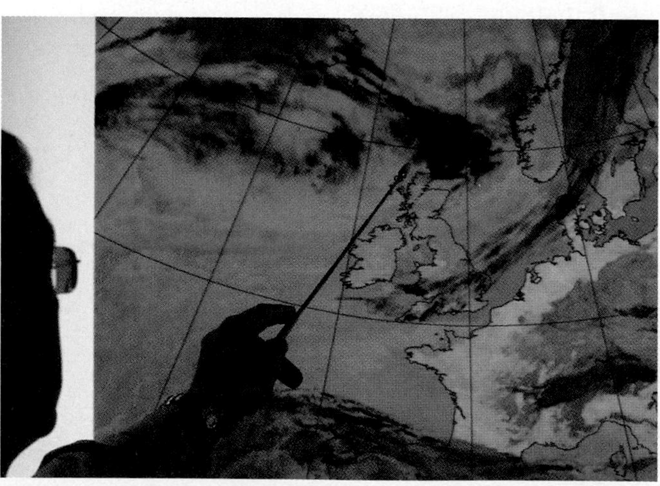
(b)

▲ FIGURE 5.29 **2010 ICELANDIC VOLCANIC ERUPTION** (a) Volcanic ash plume blanketed the Icelandic countryside following the eruption of Eyjafjallajökull Volcano beneath the Icelandic icecap. *(Arctic Images/Corbis)* (b) Meteorology scientist points to an enhanced color satellite image highlighting a volcanic ash plume moving from Iceland towards the United Kingdom on April 15, 2010; all flights in and out of Britian's airports were grounded. *(Matt Cardy/Getty Images)*

(a)

(b)

(c)

▲ FIGURE 5.30 **POISONOUS GAS FROM DORMANT VOLCANO** (a) Water in Lake Nyos, a crater lake in Cameroon, Africa, accumulated and then released immense volumes of carbon dioxide in 1986. A landslide into the lake triggered the carbon dioxide eruption and is responsible for the muddy water in this photograph. *(Thierry Orban/Corbis)* (b) Aerial view of some of the 3000 cattle that died by asphyxiation from the carbon dioxide *(Peter Turnley/Corbis)* (c) Gas being released from the bottom waters of Lake Nyos with a degassing fountain in 2001. *(University of Savoie)*

gases may be adsorbed by volcanic ash that falls onto the land. In some cases, these chemicals have contaminated the soil and have been absorbed by plants eaten by people and livestock. For example, in 1783 a "dry fog" of sulfur dioxide gas from the eruption of the Skaptar Jokull volcano in Iceland blanketed much of Europe. The cloud contained fluorine in the form of fine droplets of acidic hydrogen fluoride. These droplets contaminated pastures with fluorine and caused the death of grazing cattle in as little as 2 days.[15]

In Japan, volcanoes are monitored to detect the release of poisonous gases such as hydrogen sulfide. When releases are detected, sirens are sounded to advise people to evacuate to high ground to escape the gas.

Volcanoes can also create air pollution over larger areas. This can take the form of volcanic smog, known as *vog* (volcanic material, "v," and fog, "og"), or a haze from steam plumes created by lava flowing into the ocean. For example, the eruptions of Kilauea in Hawai'i since 1983 have at times produced vog mixed with acid rain downwind of the volcano. This combination has covered the southeastern part of the Big Island with vog, a thick, blue acidic haze that far exceeds air quality standards for sulfur dioxide. Public health warnings have been issued because small, acidic aerosol particulates and sulfur dioxide concentrations can induce asthma attacks and cause other respiratory problems. Residents and visitors have reported breathing difficulties, headaches, sore throats, watery eyes, and flulike symptoms when exposed to vog. In addition, acid rain has made the water in some shallow wells and household rainwater-collection systems undrinkable. The acidic rainwater has extracted lead from metal roofing and water pipes and may have caused elevated lead levels in the blood of some residents.[17] These eruptions have also produced nearly continuous lava flows into the ocean that result in steam explosions (Figure 5.31). Haze from the steam explosions contains corrosive hydrogen chloride gas from the vaporized seawater and tiny glass fragments that can irritate and sometimes damage eyes.[18]

DEBRIS FLOWS, MUDFLOWS, AND OTHER MASS MOVEMENTS

The most serious secondary effects of volcanic activity are **debris flows** and **mudflows**, known collectively by their Indonesian name of *lahar*. Lahars are produced when large amounts of loose volcanic ash and other tephra are saturated with water, become unstable, and suddenly move downslope. Debris flows differ from mudflows in that they are coarser; more than half of their particles are larger than

▲ **FIGURE 5.31 LAVA CREATES CORROSIVE STEAM CLOUD** Volcanoes produce a wide range of hazards including toxic gases and corrosive clouds of steam. In this picture, lava from Hawai'i's Kilauea volcano vaporizes seawater at temperatures of more than 1000°C (1800°F). Flowing from the end of a lava tube, the molten rock creates a cloud of steam mixed with corrosive hydrogen chloride gas before it cools in the Pacific Ocean. *(Richard J. Waincoat/Peter Arnold/Still Pictures/Specialist Stock)*

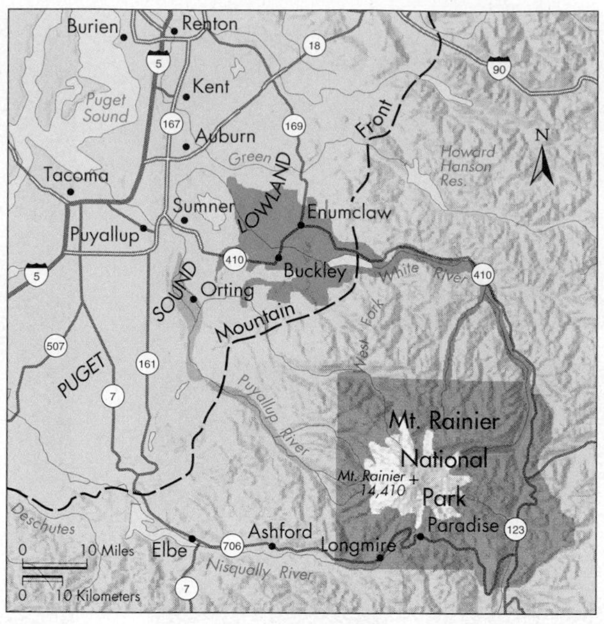

▲ **FIGURE 5.32 MUD FLOWS** Map of Mt. Rainier and vicinity showing the extent of the 5600-year-old Osceola mud flow in the White River Valley (colored orange) and the 560-year-old Electron mud flow (colored beige) in the Puyallup River Valley. *(From Crandell, D. R., and Mullineaux, D. R. U.S. Geological Survey Bulletin 1238)*

sand grains. Unlike pyroclastic flows, lahars can occur without an eruption and are generally low-temperature flows.

Debris Flows Even relatively small eruptions of hot volcanic material may quickly melt large volumes of snow and ice on a volcano. This rapid melting produces a flood of meltwater that erodes the slope of the volcano to create a debris flow. Volcanic debris flows are fast-moving mixtures of fine sediment and large rocks that have a consistency similar to wet concrete. Debris flows can travel many kilometers down valleys from the flanks of the volcano where they formed.[11] For example, early in 1990 a pyroclastic flow from the Redoubt volcano in Alaska moved across Drift Glacier. It rapidly melted snow and ice to produce voluminous amounts of water and sediment. This slurry created a debris flow with a discharge comparable to that of the Mississippi River at flood stage. Fortunately, the event was in an isolated area, so no lives were lost.[10]

Mudflows Gigantic mudflows have originated on the flanks of volcanoes in the Pacific Northwest in both historic and prehistoric times. Two of these, the Osceola mudflow and the Electron mudflow, started on Mt. Rainier (Figure 5.32). Approximately 5000 years ago, the Osceola mudflow moved 1.9 km^3 (0.5 $mi.^3$) of sediment a distance of more than 80 km (50 mi.) from the volcano. This volume would fill The Mall between the U.S. Capitol and the Washington Monument with a pile of debris more than a mile high (10 1/2 times the height of the Washington Monument). Deposits of the younger, 500-year-old Electron mudflow reached about 56 km (35 mi.) from the volcano.

Hundreds of thousands of people now live on the area covered by these old flows, and there is no guarantee that similar flows will not occur again. Many areas southeast of Puget Sound are at potential risk for debris, mud, lava, or pyroclastic flows from Mt. Rainier (Figure 5.33). An observer in the valley would see a debris or mudflow as a wall of mud the height of a ranch house moving toward them at close to 30 km (20 mi.) per hour. With the flow moving at 8.3 m (nearly 30 ft.) per second, the observer would need a car headed in the right direction toward high ground to escape being buried alive.[11]

The USGS has recently developed an automated, solar-powered lahar-detection system for several volcanoes in the United States (e.g., Mt. Rainier), Indonesia, the Philippines, Mexico, and Japan. These systems have acoustic-flow monitors that sense ground vibrations from a moving lahar and can warn people that a flow is moving down the valley. This system replaces visual sightings or cameras that are not reliable in bad weather or at night and require continual maintenance. Because dangerous lahars can occur quickly and with little warning, people in their path must be alerted immediately. Alarms from these USGS lahar-detection systems should give people sufficient time to evacuate to the safety of higher ground.[19]

Landslides Volcanic landslides are another secondary effect of volcanic activity. Like lahars, they may be triggered

CASE STUDY 5.1

Volcanic Landslides and Tsunamis

What may be the largest active landslides on Earth are located on Hawai'i. They are up to 100 km (60 mi.) wide, 10 km (6 mi.) thick, and 20 km (12 mi.) long and extend from a volcanic rift zone on land to an endpoint beneath the sea. Currently these landslides creep along at around 10 cm (4 in.) per year and contain blocks of rock the size of Manhattan Island. The fear is that they might again become a giant, fast-moving submarine debris avalanche. This avalanche would generate a huge tsunami capable of lifting marine debris hundreds of meters above sea level onto nearby islands and causing catastrophic damage around the Pacific Basin. Fortunately, such high-magnitude events are rare; the average recurrence interval seems to be about 100,000 years or so.[3]

Other huge volcano-related landslides or debris-avalanches have been documented in the Canary Islands, located in the Atlantic Ocean off the western coast of Africa. On Tenerife, the largest island, six huge landslides have occurred during the past several million years (Figure 5.A), the most recent of which took place less than 150,000 years ago. The offshore seafloor north of Tenerife is covered by 5500 km^2 (2150 $mi.^2$) of landslide deposits, an area nearly as large as the state of Delaware and more than twice the land surface of the island.[20]

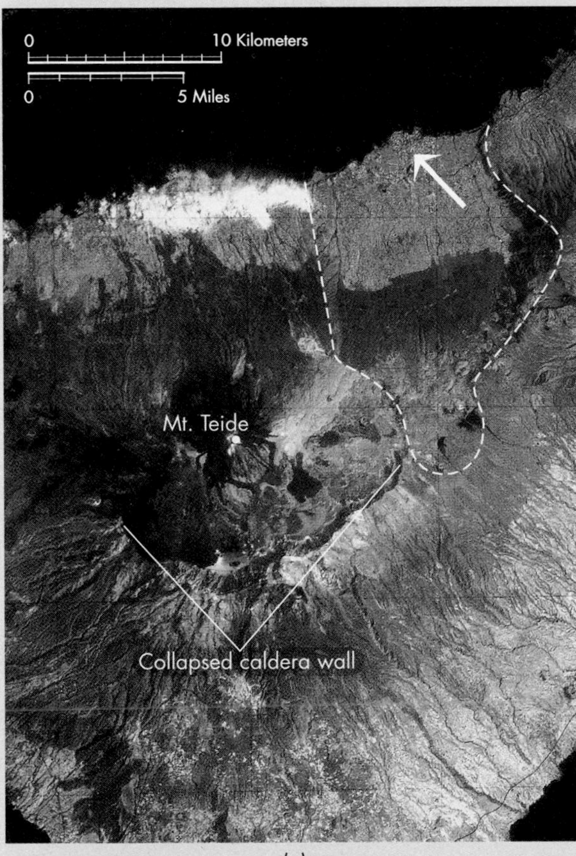

(a)

(b)

▲ FIGURE 5.A **GIANT LANDSLIDE ON VOLCANO** (a) Aerial view of part of the island of Tenerife, the largest island of the Canary Islands in the Atlantic Ocean. Labeled are the collapsed La Cañadas caldera and the Mt. Teide volcano (elevation 3.7 km [12,100 ft.]) The white dashed line outlines the extent of the Orotova landslide, one of many large landslides on the island. The white arrow in both photographs (a) and (b) points to the seaward end of this landslide. *(Espagna Instituto Geografico Nacional)* (b) Seaward end of the Orotova landslide is outlined by the white dashed line. Snow-capped Mt. Teide is in the distance. *(Jose Baerea/Espagna Instituto Geografico Nacional)*

by events other than an eruption. Large volcanic landslides may affect areas far from their source (see Case Study 5.1). For example, massive landslide deposits on the sea floor off the coasts of Hawai'i and the Canary Islands are likely to have generated huge tsunamis.[20] These large waves would cause catastrophic damage far from the islands where they originated (see Chapter 4).

MOUNT ST. HELENS 1980–2010: FROM LATERAL BLASTS TO LAVA FLOWS

The May 18, 1980, eruption of Mount St. Helens in the southwest corner of Washington State exemplifies the many types of volcanic events expected from a Cascade volcano (Figure 5.34). The eruption, like many other natural events, was unique and complex, making generalizations

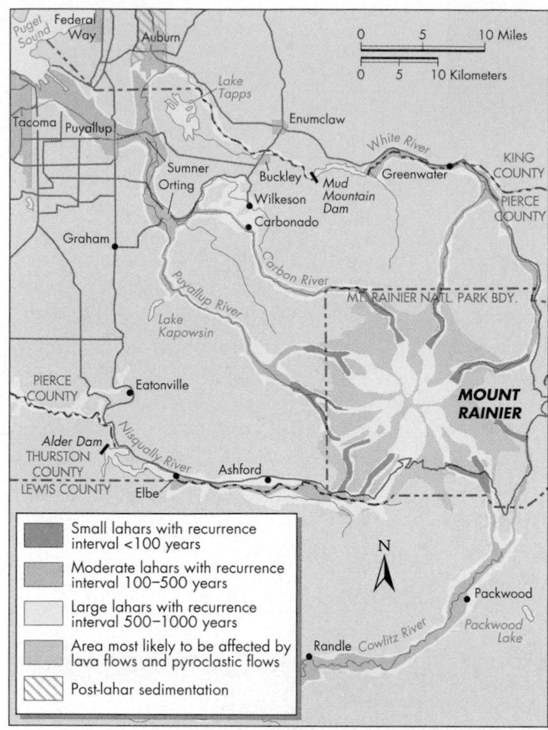

FIGURE 5.33 MUD FLOW HAZARD POTENTIAL Map of Mt. Rainier and vicinity showing potential hazards from lahars, lava flows, and pyroclastic flows. The Seattle/Tacoma suburbs of Puyallup, Sumner, Orting, and Auburn are all in potential hazard areas. *(Hoblitt and others, 1998, USGS Open-File Report 98-428)*

somewhat difficult. Nevertheless, we have learned a great deal from Mount St. Helens, and the story is not yet complete.

Mount St. Helens awoke in March 1980, after 120 years of dormancy, with seismic activity and small explosions created by the boiling of groundwater as it came in contact with hot rock. By May 1, a prominent bulge on the northern flank of the mountain could clearly be observed, and it grew at a rate of about 1.5 m (5 ft.) per day (Figure 5.35a). Only a few weeks later, at 8:32 A.M. on May 18, 1980, a magnitude 5.1 earthquake centered below the volcano triggered a large landslide/debris avalanche of approximately 2.3 km^3 (0.6 mi.^{3}) of earth material.[21] The avalanche, which involved the entire bulge area (Figure 5.34b), shot down the north flank of the mountain, displacing water in nearby Spirit Lake. It then struck and overrode a ridge 8 km (5 mi.)

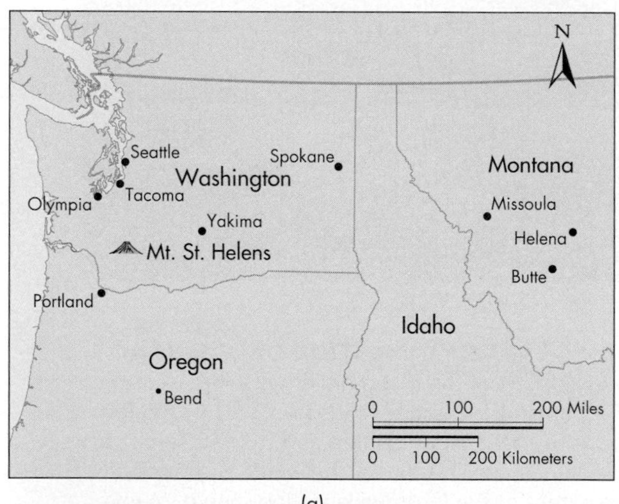

(a)

(b)

(c)

FIGURE 5.34 MOUNT ST. HELENS BEFORE AND AFTER (a) Location of Mount St. Helens, (b) Mount St. Helens before and *(Bruce Spainhower/Washington State Tourism Development)* (c) after the May 18, 1980, eruption. During the eruption, much of the northern side of the composite volcano was blown away, and the altitude of the summit was reduced by approximately 400 m (1314 ft.). The lateral blast shown in Figure 5.34 originated in the amphitheater-like area in the top center of the photograph. *(Harry Glicken/Washington State Tourism Development)*

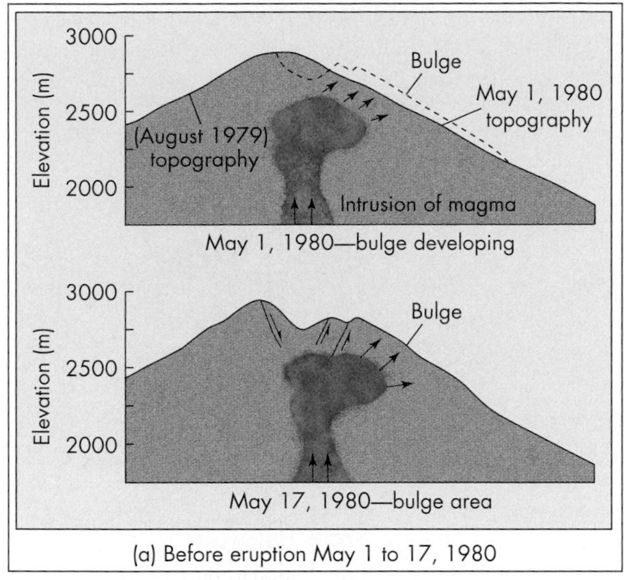

(a) Before eruption May 1 to 17, 1980

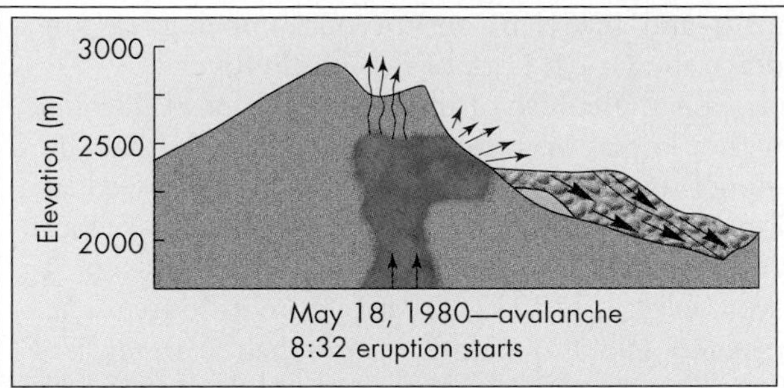

(b) Eruption starts May 18,1980

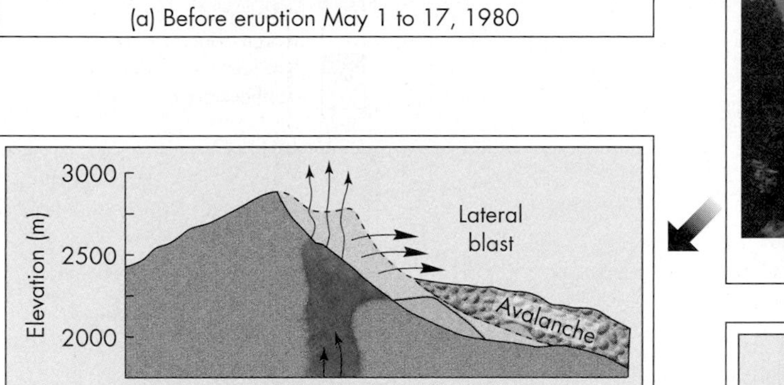

(c) Seconds after eruption starts

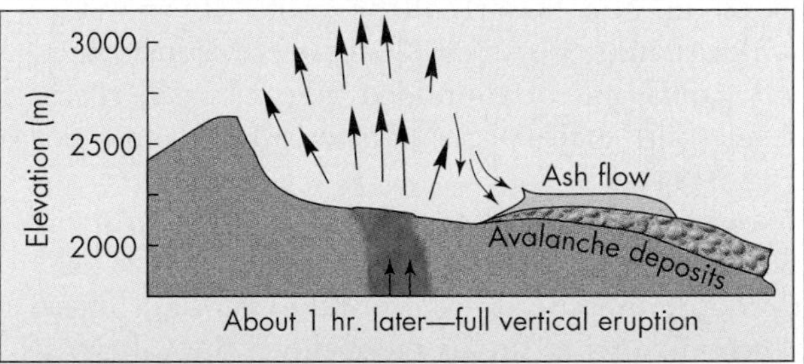

(d) About an hour after eruption starts

▲ FIGURE 5.35 **MOUNT ST. HELENS ERUPTS** Diagrams and photographs showing the sequence of events for the May 18, 1980, eruption of Mount St. Helens. Photographs (b) *(Keith B. Ronnholm)* and (c) of the lateral blast were taken less than 10 seconds apart. *(Keith B. Ronnholm)* (d) *(Roger Werth/Woodfin Camp & Associates, Inc.)*

to the north and made an abrupt turn, moving 18 km (11 mi.) down the North Fork of the Toutle River.

Seconds after the failure of the bulge, Mount St. Helens erupted with a lateral blast directly from the area that the bulge had occupied (Figure 5.35c). The blast moved at speeds of more than 480 km (300 mi.) per hour. This velocity was greater than that of the fastest bullet train. Effects of the blast were felt nearly 30 km (19 mi.) from its source. The blast devastated about 600 km^2 (more than 230 $mi.^2$).[22] Debris-avalanche deposits, blasted-down timber, scorched timber, pyroclastic flows, and mudflows from the blast covered large areas around the volcano (Figure 5.35a).

Within an hour after the lateral blast, a large vertical cloud had quickly risen to an altitude of approximately 19 km (12 mi.), extending more than 4 km (2.5 mi.) into the stratosphere (Figure 5.35d). Eruption of the vertical column continued for more than 9 hours, and large volumes of volcanic ash fell on a wide area of Washington, northern Idaho, and western and central Montana. During the 9-hour eruption, a number of pyroclastic flows swept down the northern slope of the volcano. The total amount of volcanic ash ejected was about 1 km^3 (0.26 $mi.^3$), and a large cloud of ash moved over the United States, reaching as far east as New England (Figure 5.36). In less than 3 weeks, the ash cloud had circled the Earth.

The entire northern slope of the volcano, which is the upper part of the North Fork of the Toutle River watershed, was devastated. Forested slopes were transformed into a gray landscape of mounded volcanic ash, rocks, blocks of melting glacial ice, narrow gullies, and hot steaming pits (Figure 5.37).[21]

The first of several mudflows consisted of a mixture of water, volcanic ash, rock, and organic debris, such as logs, and occurred minutes after the start of the eruption. These flows and accompanying floods raced down the valleys of both forks of the Toutle River at estimated speeds of 29 to 55 km (18 to 34 mi.) per hour, threatening the lives of people camped along the river.[21]

On the morning of May 18, 1980, two young people on a fishing trip on the North Fork of the Toutle River were sleeping about 36 km (22 mi.) downstream from Spirit Lake. They were awakened by a loud rumbling noise from the river, which was covered by felled trees. The campers attempted to run to their car, but water from the rising river poured over the road, preventing their escape. A mass of mud then crashed through the forest toward their car, and the couple climbed on its roof to escape the mud. They were safe only momentarily, however, as the mud pushed the vehicle over the bank and into the river. Leaping off the roof, they fell into the river, which was by now a rolling mass of mud, logs, collapsed train trestles, and other debris. The temperature of the water was increasing as one of the young people got trapped between logs. They disappeared several times beneath the flow but were lucky enough to emerge again. The two were carried downstream for approximately 1.5 km (0.9 mi.) before another family of campers spotted and rescued them.

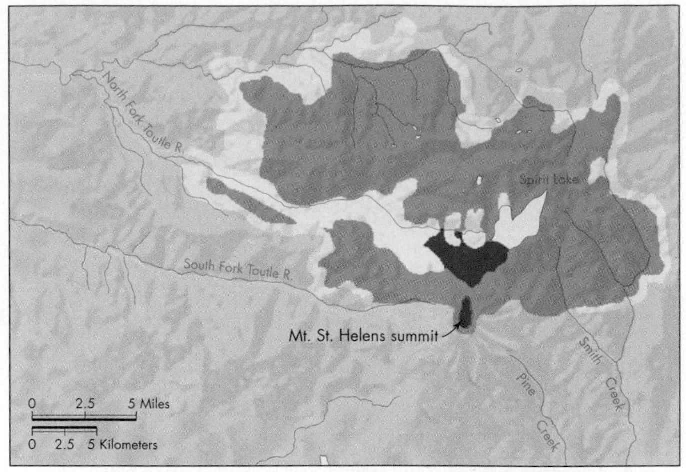

(a)

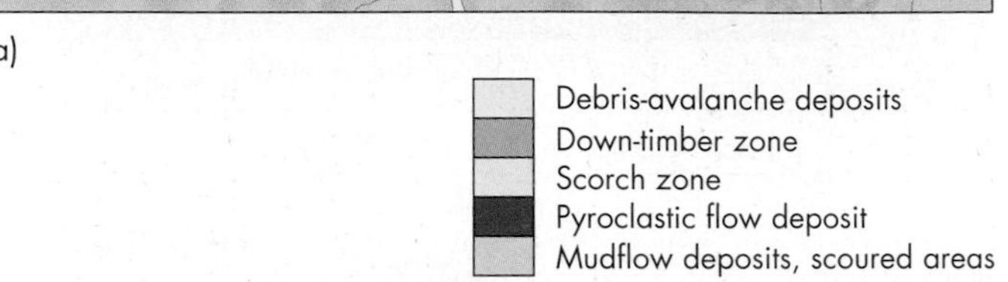

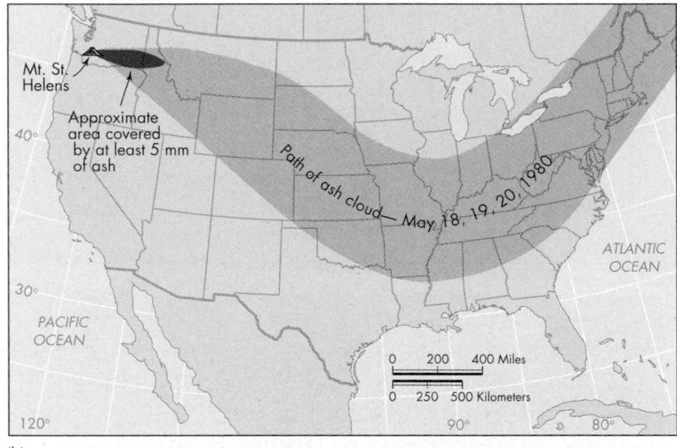

(b)

▲ FIGURE 5.36 **DEBRIS AVALANCHE AND ASH CLOUD** (a) Mount St. Helens's debris-avalanche deposits, zones of trees blown down or scorched, mud flows, and pyroclastic flow deposits associated with the May 18, 1980, eruption; and (b) path of the ash cloud (orange) from the 1980 eruption. Area covered by at least 5 mm (0.2 in.) is shown in red in the upper-left corner. *(Data from various U.S. Geological Survey publications)*

When the volcano could be viewed again following the eruption, the top of the mountain had been reduced by about 400 m (1314 ft.). What was originally a symmetrical volcano was now a huge, steep-walled amphitheater facing northward, and what was originally Spirit Lake (lower Figure 5.33b) was now filled with sediment (Figure 5.33c). The debris avalanche, horizontal blast, pyroclastic flows, and mudflows devastated an area larger than Chicago, Illinois. The eruption killed 57 people and associated flooding destroyed more than 100 homes. Approximately 800 million board feet of timber were flattened by the blast—enough wood to build 50,000 homes. Total damage from the eruption exceeded $1 billion.

◀ **FIGURE 5.37 BARREN LANDSCAPE PRODUCED BY ERUPTION** Desolate and barren landscape produced by the May 18, 1980, eruption of Mount St. Helens. A debris-avalanche and debris-flow deposit fills the North Fork of the Toutle River valley in the lower half of the photograph. Slopes in the right half of the photograph that appear to be sparsely forested are actually areas with scorched, but still standing, trees that were seared by the lateral blast. *(John S. Shelton/University of Washington Libraries)*

Both volcanic and seismic activity have continued since the deadly 1980 eruption. During the first 6 years following the eruption, 19 smaller eruptions occurred. Lava flows from these eruptions began building a lava dome above the 1980 crater floor (Figure 5.38). Between 1989 and 2004, earthquake activity increased, and the quakes were sometimes accompanied by small explosions, ash flows, and mudflows.

On September 23, 2004, Mount St. Helens came back to life when magma began moving up toward the crater floor. Construction of the lava dome resumed and continued until Jaunuary of 2008. An estimated 0.5 m^3 (0.6 $yd.^3$) of lava, about the volume of a five-drawer file cabinet, is being added to the dome every second.[23] The lava dome now rises 350 m (1155 ft.) above the crater floor, close to the height of the Empire State Building.[22]

Scientists at the USGS Cascade Volcano Observatory in Vancouver, Washington, monitor Mount St. Helens at all times. This monitoring includes a network of automated seismographs and global positioning system (GPS) satellite receivers to detect earthquakes and ground deformation, aerial gas sampling, acoustic monitors for mudflows, and surveillance with Webcams.[23]

By 1998, 18 years following the main eruption, life had returned to the mountain and its surrounding area, making many places green once more (Figure 5.39). However, the mounds of landslide deposits are still prominent, a reminder of the catastrophic event of 1980. Mount St. Helens National Monument now has two visitor centers and has attracted more than 1 million visitors.

5.4 Linkages between Volcanoes and Other Natural Hazards

Volcanoes are intimately linked to their physical environment as well as to several other natural hazards. We have stressed the relationship between volcanoes and plate tectonics, the connection between plate tectonic setting and the type of magma, and the influence that magma type has on the nature of volcanic eruptions. Volcanoes are also directly related to other natural hazards such as fire, earthquakes, landslides, and climate change.

Although volcanoes are not a major cause of fires, it isn't difficult to imagine that a linkage between these hazards exists. As molten lava pours down the sides of a volcano, or pyroclastic debris is ejected, plants and human-built structures commonly catch fire. In fact, the Hawaiian Volcano Observatory on the Big Island of Hawai'i warns tourists of the fires caused by Kilauea's active lava flows. These fires can start from explosions of methane gas that form as

◀ **FIGURE 5.38 CONTINUED GROWTH OF DOME IN CRATER** Steam and volcanic gases rise from cooling lava being added to a dome in Mount St. Helens's crater. Growth of this dome occurred from 1980 to 1986 and began again in 2004. Heat from the new growth on the dome melts snowfall. The opening in the crater wall on the lower right is where the avalanche and lateral blast occurred in 1980 (Figure 5.34) *(U S Geological Survey)*

vegetation is intensely heated by the lava. In January 2002, Africa's most destructive volcanic eruption in 25 years sent residents of the Democratic Republic of Congo fleeing raging fires ignited by lava flows (Figure 5.40). Sadly, the lava also sparked an explosion, killing 60 people.

Earthquakes commonly accompany or precede volcanic eruptions as magma rises through the Earth's crust. For example, weeks of earthquakes preceded the appearance of the Parícutin volcano 320 km (200 mi.) west of Mexico City. Some earthquakes may be large enough to do damage independent of the volcano.

Landslides are possibly the most common side effect of volcanic activity. These mass movements can be part of the slope or flank of the volcano (see Case Study 5.1) or lahars, the previously discussed volcanic debris flows and mudflows. Both types of mass wasting have the potential to do great damage and take many lives.

Last, volcanic eruptions may affect our global climate. The best-known example of climate change from a volcanic eruption occurred following the great eruption of the Tambora volcano in Indonesia in 1815. The eruption created a global cloud of sulfuric acid droplets, referred to as aerosols, in the stratosphere, which have been detected in glacial ice formed that year in both Greenland and Antarctica.[25] This cloud caused regional cooling of up to

▲ **FIGURE 5.39 THE YEARS OF RECOVERY** This field of lupine flowers on the slopes of Mount St. Helens shows that after 20 years, the landscape was recovering from the 1980 eruption. *(Gary Braasch/Woodfin Camp & Associates, Inc.)*

▲ **FIGURE 5.40 LAVA IGNITES FIRES IN AFRICA** A lava flow erupting from the Nyiragongo volcano in the Democratic Republic of Congo ignited this fire in January 2002. More than 400,000 people were displaced by fires and lava flows in the city of Goma. *(AFP/Getty Images)*

◀ **FIGURE 5.41 VOLCANO TEMPORARILY COOLS GLOBAL CLIMATE** An eruption of Mt. Pinatubo in 1991, shown here, ejected vast amounts of volcanic ash and sulfur dioxide up to about 30 km (19 mi.) into the atmosphere. Extremely fine particles, called aerosols, from the eruption remained in the upper atmosphere and circled the Earth for more than a year. This aerosol cloud temporarily lowered the average global temperature. The Mt. Pinatubo eruption was the second largest in the twentieth century. *(D. Harlow/U.S. Geological Survey, Denver)*

1°C (2°F) and resulted in 1816 being called the "year without a summer" in New England. Climate change from the eruption caused major hardships in both North America and Europe, including crop failures, famine, and disease.[26] A smaller, but still significant, climatic cooling occurred after the 1991 eruptions of Mount Pinatubo in the Philippines. The cloud of ash and sulfuric acid aerosol from the Pinatubo eruption remained in the atmosphere for more than a year (Figure 5.41). Ash particles and aerosol droplets scattered incoming sunlight and slightly cooled the global climate for two years following the eruptions.[27]

5.5 Natural Service Functions of Volcanoes

Although volcanoes pose a serious threat to those who live in their paths, like most other hazards, they also provide important natural service functions. Perhaps their greatest gift to us occurred billions of years ago when gases and water vapor released from volcanoes contributed to our atmospheric and hydrologic systems, allowing life, as we know it, to evolve. Additionally, volcanoes provide us with fertile soils, a source of power, mineral resources, and recreational opportunities, as well as the creation of new land.

VOLCANIC SOILS

From an agricultural perspective, volcanic eruptions are quite valuable, providing an excellent growth medium for plants. The nutrients produced by the weathering of volcanic rocks allow crops such as coffee, maize, pineapple, sugar cane, and grapes to thrive in volcanic soils. However, rich, fertile soils produced by volcanoes encourage people to live in hazardous areas. So although volcanic soils provide an important resource, nearby volcanic activity can make it difficult to use that resource safely.

GEOTHERMAL POWER

Another benefit provided by volcanoes is their potential for geothermal power. The internal heat associated with volcanoes may be used to create power for nearby urban areas. In fact, volcanic energy is being harnessed by geothermal power in Kilauea, Hawai'i; Santa Rosa, California; and Long Valley, California. An important benefit of geothermal energy is that it can be a renewable resource. This means that unlike fossil fuels, it can be used at a rate that doesn't outpace its replenishment. However, care must be taken that the heat and/or steam driving the system is not removed faster than it can be restored naturally, or it will become depleted, at least temporarily. In addition to using hot geothermal water to make electricity, the heat can be used for heating and industrial processes. The hot water can be pumped directly through a building or utilized with a recirculating heat-exchange system. Reykjavik, Iceland; Paris, France; and Klamath Falls, Oregon, are just a few of the many locations that use geothermal water for heating.[28]

MINERAL RESOURCES

Volcanism is the source and volcanic rocks are the host for many mineral resources. These include economic concentrations of metals such as gold, silver, platinum, copper, nickel, lead, and zinc, and nonmetallic resources, such as pumice, tuff, perlite, scoria, basalt, and volcanic clays. Although these mineral deposits form today in volcanic belts around the world, including underwater at mid-ocean ridges, most of those that we use formed in the geologic past. In the case of metallic mineral resources, most of them are found in much older Precambrian rocks. These older deposits formed at a time when the composition of some magma differed from today's volcanoes. Examples of this include gold accumulations in Precambrian volcanic rocks in Western Australia and Ontario, Canada. In addition to metals, volcanic rocks are used in a wide range of commercial products, including soap, building stone, aggregate for roads and railroads, oil and gas drilling mud, landscaping gravel, ceiling tile, cement, plaster, and cat litter.[29]

RECREATION

Besides being an energy source, the heat associated with volcanoes can also provide recreational opportunities. Many health spas and hot springs are developed in volcanic areas. Volcanoes also provide opportunities for hiking, snow sports, and education. More than 1 million tourists visit the Kilauea volcano each year, many of whom come to observe the volcano during eruption (Figure 5.42).

CREATION OF NEW LAND

In our discussion of the benefits of volcanoes, we should not forget to mention that they are responsible for creating much of the land we inhabit. In the past several decades, the residents of Hawai'i have been reminded of this as lava flows from Kilauea build deltas of land into the Pacific Ocean (see Figure 5.31). Not only are volcanic processes the major force that builds continents, but also oceanic islands such as Hawai'i and Iceland would not exist without volcanoes!

5.6 Human Interaction with Volcanoes

Unlike earthquakes, volcanoes do not lend themselves to human tinkering. That is to say, there is little we can do to affect the timing and severity of their eruptions. Whereas deep-well disposal may increase the number of earthquakes in an area, and clearing the land for agriculture may contribute to flooding, there does not appear to be any human activity that affects volcanoes. They truly are a hazard that is beyond our control, and the best we can do is attempt to minimize loss of life and property associated with eruptions.

▲ **FIGURE 5.42 ERUPTING VOLCANOES ARE TOURIST ATTRACTIONS** Tourists on the cliff to the right view a lava fountain from the Kilauea volcano on the coast of Hawai'i. *(Corbis)*

5.7 Minimizing the Volcanic Hazard

Forecasting volcanic eruptions is a major component of the goal to reduce volcanic hazards. A "forecast" for a volcanic eruption is a probabilistic statement describing the time, place, and character of an eruption. It is analogous to forecasting the weather and is not as precise a statement as a prediction.[3]

FORECASTING

It is unlikely that we will be able to accurately forecast the majority of volcanic activity in the near future, but valuable information is being gathered about phenomena that occur prior to eruptions. One problem is that most forecasting techniques require experience with actual eruptions before the mechanism is understood. Thus, we are better able to predict eruptions in the Hawaiian Islands because we have had so much experience there.

Forecasting volcanic eruptions uses information gained by

- monitoring seismic activity.
- monitoring thermal, magnetic, and hydrologic conditions.
- monitoring the land surface to detect tilting or swelling of the volcano.
- monitoring volcanic gas emissions.
- studying the geologic history of a particular volcano or volcanic center.[10,27]

Seismic Activity Our experience with volcanoes, such as Mount St. Helens and those on the Big Island of Hawai'i, suggests that earthquakes may provide the earliest warning of an impending volcanic eruption. Shallow earthquakes and tremors are produced below a volcano as upward-moving magma fractures the surrounding rock and gas bubbles in the magma form and burst. In the case of Mount St. Helens, earthquake activity started in mid-March before the eruption in May. Activity in March began suddenly with nearly continuous, shallow swarms of earthquakes. Unfortunately, there was no additional increase in earthquakes immediately preceding the catastrophic eruption on May 18. In Hawai'i, earthquakes have been used to monitor the movement of magma as it approaches the surface.

Several months prior to the 1991 Mt. Pinatubo eruptions, small steam explosions and earthquakes began.[3] Unlike Mount St. Helens, Mt. Pinatubo was an eroded ridge that did not have the classic shape of a volcano. Furthermore, it hadn't erupted in 500 years, so the majority of people living near it did not even know it was a volcano! After the initial steam explosions, scientists began monitoring earthquakes on the volcano and studying past volcanic activity, which was determined to have been explosive. Earthquakes increased in number and magnitude prior to the catastrophic eruption, and foci migrated from deep beneath the volcano to shallow depths beneath the summit.[3]

Recently, geophysicists have proposed a generalized model for seismic activity that may help in predicting eruptions.[30] This model is for explosive composite volcanoes, such as those in the Cascade Mountains, which may awaken after an extended period of inactivity (Figure 5.43). In a reawakening volcano, the magma must fracture and break previously solidified igneous rock above the magma chamber to work its way to the surface. Several weeks before reawakening, increasing pressure creates numerous fractures in the plugged volcanic conduit above the chamber. At first, the increase in seismic events will be gradual and a seismologist may need 10 days or so to confidently recognize an accelerating trend toward an eruption (Figure 5.43). However, once the trend has been recognized, there will still be several days before the eruption occurs. Unfortunately, this short warning time may be insufficient for a large-scale evacuation. Thus to forecast eruptions, it may be best to use seismic activity in concert with other eruption precursors discussed below. It is fortunate that, in contrast to earthquakes, volcanic eruptions always provide warning signs.[31]

Thermal, Magnetic, and Hydrologic Monitoring

Prior to a volcanic eruption, a large volume of magma accumulates in a holding reservoir beneath the volcano. The hot material changes the local magnetic, thermal, hydrologic, and geochemical conditions. As the surrounding rocks heat, the rise in temperature of the surficial rock may be detected by satellite remote-sensing or infrared aerial photography. Increased heat may melt snowfields or glaciers; thus, periodic remote sensing of a volcanic chain may detect new hot places related to volcanic activity. This

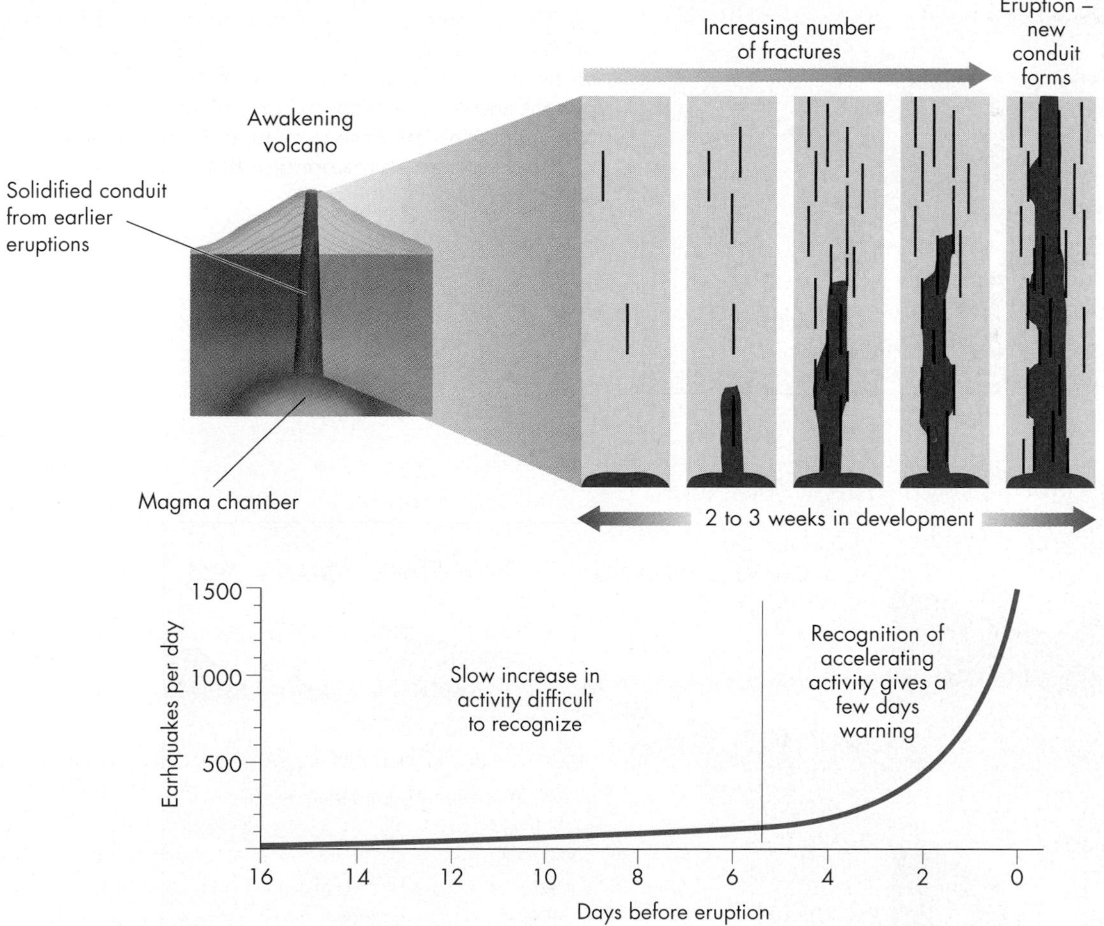

▲ **FIGURE 5.43 A VOLCANO REAWAKENS** Increased seismic activity is a good indicator of a forthcoming volcanic eruption. As a dormant volcano reawakens, rising magma fractures rock above. At first, the fracturing slowly increases the rate of seismic activity; then both the fracturing and seismic activity accelerate a few days prior to an eruption. *(Modified after Kilburn, C. R. J., and Sammonds, P. R. 2005. Maximum warning times for imminent volcanic eruptions. Geophysical Research Letters 32:L24313, doi 10.1029/2005GL024184)*

method was used with some success at Mount St. Helens prior to the main eruption on May 18, 1980.

When older volcanic rocks are heated by new magma, magnetic properties, originally imprinted when the older rocks cooled and crystallized, may change. These changes can be detailed by ground or aerial magnetic surveys.[10,32]

Land Surface Monitoring Monitoring changes in the land surface and seismic activity of volcanoes has been useful in forecasting some volcanic eruptions. Hawaiian volcanoes, especially Kilauea, have supplied most of the data. The summit of Kilauea tilts and swells prior to an eruption and subsides during the actual outbreak (Figure 5.44). Kilauea also experiences swarms of small earthquakes from the underground movement of magma shortly before an eruption. The tilting of the summit, in conjunction with earthquake swarms, was used to predict a volcanic eruption in the vicinity of the farming community of Kapoho on the flank of the volcano 45 km (28 mi.) from the summit. As a result, the inhabitants were evacuated before the event, in which lava overran and eventually destroyed most of the village.[33] Because of the characteristic swelling and earthquake activity before eruptions, scientists expect the Hawaiian volcanoes to continue to be more predictable than others. Monitoring of ground movements such as tilting, swelling, and opening of cracks or of changes in the water level of lakes on or near a volcano can identify movement that might indicate a forthcoming eruption.[10] Today, satellite-based radar and a network of GPS satellite receivers can be used to monitor change in volcanoes, including surface deformation, without sending people into a hazardous area.[34]

(a)

◀ **FIGURE 5.44 INFLATION AND TILTING BEFORE ERUPTION** (a) Idealized diagram of the Kilauea volcano in Hawai'i illustrating inflation and surface tilting as magma moves up. The red area is the underground magma chamber that fills before an eruption *(Solarfilma ehf)* (b) Graphs showing the tilting of the surface of Kilauea in two directions, the east-west component and the north-south component, from 1964 to 1966. Notice the slow increase in ground tilt before an eruption and rapid subsidence, or lowering of the land surface, during eruption. *(James R. Andrews)*

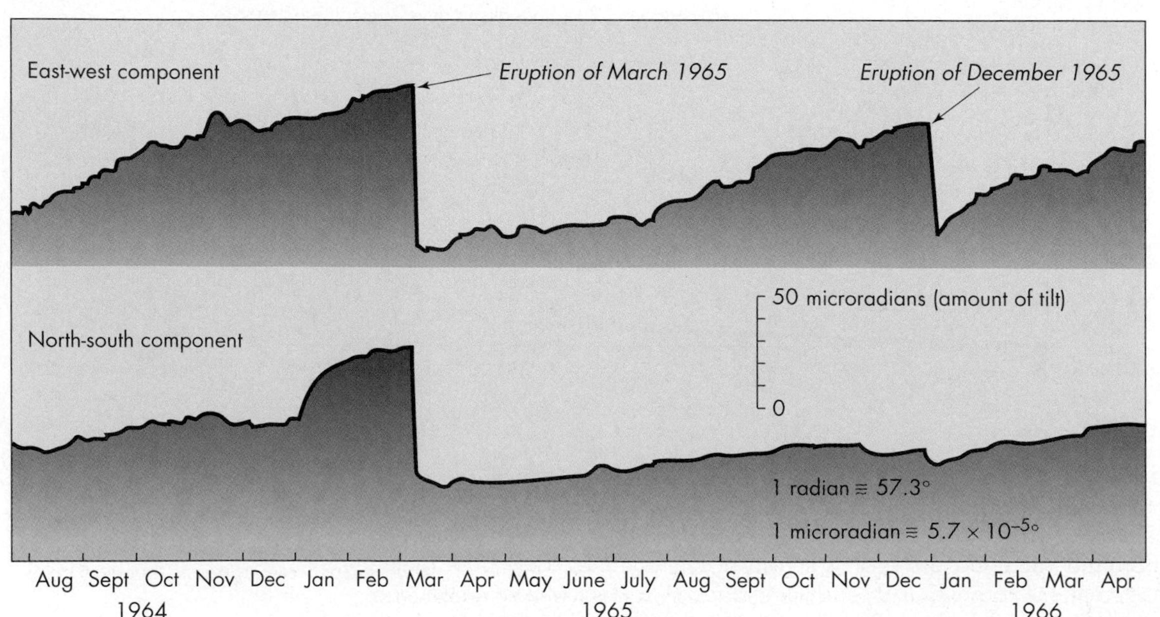

(b)

Monitoring Volcanic Gas Emissions The primary objective of monitoring volcanic gas emissions is to recognize changes in chemical composition (see Professional Profile 5.2). Changes in the relative amounts of carbon dioxide and sulfur dioxide, as well as changes in gas emission rates, are thought to correlate with subsurface volcanic processes. These changes may indicate movement of magma toward the surface. This technique was useful in studying eruptions at Mount St. Helens and Mt. Pinatubo. The volume of sulfur dioxide emitted from Mt. Pinatubo increased more than a million times 2 weeks prior to the explosive eruptions.

Geologic History An understanding of the geologic history of a volcano or volcanic system is useful as one tries to predict the types of future eruptions. The primary tool for reconstructing geologic history is geologic mapping of volcanic rocks and deposits. Attempts are made to determine the age of past lava flows and pyroclastic deposits. Geologic mapping, in conjunction with the dating of volcanic deposits at Kilauea, led to the discovery that more than 90 percent of the land surface of the volcano has been covered by lava in only the past 1500 years. The town of Kalapana, destroyed by lava flows in 1990, might never have been built if this information had been known prior to development, because the risk might have been thought too great. The real value of geologic mapping and dating of volcanic events is that it allows the preparation of hazard maps to assist in land-use planning and disaster preparedness.[10] Such maps are now available for a number of volcanoes around the world.

VOLCANIC ALERT OR WARNING

At what point should the public be alerted or warned that a volcanic eruption may occur? This question has partially been addressed by volcanologists and policy makers. The USGS recently established an alert notification system for volcanic activity for use by its five volcanic observatories.[35] This system has two components: ground-based volcano alert levels and aviation-based color code levels (Table 5.3). Each component has four levels and for most monitored volcanoes and eruptions, the alert and aviation code will be at the same level.[35] For some eruptions, however, the hazard posed to either those on the ground or in the air will differ and higher alerts or codes will be issued. For example, if a lava flow from the Kilauea volcano in Hawai'i threatens a subdivision, the Hawaiian Volcano Observatory may issue a Warning because a hazardous eruption is underway, but only an Orange aviation code because there are minor volcanic-ash emissions. Likewise, if a volcano erupts on an isolated island in Aleutians, the Alaska Volcano Observatory may issue a Watch because the eruption poses limited hazards on the ground, but a Red aviation code because of significant emissions of volcanic ash.

Although this system is a good start, the hard questions remain: When should evacuation begin? When is it safe for people to return? These are questions that public officials will have to answer when the USGS issues volcanic watches or warnings. The USGS volcanic observatories monitor approximately 50 of an estimated 170 active volcanoes in Alaska, Hawai'i, the western contiguous United States, and the Northern Mariana Islands.[35]

TABLE 5.3

U.S. Geological Survey Volcanic Alert Levels and Aviation Color Codes[a]

Ground Alert Level	Volcanic Condition	Aviation Color Code
NORMAL	(1) Typical background, noneruptive state. *Or* (2) If downgraded from higher alert level, activity has ceased and returned to a background, noneruptive state.	GREEN
ADVISORY	(1) Elevated unrest above known background level. *Or* (2) If downgraded from a higher alert level, activity has decreased significantly with close monitoring for possible renewed increase.	YELLOW
WATCH	(1) Heightened or escalating unrest with increased potential of eruption, time frame uncertain. *Or* (2) Eruption underway with limited hazards; for aviation—no or minor volcanic ash emissions.	ORANGE
WARNING	Hazardous eruption imminent, underway, or suspected; for aviation—significant emission of volcanic ash into atmosphere.	RED

[a] Note: For most eruptions, the ground alert level and aviation color code will be at the same levels; however, for some eruptions there will be a greater hazard on the ground or to aviation and different levels will be assigned for the two environments.

Source: Modified from http://volcanoes.usgs.gov/2006/warnschemes.html and U.S. Geological Survey Fact Sheet 2006-3139.

PROFESSIONAL PROFILE 5.2

Chris Eisinger, Studying Active Volcanoes

Chris Eisinger (Figure 5.B) describes volcanoes the same way he might discuss old, temperamental friends. "For the most part volcanoes are very approachable," he says, "as long as you know their cycles." As a graduate student in the Department of Geological Sciences at Arizona State University, Eisinger had 10 years under his belt of firsthand experience with volcanoes. He began as an undergraduate and spent time at the Hawaiian Volcano Observatory and in Indonesia, a region rife with volcanic activity.

But the summit of an active volcano is anything but a breezy affair. On the contrary, volcanoes emit malodorous gases, as Eisinger can attest.

"The most distinct things you notice are the fumes," he says. "You get a strong sulfur smell."

If the volcano is near enough to the ocean for lava to flow into the sea, Eisinger also says the runoff creates a steam that is highly acidic, thanks to the high chlorine content of seawater. "Your eyes will water," he said. "It's typically difficult to breathe and you end up coughing if you're not wearing a gas mask."

Rainstorms can further cloud the air, creating steam when the falling water strikes the hot surface of the lava. At times, Eisinger says, visibility for volcanologists studying at the summit was reduced to a few feet.

The heavy influx of gases doesn't affect only the eyes and nose. The taste, if one isn't wearing a mask, can be "pretty nasty," Eisinger says, adding that the smell adheres to one's clothing as well.

For volcanoes that are in a constant state of eruption, Eisinger said there are distinct patterns in the emission cycles. Volcanoes he has visited in Indonesia, for example, would emit a stream of gaseous material every 20 minutes to an hour.

Although most volcanologists are able to remain safe by paying close attention to eruption cycles, fatalities have happened. In August 2000, two volcanologists died at the summit of Semeru on the island of Java in Indonesia when the volcano erupted with no warning.

Active volcanoes also emit low, rumbling sounds, Eisinger said, due largely to subterranean explosions that tend to be muffled. Hawaiian volcanoes are unique in that they also emit an intense hissing sound, "like a jet engine," from the expulsion of gases. Eruptions are often foreshadowed by seismic activity that contributes to the rumble of the explosions. Although Eisinger himself has never witnessed a volcanic eruption of historic order, he noted that accounts of such colossal eruptions as that of Krakatau in 1883 produced reports of a boom that could be heard thousands of miles away.

—CHRIS WILSON

▲ FIGURE 5.B **VOLCANOLOGIST SAMPLES MOLTEN SULFUR** Volcanologist Chris Eisinger uses a crowbar to hit some molten sulfur on Kawah Ijen volcano in Indonesia. The 200°C (390°F) molten sulfur is still red hot and will cool to a yellow or green color. *(Chris Eisenger)*

5.8 Perception of and Adjustment to the Volcanic Hazard

PERCEPTION OF VOLCANIC HAZARD

Information concerning how people perceive a volcanic hazard is limited. People live near volcanoes for a variety of reasons: (1) They were born there and in the case of some islands, such as the Canary Islands, all land is volcanic; (2) the land is fertile and good for farming; (3) people are optimistic and believe an eruption is unlikely; and (4) they cannot choose where they live—for example, they may be limited by economics. One study of perception in Hawai'i found that a person's age and length of residence near a volcanic hazard are significant factors in his or her knowledge of the hazard and possible adjustments to it.[36] One reason the evacuation of 60,000 people prior to the 1991 eruption of Mt. Pinatubo was successful was that the government had educated people about the dangers of

SURVIVOR STORY 5.3

A Close Call with Mount St. Helens

For someone who makes his living as a reporter, Don Hamilton was awfully lucky to be nowhere near a telephone on the evening of May 17, 1980 (Figure 5.C).

The next day, at 8:32 A.M., Mount St. Helens would erupt in one of the largest eruptions in North American history, spewing monstrous proportions of ash that would eventually circle the entire planet and cover the surrounding region in hundreds of feet of debris.

But that evening, Hamilton was sitting on the porch of a lodge on Spirit Lake, no more than 10 miles from the peak of the mountain, visiting the lodge's resident, a local eccentric named Harry Truman. By this point, smaller earthquakes and seismic observations had geologists clamoring about the imminent eruption, and Truman, then 84 years old, had become a minor celebrity for his stalwart refusal to evacuate.

Hamilton was working for the now-defunct *Oregon Journal* at the time, and that night he might have stayed with Truman if he had had a telephone. Hamilton, who was 26 at the time, needed some way to file that day's story.

As it turned out, leaving was a wise decision. When the volcano erupted the very next morning, with Hamilton safely in bed in Portland, it buried Truman and his lodge in a massive landslide of debris and pyroclastic flow.

In the days and weeks leading up to the eruption, Hamilton had witnessed several violent earthquakes while visiting Truman.

"I was up there a few times in his lodge, up on his front porch, when some pretty good quakes hit," Hamilton said. "I could see the road rippling. It was the weirdest thing I'd ever seen."

By this time, the media had already caught up with Truman, who had appeared on the *Today* show and in the *New York Times*. But for all his gusto, Hamilton said, there were cracks in Truman's resolve.

"I saw some genuine fear in his face," Hamilton said. "He looked drawn. His eyes were bugging out a little bit. It made him stop and pay attention."

But in spite of all the warnings, Hamilton said most people weren't prepared for the sheer magnitude of the eruption.

"The geologists were certainly warning us that this was a very dangerous and unstable situation," he said. "But I don't think anybody really anticipated the enormity of what happened. There wasn't a lot of documented history of this kind of thing."

Natives of the Northwest certainly couldn't have been blamed for lack of first-hand experience. At the time, Mount St. Helens had been dormant for more than 120 years.

The day of the eruption, Hamilton and his brother, a photographer for the *Oregon Journal*, went up in a chartered flight and were lucky enough to have a pilot who was willing to loosely interpret the FAA restriction on the air space around the eruption.

Hamilton described the view from the plane as utterly extraordinary. Prior to the eruption, Mount St. Helens had a rounded peak that had been blown off in the blast. A massive column of volcanic ash rose straight into the cloud ceiling.

"It was unbelievable to see this emerge out of the clouds," he said. "The mountain was completely flattened off. It had been a very beautiful, conical top."

As they got closer, they saw massive mudflows, as well as cars, trucks, and bridges that had been destroyed.

"The road that I'd been going up for two weeks was completely washed out," he said.

For residents of Portland and many other areas near the blast, ash from the eruption became a regular part of life that summer.

"People were wearing surgical masks. It was piled up on the side of the road. It just stayed there all summer," Hamilton said. "For that whole summer you just lived with it."

For the 26-year-old Hamilton, the timing couldn't have been worse.

"It was the only summer I had a convertible," he said. "I kept the top up most of the time that summer."

—CHRIS WILSON

▲ **FIGURE 5.C OREGON REPORTER WHO ESCAPED DEATH ON MOUNT ST. HELENS** A fateful decision on May 17, 1980, to leave Harry Truman's lodge on the slopes of Mount St. Helens and return to Portland saved Don Hamilton's life. Don recounted his story nearly 25 years later when Mount St. Helens erupted again in March 2005. *(Don Hamilton)*

(a)

(b)

(c)

▲ FIGURE 5.45 **PEOPLE FIGHTING LAVA FLOWS** Eruptions of Mount Helgafell on the island of Heimaey, Iceland. (a) Lava fountain at night from the harbor area. (Solarfilma ehf) (b) Aerial view toward the harbor. Escaping white steam appears above the advancing black lava flow in the lower-right part of the photograph. The steam comes from water being applied to cool and slow the advance of the flow. An arching stream of water from a water cannon is also visible in the lower-right corner of the photograph. (*James R. Andrews*) (c) Aerial view showing the front of the blocky lava flow encroaching into the upper-right portion of the harbor. By fortuitous circumstances, the flow stopped at a point that has improved the protection of the harbor from storm waves. (*James R. Andrews*)

violent ash eruptions with debris flows. A video depicting these events was widely shown before the eruption, and it helped convince local officials and residents that they faced a real and immediate threat.[3]

The science of volcanoes is becoming well known. However, good science is not sufficient (see Survivor Story 5.3). Probably the greatest hazard reduction will come from an increased understanding of human and societal issues that arise during an emerging **volcanic crisis**. A volcanic crisis can develop when scientists predict that an eruption is likely in the near future. In such a crisis, improved communication among scientists, emergency managers, educators, media, and private citizens is particularly important. The goal is to prevent a volcanic crisis from becoming a disaster or catastrophe.[34]

ADJUSTMENTS TO VOLCANIC HAZARDS

Apart from the psychological adjustment to losses, the primary human adjustment to volcanic activity is evacuation. Mount Vesuvius, which devastated the cities of Pompeii and Herculaneum in A.D. 79 (see Figure 5.26), is considered by some to be one of the most hazardous volcanoes on Earth with nearly 3 million people living in the surrounding area. In an attempt to minimize the potential for a future catastrophe if another large eruption were to occur, the Italian government is offering 30,000 Euros to anyone living in the 18 towns within the immediate area who is willing to relocate.[37]

ATTEMPTS TO CONTROL LAVA FLOWS

Several methods, such as *hydraulic chilling* and wall construction, have been employed to deflect lava flows away from populated or otherwise valuable areas. These methods have had mixed success. They cannot be expected to modify large flows, and their effectiveness with smaller flows requires further evaluation.

The world's most ambitious hydraulic chilling effort was initiated in January 1973 on the Icelandic island of Heimaey. Basaltic lava flows from Mount Helgafell nearly closed the harbor of Vestmannaeyjar, the island's main town and Iceland's main fishing port. The situation prompted immediate action.

Three favorable conditions existed: (1) Slow movement of the flows allowed the time needed to initiate a control effort; (2) transport by sea and roads allowed for the delivery of pipes, pumps, and heavy equipment; and (3) water was readily available. Initially, the edges and surface of the flow were cooled with water from numerous fire hoses (Figure 5.45).

Then bulldozers were moved up on the slowly advancing flow to make a path for a large water pipe. The plastic pipe did not melt as long as water was flowing in it, and small holes in the pipe allowed the cooling of hot spots along parts of the flow. Watering had little effect the first day, but then the back edge flow began to slow down and in some cases stopped.

These actions undoubtedly had an important effect on the lava flows from Mt. Helgafell. They restricted lava movement, reduced property damage, and allowed the harbor to remain open. When the eruption stopped 5 months later, the harbor was still usable.[38] In fact, by fortuitous circumstances, the shape of the harbor was actually improved because the cooled lava provided additional protection from the sea. More recent efforts to chill and deflect lava flows on the slope of Mt. Etna in Sicily have also had some success.

REVISITING THE FUNDAMENTAL CONCEPTS

Volcanoes

1. **Hazards Are Predictable from a Scientific Evaluation**
2. **Risk Assessment Is an Important Component of Our Understanding of the Effects of Hazardous Processes**
3. **Linkages Exist between Natural Hazards**
4. **Hazardous Events That Previously Produced Disasters Are Now Producing Catastrophes**
5. **Consequences of Hazards May Be Minimized**

1) As with earthquakes, the long-term return period or frequency with which volcanoes erupt can be estimated from geologic evidence of past eruptions. However, a precise prediction of when an eruption will happen is still beyond our grasp, but significant advances in the past 3 decades have gotten us closer than ever before. The 1981 eruption of Mount St. Helens and the 1991 eruption of Pinatubo in the Philippians taught us a lot about the precursors to an eminent volcanic eruption. Together with new technological advances (seismographs, tilt meters, volatile monitoring), we may be able to narrow the potential eruption time for some events to within a few weeks. Yet, as with many natural processes, the culminating variables that lead to the actual event are numerous and often intertwined such that determination of the final causality and the precise timing of the eruption cannot be known. For example, volcanologists at Mount St. Helens were fairly certain that the eruption was eminent, but they could not have known that a large landslide would trigger the eruption nor could they have precisely predicted the timing of the earthquake that triggered the landslide.

2) The risk from volcanic eruptions is the product of the probability of the eruption times the consequences. The volcanic rocks from which the volcano is constructed give the geologist information about the frequency, type, and magnitude of previous eruptions and, consequently, provide insight into potential future events. However, it is the short-term probability based upon monitoring volcanoes and observing changes in them prior to the eruption (precursory events) that will more precisely lead to calculating the probability of eruption in the short time frame of weeks to months. Once we are able to actually predict that a volcanic eruption is likely, we may develop scenarios for consequences, and from those assess the potential cost in terms of property damage and loss of human life. Therefore, within the limits of predicting the eruption, the risk can be calculated for potential future events.

3) Volcanic eruptions are linked to many other natural hazards. Of particular importance are debris flows or mud flows (lahars), floods, and landslides. Large volcanic eruptions on high mountains will melt ice and snow, leading to the development of volcanic mud flows. Volcanic eruption that leads to lava flows that block river valleys or fill them with ash may produce temporary dams that might be overtopped, leading to flooding. Debris flow deposits may also fill valleys and place stress on rivers during times of high flows, leading to additional flood hazard. Development of steep slopes on flanks of volcanoes with loose weak materials can increase the landslide hazard, as can seismic shaking and eruption itself.

4) With ever-increasing population density, volcanic risk certainly increases. For example, in A.D. 79 when Mount Vesuvius erupted, perhaps as much as half of the populations of Pompeii and Herculenium was wiped out (approximately 2000 people). Today, there are close to 3 million people living within the vicinity of Mount Vesuvius![37] Although detailed evacuation plans for the region exist, an unexpected eruption similar to that of A.D. 79 would be catastrophic in terms of property damage and potential loss of human life.

5) In contrast to the example of Parícutin discussed in the text, the locations of most volcanoes are known and with advances in prediction and appropriate evacuation plans, the loss of life may be minimized. In general, we have been successful in recent years in encouraging people to leave once a volcanic

(continued)

warning has been issued. In areas on flanks of volcanoes that regularly might experience ash falls and so forth, buildings can be better des-igned to withstand the weight of ash fallout, and bridges can be designed to pass large volcanic debris and mud flows beneath them. When we are able to foresee the consequences of eruptions, plans may then be developed to minimize the main consequences likely to occur.

QUESTIONS TO THINK ABOUT

1. Given what you know about the effects of a volcanic eruption and the linked natural hazards, where might you build your house if you were living near a composite volcano such as Mount Rainer in Washington?
2. Do you think our current ability to monitor precursory event and predict an eminent eruption is sufficient for large populations to live on the flank of an active volcano? Why or why not?

Summary

Lava is magma that has been extruded from a volcano. Its viscosity, a characteristic related to the temperature and silica content, and gas content are important in determining the eruptive style of the different types of volcanoes. The largest volcanoes, shield volcanoes, are common at mid-ocean ridges, such as Iceland, and over mid-plate hot spots, such as the Hawaiian Islands. They are characterized by relatively nonexplosive lava flows of basalt. Most explosive volcanic eruptions are from the classic, cone-shaped composite volcanoes that occur above subduction zones, particularly around the Pacific Rim. Many of the volcanoes in the Aleutian Islands of Alaska and the Cascade Mountains of the U.S. Pacific Northwest are of this type. These volcanoes are characterized by explosive eruptions and are composed primarily of silica-rich lavas, such as andesite, and pyroclastic deposits. Volcanic domes are generally smaller, sometimes highly explosive volcanoes that occur inland of subduction zones. They include Lassen Peak in California and are composed largely of rhyolite rock.

Features of volcanoes include vents, craters, and calderas. Other features related to volcanic activity are hot springs and geysers. Large calderas are created by infrequent, huge violent eruptions. Following an explosive beginning, they often resurge and may present a volcanic hazard for a million years or longer. Recent uplift and earthquakes at the Long Valley caldera in eastern California are reminders of the potential hazard.

Volcanic activity is directly related to plate tectonics. Most volcanoes are located at plate boundaries, where magma is produced in the spreading or sinking of lithospheric plates. Two-thirds of the world's volcanoes are associated with the sinking of lithospheric plates along the Ring of Fire surrounding most of the Pacific Ocean. Specific geographic regions of North America at risk from volcanoes include the northwest coast of California; the west coasts of Oregon and Washington and parts of British Columbia and Alaska, Long Valley, and the Yellowstone National Park area.

Primary effects of volcanic activity include lava flows, pyroclastic hazards, and, occasionally, the emission of poisonous gases. Hydraulic chilling and the construction of walls have been used in attempts to control lava flows. These methods have had mixed success and require further evaluation. Pyroclastic hazards include volcanic ash falls, which may cover large areas with carpets of ash; pyroclastic flows that move as fast as 160 km (100 mi.) per hour down the side of a volcano; and lateral blasts, which can be very destructive. Secondary effects of volcanic activity include debris flows and mudflows, generated when melting snow and ice or precipitation mix with volcanic ash. These flows can devastate an area many kilometers from the volcano. All these effects have occurred in the recent history of the Cascade Range of the Pacific Northwest and will occur there in the future.

Volcanoes are linked to other natural hazards such as fire, earthquakes, landslides, and climate change. However, they also provide us with natural service functions: fertile soils, a source of power, mineral resources, recreational opportunities, and newly created land.

Efforts to meet the goal of reducing volcanic hazards are focusing on human and societal issues of communication; the objective is to prevent a volcanic crisis from becoming a disaster or catastrophe. Sufficient monitoring of seismic activity; thermal, magnetic, and hydrologic properties; and changes in the land surface, combined with knowledge of the recent geologic history of volcanoes, may eventually result in reliable forecasting of volcanic activity. Forecasts of eruptions have been successful, particularly for Hawaiian volcanoes and Mt. Pinatubo in the Philippines. Worldwide, however, it is unlikely that we will be able to accurately forecast most volcanic activity in the near future.

The U.S. Geological Survey has developed an alert notification system for volcanic activity that has four levels for hazards on land and four color codes for aviation hazards. Hard questions remain concerning when evacuation should begin and when it is safe for people to return.

Perception of the volcanic hazard is apparently a function of age and length of residency near the hazard. Some people have little choice but to live near a volcano. Community-based education plays an important role in informing people about the hazards of volcanoes. Apart from psychological adjustment to losses, the primary human adjustment to volcanic activity is evacuation. Some attempts to control lava flows once an eruption has begun have been successful.

Key Terms

ash fall (p. 140)
caldera (p. 132)
cinder cone (p. 131)
composite volcano (p. 130)
decompression melting (p. 126)
debris flow/mudflow (p. 143)
lateral blast (p. 141)
lava (p. 125)
lava flow (p. 138)
magma (p. 125)
pyroclastic flow (p. 141)
pyroclastic deposit (p. 128)
shield volcano (p. 128)
volatiles (p. 126)
volcanic crisis (p. 158)
volcanic dome (p. 131)
volcanic vent (p. 132)

Review Questions

1. What is magma and where does it come from?
2. What physical and chemical changes occur that cause rocks to melt?
3. What is viscosity, and what determines it?
4. Explain the relationship between magma composition, viscosity, and gas content.
5. List the major types of volcanoes and the type of magma associated with each.
6. Describe the major types of volcanoes and their eruption styles. Why do they erupt the way they do?
7. Explain the relationship between plate tectonics and volcanoes.
8. How do lava tubes form and move magma far from the erupting vents?
9. Explain the relationship between the Hawaiian Islands and the hot spot below the Big Island of Hawai'i.
10. How do geysers work? How can they be hazardous?
11. Explain how large caldera eruptions occur and why they are so dangerous.
12. Describe the primary and secondary effects of volcanic eruptions.
13. What methods have been attempted to control lava flows?
14. Differentiate between ash falls, lateral blasts, and pyroclastic flows.
15. What are the major gases emitted in a volcanic eruption? How can they be hazardous?
16. Explain how volcanoes can produce gigantic debris or mudflows.
17. What kinds of information help geologists forecast volcanic eruptions?
18. Explain the USGS alert notification system for volcanic eruptions.

Critical Thinking Questions

1. While looking through some old boxes in your grandparents' home, you find a sample of volcanic rock collected by your great-grandfather. No one knows where it was collected. You take it to school, and your geology professor tells you that it is a sample of andesite. What might you tell your grandparents about the type of volcano from which it probably came, its geologic environment, and the type of volcanic activity that likely produced it?
2. In our discussion of perception of and adjustment to volcanic hazards, we established that people's perceptions and what they will do in case of an eruption are associated with both their proximity to the hazard and their knowledge of volcanic processes and necessary adjustments. With this association in mind, develop a public relations program that could alert people to a potential volcanic hazard. Keep in mind that the tragedy associated with the eruption of Nevado del Ruiz was in part due to political and economic factors that influenced the apathetic attitude toward the hazard map prepared for that area. Some people were afraid that the hazard map would lower property values in high-risk areas.
3. You are going to take a scout troop to Hawai'i to see the Kilauea volcano. Some children have seen documentaries of the 1980 eruption of Mount St. Helens and are afraid; others are fearless and want to try cooking on a lava flow. What will you tell the fearful scouts? Describe the safety precautions you will follow for all the scouts.

Volcanic Hazard Assessment

THE ISSUE

Three small earthquakes have shaken Hazard City in the past two weeks. Upon learning that the earthquakes had epicenters in the Lava Mountain area, concern among the citizens ranges from indifference to alarm. The mayor realizes that information is the most important requirement for developing an appropriate response.

YOUR TASK

The mayor has hired you and two other geologists to investigate. Your job is to conduct a preliminary hazard assessment for Downtown and Ralston. A good preliminary assessment must consider all of the potentially hazardous products of a volcanic eruption, which for Lava Mountain include tephra, lahars, pyroclastic flows, lava flows, and volcanic gases. It must also include a determination of which part(s) of Hazard City could be affected by each different hazard.

This is not a trivial assignment. Expect to visit every field location, learn about the different parts of Hazard City, and use the links to learn about the different volcanic hazards. Because the geography of the deposits and the town are very important, you may want to print the maps and write notes on them while you do fieldwork and research.

6 Flooding

Mississippi River Flooding 1973–2008

n 1973, spring flooding of the Mississippi River caused the evacuation of tens of thousands of people, as thousands of square kilometers of farmland were inundated throughout the Mississippi River Valley. Fortunately, there were few deaths, but the flooding resulted in approximately $1.2 billion in property damage.[1] The 1973 floods occurred despite a tremendous investment in upstream flood-control dams on the Missouri River, the largest tributary to the Mississippi. Reservoirs behind these dams inundated some of the most valuable farmland in the Dakotas, and despite these structures, the flood near St. Louis was record breaking.[2] Impressive as this flood was at the time, it did not compare either in magnitude or in the suffering it caused with the flooding that occurred 20 and 35 years later.

During the summers of 1993 and 2008, the land around the Mississippi River and its tributaries experienced two of the largest floods in the last hundred years. There was more water, both in 1993 and 2008, than during the 1973 flood, which had a magnitude of flow expected on average once in a hundred years.

Flooding lasted from late June to early August and caused 50 deaths and more than $15 billion in property damage

The 1993 flood resulted from rain storms from April through July in the the Cedar Rapids region, where low areas produced about 900 mm (35 in.) of rain. Flooding lasted from late June to early August and caused 50 deaths and more than $15 billion in property damage. In all, about 55,000 km^2 (21,000 mi.^{2}), including numerous towns and farmlands, were inundated with water.[3–5]

The 2008 floods resulted from winter to late spring storms that in June dumped (in Indiana) as much as 255 mm (9.5 in.) of rain in a 24 hour period. Flood defenses again failed, about 24 people died, and the damage was about $9 billion.

The floodwaters in 1993 were high for a prolonged time, putting pressure on the flood defenses of the Mississippi River, particularly **levees,** which are earthen embankments constructed parallel to the river to contain floodwaters and reduce flooding (Figure 6.1). Levees in the St. Louis region on the Mississippi, Missouri, and Illinois Rivers are shown in Figure 6.2. Notice that levees completely circle some areas on the floodplains.

LEARNING OBJECTIVES

Water covers about 70 percent of Earth's surface and is critical for supporting life on the planet. However, water can also be a significant hazard to human life and property in certain situations, such as a flood. Flooding is the most universally experienced natural hazard. Floodwaters have killed more than 10,000 people in the United States since 1900. For the past decade, property damage from flooding has averaged more than $4 billion per year. Flooding is a natural process that will remain a major hazard as long as people live and work in flood-prone areas. Your goals in reading this chapter should be to

- understand basic river processes.
- understand the process of flooding and the differences between flash floods and downstream floods.
- know what geographic regions are at risk from flooding.
- know the effects of flooding and the linkages with other natural hazards.
- recognize the benefits of periodic flooding.
- understand how people interact with and affect the flood hazard.
- be familiar with adjustments we can make to minimize flood deaths and damage.

◀ **Cumberland River Floods** On Monday, May 3, 2010, after heavy weekend rains and flooding, businesses in Nashville Tennessee stand in flood water from the Cumberland River. *(Mark Humphrey/AP Images)*

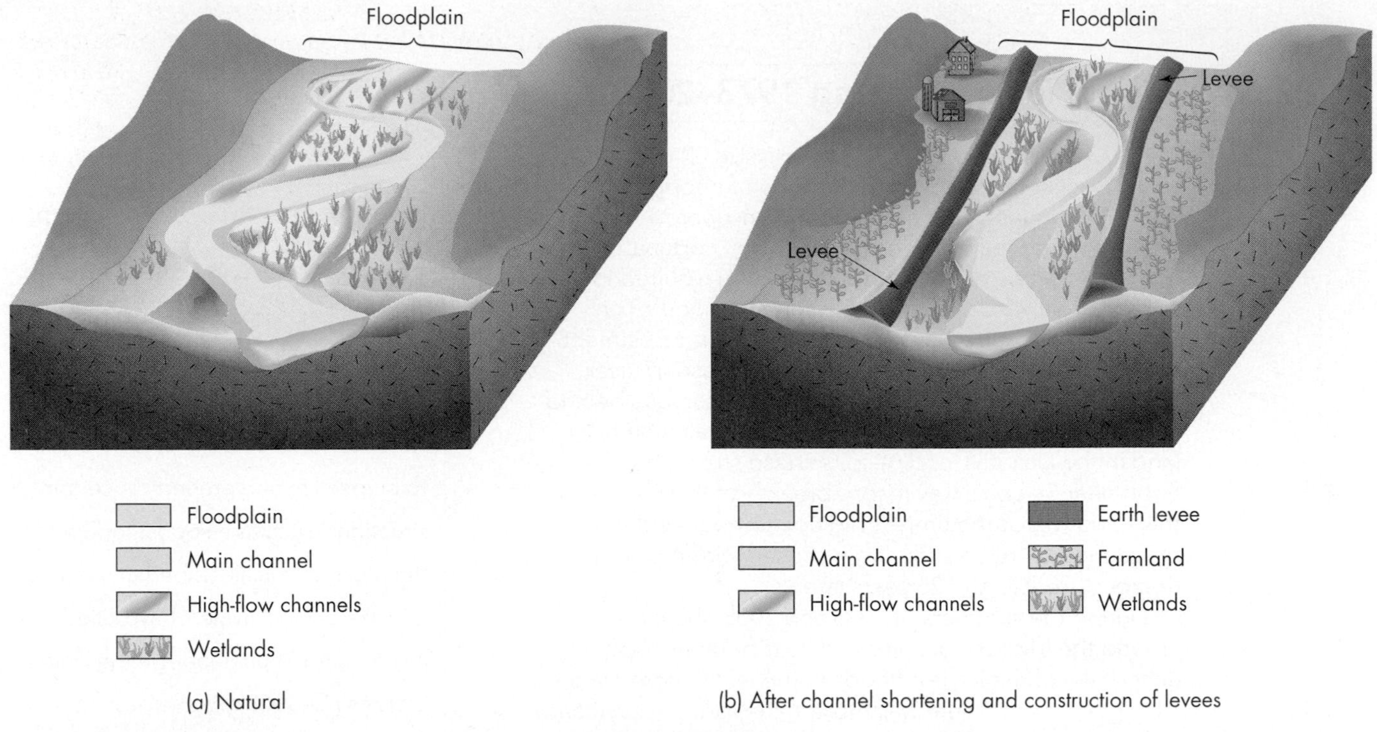

▲ FIGURE 6.1 **FLOODPLAIN WITH AND WITHOUT LEVEES** Idealized diagram of *(Edward A. Keller)* (a) natural floodplain with wetlands. (b) Floodplain after channel is shortened and levees are constructed. Land behind levees is farmed, and wetlands are generally confined between the levees.

Before construction of the levees, the Mississippi's floodplain, flat land adjacent to the river that periodically floods, was much wider and contained extensive wetlands. Since the first levees were built in 1718, approximately 60 percent of the wetlands in Wisconsin, Illinois, Iowa, Missouri, and Minnesota—all hard hit by the flooding in 1993—have been lost. In some urban areas, such as St. Louis, Missouri, levees have been replaced with floodwalls designed to protect the city against high-magnitude floods. The effects of these floodwalls can be seen in a satellite image taken in mid-July 1993 (Figure 6.3). This image shows that the river is narrow at St. Louis, where it is contained by the floodwalls, and broad upstream near Alton, Illinois, where extensive flooding occurred. The floodwalls produce a bottleneck effect—they

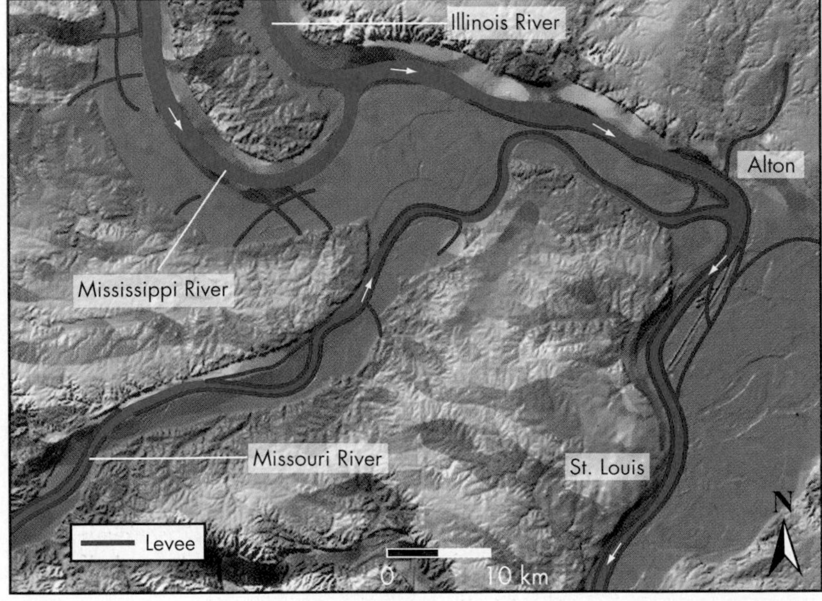

◄ FIGURE 6.2 **LEVEES NEAR ST. LOUIS** Levees almost completely enclose the Mississippi River near the cities of Alton and St. Louis, Missouri. *(Data from R.E. Criss and T. M. Kusky. 2009)*

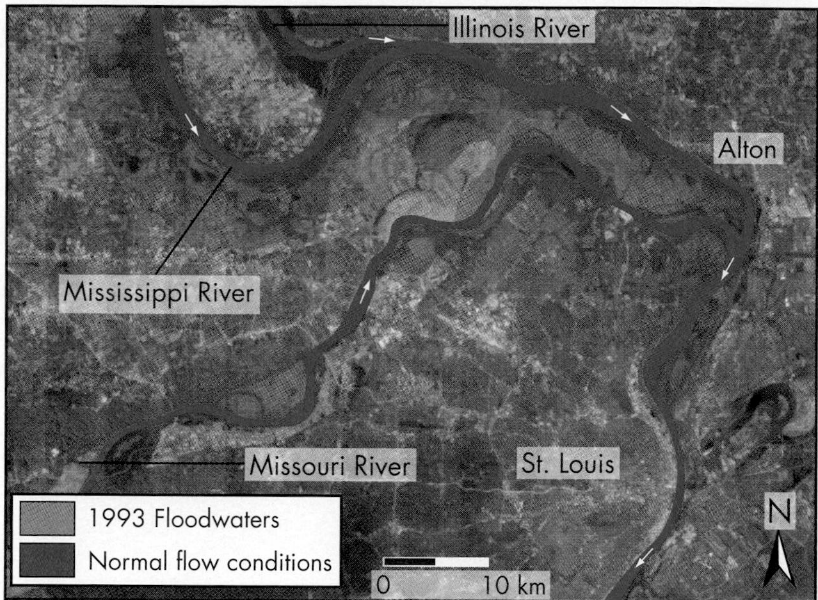

◄ FIGURE 6.3 **MISSISSIPPI RIVER FLOOD OF 1993** Satellite view of the extent of flooding in 1993 where the Illinois and Missouri Rivers join the Mississippi. Light blue is 1993 floodwaters, and dark blue is normal river flow. In the lower right, a series of floodwalls built to protect St. Louis constrict the Mississippi. This bottleneck to the flow caused widespread flooding upstream, including in the town of Alton, Illinois, on the east bank of the river near the right center edge of the image. Alton has a notorious history of flooding.

force floodwater through a narrow channel between the walls and cause it to back up waiting to get through. This effect contributed to the 1993 flooding upstream of St. Louis.

Despite the high walls constructed to prevent flooding, the rising flood peak came to within about 0.6 m (2 ft.) of overtopping the floodwalls at St. Louis. As the floodwaters in 1993 rose to record levels, they began to leak and required round-the-clock efforts to keep them from failing. Failure of levees downstream from St. Louis partially relieved the pressure, possibly saving the city from flooding. Levee failures were common during the flood event (Figure 6.4).[7] In fact, the vast majority of the private levees, that is, levees built by farmers and homeowners, along the Mississippi River and its tributaries failed.[4–6] However, most of the levees built by the federal government survived the flooding and undoubtedly saved lives and property. Unfortunately, there is no uniform building code for the levees, so some areas have levees that are higher or lower than others. Failures occurred as a result of overtopping and breaching, or rupturing, resulting in massive flooding of farmlands and towns (Figure 6.5).[4, 6]

In addition to floodwalls, other engineering structures designed to control the Mississippi have actually increased the long-term flood hazard.[8] For example, river-control structures designed to optimize the movement of loaded riverboats and barges during normal flow conditions can increase flooding at high flow. When floods overtop these structures, the floodwaters slow down and begin to back up, increasing flood height in a manner similar to the bottleneck effect created by the floodwalls in St. Louis.[9]

Of course, building houses, industrial plants, public buildings, and farms on a floodplain invites disaster; too many floodplain residents have refused to recognize the natural floodway of the river for what it is: part of the natural river system. The **floodplain,** the flat surface adjacent to the river channel that is periodically inundated by floodwater, is, in fact, produced by the process of flooding (Figure 6.6).

▲ FIGURE 6.4 **LEVEE FAILURE** A breach of this levee in Illinois during the 1993 floods of the Mississippi River caused flooding of the town of Valmeyer. Rapidly flowing whitewater can be seen rushing through the breach in the lower half of this grass-covered earthen levee. *(Comstock)*

◀ **FIGURE 6.5 DAMAGED FARMLAND** Damage to farmlands during the peak of the 1993 flood of the Mississippi River. *(Comstock)*

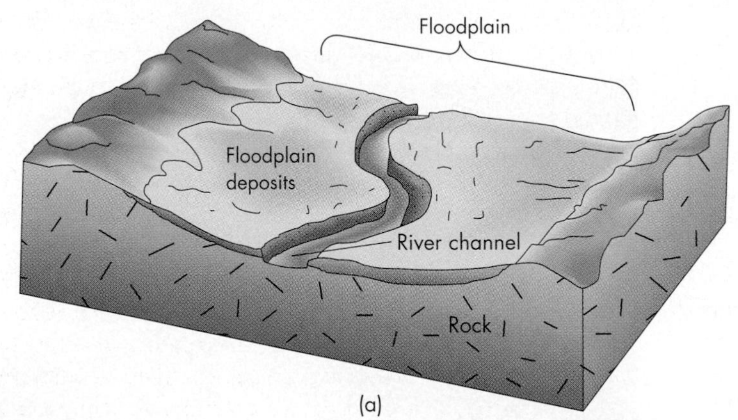

(a)

(b)

◀ **FIGURE 6.6 FLOODPLAIN** (a) Diagram illustrating the location of a river's floodplain. Most floodplains have deposits that are finer grained than those found in, and immediately adjacent to, the channel. (b) Floodplain of the Rio Grande near its headwaters in Colorado. *(Edward A. Keller)*

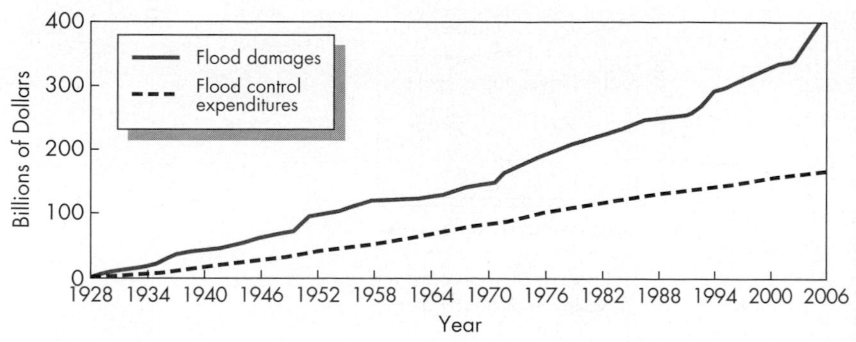

◀ **FIGURE 6.7 NATIONAL TRENDS FROM 1928 TO 2006** Increased development on river flood plains, over the past three quarters of a century, has resulted in increased cost of damages that has outpaced ever-increasing flood control cost expenditures.

If the floodplain and its relation to the river are not recognized, flood control and drainage of wetlands, including floodplains, become prime concerns.

An important lesson learned from the 1993 and 2008 floods is that construction of levees provides a false sense of security. Construction of a levee does not remove land from the floodplain, nor does it eliminate the flood risk. It is difficult to design levees to withstand extremely high-magnitude floods for a long period of time. Furthermore, by constructing a levee, the result is less floodplain space to "soak up" the floodwaters.[7] The 1993 floods caused extensive damage and loss of property; in 2008, floodwaters of the Mississippi River system inundated floodplain communities once again. National trends from 1928 to 2006 (Figure 6.7) show that flood-control expenditures and damages have increased over the 78-year period. Total flood losses have dramatically increased, in spite of the ever-increasing cost of flood-control structures.[10]

In the first 6 to 7 years after the 1993 flood, the federal government bought out many homeowners in the floodplain upstream from St. Louis.[11] Now real estate interests have reversed the trend, and local governments are allowing floodplain development behind levees.[12] Weak floodplain regulations and government subsidies have allowed the construction of more than 25,000 new homes and other buildings on the floodplain behind new, higher levees (Figure 6.8).[11] Fortunately, not all communities along the river are short sighted and have reevaluated their strategies concerning the flood hazard—they have moved to higher ground! Of course, this is exactly the adjustment that is appropriate.

6.1 An Introduction to Rivers

Streams and rivers are part of the water or hydrologic cycle, and *hydrology* is the study of this cycle. In the hydrologic cycle, water evaporates from Earth's surface, primarily from the oceans, into the atmosphere, and it returns to the oceans by flowing underground and across the land surface. Water that falls on the land as rain and snow will infiltrate into the ground, evaporate from the land surface, or drain off the land following a course determined by the local topography.

Surface drainage, referred to as *runoff*, finds its way to small streams, which may merge as *tributaries* to form a larger stream or **river.** Streams and rivers differ only in size; that is, streams are small rivers. Local usage varies as to what constitutes a creek, brook, and river. Geologists, however, commonly use the term *stream* for any body of water that flows in a channel. The region drained by a single stream or river is variously called a **drainage basin,** *watershed*, river basin, or catchment (Figure 6.9a). Thus, each stream has its own drainage basin or watershed that collects rain and other precipitation. A large

◀ **FIGURE 6.8 DEVELOPMENT INCREASES ON ST. LOUIS AREA FLOODPLAINS** This mile-and-a-half-long strip mall, one of the largest in the United States, was built on land in Chesterfield, Missouri, that was submerged under approximately 3 m (10 ft.) of Missouri River water in 1993. New levees surrounding this mall and other floodplain development will greatly reduce the area available for future floodwaters. This will increase the flood height upstream. *(Tom Gannam/AP Images)*

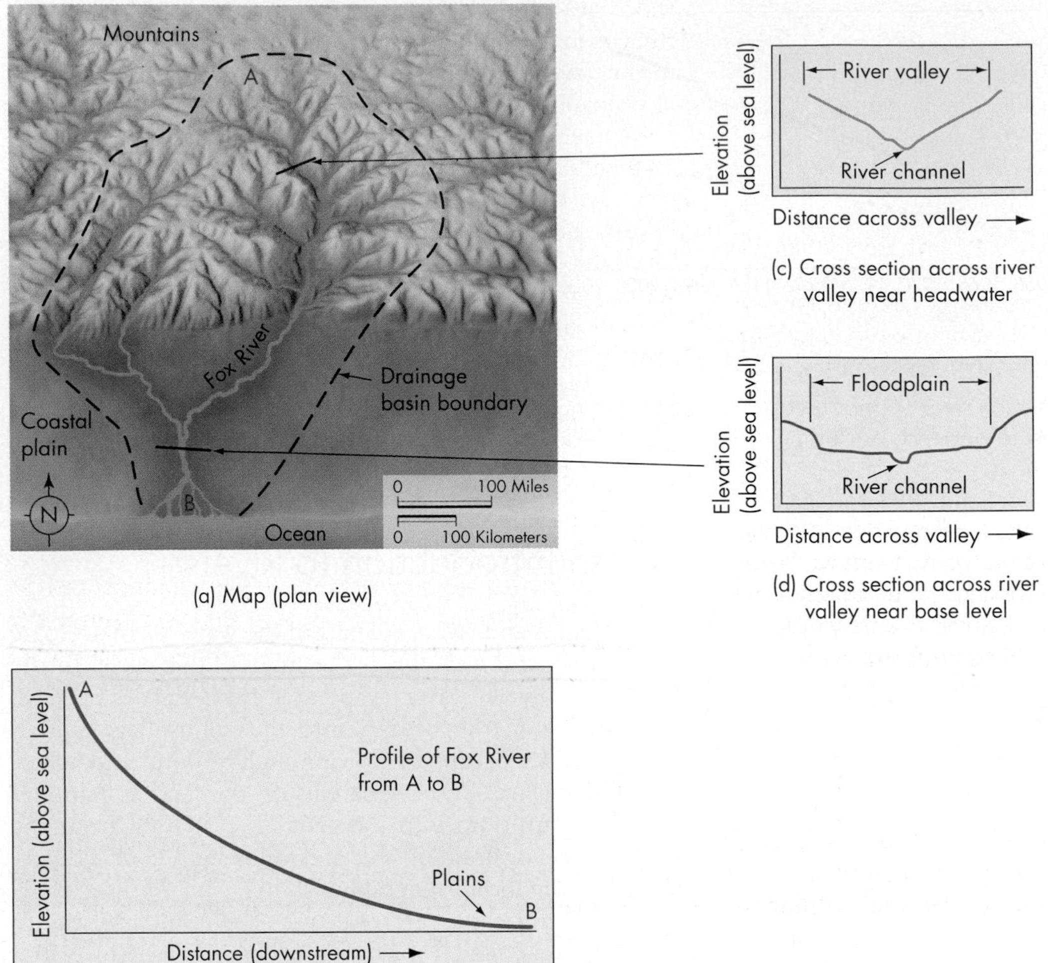

(a) Map (plan view)

(c) Cross section across river valley near headwater

(d) Cross section across river valley near base level

(b) Longitudinal profile

▲ **FIGURE 6.9 DRAINAGE BASIN AND RIVER PROFILE** Idealized diagram of the Fox River showing (a) its drainage basin outlined by the black dashed line, that is, the area where surface runoff drains into the river or its tributaries. (b) The longitudinal profile from point A at the river's head to point B at the river's mouth; note that the vertical scale on this diagram is greatly exaggerated. (c) The V-shaped cross section of the river valley near its headwaters where the valley floor is mainly the channel. (d) A cross section of the valley on the coastal plain where the valley floor is mainly the floodplain.

river basin, such as the Susquehanna River basin in the northeastern United States, is made up of hundreds of small drainage basins, commonly called watersheds, which catch the precipitation that drains into the smaller tributary streams and rivers. Except for the very few who live right next to the river, most people who live on land that drains into the Susquehanna live in a watershed of one of the tributaries. In fact, virtually everyone lives in a watershed that is a part of a larger drainage basin.

One important characteristic of a river is the slope of the land over which it flows. Referred to as the *gradient*, this slope is determined by calculating the vertical drop in elevation of the channel over some horizontal distance. Gradient is commonly given in meters per kilometer or feet per mile. In general, the gradient of a river is steepest at higher elevations in the drainage basin and levels off as the river approaches its base level. The *base level* of a river is essentially the lowest elevation to which it may erode. Most often, this elevation is at or close to sea level, although a river may have a temporary base level such as a lake. Rivers, thus, flow downhill to their base level, and a graph showing the downstream changes in a river's elevation is called a *longitudinal profile* (Figure 6.9b).

A river usually has a steeper-sided and deeper valley at high elevations near its origin, or *headwaters*, than closer to its base level where a wide floodplain may be present (Figure 6.9c and d). This happens because, at higher elevations, the steeper river gradient increases flow velocity, which, in turn, increases erosion.

EARTH MATERIAL TRANSPORTED BY RIVERS

Rivers not only move water, they also transport a tremendous amount of visible and invisible material. The quantity of this material, called the *total load*, is generally subdivided into bed load, suspended load, and dissolved load, based on how the river carries the material. The bed load of most rivers consists of sand and gravel particles that slide, roll, and bounce along the channel bottom in rapidly moving water. Bed load commonly makes up less than 10 percent of the total load. In contrast, the suspended load is composed mainly of small silt and clay particles that are carried above the streambed by the flowing water. The suspended load accounts for nearly 90 percent of the total load and makes rivers look muddy, especially during a flood. Finally, the dissolved load consists of electrically charged atoms or molecules, called *ions*, which are carried in chemical solution. Most dissolved load is derived from chemical weathering of earth materials in the drainage basin. In some places, ions from discharging underground springs, wastewater, and chemical pollution are a significant part of the dissolved load.

RIVER VELOCITY, DISCHARGE, EROSION, AND SEDIMENT DEPOSITION

Rivers are the basic transportation system of that part of the rock cycle involving erosion and deposition of sediment. They are a primary erosion agent in the sculpture of our landscape. The velocity, or speed, of water in a river varies along its course, affecting both erosion and deposition of sediment. Hydrologists combine measurements of flow velocity and water depth to determine discharge, a more useful indicator of stream flow.

Discharge (Q) is the volume of water moving through a cross section of a river per unit time. In this case, a cross section is the sideways view of the river that you would have if you took a huge knife and sliced across the valley from top to bottom at right angles to each riverbank (Figure 6.10). Once a cross section is completed with survey equipment, a hydrologist can determine the cross-sectional area (A) that is occupied by water by simply measuring the water depth. Discharge is then calculated by multiplying the cross-sectional area of the water in the channel by the velocity of flow, and it is reported in either cubic meters per second (cms) or cubic feet per second (cfs).

If there were no additions or deletions of flow along a given length of river, then its discharge would not change. Where the cross-sectional area of flow decreases, the velocity of the water must increase for discharge to remain constant. You can prove this to yourself with a garden hose. Turn on the water and observe the velocity of the water as it exits the hose. Then put your thumb partly over the end of the hose, reducing the area where the water flows out of the hose, and observe the increase in the velocity. This explains why velocity is higher where a river flows through a narrow channel in a canyon than where it spreads out and expands into a larger area downstream of the narrow channel.

Stream flow expands and slows down as a river goes from mountains onto plains or from a channel into an ocean, lake, or pond. At these points, the river commonly forms a fan-shaped deposit on land known as an *alluvial fan* (Figure 6.11), or a triangular or irregularly shaped deposit known as a *delta* if it extends into a larger body of water (Figure 6.12).

The flood hazard associated with alluvial fans and deltas is different from hazards in a river valley and its associated floodplain. Rivers entering an alluvial fan or delta often split into a system of *distributary channels*. That is, the main river divides into several channels that carry floodwaters to different parts of the fan or delta. Furthermore, these channels may change position rapidly during a single flood or from one flood to the next, creating a hazard that is difficult to predict.[13] As an example, a large recreational vehicle (RV) park on the delta of the Ventura River in Southern California flooded four times in the 1990s (see Case Study 6.1). The RV park had been constructed across a historically active distributary channel. Engineers mapping the potential flood hazards on the site before construction of the park had not recognized that the park was located on a river delta. This story emphasizes the importance of studying a river's flooding history as part of flood hazard evaluation.

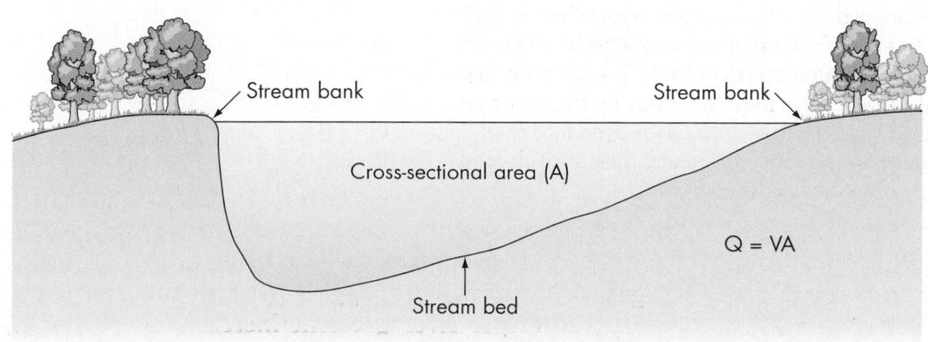

◀ **FIGURE 6.10 CALCULATING STREAM DISCHARGE** Diagram of the cross-sectional area of a stream when the entire channel is filled with water. The discharge of a stream (Q) is calculated by multiplying the velocity of the flow (V) times the cross-sectional area (A) of the water in the channel. Cross-sectional area is determined by use of a surveyed cross section of the stream channel and a measurement of the depth of water in the channel.

◀ **FIGURE 6.11 ALLUVIAL FAN IN DEATH VALLEY** Along the western foot of California's Black Mountains, this alluvial fan formed where a stream leaves a canyon and expands outward into Death Valley. The bent line in the lower left is a road that cuts across the lower part of the fan. Infrequent floods on this fan drain into the white salt flats of Death Valley, seen in the upper and lower left. *(Michael Collier)*

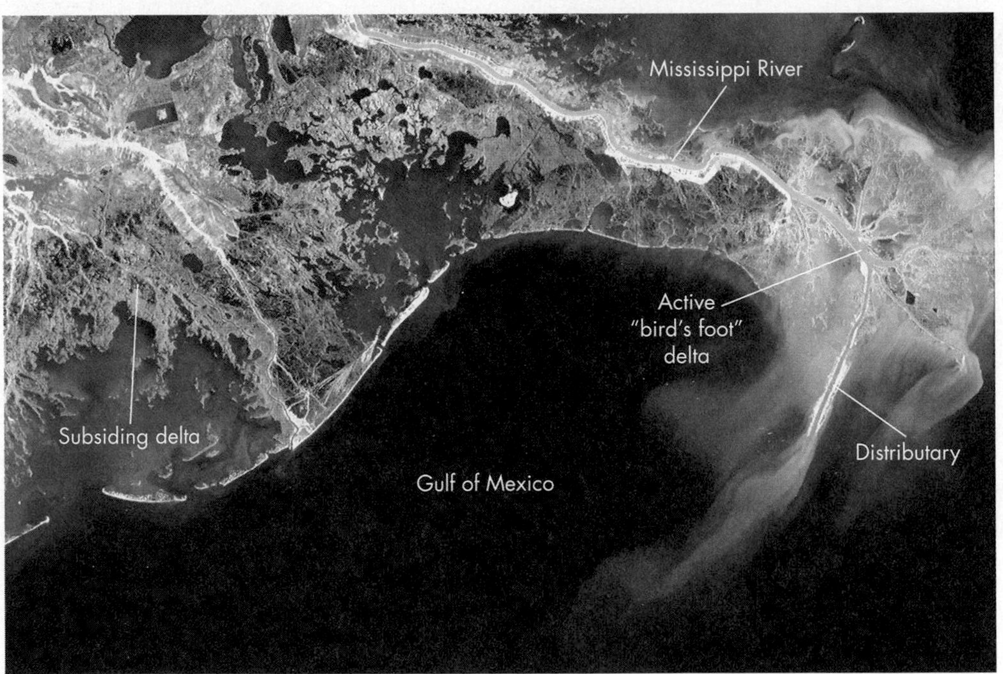

▲ **FIGURE 6.12 MISSISSIPPI DELTA** In this false-color satellite image of the Mississippi River delta, living vegetation appears red, sediment-laden waters are white or light blue, and deeper water is darker blue because it contains less suspended sediment. The main active distributary is the light-blue channel that starts in the top-left center and slopes toward the middle right. In the far-right center, the main distributary branches into several channels that make the shape of a "bird's foot." This shape indicates that river flow is stronger than the wave and current action in the Gulf of Mexico. Distance across the image is about 180 km (112 mi.). *(University of Washington Libraries)*

CHANNEL PATTERNS AND FLOODPLAIN FORMATION

Most features of rivers and floodplains result from the interaction of flowing water and moving sediment. Such features include various forms of sediment accumulation and the river channel itself. As seen from an airplane, streams and rivers develop distinctive **channel patterns.** The most common patterns are *braided*, similar to a person's hair, and *meandering*, similar to the curves of a moving snake. Braided patterns (Figure 6.13) have numerous sand and gravel bars and islands that divide and reunite the main channel, especially during low flow. Overall, braided channels tend to be wide and shallow, compared to meandering channels. One of the authors was reminded of this relationship recently on a raft trip down the braided Chilkat River in Alaska. In spite of the guide's efforts, the raft frequently ran aground on gravel bars. When that occurred, the guide yelled "shuffle" and the eight passengers bounced up and down until the raft floated back into the channel. Rivers similar to the Chilkat are likely to have a braided pattern where the stream has a steep gradient and abundant, coarse bed-load sediment. These conditions are often found in areas where tectonic processes are rapidly uplifting the land surface and where rivers receive water and sediment from melting glaciers.

Many rivers have curving channel bends called *meanders* that migrate back and forth across the floodplain over a period of years to decades (Figure 6.14). It is interesting that although the meandering behavior of rivers has been studied for nearly a century by many hydrologists, geologists, engineers, and physicists, including Albert Einstein, we don't know for sure why rivers meander. We do, however, know a great deal about how water flows in a meandering stream.

Canoeists have long known that water moves faster along the outside of a meander bend during low to relatively high flows. The fast-moving water erodes the riverbank on the outside of the bend to form a steep or near-vertical slope known as a *cutbank*. In contrast, slower water on the inside of a meander bend deposits sand and sometimes gravel to form a *point bar*. Continual erosion of the cutbank and deposition on the point bar cause each meander bend to migrate laterally in a different direction. This process is important in constructing and maintaining some floodplains (Figure 6.14). Floodplains are also built during *overbank flow*, a condition that develops when rising water spills over the river bank onto the floodplain. Flow that expands out of its channel then deposits fine sediment, such as very fine sand, silt, and clay, which builds up the floodplain. Much of the sediment transported in rivers is periodically stored by deposition in the channel and on the adjacent floodplain.

Meandering channels often contain a series of regularly spaced pools and riffles (Figure 6.15). *Pools* are deep areas produced by scour, or erosion, at high flow, and *riffles* are shallow areas formed by sediment deposited at high flow. At low flow, pools have relatively deep, slow-moving water and riffles have shallow, fast-moving water. These changes in water depth and velocity along a stream create different *habitats*, that is, environmental conditions in

(a)

(b)

▲ FIGURE 6.13 **BRAIDED RIVERS** (a) The braided pattern of the North Saskatchewan River, Alberta, Canada, is formed by shallow channels flowing around and across numerous sand and gravel bars and islands. Coarse bed-load sediment of the river comes from melting glaciers in the Canadian Rocky Mountains. *(John S. Shelton/ University of Washington Libraries)* (b) Surface view of a braided channel in Granada, southern Spain, with subdividing channels, a steep gradient, and coarse gravel. Distance across the channel in the foreground is about 7 m (23 ft.). *(Edward A. Keller)*

CASE STUDY 6.1

Flooding on the Delta of the Ventura River

In 1905, the philosopher George Santayana said, "Those who cannot remember the past are condemned to repeat it." Scholars may debate the age-old question of whether or not cycles in human history repeat themselves, but the repetitive nature of natural hazards, such as floods, is undisputed.[14] Better understanding of the historical behavior of a river is, therefore, valuable in estimating its present and future flood hazards. Consider the February 1992 Ventura River Flood in Southern California. The flood severely damaged the Ventura Beach Recreational Vehicle (RV) Resort, which had been constructed a few years earlier on an active distributary channel of the Ventura River delta. Although the recurrence interval for the 1992 flood is approximately 22 years (Figure 6.A), earlier engineering studies suggested that the RV park would not be inundated by a flood with a recurrence interval of 100 years. What went wrong?

- Planners did not recognize that the RV park was constructed on a historically active distributary channel of the Ventura River delta. In fact, early reports did not even mention a delta.
- Engineering models that predict flood inundation are inaccurate when evaluating distributary channels on river deltas where extensive channel filling and scouring, as well as lateral movement of the channel, are likely to occur.
- Historical documents, such as maps dating back to 1855, and more recent aerial photographs showed that the channels were apparently not evaluated. Maps rendered from these documents suggest that the distributary channel passing through the RV park was, in fact, present in 1855 (Figure 6.B).[15]

Clearly, the historic behavior of the river was not evaluated as part of the flood-hazard evaluation. If it had been, the site would have been recognized as unacceptable for development, given that a historically active channel was present. Nevertheless, necessary permits were issued for development of the park, and, in fact, the park was rebuilt after the flood.

Before 1992, the distributary channel carried water during 1969, 1978, and 1982. Following the 1992 flood event, the channel carried floodwaters in the winters of 1993, 1995, and 1998, again flooding the RV park. During the 1992 floods, the discharge increased from less than 25 m^3 per second (880 ft.^{3} per second) to a peak of 1,322 m^3 per second (46,670 ft.^{3} per second) in only about 4 hours! This

▲ FIGURE 6.A **FLOODING OF CALIFORNIA'S VENTURA BEACH RV RESORT IN FEBRUARY 1992** The RV park was built directly across a historically active distributary channel of the Ventura River delta. The recurrence interval of this flood is approximately 22 years. A similar flood occurred again in 1995. On the left, U.S. 101, the Pacific Coast Highway, is completely closed by the flood event. *(Mark J. Terrell/AP Images)*

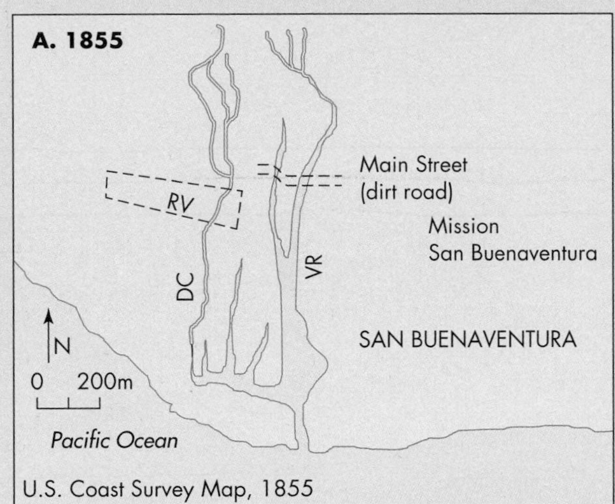

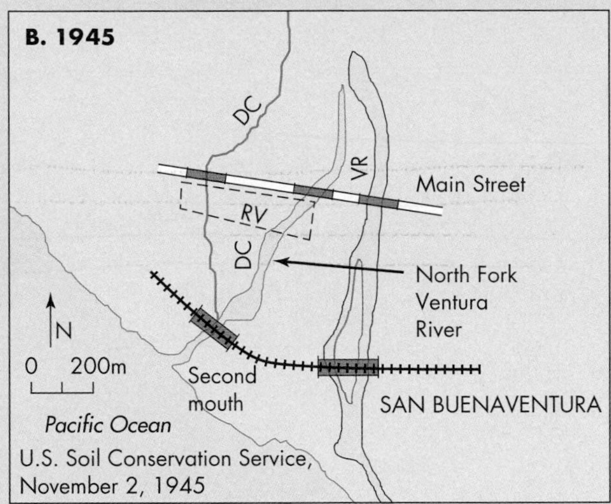

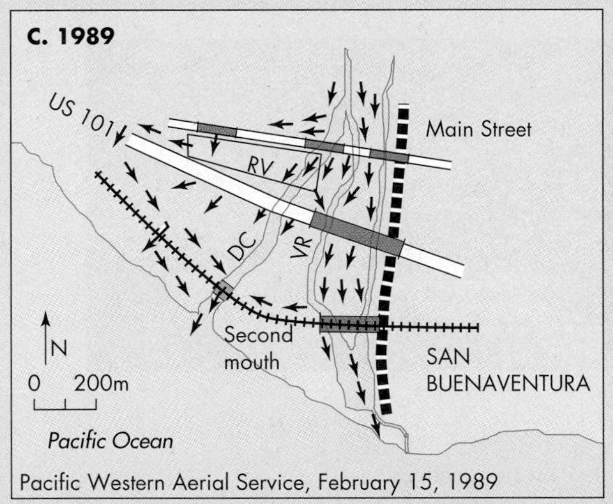

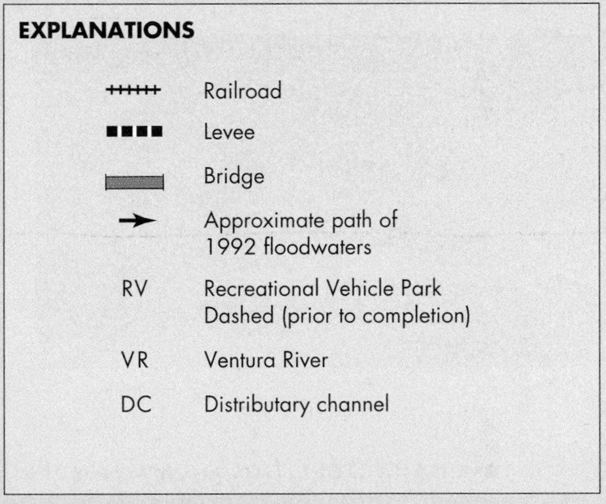

▲ FIGURE 6.B **HISTORICAL MAPS OF THE VENTURA RIVER DELTA** Distributary channel and location of the RV park are shown. (a) In 1855, a small distributary channel (DC) was flowing through the future site of the RV park (dashed box) and joined the main Ventura River channel (VR) near the river's mouth. (b) By 1945, the 1855 channel had widened to flow directly to the ocean as the North Fork of the Ventura River. A second distributary was flowing through the western part of the future RV park and joined the North Fork upstream of its mouth. (c) In 1989, around the time the RV park was built, the original 1855 channel was still active as one of the major distributaries. A levee had been built on the east side of the delta to protect San Buenaventura. In the 1992 flood (arrows), the levee and earthen fill elevating the railroad and U.S. 101 acted as dams to widen the extent of flooding on the delta. *(From Keller, E. A., and Capelli, M H. 1992. Ventura River flood of February, 1992: A lesson ignored?* Water Resources Bulletin *25(5):813–31)*

rate of flow is approximately twice as much as the daily high discharge of the Colorado River through the Grand Canyon in the summer, when river rafters navigate it. This volume is an immense discharge for a relatively small river with a drainage area of only about 585 km^2 (226 mi.^{2}). The flood occurred during daylight, and one person was killed. However, if the flood had occurred at night, the death toll would likely have been much higher.

A warning system that has been developed for the park has, so far, been effective in providing early warning of an impending flood. The park, with or without the RVs and people, is a "sitting duck." This situation was dramatically illustrated in 1995 and 1998, when winter floods again swept through the park. Although the warning system worked and the park was successfully evacuated, the facility was again severely damaged. A move is now afoot to purchase the park and restore the land to a more natural delta environment—a good move but, as of 2010, not yet a reality.

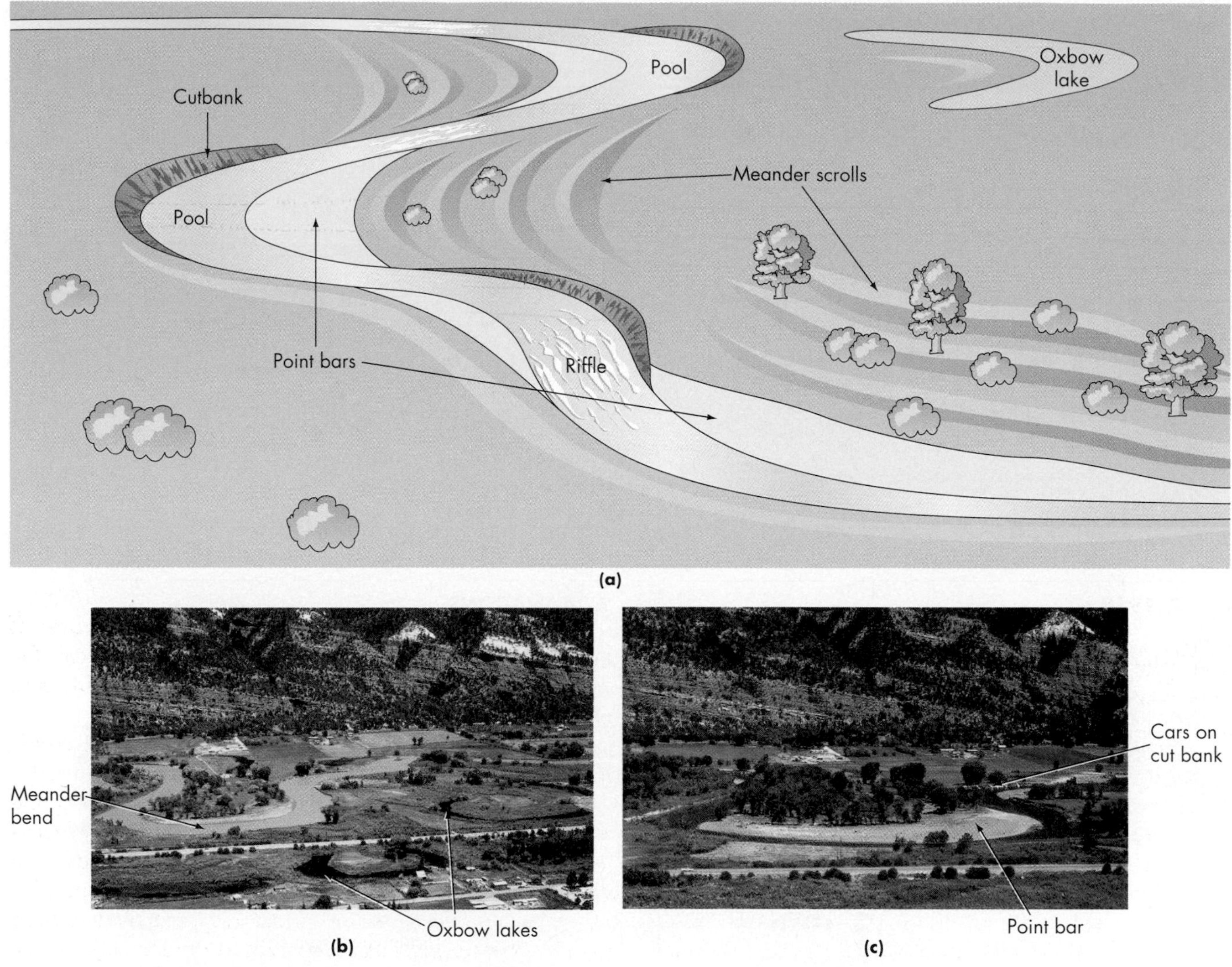

▲ **FIGURE 6.14 CHARACTERISTICS OF A MEANDERING RIVER** (a) Idealized diagram of a meandering stream showing important features. Historical migration of meander bends is commonly indicated by low, curving, vegetation-covered ridges called meander scrolls. The scrolls on the point bar in the left center of the diagram indicate that it migrated from right to left. (b) Animas River north of Durango, Colorado, at high flow with brown muddy water filling the channel. The dark-blue abandoned channels that form oxbow lakes in the lower right indicate that the river has meandered back and forth across the valley to build its floodplain. In contrast to the river, the lakes have little suspended sediment in the water. *(Robert H. Blodgett)* (c) The sandy point bar on the left side of (b) exposed at low flow. A farmer has placed junked cars on the cutbank of a meander bend in the right center to slow the erosion of his pasture. *(Robert H. Blodgett)*

which organisms live. The variety of environmental conditions found in these habitats increases the diversity of aquatic life.[16] For example, fish may feed in riffles and seek shelter in pools. Also, pools have different types of insects than are found in riffles.

Stream channels flowing through sediment, such as sand, gravel, or mud, rarely stay fixed in the same place for a long period of time. Either erosion or deposition causes the stream channel to shift position. These shifts can be continual or they can be abrupt, such as when a channel shifts to an entirely new location during a flood. In abrupt changes, a stream may abandon part of its existing channel and form a new one nearby. This process, known as *avulsion*, can have unexpected consequences for people living in a stream valley.

Having presented some of the characteristics and processes of flow of water and sediment in rivers, we will now discuss the process of flooding in greater detail.

◀ **FIGURE 6.15 POOLS AND RIFFLES** Well-developed pool-riffle sequence in Sims Creek near Blowing Rock, North Carolina. Deep pools lie under the smooth, reflective water surface in the center and lower right; shallow riffles lie under the rough, nonreflective water in the far distance and left foreground. *(Edward A. Keller)*

6.2 Flooding

The natural process of overbank flow is termed **flooding** (Figure 6.16). Most river flooding is related to the amount and distribution of precipitation in the drainage basin, the rate at which the precipitation soaks into the earth, and how quickly surface runoff from that precipitation reaches the river.

The amount of moisture in the soil at the time the precipitation starts also plays an important role in flooding. Water-saturated soil is similar to a wet sponge that cannot hold any additional moisture. If significant precipitation falls on a saturated drainage basin, flooding will occur. If the same amount of precipitation falls on a dry basin, the soil may be able to absorb considerable moisture and, thus, help prevent flooding.

In mountainous areas and in locations at higher latitudes, such as Alaska, Canada, and the northern contiguous United States, floods are common during early spring and midwinter thaws. Rain falling on frozen ground or accumulated snow, or the rapid melting of ice and snow, can cause flooding. In these areas, large masses of floating ice can create ice jams across rivers that temporarily dam their flow. Floodwaters can back up behind the ice jams or the jams can break apart abruptly, causing flooding downstream.

(a)

(b)

▲ **FIGURE 6.16 FLOODPLAIN FLOODING FROM SNOWMELT** Gaylor Creek, Yosemite National Park, California, during spring snow melt. (a) In the morning, water stays within the stream channel. *(Edward A. Keller)* (b) In the afternoon, during daily peak snowmelt, overbank flow covers the floodplain. *(Edward A. Keller)*

For example, an ice jam on the Winooski River in Vermont in 1992 threatened to flood Montpelier, the state capital, which was located upstream of the jam. In addition to natural events, human interactions with the hydrologic system, discussed in Section 6.7, can also affect river processes and flooding.

A flood can be characterized in several ways. One is the *flood discharge*, defined as the discharge of the stream at the point where water overflows the channel banks. Floods can also be defined by the height of water in the river, referred to as the *stage* of the river. A graph showing changes in stream discharge, water depth, or stage over time is called a *hydrograph*.

The term *flood stage* is frequently used to indicate that the elevation of the water surface has reached a level likely to cause damage to personal property. This definition is based on human perception, so the elevation that is considered flood stage depends on human use of the floodplain.[17] Therefore, the magnitude of a flood may or may not coincide with the extent of property damage. The relationships between the stage, discharge, and recurrence interval of floods are described and illustrated in Case Study 6.2. A **recurrence interval** of a flood is the average time between flood events that are of equal or greater magnitude.

FLASH FLOODS AND DOWNSTREAM FLOODS

Floods can further be characterized by where they occur in a drainage basin (Figure 6.17). **Flash floods** typically occur in the upper part of a drainage basin and in some small drainage basins of tributaries to a larger river. They are generally produced by intense rainfall of short duration over a relatively small area. If these floods are sudden and of relatively great volume, peak discharge can be reached in less than 10 minutes. Flash flooding is most common in arid and semiarid environments; in areas with steep topography or little vegetation; and following breaks of dams, levees, and ice jams (see Survivor Story 6.3).

Although flash floods do not generally cause flooding in the larger streams, they join downstream and can be quite severe locally. For example, a high-magnitude flash flood

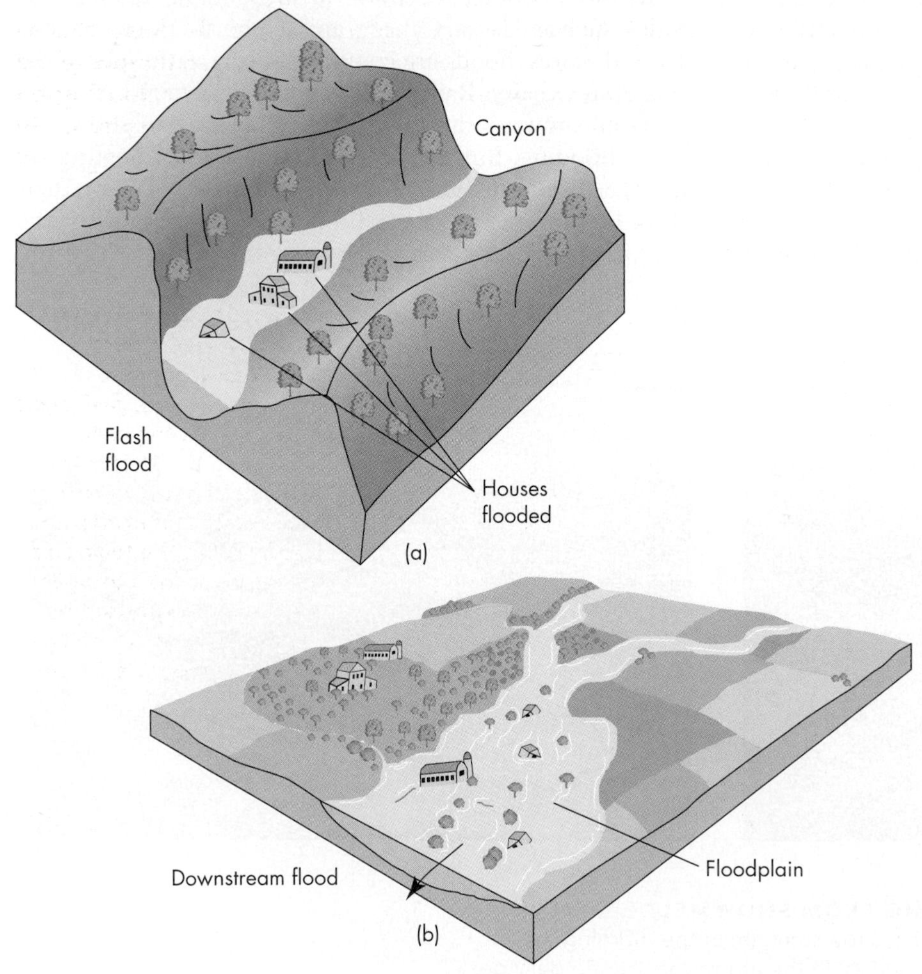

FIGURE 6.17 FLASH FLOODS AND DOWNSTREAM FLOODS Idealized diagram comparing a flash flood (a) to a downstream flood (b) Flash floods generally cover relatively small areas and are caused by intense local storms, whereas downstream floods cover wide areas and are caused by regional storms or spring runoff. *(Modified after U.S. Department of Agriculture drawing)*

▲ FIGURE 6.18 **SUBMERGED CAR** Severe storms in May of 2010 dumped heavy rain on Tennessee flooding much of the city of Nashville. A common cause of loss of life occurs when automobile drivers attempt to cross flooded patches of the road and are swept away. *(Mark Humphrey/AP Images)*

occurred in July 1976 in the Front Range of the Rocky Mountains in Colorado. This flood was caused by a complex system of thunderstorms that swept through several canyons west of Loveland and delivered up to 25 cm (9.8 in.) of rain in a few hours. The flash flooding killed 139 people and inflicted more than $35 million in damages to highways, roads, bridges, homes, and small businesses. Most damage and loss of life occurred in Big Thompson Canyon, where hundreds of residents, campers, and tourists were caught with little or no warning. Although the storms and flood were rare events in the Front Range canyons, comparable floods have occurred in the past and others can be expected in the future.[18–20]

Most people who die during flash floods are in automobiles. Deaths occur when people attempt to drive through shallow, fast-moving floodwater. A combination of buoyancy and the strong lateral force of the water sweeps automobiles off the road into deeper water, trapping people in sinking or overturned vehicles. Most automobiles will be carried away by 0.6 m (2 ft.) of water (Figure 6.18).

It is the large *downstream floods* that usually make national television and newspaper headlines. In 2004, heavy rains from a series of hurricanes and tropical storms in the eastern United States caused record and near-record flooding. Major downstream flooding occurred because the ground remained saturated as one storm after another crossed the same drainage basins. In Pennsylvania, the Susquehanna River crested 2.5 m (8 ft.) above flood stage for one of the five greatest floods in history. Downstream flooding on the Ohio River in Marietta, Ohio, was the worst in 40 years (Figure 6.19), and flooding in Atlanta, Georgia, set all-time records.

◀ FIGURE 6.19 **DOWNSTREAM FLOODING ON THE OHIO RIVER** Downstream, Marietta, Ohio, experienced its worst flooding in 40 years from the Ohio River because of heavy rains from the remnants of Hurricane Ivan in 2004. The town had to use snowplows to clear off the mud deposited by the floodwaters. *(Washington County Sheriff's Office)*

CASE STUDY 

Magnitude and Frequency of Floods

Flooding is intimately related to the amount and intensity of precipitation and runoff. Catastrophic floods reported on television and in newspapers often are produced by infrequent, large, intense storms. Smaller floods or flows may be produced by less intense storms that occur more frequently. All flow events that can be measured or estimated from stream-gauging stations (Figure 6.C) can be arranged in order of their magnitude of discharge, generally measured in cubic meters per second (Figure 6.D). The list of annual peak flows, which is the largest flow each year or the annual series so arranged, can be plotted on a discharge-frequency curve by deriving the recurrence interval R for each flow from the relationship

$$R = (N + 1) \div M$$

where R is a recurrence interval in years, N is the number of years of record, and M is the rank of the individual flow within the recorded years (Figure 6.E).[21] For example, the highest flow for 9 years of data for the Patrick River is approximately 280 m^3 (9900 $ft.^3$) per second, and that flow has a rank M equal to 1 (Figure 6.E).[22] The recurrence interval of this flood is

$$R = (N + 1) \div M = (9 + 1) \div 1 = 10$$

which means that a flood with a magnitude equal to or exceeding 280 m^3 (9900 $ft.^3$) per second can be expected about every 10 years; we call this a 10 year flood. The probability that the 10 year flood will occur in any one year is $1 \div 10$ or 0.1 (10 percent). Likewise, the probability that the **100 year flood** will occur in any year is $1 \div 100 = 0.01$ or 1 percent.

Extending, that is, extrapolating, the discharge-frequency curve is risky. The curve shouldn't be extended much beyond twice the number of years for which there are discharge records. Studies of many streams and rivers show that channels are formed and maintained by *bankfull discharge,* defined as a flow with a recurrence interval of 1.5 to 2 years. Applying this concept to the Patrick River (Figure 6.E), the bankfull discharge with an interval of 1.5 years was a discharge of 27 m^3 (950 $ft.^3$) per second. Bankfull is the flow that just fills the channel. Therefore, we can expect a stream to emerge from its banks and cover part of the floodplain with water and sediment once every year or so.

As flow records are collected, we can more accurately predict floods. However, designing structures for a 10 year, 25 year, 50 year, or even 100 year flood, or, in fact, any flow, is a calculated risk, since predicting such floods is based on a statistical probability. For many streams, the flow record is far too short to accurately predict the magnitude and frequency of large floods. In the long term, a 25 year flood happens on the average of once every 25 years, but two 25 year floods could occur in any given year, as could two 100 year floods![23] As long as we continue to build dams, highways, bridges, homes, and other structures without considering the effects on flood-prone areas, we can expect continued loss of lives and property.

◀ FIGURE 6.C **STREAM-GAUGING STATION** This gauge on San Jose Creek in Goleta, California, has (1) a water-depth sensor (lower left) that detects the amount of water in the stream by its pressure; (2) an instrument shed (upper center) that houses a computer, batteries, and a radio transmitter; and (3) a pole (top center) that has a radio antenna to send water-level data to a government agency and a solar panel to provide power for the instruments. *(Edward A. Keller)*

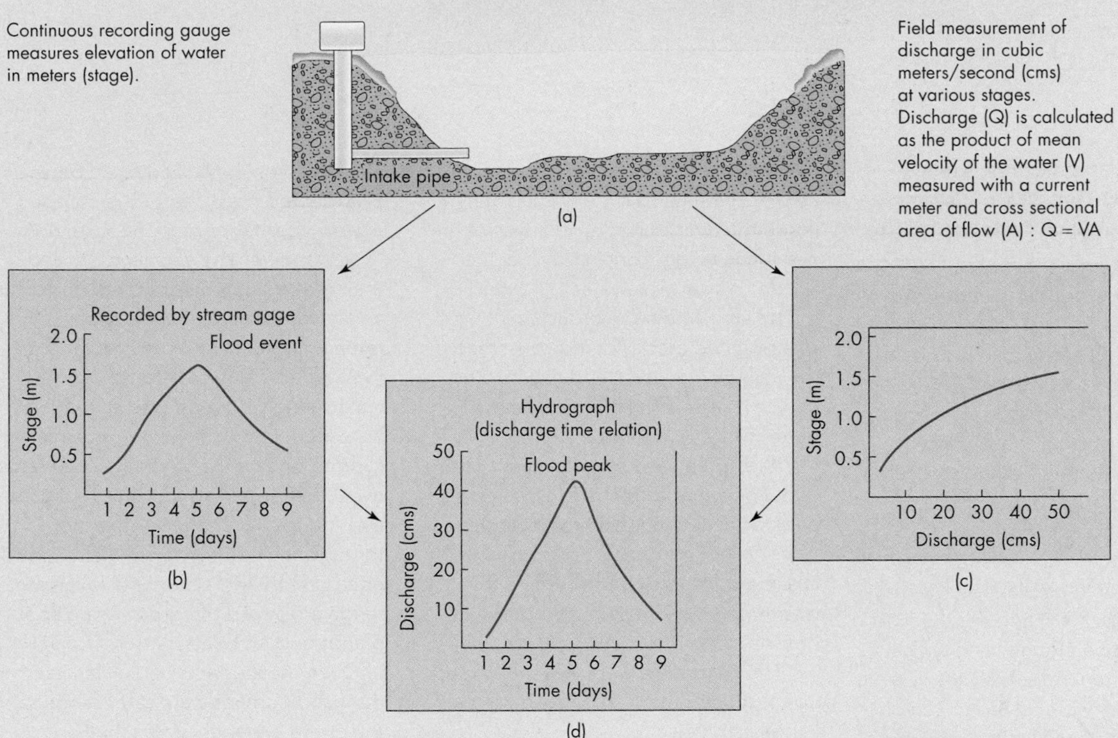

▲ FIGURE 6.D **HOW A HYDROGRAPH IS PRODUCED** To make a hydrograph of a flood, a recording gauge (a) is installed to obtain a continuous record of the water level or stage. This record is then used to produce a stage-time graph (b) Before the flood, field measurements at various flows are used to produce a stage-discharge graph (c) Graphs (b) and (c) are then combined to make the final flood hydrograph (d).

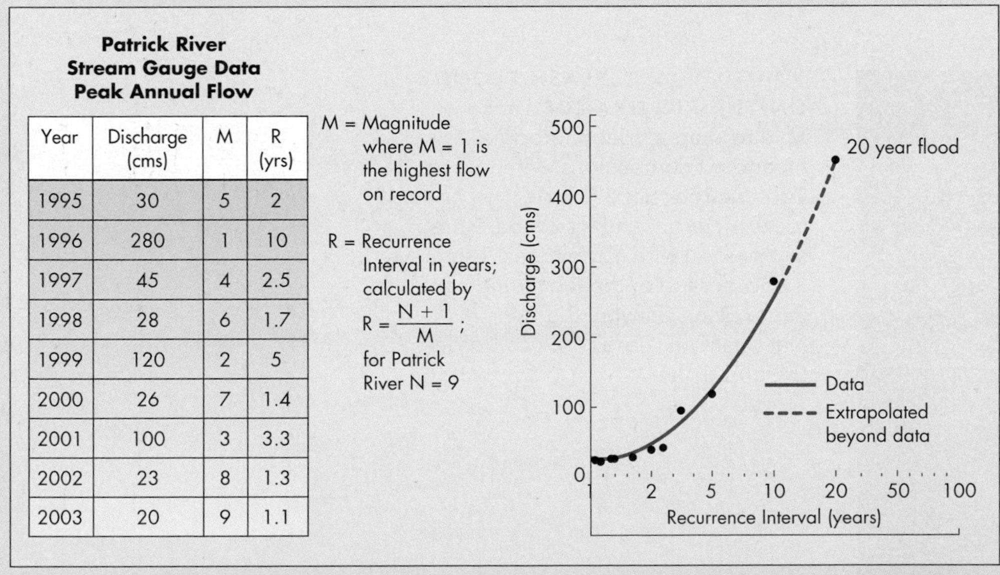

Patrick River Stream Gauge Data Peak Annual Flow

Year	Discharge (cms)	M	R (yrs)
1995	30	5	2
1996	280	1	10
1997	45	4	2.5
1998	28	6	1.7
1999	120	2	5
2000	26	7	1.4
2001	100	3	3.3
2002	23	8	1.3
2003	20	9	1.1

▲ FIGURE 6.E **EXAMPLE OF A DISCHARGE-FREQUENCY CURVE** To make a discharge-frequency graph for the Patrick River, nine years of discharge measurements in cubic meters per second (cms) were ranked (1 though 9) on the basis of the size (magnitude) of the largest flow each year. The recurrence interval, or frequency, of the largest annual flow was then calculated using the formula shown in the text. Data from the table on the left were then plotted on the graph on the right. The curve was then extended (extrapolated) to estimate the discharge of the 20 year flood, which was 450 m^3 per second. *(After Leopold, L. B. 1968. U.S. Geological Survey Circular 554)*

SURVIVOR STORY 6.3

Flash Flood

Jason Lange and His Friends Thought They Were Visiting Big Bend National Park During the Dry Season

Jason Lange and his four companions were all set for a leisurely canoe trip down Santa Elena Canyon in Big Bend National Park in southwestern Texas. And there wasn't any reason to suspect it would be anything but that. The five University of Wisconsin–Whitewater students were in Big Bend during their spring break, and as Lange says, "it was supposed to be the dry season."

They set off on their planned two-day trip on March 20, 2004, and, for the first day, the waters were quiet, as is typical for early spring. Oftentimes, the water was so shallow that they were forced to carry their boats, and Lange estimated that it never got much deeper than 1 m (3 ft).

They had stopped for lunch on the second day about a quarter of a mile from the entrance to Santa Elena Canyon when, without warning, the group saw a "wall of water coming down the river" more than 2 m (6 ft.) high.

"It literally sounded like a train coming," Lange said. "It immediately made the river explode. The river was casually running along, and all of the sudden it was whitewater."

The area where the group had stopped was quickly being consumed by the rising water level, and they had no choice but to get in their canoes and enter the canyon (Figure 6.F).

The students would later learn that the flash flood that caught them unprepared had originated in Mexico and caused a 6-m (19-ft.) depth gauge to overflow, leading park officials to estimate that it was more than 7 m (20 ft.) tall and had boasted Class V rapids.

"The canoes we were in were not set up for those kinds of rapids whatsoever," Lange said.

Once in the water, the group quickly ran into trouble. Almost immediately, one of Lange's companions, Nick Gomez, capsized in his one-person canoe. After joining Lange and one of their other companions, the five set off again down the rapids in a pair of two-person canoes.

After a second boat flipped on an unexpected drop, the group stopped at a gravel bar to regroup. Repose, however, was brief; the very land beneath their feet quickly disappeared under the rising water. Within minutes, both remaining boats had capsized and the five students were scattered.

Amazingly, Lange's original partner in his canoe, Lisa Chowdhury, managed to cling to one canoe all the way down the river. The remaining four were split into two groups, battling hypothermia through the night. A hiker finally spotted one party the next morning, and park rangers rescued the group using spotter planes and jet boats.

Once the students had all learned that their companions were safe, Lange said the entire experience was thrilling beyond belief.

"Once everyone got rescued, it was the most insane rush," he said.

And what's more, they want to go back.

—CHRIS WILSON

FIGURE 6.F FLASH FLOOD ON THE RIO GRANDE Lange was able to snap a brief photograph as he and his companions were tossed into the Rock Slide Rapids. The rocks pictured, and his canoe, were submerged within 2 minutes. This is a photograph of the Rio Grande in Santa Elena Canyon, Big Bend National Park, Texas-Mexico border. *(Jason Lange)*

Downstream floods cover a wide area and are usually produced by storms of long duration that saturate the soil and produce increased runoff. Although flooding in small tributary basins is generally limited, the combined runoff from thousands of tributary basins produces a large flood downstream. A flood of this kind is characterized by the downstream movement of the floodwaters with a large rise and fall of discharge at a particular location.[24] An example of a downstream flood comes from the Chattooga and Savannah Rivers of Georgia and South Carolina (Figure 6.20a). As a flood crest on this river system migrated 257 km (160 mi.) downstream, it took a progressively longer time for the water to rise and fall (Figure 6.20b). The hydrograph of the flood at the downstream gauging station in Clyo, Georgia, shows that the floodwaters took 5 days to reach their peak of more than 1700 m^3 (60,000 $ft.^3$) per second.[25] Another way of looking at a flood is to examine the volume of discharge per unit area of the drainage basin (Figure 6.20c). This approach eliminates the effect of downstream increases in discharge and better illustrates the shape and form of the flood peak as it moves downstream.[26]

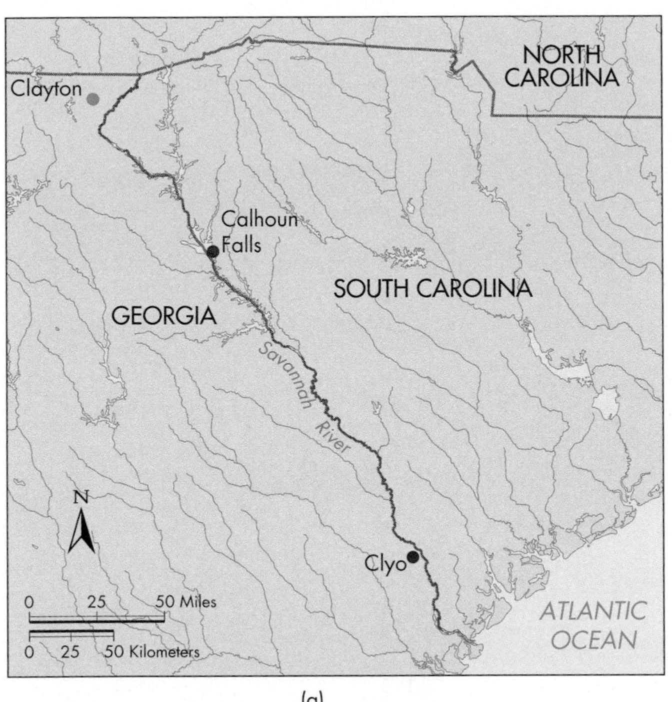

(a)

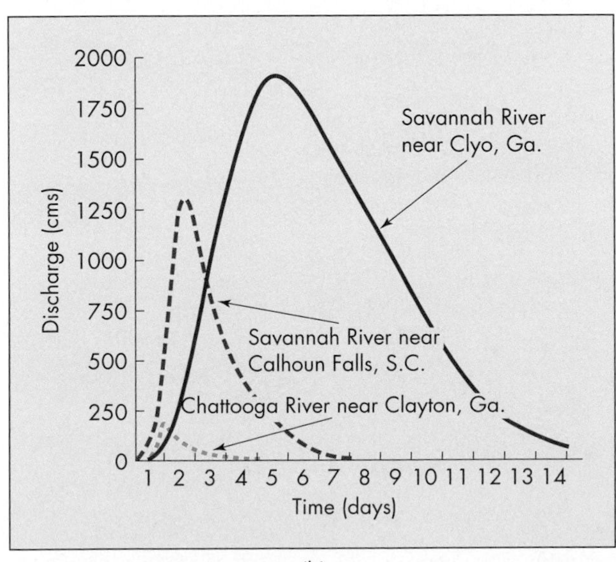

(b)

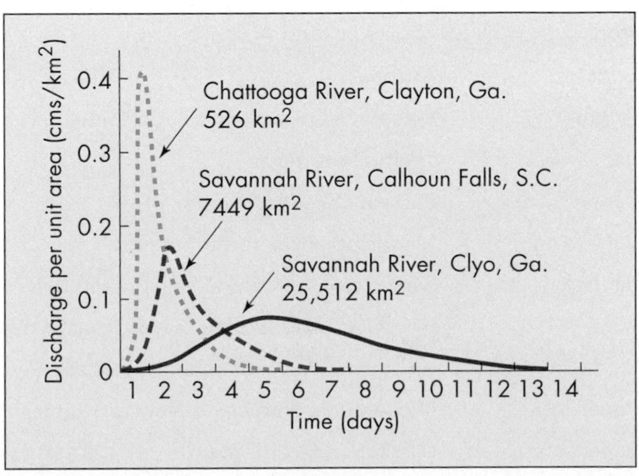

(c)

◀ **FIGURE 6.20 DOWNSTREAM MOVEMENT OF A FLOOD CREST** Floodwaters, as they moved downstream along the Savannah River on the state line between South Carolina and Georgia. The distance from Clayton to Clyo is 257 km (160 mi.). (a) Map of the area. (b) Volume of water passing Clayton, Calhoun Falls, and Clyo each second; the discharge increases as tributaries add more water. (c) Volume of water per unit area at the same points; flooding at Clio lasts much longer than flooding in Clayton because of the addition of water from thousands of small tributaries. *(After Hoyt, W. G., and Langbein, W. B. Floods. © 1955 by Princeton University Press, figure 8, p. 39. Reprinted by permission of Princeton University Press)*

6.3 Geographic Regions at Risk for Flooding

Flooding is one of the most universally experienced natural hazards. Any place that receives precipitation has the potential to flood. In the United States, floods were the number-one disaster during the twentieth century, and an average of 100 lives were lost each year from river flooding. Tragically, developing countries suffer much greater losses because of a lack of monitoring facilities, warning systems, adequate infrastructure and transportation systems, and effective disaster relief.[17, 26]

Virtually all areas of the United States and Canada are vulnerable to floods. A single flood can cause billions of dollars of property damage and hundreds of lives to be lost (Table 6.1). In as few as 5 years, large areas of the United States are affected by major flooding (Figure 6.21). Not shown on this map are the hundreds of small drainage basins that experienced flooding during this time period.

TABLE 6.1

Selected River Floods in the United States

Year	Month	Location	No. of Lives Lost	Property Damage (Millions of Dollars)
1947	May–July	Lower Missouri and middle Mississippi River basins	29	235
1937	Jan.–Feb.	Ohio and lower Mississippi River basins	137	418
1938	March	Southern California	79	25
1940	Aug.	Southern Virginia and Carolinas and eastern Tennessee	40	12
1951	June–July	Kansas and Missouri	28	923
1955	Dec.	West Coast	61	155
1963	March	Ohio River basin	26	98
1964	June	Montana	31	54
1964	Dec.	California and Oregon	40	416
1965	June	Sanderson, Texas (flash flood)	26	3
1969	Jan.–Feb.	California	60	399
1969	Aug.	James River basin, Virginia	154	116
1971	Aug.	New Jersey	3	139
1972	June	Rapid City, South Dakota (flash flood)	242	163
1972	June	Eastern United States	113	3000
1973	March–June	Mississippi River	0	1200
1976	July	Big Thompson River, Colorado (flash flood)	143	35
1977	July	Johnstown, Pennsylvania	76	330
1977	Sept.	Kansas City, Missouri, and Kansas	25	80
1979	April	Mississippi and Alabama	10	500
1983	Sept.	Arizona	13	416
1986	Winter	Western states, especially California	17	270
1990	Jan.–May	Trinity River, Texas	0	1000
1990	June	Eastern Ohio (flash flood)	21	Several
1993	June–Aug.	Mississippi River and tributaries	50	15,000
1997	January	Sierra Nevada, Central Valley, California	23	Several hundred
2001	June	Houston, Texas, Buffalo Bayou (coastal river)	22	2000
2004	Aug.–Sept.	Georgia to New York and the Appalachian Mountains	±13	>400
2006	June–July	Mid-Atlantic (Virginia to New York State)	16	1000
2008	June	Mississippi River System	24	9000
2010	May	Cumberland and other rivers (Nashville, Tennessee region)	24	several 1000s probable

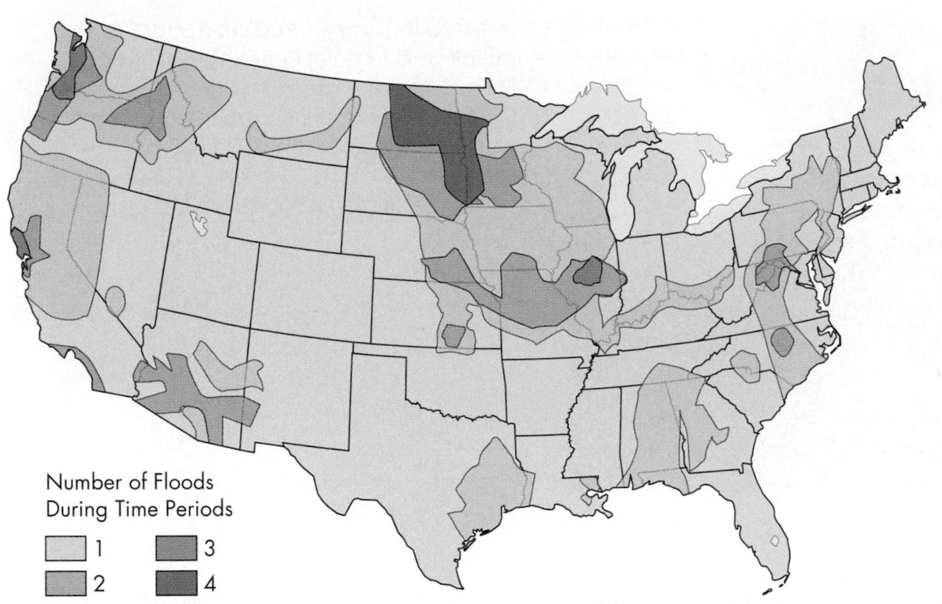

◀ **FIGURE 6.21 MAJOR FLOODING FROM 1993 TO 1997** Areas of the contiguous United States affected by major floods from 1993 to 1997. Colors show areas affected once, twice, three times, or more during this time period. This map shows only major drainage basins that were flooded. Actual flood hazard covers a much greater area. *(Modified from U.S. Geological Survey)*

For example, central Texas had six major episodes of flash flooding during the 5-year period, with damage totals exceeding $10 million.[27] When we realize that still more areas are at risk, we can see the seriousness of the flood hazard.

6.4 Effects of Flooding and Linkages Between Floods and Other Hazards

The effects of flooding may be primary, that is, directly caused by the flood, or secondary, caused by disruption and malfunction of services and systems because of the flood.[26] Primary effects include injury; loss of life; and damage caused by swift currents, debris, and sediment to farms, homes, buildings, railroads, bridges, roads, and communication systems. Erosion and deposition of sediment during a flood may also involve considerable loss of soil and vegetation.

Secondary effects may include short-term pollution of rivers, hunger and disease, and displacement of persons who have lost their homes. Failure of wastewater ponds, treatment plants, sanitary sewers, and septic systems often contaminate floodwaters with disease-causing microorganisms. For instance, in June 1998, flooding from record rainfall in southern New England caused partially treated sewage to float into Boston Harbor, and many areas of Rhode Island's Narragansett Bay were closed to swimming and shell fishing because of sewage-contaminated floodwaters.

Several factors affect the damage caused by floods:

- Land use on the floodplain
- Depth and velocity of floodwaters
- Rate of rise and duration of flooding
- Season of the year in which flooding takes place
- Quantity and type of sediment transported and deposited by floodwaters
- Effectiveness of forecasting, warning, and evacuation

In general, commercial and residential properties may experience more damage than land used for farming, ranching, or recreation. On the other hand, long-duration flooding during the growing season may completely destroy crops, whereas the same flooding of cropland during winter months may be less damaging. For downstream floods, accurate flood forecasts by the National Weather Service sometimes provide the warning time needed to build temporary levees or remove property from the hazard area.

As discussed in their respective chapters, floods can be a primary effect of hurricanes and a secondary effect of earthquakes and landslides. Although it may seem counterintuitive, floods may also cause fires in urban areas. Floodwaters may produce shorts in electric circuits and erode and break natural gas mains, resulting in dangerous fires.[26] For example, the 1997 flood in Grand Forks, North Dakota, caused a fire that burned part of the city center (Figure 6.22). Floods may also contribute to coastal erosion, especially in estuaries and deltas where high water levels may allow storm waves to reach further inland. River processes in general may also lead to landslides where stream banks are eroded.

6.5 Natural Service Functions

Although flooding is considered to be a natural hazard, and sometimes a disaster, it is important to remember that "flooding" is no more than the natural process of overbank flow. It becomes a hazard only when people live or build structures on the floodplain or try to

◀ **FIGURE 6.22 FLOODED CITY** Buildings burned in Grand Forks, North Dakota, in a fire caused by the flooding of the Red River of the North in 1997. The flood caused the evacuation of 50,000 people and inflicted nearly $4 billion in damage. *(Eric Hylden/Grand Forks Herald)*

cross a flooding river. In fact, there are many benefits to periodic flooding. Floods provide fertile sediment for farming; benefit aquatic ecosystems; and, in some cases, help keep land above sea level.

FERTILE LANDS

Floodplains themselves are built by floods. As the river overflows its banks, the velocity of flow decreases, and fine sand, silt, clay, and organic matter are deposited on the floodplain. These periodic deposits explain why floodplains are some of the most fertile and productive agricultural areas in the world. Recognizing the value of floods, the ancient Egyptians planned their farming around regular flooding of the Nile. They learned that the higher the flood, the better that year's harvest would be and even referred to flooding as "The Gift of the Nile." Unfortunately, with completion of the Aswan Dam in 1970, the river's annual floods in Egypt have effectively been stopped. Now farmers must use fertilizer to successfully grow crops on what was once naturally fertile land.

AQUATIC ECOSYSTEMS

Floods also help flush out stream channels and remove debris, such as boulders, logs, and tree branches that may have accumulated. Such events generally have a positive effect on fish and other aquatic animals. This benefit may be directly translated into a societal benefit in areas where fishing is common. Also, floods sweep nutrients and other food supplies downstream, potentially increasing the survival of aquatic organisms in this area.

SEDIMENT SUPPLY

In some cases, flooding is needed simply to keep the elevation of a landmass above sea level. For example, the Mississippi Delta in southeastern Louisiana is built up of sediment deposited as the Mississippi River repeatedly overflowed its banks through time. Construction of levees along the Mississippi has all but eliminated flood sedimentation, with the result that much of the delta is now slowly subsiding. Some areas, such as the city of New Orleans, are beneath sea level and are regularly threatened by hurricanes that cause both coastal and river flooding (see Chapter 10).

An experimental flood carried out below a dam on the Colorado River in the Grand Canyon during 1996 provides an excellent example of many of the natural service functions of floods (see Case Study 6.4).

6.6 Human Interaction with Flooding

Unlike some other natural hazards, human activity can significantly affect river processes, including the magnitude and frequency of flooding. Land-use change and dam construction may affect a stream's sediment supply, which can, in turn, alter its gradient and the shape of its channel. Urbanization, with its addition of paved areas, buildings, and storm sewers, greatly affects the flood hazard of an area.

LAND-USE CHANGES

Streams and rivers are open systems that generally maintain a rough *dynamic equilibrium*, that is, an overall balance between the work that the river does transporting sediment and the load that it receives. Sediment is supplied from tributaries and from earth material that falls down hillsides to the stream. A stream tends to have the gradient and cross-sectional shape that provides the flow velocity needed to move its sediment load.[28]

An increase or decrease in the amount of water or sediment received by a stream usually brings about changes in its gradient or cross-sectional shape, effectively changing the velocity of the water. The change in velocity may, in turn, increase or decrease the amount of sediment carried in the system. Therefore, land-use changes that affect a stream's sediment or water volume may set into motion a series of events that result in a new dynamic equilibrium.

Consider, for example, a land-use change from forest to an agricultural row crop, such as corn. Because crop lands have higher erosion rates, they provide more sediment to the stream. At first, the stream will be unable to transport the entire load and will deposit sediment, increasing the channel gradient. The new, steeper slope of the channel will increase the velocity of water and allow the stream to move more sediment. If we assume that the base level remains constant, this process will continue until the stream is flowing fast enough to carry the new load. If the notion that deposition of sediment increases channel gradient is counterintuitive to you, study the change in slope of the longitudinal profiles in Figure 6.23. A new dynamic equilibrium may be reached, provided the rate of sediment accumulation levels off and the channel gradient and shape can adjust before another land-use change occurs. Suppose the reverse situation now occurs, that is, farmland is converted to forest. The sediment supply to the stream will decrease, and less sediment will be deposited in the channel. Erosion of the channel will eventually lower the gradient, which, in turn, will lower the velocity of the water. The predominance of erosion over deposition will continue until equilibrium is again achieved between sediment supply and work done.

The sequence of change just described occurred in parts of the southeastern United States during the past two and a half centuries. On the Piedmont, an area of rolling hills between the Appalachian Mountains and the Atlantic Coastal Plain, most forests were cleared for farming by the 1800s. The land-use change from forest to farming accelerated soil erosion and subsequent deposition of sediment in local streams. This caused the channel that existed before farming to begin to fill with sediment. After 1930, the land reverted to pine forests, and this change, in conjunction with soil conservation measures, reduced the quantity of sediment delivered to streams. Thus, by 1969, formerly muddy streams choked with sediment had cleared and eroded their channels (Figure 6.24).

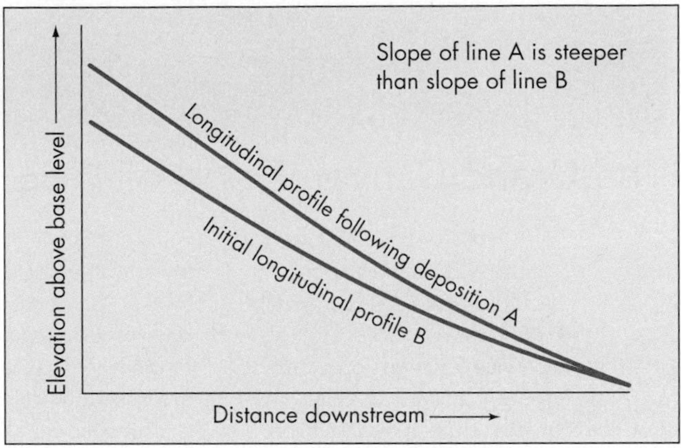

▲ **FIGURE 6.23 EFFECT OF DEPOSITION ON RIVER SLOPE** Idealized diagram illustrating that deposition in a stream channel increases the channel gradient; thus, the slope of longitudinal profile A is steeper than the slope of longitudinal profile B.

URBANIZATION AND FLOODING

Human activities increase both the magnitude and frequency of floods in small urban drainage basins of a few square kilometers. The rate of increase is determined by the percentage of the land that is covered with roofs, pavement, and cement, referred to as *impervious cover* (Figure 6.26), and the percentage of the area served by storm sewers. In most urban areas, storm sewers start at drains along the sides of streets and carry runoff to stream channels much more quickly than in natural settings. Therefore, impervious cover and storm sewers are collectively a measure of the degree of urbanization. An urban area with 40 percent impervious cover and 40 percent of its area served by storm sewers can expect to have about three times more floods of a given magnitude than before

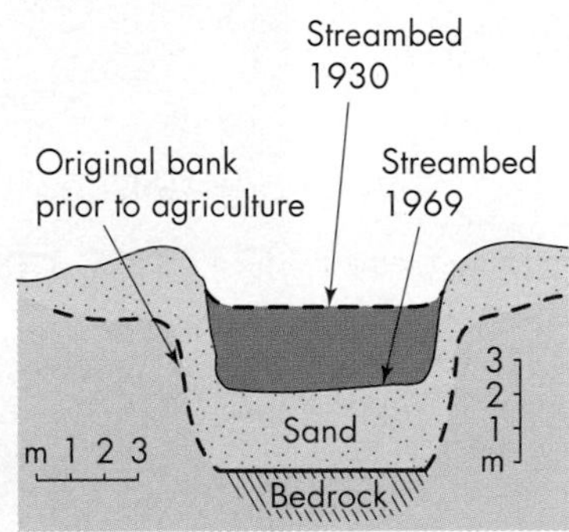

▲ **FIGURE 6.24 CHANGES IN A STREAM BED FROM CHANGES IN LAND USE** Cross section of a stream at the Mauldin Millsite, Georgia, in 1969 showing changes in channel position through time. Land use changed from natural forest to agriculture, which increased stream sedimentation until 1930, and then back to woodland, which increased stream erosion. *(After Trimble, S. W. 1969.* Culturally accelerated sedimentation on the middle Georgia Piedmont. *Master's thesis, Athens, Georgia: University of Georgia. Reproduced by permission)*

DAM CONSTRUCTION

Consider now the effect of a dam on a stream. Considerable changes will take place both upstream and downstream of the reservoir created behind the dam. Upstream, as the stream enters the reservoir, the water will slow down, deposit sediment, and form a delta. Downstream, the water coming out below the dam will have little sediment, since most of it has been trapped in the reservoir. As a result, the stream may have the capacity to transport additional sediment; if so, channel erosion rather than deposition will predominate downstream of the dam. The slope of the stream will then decrease until new equilibrium conditions are reached (Figure 6.25).

CASE STUDY 6.4

The Grand Canyon Flood of 1996

The Grand Canyon of the Colorado River (Figure 6.G) provides a good example of the natural service functions provided by periodic flooding. In 1963, the Glen Canyon Dam was built upstream from the Grand Canyon. Construction of the dam drastically altered both the natural pattern of flow and channel processes downstream—from a hydrologic viewpoint, the Colorado River was tamed.

Before the Glen Canyon Dam, the river reached a maximum flow in May or June during the spring snowmelt, and flow receded during the remainder of the year, except for occasional flash floods caused by upstream rainstorms. During periods of high discharge, the river had a tremendous capacity to transport sediment and vigorously scour its channel. This competence was evident when the river moved large boulders from the rapids. Shallow areas in the river, which become rapids, develop where the river flows over an alluvial fan or over debris flow deposits delivered from tributary canyons. As the summer low flow approached, the river was able to carry less sediment, and it deposited sand and gravel along the channel to form large bars and terraces, known as *beaches* to people who raft the river.

After the dam was built, the mean annual flood as determined by averaging the highest flow each year was reduced by about 66 percent, and the 10 year flood was reduced by about 75 percent. The dam controlled the flow to such an extent that the median, or most frequent, discharge actually increased by 66 percent. However, the flow is highly unstable because it depends on fluctuating flow for electric power. Changes in the amount of water released from the dam can cause the level of the river to vary by as much as 5 m (16 ft.) per day. The dam also greatly reduced the sediment load immediately downstream from the dam. The sediment load farther downstream is less reduced because tributary channels continue to add sediment to the channel.[29]

The Colorado River's changed flow conditions in the Grand Canyon have greatly altered both the channels and banks. Rapids may be becoming more dangerous because large floods are no longer moving many of the boulders from shallow areas of the channel. In addition, some of the large sandbars are disappearing because the river is deficient in sediment below the dam. The river is eroding this valuable habitat.

Changes in the river flow, mainly the loss of high flows, have also shifted vegetation. Before the dam was built, three nearly parallel belts of vegetation were present on the slopes above the river. Adjacent to the river and on sandbars, plant growth was scoured by yearly spring floods. Above the high waterline of the annual floods were clumps of thorned trees, such as mesquite and catclaw acacia, mixed with cactus and Apache plume. Higher yet was a belt of widely spaced brittle brush and barrel cactus.[30] For the first 20 years following its completion, the dam significantly reduced the spring floods. The absence of high, scouring flows allowed plants not formerly found in

▲ FIGURE 6.G **THE COLORADO RIVER IN THE GRAND CANYON** The sandbar "beach" in the lower left is being used by river rafters, whose numbers have increased enough to affect the canyon. The number of people allowed to raft through the canyon is now restricted. *(Craig Lovell/ Eagle Visions Photography/Alamy)*

the canyon, including tamarisk (salt cedar) and indigenous willow, to become established in a new belt along the river banks.

In June 1983, a record snowmelt in the Rocky Mountains forced the release of about three times the normal amount of water from the dam. This release was about the same volume as an average spring flood prior to construction of the dam. The resulting flood scoured sediment from the river bed and banks, which replenished the sandbars, and it scoured out or broke off some of the tamarisk and willow trees. This large release of water was, thus, beneficial to the river, and it emphasizes the importance of large floods in maintaining the system in a more natural state.

Three years later, a "test flood" was released from the dam as an experiment to redistribute the sand supply. The experimental flood formed 55 new beaches and added sand to 75 percent of the existing beaches. It also helped rejuvenate marshes and backwaters, which are important habitats to native fish and some endangered species. The experimental flood was hailed a success,[31] although a significant part of the new sand deposit was subsequently eroded away.[32]

Although the 1996 test flood distributed sand from the channel bottom and banks to sandbars, tributary streams added little new sand, since they were not in flood. The sand scoured from the river below the dam is, thus, a limited, nonrenewable source that cannot resupply sandbars on a sustainable basis. Because of this, a creative new idea has recently been suggested.[32] The plan is to use the sand delivered to the Grand Canyon by the Little Colorado River, a relatively large river with a drainage area of 67,340 km^2 (26,000 mi.^{2}) that joins the Colorado River downstream of the Glen Canyon Dam (Figure 6.H). In 1993, a flood on the Little Colorado River delivered a large volume of sand to the Colorado River in the Grand Canyon and produced a number of prominent beaches. Unfortunately, in the following year, the beaches were nearly eroded away by the Colorado River. The problem was that the beaches were not deposited high enough above the bed of the Colorado River and were, thus, vulnerable to erosion from normal flows released from the dam.

A recent study recommends that floods from Glen Canyon Dam coincide with the sand-rich spring floods of the Little Colorado River. The resulting combined flood of the two rivers would be larger, and the new sand from the Little Colorado (Figure 6.I) would be deposited higher above the channel bed in a location where it is less likely to be removed by lower flows of the Colorado. Evaluation of the hydrology of the Little Colorado River suggests that the opportunity to replenish sand on the beaches occurs, on average, once in 8 years. The proposed plan would restore or re-create river flow and sediment transport conditions as close as possible to those that existed prior to the construction of Glen Canyon Dam.[32]

One final effect of the Glen Canyon Dam has been to increase the number of people rafting through the Grand Canyon. Although rafting is now limited to 15,000 people annually, their long-range impact on canyon resources is bound to be appreciable. Prior to 1950, fewer than 100 explorers and river runners had made the trip through the canyon.

We must concede that the Colorado River is a changed river. Despite the 1983 and 1996 floods that pushed back some of the changes, river restoration efforts cannot be expected to return the river to what it was before construction of the dam.[28,29,31] On the other hand, better management of the flows and sediment transport will improve and help maintain the river ecosystem.

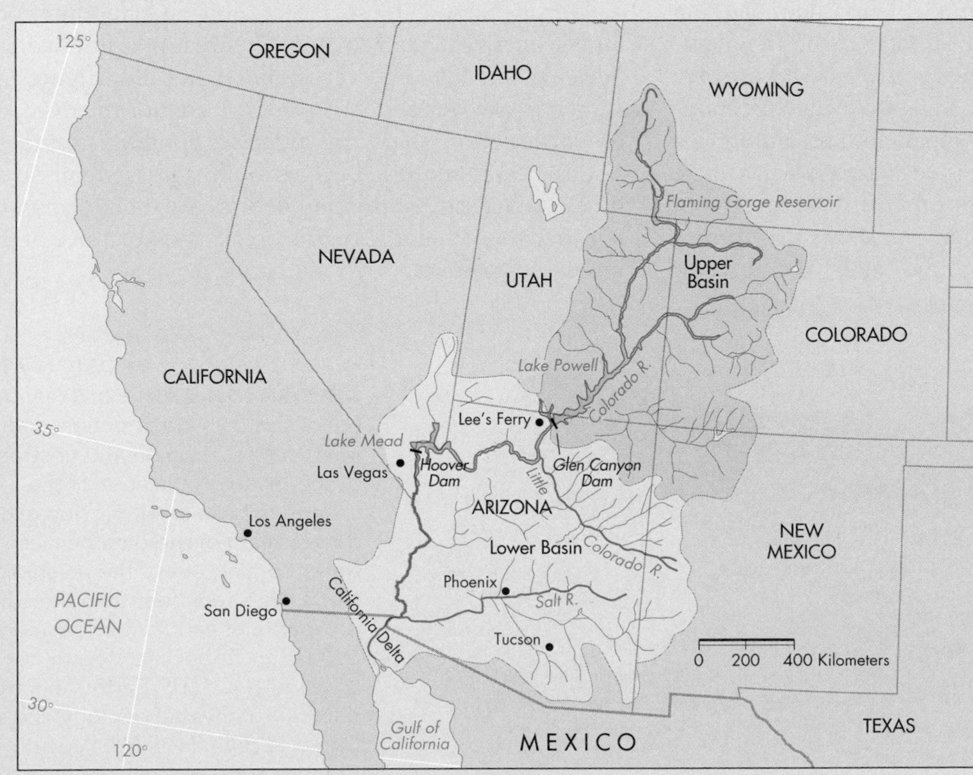

FIGURE 6.H THE COLORADO RIVER BASIN Glen Canyon Dam, just south of the Arizona-Utah state line, divides the Colorado River basin (shaded areas) in half for management purposes. Flaming Gorge Reservoir in Wyoming and Lake Powell in Utah store runoff in the Upper Basin, and Lake Mead on the Nevada-Arizona state line stores runoff in the Lower Basin. The delta, at the head of the Gulf of California, was once a large wetland area in Mexico. Today, it is severely degraded because of the diversion of Colorado River water for other uses in the United States.

FIGURE 6.1 **THE LITTLE COLORADO RIVER** The meandering channel of the Little Colorado River in the top center carries abundant sand to help maintain sandbar beaches in the Colorado River. *(Edward A. Keller)*

urbanization (Figure 6.27). This ratio applies to floods of small and intermediate frequency. However, as the size of the drainage basin increases, large floods with frequencies of approximately 50 years are less affected by urbanization.

Floods are a function of the relationship between rainfall and runoff, which is significantly changed by urbanization. One study showed that urban runoff from larger storms is nearly five times that of pre-urban conditions.[33] However, the extent of urban flooding is related not only to the peak discharge of a flood, but also to the condition of the drainage system. For example, long periods of only moderate precipitation can also cause flooding if storm drains become blocked with sediment and storm debris. In this case, water begins to pond behind a debris dam, causing flooding in low areas. An analogy is water rising in a bathtub shower when the drain becomes partly blocked by soap.

Besides increasing runoff and flooding frequency, urbanization affects how rapidly floods develop. Before urbanization, a considerable delay, or *lag time*, existed between when most rainfall occurred and the flood (Figure 6.28a). A comparison of hydrographs before and after urbanization shows that there is a significant reduction in lag time after urbanization (Figure 6.28). Short lag times, referred to as *flashy discharge*, are characterized by rapid rise and fall of floodwater. Another way that urbanization affects stream discharge is that it greatly reduces stream flow during the dry season. Normally, urban streams continue to flow during dry periods because groundwater seeps into the channel. However, because urbanization significantly reduces infiltration into the drainage basin, less groundwater is available to seep into streams. This reduced flow affects both water quality and the appearance of a stream. Very low discharge effectively concentrates pollutants in the water.[22] Some of these pollutants, such as nitrogen and phosphorus from fertilizer, can cause the growth of large amounts of algae and harm aquatic life.

During periods of surface runoff and flooding, pollutants from urban lands may enter streams as fine sediment. Sewer plants may be overwhelmed and fail (Figure 6.29). Impervious cover and storm sewers are not the only forms of construction that can increase flooding. Some flash floods occur because bridges built across small streams block the passage of floating debris that then forms a temporary dam. When the debris breaks loose, a wave of water moves downstream (see Case Study 6.5).

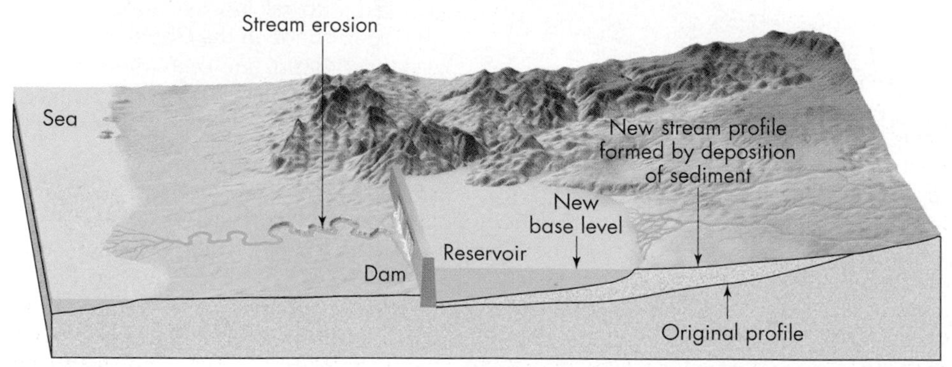

FIGURE 6.25 **EROSION AND DEPOSITION CAUSED BY A DAM** Following dam construction, sediment collects both in and upstream of the reservoir because of the change in base level. Erosion occurs downstream of the dam because water coming out of the reservoir carries less sediment than the stream is capable of carrying. *(Modified after D. Tasa, in Tarbuck, E. J., and Lutgens, F. K. 2005.* Earth: An introduction to physical geology, *8th ed. Upper Saddle River, NJ: Pearson Prentice Hall)*

◀ **FIGURE 6.26 URBANIZATION INCREASES IMPERVIOUS COVER** Aerial view of Santa Barbara, California, which, like most other U.S. cities, has much of its land surface covered by paved streets, sidewalks and parking lots, and buildings. This impervious land cover blocks the infiltration of water and increases surface runoff. *(Edward A. Keller)*

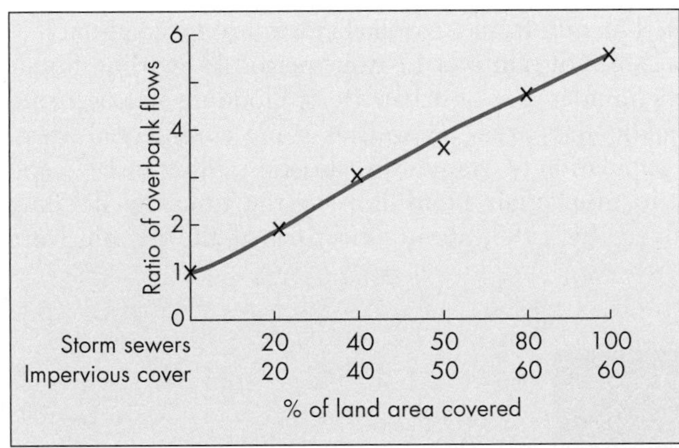

▲ **FIGURE 6.27 OVERBANK FLOODING BEFORE AND AFTER URBANIZATION** Relationship between the ratio of overbank flows (after urbanization compared to before urbanization) and the primary measures of urbanization. These measures are the percentage of area with impervious cover and storm sewers. For example, a ratio of 3 to 1, or simply 3, means that after urbanization there are three floods for every one that took place before urbanization, or that flooding is three times as common after urbanization. Note that the percentage of the land area with impervious cover typically reaches its maximum of 60 percent when about 80 percent of the urban area has storm sewers. The percentage of impervious cover then stays the same as the storm sewer system is completed for the community. This graph shows that as the degree of urbanization increases, the number of overbank flows per year also increases. *(After Leopold, L. B. 1968.* Hydrology for urban landplanning—A guidebook on the hydrologic effects of urban land use: *U.S. Geological Survey Circular 554)*

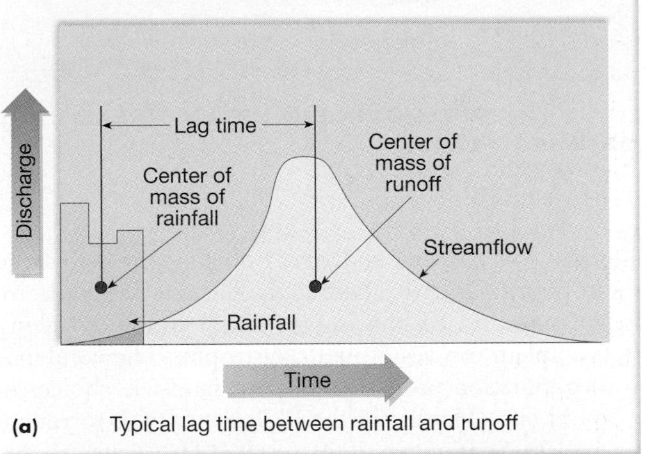

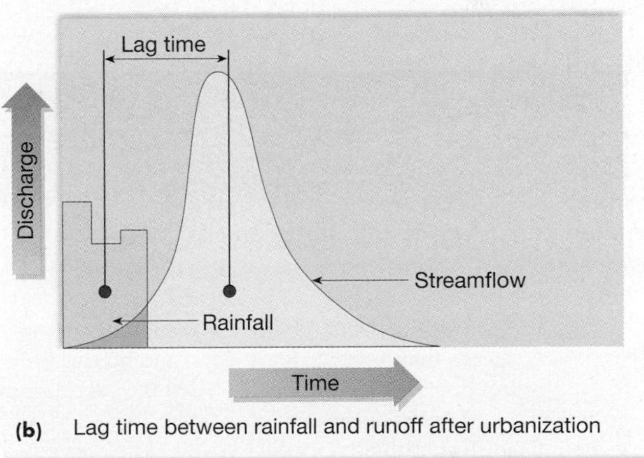

▲ **FIGURE 6.28 URBANIZATION SHORTENS LAG TIME** Generalized hydrographs. (a) Hydrograph shows the typical lag between the time when most of the rainfall occurs and the time when the stream floods. (b) Here, the hydrograph shows the decrease in lag time because of urbanization. *(After Leopold, L. B. 1968. U.S. Geological Survey Circular 559; modified after Tarbuck, E. J., and Lutgens, F. K. 2005.* Earth: An introduction to physical geology, *8th ed. Upper Saddle River, NJ: Pearson Prentice Hall)*

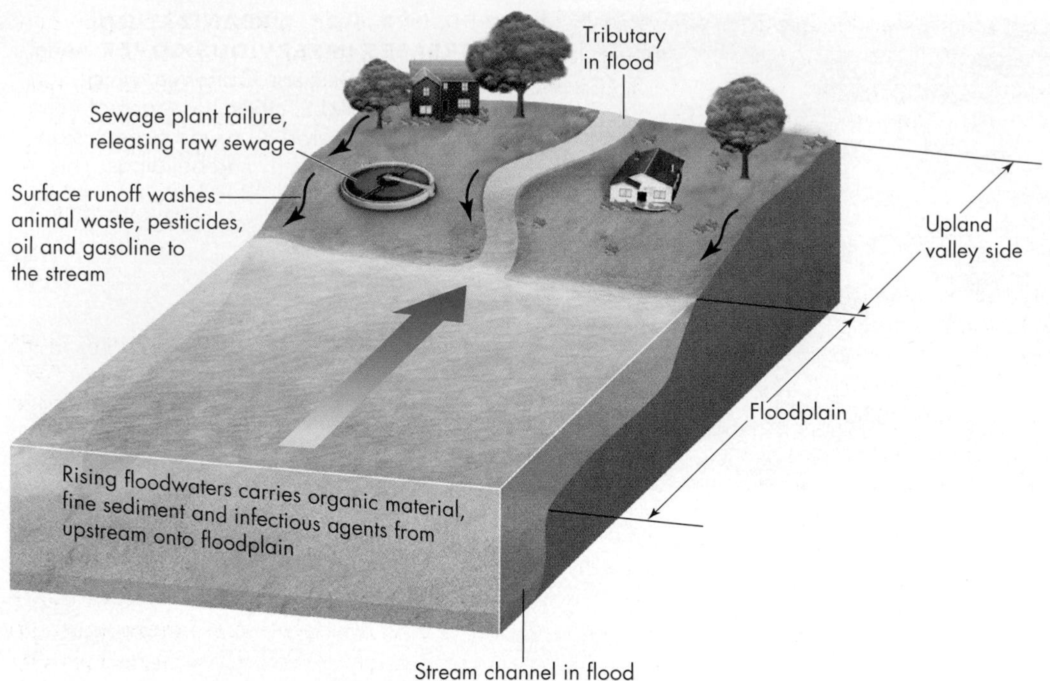

▲ FIGURE 6.29 **FLOODING IN URBAN STREAMS AND ASSOCIATED PROBLEMS/HAZARDS**

Flooding in streams and rivers flowing through large cities in poor countries where high densities of people are living in substandard housing (shanties) and encroaching on a floodplain can result in a catastrophe. The combination of population pressure and poor land-use choices in the capital city (Manila) of the Philippines with torrential rains in late September 2009 resulted in a catastrophic event. The rain from a tropical storm produced about 440 mm (18 in.) of rain in a 12-hour period.[34] Swirling floodwaters inundated poor urban areas. Flooding also occurred in middle-class areas, as well as some commercial areas. The population of Manila was about 5 million in 1975 and grew to more than 11 million by the time the flooding occurred. By 1980, about one-fifth of the people were

CASE STUDY 6.5

Flash Floods in Eastern Ohio

On Friday, June 15, 1990, more than 14 cm (5.5 in.) of precipitation fell within approximately 3 1/2 hours in parts of eastern Ohio. Two tributaries of the Ohio River, Wegee and Pipe Creeks, generated flash floods near the small town of Shadyside, killing 21 people and leaving 13 people missing and presumed dead. The floods were described as 5-m (16-ft.) high walls of water that rushed through the valley. In all, approximately 70 houses were destroyed and another 40 were damaged. Trailers and houses were seen washing down the creeks, bobbing like corks in the torrent.

The rush of water was apparently caused by the failure of debris dams that had developed across the creeks upstream of bridges. Surface runoff had washed tree trunks and other debris into the creeks from adjacent hill slopes. The debris became lodged against the bridges, creating dams. When the bridges could no longer contain the weight of the debris, the dams broke loose, sending surges of water downstream. This sequence of events has been repeated in many flash floods around the world. All too often, the supports for bridges are too close together or drainage pipes under roads, referred to as culverts, are too small to allow large debris to pass through; instead, they become temporary dams that cause flooding upstream or their failure causes flash flooding downstream.

living in crowded shantytowns on a wide floodplain with little or no infrastructure. They had no protection when the Marikina River flooded. Hundreds of people were killed, drowned in their houses and cars and in shopping malls (Figure 6.30).

6.7 Minimizing the Flood Hazard

Historically, particularly in the nineteenth century, humans have responded to floods by attempting to prevent them by modifying streams and rivers. Physical barriers, such as dams and levees, have been created, or the shape of the stream has been changed by widening, deepening, or straightening the channel. The channel shape is changed so that a stream will drain the land more efficiently. Every new flood-control project has the effect of luring more people to the floodplain in the false hope that the flood hazard is no longer significant. We have yet to build a dam or channel capable of controlling the heaviest runoff, and, when the water finally exceeds the capacity of the structure, flooding may be extensive.[26,35]

THE STRUCTURAL APPROACH

Physical Barriers Measures to prevent flooding include construction of physical barriers, such as earthen *levees* (Figure 6.31) and concrete flood walls, reservoirs to store water for later release at safe rates, and on-site storm water retention basins (Figure 6.32). Unfortunately, the potential benefits of these physical barriers are often lost because of increased development on the floodplains that they are supposed to protect. For example, the winters of 1986 and 1997 brought tremendous storms and flooding to the western states, particularly California, Nevada, and Utah. In all, damages exceeded several hundred million dollars, and several people died. During one of the floods in 1986, a levee broke on the Yuba River in California, causing more than 20,000 people to flee their homes. An important lesson learned during this flood is that levees constructed along rivers many years ago are often in poor condition and subject to failure during floods. Levee failures can also damage agricultural land by creating high-energy flows that erode topsoil and deposit a layer of sand and gravel on otherwise fertile soil in the floodplain.

▲ **FIGURE 6.30 FLOOD DAMAGE** Shantytown east of Manila in the Philippians devastated by flood waters in late September 2009. The combination of a high magnitude storm in a densely populated city with many thousands of people living on a floodplain produced a catastrophe. *(Pat Roque/Associated Press)*

The 1997 floods in California also damaged campsites and other development in Yosemite National Park. As a result, the park revised its floodplain management policy to allow the river to "run free." This new policy necessitated abandoning campsites and other facilities in the floodplain.

As described in the opening case, the Mississippi River floods caused levees to fail and the floodwalls at St. Louis, Missouri, produced a bottleneck for the river and increased flooding upstream of the city. Other communities, however, learned that strong floodplain regulations must go hand in hand with structural controls, if the hazard is to be minimized.[1,35,36,37]

Dams, like levees, are prone to failure if they are not properly maintained. Most dams in the United States are on private land. Although often constructed with government subsidies, they are poorly maintained and rarely inspected. In one recent example, more than 2000 people were evacuated in Montgomery County, Maryland, when heavy rains threatened to cause failure of a dam. Dams provide many valuable services, including flood control; however, they must be maintained and regularly inspected, or they can create an even greater flood hazard.

Channelization Straightening, deepening, widening, clearing, and/or lining existing stream channels are all methods of **channelization.** Objectives of this engineering technique include controlling floods and erosion, draining wetlands, and improving navigation.[38] Thousands of kilometers of streams in the United States have been modified without adequate consideration of the adverse effects of channelization. Thousands of additional kilometers of channelization projects are under construction or are planned.

Opponents of modifying natural streams emphasize that the practice is antithetical to the production of fish and wetland wildlife and causes extensive aesthetic degradation. Their argument is as follows:

- Drainage of wetlands adversely affects plants and animals by eliminating habitats necessary for the survival of certain species.
- Cutting trees eliminates shading and cover for fish and exposes the stream to the sun, which damages plant life and heat-sensitive aquatic organisms.
- Cutting floodplain hardwood trees eliminates many animal and bird habitats, while facilitating erosion and sedimentation of the stream.

◀ **FIGURE 6.31 MISSISSIPPI RIVER LEVEE** Earthen or concrete levees parallel the lower Mississippi River banks in Louisiana. A road on the top of this levee on the left bank of the river appears as a curving white line. This levee protects farms, homes, and businesses along the highway on the left. *(Comstock)*

- Straightening and modifying the streambed destroys both the diversity of flow patterns and feeding and breeding areas for aquatic life, while changing peak flow.
- Conversion of wetlands from a meandering stream to a straight, open ditch seriously degrades the aesthetic value of a natural area.[38] Figure 6.33 summarizes some of the differences between natural streams and those modified by channelization.

Not all channelization causes serious environmental degradation; in many cases, drainage projects are beneficial. Benefits are probably best observed in urban

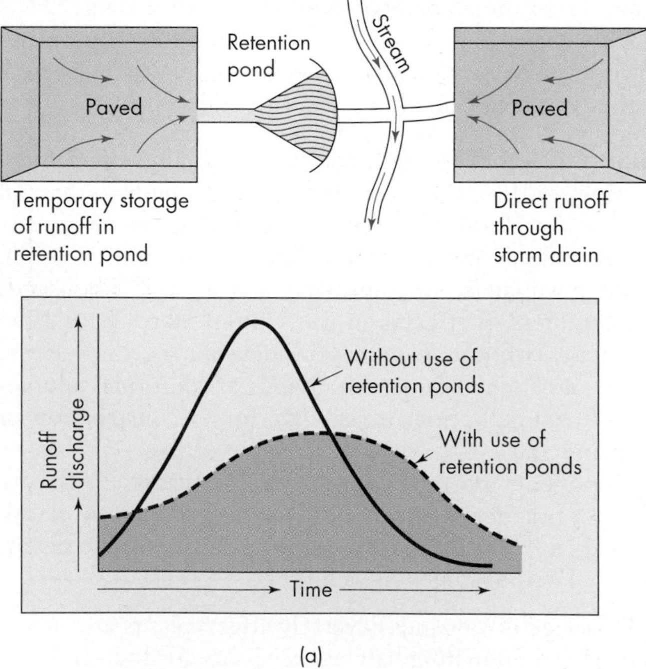

▲ **FIGURE 6.32 RETENTION PONDS REDUCE FLOOD DISCHARGE** (a) Comparison of runoff from a paved area, which goes directly through a storm drain to a stream, with runoff that is stored temporarily in a retention pond before draining to a stream. The graph shows that the use of a retention pond reduces peak discharge and the likelihood that the runoff will contribute to flooding of the stream. *(Modified after U.S. Geological Survey Professional Paper 950)* (b) A nearly dry retention pond near Santa Barbara, California; retention ponds also capture sediment that reduces pollution levels and sedimentation in streams. *(Edward A. Keller)*

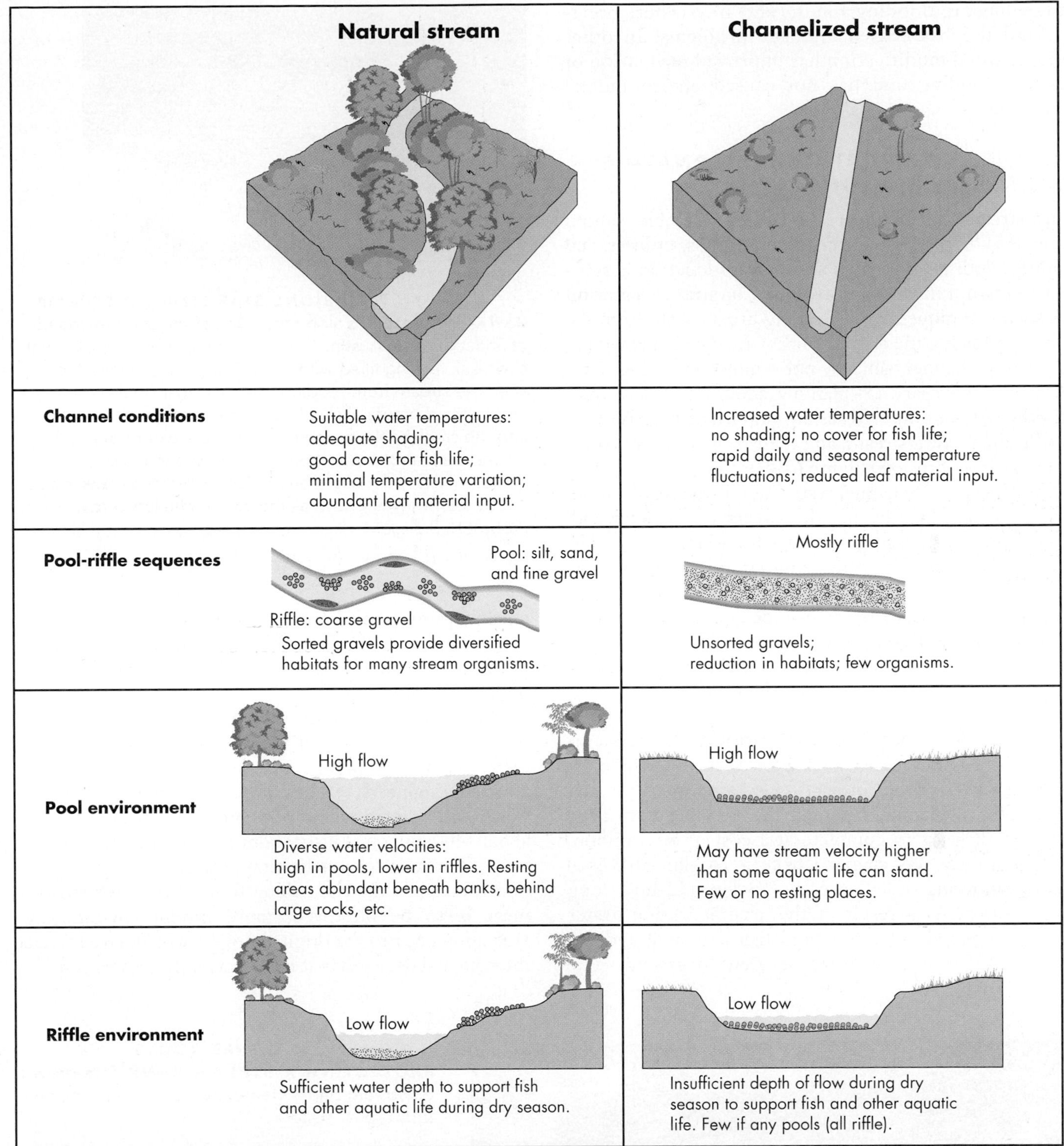

▲ FIGURE 6.33 **COMPARISON OF NATURAL AND CHANNELIZED STREAMS** Channelization of a stream significantly changes channel conditions, pool and riffle development, and environmental conditions in pools and riffles. *(Modified after Corning, Virginia Wildlife, February 1975)*

areas subject to flooding and in rural areas where previous land use has caused drainage problems. In other areas, channel modification has improved navigation or reduced flooding and has not caused environmental disruption.

CHANNEL RESTORATION: ALTERNATIVE TO CHANNELIZATION

Many streams in urban areas scarcely resemble natural channels. The process of constructing roads, utilities, and buildings with its associated sediment production is sufficient to disrupt most small streams. **Channel restoration** uses such techniques as (1) cleaning urban waste from the channel, allowing the stream to flow freely; (2) protecting the existing channel banks by not removing existing trees; and, where necessary, (3) planting additional native trees and other vegetation.[39] Trees are important because they provide shade for a stream, and their root systems protect the banks from erosion (Figure 6.34).

The objective of channel restoration is to create a more natural channel by allowing the stream to meander and, where possible, to reconstruct variable water-flow conditions with fast, shallow riffles alternating with slow, deep pools. Where lateral bank erosion must be absolutely controlled, the outsides of bends may be defended with large stones known as *riprap* or with wire baskets filled with rocks, known as *gabions* (Figure 6.35). Design criteria for channel restoration are shown on Figure 6.36.

▲ FIGURE 6.35 **GABIONS STABILIZING A STREAM BANK** These gabions along the bank of the dry streambed of Shoal Creek in Austin, Texas, are stacked baskets of metal chain-link fencing filled with rocks. They were installed to keep the stream from eroding nearby Lamar Boulevard. Their porous nature allows floodwaters and sediment to seep into the creek bank, thereby reducing the extent of flooding downstream and allowing plants to grow and partly cover the gabions. Nearly dry between rainfall events, the creek is also fed by storm sewers, such as the pipe in the left center. *(Robert H. Blodgett)*

Kissimmee River Restoration, Florida River restoration of the Kissimmee River in Florida may be the most ambitious restoration project ever attempted in the United States. Channelization of the river began in 1960 and took 10 years to complete at a cost of $32 million. The Kissimmee was changed from a 165-km (103-mi.) long meandering river into an 83-km (52-mi.) long straight ditch. As a result of this channelization, water quality decreased, waterfowl and fish declined, and the aesthetics of the area were severely degraded. The channelization drained more than 800 km^2 (310 $mi.^2$) of floodplain wetlands. Ironically, channelization increased the flood hazard because the wetlands on the floodplain no longer stored runoff.

Within a year after the channelization was completed, the state of Florida called for the river to be restored. In 1991, the U.S. Congress mandated that the U.S. Army Corps of Engineers, responsible for the original channelization, begin restoring about one-third of the river at a cost of about $400 million. This amount is more than 10 times the cost of the entire channelization project. Although restoration is certainly the right thing to do if the river environment is to be improved, more careful environmental evaluation prior to the original project would have revealed the potential damages that necessitated the restoration.[40]

◄ FIGURE 6.34 **TREE ROOTS PROTECTING A STREAM BANK** Tree growth on stream banks generally slows their erosion. *(Edward A. Keller)*

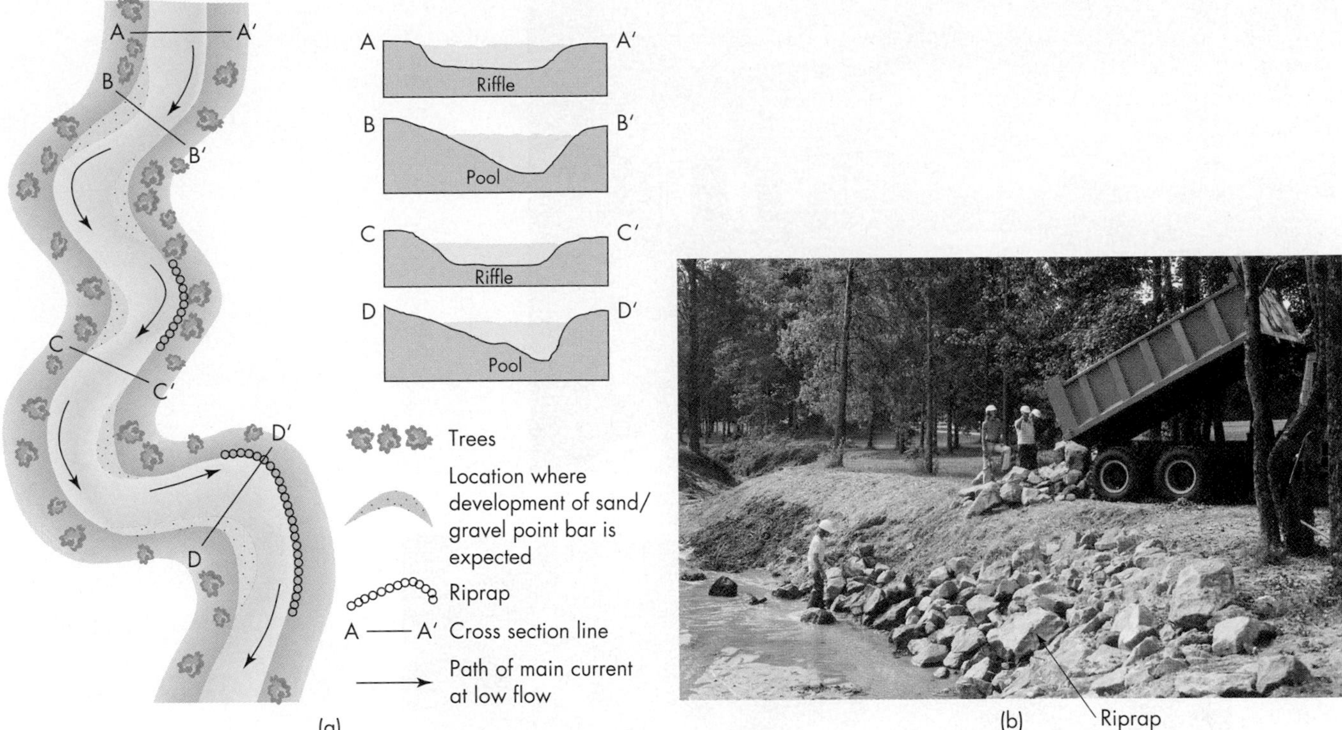

▲ FIGURE 6.36 **URBAN STREAM RESTORATION** (a) Channel restoration design criteria for urban streams, using changes in channel shape to cause scour and deposition at desired locations. Scour produces pools and deposition can create riffles. *(Modified after Keller, E. A., and Hoffman, E. K. 1977.* Urban streams: sensual blight or amenity. Journal of Soil and Water Conservation *32(5):237–240).* (b) Dump truck placing large rocks, referred to a riprap, where it is absolutely necessary to defend the bank of Briar Creek, Charlotte, North Carolina. To the left of the men, grass was planted with a straw mulch cover to help stabilize the stream bank. *(Edward A. Keller)*

In Los Angeles, California, a group called "Friends of the Los Angeles River" is working to have the Los Angeles River restored. This task will be difficult because most of the riverbed and banks are lined with concrete (Figure 6.37). However, a river park is planned for one section of the river where a more natural looking channel has reappeared since channelization (Figure 6.38).

6.8 Perception of and Adjustment to the Flood Hazard

PERCEPTION OF THE FLOOD HAZARD

Whereas most governmental agencies, planners, and policy makers have an adequate perception and understanding of flooding, many individuals do not. The public knowledge of flooding, anticipation of future flooding, and willingness to accept adjustments caused by the hazard are highly variable.

Progress at the institutional level includes the preparation of thousands of maps of flood-prone areas. The maps show areas susceptible to flooding along streams, lakes, and coastlines; areas with a flash-flood potential downstream from dams; and areas where urbanization is likely to cause problems in the near future. In addition, the federal government has encouraged states and local communities to adopt floodplain management plans.[41] Still, the idea of restricting or prohibiting development on floodplains or of relocating present development to locations off the floodplain raises its own problems; it needs further community discussion before the general population will consider accepting it. This need was tragically shown by the 2006 floods in the Mid-Atlantic United States, when severe river flooding impacted the region from Virginia to New York (Figure 6.39). More than 200,000 floodplain residents were evacuated in Wilkes-Barre, Pennsylvania, alone, and damages of approximately a billion dollars were incurred. About 16 people lost their lives as cars were swept away by floodwaters and people drowned in flood-swollen creeks and rivers. At least 70 people were rescued from rooftops. In Conklin, New York, near Binghamton, nearly three-fourths of the town was flooded as the Susquehanna River crested about 4.2 m (15 ft.) above flood level.

Increasing public awareness of the dangers of flash flooding could save many lives each year (see Professional Profile 6.6). Public safety campaigns by the National Weather Service and local governments, including the city of Boulder, Colorado; Clark County Regional Flood Control District, Nevada; and the city of Austin, Texas, have been especially

(a)

(b)

▲ FIGURE 6.37 **DIFFERENCES BETWEEN STREAM CHANNELIZATION AND RESTORATION** (a) Concrete-lined channel in the Los Angeles River system in California compared to *(Edward A. Keller)* (b) channel restoration in North Carolina. *(Edward A. Keller)*

◄ FIGURE 6.38 **STREAMS TRY TO RETURN TO NATURAL CONDITIONS** Part of Los Angeles River, California, where a more natural channel has developed since channelization. Plants have rooted in sediment deposited during flooding of the concrete channel. *(Deidra Walpole Photography)*

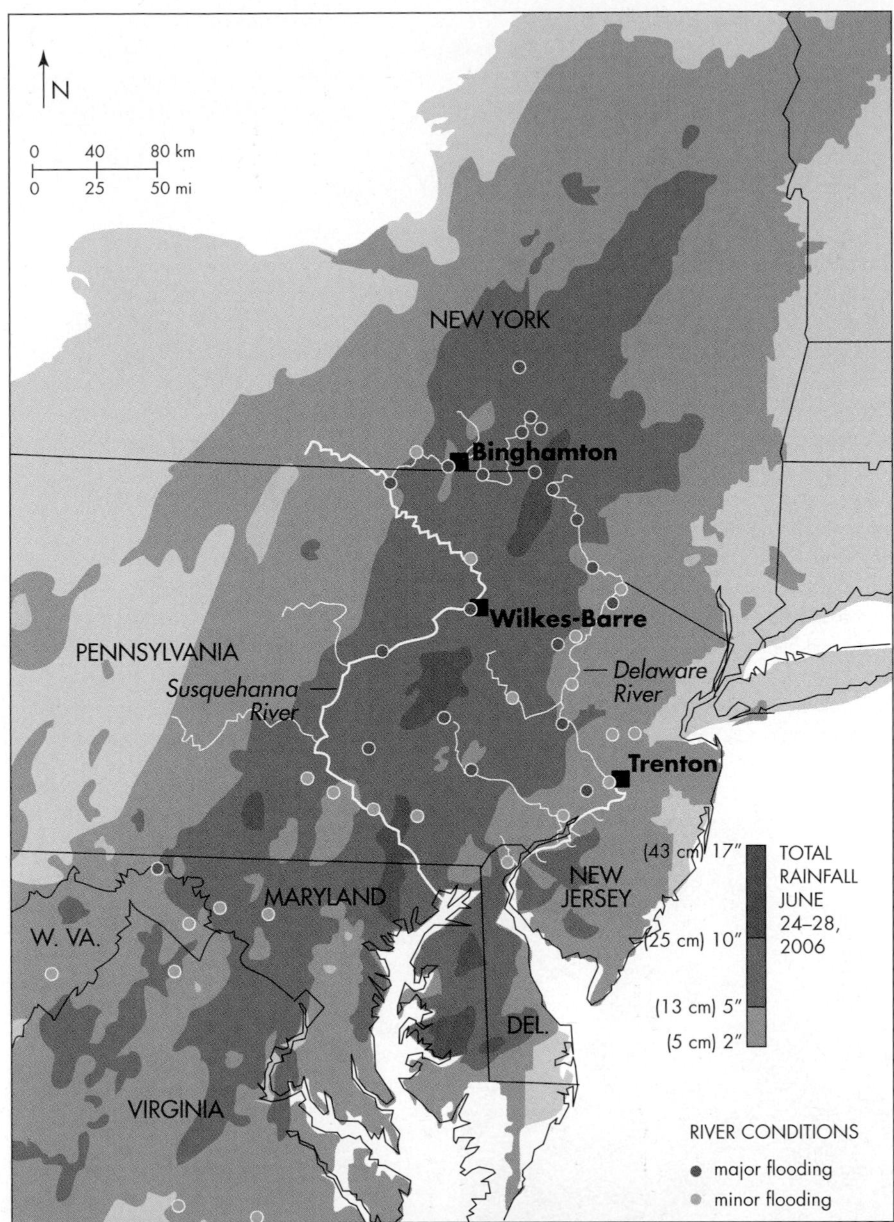

◀ **FIGURE 6.39 MID-ATLANTIC FLOODS OF JUNE AND JULY 2006** (a) Map of major and minor flooding. *(Modified from* New York Times *with data from National Weather Service)* (b) Collecting mail from flooded home in Wilkes-Barre, Pennsylvania. *(Matt Rourke/AP Images)*

PROFESSIONAL PROFILE 6.6

Professor Nicholas Pinter, Southern Illinois University

Nicholas Pinter (Figure 6.J) was one of my graduate students at the University of California, Santa Barbara, where he completed a dissertation working in the high desert of the Owens Valley and in the adjacent Sierra Nevada and White Mountains of California. After doing postdoctoral research at Yale University, Nick went to Southern Illinois University at Carbondale. Nick's research focuses on "quantifying human influences on the geology of the earth's surface." Recent research has shown that modern humans move more material each year than all the wind, running water, waves, and glacial ice on the Earth. "It's both easy and dangerous," Nick says, "to underestimate how much people have influenced the planet both locally and at the global scale."

Southern Illinois University at Carbondale is located near where the Mississippi, Ohio, Missouri, Illinois, and other rivers all come together. This central location kindled his interest in river processes, and issues of flood hazard in particular. "Some research questions are purely theoretical, but as soon as you start looking at big rivers, the science becomes relevant to people's lives and government policy really fast." Nick realized early on that some of the structures being built by the U.S. Army Corps of Engineers to facilitate shipping at low flows were behaving as large roughness elements and actually increasing the flood hazard during large floods. Over time, squeezing the river between levees and slowing down flows with structures built in the channel were leading to increasing flood stages. What that means is that, say 50 years ago, a flood of a certain discharge would inundate the banks at some depth. Today, the same discharge is associated with water many feet deeper and, hence, bringing a greater flood hazard. Although a number of outside scientists have reached similar conclusions, the U.S. Army Corps of Engineers did not receive the news that its own construction projects were actually making flooding much worse. Nick Pinter and his students have worked diligently to counter the political resistance with rock-solid science: quantifying past increases in flood stages and risk and confirming what caused those increases with hydrologic, statistical, geospatial, and modeling analyses. Nick pointed out that it is difficult to maximize for two things at the same time. That is, if you want to increase the ability of the river to transport goods through navigation at low flows, you may have one strategy. However, that strategy may not be the optimum solution for passing, say, the 100 year flood safely through the same stretch of river. "We live in the 21st century," says Nick, "in which robust science is vital to sustainable management of natural systems that are stretched to their limits in many places by the demands of modern society."

◀ **FIGURE 6.J PROFESSOR NICHOLAS PINTER** Professor Pinter is at Southern Illinois University. He is an expert on flooding of the Mississippi River and the role of human intervention through building flood defenses and structures to improve navigation. *(Nicholas Pinter.)*

successful in increasing public awareness about flash floods. With at least half of all flood deaths in automobiles, the need to educate drivers is especially critical. The National Weather Service slogan "Turn Around, Don't Drown" summarizes a life-saving safety precaution for all motorists not to drive on flooded roads. For hilly or mountainous terrain, road and trail signs can warn people to climb to higher ground if floodwaters begin to rise. In a narrow stream valley, this can mean abandoning your vehicle for the safety of higher ground if floodwaters are rising rapidly ahead of you or if you hear a roaring sound coming from upstream.

ADJUSTMENTS TO THE FLOOD HAZARD

In recent decades, we have begun to recognize the advantages of alternatives to structural flood control. These alternatives include flood insurance and controlling the land use on floodplains. Planners, policy makers, and hydrologists generally agree that no one adjustment is best in all cases. Rather, an integrated approach to minimizing the flood hazard is more effective, especially if it incorporates adjustments that are appropriate for a particular situation.

Flood Insurance In 1968, the federal government took over the flood insurance business when private companies became reluctant to continue to offer policies. Congress established the U.S. National Flood Insurance Program to make flood insurance available at subsidized rates. Administered by the Federal Emergency Management Agency (FEMA), this program requires the mapping of special flood hazard areas, defined as those areas that would be inundated by a 100 year flood. Flood hazard areas are designated along streams, rivers, lakes, alluvial fans, and deltas and along low-lying coastal areas susceptible to flooding during storms or very high tides.

New property owners in flood hazard areas must buy insurance at rates determined by the risk. Basic risk evaluation depends on identifying the area that would be inundated by the 100 year flood. The insurance program is intended to provide short-term financial aid to victims of floods and to establish long-term land-use regulations for the nation's floodplains. As part of this program, building codes are revised to limit new construction in a flood hazard area to flood-proofed buildings (Figure 6.40) and to prohibit all new construction in the area that would be inundated by the 20 year flood. For a community to join the National Flood Insurance Program, it must have FEMA prepare maps of the 100 year floodplain and adopt minimum standards of land-use regulation within the flood hazard areas. Nearly all United States communities with a significant flood risk have basic flood hazard maps and have initiated some form of floodplain regulation. Several million property owners in the United States presently have flood insurance policies.

By the early 1990s, policy makers and flood-control professionals recognized that the insurance program was in need of reform. This prompted Congress to pass the National Flood Insurance Reform Act of 1994. Provisions of the act encourage additional opportunities to mitigate flood hazards, such as flood-proofing, relocations, and buy-outs of properties likely to be frequently flooded.[42]

Flood-Proofing Several methods of flood-proofing are currently available:

- Raising the foundation of a building above the flood-hazard level using piles or columns or by extending foundation walls or earth fill.
- Constructing flood walls or earthen mounds around buildings to isolate them from floodwaters.
- Using waterproofing construction, such as waterproofed doors, basement walls, and windows.
- Installing improved drains with pumps to remove incoming floodwaters.

Other modifications to buildings are designed to minimize flood damage while allowing floodwaters to enter a building. For example, ground floors along expensive riverfront properties in some communities in Germany are designed so that they are not seriously damaged by floodwaters and may easily be cleaned for reuse following a flood.[42]

Floodplain Regulation From an environmental point of view, the best adjustment to the flood hazard in urban areas is **floodplain regulation.** The objective of floodplain regulation is to obtain the most beneficial use of floodplains while minimizing flood damage and cost of flood protection.[43] This approach is a compromise between the indiscriminate use of floodplains, which results in loss of life and tremendous property damage, and the complete abandonment of floodplains, which gives up a valuable natural resource.

There are circumstances, however, when physical barriers, reservoirs, and channelization works are required. Structural controls may be necessary to protect lives and property in areas where there is extensive development on floodplains. We need to recognize, however, that the floodplain belongs to the river system, and any encroachment that reduces the cross-sectional area of the floodplain increases flooding (Figure 6.41). An ideal solution would be discontinuing floodplain development that

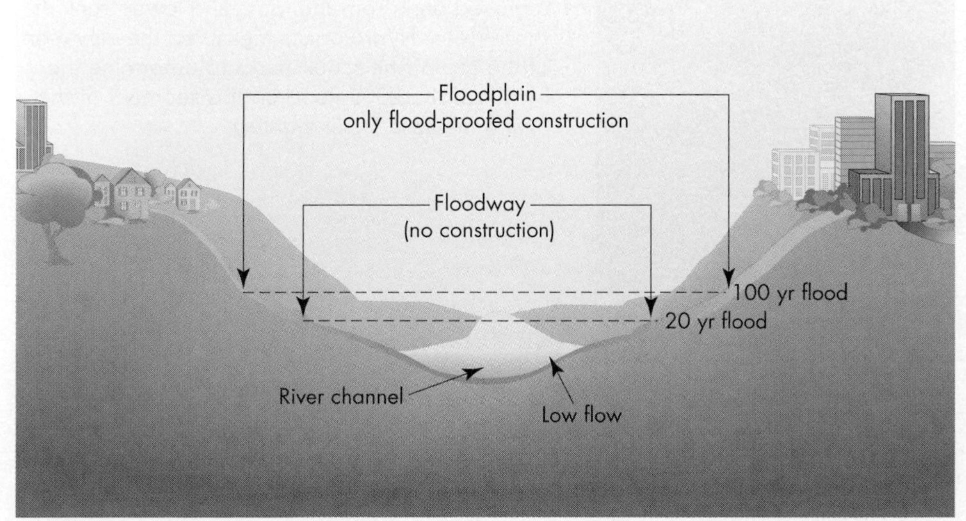

◀ **FIGURE 6.40 FLOODPLAIN REGULATION** Idealized diagram of areas inundated by the 100 and 20 year floods, the two recurrence intervals used for regulations in the U.S. National Flood Insurance Program.

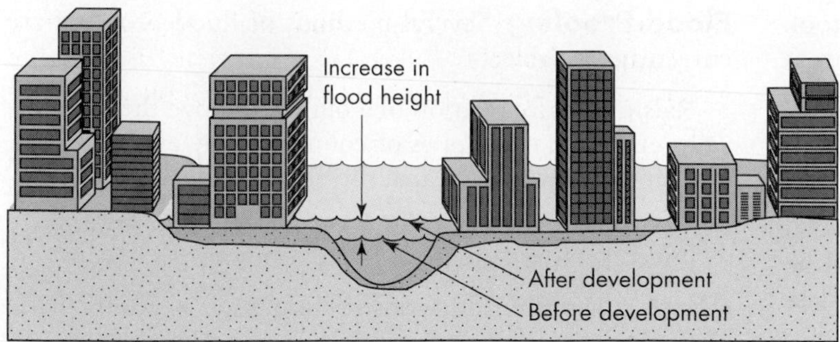

FIGURE 6.41 BUILDING ON FLOODPLAINS INCREASES HAZARD Development that encroaches on the floodplain reduces the space available for floodwater and can increase the heights of subsequent floods. *(From Water Resources Council. 1971.* Regulation of flood hazard areas, *vol. 1)*

necessitates new physical barriers. In other words, the ideal is to "design with nature." Realistically, the most effective and practical solution is a combination of physical barriers and floodplain regulations that results in less physical modification of the river system. For example, reasonable floodplain zoning in conjunction with a floodwater diversion channel or upstream reservoir may result in a smaller channel or reservoir than would be necessary without the zoning regulations.

A preliminary step toward floodplain regulation is flood-hazard mapping, which is a means of providing information about the floodplain for land-use planning.[44] Flood-hazard maps may delineate past floods or floods of a particular frequency—for example, the 100 year flood. They are useful in regulating private development, purchasing land for public use as parks and recreational facilities, and creating guidelines for future land use on floodplains.

Flood-hazard evaluation may be accomplished in a general way by direct observation and measurement of physical parameters. For example, the extensive flooding of the Mississippi River Valley during the summer of 1993 was clearly mapped using imagery from satellite and aircraft (see Figure 6.3). Flooding can also be estimated from the field measurement of high waterlines, flood deposits, scour marks (Figure 6.42), and the distribution of trapped debris on the floodplain after water has receded.[37] Once flood-hazard areas have been established, planners can modify zoning maps, regulations, and building codes (Figure 6.43).

RELOCATING PEOPLE FROM FLOODPLAINS: EXAMPLES FROM NORTH CAROLINA AND NORTH DAKOTA

For several years, local, state, and federal governments have been selectively purchasing and removing homes damaged by floodwaters to reduce future losses. In September 1999, nearly 50 cm (20 in.) of rain from Hurricane Floyd flooded many areas of North Carolina. The flooding damaged approximately 700 homes in Rocky Mount, North Carolina, population 60,000. State and federal governments subsequently decided to spend nearly $50 million to remove 430 of these homes, the largest single-home buyout ever approved. Following the purchase, the homes were demolished and the land was preserved as open space.

At Churchs Ferry, North Dakota, a wet cycle since 1993 caused nearby Devils Lake to rise approximately 8 m (26 ft.). With no natural outlet and flat land surrounding its shore, the lake more than doubled in area. The swollen

FIGURE 6.42 SCOUR MARKS INDICATE FLOOD HEIGHT Floodwaters of the San Gabriel River in Los Angeles County, California, removed bark from the roots and lower trunk of these trees. Hydrologists measured the elevation of the top of the scour marks to determine the height of the 2005 flood on this segment of the river. *(Theressa M. Carpenter)*

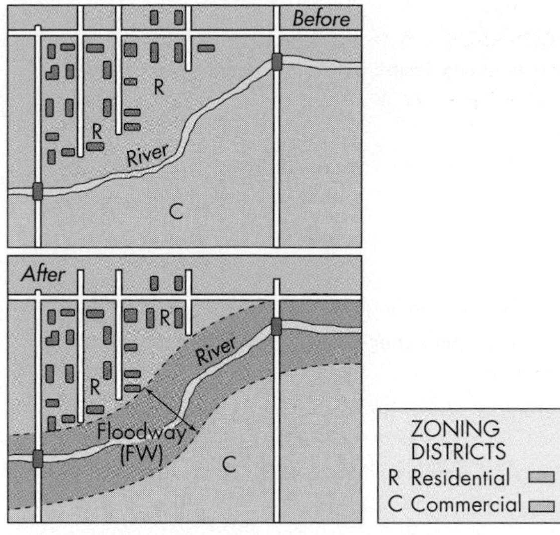

▲ **FIGURE 6.43 FLOODPLAIN ZONING** Typical zoning map before and after the implementation of flood regulations. *(From Water Resources Council. 1971. Regulation of flood hazard areas, vol. 1)*

lake inundated the land around Churchs Ferry, and by late June 2000, the town was all but deserted; the population had shrunk from approximately 100 to 7 people. Most residents took advantage of a voluntary federal buy-out plan and have moved to higher ground, many to the town of Leeds approximately 24 km (15 mi.) away. The empty houses left behind will be demolished or moved to safer ground.

This lucrative $3.5 million buyout plan seemed to be assured. Those people who participated were given the appraised value of their homes plus an incentive; most considered the offer too good to turn down. There was also recognition that the town would eventually have come to an end as a result of flooding. Nevertheless, there was some bitterness among the town's population, and not everyone participated. Among the 7 people who decided to stay were the mayor and fire chief. The buy-out program for Churchs Ferry demonstrated that the process is an emotional one; it is difficult for some people to make the decision to leave their homes, even though they know the homes are likely to be destroyed by floodwaters in the relatively near future.

PERSONAL ADJUSTMENT: WHAT TO DO AND WHAT NOT TO DO

Flooding is the most commonly experienced natural hazard. Although we can't prevent floods, individuals can be better prepared by learning what to do and what not to do before, during, and after a flood (Table 6.2).

TABLE 6.2

What to Do and What Not to Do before and After a Flood

	Preparing for a Flood
What to Do	■ Check with your local flood-control agency to see if your property is at risk from flooding. ■ If your property is at risk, purchase flood insurance if you can and be sure that you know how to file a claim. ■ Buy sandbags or flood boards to block doors. ■ Make up a flood kit, including a flashlight, blankets, raingear, battery-powered radio, first-aid kit, rubber gloves, and key personal documents. Keep it upstairs if possible. ■ Find out where to turn off your gas and electricity. If you are not sure, ask the person who checks your meter when he or she next visits. ■ Talk about possible flooding with your family or housemates. Consider writing a flood plan, and store these notes with your flood kit.
What not to Do	■ Underestimate the damage a flood can do.
	When You Learn a Flood Warning Has Been Issued
What to Do	■ Be prepared to evacuate. ■ Observe water levels and stay tuned to radio and television news and weather reports. ■ Move people, pets, and valuables upstairs or to higher ground. ■ Move your car to higher ground. It takes only 0.6 m (2 ft.) of fast-flowing water to wash your car away. ■ Check on your neighbors. Do they need help? They may not be able to escape upstairs or may need help moving furniture. ■ Do as much as you can in daylight. If the electricity fails, it will be hard to do anything. ■ Keep warm and dry. A flood can last longer than you think, and it can get cold. Take warm clothes, blankets, a Thermos, and food supplies.

(continued)

	When You Learn a Flood Warning Has Been Issued
What not to Do	▪ Walk in floodwater above knee level: It can easily knock you off your feet. Utility access holes, road works, and other hazards may be hidden beneath the water.
	After A Flood
What to Do	▪ Check home for damage; photograph any damage. ▪ If insured, file a claim for damages. ▪ Obtain professional help in removing or drying carpets and furniture as well as cleaning walls and floors. ▪ Contact gas, electricity, and water companies. You will need to have your supplies checked before you turn them back on. ▪ Open doors and windows to ventilate your home. ▪ Wash water taps and run them for a few minutes before use. Your water supply may be contaminated; check with your water supplier if you are concerned.
What not to Do	▪ Touch items that have been in contact with the water. Floodwater may be contaminated and could contain sewage. Disinfect and clean thoroughly everything that got wet.

Source: Modified after Environment Agency, United Kingdom. 2004. Floodline; Prepare for flooding.

REVISITING THE FUNDAMENTAL CONCEPTS

Flooding

1. **Hazards Are Predictable from a Scientific Evaluation**
2. **Risk Assessment Is an Important Component of Our Understanding of the Effects of Hazardous Processes**
3. **Linkages Exist between Natural Hazards**
4. **Hazardous Events That Previously Produced Disasters Are Now Producing Catastrophes**
5. **Consequences of Hazards May Be Minimized**

1) Flooding is one of the natural hazards for which predictions are fairly straightforward. Based upon flow records for rivers from the past or modeling studies, floods of various magnitudes may be predicted, as may be their average return periods. In large rivers, such as the Mississippi and others that experience regular flooding, the progress of flood waves from headwaters to the ocean may be predicted, so that people may know when flooding is likely to occur in their area. Dangerous flash floods are more difficult to predict but are based upon sustained, intense rainfall, and, when this occurs, the prediction of flash flooding is reasonably accurate.

2) Risk assessment from flooding is based upon the product of the probability of a flood occurring times the consequences. Both of these are fairly straightforward, and, as a result, we are able to estimate flood risk. This is important, since floods are the most widely experienced of the natural hazards.

3) Linkages exist between flooding and other natural hazards. For example, floodwaters may saturate banks of rivers, leading to bank failure (landslides) and other bank erosion following floods. Floodwaters may carry large amounts of sediment that are deposited on floodplains and other areas, leading to sediment pollution. Finally, floodwaters may carry biological material (dead plants, animals, chemicals, and even raw sewage) that may decay and increase the probability of the spread of diseases following flooding. For example, sewer systems may fail, leading to polluted waters that cause diseases downstream to be more likely.

4) Floods that previously produced disasters are more commonly now producing catastrophes. This results because more people are residing on floodplains, and human population along rivers and small streams is increasing. Furthermore, the activities of people during urbanization and other land-use changes, such as agriculture and deforestation, are increasing the peaks and frequency of floods.

5) Consequences of flooding may be minimized through land-use and engineering programs. The environmental solution to minimizing floods is to just say no to development on floodplains (land-use zoning). Just say no to building on the floodplain! In many parts of the world, there is significant development on floodplains from past unwise land-use decisions. As a result, the

(continued)

maintenance of levees and the building of floodwalls along rivers to protect existing development will continue into the future as a necessary endeavor. Of particular importance to flooding is disaster preparedness. We well know where floods will occur, and we can develop plans so that the impact and recovery from flooding can go smoothly.

QUESTIONS TO THINK ABOUT

1. Do you think development on flood plains should be relocated? Why or when?
2. How might you educate people living on a floodplain that has not had serious flooding in two generations that they may live in a hazard zone?
3. What are reactive policies following a flood and how do they compare to proactive steps taken before a flood? Which is better? Why?

Summary

Streams and rivers form a basic transport system of the rock cycle and are a primary erosion agent in shaping the landscape. The region drained by a stream system is called a drainage basin or watershed. Rivers carry chemicals in their dissolved load and sediment in their suspended and bed loads. Discharge refers to the volume of water moving past a particular location in a river per unit of time.

Sediment deposited by lateral migration of meanders in a stream and by the periodic overflow of the stream banks during a flood forms a floodplain. The configuration of the stream channel as seen in an aerial view is called the channel pattern. A pattern can be braided or meandering or have both characteristics in the same river.

The natural process of overbank flow is termed flooding. Flash floods in small drainage basins can be produced by intense, brief rainfall over a small area. Downstream floods in major rivers are produced by storms of long duration over a large area that saturate the soil, increasing runoff from thousands of tributary basins.

Flooding magnitude and frequency are difficult to predict on many streams because of changes in land use and limited historical records. This difficulty is especially pronounced for extreme events, such as 100 year floods. The probability that a 100 year or greater flood will take place each year is the same, regardless of when the last 100 year flood occurred.

River flooding is the most universally experienced natural hazard. Floods can occur just about anywhere there is water, and much of the United States faces the possibility of flooding. Although flooding causes many deaths and much damage, it does provide natural service functions such as the production of fertile lands, benefits to aquatic ecosystems, and maintenance of ample sediment supplies to naturally subsiding deltas such as the Mississippi.

Land-use changes, especially urbanization, have increased flooding in small drainage basins by covering much of the ground with impervious surfaces, such as buildings, parking lots, and roads, thereby increasing the runoff of storm water. Impervious surfaces are those through which water cannot penetrate easily.

Loss of life from flooding is relatively low in developed countries where adequate monitoring and warning systems are established, but property damage is much greater than in preindustrial societies because floodplains in industrialized countries are often extensively developed. Most flash flood deaths in the United States result from people attempting to drive through floodwater.

Environmentally, the best solution to minimizing flood damage is floodplain regulation, but engineering structures will still be needed to protect existing development in highly urbanized areas. These structures include physical barriers, such as levees and flood walls, and structures that regulate the release of water, such as dams and reservoirs.

Channelization is the straightening, deepening, widening, cleaning, or lining of existing streams, usually with the goal of controlling floods or improving drainage. Channelization has often caused environmental degradation, so new projects must be closely evaluated. In many places, new approaches to channel modification are using natural processes, and in some cases, channelized streams are being restored.

An adequate perception of flood hazards exists at the institutional level; however, on the individual level, more public awareness programs are needed to help people clearly perceive the hazard of living in flood-prone areas.

The best adjustments to the flood hazard include flood insurance, flood-proofing, and floodplain regulation. Floodplain regulation is critical because engineered structures tend to encourage further development of floodplains by producing a false sense of security. The first step in floodplain regulation is flood-hazard mapping, which can be difficult and expensive. Planners use flood hazard maps to zone flood-prone areas for appropriate uses. In some cases, homes in flood-prone areas have been purchased and demolished by the government and their owners relocated to safe ground.

Key Terms

100 year flood (p. 178)
channel pattern (p. 171)
channel restoration (p. 194)
channelization (p. 191)
discharge (p. 169)
drainage basin (p. 167)
flash flood (p. 176)
flooding (p. 175)
floodplain (p. 165)
floodplain regulation (p. 199)
levee (p. 163)
recurrence interval (p. 176)
river (p. 167)

Review Questions

1. Explain how levees and floodwalls can actually worsen flooding.
2. How is a drainage basin defined?
3. What are the three components that make up the total load of a stream?
4. What were the lessons learned from the 1992 flood of the Ventura River?
5. Differentiate between braided and meandering channels.
6. Differentiate between pools and riffles.
7. How do the characteristics of flash and downstream floods differ?
8. What are the primary and secondary effects of flooding?
9. What are the major factors that control damage caused by floods?
10. What is flashy discharge? How is it hazardous?
11. How does urbanization affect the flood hazard?
12. What do we mean by floodplain regulation?
13. What constitutes channelization?
14. What is channel restoration?
15. Describe the techniques that are used for flood-proofing.
16. What do we mean when we say that a 10 year flood has occurred?
17. Describe the appropriate safety precautions when walking or driving in a flooded area. What should you do in hilly or mountainous terrain if a flash flood is imminent?

Critical Thinking Questions

1. You are a planner working for a community that is expanding into the headwater portions of drainage basins. You are aware of the effects of urbanization on flooding and want to make recommendations to avoid some of these effects. Outline a plan of action.
2. You are working for a regional flood control agency that has been channelizing streams for many years. Although bulldozers are usually used to straighten and widen the channel, the agency has been criticized for causing extensive environmental damage. You have been asked to develop new plans for channel restoration to be implemented as a stream maintenance program. Devise a plan of action that would convince the official in charge of the maintenance program that your ideas will improve the urban stream environment and help reduce the potential flood hazard.
3. Does the community you live in have a flood hazard? If not, why not? If there is a hazard, what has been done or is being done to reduce the hazard? What more could be done?

Assignments in Applied Geology: Flood Insurance Rate Maps

THE ISSUE:

The Federal Emergency Management Administration has prepared flood maps for most areas of the United States. These are detailed maps that show which areas will be inundated during a 100 year flood event. Insurance rates vary widely, depending upon the map zone within which a property is located. For example, the owner of a $200,000 house located in a zone determined to be "one foot or more above base flood elevation" may pay $600 per year for flood insurance; were that same house in a zone "one foot or more below base flood elevation," the owner may pay up to $2500 per year for flood insurance.

YOUR TASK:

You are working as a geologist for an insurance company and use these maps daily. Today, your job is to assess the flooding hazard for three residential properties and determine how much the homeowners will need to pay for flood insurance. To complete your work, you will use a Flood Insurance Rate Map (FIRM), a map that shows the different flood hazard zones within a community, and detailed descriptions of each property, with historical flood information.

AVERAGE COMPLETION TIME:

1 hour

7 Mass Wasting

La Conchita: Southern California Landslide Disaster of 2005—Lessons Learned

Monday, January 10, 2005, was a disaster for the small beachside community of La Conchita, California, 80 km (50 mi.) northwest of Los Angeles. Ten people lost their lives and about 30 homes were destroyed or damaged when a fast-moving earthflow tore through the upper part of the community (see opposite page). The flow was a partial reactivation of a 1995 landslide that destroyed several homes but caused no fatalities (Figure 7.1). Although it had been raining for days prior to the 2005 slide, and the intensity of the rain was high at times, neither residents nor local officials recognized that another slide was imminent. As a result, some people were trapped in their homes and others ran for their lives (Figure 7.2).

CONTRIBUTING FACTORS

There are four primary reasons why the La Conchita has a serious landslide hazard.

1. Presence of very steep, high slopes

The slopes above La Conchita and the adjacent coastline are developed on a 140-m (450-ft.) high prehistoric sea cliff. Although the face of the slope is very steep and almost vertical in places, the top is a gently sloping to a nearly flat marine terrace planted with avocado orchards. Shells from ancient beach deposits near the top of the cliff indicate that the terrace was at sea level about 40,000 years ago. These observations and others nearby suggest that the terrace has been uplifted at a rate of 4 to 6 m (12 to 20 ft.) per thousand years. These uplift rates are some of the fastest in the world, comparable to those in Alaska or the Himalayas. The uplift at La Conchita takes place incrementally, several meters at a time, during large earthquakes. These quakes occur every few hundred to thousand years on geologic faults above La Conchita and to the south in the Santa Barbara Channel.

neither residents nor local officials recognized that another slide was imminent

2. Presence of weak rocks

Most of the La Conchita sea cliff is composed of older landslide deposits, poorly cemented claystone, siltstone,

LEARNING OBJECTIVES

Landslides, the movement of materials down a slope, constitute a serious natural hazard in many parts of North America and the world. Landslides are often linked to other hazards such as earthquakes and volcanoes. Most landslides are small and slow, but a few are large and fast. Both may cause significant loss of life and damage to property, particularly in urban areas. Your goals in reading this chapter should be to

- understand slope processes and the different types of landslides.
- know the forces that act on slopes and how they affect the stability of a slope.
- know what geographic regions are at risk from landslides.
- know the effects of landslides and their linkages with other natural hazards.
- understand how people can affect the landslide hazard.
- be familiar with adjustments we can make to avoid death and damage caused by landslides.

◀ **Reactivated Landslide Turns Deadly** For thousands of years, landslides have moved down this hillside east of Santa Barbara in Ventura County, California. This slide in 2005, a reactivation of the 1995 landslide, killed 10 people in the La Conchita subdivision at base of the hill. Shown here are dozens of rescue workers trying to locate victims in the mud and damaged houses. *(Kevork Djansezian/AP Wide World Photos)*

(a)

(b)

▲ FIGURE 7.1 **LA CONCHITA 1995 AND 2005** (a) Aerial view of the 1995 La Conchita landslide with part of the community of La Conchita in the foreground. The cliff face contains mostly older slide deposits and sediments near the top of the slope, a large slump block partially covered with vegetation on the side of the slope, and earth flow deposits at the base of the slope. This slide buried, collapsed, or damaged houses at the base of the slope. *(U S Geological Survey)* (b) Aerial view of the 2005 reactivation of the same landslide. For reference, the white arrow on the left side of each photograph points to the same building. At the base of the slope, the landslide has formed two lobe-shaped earth flows, a large one on the right and a much smaller one in the center of the photograph. Mud and debris have covered parts of four streets and damaged or buried about 30 houses. *(AP Images)*

and some sandstone. These weak sedimentary rocks have little strength to resist mass wasting on steep slopes. Historical records show that landslides have taken place on the slope above La Conchita for more than 140 years, and the geology suggests for many thousands of years![1]

3. Presence of numerous historic and prehistoric landslides

Along U.S. 101 southeast of La Conchita, landslide after landslide is visible for a distance of about 9 km (6 mi.) along the prehistoric sea cliff. Some of these slides are recent, whereas others are probably thousands of years old. Every few years, including 2005, the highway is closed for a day or so by mudflows covering the road or by landslides eroding the beachside of the road. On the same day as the 2005 La Conchita slide, mudflows closed U.S. 101 just north of La Conchita (Figure 7.3). Had homes been in the path of the flows, they would have been destroyed.

4. Periodic occurrence of prolonged and intense rainfall

Rainfall in Southern California is highly variable and the amount of precipitation fluctuates widely from year to year. Furthermore, several wet years are often followed by several dry years. Both the 1995 and 2005 La Conchita slides occurred after prolonged, and often intense, winter rains. The intensity of the rainfall was probably the major contributing factor to the 2005 landslide—more significant than land use or attempts to stabilize the slope after the 1995 landslide.

SOLUTIONS TO THE LANDSLIDE PROBLEM

Possible solutions to the recurring problem of landslides in La Conchita include the following:

1. Structural control

An engineering structure, consisting of benches and large retaining walls, could be designed for the hillslope above

◀ FIGURE 7.2 **EARTHFLOW BURIES PEOPLE, HOUSES, AND CARS** The toe of the 2005 La Conchita landslide shown here buried at least 10 people and damaged or destroyed 30 houses. When this mass of mud and debris flowed into and over these homes, it had the consistency of thick concrete pouring out of a cement truck. *(Edward A. Keller)*

La Conchita. Similar structures have been built in the Los Angeles area where a substantial landslide hazard threatens a development. However, this solution is probably cost prohibitive for La Conchita, because there are only about 100 standing structures in the community.

2. Change the land use

The homes and buildings in La Conchita could be purchased at market value and the land converted into a park. This solution would require community support and a large fund-raising effort by a nonprofit organization, such as the Land Conservancy. County, state, or federal funds would, by themselves, be insufficient to purchase the property. The drawback to this approach is that even after two major landslides in less than 10 years, some La Conchita residents like the community and want to remain, even with the knowledge that another potentially deadly landslide will occur again. Some people who died in 2005 slide earlier stated they believed the area was safer after the 1995 slide. They willingly took the risk of staying and even purchasing additional homes. Moving people from harm's way is the environmentally preferred solution to the landslide problem.

3. Install an effective warning system

Sensors to detect slope movement were installed following the 1995 slide. The 2005 slide evidently started above these sensors or the slide moved too rapidly for a warning to be issued. In hindsight, observations of patterns of the amount, duration, and intensity of rainfall might have been helpful in providing a warning. Additional research would be necessary to test this hypothesis. In Southern California, wildfire warnings, referred to as "Red Flag" days, are issued on the basis of air temperature and wind patterns. Perhaps similar warnings could be issued for landslides for La Conchita based on weather patterns. The downside to such a system would be the likelihood of false alarms. A combination of instruments to measure rainfall and slope movement might be the best approach.

◀ FIGURE 7.3 **MUDFLOW TRAPS VEHICLES NEAR LA CONCHITA** On the same day as the deadly 2005 La Conchita landslide, a separate mudflow covered about 0.5 km (0.3 mi.) of U.S. 101 just north of La Conchita. The flow broke through concrete barriers and continued onto the beach on the left. *(Dr. Grace Chang)*

4. Do nothing substantial and bear future losses
To some extent this was the course of action following the 1995 slide. Since 1995, residents were repeatedly warned of the landslide hazard; a small wall was constructed at the base of the cliff to prevent mud from small slides and hillslope erosion from covering the road, and the slope was monitored for movement. Some people moved away following the 1995 slide; however, others decided to stay, and new people bought property in the community. These were personal decisions based on individual perceptions of the hazard. One hopes that this time more will be done to reduce the continuing hazard. The question is not if another slide will occur, but when! Doing nothing and bearing future damage are not acceptable choices for protecting lives and property. However, as of August of 2010 nothing has been done to secure the lives and homes of those living in La Conchita.

LESSONS LEARNED

Several lessons can be learned from the La Conchita landslides.

1. Previous landslides are often reactivated
Many old landslides, including the 1995 and 2005 slides, lie above and adjacent to La Conchita. They will move again in the future. No part of the community is completely safe from a future landslide.[1]

2. Observation of rainfall patterns may be useful in warnings of a slide
Both the 1995 and 2005 slides occurred after periods of abundant and intense rainfall. Research might be able to establish a pattern of rainfall, which can be used as the basis for evacuating hazardous areas before a slide occurs. This approach was used in the nearby Santa Monica Mountains years ago.

3. More education about landslide hazards
Videos of the 1995 and 2005 La Conchita landslides should be shown periodically to residents in any community where similar events may happen in the future. Accompanying the videos should be discussion of how the community can adjust to the landslide hazard and develop a plan of action. The discussion and planning should involve both citizens and emergency responders such as fire, police, emergency medical services, and the Red Cross. This approach has been shown to be a valuable exercise in educating people about hazards.

7.1 An Introduction to Landslides

Mass wasting is a comprehensive term for any type of downslope movement of earth materials. In its more restricted sense, mass wasting refers to a rapid downslope movement of rock or soil as a more or less coherent mass. In this chapter, we consider landslides in the restricted sense. We will also discuss the related phenomena of earthflows and debris flows, rock falls, and avalanches. For the sake of convenience, we sometimes refer to all of these as **landslides.**

SLOPE PROCESSES

Slopes are the most common landforms on Earth. Although most slopes appear stable and static, they are actually dynamic, evolving systems. The processes that are active in these systems generally do not produce uniform slopes. Rather, most slopes are composed of several segments that are either straight or curved (Figure 7.4).

A spectacular example of these segments is on the north side of Yosemite Valley in Yosemite National Park, California (Figure 7.4a). The prominent landform there, El Capitan, has a high cliff face or *free face* of hard granite that forms a straight, nearly vertical slope segment. The free face and adjacent valley wall regularly shed pieces of rock that accumulate at the base of the cliff to form a **talus** slope (Figure 7.4a and Figure 7.5). Both the free face and the talus slope are segments of the overall slope. Frequent rock falls keep soil from developing on the free face and much of the talus slope.

In contrast to the strong granite of Yosemite Valley, gentler hillslopes develop on other rock types. For example, on Santa Cruz Island, California, slopes are developed on weaker metamorphic rocks and lack a free face. Instead, the slopes have three segments: an upper convex slope, a straight slope on the hillside, and a lower concave slope (Figure 7.4b).

As illustrated in Yosemite Valley and Santa Cruz Island, slopes are usually composed of different slope segments. The five segment types in the examples given above are sufficient to describe most slopes encountered in nature. Which slope segments are present on a particular slope depends upon the rock type and climate. Free face development is more common on strong hard rocks or in arid environments where there is little vegetation. Convex and concave slopes are more common on softer rocks or in humid environments where thick soil and vegetation are present. However, there are many exceptions to these general rules, depending upon local conditions. For example, the gentle, convex, red-colored slopes in the foreground of Figure 7.4b have formed on weak, easily eroded metamorphic rock (schist) in a semiarid climate on Santa Cruz Island, California.

Material on most slopes is constantly moving down the slope at rates that vary from an imperceptible creep of soil and rock to thundering avalanches and rock falls that move at velocities of 160 km/hr (100 mph) or more. Through time, this downslope movement is important in eroding valley walls. Slope processes are one reason that valleys are usually much wider than the streams they contain.

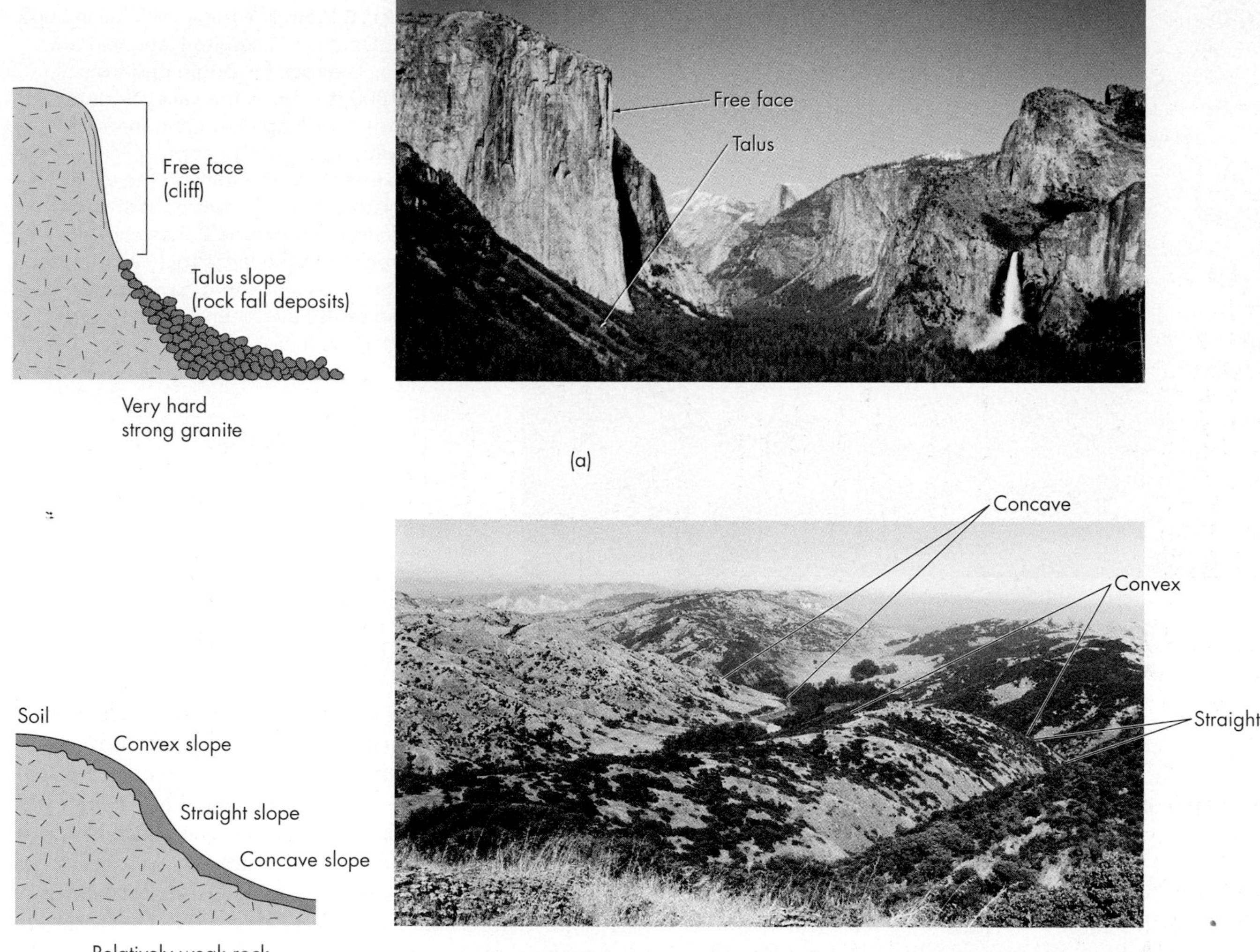

▲ **FIGURE 7.4 SLOPE SEGMENTS** (a) Slope on hard granite in Yosemite National Park, California. The cliff on the left side of the image is the several-thousand-foot-high free face of El Capitan. Below the cliff is a talus slope. *(Edward A. Keller)* (b) These slopes on Santa Cruz Island, California, formed on relatively weak metamorphic rock and have convex, straight, and concave slope segments. *(Edward A. Keller)*

TYPES OF LANDSLIDES

Earth materials on slopes may fail and move or deform in several ways (Table 7.1). **Falling** refers to the free fall of earth material, as from the free face of a cliff (Figure 7.6a). **Sliding** is the downslope movement of a coherent block of earth material along a planar *slip plane* (Figure 7.6b). **Slumping** of rock or soil is sliding along a curved slip plane producing *slump blocks* (Figure 7.6c). *Flowage*, or **flow,** is the downslope movement of unconsolidated material in which particles move about and mix within the mass, such as earthflows, debris flows, or avalanches (Figure 7.6d and Figure 7.7). Very slow flowage, called **creep,** progressive tilts telephone poles, fences, and tree trunks (Figure 7.6e). Creep occurs at such a low rate that the progressive upward growth of trees outpaces the tilting resulting in trees trunks that are curved skyward.

Many landslides are complex combinations of sliding and flowage (Figure 7.6f). For example, a landslide may start out as a *slump*, then pick up water as it moves downslope and transform into an earthflow in the lower part of the slide. Some complex landslides may form when water-saturated earth materials flow from the lower part of a slope, undermining the upper part and causing slump blocks to form (Figure 7.8).

Downslope movements are classified according to four important variables: (1) the mechanism of movement (slide, fall, flow, or complex movement), (2) type of earth material (e.g., solid rock, soft consolidated sediment, or loose unconsolidated material), (3) amount of water present, and (4) rate of movement. In general, movement is considered rapid if it can be discerned with the naked

◀ **FIGURE 7.5 ROCK FALL AND TALUS PILE** View of a huge rock fall in 2009 near Half Dome in Yosemite National Park, California. The rock fall originated from 540 m (1800 ft.) above the valley floor (out of this photo), knocking down hundreds of trees and burying a part of the popular Mirror Lake trail. The impact of the rock on the valley floor registered on seismographs around California as a magnitude 2.5 earthquake. Broken rock from the large rock fall form the light-colored talus pile that cover the ground where the trees have been bowled over. *(Duane E. DeVecchio)*

eye; otherwise, it is classified as slow. Actual rates vary from a slow creep of a few millimeters or centimeters per year to rapid at 1.5 m (5 ft.) per day, to extremely rapid at 30 m (100 ft.) or more per second.[2]

FORCES ON SLOPES

To determine the causes of landslides, we must examine the forces that influence slope stability. The stability of a slope can be assessed by determining the relationship between **driving forces** that move earth materials down a slope, and **resisting forces** that oppose such movement. The most common driving force is the downslope component of the weight of the slope material. That weight can be from anything superimposed or otherwise placed on the slope, such as vegetation, fill material, or buildings. The most common resisting force is the **shear strength** of the slope material, that is, its resistance to failure by sliding or flowing along potential slip planes (see Closer Look 7.1). Potential slip planes are geologic surfaces of weakness in

TABLE 7.1

Common Types of Landslides and Other Downslope Movements

Mechanism	Type of Mass Movement	Characteristics
Fall	Rock fall	Individual rocks fall through the air and may accumulate as talus.
	Slump	Cohesive blocks of soft earth material slide on a curved surface; also called a rotational landslide.
Slide	Soil slip	Soil and other weathered earth material slide on a tilted surface of bedrock or cohesive sediment; also called a debris slide or earth slide.
	Rock slide	Large blocks of bedrock slide on a planar surface, such as layering in sedimentary or metamorphic rocks.
	Avalanche	Granular flow of various combinations of snow, ice, organic debris, loose rocks, or soil that moves rapidly downslope.
Flow	Creep	Very slow, downslope movement of rocks and soil.
	Earthflow	Wet, partially cohesive and internally deformed mass of soil and weathered rock.
	Debris flow	Fluid mixture of rocks, sand, mud, and water that is intermediate between a landslide and a water flood; includes mudflows and lahars.
	Complex	A combination of two or more types of sliding, flowage, and occasionally falls; forms where one type of landslide changes into another as it moves downslope.

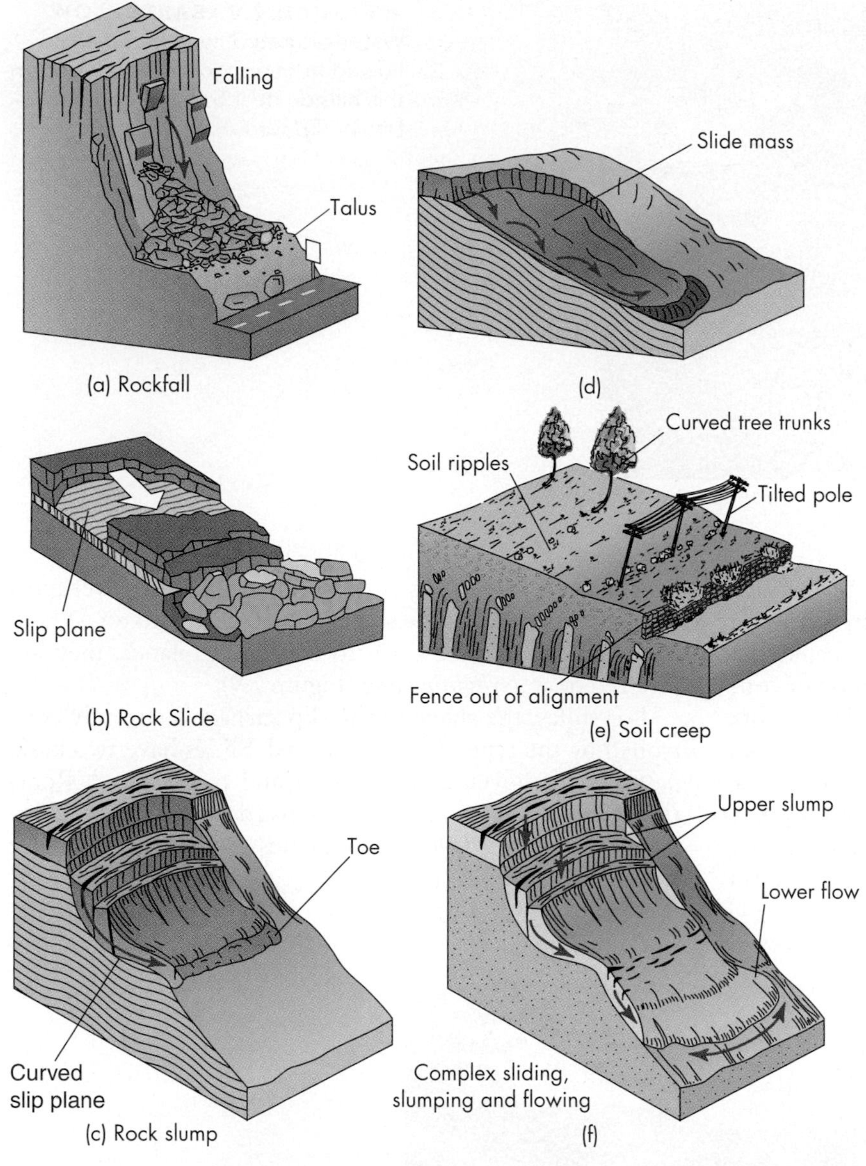

Type of Movement	Materials	
	Rock	Soil
Landslides with variable water content and rate of movement	Rotational	
	Slump(a)	Slump(b)
	Translational	
	Rock slide(c)	Soil slide (slip)(d)
Falls	Rock fall(e)	Soil fall
Flows — Slow	Rock creep	Soil creep(f)
	Unconsolidated rock and soil (saturated)	
Rapid	Earth flow(g)	
	Debris flow / mud flow(h)	
Very rapid	Debris avalanche(i)	
Complex	Combination of slides, slumps, and flows(l)	

▲ FIGURE 7.6 **TYPES OF LANDSLIDES** (a) Rock fall, in which blocks of bedrock fall through air and accumulate at the base of the cliff as a talus slope. (b) Rock slide, in which blocks of bedrock slide down a bedding plane in sedimentary rock. Soil slips are similar movements except soil, rather than rock, slides down a bedrock surface.
(c) Slump, in which cohesive blocks of soft earth material slide down a curved slip surface.
(d) Debris flow, in which mud, sand, rock, and other material are mixed with water.
(e) Soil creep causes telephone poles to tilt and trees trunks to become curved.
(f) Complex landslide, consisting of upper slump and lower flow.

the slope material, such as bedding planes in sedimentary rocks and fractures in all types of rock.

Slope stability is evaluated by computing a **safety factor** (SF), defined as the ratio of the resisting forces to the driving forces. If the safety factor is greater than 1, the resisting forces exceed the driving forces and the slope is considered stable. If the safety factor is less than 1, the driving forces exceed the resisting forces and a slope failure can be expected. Driving and resisting forces are not static; as local conditions change, these forces may change, increasing or decreasing the safety factor.

Driving and resisting forces on slopes are determined by the interrelationships of the following variables:

- type of earth materials
- slope angle and topography
- climate
- vegetation
- water
- time

◀ FIGURE 7.7 **EARTHFLOW** Water-saturated, weathered shale flowed from a scoop-shaped area on this hillside near Santa Barbara, California. *(Edward A. Keller)*

The Role of Earth Material Type The material composing a slope can affect both the type and the frequency of downslope movement. Important material characteristics include mineral composition, degree of cementation or consolidation, the presence of zones of weakness, and the ability of the earth material to transmit water. These weaknesses may be natural breaks in consistency of the earth materials, such as sedimentary and metamorphic layering, or they may be zones along which the earth has moved before, such as an old landslide slip surface or geologic fault. Weak zones can be especially hazardous if the zone or plane of weakness intersects or parallels the slope of a hill or mountain. Where the planes of weakness are rock bedding planes, they are referred to as *daylighting beds* (Figure 7.9).

For slides, the shape of the slip surface is strongly controlled by the type of earth material. Slides have two basic patterns of movement, *rotational* and *translational.* Rotational slides, or slumps, have curved slip surfaces (Figure 7.6c), whereas translational slides generally have planar slip surfaces (Figure 7.6b).

▲ FIGURE 7.8 **HOUSES BURIED BY COMPLEX LANDSLIDE** In the distance is the amphitheater-like scar in the hillslope that was the source of the 1995 La Conchita landslide. The slide started out as a slump, then liquefied into an earthflow as it moved downhill. Several homes, including the three-story home behind the orange plastic fence, were destroyed. *(Edward A. Keller)*

A CLOSER LOOK 7.1

Forces on Slopes

The fundamental driving force of mass wasting processes is gravity. Over time, the ever-constant downward pull of gravity exceeds resisting forces and causes slopes to fail, rocks to fall, soils to creep, and water-saturated earth materials to flow. On geologic time scales, gravity-driven erosional processes have sculpted Earth's surface and worn away countless ancient mountain ranges and washed them to the sea.

The downward pull of gravity exerts a quantifiable force toward the center of Earth, giving weight to potential slide masses (Figure 7.A). Yet, because Earth materials cannot be pulled straight down through the underlying rocks, the gravitational pull is divided between a downslope force and the normal force, which are at right angles to each other. The downslope force or driving force (D) is parallel to the slope of the potential slip plane, whereas the normal force (N) is perpendicular (Figure 7.A). The shear strength of the potential slip plane (resisting force) is proportional to the magnitude of the normal force, which adds frictional resistance onto the surface of the plane. The relative magnitudes of the normal and driving force are directly related to the angle of the slope. Specifically, as the slope angle (θ) increases, the normal force decreases and the driving force increases proportionally and vise versa. It follows then that the factor of safety decreases with increasing slope angle, because the factor of safety is the magnitudinal ratio of resisting forces over the driving forces.

As local conditions change naturally or as a result of human interaction with the landscape, driving and resisting forces will change. These changes may lead to an increase or decrease in the factor of safety. For example, consider the construction of a road where part of a hillside must be excavated to provide a suitable surface for automobile traffic. Removal of the material at the toe of the slope exposes a potential slip plane; a daylighting bed in this case (Figure 7.A). Although the roadcut reduces the driving forces on the slope by removing some of the weight of the potential slide mass, the cut also decrease the length of the slip plane, thereby reducing the resisting force (shear strength). If you examine Figure 7.A, you can see that only a small portion of the potential slide mass has been removed, whereas a relatively large portion of the length of the slip plane has been removed (L1 vs. L2). Therefore, the overall effect of the roadcut is to *lower the factor of safety* because the reduction of the driving force is small compared to the reduction of the resisting force.

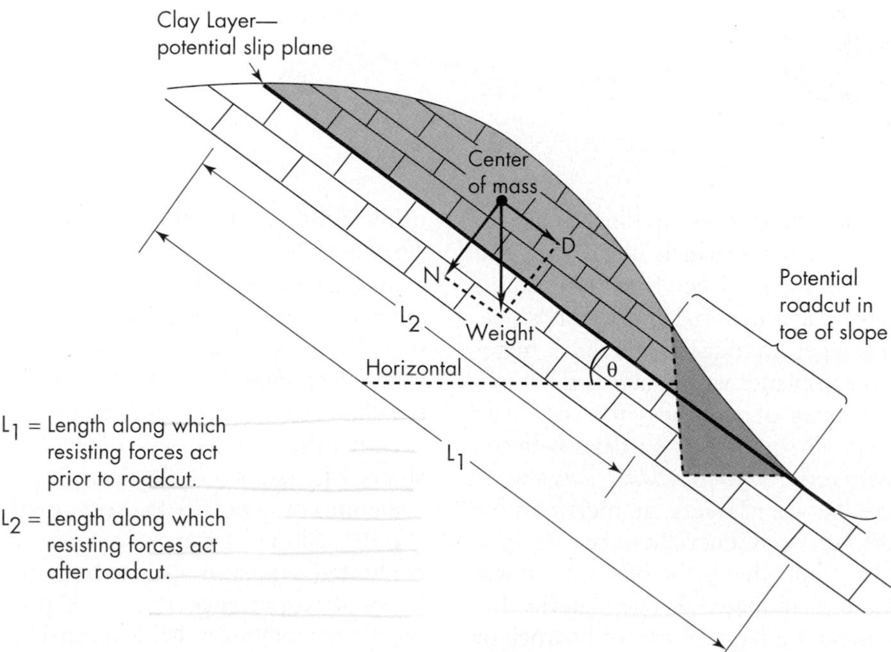

▲ **FIGURE 7.A EFFECTS ON SLOPE STABILITY** Cross section of a slope showing the principal forces acting on a potential slide mass.

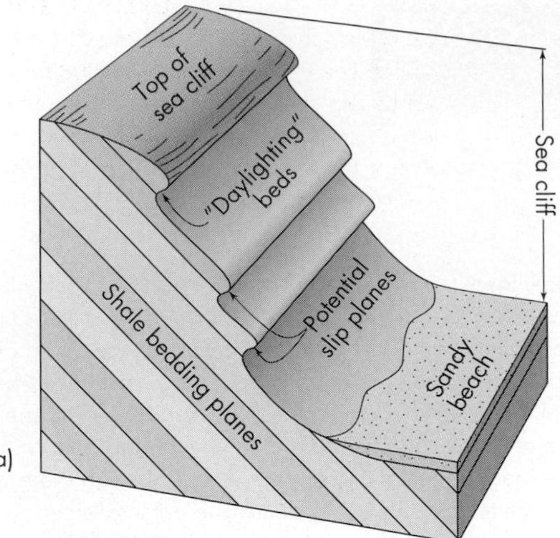

FIGURE 7.9 DAYLIGHTING BEDS (a) Bedding planes that intersect the surface of the land on a slope are said to "daylight." Such beds are potential slip planes. (b) This slide occurred in late 2003. Failure was along a daylighting bedding plane. Slide deposits cover part of the beach. *(Edward A. Keller)*

In slumps, the rotational sliding of slump blocks tends to produce small topographic benches, which may rotate and tilt in an up-slope direction (Figure 7.6c). Slumps are most common in unconsolidated earth material and in mudstone, shale, and other weak rock types. In translational slides, material moves along inclined slip planes within and parallel to a slope (Figure 7.9). The zones of weakness for these slip planes include fractures in all rock types, surfaces between layers in sedimentary rocks referred to as *bedding planes*, weak clay layers, and surfaces between layers in metamorphic rocks referred to as *foliation planes*. Generally, once a slip surface is established for a landslide, that plane of weakness will continue to exist within the earth material. The material that moves along these planes can be large blocks of bedrock or unconsolidated and weathered earth material.

A common type of translational side is a *soil slip*, a shallow slide of unconsolidated material over bedrock (Figure 7.10). The plane of weakness for soil slips is usually in **colluvium,** a mixture of weathered rock and other debris that is above the bedrock and below the soil (Figure 7.11).

Material type is also a factor in rock falls. Where resistant rock overlies weaker rock, rapid erosion of the underlying weaker rock may cause a slab failure and a subsequent rock fall (Figure 7.12).

The nature of the earth material composing a slope can greatly influence the type of slope failure. For example, on slopes of shale or weak volcanic pyroclastic material, failure commonly occurs as creep, earthflows, debris flows, or slumps. Slopes formed in resistant rock such as well-cemented sandstone, limestone, or granite are much less likely to experience the same problems. Thus, builders should assess the landslide hazard before starting construction on shale or other types of weak rock.

The ability of the earth material to transmit water, referred to as *permeability*, determines how easily soil, sediment, or rock absorbs water (see Appendix 2). As discussed later, water is involved in nearly all landslides, and permeability barriers

Soil

Shallow landslide in soil above rock

Landslide deposit

Soil

(a)

(b)

◀ **FIGURE 7.10 MULTIPLE SOIL SLIPS** (a) Cross-sectional diagram of a shallow soil slip in which the gray bands are sedimentary layers parallel to the slope. Soil slides from the slope and piles up at its base. (b) Shallow soil slips on steep slopes in Southern California. Vegetation is low brush known as chaparral. *(Edward A. Keller)*

can allow the buildup of water. For example, soil slips commonly develop because of contrasts in permeability between more permeable soil and less permeable, unweathered bedrock below the soil. Downward percolating rainwater builds up in the soil because it cannot easily infiltrate the underlying bedrock. The accumulation of water can break down cohesion between soil particles and facilitate slippage of the soil, especially on steep hillsides.

The Role of Slope and Topography Two factors are important with slope and topography: steepness of the slope and amount of topographic relief. By *slope*, we mean the slant or incline of the land surface. In general, the steeper the slope, the greater the driving force. For example, a study of landslides that occurred during two rainy seasons in California's San Francisco Bay area established that 75 percent to 85 percent of landslide activity is closely associated with urban areas on steep slopes.[3]

Topographic *relief* refers to the height of the hill or mountain above the land below. Areas of high relief are hilly or mountainous, have tens or hundreds of meters of relief, and are generally more prone to mass wasting. Within the contiguous United States and Canada, the coastal mountains of California, Oregon, Washington, and British Columbia; the Rocky Mountains; the Appalachian Mountains; and coastal cliffs and bluffs have the greatest frequency of landslides. All types of downslope movement occur on steep slopes within these areas.

Steep slopes are also associated with rock falls, avalanches, and soil slips. In Southern California, shallow soil slips are common on steep slopes when a hillside becomes saturated with water. Once soil slips move downslope, they often become earthflows or debris flows, which can be extremely hazardous (Figure 7.13). In Southern California, earthflows are common on moderate slopes, and creep can be observed on gentle slopes.

Debris flows are thick mixtures of mud, debris, and water. They range in consistency from thick mud soups to wet concrete and are capable of carrying house-size boulders. Debris flows can move either slowly or rapidly depending on conditions. They can move down established stream valleys, flow from channels filled with colluvium, or take long narrow tracks or chutes on steep hillsides. These flows can be relatively small to moderate events, transporting

◀ **FIGURE 7.11 SHALLOW SLIDES** (a) This shallow soil slip in North Carolina has removed slope vegetation. Highway engineers installed the white barrier at the base of the slide to keep debris off of the road. *(Edward A. Keller)* (b) The long, narrow area of bare soil on this Northern California hillslope was created by a debris flow. Trees and sediment carried by the flow are in a tongue-shaped pile at the bottom of the hill along the bank of the Klamath River. A logging road at the head of the flow (end of the arrow) may have helped destabilize the slope. *(Edward A. Keller)*

a few hundreds to hundreds of thousands of cubic meters of debris down a single valley, or they can be huge events transporting cubic kilometers of material from the failure of an entire flank of a mountain (see volcanic mudflows and debris flows discussed in Chapter 5).

The Role of Climate *Climate* can be defined as the weather that is typical in a place or region over a period of years or decades. A description of climate is more than just the average air temperature and amount of precipitation. It also includes the type of precipitation and its seasonal patterns. In North America, examples of type and pattern are winter rains along the west coast of the United States, summer thunderstorms in the southwestern United States, winter blizzards in the Arctic and Great Plains, heavy snows on the eastern shores of the Great Lakes, and hurricane activity on the Mexican and southeastern U.S. coasts. For landslides, the type of climate influences the amount and timing of water that infiltrates or erodes a hillslope and the type and abundance of hillslope vegetation.

In arid and semiarid climates, vegetation tends to be sparse, soils are thin, and bare rock is exposed (Figure 7.14a). Free faces and talus slopes are common, where erosion resistant bedrock is present. Resistant rock forms steep slopes (cliffs) and less resistant rock more gentle slopes.

◀ FIGURE 7.12 **ROCK FALL DAMAGES HOUSE** The large boulder embedded in the right end of this house fell 120 m (400 ft.) from a ledge of resistant sandstone in the early morning of October 18, 2001, in Rockville, Utah. Weighing an estimated 200 to 300 tons and measuring approximately 5 m (16 ft.) by 5 m (16 ft.) by 3.6 m (12 ft.), the boulder destroyed the living room, bathroom, and service area. It then entered a bedroom, narrowly missing the sleeping homeowner. The specific cause of the rock fall is unknown, but there are numerous large boulders in the area from earlier rock falls. *(Utah Geological survey)*

Rock falls, debris flows, and shallow soil slips are common types of landslides in these climatic regions.

In subhumid to humid regions of the world, abundant vegetation and thick soils cover most slopes (Figure 7.14b). These two characteristics cause the widespread development of convex and concave slope segments (see Figure 7.4). Landslide activity in these areas includes deep complex landslides, earthflows, and soil creep.

▲ FIGURE 7.13 **SHALLOW SOIL SLIPS CAN KILL** Even small, shallow landslides can be fatal. This Southern California home was destroyed in 1969 by a debris flow that originated as a soil slip. The landslide claimed two lives. *(John Shadle/Los Angeles Department of Building and Safety)*

The Role of Vegetation Vegetation has a complex role in the development of landslides and related phenomena. The nature of the vegetation in an area is a function of the climate, soil type, topography, and fire history, each of which also independently influences what happens on slopes. Vegetation is a significant factor in slope stability for three reasons:

- Vegetation provides a protective cover that cushions the impact of falling rain. This cushion allows the water to infiltrate the slope while retarding surface erosion.
- Plant roots add strength and cohesion to slope materials. They act like steel rebar reinforcements in concrete and increase the resistance of a slope to landsliding.[4]
- Vegetation also adds weight to a slope.

In some cases, this additional weight increases the probability of a landslide, especially with shallow soils on steep slopes. Such soil slips are common along the California coast where ice plant, an invasive plant imported from South Africa in the early 1900s, covers steep-cut slopes (Figure 7.15). During especially wet winter months, these shallow-rooted plants take up water and store it in their leaves. This stored water adds considerable weight to steep slopes, thereby increasing the driving forces. These plants also cause increased water infiltration into the slope, which decreases the resisting forces. When failure occurs, the plants and several centimeters of roots and soil slide to the base of the slope.

In Southern California, soil slips are also a serious problem on steep slopes covered with chaparral, a dense growth of native shrubs and small trees (see Figure 7.10b). Chaparral facilitates water infiltration into a slope and lowers its safety factor.[5]

The Role of Water Water is almost always directly or indirectly involved with landslides, so its role is particularly

(a)

(b)

▲ FIGURE 7.14 **CLIMATE AND MASS WASTING** Different climate zones around the world play and important role in, not only the appearance of the landscape, but in the processes that sculpt it. (a) Semiarid landscape in Boysen State Park, Wyioming. *(Jack Jeffers/SuperStock)* (b) Humid landscape on the Island of Taiwan *(Sean Sprague/The Image Works).*

important.[6] When studying a landslide, scientists first examine what water is doing both on and within the slope. Water affects slope stability in three basic ways: (1) Many landslides, such as shallow soil slips and debris flows, develop during rainstorms when slopes become saturated; (2) other landslides, such as slumps, develop months or even years following the deep infiltration of water into a slope; and (3) water erosion of the base or toe of a slope decreases its stability.

Erosion by flowing water along a stream bank or by wave action on the coast can remove earth material from the base of a hill or cliff. This loss of material steepens the slope, reduces the safety factor, and increases the likelihood of a landslide (Figure 7.16). Erosion is especially problematic if it removes the toe of an old landslide, thereby increasing the potential for the landslide to move again (Figure 7.17). Therefore, it is important for planners and engineers to identify old landslides before cutting into hillslopes for the construction of roads and buildings. Excavations of hillslopes must be carefully planned so that potential problems can be isolated and corrected prior to construction.

Another way that water can cause landslides is by contributing to the spontaneous liquefaction of quick or

(a)

(b)

Ice plants slipped down slope

▲ FIGURE 7.15 **ICE PLANTS ON SLOPES ARE OFTEN UNSTABLE** Ice plant was first introduced in California to protect slopes along railroads and was later widely used along highways. Instead of stabilizing slopes, ice plant often contributes to landslides because of its shallow roots and the added weight of water stored in its leaves. The plant caused these soil slips near Santa Barbara, California (a) on a steep roadside embankment and *(Edward A. Keller)* (b) at a home site. A black plastic sheet was placed over the upper part of the slide to reduce the infiltration of rain water. *(Edward A. Keller)*

expansive clay. These clays are fine-grained earth material that lose their shear strength and flow as a liquid when disturbed (see Case Study 7.4). The shaking of quick clay beneath Anchorage, Alaska, during the 1964 earthquake produced this effect and caused three extremely destructive landslides. In Quebec, Canada, several large slides associated with quick clays have destroyed numerous homes and killed more than 70 people in recent years. These slides occurred on river valley slopes when initially solid material was converted into a liquid mud as the sliding movement began.[7] The slides in Quebec are especially interesting because the liquefaction of clays occurred without earthquake shaking. In many cases, the slides were triggered by river erosion at the toe of a slope. Liquefaction first started in a small area and spread to a much larger area. Since these movements have often involved the reactivation of older slides, future problems may be avoided by mapping existing slides and restricting development on them.

The Role of Time The forces acting on slopes often change with time. For example, both driving and resisting forces may change seasonally with fluctuations in the moisture content of the slope or with changes in the position of the water table. Much of the chemical weathering of rocks, which slowly reduces their strength, is caused by the chemical action of water in contact with soil and rock near Earth's surface. Soil water is often acidic because it reacts with carbon dioxide in the atmosphere and soil to produce weak carbonic acid. This type of chemical weathering is significant in areas underlain by limestone, a rock type that is easily dissolved by carbonic acid. Changes are greater in especially wet years, as reflected by the increased frequency of landslides during or following these years. In other slopes, resisting forces may continuously diminish through time from weathering, which reduces the cohesion in slope materials, or from an increase in underground water pressure within the slope. A slope that is becoming less stable with time may have an increasing rate of creep until failure occurs.

SNOW AVALANCHES

A **snow avalanche** is the rapid downslope movement of snow and ice, sometimes with the addition of rock, soil, and vegetation. Thousands of avalanches occur every year in the western United States and Canada. As more people venture into avalanche-prone areas and more development takes place, the loss of life and property from avalanches

(a)

(b)

▲ FIGURE 7.16 **WATER ERODING THE TOE OF A SLOPE CAUSES INSTABILITY** (a) Stream-bank erosion seen in the lower left corner of this image has caused slope failure and damage to a road in the San Gabriel Mountains, California. *(Edward A. Keller)* (b) Wave erosion was a major cause of this landslide in Cove Beach, Oregon. Further movement of the slide threatens the homes above. *(Gary Braasch Photography)*

(a)

(b)

▲ **FIGURE 7.17 REACTIVATION OF A SLIDE** (a) Part of the beach (end of thin arrow) in Santa Barbara, California, has been buried by reactivation of an older landslide. Reactivation of the slide may have resulted due to erosion of the old slide toe by ocean waves. *(Donald W. Weaver)* (b) Close-up of the head of this slide where it destroyed two homes. The thick, black arrow in *(Donald W. Weaver)* (a) points to the location of this picture. *(Courtesy of Don Weaver)*

increases. Avalanche accidents in the winter of 2005–2006 killed 32 people in North America.[8] Most avalanches that kill people are triggered by the victims themselves or by members of their party.[9]

Three variables interact to create unstable conditions for snow avalanches: steepness of the slope, stability of the snowpack, and the weather. Snow avalanches generally occur on slopes steeper than about 25°.[9] The steepest angle at which snow, or any loose material, is stable is its *angle of repose*. For snow, this angle is affected by temperature, wetness, and shape of the snow grains.[10] Most snow avalanches occur on slopes between 35° and 40°.[9] On steeper slopes greater than 50° to 60°, the snow tends to continually fall off the slope. Snow-covered slopes may become unstable when the wind piles up snow on the leeward or downwind side of a ridge or hillcrest, when rapid precipitation adds weight to the slope, or when temperatures rapidly warm to make the snow very wet.

There are two common types of snow avalanches: loose-snow and slab. *Loose-snow avalanches* typically start at a point and widen as they move downslope. *Slab avalanches* start as cohesive blocks of snow and ice that move downslope. The latter type of avalanche is the most dangerous and damaging. Slab avalanches typically are triggered by the overloading of a slope or the development of zones of weakness in the snowpack. Millions of tons of snow and ice then move rapidly downslope at velocities of up to 100 km (60 mi.) per hour.

Avalanches tend to move down tracks, called *chutes*, which have previously produced avalanches (Figure 7.18). As a result, it is easier to prepare maps showing potentially hazardous areas. Avoiding these areas is obviously the preferable and least expensive adjustment to avalanches. Other adjustments include clearing excess snow with carefully placed explosives, constructing buildings and structures to divert or retard avalanches, or planting trees on slopes in avalanche-prone areas to better anchor the snow.

7.2 Geographic Regions at Risk from Landslides

As you may imagine, landslides occur everywhere that significant slopes exist and mountainous areas have a higher risk for landslides than most areas of low relief. The latter generalization is supported by a map of major landslide areas in the contiguous United States (Figure 7.19). On this map, the areas shown in red, which experience frequent landslide activity, are concentrated in the mountainous areas of the West Coast, the Rocky Mountains, and the Appalachian Mountains. The plains states are flatter and therefore relatively safe from landslides. The Rocky Mountains and coastal mountains of the East and West Coasts are at particularly high risk. Parts of central Canada, such as Saskatchewan, Manitoba, and eastern Ontario, have a relatively lower risk.

Three factors are expected to increase worldwide landslide activity in the twenty-first century:

- Urbanization and development will expand in landslide-prone areas.

(a)

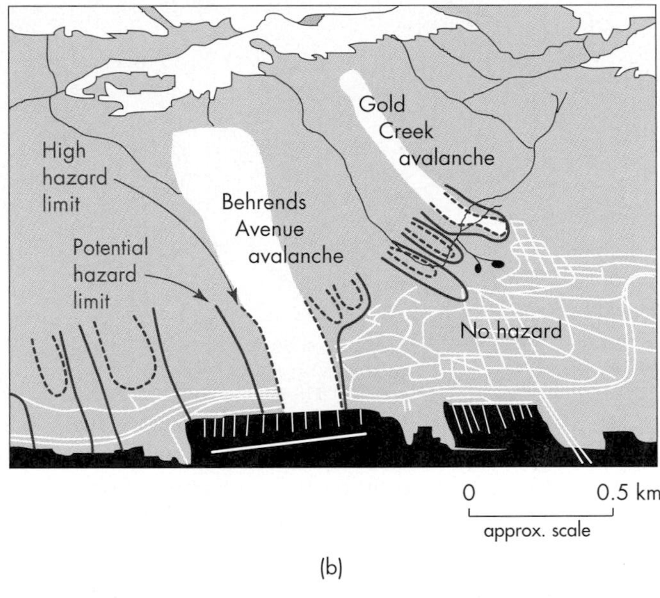

(b)

▲ FIGURE 7.18 **SNOW AVALANCHE HAZARD** (a) Snow avalanche moving down a chute in the San Juan Mountains of southwestern Colorado. This 1987 avalanche started with the failure of large slabs of snow near the ridge crest. It then fell 830 m (2720 ft.) to the valley floor and climbed up the opposite hillslope to bury 245 m (800 ft.) of U.S. 550 with 1 m (3 ft.) of snow. Known as "Battleship," this chute is a regular path for avalanches. *(Mark Rawsthorne)* (b) Map of part of Juneau, Alaska, showing areas at highest risk for snow avalanches. Most of the downtown business district is in the "no hazard" area in the lower right. The white lines in the black foreground are boat docks along the shore of Gastineau Channel. *(After Cupp, D. 1982. National Geographic 162:290–305)*

- Tree cutting will continue in landslide-prone areas.
- Changing global climate patterns will result in regional increases in precipitation.[11]

7.3 Effects of Landslides and Linkages with Other Natural Hazards

EFFECTS OF LANDSLIDES

Landslides and related phenomena have the capacity to cause substantial damage and loss of life. In the United States alone, on average 25 people are killed each year by landslides, and this number increases to between 100 and 150 if collapses of trenches and other excavations are included. The total annual cost of damages exceeds $1 billion and may be as high as $3 billion.[12]

Direct effects of landslides on people and property include being hit with or buried in falling debris (see Survivor Story 7.2). Landslides may also damage homes, roads, and utilities that have been constructed on the top or side of a hill. They regularly block roads and railroads, delaying travel for days or weeks. One massive landslide, the 1983 Thistle slide discussed later, blocked commerce for months on major highways and a transcontinental rail line. Rerouting of traffic significantly increased transportation costs in south-central Utah and resulted in the temporary or permanent closure of coal mines, oil companies, motels, and other businesses. Landslides may even

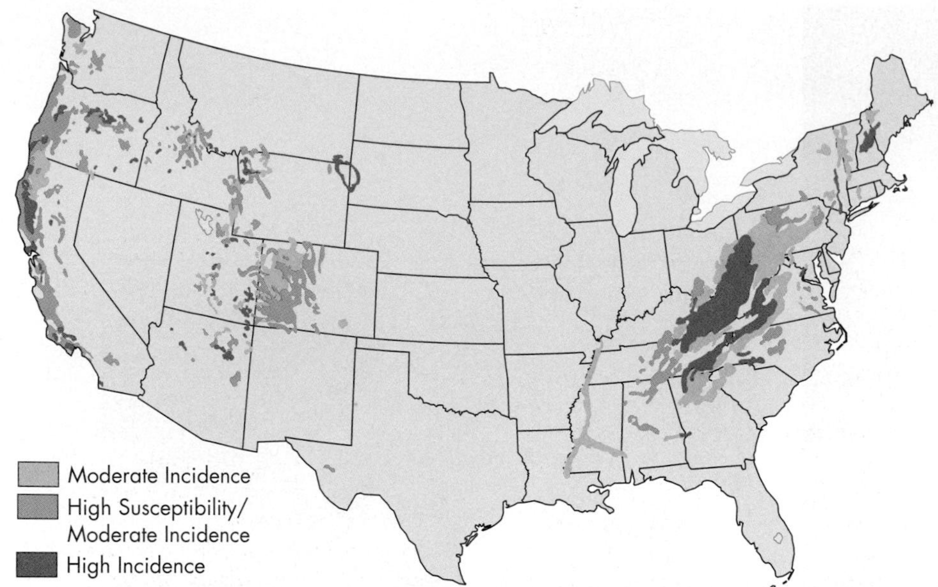

◀ **FIGURE 7.19 AREAS OF HIGH-TO-MODERATE LANDSLIDE INCIDENCE** On this map of landslide areas in the contiguous United States, the colored areas either have a high-to-moderate number of landslides or they have a moderate number of landslides and are highly susceptible to additional landslides. *(U.S. Geological Survey)*

block shipping lanes. In 1980, a debris flow from the volcano Mount St. Helens filled the Columbia River with more than 34 million m^3 (44 million $yd.^3$) of sediment. This sediment stopped cargo ships from reaching Portland, Oregon, until dredging was completed.

Indirect effects include flooding upstream from a landslide that blocks a river and the transmission of disease from fungal spores disturbed by landslides, as happened following the 1994 Northridge earthquake (see Chapter 3).

LINKAGES BETWEEN LANDSLIDES AND OTHER NATURAL HAZARDS

Landslides and other types of mass movement are intimately linked to just about every other natural hazard. Earthquakes, volcanoes, storms, and fires all have the potential to cause landslides (see Professional Profile 7.3). These relationships are discussed in their respective chapters. As described earlier, landslides themselves may be responsible for flooding if they form an earthen dam across a river. A large landslide may also cause tsunamis or widespread flooding if it displaces water out of a lake or bay. In 1963, more than 2.5 km^2 (1 $mi.^2$) of a hillside slid into the Vaiont reservoir in northeastern Italy. The landslide and resulting flood killed more than 2600 people during this tragedy.

7.4 Natural Service Functions of Landslides

Although most hazardous natural processes provide important natural service functions, it is difficult to imagine any good coming from landslides except the creation of new habitats in forests and aquatic ecosystems.

Landslides, like fire, are a major source of ecological disturbance in forests. For some old-growth forests, this disturbance can be beneficial by increasing both plant and animal diversity.[13] In aquatic environments, landslide-dammed lakes create new habitat for fish and other organisms.[14]

Mass wasting can produce deposits that become mineral resources. Weathering frees mineral grains from rocks, and mass wasting transports these minerals downslope. Heavier minerals, particularly gold and diamond, can be concentrated at the base of the slope and in adjacent streams. Although not as abundant as in stream deposits, gold and diamonds have been mined from colluvium and debris flow deposits.

7.5 Human Interaction with Landslides

Landslides and other types of ground failure are natural phenomena that would occur with or without human activity. However, in many instances, the expansion of urban areas, transportation networks, and natural-resource use has increased the number and frequency of landslides. In other settings, where the potential for landslides has been recognized, the number of events has decreased because of preventive measures. For example, grading of land surfaces for housing developments may initiate landslides on previously stable hillsides; on the other hand, building stabilizing structures and improving drainage may reduce the number of landslides on naturally sensitive slopes.

The effect of human endeavors on the magnitude and frequency of landslides varies from nearly insignificant to very significant. In cases where human activities rarely cause landslides, we may still need to learn all we can about this natural phenomenon to avoid developing in hazardous areas and to minimize damage. When human activities, such as road construction (see Closer Look 7.1), increase the number and severity of landslides, we need to learn how our practices cause slides and how we

SURVIVOR STORY 7.2

Landslide

Danny Ogg thought He was "a Goner" When the First Boulder Hit his Truck

Danny Ogg was goin' round the mountain, as they say, on Interstate 70 west of Denver, when a giant granite boulder crashed across his windshield, leaving him with only one thought: "I'm going to be buried alive."

The boulder was one of about 20 or 30 that rocketed down the steep slopes onto the interstate highway, which winds through the Rocky Mountains, early in the morning of April 8, 2004 (Figure 7.B). The largest of the rocks weighed 30 tons.

Surprisingly, though, Ogg, a 50-year-old driver from Tennessee, said he had no indication that a landslide was occurring until his semitrailer hit the first rock.

After that, it was "bang bang bang, crunch crunch," as he put it, as his truck collided with the falling debris and eventually hit the median. The vehicle's fuel tank was torn open in the process, spilling hundreds of gallons of diesel fuel.

His sense of time was skewed in the terrifying moments of the landslide, Ogg said.

"After I hit the first boulder, it took two, three minutes, I guess. It was fast but it wasn't," he said.

Luckily, Ogg suffered only an injured back and some torn cartilage and ligaments, compared to the much more serious injuries he could have incurred.

"I thought I was a goner," he said. "I don't know how long I was in there. It idled me. I was silly as a goose, as we say in Tennessee."

His sense of location was also askew: "You know you ain't dead yet, but you don't know where you are," he said. "I didn't know if I was hanging off the mountain or what."

Ogg said some of the boulders left potholes in the road a foot or two deep, and a *Rocky Mountain News* article reported that some of them bounced as high as 20 feet after impact.

Although landslides are difficult to predict and avoid, Ogg said he would be more vigilant now when driving in the mountains.

"When you go around the mountain now, you look up," he said.

—CHRIS WILSON

◀ FIGURE 7.B **BOULDERS ON I-70** These granite boulders, some as large as 4 m^3 (140 $ft.^3$), fell onto I-70 near Glenwood Springs, Colorado, on the morning of May 9, 2003. Heavy rain, combined with ice wedging in cracks within the granite, dislodged one or more of these large rocks from a free face more than 100 m (300 ft.) above the interstate. One car-sized boulder bounced across the highway and landed in the nearby Colorado River. A highway crew had to drill or blast apart the largest boulders and push them over to the shoulder so traffic could resume. The low retaining wall behind the boulders in this image was shattered in several places. *(US Geological Survey)*

PROFESSIONAL PROFILE 7.3

Bob Rasely, Mass Wasting Specialist

Bob Rasely's line of work may be the very definition of "down and dirty" (Figure 7.C). As a Utah-based geologist with the Natural Resources Conservation Service, Rasely specializes in predicting the likelihood of mass wasting in the wake of a wildfire.

"You don't see the dirt you got on you until you take your shirt off," Rasely said.

While most of the situations Rasely encounters are not emergencies, there are certainly times when the possibility of mass wasting, especially debris flows, poses an immediate threat to populated areas.

"If there's a town down below we have to rush to get ahead of the next storm," Rasely said. "In the worst-case scenario, we plant someone halfway up the mountain with a radio."

In such cases, Rasely said, they often set up evacuation plans that can remain in effect for long periods of time.

The probability and scale of a potential disaster depends on many factors. Rasely classifies most wildfires or "burns" as either "light" or "heavy." Light burns destroy only the surface growth and have a recovery span of about 1 year, whereas heavy burns destroy the roots and require 3 to 5 years for recovery.

"First, we want to know the condition of the burn, if the roots are intact," he said.

Rasely said low burns increase the erosion rate on a given surface by 3 to 5 times, whereas heavy burns increase the rate 6 to 10 times.

Other factors include the steepness and length of the slope, as well as whether established paths for water to flow already exist on the surface—a bad sign, since this means the water will run off the mountain at a much faster rate.

Pine needles and oaks also leave a waxy residue when they burn, which can further expedite the flow, Rasely said.

Once the general risk of mass wasting has been assessed, Rasely and his team have a variety of tricks they can use to minimize the potential damage, including erecting fencing made out of mesh material called "geofabric," as well as digging what they call a "debris basin" to catch falling material.

"If there's any kind of little terrace, you can stop a lot of stuff," he said.

Rasely's job has taken him to all but two of Utah's mountain ranges, and he frequently works in the many state and national parks in Utah.

And although the severity of the situation may vary widely, time is always important in this line of work.

"It's by the seat of the pants, live time," Rasely said.

—CHRIS WILSON

▲ FIGURE 7.C **GEOLOGIST EXAMINES RECENT LANDSLIDE** Bob Rasely, recently retired Utah State geologist for the Natural Resources Conservation Service, stands on the uneven, hummocky surface of the Thistle landslide shown in Figure 7.28. Irregular land surfaces such as this are common on landslides. *(Bob Rasely)*

can control or minimize their occurrence. Following is a description of several human activities that interact with landslides.

TIMBER HARVESTING AND LANDSLIDES

The possible cause-and-effect relationship between timber harvesting and erosion is a major environmental and economic issue in Northern California, Oregon, and Washington. Two controversial practices are *clearcutting*, the harvesting of all trees from large tracts of land, and road building, the construction of an extensive network of logging roads used to remove cut timber from the forest. Landslides, especially shallow soil slips, debris avalanches, and more deeply seated earthflows, are responsible for much of the erosion in these areas.

One 20-year study in the Western Cascade Range of Oregon found that shallow slides are the dominant form of erosion. This study also found that timber-harvesting activities, such as clearcutting and road building, did not significantly increase landslide-related erosion on geologically stable land; however, logging on weak, unstable slopes did increase landslide erosion by several times compared to slopes that had not been logged.[15]

Road construction in areas to be logged can be an especially serious problem because roads may interrupt surface drainage, alter subsurface movement of water, and adversely change the distribution of earth materials on a slope by cut-and-fill or grading operations.[15] As we learn more about erosional processes in forested areas, soil scientists, geologists, and foresters are developing improved management practices to minimize the adverse effects of logging. Nevertheless, landslide erosion problems continue to be associated with timber harvesting.

URBANIZATION AND LANDSLIDES

Human activities are most likely to cause landslides in urban areas where there are high densities of people, roads, and buildings. Examples from Rio de Janeiro, Brazil, and Los Angeles, California, illustrate this situation (see Case Study 7.4).

Rio de Janeiro, with a population of more than 6 million people, may have more slope-stability problems than any other city of its size.[16] Several factors contribute to a serious landslide problem in the city and surrounding area: (1) The beautiful granite peaks that spectacularly frame the city (Figure 7.20) have steep slopes of fractured rock that are covered with thin soil, (2) the area is periodically inundated by torrential rains, (3) cut-and-fill construction has seriously destabilized many slopes, and (4) vegetation has been progressively removed from the slopes.

The landslide problem started early in the city's history when many of the slopes were logged for lumber and fuel and to clear land for agriculture. Landslides associated with heavy rainfall followed the early logging activity. More recently, a lack of building sites on flat ground has led to increased urban development on slopes. The removal of additional vegetative cover and the construction of roads have led to building sites at progressively higher elevations. Excavations have cut the base of many slopes and severed the soil cover at critical points. In addition, placing fill material on slopes to expand the size of building sites has increased the load on already unstable land. Since this area periodically experiences tremendous rainstorms, it becomes apparent that Rio de Janeiro has a serious problem.

In February 1988, an intense rainstorm dumped more than 12 cm (5 in.) of rain on Rio de Janeiro in 4 hours. The storm caused flooding and debris flows that killed at least 90 people and left 3000 people homeless. Most of the deaths occurred from debris flows in hill-hugging shantytowns where housing is precarious and control of storm water runoff nonexistent. However, more affluent mountainside areas were not spared from destruction. In one area, a landslide destroyed a more affluent nursing home and killed 25 patients and staff. Restoration costs for the entire city have exceeded \$100 million. A similar story played out in April 2010, when heavy rains caused numerous landslides that destroyed 60 homes and killed more than 200 people. Much of the damage and lose of life occurred in low-income slums, where, in addition to poor construction practices, the homes were built on the steep slopes that surround the city (Figure 7.20b). If future disasters are to be avoided, Rio de Janeiro must take extensive and decisive measures to control storm runoff and increase slope stability.

Los Angeles in particular (and Southern California in general) has also experienced a high frequency of landslides associated with hillside development. Large variations in topography, rock and soil types, climate, and vegetation make interactions with the natural environment complex and notoriously unpredictable. For this reason, the area has the sometimes dubious honor of showing the ever-increasing value of studying urban geology.[18]

It took natural processes many thousands, and perhaps millions, of years to produce valleys, ridges, and hills. In a little over a century, humans have developed the machines to grade them. Nearly 40 years ago, F. B. Leighton, a geological consultant in Southern California, wrote: "With modern engineering and grading practices and appropriate financial incentive, no hillside appears too rugged for future development."[17] Thus, human activity has become a geological agent that is capable of carving the landscape at a much faster pace than glaciers and rivers. Almost overnight, we can convert steep hills into a series of flat lots and roads. The grading process, in which benches, referred to as pads, are cut into slopes for home sites, has been responsible for many landslides. Because of the extent of grading, Los Angeles has led the nation in developing codes concerning grading for development.

Grading codes that minimize the landslide hazard have been in effect in the Los Angeles area since 1963. These codes were adopted in the aftermath of destructive

Portuguese Bend, California

The Portuguese Bend landslide along the Southern California coast in Los Angeles is a famous example of how people can increase the landslide hazard. This landslide, which destroyed more than 150 homes, is part of a larger ancient slide (Figure 7.D). Road building and changes in subsurface drainage associated with urban development reactivated the ancient slide.

Aerial photography of the lower part of the reactivated landslide shows evidence of its movement (Figure 7.E). This evidence includes bare ground west of the highway where recent movement destroyed homes and roads, and a kink in the pier caused by land sliding into the ocean. Eventually the slow-moving landslide destroyed the pier and the adjacent swim club. Recent movement of the slide started in 1956 during construction of a county road. Fill dirt placed over the upper part of the ancient slide increased its instability. During subsequent litigation, the Los Angeles County was found responsible for the landslide.

From 1956 to 1978, the slide moved continually at an average rate of 0.3 to 1.3 cm (0.1 to 0.5 in.) per day. Several years of above-normal precipitation

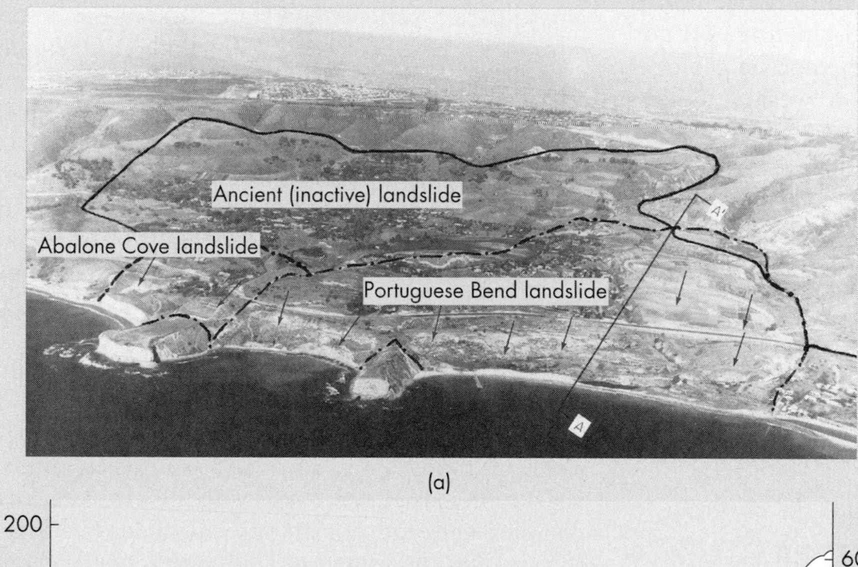

(a)

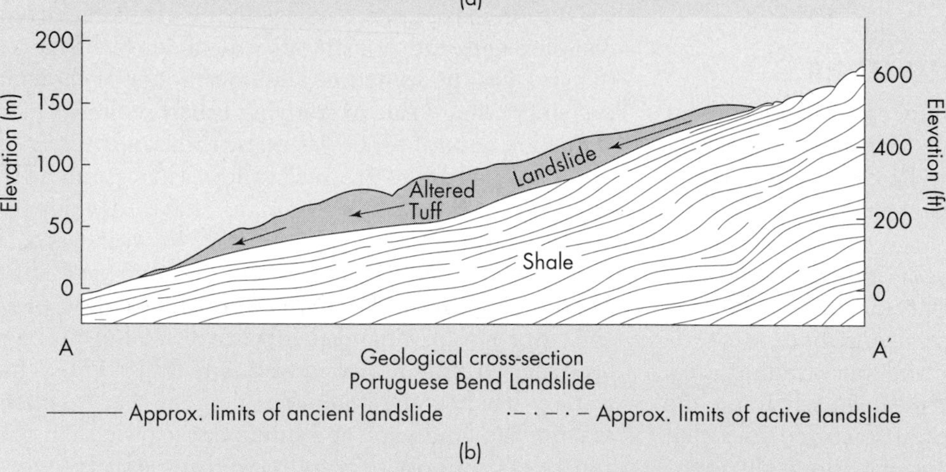

(b)

▲ FIGURE 7.D **ANCIENT LANDSLIDES CAN BE REACTIVATED** (a) Extent of ancient inactive landslide, part of which was reactivated to form the Portuguese Bend slide in the 1950s and Abalone Cove slide in the 1970s. Arrows show the direction of landslide movement toward the Pacific Ocean in the foreground. *(Los Angeles County, Department of County Engineer)* (b) Cross section on the right side of Figure 7.D passes through the Portuguese Bend slide. The landslide is underlain by a consolidated pyroclasitc volcanic tuff that has been altered to Bentonite; a type of expansive clay that deforms readily when a force is applied to it. *(Photograph and cross section courtesy of Los Angeles County, Department of County Engineer)*

◀ FIGURE 7.E **HOMES DESTROYED** The kink in the pier in the upper-left portion of this image shows initial damage from the Portuguese Bend landslide. Eventually most of the homes seen here, as well as the swim club and pier, were destroyed by the slow-moving landslide. Look for the pier in the lower center of Figure 7.D for a sense of the size of the lower part of the landslide. *(University of Washington Libraries)*

accelerated the movement to more than 2.5 cm (1 in.) per day in the late 1970s and early 1980s. Since 1956, the total displacement of the slide near the coast has been more than 200 m (660 ft.). The above-normal rainfall reactivated a second part of the ancient slide to form the Abalone Cove landslide to the north (Figure 7.D). The second slide prompted additional geologic investigation, and a successful landslide-control program was initiated. Wells were drilled into the wet rocks of the Portuguese Bend landslide in 1980 to remove groundwater from the slide mass. The pumping was an attempt to dry out and stabilize the rocks in the slide. By 1985, the slide had apparently been stabilized. However, precipitation and groundwater conditions in the future will determine the fate of the "stabilized" slides.[17]

During the two-decade period of activity, homes on the Portuguese Bend landslide continued to move. One home constantly shifted position as it moved more than 25 m (80 ft.). Other homes were not in constant motion but still moved up to 50 m (160 ft.) in the same time period. Homes remaining on Portuguese Bend during the slide's active period were adjusted every year or so with hydraulic jacks. Utility lines were placed on the surface to avoid breakage as the ground shifted. With one exception, no new homes have been constructed since the landslide began to move. The remaining occupants have elected to adjust to the landslide rather than bear total loss of their property (Figure 7.F). Nevertheless, few geologists would choose to live there now.

The story of the Portuguese Bend landslides emphasizes that science can be used to understand landslides, sometimes allowing us to at least temporarily stop their movement. However, recognizing landslides and using land-use planning to avoid building on active slides is preferable to reacting to landslide movement after homes have been constructed.

◀ FIGURE 7.F **LIVING WITH LANDSLIDES** Homes, roads, and streets were built on Portuguese Bend southwest of Los Angeles in spite of geologic maps published in the 1940s that revealed most of the area as a landslide. Palos Verdes Drive South, shown here, crosses the landslide and requires constant roadwork as it moves several feet each year toward the ocean. *(Chris Cantelmo)*

(a)

(b)

◀ **FIGURE 7.20**
LANDSLIDES ARE COMMON IN THE RIO DE JANEIRO AREA (a) Panoramic view of Rio de Janeiro, Brazil, showing the steep "sugarloaf" mountains and hills (right center of the image). A combination of steep slopes, fractured rock, shallow soils, and intense rainfall contributes to the landslide problem, as do human activities such as urbanization, logging, and agriculture. Virtually all the bare rock slopes were at one time vegetated, and that vegetation has been removed by landslides and other erosional processes *(Roger Viollet Collection/Getty Images).* (b) Aerial view of the Morro dos Prazeres slum, where several homes were destroyed by landsliding in Rio de Janeiro on April 8, 2010 *(Antonio Scorza/AFP/Getty Images/Newscom).*

and deadly landslides in the 1950s and 1960s (see Case Study 7.4). Since these codes have been in effect and detailed engineering geology studies have been required, the percentage of hillside homes damaged by landslides and floods has been greatly reduced. Although initial building costs are greater because of the strict codes, they are more than balanced by the reduction of losses in subsequent wet years.

Over-steepened slopes, increased water from lawn irrigation and septic systems, and the additional weight of fill material and buildings make formerly stable slopes unstable (Figure 7.21). As a rule, any project that steepens or saturates a slope, increases its height, or places an extra load on it may cause a landslide.[19]

Landslides related to urbanization have also been a problem in the eastern United States. In Cincinnati and surrounding Hamilton County, Ohio, most slides have taken place in colluvium, although clay-rich glacial deposits and soil developed on shale have also slipped.[18] At an average cost of more than $5 million per year, these slides are a serious problem.[12]

In Pittsburgh and surrounding Allegheny County, Pennsylvania, urban construction is estimated to be responsible for 90 percent of the landslides. Most of these slides are slow moving and take place on hillslopes of weathered mudstone or shale. In one deadly exception in an adjacent county, a rock fall crushed a bus and killed 22 passengers. Most of the landslides in Allegheny County are caused by placing fill dirt or buildings on the top of a slope, by cuts in the toe of a slope, or by the alteration of water conditions on or beneath a slope.[20] Damages from these slides average $2 million per year.

7.6 Minimizing the Landslide Hazard

To minimize the landslide hazard, it is necessary to identify areas in which landslides are likely to occur, design slopes or engineering structures to prevent them, warn people of impending slides, and control slides after they have started moving. As discussed next, the most preferable and least expensive option to minimize the landslide hazard is to avoid development on sites where landslides are occurring or are likely to occur.

IDENTIFICATION OF POTENTIAL LANDSLIDES

Recognizing areas with a high potential for landslides is the first step in minimizing this hazard. These areas can be identified where slopes are underlain by sensitive earth materials, such as clay or shale, and by a variety of surface features:

- Crescent-shaped cracks or terraces on a hillside
- A tongue-shaped area of bare soil or rock on a hillside
- Large boulders or talus piles at the base of a cliff
- A linear path of cleared or disturbed vegetation extending down a hillslope
- Exposed bedrock with layering that is parallel to the slope
- Tongue-shaped masses of sediment, especially gravel, at the base of a slope or at the mouth of a valley
- An irregular, often referred to as *hummocky*, land surface at the base of a slope

Geologists look for these features in the field and on aerial photographs. This information is used to assess the hazard and to produce several types of maps.

The first type of map is the direct result of the landslide inventory just described. It may be a reconnaissance map showing areas that have experienced slope failure or a more detailed map showing landslide deposits in terms of their relative activity (Figure 7.22a). Information concerning past landslides may be combined with land-use considerations to develop a *slope stability map* for engineering

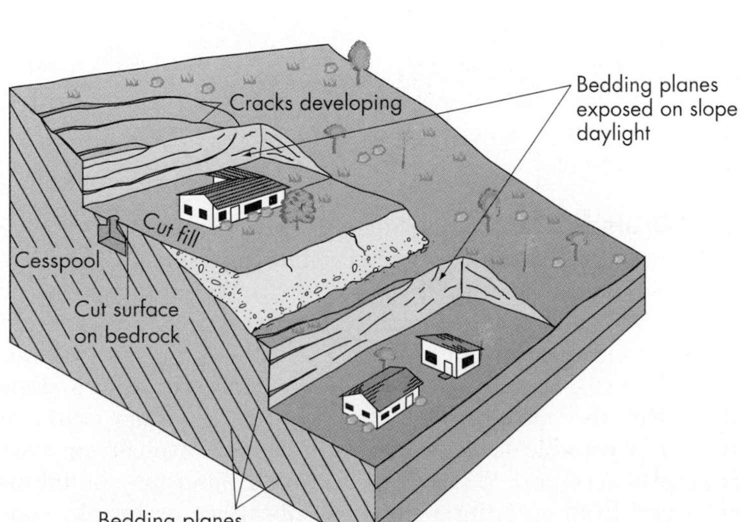

◀ **FIGURE 7.21 URBANIZATION AND LANDSLIDE POTENTIAL** This diagram shows how building on hillslopes can contribute to slope instability. The diagonal lines in the gray side of the diagram are bedding planes in sedimentary rock. Cuts into the hillside behind the houses have removed support from these bedding surfaces. The yellow fill material used to extend the flat pad for building adds weight to the slope. Cracks shown in the upper part of the diagram are an early sign that a landslide is likely to occur soon. Leaking wastewater from "cesspools" and landscape irrigation can add lubricating water to the hillslope. *(Reprinted with permission from F. Beach Leighton, Landslides and Urban Development, 1966, in Engineering Geology.)*

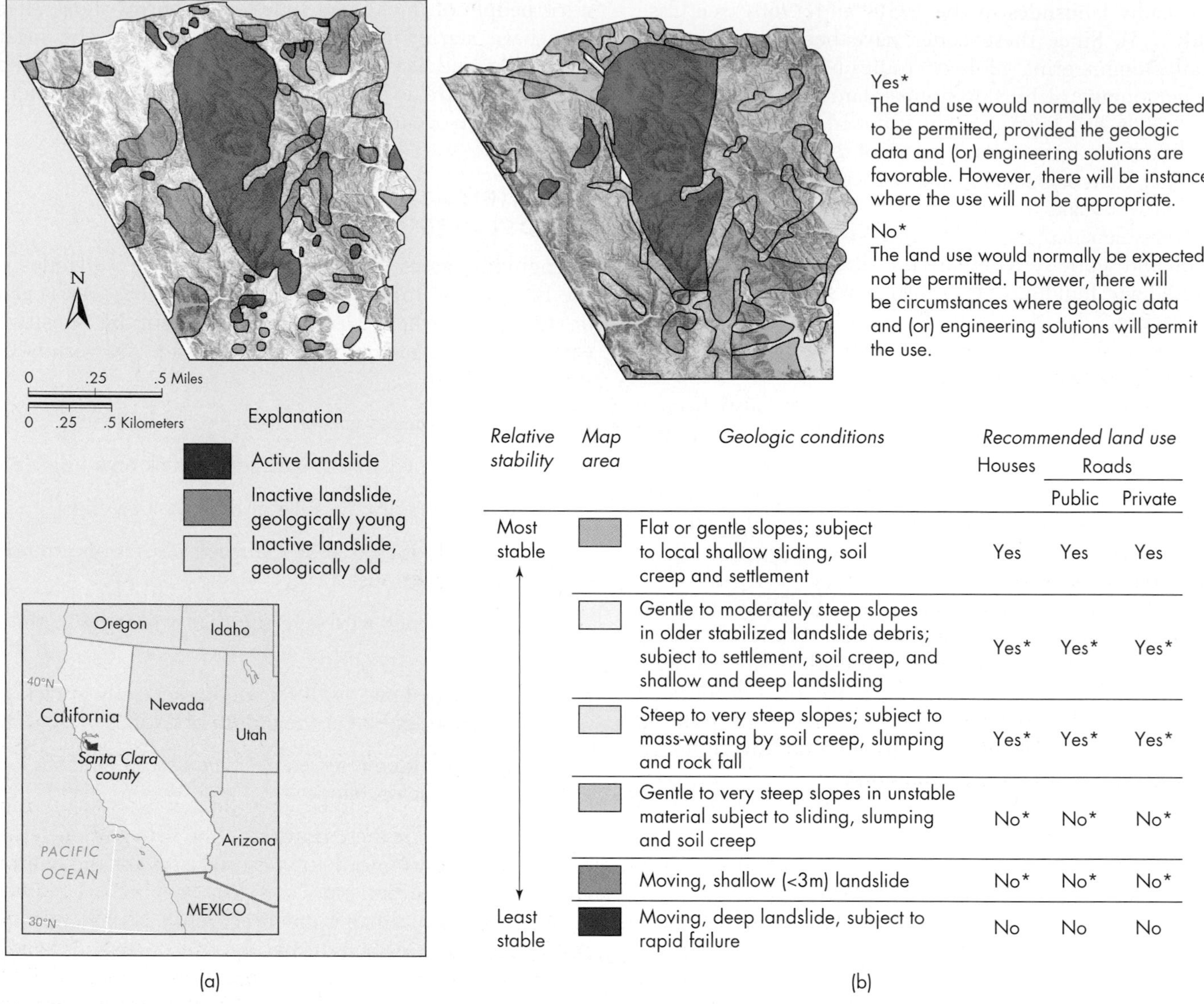

Relative stability	*Map area*	*Geologic conditions*	*Recommended land use* Houses	Roads Public	Roads Private
Most stable		Flat or gentle slopes; subject to local shallow sliding, soil creep and settlement	Yes	Yes	Yes
↑		Gentle to moderately steep slopes in older stabilized landslide debris; subject to settlement, soil creep, and shallow and deep landsliding	Yes*	Yes*	Yes*
		Steep to very steep slopes; subject to mass-wasting by soil creep, slumping and rock fall	Yes*	Yes*	Yes*
		Gentle to very steep slopes in unstable material subject to sliding, slumping and soil creep	No*	No*	No*
↓		Moving, shallow (<3m) landslide	No*	No*	No*
Least stable		Moving, deep landslide, subject to rapid failure	No	No	No

▲ **FIGURE 7.22 LANDSLIDE HAZARD MAPS** (a) Landslide inventory map and (b) a landslide risk and land-use map for part of Santa Clara County, California. *(After U.S. Geological Survey. 1982.* Goals and tasks of the landslide part of a ground-failure hazards reduction program. *U.S. Geological Survey Circular 880)*

geologists or a *landslide hazard map* with recommended land uses (Figure 7.22b) for planners. These maps do not take the place of detailed evaluation of a specific site. Preparing a *landslide risk map* is more complicated, because it involves evaluating the probability of a landside occurring and an assessment of potential losses.[21]

PREVENTION OF LANDSLIDES

Prevention of large, natural landslides is difficult, but common sense and good engineering practices can help minimize the hazard. For example, loading the top of slopes, cutting into sensitive slopes, placing fill material on slopes, or changing water conditions on slopes should be avoided or done with caution.[19] Common engineering techniques for landslide prevention include surface and subsurface drainage, removal of unstable slope materials, construction of retaining walls or other supporting structures, or some combination of these techniques.[2]

Drainage Control Surface and subsurface drainage control is usually effective in stabilizing a slope. The objective is to divert water to keep it from running across or infiltrating the slope. Surface runoff may be diverted around the slope by a series of surface drains (Figure 7.23a). The amount of water infiltrating a slope may also be controlled by covering the slope with an impermeable layer such as soil-cement, asphalt, or even plastic (Figure 7.23b). Underground water may be inhibited from entering a slope by subsurface drains. To construct a drain, a drainpipe with holes along its length is

(a)

(b)

▲ FIGURE 7.23 **TWO WAYS TO INCREASE SLOPE STABILITY** (a) Concrete drain on a roadcut that removes surface water runoff before it can infiltrate the slope. *(Edward A. Keller)* (b) Workers are covering this slope in Greece with soil-cement to reduce the infiltration of water and provide strength. Soil-cement is a blend of pulverized soil, Portland cement, and water. *(Edward A. Keller)*

surrounded with permeable gravel or crushed rock and is positioned underground to intercept and divert water away from a potentially unstable slope.[2]

Grading Although grading of slopes for development has increased the landslide hazard in many areas, carefully planned grading can increase slope stability. In a single cut-and-fill operation, material from the upper part of a slope is removed and placed near the base. The overall gradient is thus reduced, and material is removed from the upper slope, where it contributes to the driving force, and placed at the toe of the slope, where it increases the resisting force. However, this method is not practical on a high, steep slope. Instead, the slope may be cut into a series of benches or steps, each of which contains surface drains to divert runoff. The benches reduce the overall slope and are good collection sites for falling rock and small slides (Figure 7.24).[2]

Slope Supports One of the most common methods of slope stabilization is a retaining wall that supports the slope. These walls can be constructed of concrete or brick; stone-filled wire baskets called *gabions* or a series of piles of long concrete, steel, or wooden beams driven into the ground (Figure 7.25). To function effectively, the walls must be anchored well below the base of the slope, backfilled with permeable gravel or crushed rock (Figure 7.26), and provided with drain holes to reduce water pressure in the slope (Figure 7.25). With the addition of plants, these walls can be aesthetically pleasing or blend in with the natural hillside (Figure 7.27).

Preventing landslides can be expensive, but the rewards can be well worth the effort. Estimates of the benefit-to-cost ratio for landslide prevention range from 10 to 2000. That is, for every dollar spent on landslide prevention, the savings will range from \$10 to \$2000.[22]

The cost of *not* preventing a slide is illustrated by the massive Thistle landslide southeast of Salt Lake City in April 1983. This slide moved down a mountain slope and across a canyon to form a natural dam about 60 m (200 ft.) high. This dam created a lake that flooded the community of Thistle, the Denver–Rio Grande Railroad switchyard and tracks, and two major U.S. highways (Figure 7.28).[21] The total costs (direct and indirect) from the landslide and associated flooding exceeded \$400 million.[23]

The Thistle slide was a reactivated older slide, which had been known for many years to be occasionally active in response to high precipitation. It came as no surprise that those extremely high amounts of precipitation were produced by a climatic event called El Niño (see Chapter 12). In fact, a review of the landslide history suggests that this slide was recognizable, predictable, and preventable! A network of subsurface and surface drains would have prevented failure. The cost of preventing the landslide was between \$300,000 and \$500,000, a small amount compared to the damages caused by the slide.[22] Because the benefit-to-cost ratio in landslide prevention is so favorable, it seems prudent to evaluate active and potentially active landslides in areas where considerable damage may be expected and possibly prevented.

LANDSLIDE WARNING SYSTEMS

Landslide warning systems do not prevent landslides, but they can provide time to evacuate people and their

◀ **FIGURE 7.24 BENCHES ON A HIGHWAY ROADCUT** The stepped surfaces, or benches, seen in the upper-right part of this image reduce the overall steepness of the slope and provide better drainage. Benches along roadcuts into bedrock can catch rock falls before they reach a highway. *(Edward A. Keller)*

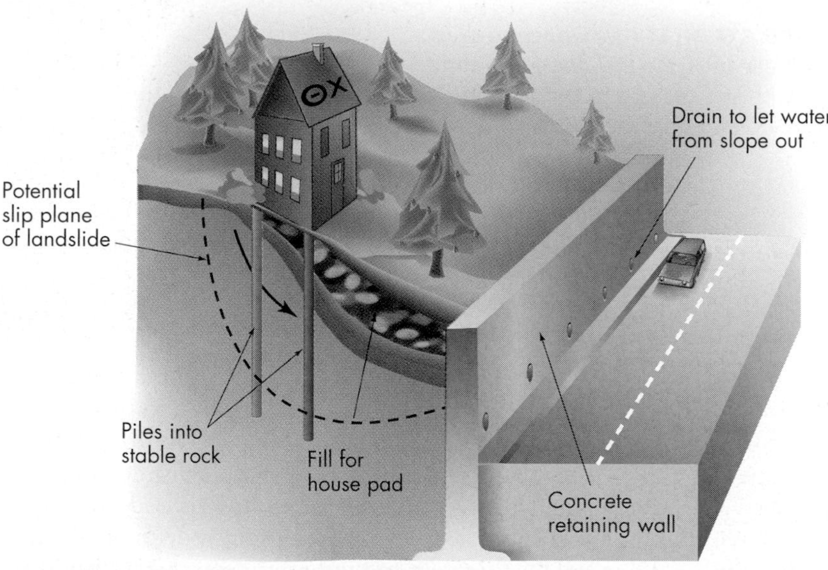

◀ **FIGURE 7.25 HOW TO SUPPORT A SLOPE** The types of slope support shown in this illustration include a roadside concrete retaining wall that is deeply anchored into the slope, concrete or steel piles sunk into stable rock, and subsurface drains that reduce water pressure in the slope.

◀ **FIGURE 7.26 RETAINING WALL** This retaining wall of concrete cribbing was installed and then backfilled to help stabilize the roadcut. *(Edward A. Keller)*

(a)

(b)

(c)

▲ **FIGURE 7.27 STEPS IN MAKING A RETAINING WALL** (a) This shallow slide occurred in the early 1990s. *(Edward A. Keller)* (b) A retaining wall of cemented stone was constructed in 1999 to correct the problem. *(Edward A. Keller)* (c) Benches behind the retaining walls were landscaped as shown by this 2001 photograph. *(Edward A. Keller)*

possessions and stop trains or reroute traffic. Surveillance provides the simplest type of warning. Hazardous areas can be visually inspected for apparent changes, and small rock falls on roads and other areas can be noted for quick removal. Human monitoring has the advantages of reliability and flexibility but becomes difficult during adverse weather and in hazardous locations.[24] For example, snow avalanche conditions in the United States have been monitored for decades by park rangers and volunteers of the National Ski Patrol.

Other warning methods include electrical systems, tiltmeters, and geophones that pick up vibrations from moving rocks. Many U.S. and Canadian railroads have slide fences on slopes above their tracks that are tied to signal systems. When a large rock hits the fence, a signal is sent to stop trains before they are in danger. This and other mitigation measures have significantly reduced the number of railroad accidents, injuries, and fatalities.[24]

A pilot landslide warning system is under development in western Washington State. The U.S. Geological Survey and Pierce County have deployed acoustic flow monitors to detect a debris flow (lahar) from Mt. Rainier.[25] A large debris flow from the mountain, similar to one that occurred about 500 years ago, would devastate more than a dozen communities and kill thousands, and possibly tens of thousands, of people.

For existing slides, shallow wells can be drilled into a slope and monitored to signal when the slide contains a dangerous amount of water. In some regions, a rain-gauge network is useful for warning when a precipitation threshold has been exceeded and shallow soil slips are probable.

7.7 Perception of and Adjustment to the Landslide Hazard

PERCEPTION OF THE LANDSLIDE HAZARD

The common reaction of homeowners who talk about landslides is, "It could happen on other hillsides, but never this one."[16] Just as flood-hazard mapping does not

◀ **FIGURE 7.28 LANDSLIDE BLOCKS A CANYON** The costliest landslide in U.S. history, this 1983 slide at Thistle, Utah, was a reactivation of an older slide. Debris from the slide blocked the Spanish Fork River and created a natural dam. This dam produced a lake that flooded the community of Thistle, the Denver–Rio Grande Railroad, and two major U.S. highways. Nearly the entire landslide moved again in 1999. *(Michael Collier)*

prevent development in flood-prone areas, landslide-hazard maps will not prevent many people from moving into hazardous areas. Prospective hillside occupants who are initially unaware of the hazard may not be swayed by technical information. The infrequency of large slides reduces awareness of the hazard, especially where evidence of past events is not readily visible. Unfortunately, it often takes catastrophic events to bring the problem to people's attention. In the meantime, people in many parts of the Rocky Mountains, Appalachian Mountains, California, and other areas continue to build homes in areas subject to future landslides.

ADJUSTMENTS TO THE LANDSLIDE HAZARD

Although the most reasonable adjustment to the landslide hazard is simply not to build in landslide-prone areas, many people still go ahead. As long as people continue to build and buy expensive homes "with a view," we will need to find other adjustments to avoid death and damage from landslides. Such adjustments can include locating critical facilities away from landslide-prone areas and landslide correction.

Siting of Critical Facilities As with earthquakes (Chapter 3), safely siting critical facilities such as hospitals, schools, and police stations is crucial. Ensuring that these buildings are not located on or directly below hillsides is a simple way to guarantee that they will remain functional in the event of a landslide. In urban areas that have been built up against mountains, such as Juneau, Alaska, such selective siting can be difficult.

Landslide Correction After a slide has begun, the best way to stop it is to attack the process that started the slide. In most cases, the cause is an increase in underground water pressure within and below the slide. The pressure may be reduced by an effective drainage program. This program often includes surface drains at the head of the slide to reduce surface water infiltration and subsurface drainpipes or wells to remove groundwater and lower the water pressure. Draining tends to increase the resisting force of the slope material, thereby stabilizing the slope.[6]

PERSONAL ADJUSTMENTS: WHAT YOU CAN DO TO MINIMIZE YOUR LANDSLIDE HAZARD

Consider the following advice if purchasing property on a slope:

- Landslides often develop in areas of complex geology, and a geologic evaluation by a professional geologist of any property on a slope is recommended.
- Avoid homes at the mouth of a valley or canyon, even a small one, where debris flows may originate from upstream slopes and travel down the channel.
- Consult local agencies such as city or county engineering departments that may be aware of landslides in your area.
- Watch out for "little landslides" in the corner of the property—they usually get larger with time.
- If purchasing a home, look for cracks in house walls and look for retaining walls that lean or are cracked. Be wary of doors or windows that stick or floors that are uneven. Foundations should be checked for cracks or tilting. If cracks in walls of the house or foundation can be followed to the land outside the house, be especially concerned that a landslide may be present.
- Be wary of leaks in a swimming pool or septic tank, trees or fences tilted downslope, or utility wires that are taut or sagging.
- Be wary if small springs are present, because landslides tend to leak water. Look for especially "green" areas, other than a septic system leach field, where more subsurface water is present.
- Walk the property and surrounding property, if possible, looking for linear or curved cracks (even small ones) that might indicate instability of the land surface.
- Look for the surface features listed earlier that geologists use to identify potential landslides.
- Although correcting a potential landslide problem is often cost effective, it can still be expensive; much of the fix is below ground where you will never see the improvement. Overall it is better not to purchase land with a potential landslide hazard.
- The presence of one or more of these features does not prove that a landslide is present or will occur. For example, cracks in walls and foundations, doors or windows that stick, or floors that are uneven can be caused by soils that shrink and swell. However, further investigation is warranted if these features are present.

If you enjoy snow sports, hike, climb, live, or travel in areas that have snow avalanches, you should be aware of the following:

- Most avalanches that involve people are triggered by the victims themselves or by others in their party.[9]
- Obtain forecasts from the nearest avalanche center. Remember that changing weather conditions can affect the avalanche hazard.
- Most people who survive avalanches are rescued by the other members in their party.[9]
- Learn avalanche safety procedures and how to evaluate snow conditions before traveling in an area that has avalanches.

REVISITING THE FUNDAMENTAL CONCEPTS

Mass Wasting

1. **Hazards Are Predictable from a Scientific Evaluation**
2. **Risk Assessment Is an Important Component of Our Understanding of the Effects of Hazardous Processes**
3. **Linkages Exist between Natural Hazards**
4. **Hazardous Events That Previously Produced Disasters Are Now Producing Catastrophes**
5. **Consequences of Hazards May Be Minimized**

1) As a result of the role of water in most mass wasting processes, heavy precipitation is often correlated with landslide events. However, because precipitation is not the only driving force that causes slopes to fail, prediction is limited when no data of other controlling factors are available, such as the presence of crescent-shaped cracks, quick clays, recent fire, or over-steepened slopes or evidence of older landslide events. In such areas, mass wasting

(continued)

may be anticipated especially during or following intense precipitation.

2) On most modern building sites, landslide hazard assessment is conducted before construction begins, and, therefore, mass wasting processes are amenable to risk analysis.

3) Landslides are linked to severe weather, earthquakes, volcanoes, coastal erosion, wildfires, and tsunamis. In most cases, landslides are the result and not the cause of other natural hazards. Although, as the example of Mount St. Helens in Chapter 5 and the Papua New Guinea tsunami (see Section 4.1) demonstrate, large volcanic eruptions and deadly tsunamis can be triggered by landslides.

4) With ever-increasing population, suitable construction sites on low relief areas become less available. This leads to increased development in areas where slopes are steeper and inherently less stable. As the loss of life and homes in Rio de Janeiro over the past few decades exemplifies (see Section 7.5), rapid population growth has led to increased construction on the steep slopes outside the city limits. Had people not been forced to build on these steep slopes, it is likely that the loss of 200 lives in April 2010 from landslides could have been avoided.

5) It is clear from construction practices in most of the developed countries in the world that the consequences of mass wasting processes can be minimized through proper land use and engineering. Evaluation of construction sites for potential landslides and remediation if evidence is documented before construction begins is standard practice in the United States. Surface runoff can be channelized and transported off steep slopes to avoid saturation of earth materials, thereby minimizing the driving force that water provides to the mass wasting process. Slopes can be graded to decrease the slope angle, and retaining walls and debris basins can be constructed to divert or catch landslide materials where anticipated.

Summary

The most common landforms are slopes—dynamic, evolving systems in which surficial material is constantly moving downslope, or mass wasting at rates ranging from imperceptible creep to thundering avalanches. Slope failure may involve *flowage*, *sliding*, or *falling* of earth materials; landslides are often complex combinations of sliding and flowage.

The forces that produce landslides are determined by the interactions of several variables: the type of earth material on the slope, topography and slope angle, climate, vegetation, water, and time. Determining the cause of most landslides can be accomplished by examining the relations between forces that tend to make earth materials slide, the *driving* forces, and forces that tend to oppose movement, the *resisting* forces. The most common driving force is the weight of the slope materials, and the most common resisting force is the *shear strength* of the slope materials. Geologists and engineers determine the *safety factor* of a slope by calculating the ratio of resisting forces to driving forces; a ratio greater than one means that the slope is stable; a ratio less than one indicates potential slope failure. The type of rock or soil on a slope influences both the type and the frequency of landslides.

Water has an especially significant role in producing landslides. Moving water in streams, lakes, or oceans erodes the base of slopes, increasing the driving forces. Excess water within a slope increases both the weight and underground water pressure of the earth material, which in turn decreases the resisting forces on the slope.

Snow avalanches present a serious hazard on snow-covered, steep slopes. Loss of human life from avalanches is increasing as more people venture into mountain areas for winter recreation.

Landslides may occur just about anywhere that slopes exist. The areas of greatest hazard in the United States include the mountainous areas of the West Coast and Alaska, the Rocky Mountains, and the Appalachian Mountains. Where landslides do occur, they may cause significant damage and loss of life. They are also linked to other hazards, especially floods, earthquakes, and wildfires.

The effects of land use on the magnitude and frequency of landslides range from insignificant to very significant. Where landslides occur independently of human activity, we need to avoid development or provide protective measures. In other cases, when land use has increased the number and severity of landslides, we need to learn how to minimize their recurrence. In some cases, filling large water reservoirs has altered groundwater conditions along their shores and caused slope failure. Logging operations on weak, unstable slopes have increased landslide erosion. Grading of slopes for development has created or increased landslide problems in many urbanized areas of the world.

To minimize the landslide hazard, it is necessary to establish identification, prevention, and correction procedures. Monitoring and mapping techniques help identify hazardous sites. Identification of potential landslides has been used to establish grading codes, and these codes, in turn, have reduced landslide damage. Prevention of large natural slides is difficult, but careful engineering practices can minimize the hazard where they cannot be avoided. Engineering techniques for landslide prevention include drainage control, proper grading, and construction of supports such as retaining walls. Efforts to stop or slow existing landslides must attack the processes that started the slide—usually by initiating a drainage program that lowers water pressure in the slope. Even with these improvements in the recognition, prediction, and mitigation of landslides, the incidence of landslides is expected to increase in the twenty-first century.

Most people perceive the landslide hazard as minimal, unless they have prior experience. Furthermore, hillside residents, like floodplain occupants, are not easily swayed by technical information. Nevertheless, the wise person will have a geologist inspect property on a slope before purchasing.

Key Terms

colluvium (p. 216)
creep (p. 211)
debris flow (p. 217)
driving forces (p. 212)
falling (p. 211)
flow (p. 211)
landslide (p. 210)
mass wasting (p. 210)
resisting forces (p. 212)
safety factor (p. 213)
shear strength (p. 212)
sliding (p. 211)
slumping (p. 211)
snow avalanche (p. 221)
talus (p. 210)

Review Questions

1. What is a landslide?
2. What are slope segments?
3. What are the common types of slope segments and how do they differ?
4. What are the three main ways that materials on a slope may fail?
5. What is the safety factor and how is it defined?
6. How do slumps (rotational slides) differ from soil slips and rock slides (translational slides)?
7. How does the slope angle affect the incidence of landslides?
8. How and where do debris flows occur?
9. What are the three ways that vegetation is important in slope stability?
10. Why does time play an important role in landslides?
11. What is the relationship between the downslope force and normal force?
12. What variables interact to cause snow avalanches?
13. What is the angle of repose?
14. What are the two types of snow avalanches and how do they differ?
15. How might processes involved in urbanization increase or decrease the stability of slopes?
16. What types of surface features are associated with landslides?
17. What are the main steps that can be taken to prevent landslides?

Critical Thinking Questions

1. Your consulting company is hired by the national park department in your region to estimate the future risk from landsliding. Develop a plan of attack that outlines what must be done to achieve this objective.
2. Why do you think that few people are easily swayed by technical information concerning hazards such as landslides? Assume you have been hired by a municipality to make its citizens more aware of the landslide hazard on the steep slopes in the community. Outline a plan of action and defend it.
3. The Wasatch Front in central Utah frequently experiences wildfires followed by debris flows that exit mountain canyons and flood parts of communities built next to the mountain front. You have been hired by the state emergency management office to establish a warning system for subdivisions, businesses, and highways in this area. How would you design a warning system that will alert citizens to evacuate hazardous areas?

Selected Web Resources

Landslide Hazard Program
landsslides.usgs.gov—information on landslides from the U.S. Geological Survey

Landslide and Debris Flow (Mudslide)
www.fema.gov/hazard/landslide/—information on landslide hazards and mitigation from the Federal Emergency Management Agency

Landslides
www.consrv.ca.gov/cgs/geologic_hazards/landslides/—information on landslides and landslide maps from the California Geological Survey

Landslides Hazards in Oregon
www.oregongeology.com/sub/Landslide/Landslide-home.htm—information about debris flows and other landslides from the Oregon Department of Geology and Mineral Industries

Hazards
geosurvey.state.co.us/Default.aspx?tabid=35—information on landslides, debris flows, mudslides, snow avalanches, and rock falls from the Colorado Geological Survey

Puget Sound Landslides
www.ecy.wa.gov/programs/sea/landslides/—information on landslides and landslide prevention from the Washington Department of Ecology

Landslides
www.geology.enr.state.nc.us/Landslide_Info/Landslides_main.htm—information on landslides from the North Carolina Geological Survey

Landslides
landslides.nrcan.gc.ca—comprehensive site on landslides from Natural Resources Canada

avalanche.org
www.avalanche.org/—information on snow avalanches and links to North American avalanche forecast centers from the American Avalanche Association

Forest Service National Avalanche Center
www.fsavalanche.org/—information on snow avalanches and avalanche safety from the U.S. Forest Service

Assignments in Applied Geology: Landslide Hazard Assessment

THE ISSUE

Mass movements, including landslides, are a significant source of property damage. Damage or destruction due to landslides can be nearly instantaneous (such as in the La Conchita landslide of January 2005). Conversely, slow-moving landslides can take place over many years, gradually tearing apart buildings in their path. In either case, the conditions conducive to landslides are often readily identifiable. As part of choosing a site for residential and commercial development, a contractor commissions (or should commission) a landslide-hazard assessment. Determining the potential for landslides is a common job done by geologists. They use maps, field surveys, soil reports, and aerial photographs to accomplish this work.

YOUR TASK

Your client, a housing contractor, is considering five sites for purchase and development for single-family residences. Your job is to determine the landslide risk for each of the five sites. After that is done, you will determine the suitability of these five sites for development using other information that can be found in a soil survey. You must be careful not to recommend any sites that will put your client at risk. Sites with even a small amount of risk should be avoided.

AVERAGE COMPLETION TIME:

1 hour

8 Subsidence and Soils

Venice Is Sinking

Italy's beautiful and famous city of Venice faces a serious geologic problem. The city is sinking, or subsiding, up to 2 mm (0.08 in.) per year in some areas.[1] Venice is built on 17 small islands in a coastal lagoon. The islands are connected by more than 400 bridges. Although its coastal location and numerous canals are part of Venice's draw as a tourist destination, the presence of so much water surrounding a subsiding city makes Venice extremely prone to flooding.

The land on which Venice is built is often only a few centimeters above sea level, and many buildings are flooded repeatedly. Although subsidence (vertical deformation, in this case sinking) has been occurring naturally for millions of years, the overpumping of groundwater from the 1930s to the 1960s significantly increased the rate at which Venice is sinking.[2] This subsidence has resulted in numerous floods from the sea, the response to which has been to raise buildings and streets above the flood level. The overpumping, and thus most of the human contribution to this natural hazard, ended in the 1970s. Unfortunately, natural subsidence is still occurring and floods are increasingly common.

As worldwide sea level continues to rise, the future of Venice is uncertain

To combat the flooding, the Italian government has started a $2.6 billion project to construct 78 hinged, steel floodgates across the three tidal inlets to the Venice lagoon.[3] These floodgates will be designed to swing upward when a storm warning is received to prevent the wind-driven surge of seawater from the Adriatic Sea from entering the lagoon. However, these gates will neither slow the subsidence of the city nor completely stop flooding, and their long-term utility has been questioned.[4] As worldwide sea levels continue to rise, the future of Venice is uncertain.

The story of the subsidence of Venice involves linkages between soil, rock, compaction of sediment, tectonics, hydrology, and human use of the land. Soils are a very common factor in subsidence and land use because the first earth material we encounter when building anything at the surface of the Earth (homes and roads, for example) is soil.

In this chapter, we will focus mainly on subsidence, but soils are linked to several aspects of natural hazards. Therefore, we will begin with a general discussion of soil form and process.

LEARNING OBJECTIVES

Subsidence, the sinking of the land, and expansion and contraction of the soil are important geologic processes capable of causing extensive damage in some areas of the world. Your goals in reading this chapter should be to

- know what a soil is and the processes that form and maintain soils.
- understand the causes and effects of subsidence and volume changes in the soil.
- know the geographic regions at risk for subsidence and volume changes in the soil.
- understand the hazards associated with karst regions.
- recognize linkages between subsidence, soil expansion and contraction, and other hazards, as well as natural service functions of karst.
- understand how humans interact with subsidence and soil hazards.
- know what can be done to minimize the hazard from subsidence and volume changes in the soil.

◀ **Flooding in St. Mark's Square, Venice, Italy** An exceptional 137-cm (54-in) high tide on this day, November 1, 2004, caused the flooding of 80 percent of Venice in nothern Italy. Flooding is becoming more common as Venice subsides and sea level rises. *(AP Images)*

8.1 Soil and Hazards

Soil may be defined in several ways. To soil scientists, it is solid earth material that has been altered by physical, chemical, and organic processes such that it can support rooted plant life. To engineers, on the other hand, soil is any solid earth material that can be removed without blasting. Both of these definitions are important in earth science.

Study of soils helps in the evaluation of natural hazards, including floods, landslides, and earthquakes, since floodplain soils differ from upland soils, consideration of soil properties helps delineate natural floodplains. Evaluation of the relative ages of soils on landslide deposits may provide an estimate of the frequency of slides and thus assist in planning to minimize their impact. The study of soils has also been a powerful tool in establishing the chronology of earth materials deformed by faulting, which has led to better calculations of the recurrence intervals of earthquakes at particular sites.

The development of a soil from inorganic and organic materials is a complex process. Intimate interactions of the rock and hydrologic cycles produce the weathered rock materials that are basic ingredients of soils. **Weathering** is the physical and chemical breakdown of rocks and the first step in soil development. Weathered rock is further modified by the activity of soil organisms into soil, which is called either *residual* or *transported*, depending on where and when it has been modified. The more insoluble weathered material may remain essentially in place and be modified to form a residual soil, such as the red soils of the Piedmont in the southeastern United States. If weathered material is transported by water, wind, or glaciers and then modified in its new location, it forms a transported soil, such as the fertile soils formed from glacial deposits in the American Midwest.

A soil can be considered an open system that interacts with other components of the geologic cycle. The characteristics of a particular soil are a function of *climate, topography, parent material* (the rock or alluvium from which the soil is formed), *time* (age of the soil), and *organic processes* (activity of soil organisms). Many of the differences we see in soils are effects of climate and topography, but the type of parent rock, the organic processes, and the length of time the soil-forming processes have operated are also important.

SOIL HORIZONS

Vertical and horizontal movements of the materials in a soil system create a distinct layering, parallel to the surface, collectively called a *soil profile*. The layers are called zones or **soil horizons.** Our discussion of soil profiles will mention only the horizons most commonly present in soils. Additional information is available from detailed soils texts.[5,6]

Figure 8.1a shows the common master (prominent) soil horizons. The ***O* horizon** and ***A* horizon** contain highly concentrated organic material; the differences between these two layers reflect the amount of organic material present in each. Generally, the *O* horizon consists entirely of plant litter and other organic material, whereas the underlying *A* horizon contains a good deal of both organic and mineral material. Below the *O* or *A* horizon, some soils have an ***E* horizon,** or *zone of leaching*, a light-colored layer that is leached of iron-bearing components. This horizon is light in color because it contains less organic material than the *O* and *A* horizons and little inorganic coloring material such as iron oxides.

The ***B* horizon,** or *zone of accumulation*, underlies the *O*, *A*, or *E* horizon and consists of a variety of materials translocated downward from the overlying horizons. Several types of *B* horizon have been recognized. Probably the most important type is the *argillic B*, or B_t horizon. A B_t horizon is enriched in clay minerals that have been translocated downward by soil-forming processes. Another type of *B* horizon of interest to environmental geologists is the B_k horizon, characterized by the accumulation of calcium carbonate. The carbonate coats individual soil particles in the soils and may fill some pore spaces (the spaces between soil particles), but it does not dominate the morphology (structure) of the horizon. A soil horizon that is so impregnated with calcium carbonate that its morphology is dominated by the carbonate is designated a ***K* horizon** (Figure 8.1b). Carbonate completely fills the pore spaces in *K* horizons, and the carbonate often forms in layers parallel to the surface. The term *caliche* is often used for irregular accumulation or layers of calcium carbonate in soils.

The ***C* horizon** lies directly over the unaltered parent material and consists of parent material partially altered by weathering processes. The ***R* horizon,** or unaltered parent material, is the consolidated bedrock that underlies the soil. However, some of the fractures and other pore spaces in the bedrock may contain clay that has been translocated downward.[5]

The term *hardpan* is often used in the literature on soils. A hardpan soil horizon is defined as a hard (compacted) soil horizon. Hardpan is often composed of compacted and/or cemented clay with calcium carbonate, iron oxide, or silica. Hardpan horizons are nearly impermeable and, thus, restrict the downward movement of soil water.

SOIL COLOR

One of the first things we notice about a soil is its color, or the colors of its horizons. The *O* and *A* horizons tend to be dark because of their abundant organic material. The *E* horizon, if present, may be almost white, owing to the leaching of iron and aluminum oxides. The *B* horizon shows the most dramatic differences in color, varying from yellow-brown to light red-brown to dark red, depending

O. Horizon is composed mostly of organic materials, including decomposed or decomposing leaves, twigs, etc. The color of the horizon is often dark brown or black.

A. Horizon is composed of both mineral and organic materials. The color is often light black to brown. Leaching, defined as the process of dissolving, washing, or draining earth materials by percolation of groundwater or other liquids, occurs in the A horizon and moves clay and other material such as iron and calcium to the B horizon.

E. Horizon is composed of light-colored materials resulting from leaching of clay, calcium, magnesium, and iron to lower horizons. The A and E horizons together comprise the zone of leaching.

B. Horizon is enriched in clay, iron oxides, silica, carbonate or other material leached from overlying horizons. This horizon is known as the zone of accumulation.

C. Horizon is composed of partially altered (weathered) parent material; rock as shown here but the material could also be alluvial in nature, such as river gravels in other environments. The horizon may be stained red with iron oxides.

R. Unweathered (unaltered) parent material.

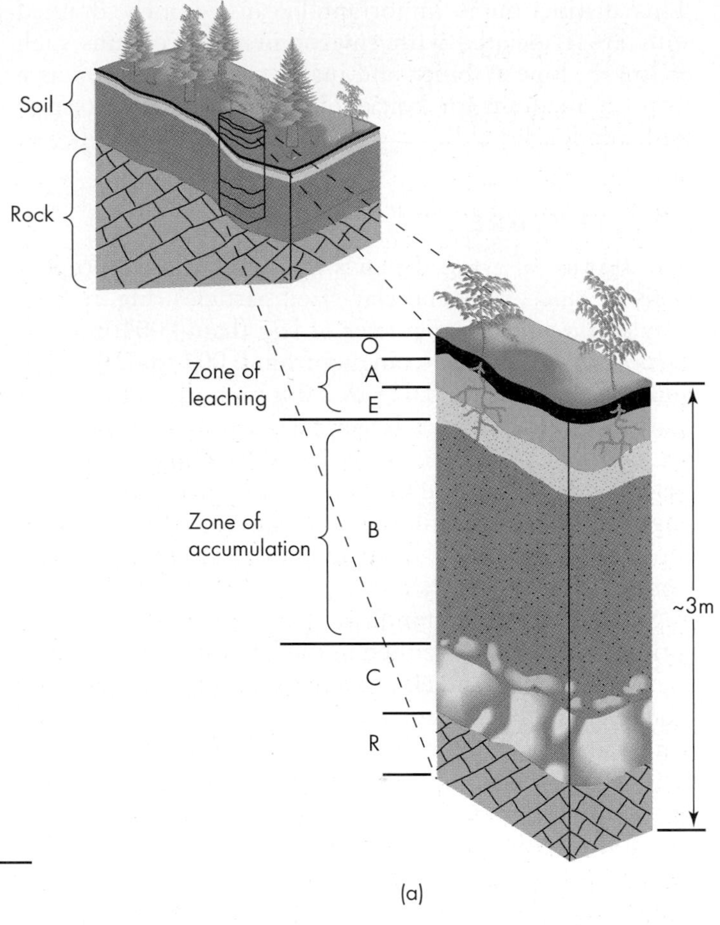

(a)

(b)

▲ **FIGURE 8.1 SOIL PROFILES** (a) Idealized diagram showing a soil profile with soil horizons. (b) Soil profile showing a dark brown *A* horizon, a light-red *B* horizon, a white *Bk* horizon rich in pedogenic calcium carbonate, and a lighter colored C horizon. *(Edward A. Keller)*

upon the presence of clay minerals and iron oxides. The B_k horizons may be light-colored due to their carbonates, but they are sometimes reddish as a result of iron oxide accumulation. If a true *K* horizon has developed, it may be almost white because of its great abundance of calcium carbonate. Although soil color can be an important diagnostic tool for analyzing a soil profile, one must be cautious about calling a red layer a *B* horizon. The original parent material, if rich in iron, may produce a very red soil even when there has been relatively little soil profile development.

Soil color may be an important indicator of how well drained a soil is. Well-drained soils are well aerated (oxidizing conditions), and iron oxidizes to a red color. Poorly drained soils are wet, and iron is reduced rather than oxidized. The color of such a soil is often yellow.

This distinction is important because poorly drained soils are associated with environmental problems such as lower slope stability and inability to be utilized as a disposal medium for household sewage systems (septic tank and leach field).

SOIL TEXTURE

The texture of a soil depends upon the relative proportions of sand-, silt-, and clay-sized particles (Figure 8.2). *Clay particles* have a diameter of less than 0.004 mm, *silt particles* have diameters ranging from 0.004 to 0.074 mm, and *sand particles* are 0.074 to 2.0 mm in diameter. Earth materials with particles larger than 2.0 mm in diameter are called *gravel*, *cobbles*, or *boulders*, depending on the particle size. Note that the sizes of particles given here are for engineering classification and are slightly different from those used by the U.S. Department of Agriculture for soil classification.

Soil texture is commonly identified in the field by estimation and then refined in the laboratory by separating the sand, silt, and clay and determining their proportions. A useful field technique for estimating the size of sand-sized or smaller soil particles is as follows: It is sand if you can see individual grains, silt if you can see the grains with a 10× hand lens, and clay if you cannot see grains with such a hand lens. Another method is to feel the soil: Sand is gritty (crunches between the teeth), silt feels like baking flour, and clay is cohesive. When mixed with water, smeared on the back of the hand, and allowed to dry, clay cannot be dusted off easily, whereas silt or sand can.

Soil particles often cling together in aggregates, called *peds*, that are classified according to shape into several types. Figure 8.3 shows some of the common structures of peds found in soils. The type of structure present is related to soil-forming processes, but some of these processes are poorly understood.[5] For example, *granular structure* is fairly common in *A* horizons, whereas *blocky* and *prismatic structures* are most likely to be found in *B* horizons. Soil structure is an important diagnostic tool in helping to evaluate the development and approximate age of soil profiles. In general, as the profile develops with time, structure becomes more complex and may go from granular to blocky to prismatic as the clay content in the *B* horizon increases.

RELATIVE SOIL PROFILE DEVELOPMENT

Most geologists will not have occasion to make detailed soil descriptions and analyses of soil data. However, it is important for geologists to recognize differences among weakly developed, moderately developed, and well-developed soils, that is, to recognize their relative profile development. These distinctions are useful in the preliminary evaluation of soil properties and help determine whether the opinion of a soil scientist is necessary in a particular project:

- *A weakly developed soil profile* is generally characterized by an *A* horizon directly over a *C* horizon (there is no *B* horizon or it is weakly developed). The *C* horizon may be oxidized. Such soils tend to be only a few hundred years old in most areas, but they may be several thousand years old.

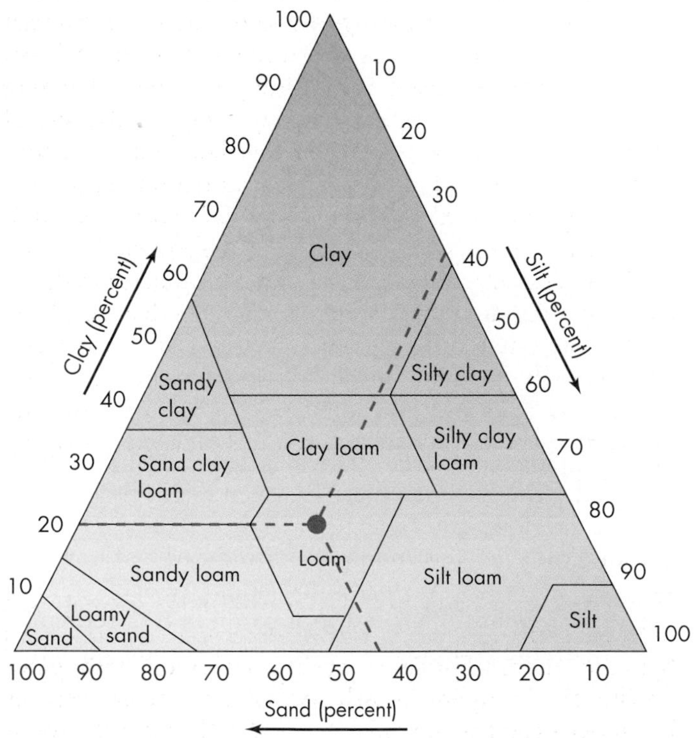

FIGURE 8.2 SOIL TEXTURAL CLASSES The classes are defined according to the percentage of clay-, silt-, and sand-sized particles in the soil sample. The point connected by dashed lines represents a soil composed of 40 percent sand, 40 percent silt, and 20 percent clay, which is classified as loam. *(U.S. Department of Agriculture standard textural triangle)*

Type	Typical size range	Horizon usually found in	Comments
Granular	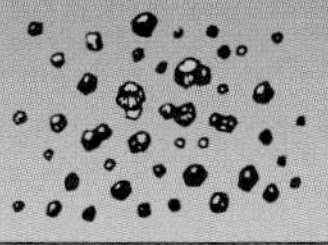 1–10 mm	A	Can also be found in B and C horizons
Blocky	 5–50 mm	B_t	Are usually designated as angular or subangular
Prismatic	 10–100 mm	B_t	If columns have rounded tops, structure is called *columnar*
Platy	 1–10 mm	E	May also occur in some B horizons

◀ **FIGURE 8.3 CHART OF DIFFERENT SOIL STRUCTURES (PEDS)**

- *A moderately developed soil profile* may consist of an *A* horizon overlying an argillic B_t horizon that overlies the *C* horizon. A carbonate B_k horizon may also be present but is not necessary for a soil to be considered moderately developed. These soils have a *B* horizon with translocated changes, a better-developed texture, and redder colors than those that are weakly developed. Moderately developed soils often date from at least the Pleistocene (more than 10,000 years old).
- *A well-developed soil profile* is characterized by redder colors in the B_t horizon, more translocation of clay to the B_t horizon, and stronger structure. A *K* horizon may also be present but is not necessary for a soil to be considered strongly developed. Well-developed soils vary widely in age, with typical ranges between 40,000 and several hundred thousand years and older.

A *soil chronosequence* is a series of soils arranged from youngest to oldest on the basis of their relative profile development. Such a sequence is valuable in hazards work, because it provides information about the recent history of a landscape, allowing us to evaluate site stability when locating such critical facilities as a waste disposal operation or a large power plant. A chronosequence combined with numerical dating (applying a variety of dating techniques, such as radiocarbon ^{14}C, to obtain a date in years before the present time of the soil) may provide the data necessary to make such inferential statements as, "There is no evidence of ground rupture due to earthquakes in the past 1000 years," or "The last mudflow was at least 30,000 years ago." It takes a lot of work to establish a chronosequence in soils in a particular area. However, once such a chronosequence is developed and dated, it may be applied to a specific problem.

Consider, for example, the landscape shown in Figure 8.4, an offset alluvial fan along the San Andreas fault in the Indio Hills of Southern California. The fan is offset about 0.6 km (60,000 cm). Soil pits excavated in the alluvial fan suggest that it is at least 20,000 years old, but younger than 45,000 years. The age was estimated on the basis of correlation with a soil chronosequence in the nearby Mojave Desert, where numerical dates for similar soils are available. Soil development on the offset alluvial fan allowed the age of the fan to be estimated. This allowed the slip rate (the amount of offset of the fan divided by the age of the fan—that is, 60,000 cm ÷ 20,000 years) for this part of the San Andreas fault to be estimated at about 3 cm annually.[7,8] More recent work, using a numerical dating technique known as exposure dating, suggests the age of 35,500 ± 2500 years and a total displacement of about 570 m. This new and more accurate estimation of age provides a maximum slip rate of about 1.7 cm/yr (0.7 in.). Thus, the earlier soil date has been improved by more recent technology. As numerical dating has improved, the use of soil development as a dating tool has decreased.

The slip rate for this segment of the fault was not previously known. The rate is significant because it is a necessary ingredient in the eventual estimation of the probability and the recurrence interval of large, damaging earthquakes.

WATER IN SOILS

If you analyze a block of soil, you will find it is composed of bits of solid mineral and organic matter with pore spaces between them. The pore spaces are filled with gases (mostly air) or liquids (mostly water). If all the pore spaces in a block of soil are completely filled with water, the soil is

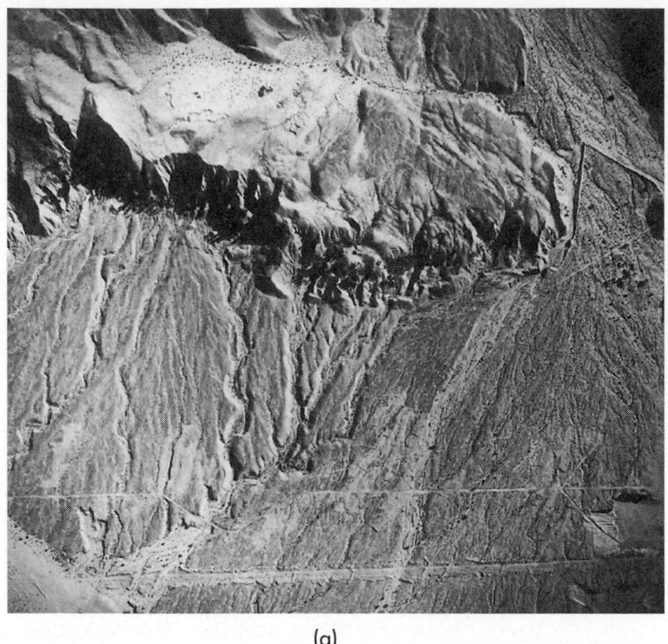

(a)

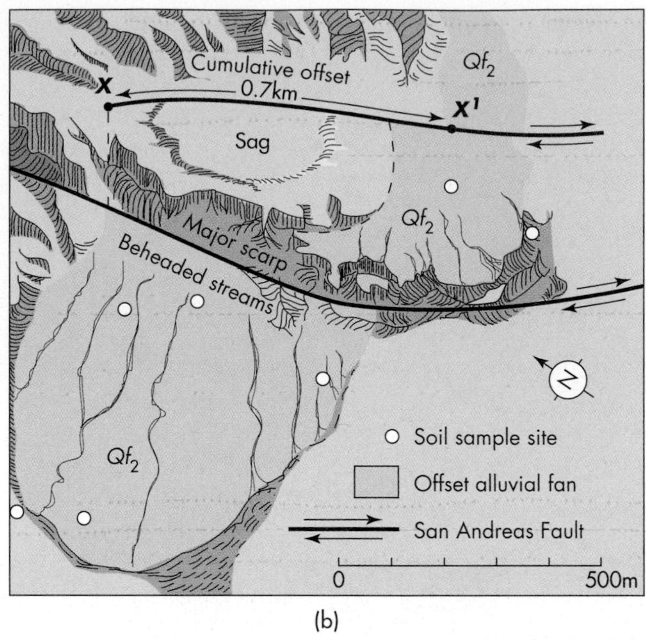

(b)

▲ **FIGURE 8.4 OFFSET ALLUVIAL FAN ALONG THE SAN ANDREAS FAULT NEAR INDIO, CALIFORNIA** (a) Aerial photograph. *(Woodward-Clyde Consultants)* (b) Sketch map.

said to be in a *saturated condition;* otherwise, it is said to be *unsaturated.* Soils in swampy areas may be saturated year-round, whereas soils in arid regions may be saturated only occasionally (Figure 8.5).

The amount of water in a soil, called its *water content* or its *moisture content*, can be important in determining engineering properties such as the strength of a soil and its potential to shrink and swell. If you have ever built a sand castle at the beach, you know that dry sand is impossible to work with, but that moist sand will stand vertically, producing walls for your castle. Differences between wet and dry soils are also apparent to anyone who lives in or has visited areas with dirt roads that cross clay-rich soils. When the soil is dry, driving conditions are excellent, but, following rainstorms, the same roads become mud pits and nearly impassable.[9]

▲ **FIGURE 8.5 PARTLY SATURATED SOIL SHOWING PARTICLE-WATER-AIR RELATIONSHIPS** Particle size is greatly magnified. Attraction between the water and soil particles (surface tension) develops a stress that holds the grains together. This apparent cohesion is destroyed if the soil dries out or becomes completely saturated. *(After R. Pestrong. 1974.* Slope Stability. *American Geological Institute)*

Water in soils may flow laterally or vertically through soil pores, which are the void spaces between grains, or in fractures produced as a result of soil structure. The flow is termed *saturated flow* if all the pores are filled with water and *unsaturated flow* when, as is more common, only part of the pores are filled with water.

In unsaturated flow, movement of water is related to processes such as thinning or thickening of films of water in pores and on the surrounding soil grains.[10] The water molecules closest to the surface of a particle are held the tightest, and, as the films thicken, the water content increases and the outer layers of water may begin to move. Flow is, therefore, fastest in the center of pores and slower near the edges.

The study of soil moisture relations and movement of water and other liquids in soils, along with how to monitor movement of liquids, is an important research topic. It is related to many processes including slope stability and subsidence of earth materials.

SOIL CLASSIFICATION

Both the terminology and the classification of soils are a problem in environmental studies because we are often interested in both soil processes and the human use of soil.

A *taxonomy* (classification system) that includes engineering as well as physical and chemical properties would be most appropriate—but none exists. We must, therefore, be familiar with two separate systems of soil classification: *soil taxonomy*, used by soil scientists, and the *engineering classification*, which groups soils by material types and engineering properties.

Soil Taxonomy Soil scientists have developed a comprehensive and systematic classification of soils known as **soil taxonomy**, which emphasizes the physical and chemical properties of the soil profile. This classification is a sixfold hierarchy, with soils grouped into Orders, Suborders, Great Groups, Subgroups, Families, and Series. The eleven Orders (Table 8.1) are mostly based on gross soil morphology (number and types of horizons present), nutrient status, organic content (plant debris, etc.), color (red, yellow, brown, white, etc.), and general climatic considerations (amount of precipitation, average temperature, etc.). With each step down the hierarchy, more information about a specific soil becomes known.

Soil taxonomy is especially useful for agricultural and related land-use purposes. It has been criticized for being too complex and for lacking sufficient textural and engineering information to be of optimal use in site evaluation for engineering purposes. Nevertheless, the serious earth scientist must have knowledge of this classification because it is commonly used by soil scientists and Quaternary geologists, those who study earth materials and the processes of recent (past 1.8 million years) earth history.

Engineering Classification of Soils The **unified soil classification system,** widely used in engineering practice and thus the evaluation of hazards, is shown in Table 8.2. Because all natural soils are mixtures of coarse particles (gravel and sand), fine particles (silt and clay), and organic material, the major divisions of this system are *coarse-grained soils*, *fine-grained soils*, and *organic soils*. Each group is based on the predominant particle size or the abundance of organic material. Coarse soils are those in which more than 50 percent of the particles (by weight) are larger than 0.074 mm in diameter. Fine soils are those with less than 50 percent of the particles greater than 0.074 mm.[11] Organic soils have a high organic content and are identified by their black or gray color and sometimes by an odor of hydrogen sulfide, which smells like rotten eggs. With this introduction to soils behind us, we will consider more specific aspects of subsidence.

TABLE 8.1

General Properties of Soil Order Used with Soil Taxonomy by Soil Scientists

Order	General Properties
Entisols	No horizon development; many are recent alluvium; synthetic soils are included; are often young soils.
Vertisols	Include swelling clays (greater than 35 percent) that expand and contract with changing moisture content. Generally form in regions with a pronounced wet and dry season.
Inceptisols	One or more horizons have developed quickly; horizons are often difficult to differentiate; most often found in young but not recent land surfaces, have appreciable accumulation of organic material; most common in humid climates but range from the Arctic to the tropics; native vegetation is most often forest.
Aridisols	Desert soils; soils of dry places; low organic accumulation; have subsoil horizon where gypsum, caliche (calcium carbonate), salt, or other materials may accumulate.
Mollisols	Soils characterized by black, organic-rich *A* horizon (prairie soils); surface horizons are also rich in bases. Commonly found in semiarid or subhumid regions.
Andisols	Soils derived primarily from volcanic materials; relatively rich in chemically active minerals that rapidly take up important biologic elements such as carbon and phosphorus.
Spodosols	Soils characterized by ash-colored sands over subsoil, accumulations of amorphous iron-aluminum sesquioxides and humus. They are acid soils that commonly form in sandy parent materials; found principally under forests in humid regions.
Alfisols	Soils characterized by a brown or gray-brown surface horizon and an argillic (clay-rich) subsoil accumulation with an intermediate to high base saturation (greater than 35 percent as measured by the sum of cations, such as calcium, sodium, magnesium, etc.). Commonly form under forests in humid regions of the mid-latitudes.
Ulfisols	Soils characterized by an argillic horizon with low base saturation (less than 35 percent as measured by the sum of cations); often have a red-yellow or reddish-brown color; restricted to humid climates and generally form on older landforms or younger, highly weathered parent materials.
Oxisols	Relatively featureless, often deep soils, leached of bases, hydrated, containing oxides of iron and aluminum (laterite) as well as kaolinite clay. Primarily restricted to tropical and subtropical regions.
Histosols	Organic soils (peat, muck, bog).

Source: After Soil Survey Staff. 1994. Keys to Soil Taxonomy, 6th ed. Soil Conservation Service, U.S. Department of Agriculture.

TABLE 8.2

Unified Soil Classification System Used by Engineers

Major Divisions				Group Symbols	Soil Group Names
COARSE-GRAINED SOILS (Over half of material larger than 0.074 mm)	Gravels	Clean Gravels	Less than 5% fines	GW	Well-graded gravel
				GP	Poorly graded gravel
		Dirty Gravels	More than 12% fines	GM	Silty gravel
				GC	Clayey gravel
	Sands	Clean Sands	Less than 5% fines	SW	Well-graded sand
				SP	Poorly graded sand
		Dirty Sands	More than 12% fines	SM	Silty sand
				SC	Clayey sand
FINE-GRAINED SOILS (Over half of material smaller than 0.074 mm)	Silts Nonplastic			ML	Silt
				MH	Micaceous silt
				OL	Organic silt
	Clays Plastic			CL	Silty clay
				CH	High plastic clay
				OH	Organic clay
Predominantly Organics				PT	Peat and Muck

8.2 Introduction to Subsidence and Soil Volume Change

In contrast to mass wasting (Chapter 7), which originates on slopes, subsidence and changes in the volume of the soil occur on both slopes and flat ground. **Subsidence** is a type of ground failure characterized by nearly vertical deformation, or the downward sinking of earth materials. It often produces circular surface pits, but it may produce linear or irregular patterns of failure. Volume changes in the soil result from various natural processes. These processes include shrinking and swelling caused by changes in the water content of the soil and frost heaving related to the freezing of water in the soil. The effects of shrink/swell and freeze/thaw movement in soil are generally not as dramatic as some subsidence—you are unlikely to see a headline "Dozens Die in Denver as Soil Expands"; however, volume changes in the soil are one of the most widespread and costly of natural hazards.

Subsidence is commonly associated with the dissolution of soluble rocks, such as limestone, beneath the surface. The resultant landscape has closed depressions and is known as **karst topography.** Other major causes of subsidence include the thawing of frozen ground and compaction of recently deposited sediment. To a lesser degree, earthquakes and the underground drainage of magma are also responsible for causing subsidence. Human-induced subsidence, discussed in Section 8.7, can result from the withdrawal of underground fluids, from the collapse of soil and rock over underground mines, and from the draining of marshes and swamps.

KARST

Although the term *karst* is not in most people's vocabulary, it is the name given to a common type of landscape in the United States and in many other parts of the world. Parts of Austin, Texas; Birmingham, Alabama; Nashville, Tennessee; and St. Louis, Missouri, are built on karst topography.[11] In this type of topography, subsidence results from dissolution of rocks beneath the land surface. Dissolution occurs as percolating surface water or groundwater moves through rock that is easily dissolved. Four common sedimentary rocks—rock salt and rock gypsum, which are evaporites; limestone and dolostone, which are carbonates; and one common metamorphic rock, marble—are easily dissolved. Rock salt

◀ FIGURE 8.6 **SMALL SINKHOLE** Not all sinkholes are large subsidence features. Small collapse sinkholes, such as this one in Boyle County, Kentucky, are common. *(Kentucky Geological Survey)*

and rock gypsum dissolve rapidly as fresh surface water flows through holes in the rock, whereas limestone, dolostone, and marble will dissolve if the percolating water is acidic. Although rock salt is approximately 7500 times more soluble and rock gypsum 150 times more soluble than limestone, most karst topography is underlain by limestone because of its greater abundance in near-surface rocks.[12]

Percolating water may become acidic when carbon dioxide is dissolved in it. This acidification generally occurs in the soil where carbon dioxide is produced by bacterial decomposition of dead plants and animals. Most soil bacteria "breathe" like humans; that is, their respiration takes in oxygen and releases carbon dioxide. Dissolving carbon dioxide in water produces carbonic acid, the same weak acid that is present in carbonated soft drinks such as Coke and Pepsi. Where acidified water comes in contact with limestone, it dissolves the bedrock and lowers the land surface on average 10 cm (0.4 in.) per century.[13] Carbonic acid dissolves limestone more readily than dolostone; thus, land underlain by limestone is more susceptible to dissolution and more likely to become karst topography.

Areas underlain by dense, thin-bedded, fractured, or well-jointed crystalline limestone are especially vulnerable to dissolution. In such areas, surface waters are easily diverted to subterranean routes along irregular fractures or to planar cracks between sedimentary layers. If the percolating surface water has become acidic, it can enlarge these openings in the rock to produce underground holes, or voids, of various shapes and sizes. Where the void space is relatively close to the surface, pits known as **sinkholes** may develop.

Sinkholes may exist individually (Figure 8.6) or develop in large numbers to form a pockmarked surface known as a *karst plain*. The limestone plateau in southern Indiana and central Kentucky and Tennessee is an example a karst plain (Figure 8.7).

In addition to sinkholes, many karst areas have other features developed by the chemical weathering of bedrock (Figure 8.8). In humid temperate climates, karst landscapes

▲ FIGURE 8.7 **KARST TOPOGRAPHY** This rolling landscape in west-central Kentucky is typical of karst topography in a humid temperate climate. *(Graphic Berdeaux)*

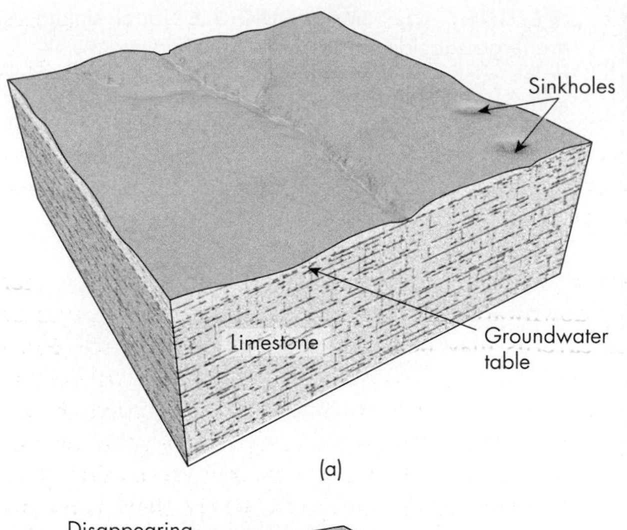

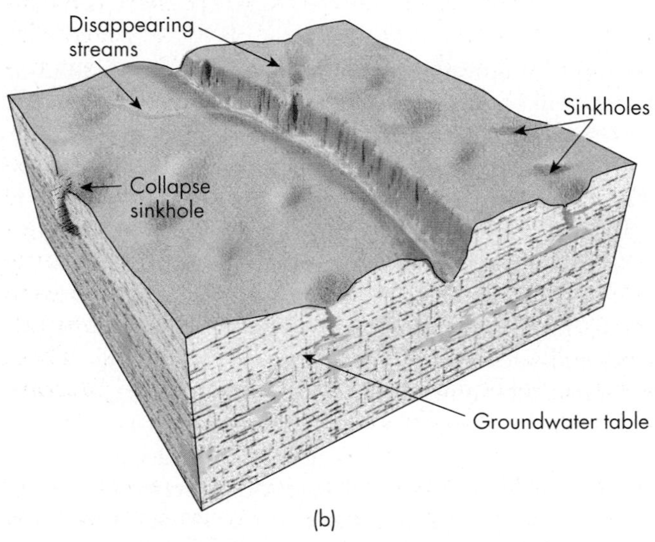

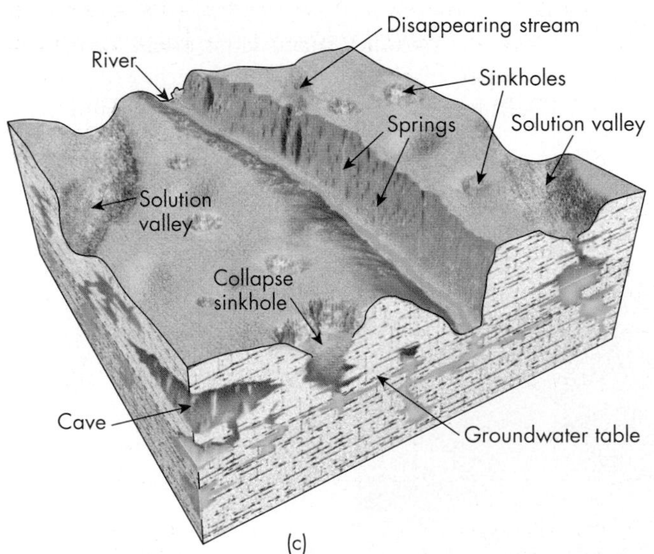

◀ FIGURE 8.8 **DEVELOPMENT OF KARST TOPOGRAPHY** (a) In the early stage of karst formation in a limestone terrain, water from the surface seeps through fractures and along layering in the soluble rock. Weak acid in the water then dissolves the rock. (b) As a river erodes more deeply into the land surface, the groundwater table drops. Caves begin to form and collapse to become sinkholes. Some surface streams disappear underground to become groundwater. (c) In later stages of karst development, a downward-eroding river continues to lower the groundwater table. Large caverns and sinkholes develop and eventually merge to form solution valleys that form without surface streams. In humid tropical climates, intense dissolution removes nearly all of the rock, leaving behind pillars of limestone referred to as *karst towers*. *(Modified from illustration by D. Tasa in Tarbuck, E. J. and Lutgens, F. K. 2005.* Earth: An introduction to physical geology, *8th ed. Upper Saddle River, NJ, Pearson Prentice Hall)*

are characterized by beautiful rolling hills with alternating areas of subsidence and undisturbed land. These areas may be underlain by large natural caves, commonly referred to as *caverns*, if the voids excavated by dissolution are at sufficient depth below the surface. Cave openings can be the site of disappearing streams where surface water goes underground, or the place where groundwater comes out at the surface to form springs. In humid tropical regions, extensive dissolution removes most of the soluble bedrock, leaving behind a landscape of steep hills known as *tower karst*.

Sinkholes In karst areas, sinkholes vary in size from one to several hundred meters in diameter and can open up extremely rapidly. There are two basic types of sinkholes (Figure 8.9):

1. **Solutional sinkholes**
   Solutional sinkholes form by dissolution on the top of a buried bedrock surface and are the most common type of sinkhole. Dissolution occurs where the downward infiltration of acidic groundwater becomes concentrated in holes created by joints and fractures. In the formation of these sinkholes, groundwater is typically drawn into a cone above a hole in the limestone, like water being drawn into a sink drain.
2. **Collapse sinkholes**
   Collapse sinkholes develop by the collapse of surface or near-surface material into an underground cavern. As subsidence features, these sinkholes can develop into spectacular collapse structures.

Some sinkholes have openings into subterranean passages that allow water to escape during a rainstorm. Most, however, are filled with rubble that blocks any underground passages. Blocked sinkholes usually fill up with water, forming small lakes. Most of these lakes eventually drain when the water is able to filter through the debris.[14] Similar drainage problems can result when artificial ponds and lakes are constructed over sinkholes (see Survivor Story 8.1).

Cave Systems As solutional pits enlarge and water moves downward through limestone, a series of caves or larger caverns may be produced. Mammoth Cave, Kentucky, and Carlsbad Caverns, New Mexico, are two of the famous caverns, or **cave systems,** in the United States. The primary mechanism for forming caves is groundwater moving through rock. Cave systems tend to develop at or near the present groundwater table where there is a continuous replenishment of water that is not saturated with the weathering products of the limestone. Many cave systems have passages and underground rooms on a number of levels. In these systems, each level may represent a different time of cave formation that is related to a fluctuating groundwater table. Fundamentally, caves are enlarged as groundwater moves through limestone along fractures or along planes between sedimentary layers, eventually forming a cavern. Later, if the groundwater table moves to a lower level, water seeping into the cave will deposit calcium carbonate on the sides, floor, and ceiling. These deposits form beautiful cave formations such as *flowstone*, *stalagmites*, and *stalactites* (Figure 8.10).

Tower Karst Large, steep limestone "towers" rising above the surrounding landscape are known as *tower karst*. Most common in humid tropical regions, karst towers are residual landforms of a highly eroded karst landscape.

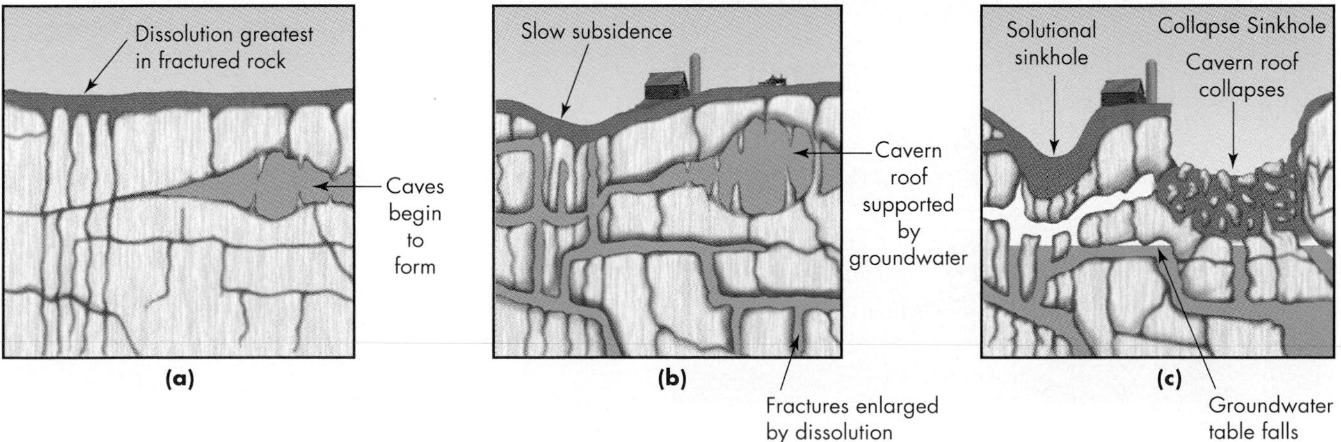

▲ FIGURE 8.9 **FORMATION OF SINKHOLES** (a) Initial dissolution of soluble bedrock takes place along vertical and horizontal fractures, and some fractures become small caves. (b) Over geologic time, dissolution continues to enlarge fractures and some caves become caverns with roofs partly supported by groundwater. (c) Subsequent lowering of the groundwater table removes the cavern roof support, and it falls into the cave to form a *collapse sinkhole*. Collapse may occur only a few years after the groundwater table has lowered. The heavily fractured area of bedrock is dissolved to create a *solutional sinkhole*.

SURVIVOR STORY 8.1

Sinkhole Drains Lake

The shores of Scott Lake, in central Florida, have long been a coveted address for residents of central Florida. Influential executives have built multimillion-dollar mansions alongside the 285-acre private waterway. Scott Lake's pristine waters attracted birds; otters; alligators; and people who fish for the teeming populations of largemouth bass, grass carp, bluegill, and catfish.

Then, within the space of a week, virtually the entire lake disappeared.

The story began on June 13, 2006, when the lake received more than eight inches of rain. But instead of rising, the water level of the lake began dropping.

Over the next few hours, the water began to drain in earnest. Residents on the lake's southeastern shore watched the lake's normally placid waters begin to bubble, then whirlpool around a single point. Boathouses collapsed. Docks across the lake were left standing high and dry (Figure 8.A).

Dave Curry, who has lived on the lakeshore for more than 30 years, guessed the cause immediately: a sinkhole.

As in most of Florida, the soil around Scott Lake is shallow, just a few tens of feet deep. Beneath the soil are a porous layer of limestone and the water-saturated rock of the great Floridan aquifer.

Several seasons of drought had lowered the water level in the aquifer, undermining support for the soil above. With the sudden weight of so much rain, the lake bottom collapsed into a hole more than 1.5 m (5 ft.) in diameter and 16m (52 ft.) deep.

"This happens all the time in Florida," Curry says. He had watched the lake drain once before, in 1969, likely due to a previous sinkhole.

This time, Scott Lake lost more than a billion gallons of water, more than 95 percent of the lake's volume. The event was catastrophic for lake wildlife. "I counted seven alligators down in that one hole. They couldn't get out; they ended up being washed into the Floridan aquifer along with the fish and the turtles," Curry says.

Of the estimated 85,000 pounds of fish in the lake, those not sucked into the sinkhole were left gasping on the muddy bottom. "There were huge dead fish laying all around the banks. But the smell didn't last too long because thousands of buzzards and all kinds of other bird life I'd never seen before came in and cleaned it up in a few days. Otters were catching some of the live fish in distress," Curry says.

On shore, houses nearest the sinkhole fared the worst. On two houses, the foundations cracked, windows shattered, and fissures opened up in lawns and swimming pools. The collapse of the soil had steepened the shoreline, causing the houses to slip. "It looked like a California mudslide," Curry says.

Since then, Scott Lake has steadily begun to refill. Mud and other debris seem to have formed a natural plug. But local residents aren't taking any chances. They plan to seal the hole with concrete, which should guarantee the lake's return, at least until Mother Nature strikes again.

—KATHLEEN WONG

◀ FIGURE 8.A **DAVE CURRY VIEWS DROP IN LAKE LEVEL** Property owner Dave Curry walks out on a boat dock that once served Scott Lake in south-central Florida. The lake level was lowered drastically when a sinkhole opened in the lake bottom. *(Dave Curry)*

◀ FIGURE 8.10 **CAVE FORMATIONS** Carlsbad Caverns, New Mexico, contains stalactites, which hang from the ceiling; stalagmites, which grow up from the cave floor; and flowstone, which forms as water flows slowly down the walls or across an incline. Recall that a stalactite hangs "tight" to the ceiling and a stalagmite "might" grow up from the cave floor to meet the stalactite. *(Bruce Roberts/ Photo Researchers, Inc.)*

Cuba and Puerto Rico exhibit tower karst, as does much of Southeast Asia. The most spectacular examples come from Guilin, China, where they have been painted by artists since the first century (Figure 8.11).

Disappearing Streams Karst regions are often underlain by a complex groundwater system that occasionally intersects with the land surface. In these areas, surface streams may suddenly disappear into cave openings. Such **disappearing streams** do not actually disappear; rather, they flow directly into the groundwater system and continue along a subterranean route (see Figure 8.8b).

Springs Areas where groundwater naturally discharges at the land surface are known as **springs.** Most springs in karst areas are highly productive, especially during rainy periods. Karst springs are an important resource, but many are drying up as a result of overpumping of groundwater.

THERMOKARST

Some karst terrains develop from processes other than the groundwater dissolution of rock. In polar regions, and at high altitude, much of the soil and underlying sediment remain frozen throughout the year. This natural condition, called **permafrost,** may be continuous throughout the soil or, in slightly warmer climates, may exist as discontinuous patches or thin layers. Permafrost consists of particles of soil or sediment that remain cemented with ice for at least two years. The melting of this frozen earth material can produce subsidence, especially if it contains a large amount of ice. Land subsidence of several meters or more is possible through the melting of permafrost.[15]

If undisturbed by human activity, the thawing of permafrost is generally limited to the upper few meters of earth material. This forms an active layer directly below the land surface, in which the permafrost thaws during summer months and the water in the soil refreezes in the winter. However, more extensive thawing of permafrost can produce an irregular land surface known as *thermokarst.* Climatic warming over the past four decades has thawed large areas of Arctic permafrost and formed thermokarst. In some areas, the uppermost layers of permafrost are thawing at rates of close to 20 cm (8 in.) per year.[16]

SEDIMENT AND SOIL COMPACTION

Rapidly deposited fine sediment, soil and sediment cemented with soluble minerals, and organic-rich soil are all susceptible to subsidence. This subsidence may occur as the rapidly deposited sediment compacts; as sediment and soil collapse when the binding material is loosened; or, in organic soils, as water is drained from the soil. Compaction of sediment and soil can occur naturally or as the result of human activities.

Fine Sediment Rapidly deposited fine sand and mud contain a great deal of water-filled space between the sediment particles. With time, the amount of water between the particles is reduced and the sediment compacts. Rapid deposition and compaction are especially common in river deltas. In deltas, natural subsidence has to be balanced by additional sedimentation to keep the land surface of the delta, called the *delta plain,* from sinking below sea level. Episodic events such as floods and earthquakes can cause deltaic sediment to remobilize and subside. Historically, this subsidence has submerged coastal cities, including Eastern Canopus, Herakleion, and part of Alexandria on the Nile Delta.[11]

▲ FIGURE 8.11 **TOWER KARST** (a) Painting of Chinese tower karst, "Peach Garden Land of Immortals" by Qiu Ying. *(Asian Art & Archaeology/Corbis* (b) Tower karst south of Guilin in South China's Guangxi Zhuang Autonomous Region. *(A C Waltham/ Robert Harding)*

The natural tendency for deltas to subside can be amplified if sedimentation on the delta plain is slowed or stopped. Sedimentation has been sharply reduced on both the Mississippi and Nile Deltas during the past 125 years[17]. On the Mississippi, sedimentation on the delta plain was slowed and stopped by levees built on both sides of the river. These levees have protected communities such as New Orleans from river flooding, but they have also kept new sediment from being added to the delta plain (see Case Study 8.2). In the case of the Nile Delta, construction of the Aswan dams upstream and the diversion of two-thirds of the river water into canals have stopped sediment from reaching much of the delta plain.[18] Part of the subsidence in both the Mississippi and Nile Deltas is also from the compaction of organic soils.

Collapsible Soils Some windblown dust deposits, referred to as *loess*, and stream deposits in arid regions are loosely bound with clay particles and water-soluble minerals. These deposits dry out rapidly after they form and may remain essentially dry for a long time. This dry condition can change if water ponds on the surface long enough for it to percolate downward through the entire deposit. Percolating water weakens the bonds of clay particles and dissolves minerals that hold the soil together. The entire deposit can then collapse and lower the land surface, sometimes more than a meter (3 ft.).[19] Soil and sediment that are prone to this behavior are referred to as **collapsible soil.**[20]

Organic Soils Some wetland soils forming in marshes, bogs, and swamps contain large amounts of organic matter. Called **organic soils,** these earth materials consist of partially decayed leaves, stems, roots, and in colder regions, moss, which soak up water like a sponge. When water is drained from these soils, they dry out, compact, and are exposed to processes that cause them to disappear.

Bacterial decomposition is the primary process causing subsidence in drained organic soil. This process converts organic carbon compounds to carbon dioxide gas and water. Other destructive processes include water and wind erosion and the combustion of peat in prescribed burns and wildfires. The decomposition, erosion, and burning of organic soils cause the irreversible subsidence of drained wetlands. Part of the subsidence of the city of New Orleans can be attributed to compaction of organic soil (see Case Study 8.2).[21]

One of the most dramatic examples of subsidence in organic soils has occurred in the Florida Everglades. Land drainage, primarily for agriculture and urban development, has combined with droughts to cause more than half of the freshwater Everglades to subside from 0.3 to 3 m (1 to 9 ft.) during the twentieth century.[22,23]

EARTHQUAKES

Although we commonly think of earthquakes as associated with uplift of the ground surface, they may also cause subsidence. As mentioned in Chapter 3, the 1964 Alaskan earthquake (**M** 9.2) caused extensive subsidence, resulting in flooding of some communities. In the Pacific Northwest, the geologic record contains evidence for repeated episodes of subsidence along the coasts of British Columbia, Washington, and Oregon. These subsidence episodes are believed to indicate episodic great earthquakes along the Cascadia subduction zone. In these megathrust earthquakes, strain builds up along "locked" segments of the subduction zone. Through time, this strain causes the western edge of the North American tectonic plate to buckle. This buckling drags the underwater seaward edge of the continent downward and produces an upward bulge along the coast (Figure 8.12). When strain is released during a great earthquake, the buckled edge of the North American plate bounces back to its undeformed position. This realignment results in the underwater uplift of the previously downwarped area and subsidence of the coastal bulge area. A similar cycle has been observed along subduction zones in Japan and Indonesia.

UNDERGROUND DRAINAGE OF MAGMA

Subsidence can also occur from volcanic activity, related to either the central magma chamber of the volcano or shallow tunnels on the flank of the volcano. As magma moves upward into underground chambers below a volcano, the surface of the volcano may be forced upward. When it erupts, the volume of magma in the underground chamber is reduced and the land initially uplifted by the magma will subside (see Figure 5.15). As mentioned in Chapter 5, cycles of uplift and subsidence are useful in predicting volcanic eruptions.

When basaltic lava flows down the flank of a volcano, the surface of the flow may cool as molten lava continues to move underground through a tunnel called a *lava tube*. Lava eventually drains from the tube, leaving a void near the surface that is susceptible to collapse. Unfortunately, some unsuspecting home buyers on the Big Island of Hawai'i have discovered too late that it is not advisable to build on a lava tube.

EXPANSIVE SOILS

Changes in moisture conditions can produce substantial movement in some clay-rich soils. Referred to as **expansive soils,** these soils shrink significantly during dry periods and expand or swell during wet periods. Most of the swelling is caused by the chemical attraction of water molecules to the surface of very fine particles of clay (Figure 8.13a).[30] Swelling can also be caused by the chemical attraction of water molecules to layers within the crystal structure of some clay minerals.

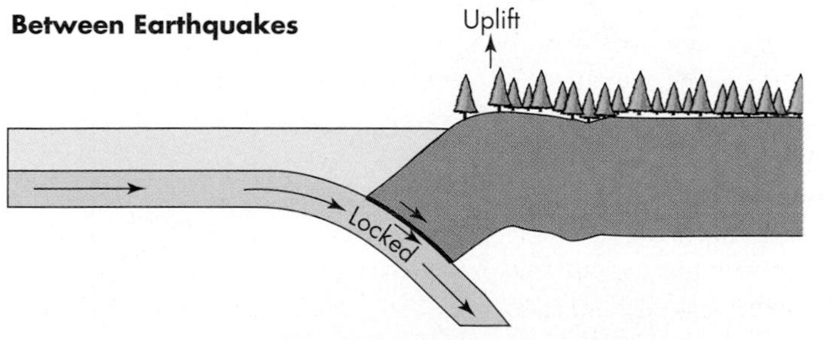

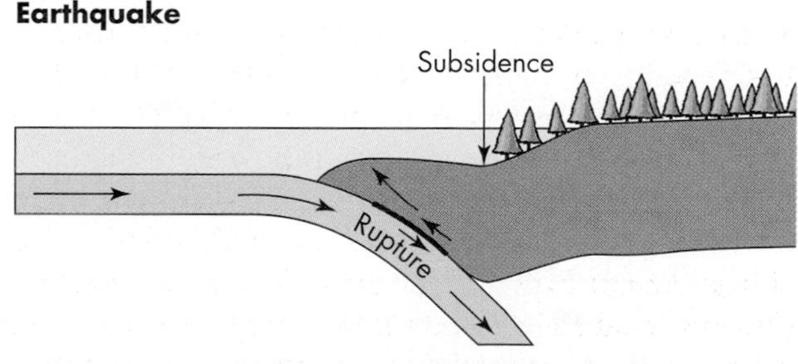

▲ FIGURE 8.12 **SUBSIDENCE CAUSED BY EARTHQUAKES** A proposed model for the cycle of uplift and subsidence associated with subduction along the Pacific Northwest coast. The Juan de Fuca oceanic plate on the left is being subducted below the North American continental plate on the right. (a) Between earthquakes, the two tectonic plates are locked together along the subduction zone and elastic strain builds up. The seaward edge of the North American plate is dragged downward, causing the coastline to be uplifted. (b) When a major earthquake (**M** 7) occurs in the locked portion of the subduction zone, strain is released and the deformed seaward edge of the North American plate "snaps" back into place. This may cause uplift of the seafloor and subsidence along the coastline. *(After Dragert, H., Hyndman, R. D. Roger, G. C. and Wang, K. 1994. Current deformation and width of the seismogenic zone of the northern Cascadia subduction thrust,* Journal of Geophysical Research *99(B1):653–68)*

CASE STUDY 8.2

Subsidence of the

Home to nearly 40 percent of the coastal marsh and swamp wetlands in the United States, the Mississippi Delta is the major land feature in southern Louisiana.[15] Geologically young, the present coastline of the delta was built by sedimentation from both the Mississippi and Atchafalaya Rivers during the past 7500 years.[24] As these rivers reach the sea, their channels subdivide into multiple distributaries. Actively flowing distributary channels build out a lobe-shaped pile of sand and mud into the ocean. This outbuilding continues until floodwaters break through a channel bank and the river finds a shortcut to the sea. If the break is permanent, then the river abandons its old distributaries and forms a new lobe of deltaic sediment. For the Mississippi Delta, this has occurred, on average, once each millennium (Figure 8.B).

As the river shifts its distributaries, two natural processes keep the land surface, the delta plain, above sea level. These processes are the deposition of sand and mud by distributary floodwaters and the accumulation of organic soil from dead marsh and swamp vegetation. The resulting sedimentary deposit is loosely packed together and saturated with water like a wet sponge. Over time, this deposit naturally compacts, and the land surface will subside unless new sediment is added to the delta plain.

Nearly three centuries ago, French settlers in New Orleans began to alter the natural balance between sedimentation

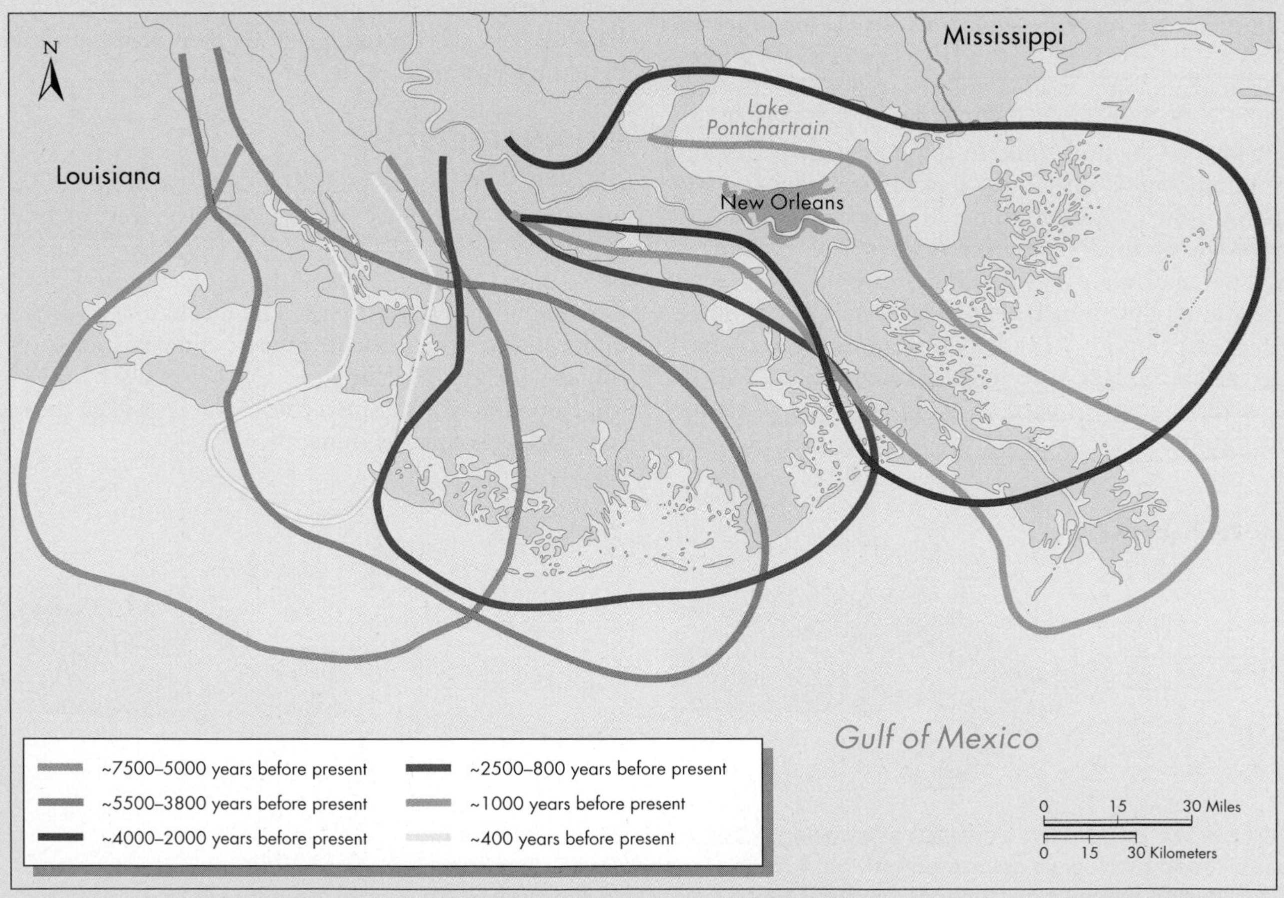

▲ FIGURE 8.B **MISSISSIPPI DELTA BUILT IN STAGES** During the past 7500 years, the Mississippi and Achafalaya Rivers have built the Mississippi Delta plain in a series of stages. In each stage, the rivers built a complex lobe of sand and mud into the Gulf of Mexico. Each lobe became inactive when the river breeched its levee during a flood and shifted its flow to a shorter channel to the sea. *(After Roberts, H. H. 1997. Dynamic changes of the Holocene Mississippi River delta plain: The delta cycle,* Journal of Coastal Research *13(3), p. 605-627)*

and subsidence with the construction of levees along the Mississippi. By the early twentieth century, the U.S. Army Corps of Engineers had completed levees on both sides of the river and essentially stopped the Mississippi from flooding the delta plain. Without river flooding, the delta plain lost its regular supply of freshwater, nutrients, and sediment, and near-surface subsidence increased.

Starving the delta plain of new sediment and nutrients has not been the only cause of subsidence. Withdrawal of large amounts of groundwater, oil, and natural gas from below the delta surface appears to have made a significant contribution to subsidence and a loss of wetlands.[25] Other scientists have hypothesized that tectonic subsidence, the sinking of the Earth's crust below the delta, has contributed to lowering the delta surface. Recent investigations testing this hypothesis have yielded conflicting results and further study will have to be done.[26,27]

A major consequence of the increased subsidence of the Mississippi Delta plain has been a significant loss of wetlands. Marshes and swamps in coastal Louisiana are disappearing at an average rate of a football-field-sized area every 30 minutes.[28] This rate translates to the disappearance of 50 to 80 km^2 (20 to 30 $mi.^2$) of land each year.[28] Wetland loss makes the coastline, including New Orleans, increasingly vulnerable to hurricane storm surges and has resulted in major economic losses for fisheries.

In addition to subsidence, rising sea levels and the construction of numerous navigation and drainage canals have contributed significantly to the loss of wetlands. The canals, built primarily for oil and gas operations, have exposed the delta plain to erosion and brought in saltwater that has killed wetland plants (Figure 8.C).[29]

An estimated 25 percent of the original coastal wetlands in Louisiana disappeared in the past century through subsidence, vegetation loss, and erosion. What will happen to the remaining 14,000 km^2 (5500 $mi.^2$) of wetlands? Nearly a decade ago, scientists, engineers, government officials, and business leaders in Louisiana determined that a massive wetland and barrier island restoration effort was needed to combat the loss of coastal marshes and swamps (Figure 8.D).[28] Although this effort is projected to cost more than $14 billion over the next 40 to 50 years, it is substantially less than the $135 billion in property damage from Hurricane Katrina and other large consumer and federal expenditures. Dealing with coastal land loss, such as that on the Mississippi Delta, will be one of the major environmental and disaster-planning challenges for the United States in the twenty-first century.

▲ FIGURE 8.C **WETLAND LOSS ON THE MISSISSIPPI DELTA** Excavation of numerous canals for oil and gas drilling, such as these near Leesville, Louisiana, have contributed significantly to wetland loss in the Mississippi Delta. Each canal drains the adjacent wetland and exposes organic soil to decomposition and erosion. Loss of this soil causes the land surface to subside. In many areas, the only land remaining is the spoil piles of sediment that were dredged to make the canals. *(Philip Gould/Corbis)*

(continued)

▲ FIGURE 8.D **CUTTING A LEVEE TO RESTORE A WETLAND** The floating dredge *California* in the lower center of the photograph is cutting a channel through the levee along the west bank of the Mississippi River south of New Orleans. The channel, part of the $22 million U.S. Army Corps of Engineers West Bay Sediment Diversion Project, was completed in 2003. This project was designed to divert 1400 m^3 (50,000 $ft.^3$) of river water per second to build a small, deltaic wetland in an adjacent bay. *(U.S. Army Corps of Engineers, New Orleans District)*

The *smectite* group of clay minerals, including the mineral montmorillonite, typically has the smallest clay crystals and thus, in bulk, has the greatest surface area to attract water molecules. Smectites are abundant in many clay and shale deposits derived from the weathering of volcanic rock, and they are the primary mineral in bentonite, a rock that forms from the alteration of volcanic ash (see Case Study 7.4). Clay, shale, and clay-rich soil containing smectite have the greatest potential for shrinking and swelling.

The presence of *shrink-swell clays*, such as smectites, can often be recognized from features on the land surface. These features include deep cracks produced by the drying of the soil (Figure 8.13c); a popcorn-like weathering texture on bare patches of clay; an alternating pattern of small mounds and depressions in the land surface; a series of wavy bumps in asphalt pavement (Figure 8.13d); the tilting and cracking of blocks of concrete and bricks in sidewalks and foundations (Figure 8.14); and the random tilting of utility poles and gravestones (Figure 8.15).

Shrinking and swelling of expansive soils cause structural damage to building foundations and the shifting and breaking of pavement and utility lines. This decrease or increase in the overall volume of the soil results from changes in its moisture content. Factors that affect the moisture content include climate, vegetation, topography, and drainage.[31] Regions that have a pronounced wet season followed by a dry season often have shrink-swell soils. These regions, such as the southwestern United States, are more likely to experience an expansive soil problem than are regions where precipitation is more evenly distributed throughout the year. Vegetation can also cause changes in the moisture content of a soil. Large trees draw and use a lot of local soil moisture, especially during a dry season. This withdrawal of water may produce soil shrinkage (Figure 8.13b).

Expansive soils swell when they become wet and can produce an expansion pressure in excess of 500 kPa (10,000 lbs. per $ft.^2$).[32] This intense pressure is strong

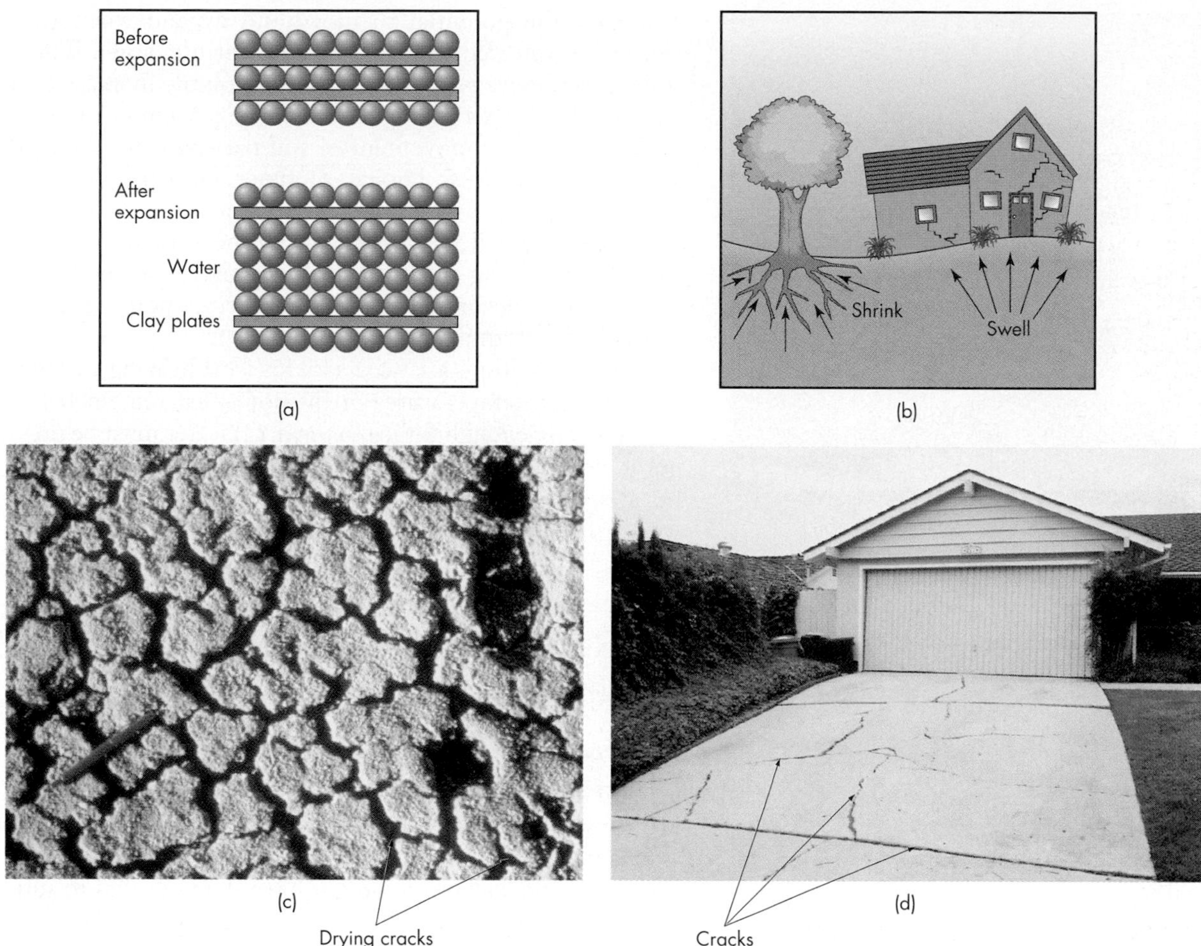

▲ FIGURE 8.13 **EXPANSIVE SOILS** (a) Smectite clay expands as water molecules are added onto and within the clay particles. (b) Effects of soil's shrinking and swelling at a home site. *(After Mathewson, C. C., and J. P. Castleberry, II.* Expansive soils: Their engineering geology. *Texas A&M University)* (c) Drying of an expansive soil produces this popcorn-like surface and a network of polygonal desiccation cracks. Pen in lower left for scale. *(U.S. Geological Survey Department of Geology)* (d) Shrinking and swelling of expansive soil cracked the concrete in this driveway. *(Edward A. Keller)*

◀ FIGURE 8.14 **DAMAGE FROM EXPANSIVE SOIL** Uneven shrinking and swelling of expansive clays in layers of steeply dipping bedrock produced the rolling surface of this road and sidewalk in Colorado. *(David C. Noe)*

▲ FIGURE 8.15 **SURFACE MOVEMENT IN EXPANSIVE SOIL** Repeated shrinking and swelling of a clay-rich soil caused the tilting of these gravestones in a cemetery in Manor, Texas. *(Robert H. Blodgett)*

enough to move the foundation and break open walls of multistory buildings, cause building support pilings to rise up out of the ground, and rupture major underground pipelines.

FROST-SUSCEPTIBLE SOILS

Freezing temperatures over a sufficient period of time can produce a great deal of expansion in silty soil. Silt is gritty sediment that is finer than sand, but coarser than clay. Soils that have the potential to move and expand when they freeze are referred to as **frost-susceptible soils.** These soils expand because of the 9 percent volume increase that occurs when water changes to ice. This volume increase causes an upward movement of soil particles and the land surface in a process known as **frost heaving.** Upward movement is generally much greater than 9 percent because additional water is drawn to the zone of freezing, thereby increasing the amount of ice that is formed. The additional ice accumulates to form discrete pods or lenses of ice in the earth material (Figure 8.16).

Three conditions are necessary for frost heaving to occur in a soil: (1) Temperatures in the soil must remain below freezing long enough for ice to form, (2) water must be present to create a substantial amount of ice, and (3) there must be a significant proportion of silt-size particles to allow water to be drawn through and retained in the soil.

8.3 Regions at Risk for Subsidence and Soil Volume Change

The type of earth material and climate are important factors in determining the geographic regions that are most at risk for subsidence or soil volume change. Landscapes underlain by soluble rocks, permafrost, or easily compacted soil and sediment have the potential to subside. Soils containing a high percentage of smectite clay

▲ FIGURE 8.16 **ICE FORMATION IN SOIL CAUSES FROST HEAVING** In cold regions, pods or lenses of ice can form in the soil and cause the land surface to heave upward. This photo of Arctic tundra in northern Alaska shows ice lenses below mounds of heaved soil. *(Anton Largiader)*

minerals are susceptible to shrinking and swelling, and soils containing an abundance of silt are especially susceptible to frost heaving.

Geology, alone, does not determine if a sinkhole will form or if the soil will shift below a building. Climate, primarily the amount and timing of rainfall and the duration of freezing temperatures, can also be a determinant. Active sinkhole formation is most common in humid climates. Expansive soils are most common in regions that have distinct wet and dry seasons, and collapsible soils are found primarily in arid and semiarid climates. Those parts of the northern United States and Canada that experience extended periods of below-freezing temperatures are likely to experience frost heaving.

Dissolution of soluble rocks and permafrost and compaction of sediment and soil are the most common causes of subsidence. In particular, the dissolution of limestone produces karst topography, which is estimated to compose up to 10 percent of Earth's surface. In the contiguous United States, approximately 25 percent of the land surface is underlain by limestone or evaporites that have some degree of karst development (Figure 8.17). Major belts of karst topography include (1) a region extending through the states of Tennessee, Virginia, Maryland, and Pennsylvania; (2) south-central Indiana and west-central Kentucky; (3) the Salem-Springfield plateaus of Missouri and northernmost Arkansas; (4) the Edwards Plateau of central Texas; (5) most of central Florida; and (6) Puerto Rico. Subsidence and other karst-related phenomena are a major problem in these areas. The most extensive development of karst in evaporite rocks in the United States is in the Permian basin of eastern New Mexico and western Texas.[33]

Permafrost covers more than 20 percent of the world's land surface.[16] Most of Alaska and more than half of Canada and Russia are underlain by permafrost. Hazard maps show that many towns and smaller settlements are threatened by thawing permafrost, including Barrow, Alaska; Inuvik, Northwest Territories, Canada; and Yakutsk, Norilsk, and Vorkuta in Russia.[16]

Subsidence caused by the compaction of sediment is most pronounced in areas where it was rapidly deposited or where it contains abundant organic matter. These areas include many of the world's marine deltas, such as those of the Mississippi River, the Sacramento–San Joaquin Rivers in California, and the Nile River in Egypt.

Organic-rich soils susceptible to subsidence are common in cold-region wetlands of New England, New York State, the Upper Midwest, Washington State, Alaska, and Canada, where the wetlands are variously called "bogs," "fens," "moors," and "muskeg." Coastal wetlands underlain by peat or muck deposits are also susceptible to subsidence in the Florida Everglades, the Sacramento–San Joaquin Delta of California, and coastal Louisiana and North Carolina.

In North America, expansive soils are a problem primarily in the western United States and Canada. Soils with the highest swelling potential are found in the blackland prairies from Texas to Alabama and in parts of the High Plains from eastern Colorado north to Saskatchewan and Alberta, Canada, although swelling clays are found in soils in all 50 states of the United States, Puerto Rico, the Virgin Islands, and in most Canadian provinces and territories.[34]

Frost heaving occurs in regions underlain by permafrost, as well as in colder parts of the United States and Canada that do not have permanently frozen ground. Frost-susceptible soils occur throughout Canada, Alaska, the northern contiguous United States, and at higher altitudes in the Appalachians, Rocky Mountains, Sierra Nevada, and Cascades.

Large-scale, seismic-related subsidence is a risk for areas of Alaska, Hawai'i, and the Pacific Northwest in both Canada and the United States. Deflation of a magma chamber can cause subsidence in any volcanic area, and subsidence over lava tubes occurs in the Hawaiian Islands and in some lava flows of western North America.

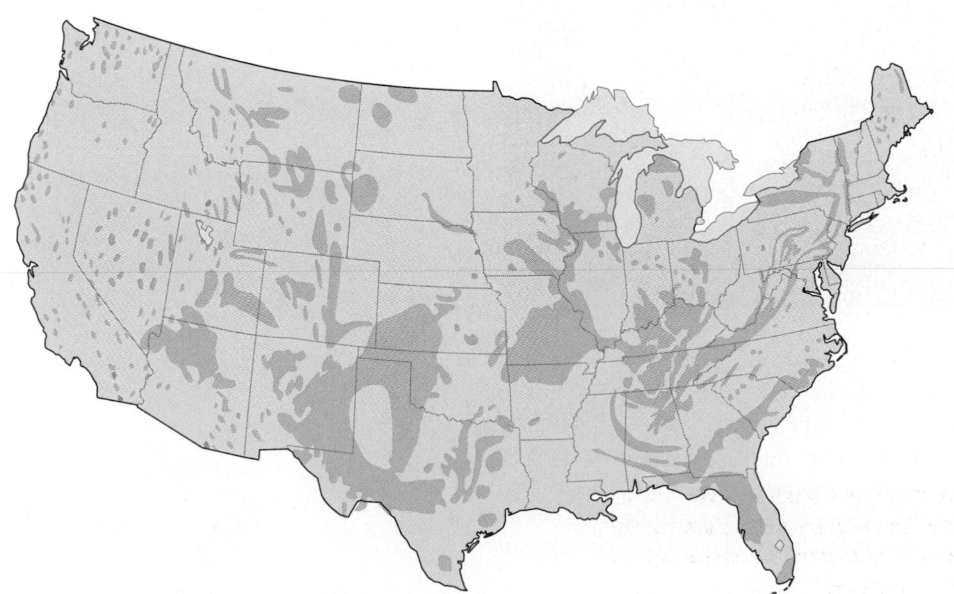

◀ **FIGURE 8.17 KARST MAP** Distribution of karst topography in the contiguous states of the United States. Approximately 40 percent of the land in the eastern half of this map is karst topography. Alaska, Puerto Rico, Hawai'i, and Canadian provinces and territories also have areas of karst. *(After White, W. B., Culver, D. C., Herman, J. S., Kane, T. C., and Mylroie, J. E. 1995. Karst lands. American Scientist 83:450–59)*

8.4 Effects of Subsidence and Soil Volume Change

Both karst formation and soil expansion and contraction cause significant economic damage each year. Karst regions are home to a host of problems such as sinkhole formation, groundwater pollution, and variable water supply. Heaving in expansive and frost-susceptible soils often damages highways, buildings, pipelines, and other structures. Additional damage from subsidence occurs in marine deltas, drained wetlands, collapsible soils, and in many areas underlain by permafrost.

SINKHOLE FORMATION

Sinkholes have caused considerable damage to highways, homes, sewage facilities, and other structures. Both natural and artificial fluctuations in the water table are probably the trigger mechanism for the formation of collapse sinkholes. High groundwater table conditions enlarge caverns closer to the surface of Earth by dissolving the roof and side of the caves. As long as a cavern remains filled with water, the buoyancy of the water helps support the weight of the overlying earth material. Lowering of the groundwater table eliminates some of the buoyant support, and the cave roof may collapse. This situation was dramatically illustrated in Winter Park, Florida, on May 8, 1981, when a large collapse sinkhole began developing. The sinkhole grew rapidly and within 24 hours had swallowed a house, part of a community swimming pool, half of a four-lane street, parts of three businesses, and parking lots containing several Porsches and a pick-up camper (Figure 8.18).[35] Damage caused by this sinkhole exceeded $2 million.

Sinkholes form nearly every year in central Florida when the groundwater level is lowest. Although exact positions cannot be predicted, more sinkholes form during droughts. The Winter Park sinkhole and several smaller sinkholes developed during the 1981 drought when groundwater levels were at a record low.

In urban and some rural areas the evidence of a subsidence hazard is sometimes masked by human activities. For example, a sinkhole in the Lehigh Valley in eastern Pennsylvania was first identified in the 1940s and was then completely filled with dead stumps, blocks of asphalt, and other trash and covered with a cornfield. Although no longer recognizable as a sinkhole, the filled depression collected urban runoff from nearby pavement and buildings.

▲ FIGURE 8.18 **SINKHOLE SWALLOWS PART OF TOWN** This sinkhole in Winter Park, Florida, grew rapidly for three days in 1981, swallowing part of a community swimming pool as well as several businesses, houses, and vehicles. The sides of the sinkhole have since been stabilized and landscaped, and it is now a park with a small lake. *(Leif Skoogfors/ Woodfin Camp and Associates, Inc.)*

The runoff loosened the plug of soil and trash and continued to dissolve the underlying limestone. An increased demand for groundwater in the area also contributed to a lowering of the groundwater table. These factors combined to cause a catastrophic collapse of the land surface in 1986. Within only a few minutes, the collapse produced a pit that was at least 30 m (100 ft.) in diameter and 14 m (46 ft.) deep. Although the damage was confined to a street, parking lots, sidewalks, and utility lines, the subsequent stabilization and repair costs were nearly $500,000.

GROUNDWATER CONDITIONS

Karst development creates a geologic environment in which groundwater is intensively used by humans and wildlife and is also easily polluted. Sinkholes, caves, and related karst features can form direct connections between surface water and groundwater (Figure 8.19). Such connections make the groundwater vulnerable to pollution and to groundwater table fluctuation during droughts. These vulnerabilities are a major concern for both public and private sources of drinking water.

One common source of pollution comes from sinkholes that have been used for waste disposal, especially where the bottom of the depression is near the groundwater table. Groundwater can also be contaminated where polluted water from surface streams flows into caves and fractures. This water can reach the groundwater table without natural filtration by soil or sand.

Groundwater-table fluctuations in karst areas affect humans, plants, and wildlife. For example, groundwater from karst limestone is heavily used along the edge of the Edwards Plateau in central Texas. In this area, frequent drought and heavy groundwater use can rapidly lower the groundwater table and cause springs to reduce or stop their flow. Diminished groundwater flow threatens communities that rely on it as a source of water, such as the city of San Antonio, and a number of unique plants and animals that are found only in the springs.

DAMAGE CAUSED BY MELTING PERMAFROST

Early settlers in the Arctic built their homes directly on permafrost only to find that heat radiated from the buildings thawed the soil. By the middle of the twentieth century, most structures were built on posts or pillars sunk into the permafrost. This technique elevates the floor of a building above the land surface and keeps its heat from melting the underlying soil. However, this use of pilings assumed that the permafrost would remain frozen. For the past several decades, that assumption has not always been the case.

Recent thawing of permafrost has caused roads to cave in; airport runways to fracture; railroad tracks to buckle; and buildings to crack, tilt, or collapse (Figure 8.20).[16] In two Siberian cities alone, an estimated 300 apartment buildings have been damaged. The state of Alaska now spends around 4 percent of its annual budget repairing permafrost damage.[36]

COASTAL FLOODING AND LOSS OF WETLANDS

Flooding of coastal areas and destruction of wetlands are two major effects of subsidence in marine deltas and bays. Subsidence of the Mississippi Delta during the past century has contributed to wetland loss and the sinking of New Orleans (see Case Study 8.2). The city of New Orleans and adjacent suburbs are subsiding relative to sea level at an

◀ **FIGURE 8.19 WATER FROM A SUBTERRANEAN STREAM** In karst topography, groundwater and surface water may be directly connected. This waterfall develops from a subterranean stream flowing from a cave. *(Kenneth Murray/Photo Researchers, Inc.)*

◀ FIGURE 8.20 **MELTING PERMAFROST DESTROYS BUILDING** The foundation of this apartment building in Cherskii in eastern Siberia was undercut by thawing permafrost. Structural damage from melting permafrost is becoming common in Russia, Alaska, and Canada. *(Professor V. E. Romanovsky/ University of Alaska Fairbanks)*

average rate of 8 mm (0.3 in.) per year (Figure 8.21). As a result, most of New Orleans is below sea level and parts of east New Orleans have subsided to 3 to 5 m (10 to 16 ft.) below sea level.[37] Rings of levees and floodwalls surrounding New Orleans and many of its suburbs were established to keep urban areas from flooding by the Gulf of Mexico, Lake Pontchartrain, and the Mississippi River.

Outside the levees and floodwalls are wetlands that help protect the city and its suburbs from ocean waves and storm surges. Without intervention, this marsh and swampland will subside or erode by 2090 and New Orleans will be directly on the sea.[21]

▲ FIGURE 8.21 **SUBSIDENCE IN NEW ORLEANS** Land surrounding many New Orleans' buildings has subsided since they were built. The land surface around this house in the Lakeview area appears to have subsided 0.25 to 0.5 m (10 to 19 in.) since 1956. *(Professor J. David Rogers)*

DAMAGE CAUSED BY SOIL VOLUME CHANGE

Expansive and frost-susceptible soils cause significant environmental problems. As one of our most costly natural hazards, soil volume change is responsible for many billions of dollars in damages annually to highways, buildings, and other structures. Annual costs from soil expansion and contraction exceed $15 billion in the United States and are substantial in Canada and Mexico as well.[38] Frost action on roads and streets in the United States alone costs an estimated $2 billion each year in rebuilding thousands of miles of pavement.[39] In many years, damage caused by soil volume change exceeds the cost for all other natural hazards combined.

Every year, more than 250,000 new houses are constructed on expansive soils. Of these, about 60 percent will experience some minor damage, such as cracks in the foundation,

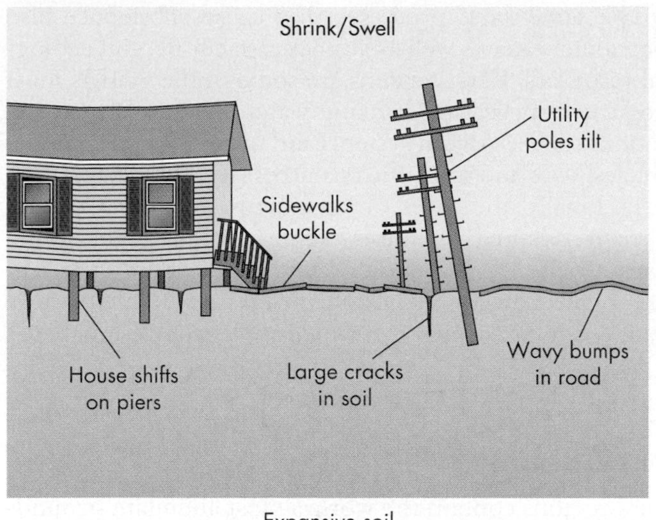

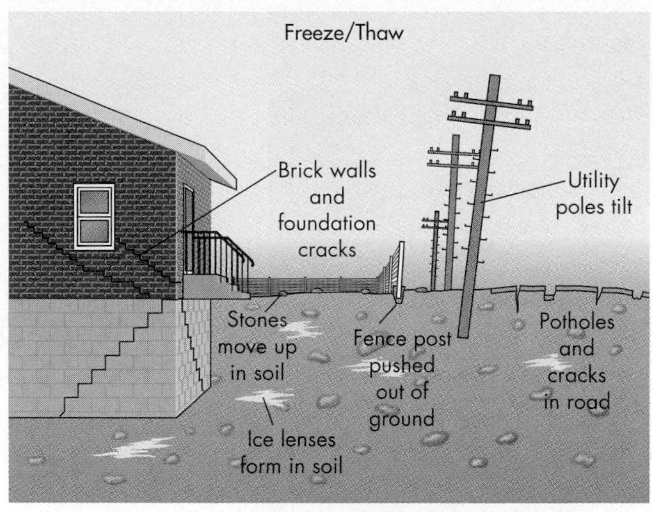

▲ FIGURE 8.22 **SOIL MOVEMENT EFFECTS** Shrinking and swelling of expansive soils (a) and freezing and thawing of frost-susceptible soils (b) have similar effects. Both types of soil movement crack and shift foundations and pavement, and tilt utility poles, road signs, and fence posts. Shrinking of the soil can produce deep cracks in the ground, and swelling of the soil can create small mounds on the land surface. Freeze and thaw cycles move buried objects, such as stones, upward in the soil to the surface. These cycles also contribute to the formation of potholes in asphalt pavement.

walls, driveway, or walkway, and 10 percent will be seriously damaged, some beyond repair (see Figures 8.13d and 8.14).[40] Underground water lines in expansive soils may rupture when there is a significant change in soil moisture. This rupture can result in a loss of water pressure, which makes the line vulnerable to contamination and requires customers to boil their water before using it.

In frost-susceptible soils, frost heaving has some of the same effects as swelling in expansive soils (Figure 8.22). Cracking occurs in foundations and asphalt and concrete pavement, and pipelines break. Frost heaving also moves larger particles in the soil upward at a faster rate than smaller particles. This upward movement, as much as 5 cm (2 in.) per year, can simply be an annoyance, such as the continual appearance of large stones in a garden or farmer's field each spring, or it can be destructive, such as the lifting of fence posts and utility poles out of the ground. Even coffins have been known to rise to the surface by frost heaving![41]

8.5 Linkages Between Subsidence, Soil Volume Change, and Other Natural Hazards

As mentioned previously, subsidence can be a side effect of earthquakes, volcanoes, and climate change. However, subsidence may also cause other natural hazards to occur. As described earlier for the Mississippi Delta, the link between subsidence and flooding is a common one. Soil expansion and contraction are also linked to increased rates of mass wasting on slopes.

In areas undergoing rapid subsidence, especially those that are also subsiding because of the overpumping of groundwater, flooding can be a major problem. As the land is lowered relative to surrounding bodies of water, the incidence of flooding increases (see opening Case Study). In many growing cities, there is a high demand for clean drinking water. Unfortunately, this demand leads to the mining of groundwater, often depleting the resource faster than it can be replenished. Because of groundwater mining, coastal and river floods have become more common and severe in low-lying coastal cities such as Bangkok, Thailand, and urban flooding has become more common in Mexico City.[42]

On sloping land surfaces, frost heaving and the shrinking and swelling of clay interact with mass wasting. This interaction can increase the rate of gravity-driven creep and produce an uneven land surface.

Subsidence also has direct linkages to climate change. This linkage is apparent in hot deserts, cold Arctic regions, and coastal areas. In hot arid areas, drought conditions commonly lower the groundwater table. The withdrawal of groundwater can further compact and shrink unconsolidated earth materials. In the desert of the southwestern United States, the drying of earth materials is contributing to regional subsidence and to the formation of large, polygonal *desiccation cracks* (Figure 8.23). These cracks are similar in shape to the cracks that you see in mud after it has dried out. Except for their polygonal form, the large

▲ FIGURE 8.23 **LARGE DESICCATION CRACK** This crack in the road is part of a polygonal network of large desiccation (drying) cracks in Graham County, southeastern Arizona. Unlike similar earth fissures that result from over pumping, these cracks appear to be the result of the natural lowering of the groundwater table in a drought. *(Mike Bryce/Graham County Highway Department)*

desiccation cracks are the same size and depth as large linear cracks, called *earth fissures*, which are produced by the overpumping of groundwater.[43,44] In cold Arctic areas, global warming is the primary cause of the melting of permafrost. Global warming is also the primary cause for the increased rate of sea-level rise. In coastal areas, such as Venice, New Orleans, Galveston Bay near Houston, and the Nile Delta, subsidence increases the local rate of sea-level rise and the resulting loss of land.

8.6 Natural Service Functions of Subsidence and Soil Volume Change

Although subsidence and soil volume change cause many environmental and economic problems, there are also benefits from these processes. Expansion and contraction of the soil form natural coherent blocks of soil material called *peds* that are similar to clods of soil produced by plowing or digging. The creation of peds contributes to soil productivity, especially in clay-rich soils, by facilitating root growth and soil drainage.

The same karst processes that cause subsidence also contribute water as well as aesthetic, scientific, and ecological resources. Karst terrains are some of the world's most productive sources of drinking water. Beautiful karst formations such as cavern systems and tower karst are important aesthetic and scientific resources (see Figures 8.10 and 8.11). Finally, the caves of karst areas are home to rare, specially adapted creatures, some of which are found nowhere else. For example, caverns and other karst features in the Edwards Limestone in Texas are home to more than 40 unique species, 8 of which are legally designated as endangered.

WATER SUPPLY

Karst regions contain the world's most abundant groundwater supply, thus providing a critical world resource. About 25 percent of the world's population either lives on or gets its drinking water from karst formations. In the United States, an estimated 40 percent of the groundwater used for drinking comes from karst terrains.[45] For example, the Edwards Formation, discussed previously, provides drinking water for more than 2 million people in Texas.

AESTHETIC AND SCIENTIFIC RESOURCES

Rolling hills, extensive cave systems, and beautiful formations of tower karst are among the aesthetic resources found in karst areas. Unique landscapes, such as the tower karst regions of China, are areas of unparalleled beauty that offer stunning vistas. Caves, too, have proved to be a popular destination for both spelunkers and tourists. Mammoth Cave National Park, Kentucky, contains the world's longest cave system and attracts droves of visitors each year. Aesthetics aside, karst regions also provide scientists with a natural laboratory in which to study the record of climate change contained in cave formations. Caves also provide an ideal environment for preserving animal remains, making them important resources for paleontologists and archaeologists.

UNIQUE ECOSYSTEMS

Caves are home to rare creatures specially adapted to live in the karst environment. Karst-dependent species known as *troglobites* have evolved to live in the total darkness of caves. Such species include eyeless fish, shrimp, salamanders, flatworms, and beetles (Figure 8.24). Other animals, such as bats, rely on caves to provide shelter. Karst areas generally support a diverse cross section of species; preserving these areas, and hence these organisms, for the future should be a societal objective.

◀ FIGURE 8.24 **CAVES ARE UNIQUE ECOSYSTEMS** Unique species, such as this endangered Texas blind salamander from Ezell's Cave National Natural Landmark in central Texas, have evolved to live in the total darkness of caves. *(Robert & Linda Mitchell Photography)*

8.7 Human Interaction with Subsidence and Soil Volume Change

As discussed earlier, subsidence can both cause problems and provide benefits. Commonly, when human beings live in areas underlain by karst, compacting sediment and soil, permafrost, and expansive or frost-susceptible soil, previously existing problems are exacerbated and new problems arise. Human beings contribute to problems associated with subsidence and soil volume change by withdrawing subsurface fluids, excavating underground mines, thawing frozen ground, restricting deltaic sedimentation, altering surface drainage, and using poor landscaping practices.

WITHDRAWAL OF FLUIDS

The withdrawal of subsurface fluids, such as oil with associated natural gas and water, groundwater, and mixtures of steam and water for geothermal power, have all caused subsidence.[46] In each case, the high fluid pressure in the sediment or rock helped support the earth material above. For this reason, a large rock at the bottom of a swimming pool seems lighter: Buoyancy produced by the liquid tends to lift the rock. Support or buoyancy by fluid pressure can be especially important in shallow or rapidly deposited earth material. In these instances, pumping out the fluid reduces support and causes surface subsidence.

A classic example can be found in the central valley of California where thousands of square kilometers have subsided from overpumping groundwater for irrigation and other uses (Figure 8.25a). One of the areas of greatest subsidence has been the Los Banos–Kettleman City area where most of the land surface has subsided more than 0.3 m (1 ft.), and locally up to 9 m (30 ft.) (Figure 8.25b). As the groundwater was mined, water pressure was reduced, sedimentary grains compacted, and the surface subsided (Figure 8.26).[47,48] Similar examples of subsidence caused by overpumping have been documented near Phoenix, Arizona; Las Vegas, Nevada; the Houston-Galveston area in Texas; San Jose, California; and Mexico City, Mexico. Such subsidence has reactivated geologic faults and formed extremely long earth fissures in unconsolidated sediment (Figure 8.27).[48]

UNDERGROUND MINING

Serious subsidence events have been associated with the underground mining of coal and salt. Most subsidence in coal mining is caused by the failure of pillars of coal that have been left behind to support the mine roof. With time, these pillars weather, weaken, and collapse, causing the roof to cave in and the land surface above the mine to subside (see Professional Profile 8.3). In the United States, more than 8000 km^2 (3100 mi.^{2}) of land, an area twice the size of Rhode Island, has subsided because of underground coal mining. This subsidence continues today, long after mining has ended. In 1995, a coal mine that was last operated in the 1930s collapsed beneath a 600 m (2000 ft.) length of I-70 in Ohio; repairs of the highway took 3 months.[49] Although coal mine subsidence most often affects farmland and rangeland, it has also damaged buildings and other structures in urban areas such as Scranton, Wilkes-Barre, and Pittsburgh, Pennsylvania; Youngstown, Ohio; and Farmington, West Virginia.[49,50]

In the case of salt, subsidence has taken place over both solution and open-shaft mines. Solution mines, the source

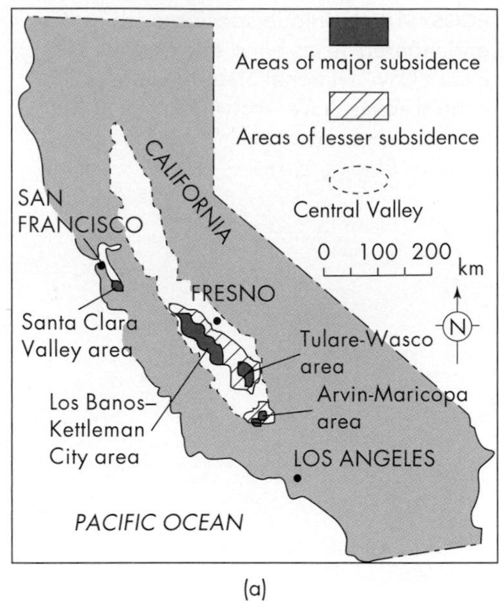

(a)

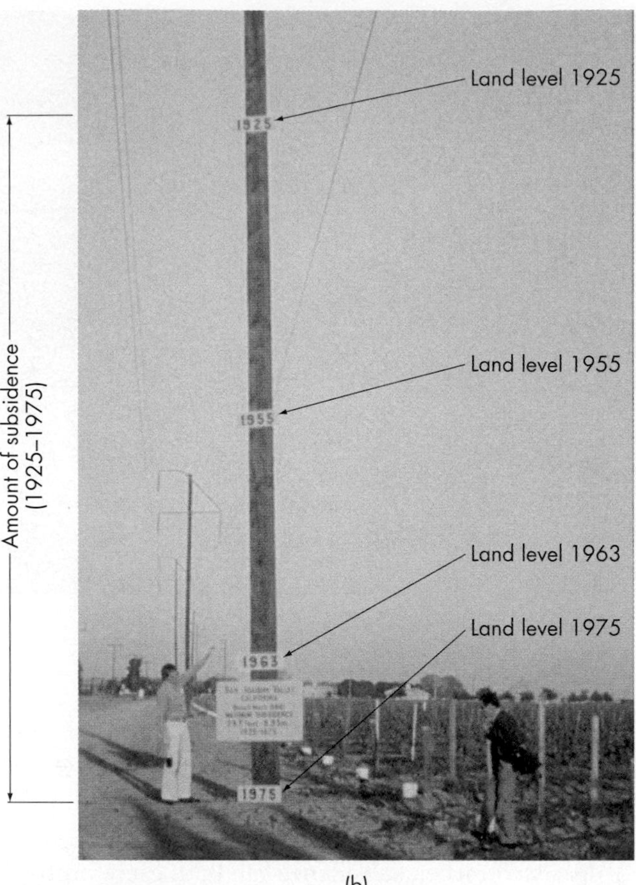

(b)

▲ **FIGURE 8.25 LAND SUBSIDENCE FROM GROUNDWATER EXTRACTION** (a) Principal areas of land subsidence in California resulting from groundwater withdrawal. *(After Bull, W. B. 1973. Geologic factors affecting compaction of deposits in a land-subsidence area,* Geological Society of America Bulletin *84; reprinted by permission)* (b) Photograph illustrating the amount of subsidence in the San Joaquin Valley, California. The marks on the telephone pole are the positions of the ground surface in recent decades. The photo shows nearly 8 m (26 ft.) of subsidence. *(Ray Kenny Ph.D.,PG)*

for most of our table salt, use wells to inject fresh water into salt deposits. The dissolved salt is then pumped out of the well, leaving a cavity behind. Collapse of this cavity and subsequent surface subsidence have taken place in solution mines in Kansas, Michigan, New York, and Texas.[33]

Open-shaft mining is used for extracting rock salt from the Earth and is taking place underground below Detroit, Michigan, and Cleveland, Ohio. In the past three decades, the catastrophic flooding of two underground salt mines has caused surface damage and subsidence. One, the Retsof Mine near Geneseo, New York, was the largest salt mine in the world. Collapse of its roof in 1994 allowed groundwater to flood the mine. Flooding produced two large sinkholes and subsidence damage to roads, utilities, and buildings.[51]

The second flooding event occurred on the Jefferson Island salt dome in southern Louisiana. In 1980, an oil rig drilling for natural gas accidentally penetrated an underground salt mine (Figure 8.28). The rig was mounted on a floating barge in a small lake above the salt dome. After drilling into the mine shaft, the rig toppled over into the lake and disappeared, with torrents of water flooding the mine. Fortunately, 50 miners and 7 people on the drilling rig escaped injury. The mine was a total loss, and buildings and gardens were damaged. Within 3 hours, the entire lake had drained and a 90 m (300 ft.) deep and 0.8 km (0.5 mi.) wide subsidence crater formed above the flooded mine. The structural integrity of underground salt mines is of particular concern because the U.S. Strategic Petroleum Reserve is stored in four Gulf Coast salt mines. A fifth storage site below Weeks Island, Louisiana, was emptied of oil in 1999 because of groundwater seepage through a sinkhole.

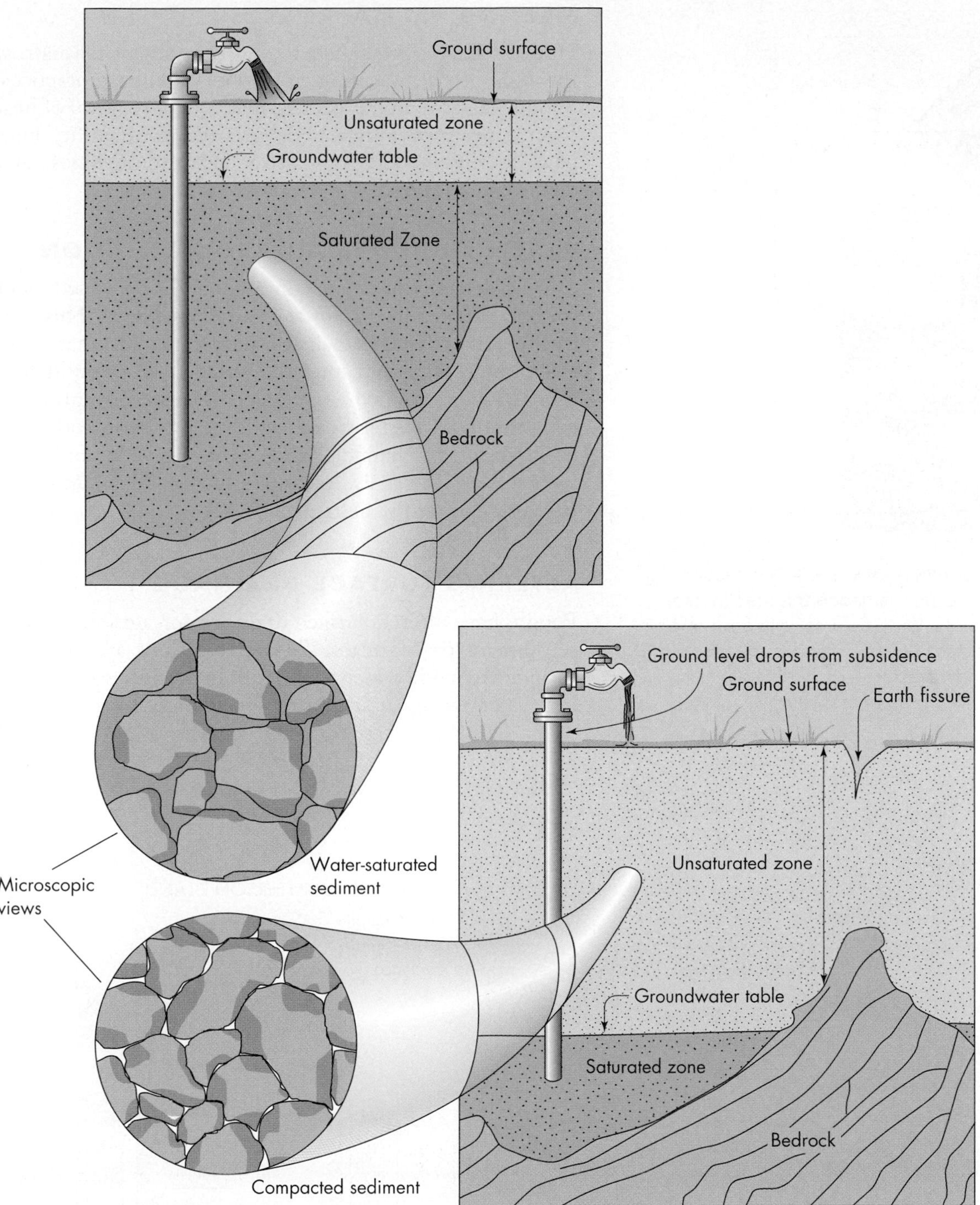

▲ FIGURE 8.26 **SUBSIDENCE FROM PUMPING OF FLUIDS** Idealized diagram showing how surface subsidence results from pumping groundwater. The unsaturated zone is the area above the groundwater table where the holes (pores) between grains contain both air and water. In the saturated zone beneath the groundwater table, the pores are completely filled with water. When groundwater is pumped out, the pores collapse and the surrounding earth material compacts. This compaction can cause the land surface to subside. *(Modifed from Kenny, R. 1992. Fissures.* Earth *2(3):34–41)*

▲ FIGURE 8.27 **EARTH FISSURES DAMAGE ROAD** A caution sign was erected after an earth fissure damaged this road in rural Pima County, Arizona, in 1981. Earth fissures have developed in the desert of the southwestern United States where over pumping has lowered the groundwater table. *(S. R. Anderson/U.S. Geological Survey Survey Department of Geology)*

MELTING PERMAFROST

Humans have contributed to the thawing of permafrost through global warming and poor building practices. Shoddy construction practices, inadequate removal of heat from beneath buildings, and burial of warm utility lines have locally melted permafrost, broken pipelines, and damaged buildings.[16,52]

RESTRICTING DELTAIC SEDIMENTATION

Marine deltas require the continual addition of sediment to their surfaces to remain at or above sea level. This sediment comes from the distributary channels that carry river water, sand, and mud to an ocean. In many deltas, humans have stopped or reduced this sedimentation by constructing dams upstream, building levees on both sides of distributary channels, and diverting sediment-laden river water into canals. All these practices may contribute to subsidence of the delta plain.

ALTERING SURFACE DRAINAGE

People have altered surface drainage for agriculture and settlement for centuries. These alterations are designed to remove or add water to the soil. In organic soils, draining water causes or increases subsidence. In the United

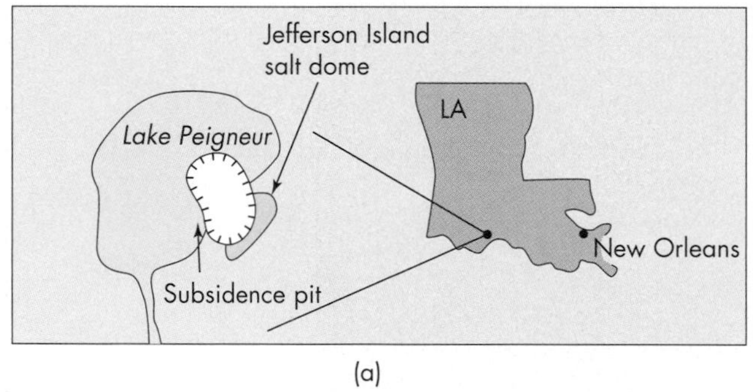

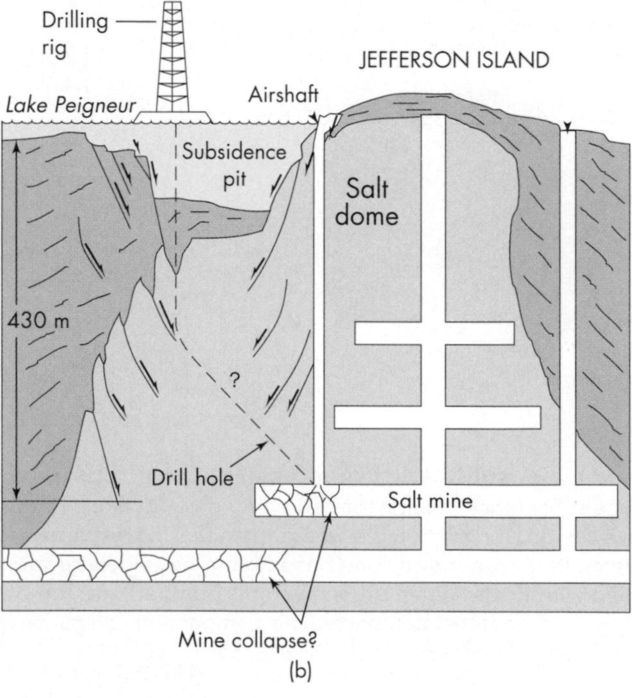

▲ FIGURE 8.28 **SUBSIDENCE FROM SALT DISSOLUTION** (a) Location of Lake Peigneur, Louisiana. (b) Idealized diagram showing the Jefferson Island salt dome collapse that caused a large subsidence pit to form in the bottom of the lake. An estimated 15 million m^3 (530 million $ft.^3$) of water flooded the salt mine after a natural gas well penetrated the mine shaft.

PROFESSIONAL PROFILE 8.3

Helen Delano, Environmental Geologist

For anyone looking to build a house in Pennsylvania, any number of natural hazards need to be considered. "We have something for everyone," says Helen Delano, from the Pennsylvania Bureau of Topographic and Geologic Survey. "As a geologist working for the state of Pennsylvania, one of my tasks is looking at where hazards may occur" (Figure 8.E).

Some of those dangers include deadly gases, such as methane and radon, which can migrate up from buried organic or radioactive material, respectively, but one of the more common problems is the potential for sinkholes and subsidence, particularly given the many abandoned coal mines in the area. Delano said the southwestern region of the state has been mined for coal since the mid-1700s, and that the shallower mines are only 15 to 30 m (50 to 100 ft.) below the surface. Although the technology does exist to empirically test an area, Delano said a lot of information could be gained simply by consulting coal mining maps.

But when the maps do not allow for a complete picture, relatively simple and inexpensive technology can help geologists determine whether there are any major cavities in the ground. Delano said geologists sometimes use what's called a "television borehole camera": They literally drill a hole and lower a small camera into the ground to examine the underground contents. This technology, however, is available for only relatively shallow depths. Delano said geophysical methods for deeper depths are far more expensive.

In addition to old coal mines, shallow caves, which have the potential to form sinkholes, can be detected with ground-penetrating radar—useful information for anyone considering a construction project in several parts of the state. "In eastern Pennsylvania, we get sinkholes from limestone subsidence," Delano said.

Delano began her career examining trouble from natural hazards in the Pittsburgh area. "When I first started working I would go out and look at people's backyard problems and help them understand what was going on and collect data for our use, trying to keep track of the scope of the problems for the state," she said.

Now she deals also with municipalities, advising them when natural hazards, including subsidence, pose problems for building projects.

"I have a folder on my desk right now for a rural community that has a proposal for a housing project over eighteenth- and nineteenth-century iron mines," Delano said.

When a major construction project is slated to be built over a region where subsidence is a risk, there are several options, Delano said. "Usually the cheapest thing to do is build your house somewhere else," she said. But when the location is just too good to pass up, there are engineering solutions, such as digging a deep foundation in firm bedrock.

The key, Delano said, is in knowing the risks ahead of time: "It's much, much cheaper to factor in those costs at the beginning." But for the most part, a close study of geological hazards, such as subsidence, is not yet routine in Pennsylvania when it comes to small construction projects.

"Most residential construction doesn't take into account geological hazards," Delano said, adding that, as more and more of the valuable land is occupied, it will become increasingly necessary to consider such dangers.

—CHRIS WILSON

▲ FIGURE 8.E **FIELDWORK IS REQUIRED FOR NATURAL HAZARD ASSESSMENTS** Environmental geologist Helen Delano is the Pennsylvania Geological Survey contact for local governments in southeastern Pennsylvania. Here she examines an exposure of the Gettysburg Formation in her area of responsibility. *(James Shaulis/Pennsylvania Geological Survey)*

States, draining organic soils has caused or increased subsidence in the Mississippi Delta, the Florida Everglades, and the Sacramento–San Joaquin Delta. Subsidence in the Sacramento–San Joaquin River Delta threatens the integrity of levees that protect cropland from flooding and keep saltwater from intruding into California's massive water supply system.[53] Adding water to the soil, such as through irrigation, can cause land subsidence on collapsible soils, and as described next, increase the swelling of expansive soils.

POOR LANDSCAPING PRACTICES

The shrinking and swelling of expansive soils are often amplified by poor landscaping practices. Planting trees and large shrubs close to foundations may cause damage from soil shrinkage during dry periods as plant roots pull moisture from the soil.[31] At the other extreme, planting a garden or grass that needs frequent watering close to foundations can cause damage from soil swelling. Rather than keeping the soil at a constant moisture level, irrigation systems commonly leave excess water in the soil. Excess water is considered the most significant cause of damage from swelling soil.[25]

8.8 Minimizing Subsidence and Soil Volume Change

Minimizing the hazards from, and related to, subsidence and the expansion and contraction of soil requires an understanding of the landscape from a geologic perspective. Even with this understanding, it is difficult to prevent natural subsidence and changes in soil volume. However, there are steps that can be taken to minimize the damage associated with this hazard.[11]

Artificial Fluid Withdrawal We will always be plagued with subsidence problems in areas where bedrock is aggressively being dissolved or where groundwater levels are continuing to fall because of drought. Natural sinkholes will continue to open up. We can, however, prevent some human-caused subsidence associated with the mining of groundwater or pumping of oil and gas.

Groundwater mining occurs when the amount of water coming to the surface in wells and springs exceeds the volume that is being replenished through the percolation of rain and surface water. Where groundwater, oil, or gas removal is causing the land surface to subside, it is possible to prevent or minimize further subsidence.

For example, from the early 1900s to the mid-1970s, groundwater mining in the Houston-Galveston area of Texas was the primary cause for up to 3 m (10 ft.) of subsidence over an 8300 km^2 (3200 mi.^{2}) area.[54] This prompted the 1975 Texas legislature to create a regulatory district to issue well permits. Since creation of this district, subsidence has essentially stopped in areas where groundwater pumping has decreased. This has not been the case in parts of Florida, Arizona, and Nevada, where increasing use of groundwater continues to cause sinkholes or earth fissures.

Installation of *injection wells* is often suggested as a way to minimize or stop subsidence from fluid withdrawal. This technique was used with some success in the 1950s when water was injected at the same time that oil was being pumped from the Long Beach, California, oil field. However, this method does not work for most groundwater mining because irreplaceable water has been extracted from layers of fine earth material. Once fine earth material has been compacted, it is not possible to push the particles back apart.[32]

Regulating Mining The best way to prevent damage from subsidence caused by mining activities is to prevent mining in urban areas. Although such laws are currently in place in many countries, the threat from older mines still exists.

Prevention of Damage from Thawing Permafrost Most existing engineering practices for building on permafrost have assumed that the permafrost will remain frozen if heat from a building or pipeline does not radiate into the ground. With the recent widespread thawing of permafrost, new and more costly practices are being developed, such as putting buildings on adjustable screw jacks or latticelike foundations to allow for the freezing and thawing of permafrost.[16]

Reducing Damage from Deltaic Subsidence Completely stopping further subsidence of human settlements on delta plains and restoring the deltas to natural conditions are unrealistic. Levees must continue to be elevated to protect urbanized areas, and adequate pumping systems must be maintained to remove excess surface water from the levee-protected enclosures. However, in undeveloped areas, levees could be breached to restore the sediment and freshwater supply necessary to rebuild marshes. The restored marshes would again help protect subsided urban areas from storms and rising sea levels.

Managing Drainage of Organic and Collapsible Soils As with most groundwater-induced subsidence, restoration of drained organic soils or collapsed soils is not possible. Only proper water management of existing drainage to marshes and swamps will minimize future subsidence from organic soils. For collapsible soils, it is especially important to limit irrigation and modify the land surface so that water does not pond near buildings.

Prevention of Damage from Expansive Soils Proper design of subsurface drains, rain gutters, and reinforced foundations can minimize the damage from expansive soil. These techniques improve drainage and allow the foundation to accommodate some shrinking and swelling of the soil.[31] Another preventive method is to construct buildings on a layer of compacted fill that acts as a barrier between the structure above and the expansive soil below. This method helps control the moisture level in the soil and provides a stable base upon which to build. For larger buildings, roads, and airports, it may be cost effective to excavate and redeposit the upper part of an expansive soil or to mix in calcium-based stabilizers, such as quicklime, to bind soil particles together.

8.9 Perception of and Adjustment to Subsidence and Soil Volume Change

PERCEPTION OF SUBSIDENCE AND SOIL VOLUME CHANGE

Subsidence and expansion and contraction of the soil are natural hazards that get little media coverage. Few people living in the United States are concerned about subsidence or soil volume change hazards. However, people living in areas directly affected, such as those in areas of expansive soils, permafrost, or rapid groundwater withdrawal, are more likely to understand the hazard. Furthermore, people living in regions where sinkholes commonly develop are generally familiar with the hazard and perceive it to pose a real risk to property.

ADJUSTMENT TO SUBSIDENCE AND SOIL VOLUME CHANGE

The most appropriate adjustment to subsidence and soil volume change is to avoid building in areas prone to these hazards. Clearly this is not always possible, because a significant portion of the eastern United States is underlain by karst, large areas of the western United States are underlain by swelling soil, and areas of the northern United States and Canada have frost-susceptible or organic soils or permafrost. The best we can do is to identify high-risk areas in which construction should be prohibited or limited. Unfortunately, in areas of natural subsidence, such as karst regions, subsidence may be difficult or impossible to predict. The groundwater system is constantly altering the subsurface rock, and sinkholes are common. There are, however, some methods that can help identify areas of potential subsidence and soil volume change.

Geologic and Soil Mapping Detailed geologic and soil maps can be made to identify as accurately as possible the hazards present. An understanding of the geology and soil, coupled with the surface and groundwater systems in an area, will greatly aid in the prediction and avoidance of subsidence and soil expansion and contraction.

Surface Features In areas underlain by limestone, surface features such as cracking of the land surface should be noted. The appearance of cracks in the ground may indicate that sinkhole collapse is imminent, and appropriate steps should be taken to avoid damage and injury. In the western United States, cracks in the ground may indicate expansive soils or, in some desert areas, the falling of the groundwater table.

Subsurface Surveys When planners must make decisions about where to build structures in karst regions, knowledge of the subsurface environment is critical. Subsurface exploration with *ground penetrating radar* (GPR) and drilling boreholes to examine the subsurface geology are often desirable before construction begins. These techniques may help prevent structures from being built above shallow caves. Additional geologic surveys may be needed to evaluate high-risk areas encountered during construction. In areas of expansive and frost-susceptible soils or areas underlain by permafrost, geotechnical borings and soil testing may be needed to properly design foundations.

Some states, such as Colorado, require disclosure of the presence of expansive soils when houses are sold. Disclosure requirements apply to new-home builders, homeowners, and real estate brokers.[55] A recent lawsuit for construction defects on expansive soils brought by owners of a 246-unit condominium complex in suburban Denver, Colorado, resulted in a settlement of more than $39 million. Homeowners who live in areas where subsidence hazards are present should check the hazard coverage in their insurance policies. For example, in many areas, neither sinkholes nor mine subsidence is covered in standard homeowners' policies and they require additional coverage.

REVISITING THE FUNDAMENTAL CONCEPTS

Soil and Subsidence

1. **Hazards Are Predictable from a Scientific Evaluation**
2. **Risk Assessment Is an Important Component of Our Understanding of the Effects of Hazardous Processes**
3. **Linkages Exist between Natural Hazards**
4. **Hazardous Events That Previously Produced Disasters Are Now Producing Catastrophes**
5. **Consequences of Hazards May Be Minimized**

1. Subsidence is a natural hazard for which predictions are difficult. We can predict based on geology and hydrology where subsidence and soil movement (from expansive soil) are likely to occur. Subsurface geophysical exploration may be used to identify large subsurface openings that might initiate subsidence (collapse sinkhole, for example). Predicting when a subsidence event might actually happen is not possible. Chronic problems such as expansion and contraction of clay-rich soils can be predicted.
2. Risk assessment from subsidence is based upon the product of the probability of a subsidence event occurring times the consequences. Both of these are fairly straightforward, and, as a result, it is theoretically possible to estimate the risk from subsidence. Risk assessment from expansive soil is straightforward and is standard engineering practice before construction.
3. Linkages exist between subsidence and other hazards. For example, subsidence pits (collapse sinkholes) in limestone areas may be associated with a water pollution hazard if the pit is used as a place to dump trash.
4. As a result of urbanization population growth with more development in areas with soil and subsidence hazard, the consequences of these hazards has increased. In areas with abundant limestone such as the southeastern USA sinkhole collapses damage property because there are more people and structures and more groundwater is being withdrawn. In the southwestern USA where hazards associated with expansive soil are more common as more people move into areas where soil stability is a potential problem.
5. Consequences of expansive soils that cause billions of dollars of damage every year may be minimized through land-use and engineering programs. We know where damage from expansive soil will occur, and we can develop plans so that the impact can be eliminated.

QUESTIONS TO THINK ABOUT

1. Your city is in an area with abundant limestone. Collapsed sinkholes have occurred in the past that have damaged roads and buildings. What factors would you consider in developing a plan to minimize subsidence events. Consider physical, hydrological, cultural and economic factors.
2. What are reactive policies following a large (200-m diameter 30-m deep) subsidence due to a collapse sinkhole in a housing area suddenly forming and how might they compare to proactive steps taken before the event? Which is better? Why?

Summary

Engineers define soil as earth material that may be removed without blasting, whereas, to a soil scientist, soil is solid earth material that can support rooted plant life. A basic understanding of soils and their properties is becoming crucial in several areas of environmental geology, including land-use planning, waste disposal, and evaluation of natural hazards such as flooding, landslides, and earthquakes.

Soils result from interactions of the rock and hydrologic cycles. As open systems, they are affected by variables such as climate, topography, parent material, time, and organic activity. Soil-forming processes tend to produce distinctive soil layers, or horizons, defined by the processes that formed them and the type of materials present. Of particular importance are the processes of leaching, oxidation, and accumulation of materials in various soil horizons. Development of the argillic *B* horizon, for example, depends on the translocation of clay minerals from upper to lower horizons. Three important properties of soils are color, texture (particle size), and structure (aggregation of particles).

An important concept in studying soils is relative profile development. Young soils are weakly developed. Soils older than 10,000 years tend to show moderate development, characterized by stronger development of soil structure, redder soil color, and more translocated clay in the *B* horizon. Strongly developed soils are similar to those of moderate development, but the properties of the *B* soil horizon tend to be better developed. Such soils can range in age from several tens of thousands of years to several hundred thousand years or older. A soil chronosequence is a series of soils arranged from youngest to oldest in terms of relative soil profile development. Establishment of a soil chronosequence in a region is useful in evaluating rates of processes and recurrence of hazardous events such as earthquakes and landslides.

A soil may be considered as a complex ecosystem in which many types of living things convert soil nutrients into forms that plants can use. *Soil fertility* refers to the capacity of the soil to supply nutrients needed for plant growth.

Soil has a solid phase consisting of mineral and organic matter; a gas phase, mostly air; and a liquid phase, mostly water. Water may flow vertically or laterally through the pores (spaces between grains) of a soil. The flow is either saturated (all pore space filled with water) or, more commonly, unsaturated (pore space partially filled with water). The study of soil moisture and how water moves through soils is becoming an important topic in environmental geology.

Several types of soil classification exist, but none of them integrate both engineering properties and soil processes. Environmental geologists must be aware of both the soil science classification (soil taxonomy) and the engineering classification (unified soil classification).

Subsidence is a type of ground failure characterized by nearly vertical deformation, or the downward sinking, of earth materials. This failure may be caused by natural processes, human activities, or a combination of the two. Most subsidence is caused by the underground dissolution of soluble rocks such as limestone, dolostone, marble, rock salt, and rock gypsum. Other causes of subsidence are the lowering of groundwater levels and fluid pressures in sediment, the thawing of permafrost, reduced sedimentation on delta plains, the flooding of collapsible soils, and the drainage of organic soils. Earthquakes, the deflation of magma chambers, and the drainage of lava tubes also may cause some subsidence.

Underground dissolution of limestone by acidic groundwater creates a landscape of caves and sinkholes known as karst topography. Other karst features include disappearing streams, springs, and tower karst. Most sinkholes form by the slow dissolution of limestone. Other sinkholes form from the collapse of cave roofs. This collapse is often caused by a falling groundwater table during a drought, or by an increase in pumping of water wells. Karst topography also develops where layers of highly soluble rock salt or rock gypsum are near the surface.

During the past several decades, the thawing of permafrost has become a major hazard in Arctic and near-Arctic regions. Most of this melting is a direct result of climatic warming. Thawing permafrost causes subsidence and structural damage, as well as the formation of thermokarst, a terrain consisting of uneven ground with sinkholes, mounds, ponds, and caves.

Loosely compacted fine sediment subsides where the groundwater table has fallen or fluid pressure has been reduced. Groundwater changes may be natural or result from human activities, such as groundwater mining. This subsidence is often irreversible because of the drying of underground layers of very fine sediment. Surface features associated with this compaction include large earth fissures or desiccation cracks.

Marine deltas are areas of natural compaction and subsidence. In these areas, the continual deposition of sediment on the delta plain generally keeps up with the compaction of sediment underground. Reducing or stopping sedimentation on the delta plain by the construction of dams, levees, and canals causes the delta surface to subside below the sea. This subsidence destroys wetlands and can produce flooding in urban areas such as New Orleans.

Soil expansion and contraction can cause the land surface to either heave upward or sink downward. Both heaving and sinking occur in expansive and frost-susceptible soils, and sinking takes place in collapsible soils.

Expansive soils are those that swell when they become wet and shrink when they dry, commonly a result of changes in the amount of water clinging onto the surface of fine smectite clay. Wetting and drying of this clay cause the expansion and contraction of the soil. These volume changes can cause extensive structural damage. Factors that affect the moisture content of an expansive soil include climate, vegetation, topography, and drainage.

Frost-susceptible soils are those that are likely to accumulate ice in pockets or lenses between silty earth material. Growth of these ice accumulations displaces the surrounding soil and produces frost heaving. The upward heaving and later thawing of the soil cause structural and pavement damage in areas underlain by permafrost and in soils that are only intermittantly frozen.

Collapsible soils are normally dry with particles loosely packed or weakly cemented together. These soils are susceptible to subsidence when water ponds on the surface.

Subsidence is a hazard in more than 45 states in the United States and most Canadian provinces. In areas of karst topography, hazards include sinkhole collapse, groundwater pollution, and unreliable water supplies. Large areas of karst topography are found in (1) a region extending through the states of Tennessee, Virginia, Maryland, and Pennsylvania; (2) south-central Indiana and west-central Kentucky; (3) the Salem-Springfield plateaus of Missouri and Arkansas; (4) central Texas; (5) central Florida; and (6) Puerto Rico.

Soil volume change is also a common hazard in many areas. Permafrost and frost-susceptible and organic soils are common in Alaska, Canada, and Russia. Organic soils are also abundant in the Upper Midwest, the Pacific Northwest, and the Gulf and Atlantic Coasts in the United States. Frost-susceptible soils also occur in areas of the northern contiguous United States and at high altitudes in mountainous areas. Collapsible soils occur in arid and semiarid regions, such as the southwestern United States. Expansive soils are a problem primarily in the western United States and Canada. These soils are responsible for significant economic damage each year, mostly from damage to highways, buildings, pipelines, and other structures.

Although subsidence causes numerous problems, it has benefits in the development of karst topography. About 25 percent of the world's population gets its drinking water from karst formations. Karst regions offer important aesthetic and scientific resources. They are home to rare creatures that are specially adapted to live underground.

Human beings exacerbate subsidence by removal of subsurface fluids, underground mining, melting permafrost, reduction of sediment accumulation on deltas, and the draining of organic soils. The effects of soil volume change can be intensified by using poor landscaping and drainage practices on expansive and collapsible soils. Natural subsidence and changes in the volume of the soil are difficult to prevent, but human-induced subsidence may be avoided or minimized. Methods for limiting human-induced subsidence include injecting water during crude-oil production and regulating groundwater pumping and underground mining. Problems with soil volume change may be minimized with sound construction and landscaping techniques. An understanding of the local geologic and hydrologic systems can help prevent water pollution in karst areas.

Adjustments to the subsidence and soil volume change hazard include identification of problem areas through geologic, soil, and subsurface mapping. Homeowners can protect themselves with insurance that covers the subsidence and soil volume change hazards in their area.

Key Terms

A horizon (p. 244)
B horizon (p. 244)
C horizon (p. 244)
cave system (p. 253)
collapsible soil (p. 256)
disappearing stream (p. 255)
E horizon (p. 244)
expansive soil (p. 257)
frost heaving (p. 262)
frost-susceptible soil (p. 262)
groundwater mining (p. 274)
karst topography (p. 250)
K horizon (p. 244)
O horizon (p. 244)
organic soil (p. 256)
permafrost (p. 255)
R horizon (p. 244)
sinkhole (p. 251)
soil (p. 244)
soil horizon (p. 244)
soil taxonomy (p. 249)
spring (p. 255)
subsidence (p. 250)
unified soil classification system (p. 249)
weathering (p. 244)

Review Questions

1. What is a soil, and soil profile?
2. How does limestone bedrock dissolve?
3. Which rock types are espcially susceptible to dissolution?
4. What features are found in karst areas?
5. Describe the two processes by which sinkholes form.
6. How do caves and caverns form?
7. Describe the natural cycle of permafrost thawing and how climate change has influenced this cycle.
8. What keeps a delta plain from subsiding?
9. What happens to organic soils when they are drained of water?
10. Explain how expansive soils shrink and swell, how frost heaving occurs, and how collapsible soils subside.
11. What natural and artificial features might indicate the presence of expansive soils?
12. What factors influence the moisture content of expansive soils?
13. Explain how subsidence might be connected to earthquakes in subduction zones.
14. How can subsidence occur on a volcano?
15. Identify the types of subsidence and soil volume change hazards that are likely to be found in the
    a. Eastern United States
    b. Western United States
    c. Alaska and Canada
16. Which subsidence or soil volume change hazards cause the most economic damage? Why are these hazards so costly?
17. What factors contribute to the formation of sinkholes?
18. Why is groundwater sometimes polluted in karst terrains?
19. What are natural and anthropogenic causes for the subsidence of the Mississippi Delta and New Orleans?
20. What types of damage are caused by the thawing of permafrost, shrinking and swelling of soil, and frost heaving?
21. How are subsidence and soil volume change hazards linked to changes in climate?
22. What are the natural service functions of subsidence?
23. Explain how fluid withdrawal and mining can increase subsidence.
24. How can we minimize or adjust to subsidence and soil volume change hazards?

Critical Thinking Questions

1. You are considering building a home in rural Kentucky. You know the area is underlain by limestone, and you are concerned about possible karst hazards. What are some of your concerns? What might you do (or have done) to determine where to build your home?
2. You work in the planning department in one of the parishes (counties) close to New Orleans. What would you advocate in the long term and in the short term to protect your community from subsidence and flooding? Consider both regional and local solutions to the problem.
3. You have inherited a ranch house built on a concrete slab on clay soil in a suburb east of Denver, Colorado. What would you look for, or do, to determine if the house is on expansive soil? If you found that the soil was expansive, how would you minimize damage from the shrinking and swelling of the soil?
4. You, or your parents, would like to build a retirement home in a desert development in Arizona or New Mexico. What would you look for to determine if subsidence or soil volume change is a potential problem? What could you do to protect your investment?
5. As a town council representative in a small village in New England or Ontario, Canada, you have been asked to approve a building permit on property that is partly underlain by silty glacial deposits and partly by a marsh. What questions should you ask the permit applicant regarding his or her planned construction on the silty soil, and his or her proposal to drain and build on the wetland?

9 Atmosphere and Severe Weather

Tri-State Tornado, March 18, 1925

Although the majority of weather-related deaths are caused by blizzards and heatwaves, it is the tornado that strikes fear in the hearts of many people living in the midcontinent area of the United States. This fear is well founded—a **tornado,** simply defined as a violently rotating column of air associated with extreme horizontal winds, can cause tremendous property damage and loss of life.

In just 3 1/2 hours on March 18, 1925, the Tri-State Tornado killed more people and destroyed more property than any other tornado in historic time. The tornado was unique in several ways. First, it was in contact with the ground for 349 km (217 mi.), a distance of 183 km (114 mi.) longer than any other tornado to date.[1] Second, the average width of the tornado was 3/4 of a mile and at times it reached 1 mile in width. Only 73 of the more than 35,000 reported tornadoes have been as wide as or wider than the Tri-State tornado.[1] This width produced extreme damage over at least 425 km^2 (164 $mi.^2$) (Figure 9.1). Third, the tornado's track was a straight line over much of its path through the states of Missouri, Illinois, and Indiana. This track contrasts with the curved track of most tornadoes. Finally, the tornado's average forward speed of 100 km (62 mi.) per hour was one of the fastest ever reported. Most tornadoes travel at an average speed of 50 km (30 mi.) per hour.[2] People seeing the storm approach assumed it was a thunderstorm because it was so broad and, in fact, there were associated thunderstorms. The tornado's violent winds inflicted unprecedented damage that included the deaths of at least 695 people and more than 2000 injuries. Total damages, expressed in year 2000 dollars, were about $170 million.

The tornado was unique in several ways.

As the Tri-State Tornado moved across the landscape, it traversed hills as high as 425 m (1400 ft.) and crossed valleys and lowlands; the topography had essentially no effect on the tornado. Although much of the death and damage resulted from the collapse of buildings, flying debris was also responsible for significant destruction (Figure 9.2). As the storm moved northeast, the damaging rotational winds had velocities of up to 290 km (180 mi.) per hour.

The extensive death and damage from the Tri-State Tornado resulted from several factors: (1) There were no tornado forecasts or warnings because the technology for

LEARNING OBJECTIVES

Atmospheric processes and energy exchanges are driven by Earth's energy balance and linked to climate and weather. Hurricanes, thunderstorms, tornadoes, blizzards, ice storms, dust storms, and heatwaves, as well as flash flooding resulting from intense precipitation, are all natural processes that are hazardous to people. These severe hazards affect considerable portions of North America and are responsible for causing significant death and destruction each year. Your goals in reading this chapter should be to

- understand Earth's energy balance and energy exchanges that produce climate and weather.
- know the different types of severe weather events.
- know the main effects of severe weather events, as well as their linkages to other natural hazards.
- recognize some natural service functions of severe weather.
- understand how human beings interact with severe weather hazards and how we can minimize the effects of these hazards.

◀ **Tornado in Manchester South Dakota** A violent tornado blocks passage down a road on June 24th, 2003 in South Dakota. This storm was part of the largest single-day outbreak of tornadoes in the state's history. *(Gene & Karen Rhoden/Peter Arnold/Still Pictures/Specialist Stock)*

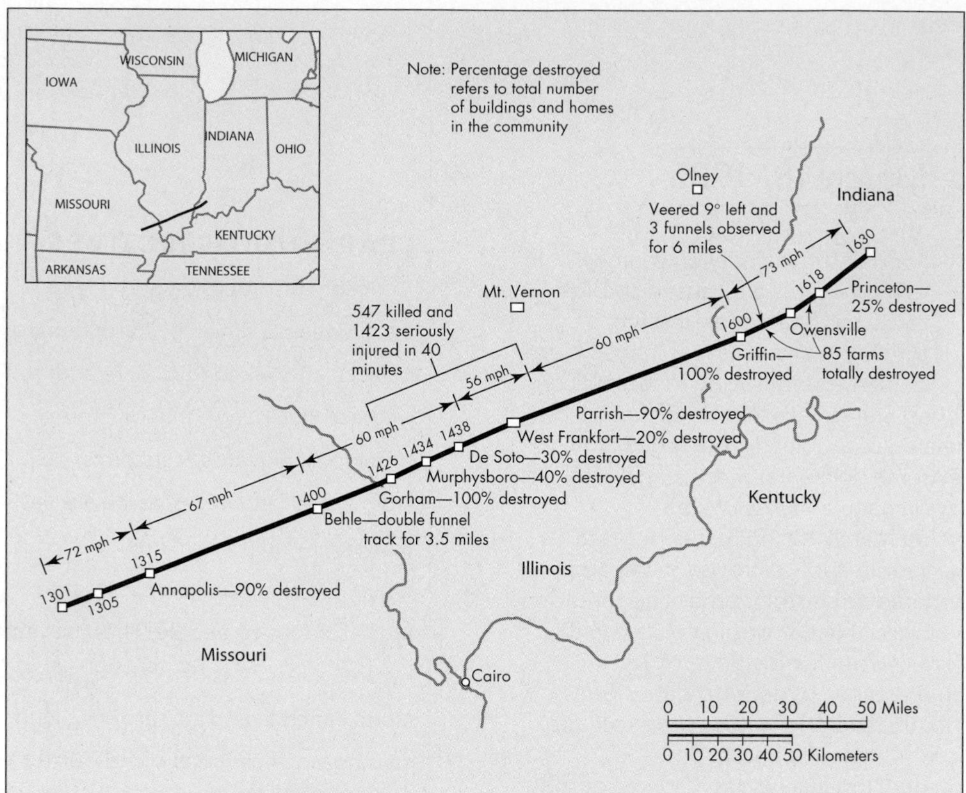

▲ **FIGURE 9.1 TORNADOES CAN TRAVEL LONG DISTANCES** In a span of 3 1/2 hours, the Tri-State Tornado of March 18, 1925, traveled 350 km (217 mi.) from Reynolds Co., Missouri, across Illinois to near Princeton, Indiana. Traveling at an average speed of 100 km (62 mi.) per hour, this tornado varied from 1 to 1.5 km (0.5 to 1 mi.) in width. The numbers from 1301 to 1630 along the tornado track refer to the time of day.*(After Wilson, J. W., and Changnon Jr., S. A. 1971.* Illinois tornadoes. *Illinois State Water Survey Circular 103. Urbana, IL)*

◄ **FIGURE 9.2 MOST DESTRUCTIVE TORNADO IN HISTORY** Ruins of the Longfellow School in Murphysboro, Illinois, where 17 children were killed by the Tri-State Tornado. The tornado struck the school around 2:34 P.M. on March 19, 1925. Most of the children were killed by the collapse of the unreinforced brick walls. Trees surrounding the school were also destroyed. A total of 234 people were killed in Murphysboro, the largest number of tornado deaths in a single town in U.S. history. *(NOAA)*

them did not exist in 1925, (2) the storm destroyed telephone lines that could have been used to warn people in the projected path of the tornado, (3) the tornado was exceptionally large and strong, (4) massive amounts of flying debris and dust masked any funnel shape and made it hard to recognize the storm as a tornado, and (5) many of the homes and farms were poorly constructed and unable to withstand the strong winds.[3] Should such a tornado occur in the future, successful forecasting and warning, along with better construction techniques, would likely save countless lives.

9.1 Energy

The concept of *energy* is fundamental to an understanding of severe weather. Energy is an abstract concept, because we cannot see or feel it. We can, however, experience the part of energy referred to as a *force*. Many people experience force by either pushing or pulling an object. For example, when we pull a box along the floor or push our car when it has stalled, we are exerting a force in a specific direction. The strength or magnitude of this force can be measured by how much the force accelerates the motion of the box or car. In the metric system, force is measured in Newtons. A *Newton* (N) is defined as the force necessary to accelerate a 1 kg (2.2 lb) mass 1 m (3.3 ft.) per second each second that it is in motion.

Another important concept in understanding energy is *work*. Work is done when energy is expended. In physics, work is done when a force is applied to an object and that object moves a given distance in the direction of the applied force. Work is thus calculated by multiplying the force times the distance over which it is applied. In the metric system, work is measured in joules. A *joule* is defined as a force of 1 Newton applied over a distance of 1 m (3.3 ft.). To relate this concept to weather, the amount of work that is taking place in a typical thunderstorm is approximately 10 trillion joules, whereas an average hurricane expends approximately 100,000 times that amount of work.

The rate at which work is done is its *power*. Another way of stating this is that power is energy divided by time. In the metric system, power is expressed as joules per second, or *watts* (W), and 1 joule per second is equal to 1 W. Many people associate the latter unit of measurement with the power used by appliances and lightbulbs.

When dealing with atmospheric processes, we are often concerned with large amounts of energy. For example, when we discuss global energy consumption, the amounts are so large that they are often expressed in exajoules (EJ). One EJ is one quintillion (10^{18}) joules. U.S. Department of Energy statistics indicate that global energy consumption is more than 440 EJ per year, a little more that 1.2 EJ each day. For comparison, the internal heat energy of Earth that reaches the surface is 3 times human global energy consumption, whereas solar energy reaching the surface of the Earth is nearly 10,000 times more energy than humans consume each year.[4] Thus it is primarily solar power that heats the surface of our planet, evaporates water, and produces the differential heating that causes air masses to move across our landscape.

TYPES OF ENERGY

We must expend energy to do work; another way of stating this concept is to define energy as the "ability to do work." The three main types of energy are potential energy, kinetic energy, and heat energy. *Potential energy* is stored energy. For example, the water held behind a dam contains gravitational potential energy that if released would flow downward. *Kinetic energy* is the energy of motion. From the previous example, the water flowing from the dam loses potential energy as it accelerates downward, while gaining kinetic energy. In the case of a hydroelectric dam, the kinetic energy of the flowing water is used to turn turbines (doing *work*) to generate electricity.

Heat energy is the energy of random motion of atoms and molecules. Heat itself may then be defined as the kinetic energy of atoms or molecules within a substance. Heat may also be thought of as the energy that is transferred from one body to another as a result of a difference in temperature.[5] The two types of heat that are important in atmospheric processes are sensible heat and latent heat. As the name suggests, *sensible heat* is heat that may be sensed or monitored by a thermometer. On a warm day, it is the sensible heat that we feel in the air. *Latent heat* is a more difficult concept to comprehend; it is the amount of heat that is either absorbed or released when a substance changes phase (from solid to liquid, for example). In the atmosphere, latent heat is related to the three phases of water: ice, liquid water, and water vapor. For example, the evaporation of water involves a phase change from liquid to water vapor, a gas. The energy required for this transformation is known as the *latent heat of vaporization*. This energy is recoverable when water vapor changes back to a liquid through condensation in the atmosphere when it rains.[5]

HEAT TRANSFER

To complete our discussion of energy, work, and power, we need to consider how heat energy is transferred in the atmosphere. The three major heat-transfer processes are conduction, convection, and radiation. You can see these processes when you have a pan of boiling water on an electric range (Figure 9.3).

Conduction is the transfer of heat through a substance by means of atomic or molecular interactions. The process of conduction relies on temperature differences, causing heat to flow through a substance from an area of greater temperature to an area of lesser temperature. In our example, conduction of heat through the metal pan causes the handle to heat up. Although conduction also occurs in the atmosphere, on land, and in bodies of water such as the

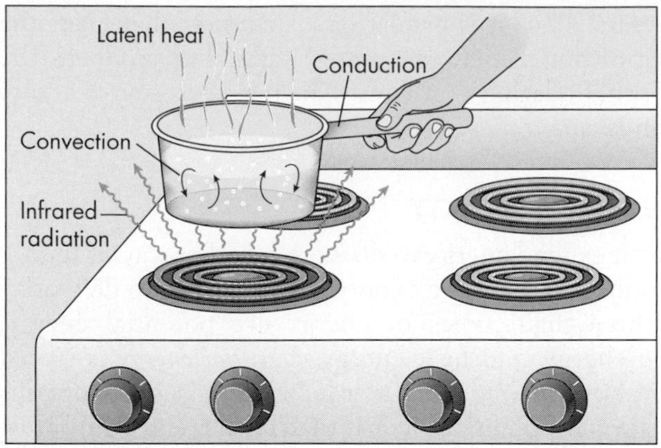

▲ FIGURE 9.3 **HEAT TRANSFER PROCESSES** Idealized heat transfer for a pan of boiling water on an electric range. Conduction is taking place in the metal of the pan and its handle; circulating water is transfering heat by convection, and visible and invisible electromagnetic radiation is being transferred through the air from the glowing coil. Latent heat is being absorbed by water vapor as liquid water evaporates. *(After Christopherson, R. W. 2006.* Geosystems: An introduction to physical geography, *6th ed. Upper Saddle River, NJ: Pearson Prentice Hall)*

ocean, it is least important in the atmosphere. Air is actually a poor conductor of heat. This is why trapped air is used to increase the insulation value of goose down comforters, double-paned windows, and plastic foam. In the atmosphere, significant conduction is generally limited to very thin layers of air in contact with Earth's surface.

Convection is the transfer of heat by the mass movement of a fluid, such as water or air. In our example (Figure 9.3), water in the bottom of the pan is heated and rises upward to displace the cooler water at the surface. The cooler water then sinks downward to the bottom of the pan. This physically mixes the water by moving the heat energy and creates a circulation loop known as a *convection cell.* Convection is an important process for the transfer of atmospheric heat in thunderstorms, and in the large-scale circulation of air away from the Equator.

Finally, *radiation* refers to wavelike energy that is emitted by any substance that possesses heat. The transfer of energy by radiation occurs by oscillations in an electric field and a magnetic field, and thus the waves are generally called *electromagnetic waves.* In our example, the red glow from the heating element of the electric range transmits visible electromagnetic radiation as red light and nonvisible infrared radiation. Most electromagnetic waves are not visible to us.

Our example is summarized as follows: The heat transfer from the electric range includes electromagnetic radiation from the glowing heating coil on the stove, conduction through the solid metal pan, and finally convection bringing warm water bubbling upward from the bottom to the top of the pan.

9.2 Earth's Energy Balance

Earth receives energy from the sun, and this energy affects the atmosphere, oceans, land, and living things before being radiated back into space. This process creates *Earth's energy balance*—a general equilibrium between incoming and outgoing energy (Figure 9.4). In the shift from incoming to outgoing energy, some of the energy changes form, but as stated in the First Law of Thermodynamics, it is neither created nor destroyed.

Although Earth intercepts only a tiny fraction of the total energy emitted by the sun, the intercepted energy is adequate to sustain life. The sun's energy also drives the hydrologic cycle, ocean waves and currents, and global atmospheric circulation. Although Earth's energy balance or budget contains several important components, nearly all of the energy that is available at Earth's surface comes from the sun (Figure 9.4).

ELECTROMAGNETIC ENERGY

Much of the energy emitted from the sun is *electromagnetic energy.* This energy, a type of radiation, travels through the vacuum of space at the speed of light, a velocity close to 300,000 km (186,000 mi.) per second. Electromagnetic radiation is commonly described as a wave, and the distance between the tops of two successive waves is referred to as the *wavelength.* The various types of electromagnetic radiation are distinguished by their wavelengths, and the collection of all possible wavelengths is known as the *electromagnetic spectrum* (Figure 9.5). Longer wavelengths, those greater than 1 m (3.3 ft.), include radio waves and microwaves, whereas the shortest wavelengths are X-rays and gamma rays. Visible electromagnetic radiation, referred to as light, makes up only a very small fraction of the total electromagnetic spectrum. Other types of electromagnetic radiation with environmental significance include infrared (IR) and ultraviolet (UV) radiation. Infrared radiation is involved in global warming, and levels of UV radiation at the Earth's surface are influenced by the depletion of ozone in the upper atmosphere (see Chapter 12).

ENERGY BEHAVIOR

Once electromagnetic energy from the sun reaches Earth, it is redirected, transmitted, or absorbed by the atmosphere, ocean, or land. In redirection, the energy is either reflected like a light bouncing off of a mirror or scattered in different directions. Reflection—from the tops of clouds, the water, and the land—is one of two ways that solar energy returns to outer space (see Chapter 12). Scattering disperses the energy in many directions, with scattered energy generally weaker than reflected or transmitted energy. Transmission involves the energy passing through the atmosphere, like light passing directly through window glass. On a clear day, the atmosphere transmits most

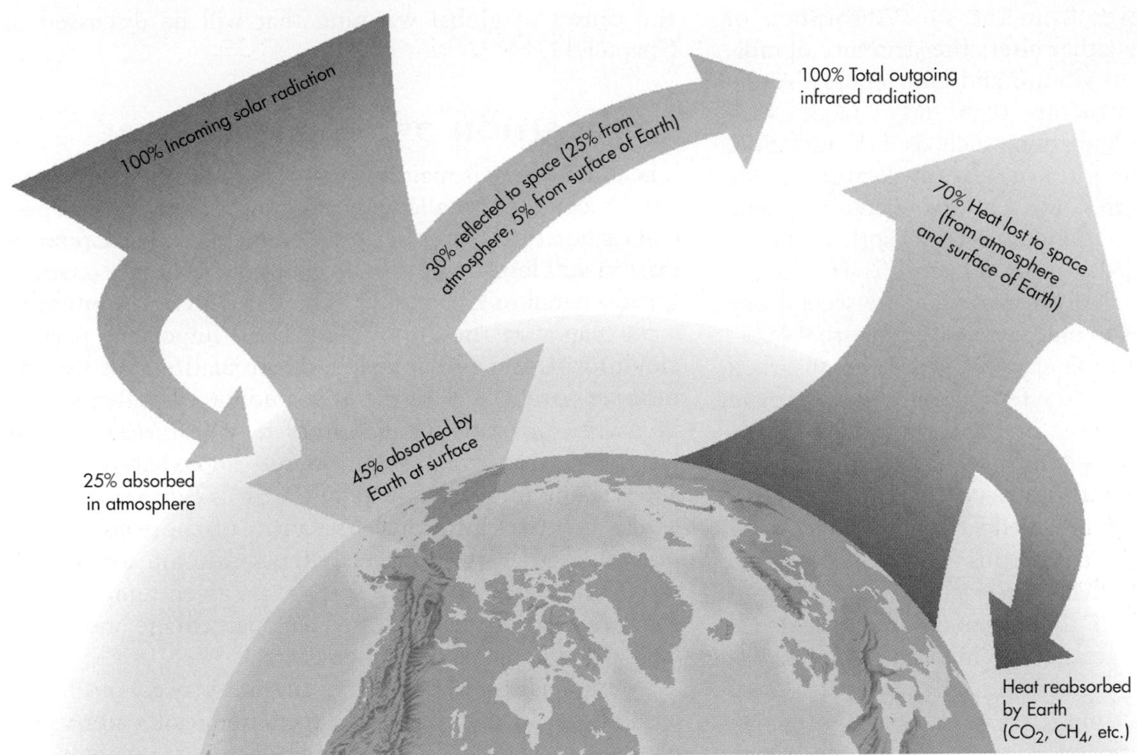

▲ **FIGURE 9.4 EARTH'S ENERGY BALANCE** Most of the annual energy flow to the Earth from the sun is either reflected or reradiated back into outer space. Only a small component of Earth's heat is actually coming from its interior. *(After Pruitt, N. L., Underwood, L. S., and Surver, W. 1999.* Bioinquiry, Learning system 1.0, making connections in biology. *New York: John Wiley & Sons)*

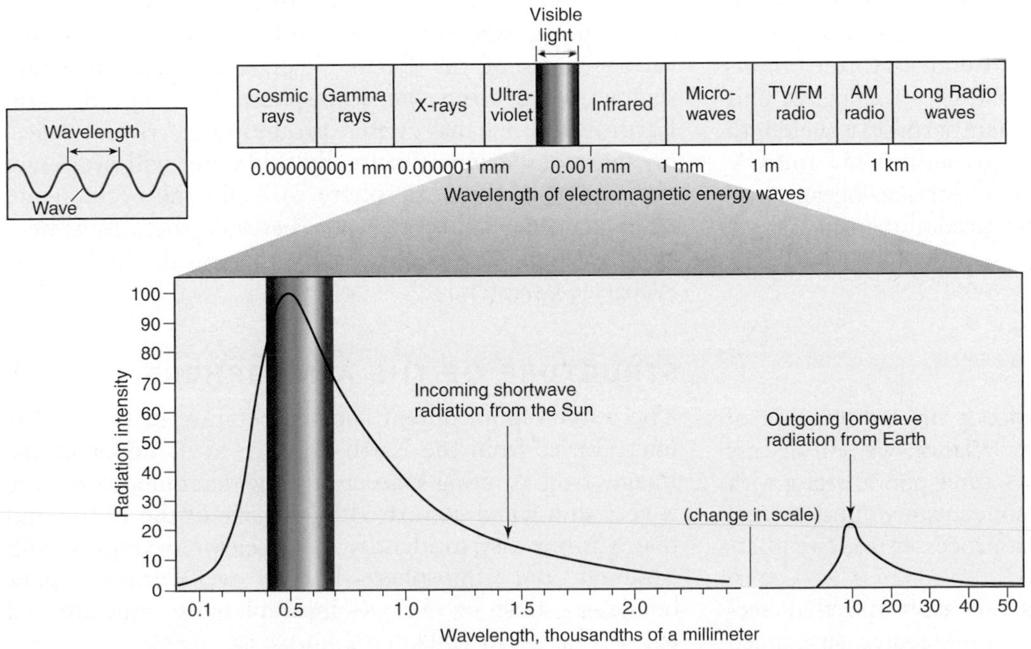

▲ **FIGURE 9.5 THE ELECTROMAGNETIC SPECTRUM** Wavelength is the distance between one wave crest and the next. Values for wavelength in the electromagnetic spectrum have an enormous range from billionths of a meter for X-rays to thousands of meters for long radio waves. The incoming energy from the sun has shorter wavelengths than the radiation emitted from the Earth. Only a small amount of the electromagnetic spectrum is visible light, which can be seen by the human eye.

of the energy it receives from the sun. Absorption of electromagnetic energy either alters the structure of molecules or causes them to vibrate and emit energy. Alteration of molecular structure can take place with penetrating short-wave radiation, such as UV radiation. Emitted energy is in the form of heat or electromagnetic radiation. For many gases, the emitted electromagnetic energy is at the same wavelength as the absorbed energy, whereas in many solids, the emitted energy is of a longer wavelength, such as IR energy. Emission is the second way that solar energy returns to outer space (see Figure 9.4).

The relative amount of energy reflected and absorbed is important in determining the temperature of the air and land. For example, the color of the land surface determines its reflectivity to visible light. This reflectivity is referred to as the *albedo* of the surface. In general, darker-colored surfaces have lower albedo than lighter-colored surfaces. Darker-colored coniferous woodland, such as pine or spruce forests, reflects 5 percent to 15 percent of solar radiation compared to lighter-colored grasslands that reflect up to 25 percent of the incoming solar radiation.[6] Much of the energy that is not reflected is absorbed, causing darker surfaces to heat up more than lighter surfaces. Many atmospheric gases are selective absorbers; that is, they absorb some wavelengths of energy and are transparent to others. For example, carbon dioxide and water vapor selectively absorb IR wavelengths and contribute to the warming of the lowest layer of the atmosphere (see Chapter 12).

Absorbed energy that is not emitted as heat is emitted as electromagnetic radiation. The temperature of the object emitting the radiation also affects the wavelength of electromagnetic radiation that it emits. Hotter objects radiate energy more rapidly and at shorter wavelengths. This fact explains why the sun emits mainly short-wavelength radiation, such as gamma and X-rays, and visible and UV light. In contrast, the Earth's land surface, oceans, and clouds are so cool that they emit predominantly longer-wavelength IR radiation.

9.3 The Atmosphere

Now that we have finished our brief discussion of energy and Earth's energy balance, we will discuss the various components of Earth's atmosphere, along with atmospheric circulation. These concepts are fundamental to the understanding of weather processes and weather-related hazards.

The **atmosphere** is the gaseous envelope that surrounds Earth. It is made up of gas molecules, suspended particles of solid and liquid, and falling precipitation. The atmosphere causes the weather we experience every day and is responsible for trapping the heat that keeps the Earth warm enough to be habitable. Knowledge of the structure and dynamics of the atmosphere is critical to understanding severe weather, as well as the mechanism and causes of global warming that will be discussed in Chapter 12.

COMPOSITION OF THE ATMOSPHERE

The atmosphere is mainly composed of nitrogen and oxygen; it contains smaller amounts of argon, water vapor, and carbon dioxide. Other trace elements and compounds exist in still lesser amounts. We will discuss these gases in greater detail in Chapter 12. The behavior and content of water vapor in the atmosphere is an important part of cloud formation and atmospheric circulation. We use the term *humidity* to describe the amount of water vapor. Humidity is largely a function of temperature; warm air has the capacity to hold more water vapor than does cold air. The amount of moisture in the air is commonly given as the **relative humidity,** the ratio of the water vapor present in the atmosphere to the maximum amount of water vapor that could be there for a given temperature. Relative humidity is expressed as a percentage and varies from a few percent to 100 percent.

Natural changes in relative humidity occur each day without significant changes in the amount of water vapor in the atmosphere. Relative humidity increases at night because of the cooler temperature and decreases with daytime heating. Changes in the actual water vapor content of the atmosphere take place where water evaporates from the surface or where air masses mix, either vertically or horizontally.

Virtually all the water vapor in the air is derived from evaporation of water from Earth's surface. Water is constantly being exchanged between the atmosphere and the various parts of the Earth. Sleet, snow, hail, and rain remove water from the atmosphere and deposit it on Earth where it may enter groundwater, rivers, lakes, oceans, and glaciers. Eventually this water will evaporate and return to the atmosphere to begin the cycle again. This constant cycling of water between the atmosphere and the Earth's surface is a major part of the hydrologic cycle (see Chapter 1).

STRUCTURE OF THE ATMOSPHERE

The water vapor content and temperature of the atmosphere varies from the Earth's surface to its upper limits. Images from orbiting spacecraft show that our atmosphere is very thin when compared to the size of the Earth, and that it is not easy to identify its upper limits (Figure 9.6). Although the atmosphere has no well-defined upper boundary, most of the gas molecules are concentrated below a height of 100 km (62 mi.).

Earth's atmosphere has a structure consisting of four major layers or spheres (Figure 9.7). The lowest layer, the **troposphere,** extends about 8 to 16 km (5 to 10 mi.) above the surface of Earth. With the exception of some jet airplane travel, we spend our entire lives within the troposphere. Not even the highest mountains breach the upper boundary of

◀ **FIGURE 9.6 EARTH'S THIN ATMOSPHERE** Viewed from space, the atmosphere appears as a thin layer surrounding Earth. *(NASA)*

the troposphere known as the *tropopause*. The defining characteristic of the troposphere is a rapid upward decrease in temperature that results from decreasing air pressure with increasing altitude. However, the most visible characteristic of the troposphere is abundant condensed water vapor in the form of clouds. Most of the clouds and weather that directly affect us are found in the troposphere.

The formation and development of clouds are particularly important. Clouds develop when very small water droplets or ice crystals condense from the atmosphere. Without clouds there would be no rain, snow, thunder, lightning, or rainbows. You are probably familiar with two of the most common types of clouds—puffy, fair-weather *cumulus* clouds that may look like pieces of floating cotton,

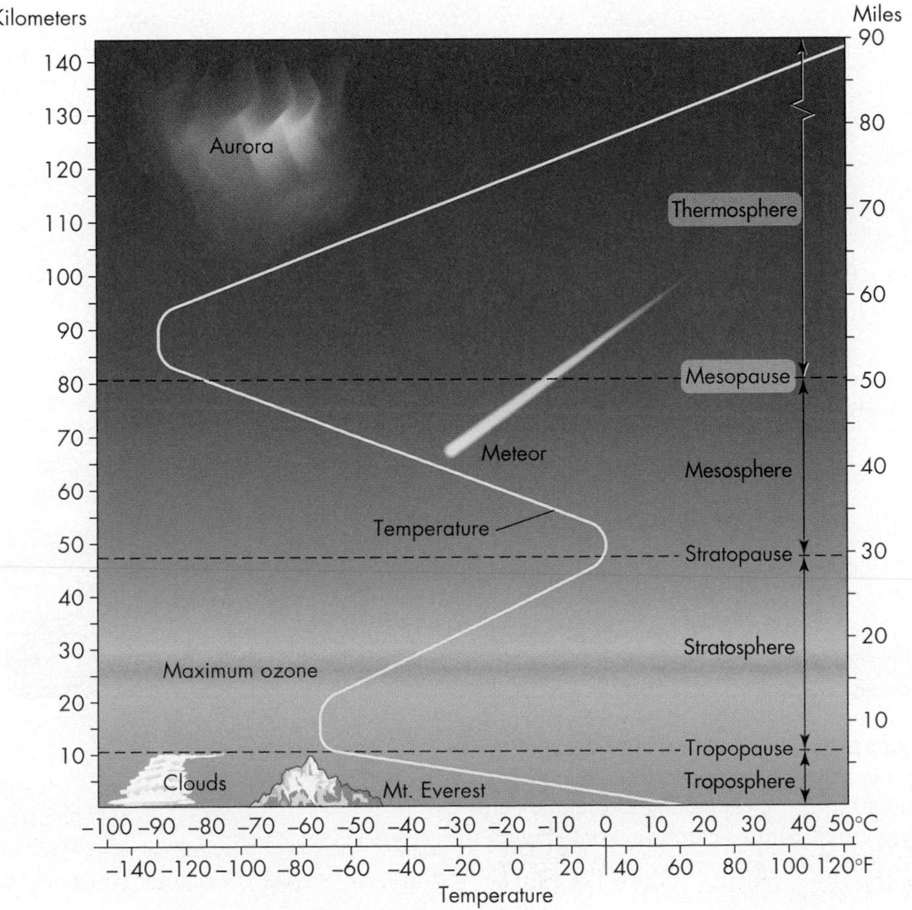

◀ **FIGURE 9.7 ATMOSPHERIC STRUCTURE** Earth's atmosphere has a structure based on changes in air temperature from the surface upward. The curving yellow line shows the average change in air temperature with height. Weather develops in the lowest layer, the troposphere. *(After Lutgens, F. K., and Tarbuck, E. J. 2007. The atmosphere: An introduction to meteorology, 10th ed. Upper Saddle River, NJ: Pearson Prentice Hall)*

and the towering *cumulonimbus* thunderstorm clouds that release tremendous heat energy by condensation of water vapor (Figure 9.8).

Most of the atmospheric water vapor condenses in the troposphere, leaving little water in the upper layers of the atmosphere. In addition to water vapor, the troposphere contains most of the atmospheric carbon dioxide and methane, two gases that are important in global warming (Chapter 12). Only ozone, a gas composed of oxygen atoms (O_3), is significantly less abundant in the troposphere when compared with the upper atmosphere (see Figure 9.7).

9.4 Weather Processes

A complete discussion of the atmospheric conditions and processes associated with severe weather is beyond the scope of this book. Students who find this topic especially interesting should pursue coursework in meteorology, the scientific study of weather. Instead, we will focus on four aspects of the atmosphere that are directly related to severe weather: atmospheric pressure and circulation patterns, the vertical stability of the atmosphere, the Coriolis effect (see A Closer Look 9.1), and the interaction of different air masses.

ATMOSPHERIC PRESSURE AND CIRCULATION

Weather forecasters often talk about areas of low or high pressure. The pressure that they are referring to is *atmospheric pressure*. Also called *barometric pressure*, atmospheric pressure is the weight of the column of air that is above any given point (Figure 9.9). This point may be on the Earth's surface or above it, such as in an airplane. Atmospheric pressure can also be thought of as the force exerted by the gas molecules on a surface (Figure 9.10a). As you might expect, atmospheric pressure is greater at sea level than at the top of a high mountain where there is less air above the surface (Figure 9.10b). Nearly all of the weight of the atmosphere and, thus, nearly all the pressure, is in the lower atmosphere below an elevation of 50 km (31 mi.) (Figure 9.10b). Atmospheric density and pressure decrease rapidly as one goes to higher elevations. You may have noticed this pressure change if your ears have ever "popped" during a drive up or down a mountain or in an airplane that was changing altitude.

Changes in air temperature and air movement are responsible for most of the horizontal variation in atmospheric pressure. Temperature change influences pressure because cold air is denser than warm air and exerts a higher pressure on the underlying surface. The density of cold air is higher because its gas molecules have lower kinetic

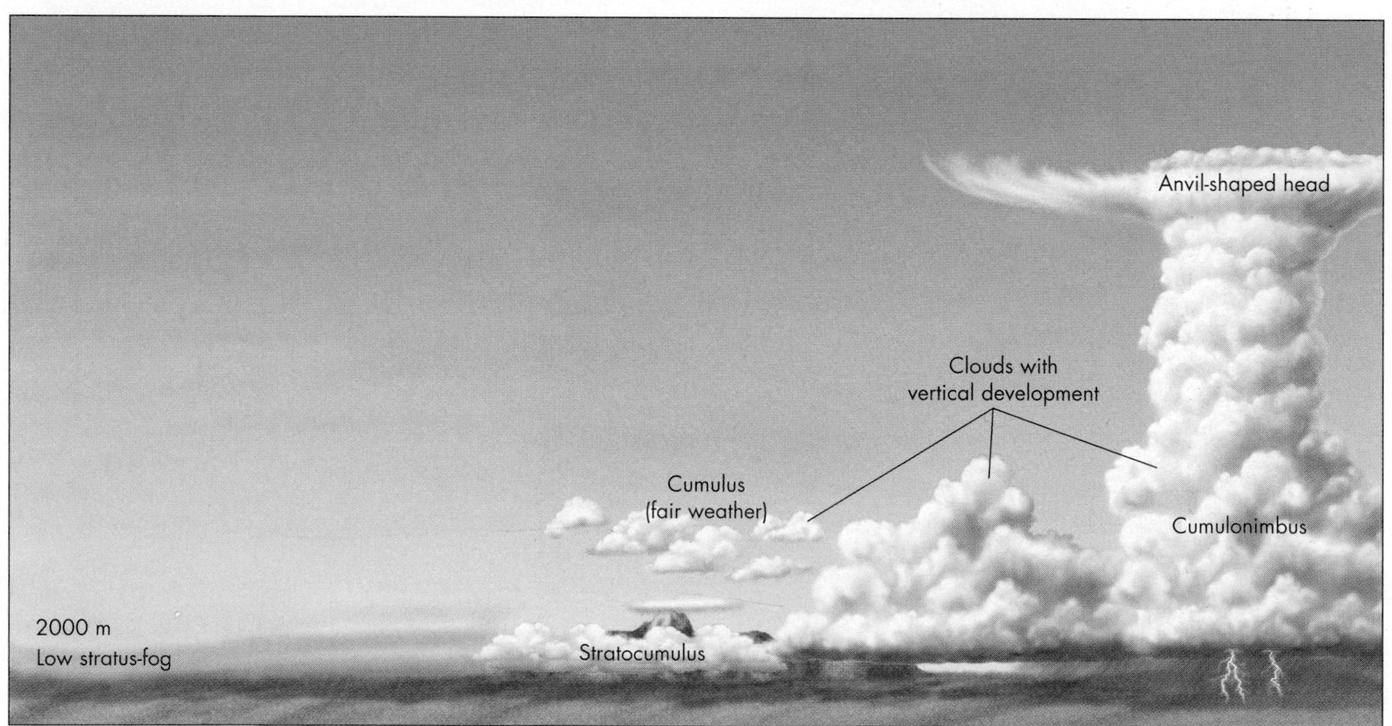

▲ FIGURE 9.8 **CLOUD TYPES ASSOCIATED WITH SEVERE WEATHER** Clouds consist of small water droplets and are classified on the basis of altitude (low, middle, and high) and form (cirriform, cumuliform, and stratiform). Cumulonimbus clouds are associated with severe thunderstorms and tornadoes. Stratus clouds in contact with the ground form fog. *(After Christopherson, R. W. 2006. Geosystems: An introduction to physical geography, 6th ed. Upper Saddle River, NJ: Pearson Prentice Hall)*

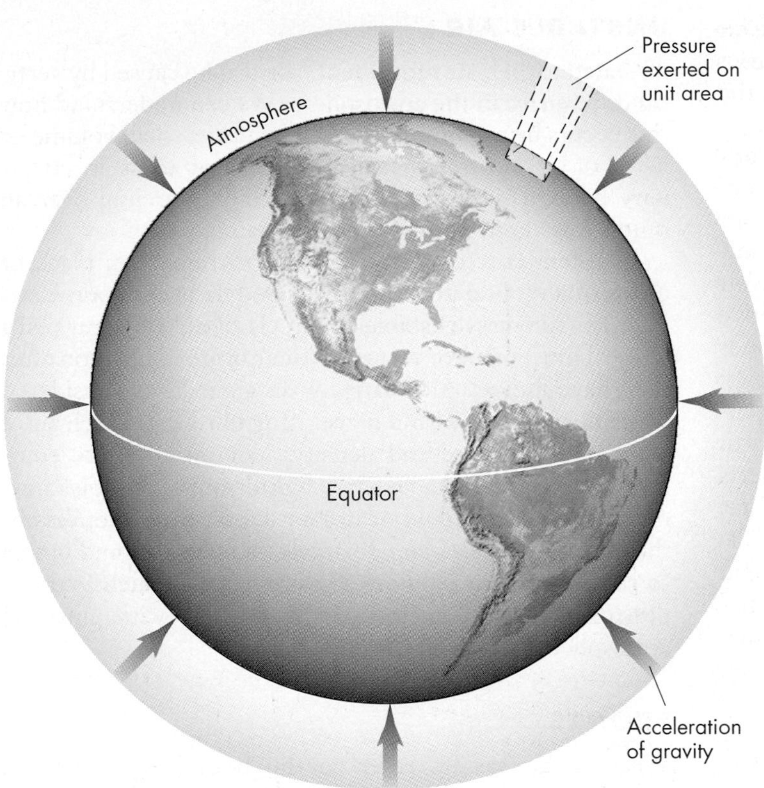

◀ **FIGURE 9.9 ATMOSPHERIC FORCE REMAINS CONSTANT AS AIR PRESSURE CHANGES** Idealized diagram showing Earth and its atmosphere. The total force of the entire atmosphere is the product of the mass of the atmosphere and the acceleration of gravity. Both the mass and acceleration are constant; therefore, the total force from the atmosphere is constant. However, the air pressure at points on the Earth does vary because the mass of overlying air varies from place to place. Variations in the mass of overlying air are caused by differences in the temperature and density of air masses. Air pressure can be expressed as atmospheric force per unit area. *(After Aguado, E., and Burt, J. E. 2007. Understanding weather and climate, 4th ed. Upper Saddle River, NJ: Pearson Prentice Hall)*

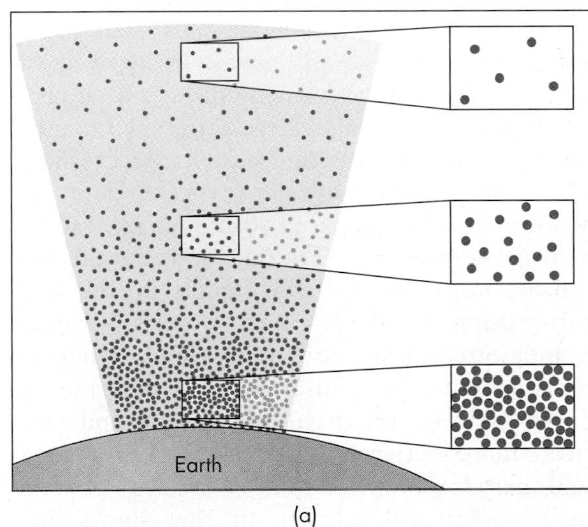

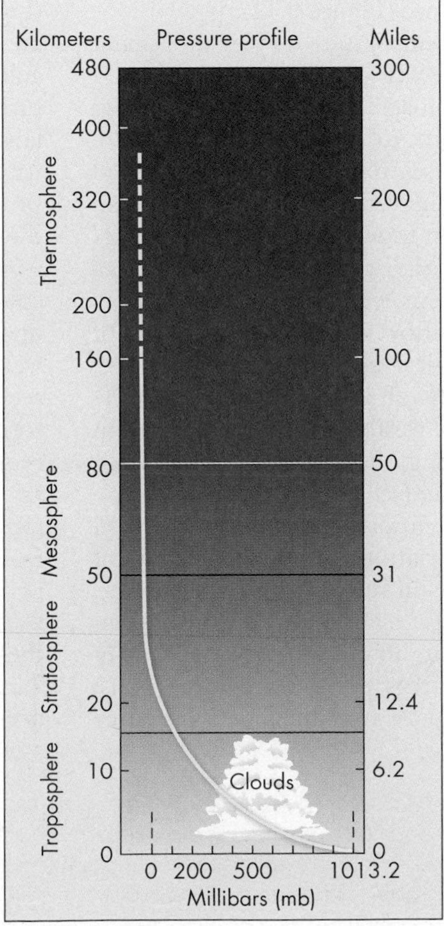

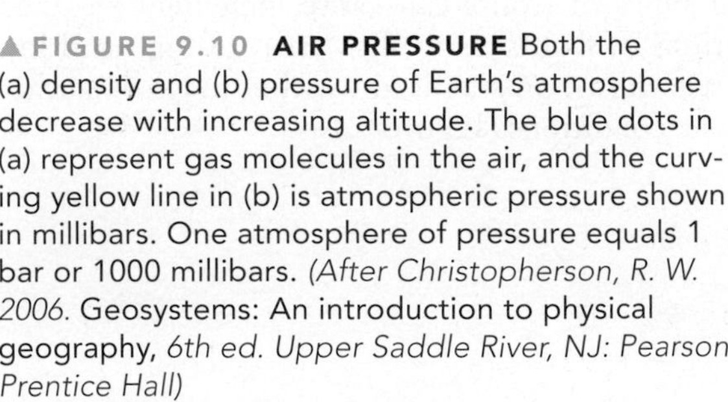

▲ **FIGURE 9.10 AIR PRESSURE** Both the (a) density and (b) pressure of Earth's atmosphere decrease with increasing altitude. The blue dots in (a) represent gas molecules in the air, and the curving yellow line in (b) is atmospheric pressure shown in millibars. One atmosphere of pressure equals 1 bar or 1000 millibars. *(After Christopherson, R. W. 2006.* Geosystems: An introduction to physical geography, *6th ed. Upper Saddle River, NJ: Pearson Prentice Hall)*

energy and stay more closely packed together. Because air pressure and temperature are related, air pressure varies geographically, and this variation has a strong effect on the weather and global atmospheric circulation patterns.

Solar heating and evaporation of seawater at low latitudes near the equator creates warm air masses that have high relative humidity. The low density of the heated air creates low-pressure zones at equatorial latitudes that convect upward in the atmosphere (Figure 9.11). The air masses cool as they rise, causing an increase in the relative humidity and condensation of water vapor to form clouds and the abundant precipitation characteristic of the tropics. Having dropped the rain, the now dry, cold, and dense air in the upper troposphere spreads out and sinks at 30° north and south latitude. The subtropical middle latitude deserts of the world, such as the Sahara, Arabian, and Mojave, are, in fact, the result of these dry descending air masses. Similar vertical circulation cells are observed at middle and higher latitudes, where low-pressure zones at 60° north and south latitude are characterized by rising humid air that spreads out at the base of the stratosphere or *tropopause* and descends at the poles and 30° latitude (Figure 9.11).

Air movement can also cause changes in atmospheric pressure. If there is an overall flow of air into a region, *convergence* of the air occurs, and the air piles up to increase atmospheric pressure. In contrast, if there is an overall flow of air out of a region, then *divergence* occurs, and the loss of air lowers atmospheric pressure (Figure 9.12).

The combined effects of temperature and air movement produce the *low-pressure centers* (L) and *high-pressure centers* (H) that you see on weather maps. At the surface, air flows from areas of high pressure to areas of low pressure (Figure 9.12). Less dense air rises in areas of low pressure and diverges in the upper troposphere. This generally creates an area of high pressure aloft on top of the surface low-pressure center. Air flows from a high-pressure center aloft to a low-pressure center aloft where it, in turn, sinks to create a surface high-pressure center. Low pressures near the top of the troposphere, where vertical circulation cells diverge, cause narrow fast-flowing jets of air (see Figure 9.11). *Jet streams*, as they are called, are westerly winds (flowing from west to east) that encircle the globe and play an important role in creating severe weather (see Chapter 10).

As mentioned previously, differences in the absorption of solar radiation produce variations in the temperature of the Earth's surface, as well as in the air above the surface. These temperature variations create differences in atmospheric pressure and cause air to flow horizontally from areas of high pressure to areas of low pressure. Thus, changes in atmospheric pressure are a major driving force for wind and weather.

UNSTABLE AIR

As stated earlier, air movement can also be caused by vertical differences in the atmosphere. We can understand how this occurs by examining the behavior of a small volume or *parcel* of air. You can visualize a parcel of air as an imaginary balloon similar to a small dirigible circling over an outdoor stadium during a football game.

The tendency of a parcel of air to remain in place or change its vertical position is referred to as *atmospheric stability*. An air mass is stable if parcels of air within it resist vertical movement or return to their original position after they have moved. Alternatively, an air mass is considered unstable if parcels within it are rising until they reach air of similar temperature and density.[7] The atmosphere commonly becomes unstable when lighter warm or moist air is overlain by denser cold or dry air. Under these conditions the instability causes some parcels of air to sink and others to rise like hot air balloons. Severe weather, such as thunderstorms and tornadoes, is associated with unstable atmospheric conditions.

FRONTS

Weather forecasters refer to the boundary between a cooler and warmer air mass as a *front*. The boundary is called a cold front when cold air is moving into a mass of warm air, and a warm front when the opposite is true (Figure 9.13). Regardless of which air mass is advancing on the other, the warmer air will always be lifted by the colder air mass. In addition to having different temperatures, the air masses may also have different humidity levels, densities, wind patterns, and stability. As a result of friction with the ground, high-density cold fronts are slowed at lower levels, compared to higher up, causing the front to become steeper (Figure 9.13a). As the cold front advances, warm air is driven rapidly up the steep cold front causing localized cloud formation, strong updrafts, and, often, heavy precipitation. In contrast, warm fronts tend to be gentler as they override the receding cold air mass, and, therefore, updrafts and cloud formation are more scattered (Fig-ure 9.13b). Fronts may remain stationary for a few hours or even days. A boundary between cooler and warmer air that shows little movement is called a *stationary front*. A fourth type of front, called an *occluded front*, develops where rapidly moving cooler air overtakes another cooler air mass and warm air is wedged above the frontal boundary. Each of these four types of fronts can cause inclement weather. The positions of fronts, as well as areas of high and low pressure, are shown on standard weather maps of surface atmospheric conditions (Figure 9.14).

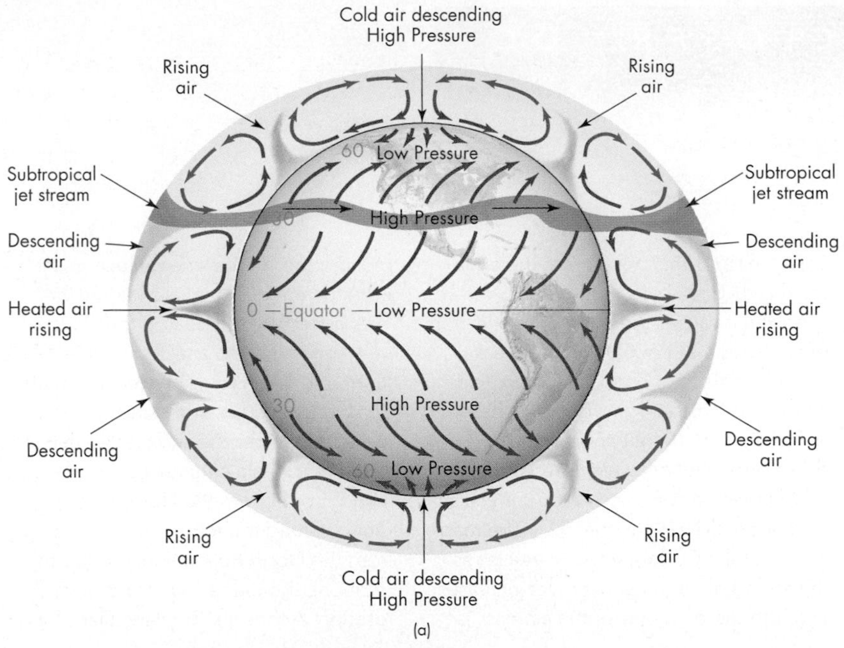

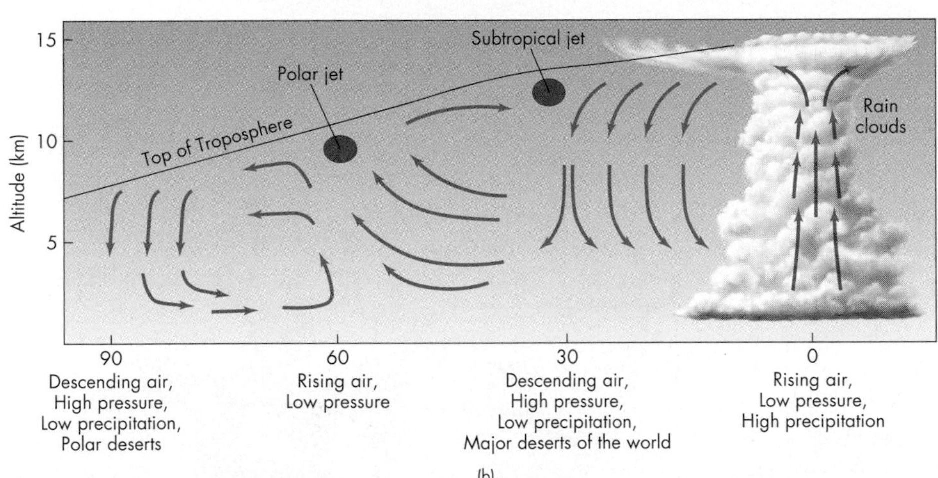

FIGURE 9.11 ATMOSPHERIC CIRCULATION (a) Global circulation of the lower atmosphere showing zones of rising and descending air masses and corresponding zones of low and high air pressure. (b) Idealized diagram showing atmospheric conditions from the equator to a pole illustrating zones of rising and descending air masses and the location of jet streams.

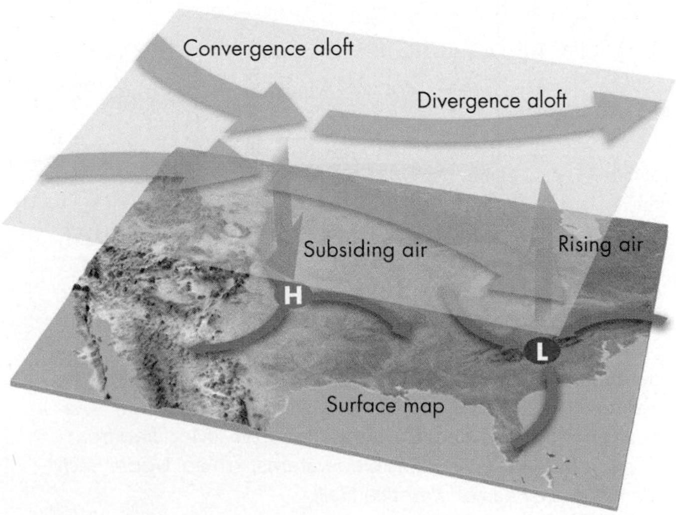

FIGURE 9.12 LOW- AND HIGH-PRESSURE CENTERS This diagram illustrates airflow in low- and high-pressure centers in the Northern Hemisphere. Horizontal changes in atmospheric pressure and temperature variations cause air to circulate in the troposphere. Airflow at the surface converges (purple arrows point inward) in low-pressure centers and diverges (purple arrows point outward) in high-pressure centers. *(After Lutgens, F. K., and Tarbuck, E. J. 2007.* The atmosphere: An introduction to meteorology, *10th ed. Upper Saddle River, NJ: Pearson Prentice Hall)*

A CLOSER LOOK 9.1

Coriolis Effect

The unequal distribution of solar energy that reaches the surface of Earth inevitably leads to temperature and pressure gradients that drive atmospheric circulation. You might expect air moving from a high-pressure area to a low-pressure to follow a straight path; yet, when wind patterns are traced across Earth's surface, they appear to curve (Figure 9.A). This is because the Earth, our frame of reference, rotates beneath the flowing air masses, causing a deflection of the wind to the right or to the left. This apparent change in motion or deflection is known as the **Coriolis effect,** named for Gaspard-Gustave Coriolis, who first published a paper on the effects of a rotating reference frame in 1835. As you will see in this chapter and Chapter 10, the Corriolis effect is important in controlling prevailing wind patterns on Earth and plays an important role in the formation and movement of hurricanes.

In order to understand the Coriolis effect, we must first step off our planet and get some perspective on our rotating home. As Figure 9.B. shows, the Earth rotates from west to east, and when viewed from the Northern Hemisphere rotates in a counterclockwise manner. Conversely and perhaps surprisingly, when viewed from above the South Pole, the west to east rotation of Earth results in a clockwise rotation. We recommend trying this with a globe or basketball. Continuously rotate the ball in one direction and view it from above and from below. What you will see is two different rotation directions as described above. Because of the different rotation directions, the Coriolis effect behaves differently in the Northern Hemisphere than in the Southern Hemisphere. Specifically in the Northern Hemisphere, the deflection will be to the right and in the Southern Hemisphere the deflection will be to the left (Figure 9.A).

Let us now examine what causes the Coriolis deflection. Notice in (Figure 9.B), the rotation speed of Earth varies with latitude increasing from zero at the poles to 1675 km (1041 mi.) per hour at the equator. It is this variability in the rotation speed that causes the apparent deflection of anything that flows or flies above the rotating surface of Earth, including airplanes, oceans, and air masses. The Coriolis force acts at a right angle to the direction of motion, which causes the path of an object moving above the surface to have a curved path (Figure 9.C). The magnitude of the Coriolis effect increases as the speed of the moving object increases, and, therefore, the faster the air mass is moving the greater the deflection. At the equator, the effect is zero (Figure 9.A).

A good way to visualize the Coriolis effects is to use an analogy of an airplane flight to explain how the rotation of the Earth could cause air masses to be deflected across the surface of the Earth. The pilot plots a course due south along a straight path from the North Pole to Quito, Ecuador (Figure 9.B). Departing from the North Pole (the axis of Earth's rotation), the plane has zero eastward rotation. Although the plane flies along a

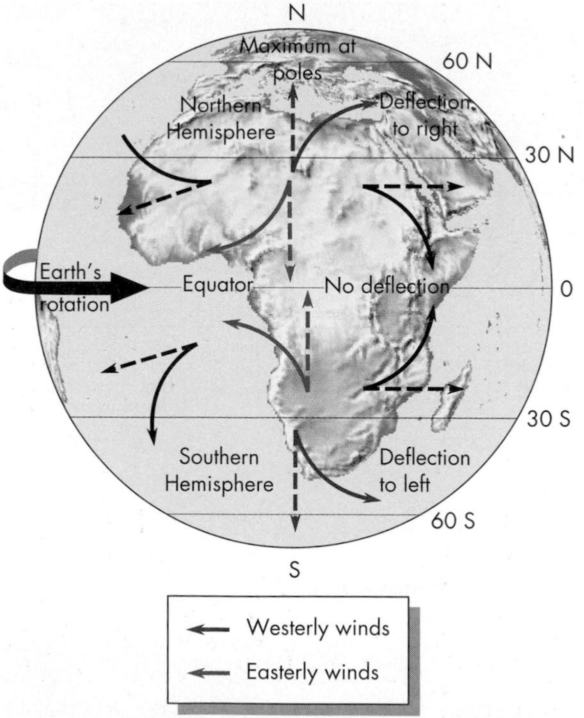

▲ **FIGURE 9.A SURFACE WIND PATTERNS THAT RESULT FROM THE CORIOLIS EFFECT** Air masses moving across the surface of Earth will be deflected to the right in the Northern Hemisphere and to the left in the Southern Hemisphere due to the Coriolis effect. The deflection influences the prevailing wind directions creating westerly (red arrows) and easterly winds (blue arrows) flowing from mid-latitude high-pressure zones. *(Modified from Christopherson, R. W., 2009.* Geosystems, *7th ed. Upper Saddle River, NJ: Pearson Prentice Hall)*

(continued)

straight path toward Quito, the greater rotation speed of Earth near the equator carries the city eastward. Upon arriving at the latitude of Quito, the pilot would find herself over the Pacific Ocean (Figure 9.B). To an observer in Quito watching the airplane, the land surface would feel stationary, yet the airplane would appear to turn toward the west and head out over the sea, even though its path was straight. On the return flight from Quito, the greater eastward rotation at the equator compared more northern latitudes would cause the airplane to be deflected toward the east. In both cases, when viewed from above, the plane appears to have been deflected to the right.

Let's now look at an example of how the Coriolis effect can be used to explain the predominant wind directions on Earth. Wind traveling southward from the Northern Hemisphere's subtropical high-pressure zone (30°N latitude) will be deflected toward the west, producing easterly winds, (Figure 9.A). Whereas winds moving toward the north will produce westerly winds. In contrast, air moving southward from the Southern Hemisphere's subtropical high-pressure zone (30°S latitude) will be deflected toward the west, producing easterly winds, whereas wind moving toward the north will produce westerly winds (see Figure 9.A).

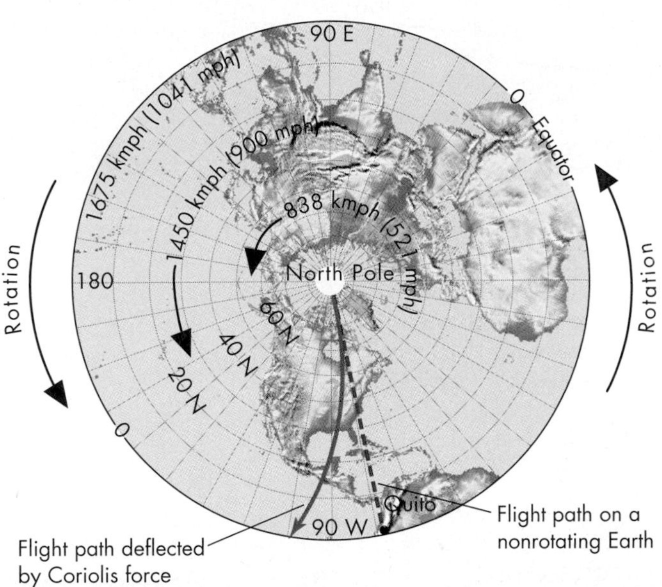

FIGURE 9.B EARTH'S ROTATION The example here illustrates the Coriolis effect on an airplane traveling along a straight path between the North Pole and Quito, Ecuador, which is on the equator. *(Modified from Christopherson,R. W., 2009.* Geosystems, *7th ed. Upper Saddle River, NJ: Pearson Prentice Hall)*

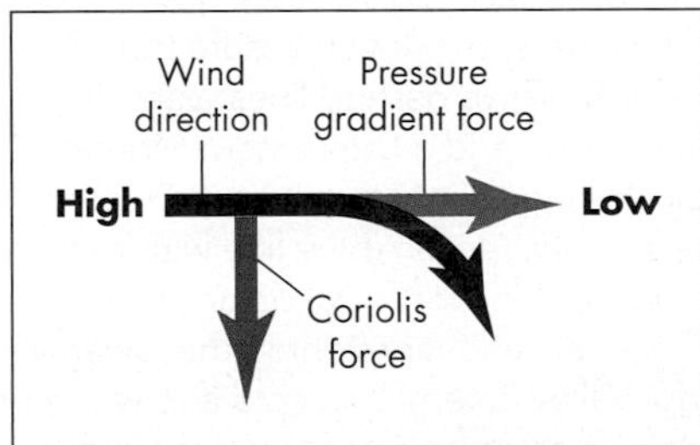

FIGURE 9.C FORCES INVOLVED IN WIND PATTERNS The path of an air mass is controlled, in part, by the pressure gradient and the Coriolis forces. On a nonrotating Earth, air masses would flow along the pressure gradient from areas of high pressure to areas of low pressure (blue arrow). However on a rotating Earth, the Coriolis force (red arrow) exerts a force perpendicular to the horizontal wind direction, which causes the path of an air mass to curve (black arrow). This cartoon shows a wind deflection to the right consistent with the Northern Hemisphere.

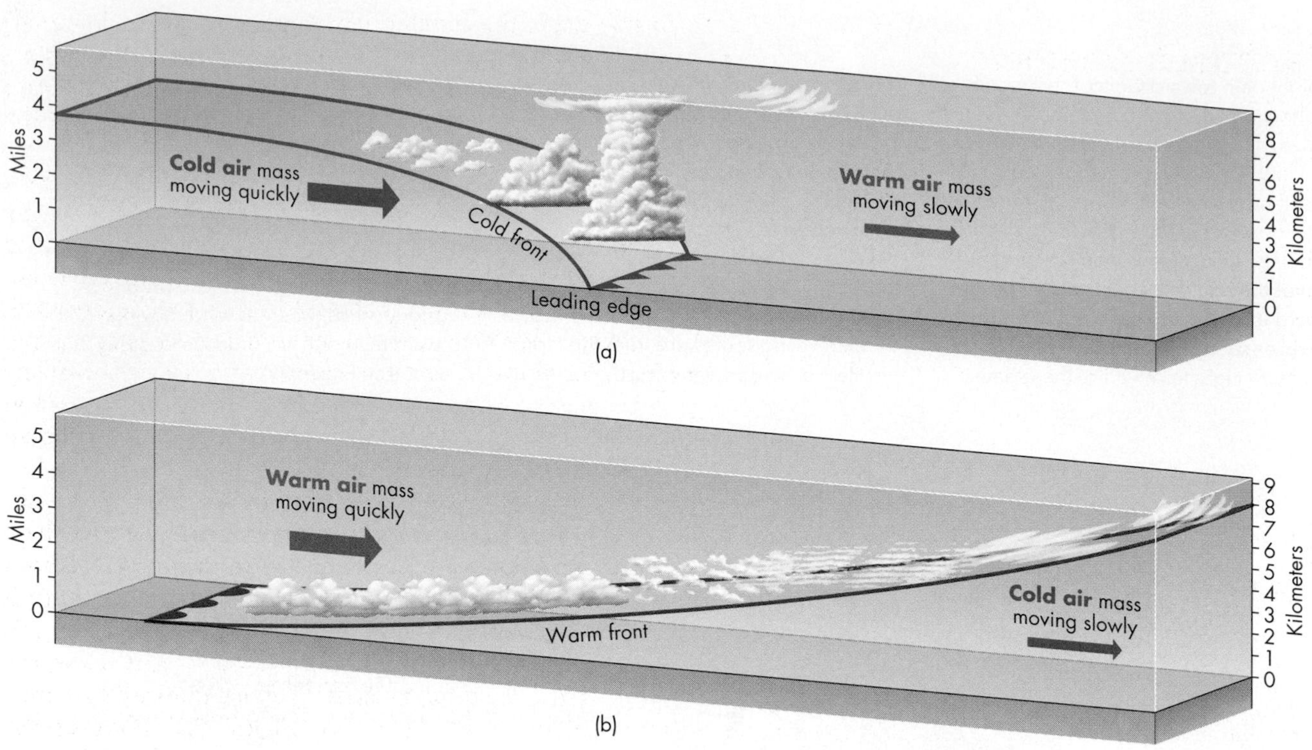

▲ FIGURE 9.13 **FRONTS** Weather fronts mark the boundary of air masses that have different densities, generally as the result of differences in temperature. (a) Advancing cold front that forces warm air upward. The rising warm air can create clouds and heavy precipitation. (b) Advancing warm front which forces warm air to rise over cooler air. Clouds and precipitation may develop. *(After McKnight, T. L., and Hess, D. 2004.* Physical geography, *8th ed. Upper Saddle River, NJ: Pearson Prentice Hall)*

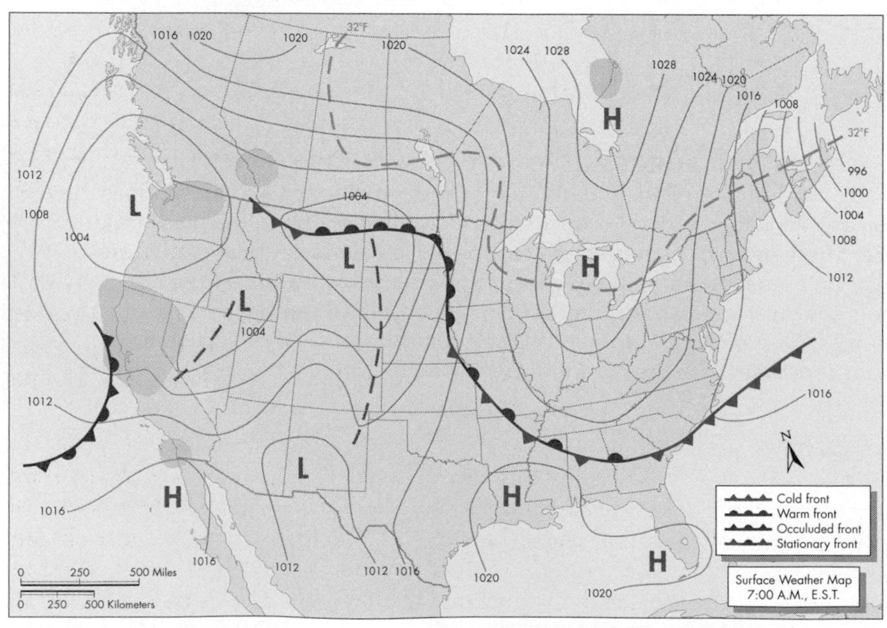

◀ FIGURE 9.14 **WEATHER MAP** The weather showing surface atmospheric conditions for the United States and southern Canada on April 13, 2003. Map shows positions of surface high- (H) and low- (L) pressure centers, two low-pressure troughs (reddish brown dashed lines extending from a L in Idaho and a L in eastern Montana), a cold front extending east from Georgia into the Atlantic Ocean (blue line with triangles), a warm front in North Dakota (red line with half circles), a stationary front in the lower Mississippi Valley (alternating cold and warm front symbols on opposite sides of the same line), an occluded front off the California coast (alternating cold and warm front symbols on the same side of a purple line), the line marking freezing temperature (curving dashed blue line), and areas of precipitation (solid green). The green contour lines show atmospheric pressure in millibars (e.g., 1016). *(Courtesy of the National Weather Service)*

9.5 Hazardous Weather

The basic principles of atmospheric physics described earlier can help us understand severe weather and its associated hazards. Severe weather refers to events such as thunderstorms, tornadoes, hurricanes (see Chapter 10), blizzards, ice storms, mountain windstorms, heatwaves, and dust storms. These events are considered hazardous because of the energy they release and damage they are capable of causing.

THUNDERSTORMS

At any one time, thousands of thunderstorms are in progress on Earth. Most occur in the equatorial regions. For example, the city of Kampala, Uganda, near the equator in East Africa, holds the world record for thunderstorm frequency; it has thunderstorms nearly 7 out of every 10 days. In North America, the regions with the highest number of days each year with thunderstorms are in an area along the Front Range of Rocky Mountains in Colorado and New Mexico and in a belt that encompasses all of Florida and the southern parts of Georgia, Alabama, Mississippi, and Louisiana (Figure 9.15).

Most readers are likely to have experienced at least one thunderstorm, because they occur in virtually every part of the United States and Canada. Although they can form any time, thunderstorms are most common during afternoon and evening hours, especially in the spring and summer. Their abundance at this time of maximum daytime heating is related to the special set of atmospheric conditions that are required to produce a thunderstorm.

Three basic atmospheric conditions are necessary for a thunderstorm to form:

1. Warm humid air must be available in the lower atmosphere to feed clouds and precipitation and provide energy to the storm as it develops.
2. A steep vertical temperature gradient must exist in the environment such that the rising air is warmer than the air through which it is moving. This gradient places colder air over warmer, moist air.
3. An updraft must force moist air up to colder levels of the atmosphere.

These three conditions are common at cold fronts where the warm humid air is forced upward along the relatively steep front of the cold air mass (see Figure 9.13a).

Thunderstorm formation starts as moist air is forced upward and cools, and water vapor condenses to form a puffy cumulus cloud. Initially the cloud will form and then evaporate with little increase in height. If the moisture supply and updraft continue, the relative humidity increases in the air surrounding the cloud and it grows in size instead of evaporating. The upward growth in size of a cumulus cloud begins the *cumulus stage* (Figure 9.16) of thunderstorm development. In this stage, the cumulus cloud becomes a cumulonimbus cloud with the growth of domes and towers that look like a head of cauliflower. This growth requires a continuous release of latent heat from water vapor condensation to warm the surrounding air and cause the air to rise farther.

As the domes and towers grow upward, precipitation starts by one of two mechanisms. First, growth of the cloud into colder air causes water droplets to freeze into ice crystals and snowflakes. The larger snowflakes fall until they enter air that is above freezing and melt to form raindrops. Second, in warm air in the lower part of the cloud, large cloud droplets collide with smaller droplets and coalesce to become raindrops. Once raindrops are too large to be supported by updrafts in the cloud, they begin to fall, creating a downdraft.

The *mature stage* of thunderstorm development begins when the downdraft and falling precipitation leave the base of the cloud (Figure 9.16). At this stage, the storm has both updrafts and downdrafts, and it continues to grow until it reaches the top of the unstable atmosphere. Commonly this upper limit of growth is the tropopause. At this point, the updrafts may continue to build the cloud outward to form a characteristic anvil shape (see Figure 9.8). During the mature stage, the storm produces heavy rain, lightning and thunder, and occasionally hail (see Case Study 9.2)

The final or *dissipative* stage begins when the upward supply of moist air is blocked by downdrafts at the lower levels of the cloud (Figure 9.16). Downdrafts incorporate cool, dry air surrounding the cloud and cause some of the falling precipitation to evaporate. This evaporation further cools the downdraft and limits the updraft of warm humid air. Deprived of moisture, the thunderstorm weakens, precipitation decreases, and the cloud dissipates. Most individual thunderstorms, sometimes called *air mass thunderstorms*, last less than an hour and do little damage.

Severe Thunderstorms The scenario just described is typical of most thunderstorms, which never reach severe levels. However, under the right conditions, these storms can become severe. In the United States, the National Weather Service classifies a thunderstorm as *severe* if it has wind speeds in excess of 93 km (58 mi.) per hour, or hailstones larger than 1.9 cm (0.75 in.), or generates a tornado.[2] Severe thunderstorms, which rely on favorable atmospheric conditions over a large area, are able to self-perpetuate. They often appear in groups and can last from several hours to several days.

Conditions necessary for the formation of a severe thunderstorm include large changes in wind shear, high water-vapor content in the lower troposphere, updraft of air, and the existence of a dry air mass above a moist air mass.[2] Of these four conditions, vertical wind shear is especially important. *Vertical wind shear* is produced by an upward increase in the velocity or change in the horizontal direction of wind (Figure 9.17). In general, the greater the vertical wind shear, the more severe a thunderstorm will become.[8]

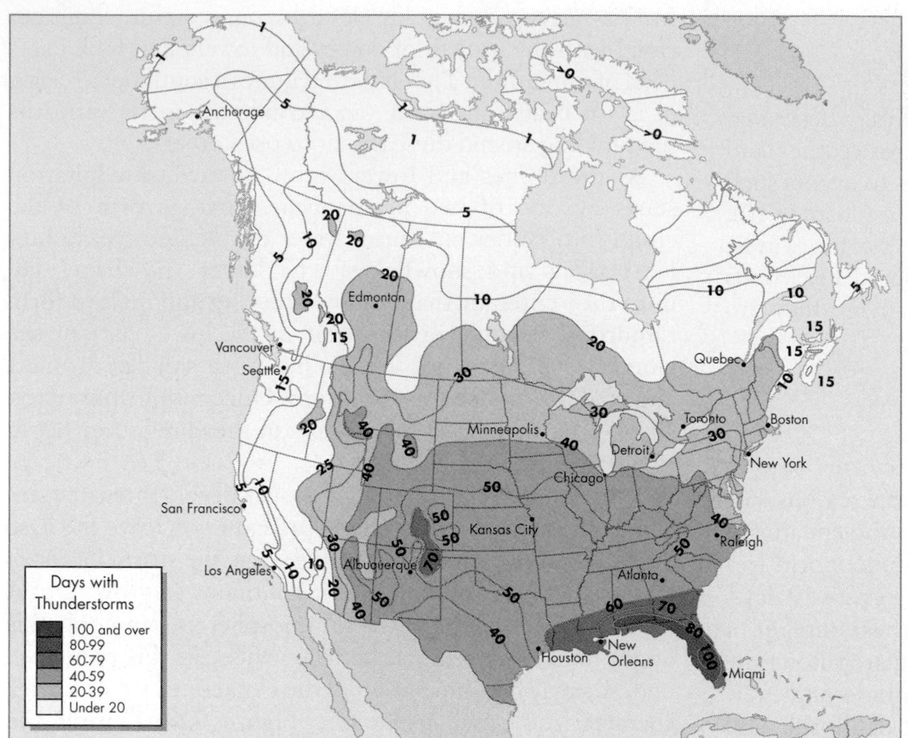

FIGURE 9.15 THUNDERSTORM OCCURRENCE IN NORTH AMERICA Average number of days per year with thunderstorms in the United States and Canada. South-central Florida has the greatest number of days with thunderstorms and Baffin Island in the Canadian Arctic the least. *(From Christopherson, R. W. 2006. Geosystems: An introduction to physical geography, 6th ed. Upper Saddle River, NJ: Prentice Hall, with data courtesy of the National Weather Service and Map Series 3,* Climatic atlas of Canada*)*

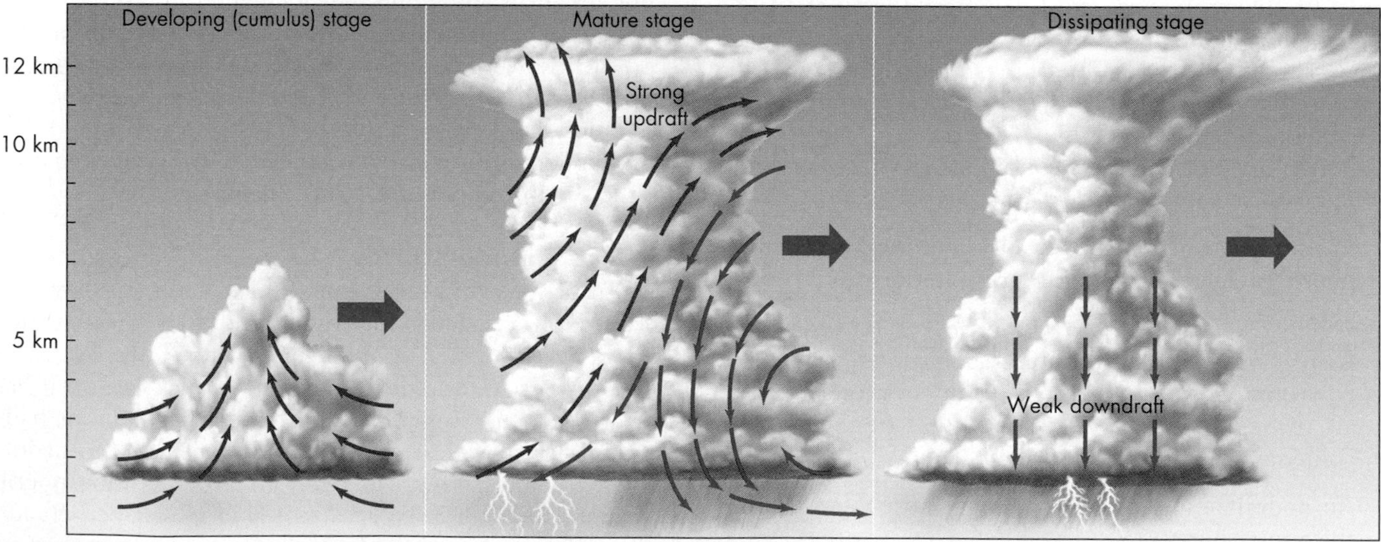

FIGURE 9.16 LIFE CYCLE OF A THUNDERSTORM Idealized diagram showing stages of development and dissipation (disappearance) of a thunderstorm. Red arrows show updrafts of warm air and thin blue arrows show downdrafts of cold air. The large blue arrows show the direction that the storm is moving. *(After Aguado, E., and Burt, J. E. 2007.* Understanding weather and climate, *4th ed. Upper Saddle River, NJ: Pearson Prentice Hall)*

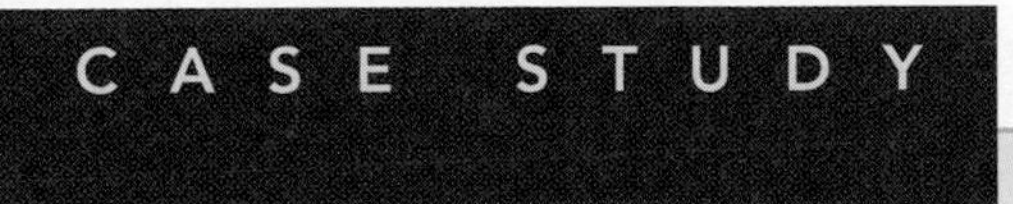

CASE STUDY 9.2

Lightning

A common occurrence during thunderstorms, **lightning,** consists of flashes of light (Figure 9.D) produced by the discharge of millions of joules of electricity. These discharges can heat the air in its path to as high as 30,000°C (54,000°F)—five times hotter than the surface of the sun.[2] This rapid heating causes the surrounding air to expand rapidly producing *thunder.*

Most lightning bolts are cloud-to-cloud; that is, they start and end within the thunderstorm or end in nearby clouds. Although cloud-to-ground lightning is less common, an estimated 25 million lightning bolts strike in the United States each year.[9] Cloud-to-ground lightning is more complex than it appears to an observer (Figure 9.D).

Most lightning comes from cumulonimbus clouds that have grown high enough for ice crystals and pellets to form. Interactions between rising ice crystals and falling ice pellets are thought to create an electrical field strong enough to produce lightning.[10] Electrons moving between the two forms of ice build up positive electrical charge in the upper part of the cloud and negative charge in the lower part (Figure 9.Ea). Since like electrical charges repel, the increased negative charge in the base of the cloud drives away the negative charge on the Earth surface below. This leaves the ground beneath the thunderstorm with a net positive charge. A strong difference in electrical charge between the cloud and the ground is required for lightning to overcome the natural insulating property of air.

The majority of lightning strikes begin as a narrow column of high-speed electrons moving downward from the base of a cloud toward a pocket of positive charge on the Earth. Within milliseconds this column, called a *stepped leader,* branches downward until it is close to the ground (Figure 9.Eb). As the stepped leader approaches, a spark jumps from a tall object, such as a tree or building and attaches to the leader (Figure 9.Ec). The spark completes a conductive channel of ionized air molecules for an upward surge of positive electrical charge. This surge heats the air to create a brilliant *return stroke* of lightning to the cloud (Figure 9.Ed). Within milliseconds, additional leaders and return strokes move along the same path and often cause the lightning flash to appear to flicker to the human eye.

Lightning strikes constitute a serious natural hazard even though the number of annual lightning deaths in the United States has decreased as more people

▲ **FIGURE 9.D CLOUD-TO-GROUND AND CLOUD-TO-CLOUD LIGHTNING** Nighttime photograph of cloud-to-ground lighting strokes in the left half of the photograph and cloud-to-cloud lightning strokes in the upper right part of the photograph. Although most lightning that we see goes from cloud-to-ground or cloud-to-cloud, more than half of all lightning occurs within thunderstorm clouds. *(NOAA)*

(continued)

leave rural areas to live in cities. The chances of being struck by lightning in the United States are estimated to be 1 in 240,000 each year.[11] For an 80-year lifespan, this becomes 1 in 3000.[11] Your risk, however, depends on where you live and work. Nevertheless, lightning is a serious weather-related hazard that kills an estimated 100 people and injures more than 300 per year in the United States and kills 7 people and injures 60 to 70 each year in Canada.[10,11,12] Of those who survive a lightning strike, 70 percent suffer serious long-term health effects (see Survivor Story 9.3).[13]

(a)

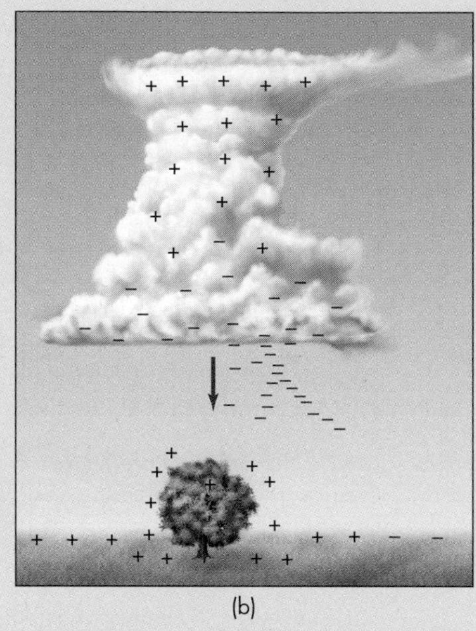
(b)

(c)

(d)

◀ **FIGURE 9.E**
DEVELOPMENT OF CLOUD-TO-GROUND LIGHTNING
(a) Electrical charge separation occurs in a cumulonimbus cloud, most likely by the interaction between rising ice crystals and falling ice pellets. The buildup of a negative electrical charge at the base of the cloud locally induces a positive electrical charge below the thunderstorm. (b) A nearly invisible, stepped leader forms a column of electrically charged air that branches downward toward the ground. (c) When the stepped leader gets close to the ground, a spark jumps from a tall object and attaches to the leader; this completes a narrow channel for the rapid flow of electrons. (d) Flowing electrons heat the channel to extremely high temperature and create a return stroke back to the cloud. The process of downward leader followed by upward return stroke repeats itself multiple times in a single lightning flash.

Three types of severe thunderstorms have been identified on the basis of their organization, shape, and size. They include roughly circular clusters of storm cells called *mesoscale convective systems* (MCSs), linear belts of thunderstorms called *squall lines*, and large cells with single updrafts called *supercells*.

MCSs are the most common of the three types. They are large clusters of self-propagating storms in which the downdraft of one cell leads to the formation of a new cell nearby. Unlike many single-cell, air-mass thunderstorms that last for less than an hour, these complexes of storms can continue to grow and move for periods of 12 hours or more. The downdrafts of these storms can come together to form *outflow boundaries*, curved lines of thunderstorms that may travel long distances.

Squall lines, which average 500 km (310 mi.) in length, are long lines of individual storm cells.[2] These lines commonly develop parallel to cold fronts at a distance of 300 to

9.3

Struck by Lightning: Michael Utley came within inches of death when lightning struck him on a Cape Cod golf course.

Like most of us, Michael Utley didn't worry much about being struck by lightning. The odds, after all, are second to none.

And so, at a charity golf tournament to benefit a local YMCA near his Cape Cod home, he was not overly apprehensive about the looming threat of a thunderstorm.

"I didn't pay attention to it," he said.

Four holes into the game, the warning horn blasted, urging golfers to seek shelter. Utley replaced the flag in the hole and was several yards behind his three companions when he was struck by lightning.

Hearing a thunderous crack behind them, Utley's friends turned around. Utley says they reported smoke coming from his body, his shoes torn from his feet, and his zipper blown open.

Luckily, Utley remembers none of this.

In fact, he remembers nothing from the next 38 days, which he spent in an intensive care unit.

Fortunately for Utley, one of his companions had just recently been retrained in CPR, which may very well have saved Utley's life while the EMS vehicles were on their way.

As for the various images we commonly associate with those struck by lightning, Utley said some are accurate and some are not.

"The hair standing up, that's real. When that happens, you're pretty close to being dead," he said. But "I didn't see a white light at the end of the tunnel. I don't burn clocks off the wall when I touch them."

Utley's reaction, like that of many victims of lightning, was filled, in his own words, "with a variety of ups and downs." His wife, Tamara, recalls that he did not slip into a comalike state for several days.

"It was very, very strange," she said. "When he first got to the ICU, he was able to move and he was very lucid." But within a few days, he began to lose consciousness for extended periods of time. Michael remembers none of this either.

The road to recovery is hardly over for Michael. He still bumps into walls when he walks and has had to relearn many basic activities. He has also spent a great deal of time working to educate the community about the perils of lightning, including working with the PGA on the danger electrical storms pose to golfers (Figure 9.F).

He sums up his message to golfers in one easy slogan: "If you see it, flee it; if you hear it, clear it."

—CHRIS WILSON

▲ **FIGURE 9.F LIGHTNING SAFETY** This National Weather Service poster of professional golfer Vijay Singh is a reminder that more people are killed and injured by lighting strikes in outdoor recreation than in any other activity. *(NOAA)*

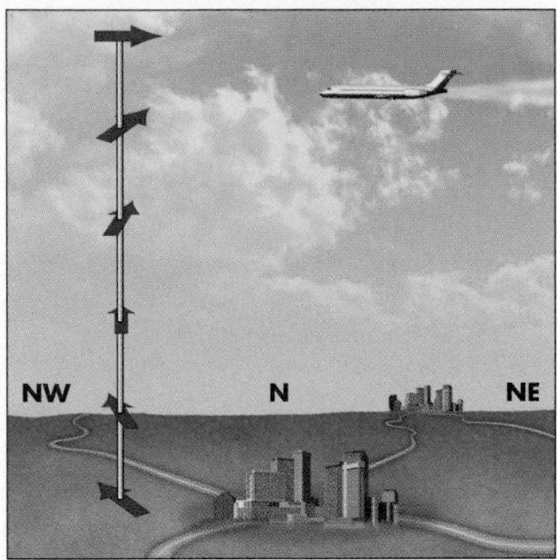

(a)

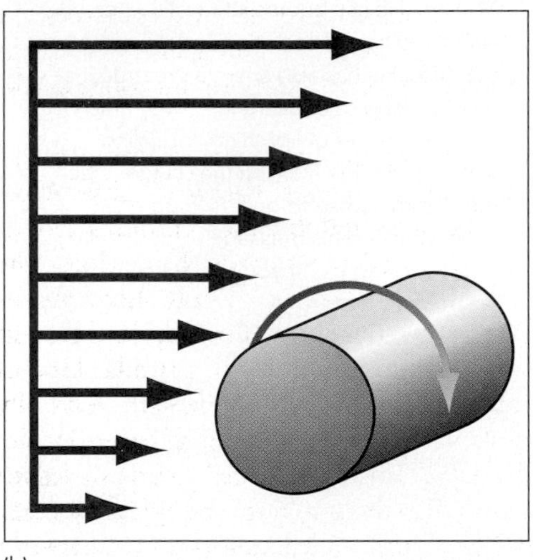

(b)

▲ **FIGURE 9.17 VERTICAL WIND SHEAR** (a) Vertical wind shear resulting from changes in the wind direction with increasing altitude. (b) Vertical wind shear resulting from changes in the velocity of wind with increasing altitude.

500 km (180 to 310 mi.) ahead of the front.[2] Updrafts in the advancing line of storms typically form anvil-shaped clouds whose tops extend ahead of the line. Downdrafts originating on the back side of the storms often surge forward as a *gust front* of cold air in advance of precipitation. Squall lines can also develop along *drylines*, an air mass boundary similar to a front, but in which the air masses differ in moisture content rather than air temperature. Drylines develop in the southwestern United States during the spring and summer, sometimes producing daily squall lines.

The most damaging of all severe thunderstorms is the *supercell storm*. Although smaller than MCSs and squall lines, supercell storms are extremely violent and are the breeding ground for most large tornadoes. They usually range from 20 to 50 km (12 to 30 mi.) in diameter, last from 2 to 4 hours, and are capable of doing significant damage in that time.

Downbursts of air from severe thunderstorms, especially MCSs, can also generate strong, straight-line windstorms. These windstorms vary in size, with the largest ones, called *derechos*, producing severe, tornado-strength wind gusts along a line that is at least 400 km (250 mi.) in length.[9] In general, derecho winds exceed 90 km (57 mi.) per hour and can cause numerous trees to fall, widespread power outages, serious injuries, and multiple fatalities. Strong derechos generate winds approaching 210 km (130 mi.) per hour and produce the same damage as a medium-sized tornado.[15] More than a dozen derechos strike North America each year, with most occurring in the eastern two-thirds of the contiguous United States and southern Canada.[9] Smaller thunderstorm downbursts, called *microbursts*, are more common than derechos and are a serious hazard to aviation.

Hail Although many large thunderstorms can produce hard, rounded, or irregular pieces of ice called *hailstones*, it is large hail from severe thunderstorms that is the greatest hazard. Evidence for the origin of hailstones can be found by cutting a stone in half to reveal a bull's-eye pattern of concentric rings of ice. These rings form as a hailstone is moved up and down in a thunderstorm. Starting with a small ice pellet as a nucleus, a hailstone gets a coating of liquid water in the lower part of the storm, and that coating freezes when a strong updraft carries the stone upward into cold air. This process is repeated many times to form a large piece of hail. The largest authenticated hailstone in North America fell from a severe thunderstorm in Aurora, Nebraska, in June 2003. Nearly as large as a volleyball, the hailstone measured 18 cm (7 in.) in diameter, weighted 0.8 kg (1.7 lbs), and is estimated to have hit the ground at a velocity of more than 160 km (100 mi.) per hour![16]

Hailstorms generally cause more property damage than casualties. The damage caused by hail in the United States alone averages $1 billion per year.[16] In North America, damaging hailstorms are most common in the Great Plains, particularly in northeastern Colorado and southeastern Wyoming, and in the Calgary area of Alberta, Canada. North-central India, Bangladesh, Kenya, and Australia also frequently experience damaging hailstorms. Although rare in North America, deaths from hailstones are not uncommon in Bangladesh and India in heavily populated areas with poorly constructed dwellings.[16]

TORNADOES

Usually spawned by severe thunderstorms, a tornado is one of nature's most violent natural processes. From 1992 to 2002, tornadoes in the United States killed an average of 57 people per year. These spinning columns of wind can take on a variety of shapes, including a rope, funnel, cylinder, and wedge. Other names given to tornadoes include "twisters" and "cyclones," although cyclones also include large spinning columns of wind within clouds, hurricanes, and very large atmospheric low-pressure systems. To be called a tornado, a spinning column of wind or a *vortex* must extend downward from a cloud and touch the ground. In small or newly developed tornadoes, the vortex can be almost invisible until a condensation cloud forms or the tornado picks up dust and debris. Funnel-shaped vortices that have not touched the ground are called *funnel clouds* (Figure 9.18a). Undoubtedly the number of funnel clouds that develop each year far exceeds the number of tornadoes.

Tornadoes form where there are large differences in atmospheric pressure over short distances, as often results during a major storm, such as a supercell thunderstorm. Although meteorologists do not completely understand precisely how tornadoes form, they have recognized that most tornadoes go through similar stages in development.

In the initial *organizational stage*, vertical wind shear causes rotation to develop within the thunderstorm (Figure 9.19a). Strong updrafts developed in front of the advancing cold front and upwelling due to latent heat of condensation tilts the horizontally rotating air vertically (Figure 9.19b). This rotation may be detected on weather radar as part of a large, upward rotating column of air known as a *mesocyclone* (Figure 9.19c). Major updrafts, often in the rear or southwestern part of the storm, lower a portion of the cumulonimbus cloud to form a *wall cloud* (Figure 9.18b). This wall cloud may begin to slowly rotate and a short funnel cloud may descend (Figure 9.18a). A tornado has formed if dust and debris on the ground begin to swirl below the funnel (Figure 9.18c). Not all wall clouds and mesocyclones produce a tornado, and a tornado can develop without the formation of a wall cloud or mesocyclone.

In the second, *mature* stage, a visible condensation funnel extends from the thunderstorm cloud to the ground as moist air is drawn upward (Figure 9.18c). In stronger tornadoes, smaller intense whirls, called *suction vortices*, may form within the larger tornado . The suction vortices orbit the center of the large tornado vortex and appear to be responsible for its greatest damage.[1]

When the supply of warm moist air is reduced, the tornado enters the *shrinking stage*. In this stage, it thins and begins to tilt. As the width of the funnel decreases, the winds can increase, making the tornado more dangerous.

In the final decaying, or *rope stage* (Figure 9.18d), the upward-spiraling air comes in contact with downdrafts and the tornado begins to move erratically. Although this is the beginning of the end for the tornado, it can still be extremely dangerous at this point. Tornadoes may go through all the stages just described or they may skip stages. Like other types of clouds, new tornadoes can form nearby as older tornadoes disappear.

As tornadoes move along the surface of Earth, they pick up dirt and debris. This debris gives the tornado cloud its characteristic dark color (Figure 9.18d). Tornadoes typically have diameters measured in tens of meters and wind speeds of 65 km (40 mi.) to more than 450 km (280 mi.) per hour.[2] Once they touch down, tornadoes usually travel 6 to 8 km (4 to 5 mi.) and last only a few minutes before weakening and disappearing. However, as we discussed in the case history that opened this chapter, the largest, most damaging tornadoes may move at speeds of close to 100 km (60 mi.) per hour along a path several hundreds of kilometers long.

Classification of Tornadoes Tornadoes are classified by the most intense damage that they have produced along their path. Each tornado can be assigned a value on the *Enhanced Fujita* or **EF Scale** (Table 9.1) based on a post-storm damage survey. This survey determines the levels of damage experienced by 26 types of buildings, towers, and poles and by hardwood and softwood trees. Values on the EF Scale estimate the maximum 3-second wind gust in the tornado and range from 105 to 137 km (65 to 85 mi.) per hour for an EF0 tornado, to more than 322 km (200 mi.) per hour for an EF5 tornado.[17] The EF Scale replaces the F-scale developed by T. Theodore Fujita in 1971 and is similar to the Modified Mercalli Scale for assessing earthquake intensity.

Tornadoes that form over water are generally called *waterspouts* rather than tornadoes (Figure 9.20). Most *waterspouts* develop beneath fair-weather cumulus clouds and appear to be associated with wind shear along the boundary of contrasting air masses.[1] Wind shear means that the winds along a boundary are blowing in different directions and velocities. You can produce a similar kind of shear-induced rotation by holding a pencil vertically between your hands and moving your hands in opposite directions. Small, wind-shear–produced tornadoes can develop on land and, like waterspouts, are typically weak EF1 or EF0 tornadoes.[1] Although not as powerful as most tornadoes in supercell storms, waterspouts and related weak tornadoes on land can cause considerable damage and should be respected.

Occurrence of Tornadoes Although tornadoes are found throughout the world, they are much more common in the United States than in any other location on Earth. The United States has just the right combination of weather, topography, and geographic location to make it the perfect spawning ground for tornadoes.[1] Most U.S. tornadoes occur in the Plains states, between the Rocky Mountains and the Appalachians. As with severe thunderstorms, spring and summer are the most common time for tornadoes, and most develop in the late afternoon and evening. The highest risk for tornadoes occurs along what is called "Tornado Alley," a belt that stretches from north to south through the central United States (Figure 9.21). This is where all the ingredients for the formation of a tornado collide. Areas of high annual tornado occurrence include a huge region from Florida to Texas and north to the Dakotas, Indiana, and Ohio. Although Canada experiences far fewer tornadoes than the United States, it has several tornado-prone regions, such as Alberta, southern Ontario, and southeastern Quebec, as well as an area running from

▲ FIGURE 9.18 **STAGES IN TORNADO DEVELOPMENT** (a) Airborne funnel clouds, like this one over the state capitol in Austin, Texas, are not considered tornadoes until they touch the ground. *(Austin Public Library)* (b) This wall cloud near Miami, Texas, is the lighter gray cloud to the left of the lightning bolt. Wall clouds hang downward from a severe thunderstorm and are often where tornadoes form *(NOAA)* (c) A tornado in the mature stage of development near Seymour, Texas. *(NOAA)* (d) A tornado in the decaying rope stage of development near Cordell, Oklahoma. A tornado can still be dangerous at this stage. *(NOAA)*

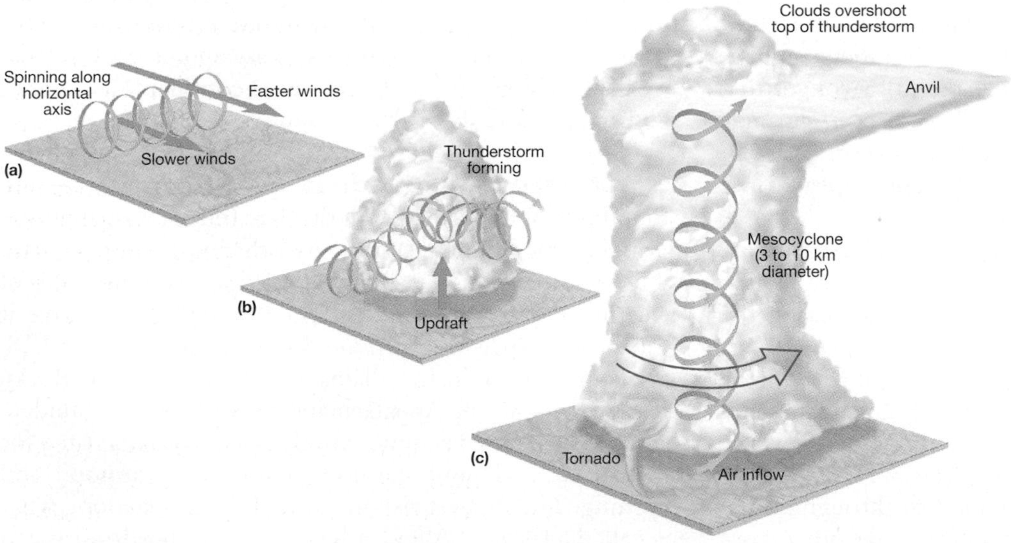

▲ FIGURE 9.19 **TORNADO FORMATION** (a) Vertical wind shear creates a horizontal axis rotation or rolling of air. (b) Updrafts produced by rising warm air lift the horizontally rolling air toward the vertical. (c) Development of a vertical rotating mesocyclone is developed. If conditions are right, a tornado will develop from the slowly rotating wall cloud toward the rear of the storm *(After McKnight, T. L., and Hess, D. 2004.* Physical Geography, *10th ed. Upper Saddle River, NJ: Pearson Prentice Hall)*

TABLE 9.1

Enhanced Fujita Scale of Tornado Intensity

Scale	Wind Estimate[a]	Typical Damage[b]
EF0	105–137 km/hr (65–85 mph)	*Light damage.* Some damage to chimneys; branches broken off trees; shallow-rooted trees pushed over; sign boards damaged.
EF1	138–177 km/hr (86–110 mph)	*Moderate damage.* Peels surface off roofs; mobile homes pushed off foundations or overturned; moving autos blown off roads.
EF2	178–218 km/hr (111–135 mph)	*Considerable damage.* Roofs torn off frame houses; mobile homes demolished; boxcars overturned; large trees snapped or uprooted; light-object missiles generated; cars lifted off ground.
EF3	219–266 km/hr (136–165 mph)	*Severe damage.* Roofs and some walls torn off well-constructed houses; trains overturned; most trees in forest uprooted; heavy cars lifted off the ground and thrown.
EF4	267–322 km/hr (166–200 mph)	*Devastating damage.* Well-constructed houses leveled; structures with weak foundations blown away some distance; cars thrown and large missiles generated.
EF5	More than 322 km/hr (More than 200 mph)	*Incredible damage.* Strong frame houses leveled off foundations and swept away; automobile-sized missiles fly through the air in excess of 100 m (110 yds.); trees debarked; incredible phenomena will occur.

[a] Wind speeds are the maximum estimated for a 3-second gust.

[b] Accurate placement on this scale involves expert assessment of the degree of damage to 28 indicators including homes, buildings, towers, poles, and trees.

Source: Wind Science Engineering Center. 2006. A recommendation for an Enhanced Fugita Scale (EF-Scale)*. Lubbock, TX: Texas Tech University. www.wind.ttu.edu/EFScale.pdf.Accessed1/16/06.*

▲ FIGURE 9.20 **WATERSPOUTS** Two waterspouts over Lake Winnipeg, Manitoba, Canada. Most waterspouts are relatively weak tornadoes that form over the ocean or large lakes. In North America, they are most common in southern Florida. *(Dr. Scott Norquay/Tom Stack & Associates, Inc.)*

southern Saskatchewan and Manitoba to Thunder Bay, Ontario.[18]

Possibly with the exception of Bangladesh, violent tornadoes (EF4 or EF5) are rare or nonexistent outside of the United States and Canada.[1] Waterspouts are both less hazardous and more common than tornadoes on land. Most take place in tropical and subtropical waters, but they have been reported off the New England and California coasts as well as on the Great Lakes.[1] They are especially common along the Gulf Coast, Caribbean Sea, Bay of Bengal in the Indian Ocean, and the South Atlantic. A study conducted in the Florida Keys counted 390 waterspouts within 80 km (50 mi.) of Key West during a 5-month period.[1]

BLIZZARDS AND ICE STORMS

Blizzards are severe winter storms in which large amounts of falling or blowing snow are driven by high winds to create low visibilities for an extended period of time. Extremely low visibility, commonly referred to as a *whiteout*, cold temperatures, and high winds make a blizzard dangerous. Official thresholds for blizzard conditions differ in the United States and Canada. In the United States, winds must exceed 56 km (35 mi.) per hour with visibilities of less than 0.4 km (0.25 mi.) for at least 3 hours, whereas in Canada winds must exceed 40 km (25 mi.) per

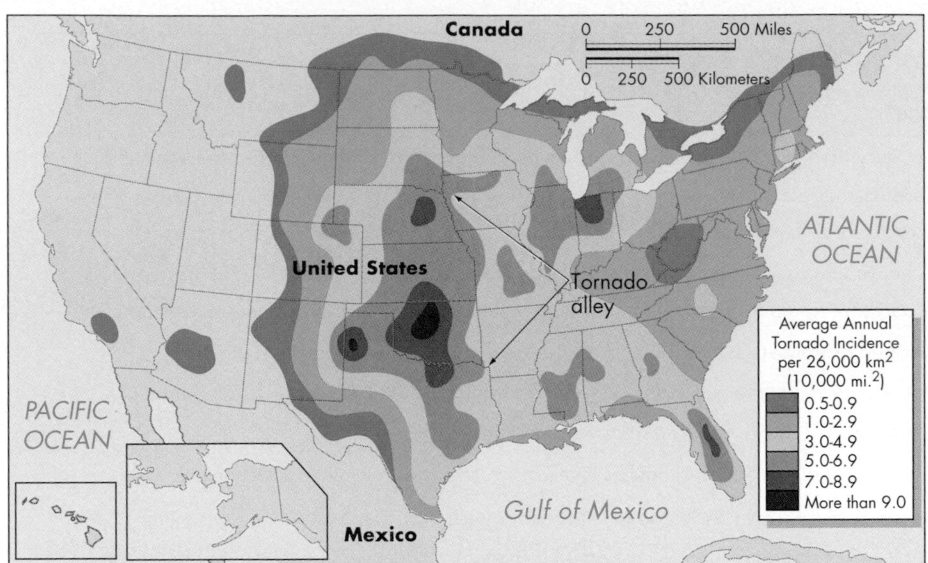

FIGURE 9.21 TORNADO OCCURRENCE IN THE UNITED STATES Average occurrence of tornadoes per 26,000 km^2 (10,000 mi.^{2}) in the United States and southernmost Canada from 1950 to 2000. *(After Christopherson, R. W. 2006. Geosystems: An introduction to physical geography, 6th ed. Upper Saddle River, NJ: Pearson Prentice Hall. Data from National Severe Storm Forecast Center, National Weather Service)*

hour with visibilities of less than 1 km (1.6 mi.) for at least 4 hours.[19,20] In either case, a blizzard may involve no additional snowfall. High winds picking up previously fallen snow create a *ground blizzard* in which visibility can drop to a few meters or less for many hours. Ground blizzards develop numerous times each winter in Antarctica, Alaska, Canada, and on the Great Plains.

In North America, blizzards associated with heavy snowfall are most common in the Great Plains, the Great Lakes area, the U.S. Northeast, the Canadian Atlantic provinces, and the eastern Arctic. Storms producing heavy snowfall and blizzards form from an interaction between upper-level winds associated with a low-pressure trough and a surface low-pressure system.[21] In North America, blizzard-producing storms generally form in eastern Colorado; central Alberta, Canada; or along the coast of North Carolina or the northern Gulf of Mexico (Figure 9.22). Storms in Colorado and along the coast that produce heavy snowfall are derived from moist air originating over the Atlantic Ocean. In contrast, fast-moving storms forming east of the Canadian Rockies, called *Alberta Clippers*, are generally drier with less snow and extremely cold temperatures.

Blizzards can produce lengthy periods of heavy snowfall, wind damage, and large snowdrifts. For example, the Saskatchewan blizzard of 1947 lasted for 10 days and buried an entire train in a 1-km (0.6-mi.) long snowdrift that was 8 m (25 ft.) high.[22] Another famous storm, the "Blizzard of 1888," killed more than 400 people and paralyzed the northeastern United States for 3 days with snow drifts that reportedly were as high as the second floor of buildings.

Storms that move along the east coast of the United States and Canada, called *nor'easters*, can produce blizzards (see Chapter 10). Nor'easters wreak havoc with hurricane-force winds, heavy snows, intense precipitation, and high waves that damage coastal areas. These storms are most common between September and April and often create blizzard conditions in large cities such as New York and Boston. In March 1993, a severe nor'easter paralyzed the East Coast, causing snow, tornadoes, and flooding from Alabama to Maine. Damages from this storm, known as the "Blizzard of '93," were more than a billion dollars, and more than 240 people were killed.

Less than three years later, in January 1996, a strong winter storm brought another massive blizzard to the East Coast. The storm crippled the eastern United States for several days, produced record-breaking snowfall in Philadelphia and parts of New Jersey, and dropped 51 cm (20 in.) of snow in New York City's Central Park (Figure 9.23). In all, the blizzard killed at least 100 people and caused an estimated $2 billion in damages.

One reason that blizzards are generally more dangerous than other snowstorms is the **wind chill** effect. Moving air rapidly cools exposed skin by evaporating moisture and removing warm air from next to the body. This chilling reduces the time it takes for *frostbite* to form. Thus in blizzards the lower wind chill temperature is more important than the air temperature.

Ice storms, prolonged periods of freezing rain, can be more damaging than, and just as dangerous as, blizzards. Ice accumulates on all cold surfaces and is especially harmful to utility lines and trees and to surface travel (see Case Study 9.5). These storms typically develop during the winter in a belt on the north side of a stationary or warm front. In this setting, a combination of three conditions leads to freezing rain: (1) an ample source of moisture in the warm air mass south of the front, (2) warm air uplifted over a shallow layer of cold air, and (3) objects on the land surface at or very close to freezing. Under these conditions, snow begins to fall from the cooled top of the warm air mass. The snow melts as it passes through the warm air and the resulting raindrops become supercooled when they hit the cold air at the surface (Figure 9.24). Upon contact with cold objects, such as roads, trees, and utility lines, the rain immediately freezes to form a coating of ice. Coatings of

PROFESSIONAL PROFILE 9.4

Sarah Tessendorf–Severe Storm Meteorologist

Sarah Tessendorf (Figure 9.G) has been a storm watcher for as long as she can remember. "My favorite thing as a child was being woken up in the middle of the night by a really bright flash of lightning. I was scared to death of it—but it got my blood running every time," she said.

Tessendorf's fascination with extreme weather has never faltered. While still an undergraduate, she began studying tornadoes at SOARS (Significant Opportunities in Atmospheric Research and Science), a summer internship program at the National Center for Atmospheric Research (NCAR) in Boulder, Colorado.

Working with Jeffrey Trapp of Purdue University, Tessendorf studied two types of storms known to trigger tornadoes. Supercells—severe thunderstorms containing rotating updrafts—were considered responsible for the most severe tornadoes. Squall lines—storms that travel in one long line—were thought to spawn only weak twisters. Using radar data, Tessendorf and Trapp discovered that squall lines birth up to 20 percent of tornadoes, some of which can be violent. Their results have sent atmospheric scientists scrambling to learn how to spot squall line tornadoes in time to warn the public.

"It was exciting to have my first project lay the groundwork that there is a threat from this type of storm," Tessendorf said.

Building on this expertise, Tessendorf went on to study lightning for her Ph.D. The high plains of Colorado are struck by unusual, positive cloud-to-ground lightning. To find out why, dozens of scientists from the federal government, universities, and private industry combined forces in 2000 to observe high plains storms with three radars, instrumented balloons, even a chase plane. "We wanted to figure out what made these storms unique," Tessendorf said.

Tessendorf and her colleagues confirmed that these odd storms have what is known as inverted charge structures—a midlevel positive layer of charge sandwiched by negative charge layers. She then hypothesized how such a charge structure could develop. Scientists believe that storms become electrified after small ice crystals collide and transfer charge with small hail-like particles. Updrafts separate the charge by carrying the small crystals skyward, leaving the graupel at lower elevations. Tessendorf found that unusually strong and large updrafts were common features of Colorado storms with inverted charge structures and could aid in altering the charge transfer process.

Now a postdoctoral researcher at the University of Colorado at Boulder, Tessendorf uses computer models to study precipitation processes in storms. In her spare time, she enjoys storm chasing. "Out in front of the storm I can feel warm humid air being sucked into the clouds. Then it suddenly gets really still, and cool wind flows outward from the storm. That's the key to get moving again," Tessendorf said. "To feel and experience that calm is both eerie and amazing at the same time."

Tessendorf hopes to continue working in meteorology while sharing her love of storms with others. "The weather is something we all experience in common. I'd like to use it to help convey the excitement of science to the public," she said.

— KATHLEEN WONG

▲ **FIGURE 9.G DR. SARAH TESSENDORF—SEVERE STORM METEOROLOGIST** Dr. Tessendorf's smile comes from a successful day of storm observations on the High Plains in eastern Colorado. *(Brenda Cabell/Sarah Tessendorf)*

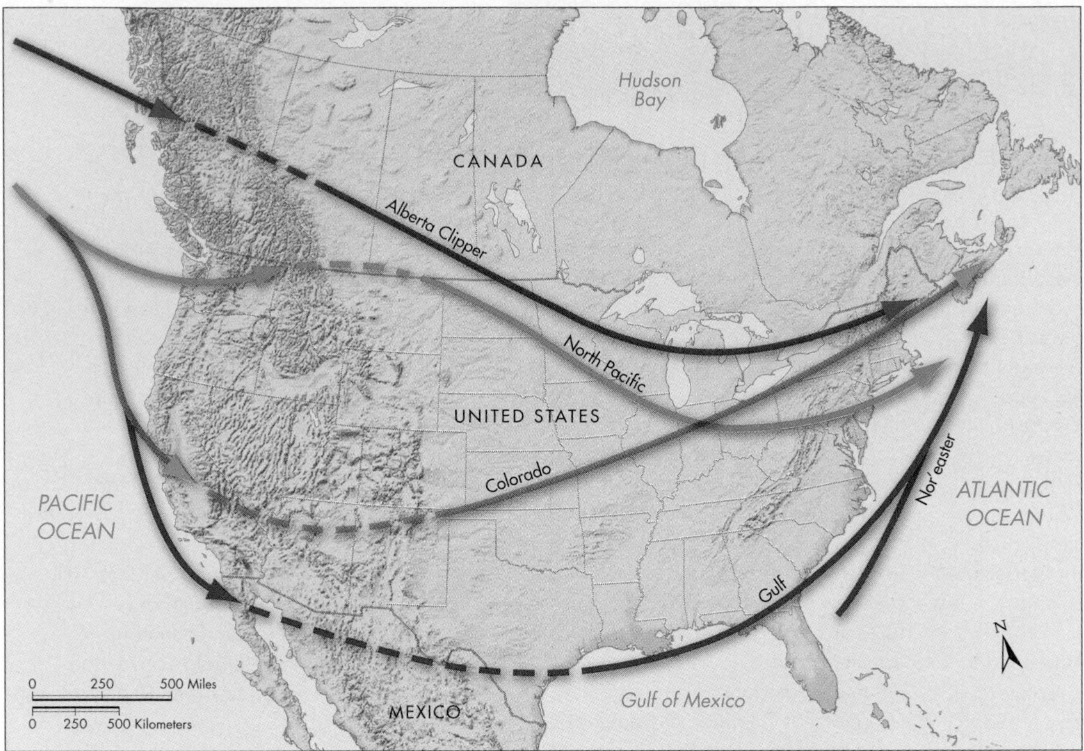

▲ FIGURE 9.22 **WINTER STORM TRACK IN THE UNITED STATES AND SOUTHERN CANADA** In winter months, low-pressure centers and associated storms develop over the northern Pacific Ocean and come onshore along the west coast of the United States, Canada, and Mexico. The upper-level remnants of these storms (dashed lines) cross the western mountains and low-pressure centers and associated storms redevelop on the Great Plains. Low-pressure centers and storms also develop along the Gulf and Carolina coasts. Storms tracks are named for geographic origin (e.g., Alberta Clippers, Colorado) or the direction of their prevailing high winds (e.g., Nor'easter). *(After Keen, R. A. 2004. Skywatch west: The complete weather guide, revised ed. Golden, CO: Fulcrum Publishing)*

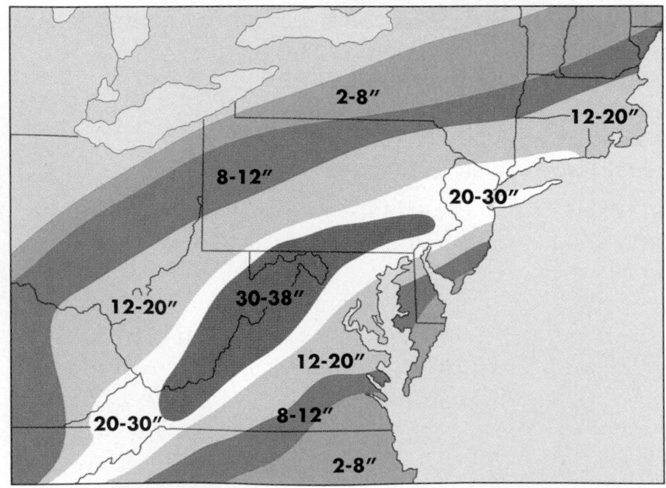

▲ FIGURE 9.23 **SNOWFALL FROM THE BLIZZARD OF 1996** The greatest amount of snow in this storm fell along the Appalachian Mountains. Blizzard-force winds created snow drifts much deeper than the snow depths indicated on this map.

15 to 20 cm (6 to 8 in.) of ice have been produced by prolonged ice storms in Idaho, Texas, and New York. The regions most prone to ice storms include the Columbia River Valley in the Pacific Northwest, the south-central Great Plains, the Ohio River Valley, the Mid-Atlantic and New England states, and the Atlantic provinces and St. Lawrence River Valley in eastern Canada.[16]

FOG

Like many other natural phenomena, there is nothing innately hazardous about **fog;** it is simply a cloud in contact with the ground. Fog can become hazardous when it obscures visibility for travel and other human activities, and when air pollutants are added to form smog or vog (see Chapter 3).

Clouds, and thus fog, form either by air cooling until condensation occurs or by evaporation adding water vapor to already cool air. Cooling can produce fog as heat radiates from the land at night, as warm moist air blows over cold coastal water, and as humid air rises up a mountainside. Fog can also

▲ FIGURE 9.24 **THICKNESS OF COLD AIR DETERMINES TYPE OF PRECIPITATION** In winter storms, the thickness of cold air at the surface determines the type of precipitation. (a) Most precipitation begins falling as snowflakes and remains as snow if it does not pass through warm air before reaching the ground. (b) Falling snowflakes melt if they encounter warm air aloft, and the resulting rain refreezes to form ice pellets (sleet) if there is sufficient time and cold air before they reach the ground. (c) If there is only a thin layer of cold air at the surface, raindrops become supercooled and then freeze immediately upon contact with a cold surface such as a tree limb, house, or power line. (d) Snowflakes will melt and rain will fall if a thick warm air layer is at the surface. *(After Linda Scott/*Austin American-Statesman*)*

develop by evaporation when cold air flows over river, lake, or coastal water that is warmer than the air, and where warm rain falls through cool air along a frontal boundary.

In North America, the foggiest areas are the Pacific, New England, and Atlantic Canada coasts, and the valleys and hills of the Appalachians. Fog contributed to the worst aircraft accident ever, when 583 people died in the 1977 collision of two Boeing 747 airliners on a runway in the Canary Islands.[31] Dense fog was also responsible for the worst maritime accident in Canadian history, when 1014 people died in the 1914 collision of two ships on the St. Lawrence River.[32] Although technological advancements make collisions in the fog less likely, dense fog still contributes to numerous accidents, injuries, and deaths each year.

DROUGHT

Drought is defined as an extended period of unusually low precipitation that produces a temporary shortage of water for people, other animals, and plants. More than 1 billion people live in semiarid regions where droughts are common, and more than 100 million people are threatened with malnutrition or death if drought causes their crops to fail. Droughts commonly contribute to regional food shortages, but today the worldwide food distribution system is usually able to prevent drought from causing widespread famine.

In the United States, drought affects more people than any other natural hazard and causes losses of $6 to 8 billion each year.[33] Drought in the United States continues to cause serious water and power shortages as well as agricultural problems. For example, a drought in 1977 brought crop failures, domestic water shortages, and a decrease in industrial productivity from a loss of hydroelectric power.[31] In mid-2004, drought conditions in the western United States were again causing water shortages and agricultural problems (Figure 9.25). After more than 5 years of low precipitation, water levels in major reservoirs from Montana to Arizona and New Mexico were at their lowest points in decades.

MOUNTAIN WINDSTORMS

Destructive windstorms develop seasonally on the downwind or lee side of mountain ranges and downslope of large glacial ice fields. Mountains act as barriers to prevailing winds and can, under specific temperature and pressure conditions, cause winds to roar down their slopes at speeds that can exceed 185 km (115 mi.) per hour. Referred to as Chinooks east of the Rocky Mountains and Santa Ana winds west of the Transverse Ranges in Southern California, these windstorms also occur downwind of the Sierra Nevada, Cascades, Alps, and other mountain ranges. Mountain windstorms sometimes persist for a day or more, cause widespread roof and tree damage, can blow vehicles off highways, and contribute to large wildfires (see Chapter 13).

DUST AND SAND STORMS

Dust storms are strong windstorms in which suspended dust that is carried by the wind reduces visibility for a significant period of time. Wind velocities in these storms exceed 48 km (30 mi.) per hour and visibility is reduced to less than 0.8 km (0.5 mi.).[5] A typical dust storm is several hundred kilometers in diameter and may carry more than 100 million tons of dust. Most natural dust particles are pieces of minerals less than 0.05 mm (0.002 in.) in diameter. Natural dust also contains minor amounts of biological particles, such as spores and pollen. In addition to being a safety hazard for travel, airborne dust particles can affect climate and human health. Once suspended in the air, fine dust particles less than 0.01 mm (0.0004 in.) can travel long distances in the upper atmosphere. Satellite images show dust storms from West Africa crossing the Atlantic Ocean to Florida. It

CASE STUDY 9.5

The Great Northeastern Ice Storm of 1998

One of the most destructive and long-lasting ice storms in the history of Canada and the northeastern United States struck U.S./Canadian border states and provinces in January 1998. In most areas, the storm started late on January 4 and continued for 6 days. As least 45 people died in the United States and Canada as electric power was cut off to more than 5.2 million people.[23] The power loss contributed to most deaths, primarily from the loss of heat for an extended time. Causes of death ranged from carbon monoxide poisoning produced by improperly vented heat sources to hypothermia from cold temperatures.[23]

Although workers from 14 power companies and military personnel worked long hours to restore power, more than 700,000 people were still without power in the third week after the onset of the storm.[24] The provinces of Quebec, Ontario, and New Brunswick and the states of New York and Maine were the worst hit, although damage was also considerable in Vermont, New Hampshire, and Nova Scotia (Figure 9.H). In addition to substantial damage to the power and telecommunications system, forestry, dairy farming, maple syrup production, and homeowners were severely affected by the storm. Overall losses exceeded $6.2 billion, with more than 80 percent of the losses occuring in Canada.[25] It was Canada's costliest disaster in history.

◀ **FIGURE 9.H ICE GLAZES POWER LINES AND TREES** Freezing rain turns to clear ice once it falls on a surface that is at or below freezing. Many tree limbs and utility lines, like these in the January 1998 ice storm in Watertown, New York, bend and break with the weight of accumulated ice. *(Syracuse Newspaper/The Image Works)*

Meteorological Conditions

In many ways, the meteorological conditions that caused the Ice Storm of 1998 were typical for ice storms. A persistent flow of warm moist air from the Gulf of Mexico (Figure 9.I) rose up over a thin wedge of cold air to the north of a stationary front. Cold air was driven south by northeasterly winds circulating around an Arctic high-pressure center over Hudson Bay. Much of the cold air settled in river and mountain valleys, including the St. Lawrence and Ottawa River Valleys and the Lake Champlain Valley. Warm air was driven northward by a high-pressure center over Bermuda and a low-pressure trough over the lower Mississippi Valley.[26]

Weather conditions in the region were essentially unchanged for almost a week. The flow of moist air from the Gulf of Mexico acted like a conveyor belt to constantly deliver water vapor for precipitation both north and south of the stationary front. South of the front, heavy rains caused severe flooding in the mountains of North Carolina and Tennessee, killing nine people.[27] North of the front, most of the precipitation took the form of freezing rain with total liquid-equivalent accumulations of 20 to 100 mm (0.8 to 4 in.).[24] Ice accumulations of more than 250 mm (1 in.) were reported from the freezing rain at stations in Canada and the United States.[24,28]

Lessons Learned

Why was this storm so catastrophic? As with most other modern catastrophes, both natural processes and humans contributed to the losses. For many Canadian communities, including Ottawa and Montreal, and some American towns, such as Burlington, Vermont, and Massena, New York, the Ice Storm of 1998 was the most prolonged period of freezing precipitation on record.[24,26] Historically most episodes of freezing rain in Quebec, Ontario, and the northeastern United States last only a few hours at a time; however, the 1998 Ice Storm brought a total of more than 80 hours of freezing rain and drizzle over a 6-day period, twice the annual average.[26]

On the human side, the ice storm demonstrated how dependent our society has become on electricity, especially the electrical grid. The storm destroyed more than 1000 high-voltage towers, 35,000 telephone poles, and more than 120,000 km (75,000 mi.) of power lines (Figure 9.J).[23] Montreal lost its water supply after treatment plants and pumping stations experienced a power failure on January 9. In many places in Canada and the United States, emergency backup electrical generators were either absent, failed, or eventually ran out of fuel.

Also apparent was an increasing reliance on "just-in-time" delivery for many items. It became difficult, in some cases, to repair the electrical grid when needed supplies, such as telephone poles, had to be shipped long distances when transportation systems were shut down. This was repeated recently in

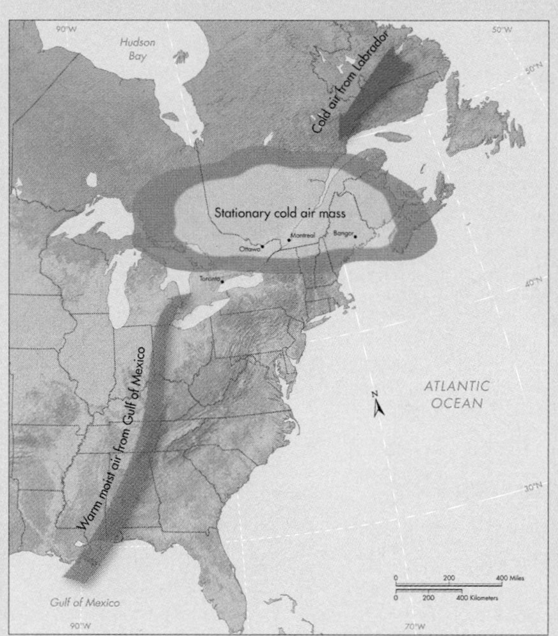

◀ **FIGURE 9.I WEATHER CONDITIONS THAT CAUSED THE ICE STORM OF 1998** A conveyor-belt–like flow of warm, moist air from the Gulf of Mexico overrode a stationary cold air mass in Quebec, eastern Ontario, Nova Scotia, Prince Edward Island, and the northeastern United States to cause the Ice Storm of 1998. This system remained in place for nearly a week, causing the worst disaster in Canadian history. *(After Environment Canada. http://www.weatheroffice.gc.ca/canada_e.html)*

◀ **FIGURE 9.J TRANSMISSION TOWERS COLLAPSE UNDER WEIGHT OF ICE** A series of Hydro-Quebec high-voltage power transmission towers collapsed under the weight of ice accumulations during the January 1998 ice storm near Bruno, Quebec, south of Montreal. More than 5.2 million people lost power in the ice storm and some power outages lasted more than 3 weeks. *(AP Images)*

Austin, Texas, when an ice storm depleted deicing fluid for several airlines at the city's airport, resulting in the cancellation of more than 100 flights.[29] The airline's reliance on ground delivery, rather than adequate on-site storage, caused the transportation disruption.

Sometimes the lessons learned from a disaster or catastrophe are not conclusions drawn from scientific observations, technological malfunctions, or policy failures. They are simple observations about human interactions. For the Ice Storm of 1998, several simple lessons were summarized in responses from Maine schoolchildren to the question "What did you learn from the Ice Storm of '98?"[30]

> "I learned that we take things like power, heat, and water for granted." (student at Readfield Elementary School)
>
> "People must have been together as families a lot more before there was electricity." (13-year-old student at South Meadow School)
>
> "I learned that if you are in a big storm it helps if you help others." (12-year-old student at Woodside Elementary School)
>
> "I learned that ice can cripple a whole state." (12-year-old student at Woodside Elementary School)

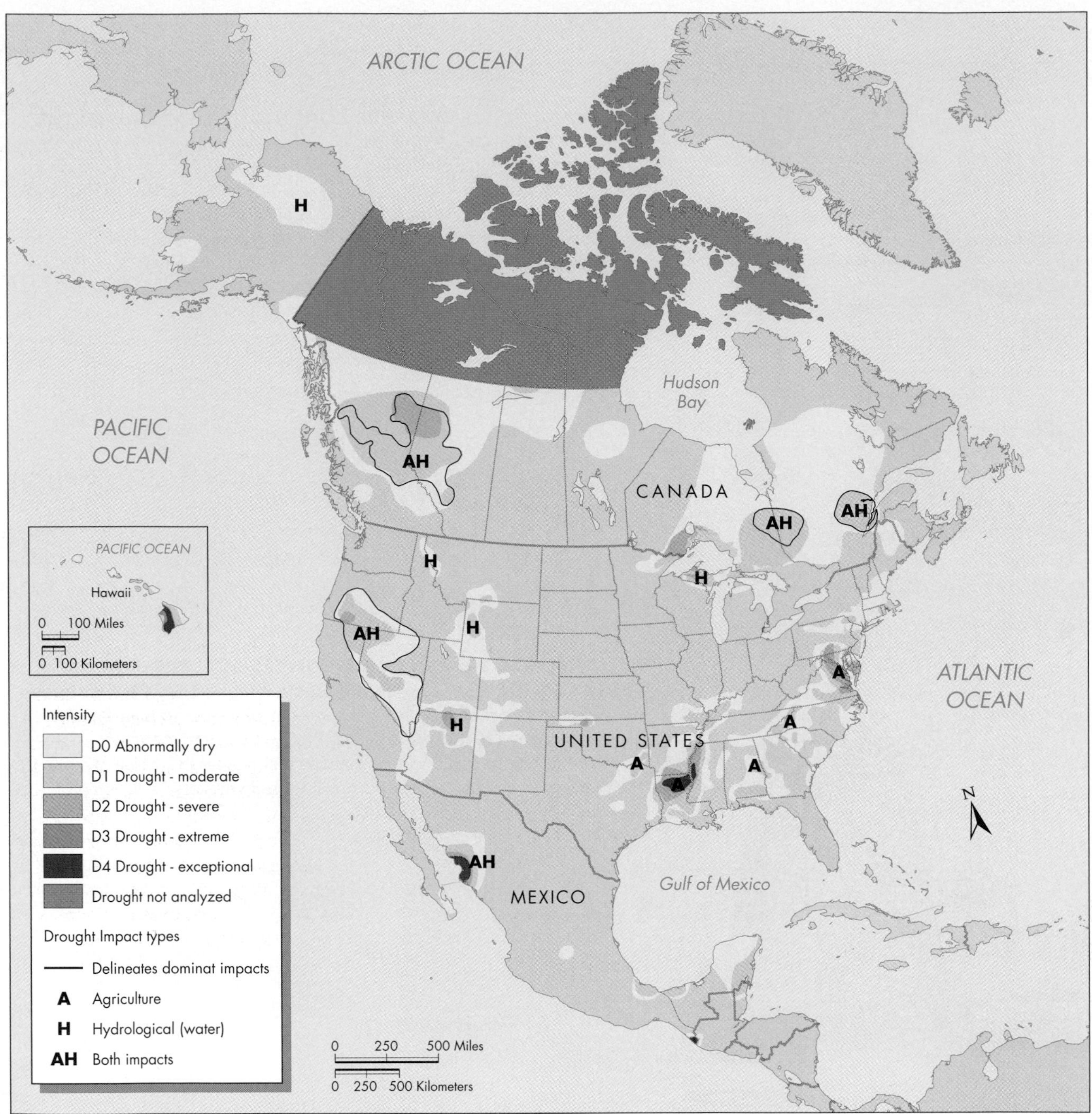

▲ **FIGURE 9.25 NORTH AMERICAN DROUGHT CONDITIONS AND EFFECTS** Intensity and effects of drought in the United States, Canada, and Mexico as of October 2004. Record-setting rainfall and snowfall in the late fall of 2004 began to reduce the intensity and effects of the 5-year drought in the western United States. *(Courtesy NOAA National Climate Data Center)*

is not uncommon to hear television weather reporters mistakenly call these events "sandstorms."

Unlike dust storms, *sandstorms* are almost exclusively a desert phenomenon in which sand is transported in a cloud of bouncing grains that rarely extends more that 2 m (6.5 ft.) above the land surface.[34] Blowing sand is abrasive. One of your authors learned this firsthand when his windshield was frosted by blowing sand while driving in Colorado.

Dust storms and sandstorms occur mostly in midlatitude, semiarid, and arid regions. In the United States, huge dust storms in the southern High Plains during the 1930s produced conditions known as the "Dust Bowl." A combination of drought and poor agricultural practices during the Great Depression caused severe soil erosion in parts of five states centered on the Oklahoma panhandle. Frequent, sometimes daily, dust storms originated in this area and destroyed crops and pastureland (Figure 9.26).

HEATWAVES

All of North America, and much of the world, is vulnerable to the effects of heatwaves. In most areas, **heatwaves** are considered to be prolonged periods of extreme heat that are both longer and hotter than normal. From 1992 to 2002, heatwaves in the United States killed an average of 200 people per year, which is about equivalent to deaths from flooding, lightning, tornadoes, and hurricanes combined during the same 10-year period. Heatwaves can be especially deadly in urban areas in cooler climates where residents are not acclimated to periods of prolonged heat and humidity. This was apparent in heatwaves in 1995, when there were more 500 heat-related deaths in the Chicago area, and in August 2003, when there were an estimated 14,800 heat-related deaths in France (see Case Study 9.6).[35]

Heatwaves in the eastern United States and Canada are commonly associated with elongated areas of high pressure, called *ridges*. Wet conditions are generally found to the west of the ridge, whereas sunny dry conditions prevail to the east. If such a ridge stays in place for several days, air temperatures below the ridge will rise to above-normal levels and cause a heatwave.

Heatwaves can accompany either severe humidity or extreme dryness. In either case, it is important to monitor the **heat index** (Figure 9.27). This index measures the body's perception of air temperature, which is greatly influenced by humidity. For example, a temperature of 35°C (95°F) will feel significantly hotter in parts of Florida when the relative humidity is 75 percent. In this example, the combination of high temperature and high humidity produces a heat index of 128, which is extremely dangerous. In addition to humidity, the length of time that one is exposed to direct sunlight, the wind speed, and an individual's health will affect heat stress.

9.6 Human Interaction with Weather

Many natural hazards are clearly and significantly altered by human activities. For example, we have discussed how changes in land use affect flooding

(a)

(b)

FIGURE 9.26 U.S. "DUST BOWL" (a) Dust storm approaching Elkhart, Kansas, on May 21, 1937. Sometimes called "black blizzards" because of their color, these storms eroded topsoil from cropland in the southern High Plains. *(Getty Images)* (b) Farmer Arthur Coble and sons walking into the face of a dust storm in Cimmaron County, Oklahoma, in April 1936. The house shown here is partially buried in dust. *(AP Images)*

Air Temperature (°F)	Relative Humidity (%) 40	45	50	55	60	65	70	75	80	85	90	95	100
110	136												
108	130	137											
106	124	130	137										
104	119	124	131	137									
102	114	119	124	130	137								
100	109	114	118	124	129	136							
98	105	109	113	117	123	128	134						
96	101	104	108	112	116	121	126	132					
94	97	100	102	106	110	114	119	124	129	135			
92	94	96	99	101	105	108	112	116	121	126	131		
90	91	93	95	97	100	103	106	109	113	117	122	127	132
88	88	89	91	93	95	98	100	103	106	110	113	117	121
86	85	87	88	89	91	93	95	97	100	102	105	108	112
84	83	84	85	86	88	89	90	92	94	96	98	100	103
82	81	82	83	84	84	85	86	88	89	90	91	93	95
80	80	80	81	81	82	82	83	84	84	85	86	86	87

With prolonged exposure and/or physical activity

Extreme danger
Heat stroke or sunstroke highly likely

Danger
Sunstroke, muscle cramps, and/or heat exhaustion likely

Extreme caution
Sunstroke, muscle cramps, and/or heat exhaustion possible

Caution
Fatigue possible

▲ FIGURE 9.27 **HEAT INDEX CHART** This National Weather Service chart is used to calculate the heat index by combining information about air temperature and relative humidity. A similar chart, the Humidex Chart, is used by the Meteorological Service of Canada. *(See http://www.msc.ec.gc.ca/cd/brochures/humidex_table_e.cfm)*

and landslides, and how deep-well disposal and the filling of large water reservoirs may contribute to earthquakes. Land-use practices may also increase the effects of weather events. For example, the farming practice of plowing cropland after fall harvest and leaving the topsoil exposed to wind erosion during the winter significantly increased the size of dust storms in the Dust Bowl of the 1930s. Also, locating mobile homes in areas subject to frequent high winds and tornadoes greatly increases damages and loss of life from this type of severe weather. Finally, land-use practices in cities can intensify the effects of heatwaves. Large areas of pavement, "big box" one-story buildings, and sparse park land contribute to the **urban heat island effect,** a local climatic condition in which a metropolitan area may become as much as 12°C (22°F) warmer than the surrounding countryside (see Case Study 9.6).[2]

On a larger scale, human interaction with severe weather is taking place through global warming. On the basis of computer models, atmospheric scientists conclude that global warming is likely to increase the heat index and number of heatwaves over land and the intensity of precipitation events in most areas.[44] Computer models also indicate that global warming is likely to increase the risk of drought in midlatitude continental interiors and increase wind and precipitation intensities in hurricanes, typhoons, and other tropical cyclones.[45] The effect of global warming on small-scale events such as tornadoes, thunderstorms, hail, and lightning is still being studied. Overall, global warming will likely increase the incidence of severe weather.[45]

9.7 Linkages with Other Hazards

Severe weather is directly linked to short-term events such as flooding, mass movements, and wildfires, and to long-term changes in global climate. Linkages related to large tropical and extratropical cyclones are discussed in Chapter 10.

Flooding, often flash flooding, can be produced by one or more intense, slow-moving thunderstorms that produce a large amount of rain in a relatively short time. Thunderstorms move slowly if prevailing winds are relatively stagnant. Light prevailing winds can cause essentially the same storm, such as a mesoscale convective system, to remain fixed over a geographic area, or it can cause stagnation of a frontal boundary, such as a stationary front. Since thunderstorms commonly move parallel to a stationary front, the stagnation of a front can cause storms to track over the same area, dropping large amounts of rainfall in what is sometimes called a "training" effect. Stagnation of thunderstorms can also occur over mountain foothills and cause deadly flash floods, like those in Rapid City, South Dakota, in 1972; Big Thompson Canyon west of Longmont, Colorado, in 1976; and Fort Collins, Colorado, in 1997.

Thunderstorms are also produce lightning, the primary, natural ignition source for wildfires (see Chapter 13). However, not all lightning strikes in undeveloped areas cause wildfires. Only an estimated one out of every four cloud-to-ground lightning strikes has the continuous current necessary to start a wildfire, and then the fire may spread only if moisture conditions are low.[46]

Drought, dust/sandstorms, and heatwaves are all hazardous weather events that may become more prevalent in some regions with global warming (see Chapter 12). Heatwaves in the Arctic have already contributed to the melting of permafrost, which in turn has caused subsidence (see Chapter 7) and the release of additional greenhouse gases (see Chapter 12).

9.8 Natural Service Functions of Severe Weather

For many of us, it may be difficult to envision any benefits from severe weather; this is probably because most of the natural service functions are long term and not obvious. For example, lightning is the primary ignition source for natural wildfires, and wildfires are a vital process in prairie, tundra, and forest ecosystems (see Chapter 13). Windstorms also help maintain the health of forests. These storms topple dead and diseased trees, which then are recycled in the soil. Fallen trees also create clearings that become new habitats for diverse plants and animals. Even ice storms, which appear to be so destructive to trees, are part of a natural ecological cycle that increases plant and animal diversity in the forest.

In the hydrologic cycle, blizzards and other snowstorms, thunderstorms, and tropical storms are important, and in some areas the primary source of water. A continual supply of water from snowmelt and seasonal rainfall reduces a region's vulnerability to drought.

Finally, there are some uniquely human benefits to severe weather. Snowfall, cloud formations, and lightning displays have long had an aesthetic value. Severe weather also excites many people. Thanks to movies and television, tornado chasing has become a popular avocation and a new form of tourism. Guided expeditions in specially equipped vehicles regularly drive into "Tornado Alley" to chase and photograph tornadoes. Tornado chasing, however, can be extremely dangerous—serious injury, and even death, can occur if a vehicle encounters a tornado.

9.9 Minimizing Severe Weather Hazards

Thunderstorms, tornadoes, mountain windstorms, ice storms, blizzards, and heatwaves will continue to threaten human lives and property. As long as people live in the path of such hazards, we must take steps to minimize the damage and loss of life associated with them. We must be able to predict these events accurately in order to reduce their hazard.

FORECASTING AND PREDICTING WEATHER HAZARDS

Timely and accurate prediction of severe weather events is extremely important if human lives are to be spared. Even with improvements in satellite sensors and computer modeling, severe weather events are still difficult to forecast, and their behavior is unpredictable. A network of *Doppler radar* stations across North America has significantly improved our ability to predict the path of severe storms. Doppler radar antennas send out electromagnetic radiation that has a wavelength a little longer than microwaves (see Figure 9.5). Clouds, raindrops, ice particles, and other objects in the sky reflect these electromagnetic waves back to a receiving antenna. The wavelength of the reflected waves changes depending on whether the objects are moving toward or away from the antenna. This change in wavelength, called the *Doppler effect*, is similar to the difference in pitch of sound waves as an ambulance siren approaches you and then goes away from you. The changes in radar wavelength are analyzed and can be used to make short-term predictions about the weather, on the scale of hours. For example, Doppler radar can detect a mesocyclone within a thunderstorm and allow meteorologists to issue some tornado warnings up to 30 minutes in advance.

Watches and Warnings You may have heard on the news that a tornado **watch** has been issued for a given area. A tornado watch warns the public of the possibility of a tornado, or tornadoes, developing in the near future. A typical tornado watch might include an area of 52,000 to 104,000 km^2 (20,000 to 40,000 mi.^{2}) and last from 4 to 6 hours.[47] A watch does not guarantee that the event will occur; rather, it alerts the public to the possibility of a severe weather event and suggests that they monitor local weather conditions and listen to radio or television stations for more information.

When a tornado has actually been sighted or detected by weather radar, the watch is upgraded to a **warning.** A warning indicates that the area affected is in danger, and people should take immediate action to protect themselves and others. Watches may be upgraded to warnings, or warnings may be issued for an area not previously under a watch. Both watches and warnings may be issued for any type of severe weather—severe thunderstorms, tornadoes, tropical storms, hurricanes, heatwaves, blizzards, and others, with some variation in the area covered and duration of the watch or warning.

People's perception of the risk of severe weather hazards differs according to their experience. Someone who has survived a tornado is more likely to perceive the hazard as real than someone who has lived in a region at risk for

CASE STUDY 9.6

Europe's Hottest Summer in More Than 500 Years

Heatwaves are one of the deadliest of natural hazards, yet they often get little publicity and are not long remembered. Many Americans have heard of the hurricane that destroyed Galveston, Texas, in 1900 and killed more than 6000 people; yet, few people know that a year later a Midwestern July heatwave killed more than 9500 people.[12] Even today, with rapid global communication, heatwaves do not get as much attention as other less deadly natural hazards. This was even true for the 2003 European heatwave that killed more than 30,000 people, more than any other heatwave in recorded history.

Like most heatwaves, the severe 2003 heatwave in Europe started gradually, many weeks before it peaked. Late spring and summer drought conditions in Europe set the stage for a long, hot summer. In June, record-setting temperatures began in Switzerland, breaking a 250-year record for the hottest month ever reported.[36] Extreme daily high temperatures of 35° to 40°C (95° to 104°F) continued through July in central and southern Europe (Figure 9.K). The heatwave peaked in early August with August 10

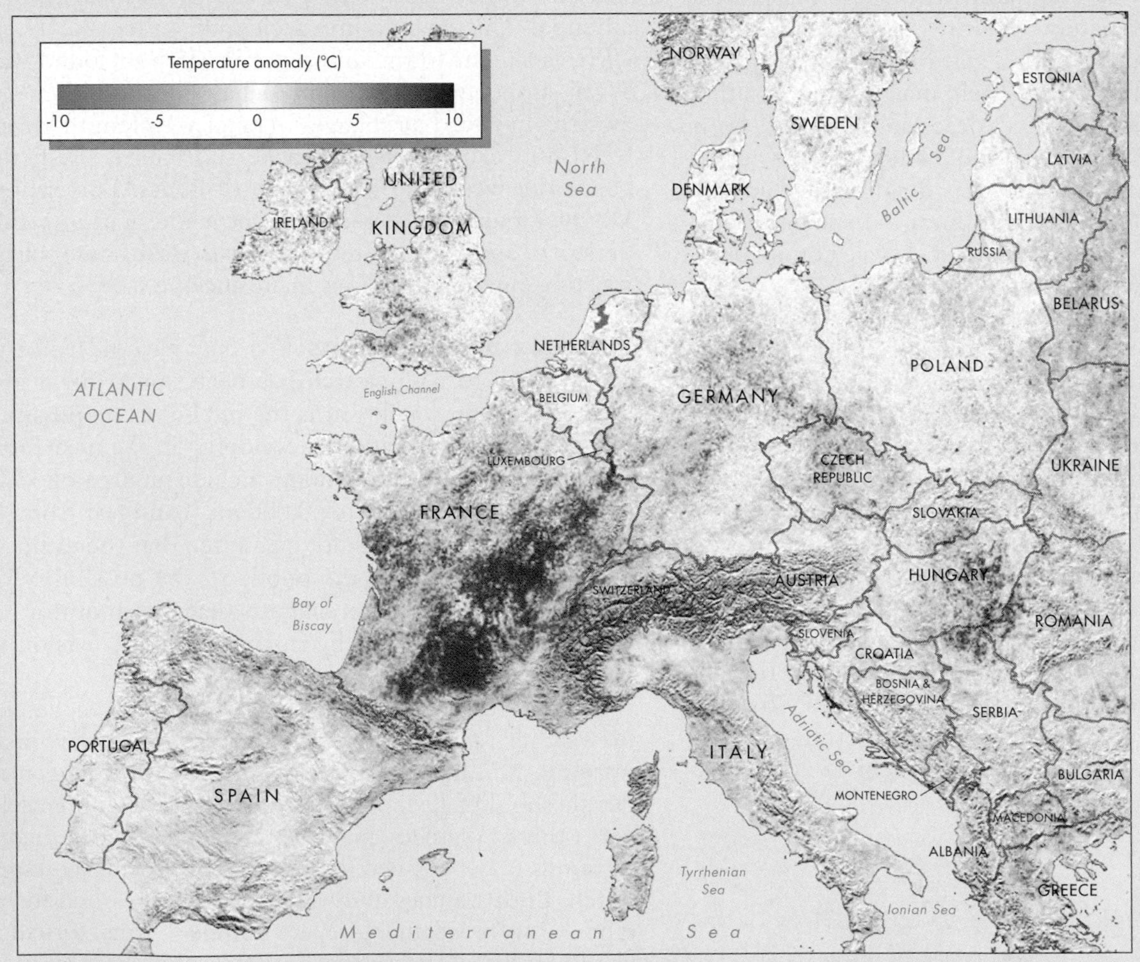

▲ **FIGURE 9.K LAND SURFACE HEATS UP IN 2003 EUROPEAN HEATWAVE** Difference between the daytime temperature of the land surface in July 2003 and the cooler, more normal temperatures of July 2001. Temperatures were measured by the Moderate Resolution Imaging Spectroradiometer (MODIS) on NASA's Terra satellite. The dark red areas were at least 10°C (18°F) hotter in 2003 than in 2001. *(Reto Stocki/Robert Simmon/NASA Earth Observing System)*

the hottest day ever recorded in Great Britain—a high temperature of 38.1°C (100.2°F) was measured in Gravesend, Kent, breaking records dating back to 1875.[37] That same day, Belgium had a high of 40°C (104°F). Overall, climatologists have concluded that the summer of 2003 was probably the hottest in Europe in more than 500 years.[38]

Paris—An Urban Heat Island

For several reasons, the greatest number of deaths occurred in Paris and surrounding French communities. First, heatwaves are generally worse in cities because of the *urban heat island* effect. Environmental conditions associated with cities often make them 5.6°C (10°F) warmer than surrounding natural areas (Figure 9.L).[39] The urban heat island effect is caused primarily by waste heat from vehicles, factories, and air conditioners; reduced airflow between buildings that traps heat; and less vegetation in urban areas. In contrast, evaporation from plant leaves and the soil cools the air in natural areas and mitigates processes that heat up urban areas. For Paris, the heat island effect during the first 6 days of August contributed to raising the air temperatures 7°C (13°F) above normal (Figure 9.M). A second reason that Paris and its suburbs experienced numerous casualties is that air pollution levels generally rise in cities during a heatwave. Stagnant air circulation associated with most heatwaves allows air pollutants to accumulate and form photochemical smog. The smog contains ozone and other gases that cause or amplify respiratory problems. Finally, for Paris and other European cities, August is customarily a holiday month. With many people on vacation, health services were short staffed and unable to handle the thousands of heat-related illnesses. Even morgues were unable to keep pace with the increased number of deaths.

Effects Across the Continent

In human terms, the 2003 heat wave was a disaster of major proportions. France was hit the hardest with more than 14,800 deaths, followed by 7000 deaths in Germany, 4200 deaths in Spain, and 4000 deaths in Italy.[36] Drought conditions severely affected agriculture, and Europe experienced more than 25,000 fires burning an estimated 647,069 hectars (1,598,910 acres) of forests.[36] The global impact of the European drought and forest fires has been estimated at more than $17 billion.[36] Drought also had a major impact on power production: France, Europe's major energy producer, had to cut its power exports in half because of a lack of cooling water for its power plants.

European ecosystems were also significantly altered by the 2003 heat wave. Alpine glaciers lost up to 10 percent of their volume, the greatest loss of ice in any year on record.[36] Ice also melted in the soil, as Alpine permafrost thawed to depths of up to 2 m. (6 ft.).[36] With permafrost ice no longer present to bind weathered rock, there were numerous rock falls in the Alps. A rock fall on the Matterhorn, Switzerland's highest peak, forced the evacuation of 90 climbers.

Contributing Factors

As with many other heat waves, the 2003 event in Europe was enhanced by a stagnant pattern of atmospheric circulation. Stagnant air circulation centered around an anticyclone, a ridge of high atmospheric pressure, that remained nearly stationary over Western Europe for more than 20 days. The anticyclone produced nearly cloudless skies and a downward flow of dry air and also blocked rainfall from storms in the North Atlantic. Exceptionally dry soil moisture and clear

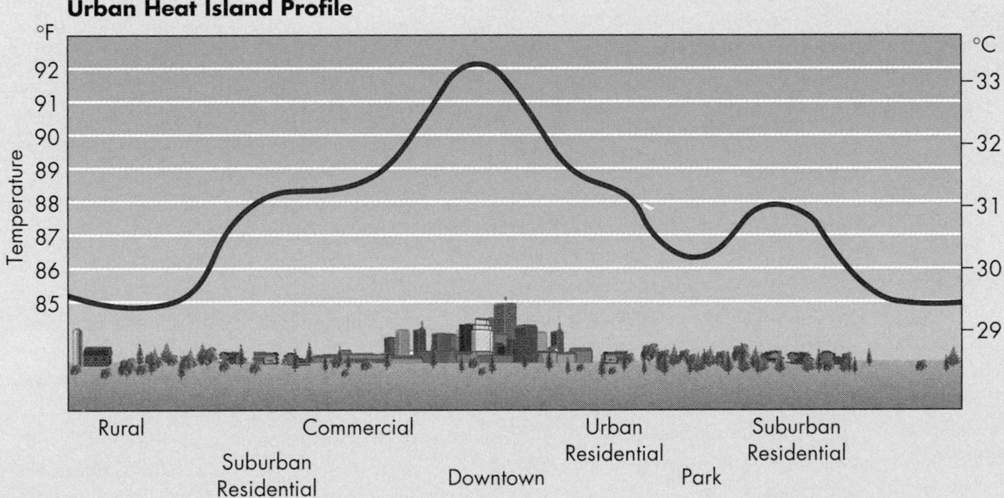

▲ FIGURE 9.L **URBAN HEAT ISLAND EFFECT** Late afternoon temperature profile across urban, suburban, and rural areas on a summer day. The urban and suburban areas are significantly warmer than nearby rural areas, a phenomenon known as the urban heat island effect. This effect is greatest on calm, clear evenings when a city is often 5.6°C (10°F) warmer than the adjacent countryside. *(After U.S. Environmental Protection Agency. 2006.* Heat Island Effect. *http://www.epa.gov/heatisland/about)*

(continued)

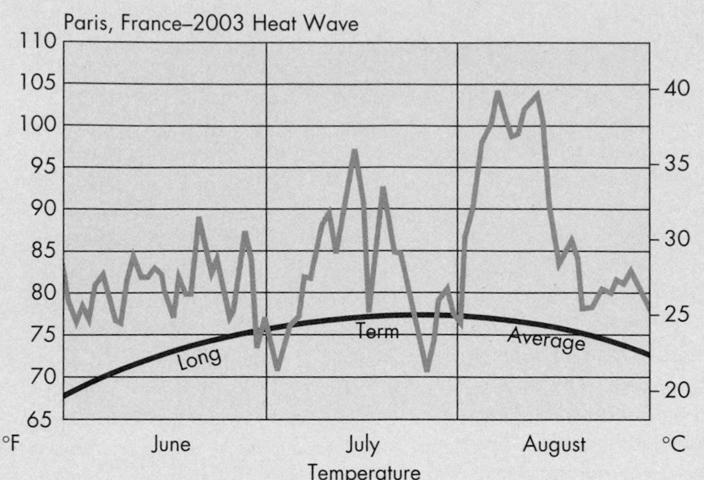

FIGURE 9.M A HOT TIME IN PARIS 2003 (orange line) and long-term average (purple line) daily high temperatures in Paris, France. Although the June and July daily high temperatures are not exceptional for many U.S. cities, they were a tremendous increase above the long-term average for Parisians. *(After U.S. Environmental Protection Agency. 2006.* Excessive heat events guidebook. *Publication EPA 430-B-06-005)*

atmospheric conditions lasted from May through August and contributed significantly to the heatwave.[40] With clear skies, solar radiation heated the ground day after day and dried out the soil. As soil moisture decreased, there was less water to evaporate and cool the air. Heat built up in the soil and then radiated back into the atmosphere each night to significantly warm nighttime temperatures. Hot days followed by hot nights greatly increased heat stress for many people.

Many Europeans wanted to know if the heatwave was caused by global warming. Meteorologists cannot tell if a single weather event is the result of changes that humans have made to the atmosphere. They can, however, estimate the probability that human influence has increased the risk of a heatwave exceeding a given magnitude. For the 2003 European heat wave, computer climate models indicate that the greenhouse gases that humans have added to the troposphere (see Chapter 12) have doubled the probability of an event of that magnitude.[41]

Lessons Learned

Like heatwaves themselves, warning and mitigation programs have been slow to develop. Philadelphia, Pennsylvania, and Toronto, Ontario, have established model programs, and Lisbon, Portugal, and Rome, Italy, had systems in place before the 2003 heatwave.

Warning Systems

At the time of the 2003 heat wave, most areas of Europe did not have heat health warning systems.[42] A similar situation existed in North America prior to the deadly 1995 heat wave in Chicago, Illinois. Today, a number of European cities and countries and more than 15 major metropolitan areas in the United States have heat health warning systems. These systems establish different criteria for warnings based on local meteorological conditions and the adaptation of residents to high heat conditions. The conditions required to trigger excessive heat warnings in Phoenix, Arizona, which regularly experiences temperatures higher than 38°C (100°F), differ from those in normally cooler Toronto, Ontario.

Vulnerable Populations

The elderly, infants and young children, people with mental illness or mobility impairments, the homeless and poor, people engaged in vigorous outdoor exercise or work, and those who are socially isolated or physically ill are at greatest risk from periods of prolonged heat. Most of those who died in France in the 2003 heatwave were the elderly, although an unusually high number of heat-related deaths occurred among younger people.[43] Vulnerable individuals need to be monitored regularly by caretakers or other concerned citizens for signs of heat exhaustion or heat stroke.

Cooling Systems

Heatwaves, including the 2003 European event, are often accompanied by power failures. Loss of air conditioning and power for fans and refrigeration is especially problematic for people who live in urban areas, especially in high-rise buildings (Figure 9.N). Spending a few hours every day in air conditioning can lower heat stress. Mobile cooling centers, such as city buses, can be routinely dispatched to publicized locations to provide heat relief to urban residents.

Public Education

Government officials and the mass media must provide the public with timely and consistent information about heat conditions and mitigation actions. Use of a heat index, which combines temperature and humidity (see Figure 9.27) is critical to communicating conditions that could result in heat-related illness. Individuals can take a number of actions (Table 9.A) to reduce their heat stress and prevent medical conditions caused by excessive heat exposure, such as heat cramps, heat exhaustion, and heat stroke (aka sunstroke).

Prospects for the Future

Climate studies indicate that additional extreme heat events similar to the European heatwave of 2003 are likely to occur in the twenty-first century. Atmospheric scientists have calculated that because of continued global warming, it is 100 times more likely that an event similar to the 2003 heatwave will take place in the next 40 years.[41] Environmental and public health agencies in the United States, Canada, and elsewhere have established programs to prepare for extreme heat events. In particular, these agencies advocate long-term programs to reduce the urban heat island effect, which intensifies heatwaves. Methods to reduce the effect include increasing the reflectivity of urban surfaces, such as by installing light-colored rooftops; increasing the amount of pervious pavement to provide the air access to moisture in the soil; and adding vegetation to urban areas, even to the flat roofs of buildings.

◀FIGURE 9.N **PARISIANS COOL OFF IN RECORD HEAT** Children and adults cool off in the Trocadero Fountains near the Eiffel Tower in Paris on August 12, 2003. Temperatures in Paris reached 40°C (104°F). *(Franck Prevel/AP Images)*

TABLE 9.A

Personal Adjustments for Heatwaves

Do
• Minimize direct exposure to the sun
• Use air conditioners or spend time in air-conditioned places
• Use portable fans to exhaust hot air from or draw cooler air into rooms
• Take a cool bath or shower
• Stay hydrated by regularly drinking water or other nonalcoholic beverages
• Eat light, cool, and easy-to-digest foods such as fruit or salads
• Wear loose-fitting, light-colored clothes
• Check on older, sick, or frail people who may need help adjusting to the heat
• Know the symptoms of heat exhaustion and heatstroke, and the appropriate responses
Don't
• Direct air from fans toward yourself when room temperature is hotter than 32°C (90°F)
• Leave children and pets alone in cars for any amount of time
• Drink alcoholic beverages to try to stay cool
• Eat heavy, hot, or hard-to-digest foods
• Wear heavy or dark clothing

Source: After U.S. Environmental Protection Agency. 2006. Excessive heat events guidebook. Publication EPA 430-B-06-005.

tornadoes yet has never experienced one. Incorrect predictions of where or when a hazard will strike may also lower risk perception. For example, if people are repeatedly warned of severe thunderstorms that never arrive, they may become complacent and ignore future warnings. As with any other hazard, accurate risk perception by planners and the public alike is key to reducing the threats associated with severe weather events.

Finally, our current theoretical understanding and monitoring technology do not always allow accurate prediction of the intensity and extent of many severe weather events. For example, even though less than 1 percent of tornadoes, the severe EF4 and EF5 storms, are responsible for half of tornado-caused deaths, meteorologists are not yet able to include predictions of intensity in official tornado watches and warnings.[48] Likewise, forecasting of the amount of icing and depths of snow accumulation in winter storms is often difficult. However, the National Weather Service and private sector meteorologists are developing computer-based techniques for making real-time predictions of the intensity of some severe weather events.[49] Referred to as *nowcasting*, these techniques use information from weather radar, satellites, and automated weather stations to predict the path and development of severe storms once they have formed.

ADJUSTMENT TO THE SEVERE WEATHER HAZARD

Although we cannot control Earth's atmospheric system to prevent severe weather, we can take a number of steps to reduce the associated death and damage. These actions include both long-term changes to community infrastructure and plans or procedures to be implemented for severe weather. Long-term actions to prevent or minimize death, injuries, and damage are considered *mitigation*. Mitigation activities include the safety-conscious engineering and building of structures, the installation of warning systems, and the establishment of hazard insurance. Establishing community and individual plans and procedures to deal with an impending natural hazard is considered *preparedness*.[50]

Mitigation Although mitigation techniques differ for each weather hazard, some general statements can be made. Building new structures and modifying existing buildings can save lives and protect property from weather hazards. For example, windproofing buildings may significantly reduce damage from severe storms such as tornadoes, derechos, mountain windstorms, and hurricanes. In the United States, the Federal Emergency Management Agency (FEMA) offers grants and architectural plans to establish community shelters and safe rooms in buildings for tornado protection.[51]

Ensuring that electric, gas, water, wastewater, and communications systems continue to function following storms or severe winter weather is a continuing challenge. For example, electrical transmission lines are generally designed to survive ice storms that have a 50-year recurrence interval in the area that they serve.[28] However in wooded areas, winds can combine with ice storms to cause power outages for weeks because ice-loaded branches fall on low-voltage power lines. For many water and wastewater systems, the use of pump stations makes them vulnerable to power interruptions caused by lightning, wind, and ice storms. Without backup sources that can provide power for days until a power grid is restored, these systems remain vulnerable to weather-related interruption. Finally, communication systems have the greatest potential to survive severe weather events if they have redundancy. A combination of landline telephone, cell phone, voice-over-Internet, multifrequency radio, and satellite communication links has the best potential for surviving various combinations of wind, water, ice, and lightning damage.

Other mitigation techniques include developing and installing warning systems and ensuring that universal hazard insurance is available. The goal of warning systems is to give the public the earliest possible notification of impending severe weather. Announcements can be made by commercial radio, television, and the Internet; by U.S. and Canadian government weather radio broadcasts; and by local warning sirens. Some local U.S. and Canadian governments have recently installed community notification systems that call telephone subscribers, including registered mobile/Internet phone users, and warn them of emergencies.

Finally, insurance policies should be available to property holders living in regions at risk for weather disasters. Basic policies cover damage from water and wind but often exclude specific types of hazards such as flooding. Residents of risk areas should determine whether extra coverage is required for a severe storm, tornado, hurricane, blizzard, or other natural hazards.

Preparedness and Personal Adjustments Individuals can take several steps to prepare for severe weather. Many of these steps can and should be carried out before a watch or warning is issued, whereas others are more appropriate when the danger is imminent. In areas prone to severe weather, people should be aware of the times of year that are most hazardous and make adequate preparations for themselves and their home. Information about how to prepare for various weather-related disasters is available from the U.S. National Oceanic and Atmospheric Administration (NOAA) and its subsidiary, the National Weather Service; from FEMA; and from Environment Canada and its subsidiary, the Meteorological Service of Canada.

Wearing proper clothing and modifying travel plans are prudent adjustments to many types of severe weather. Protection from direct sunlight in heatwaves and the need for insulated clothing in snowstorms and blizzards are intuitive for most adults. Where people often are not prepared is in cool (5° to 15°C [40° to 60°F]), wet, and windy weather during which the human body can loose heat rapidly by evaporation from the skin and lungs and by contact with cold air and water. Rapid loss of core body heat causes intense shivering, loss of muscle coordination, mental sluggishness, and confusion, a medical condition known as *hypothermia*. In cool, wet, and windy weather, an improperly dressed person can show symptoms of hypothermia within several hours. People can become hypothermic immersed in 25°C (77°F) water, or if improperly dressed and wet, in 20°C (70°F) air temperature.[52,53] Unless body heat is restored, the continued loss of heat from a hypothermic person will cause unconsciousness and eventually death.

Low visibility conditions can develop rapidly in dust storms, blizzards, and fog and are especially hazardous for travel on freeways and other roads where vehicles normally travel at high speed. For example, on average in the United States 40,000 highway accidents each year occur in fog conditions, resulting in nearly 640 deaths and more than 16,000 injuries.[54] When traveling by automobile, "Onward Through the Fog" in not the best course of action.

REVISITING THE FUNDAMENTAL CONCEPTS

Atmosphere and Severe Weather

The chapter on earthquakes is the first of several in this book that directly discuss a specific natural hazard. For those chapters, we will have, following the summary, a short discussion of how that chapter relates to the fundamental concepts introduced in Chapter 1. Those concepts are

1. **Hazards Are Predictable from a Scientific Evaluation**
2. **Risk Assessment Is an Important Component of Our Understanding of the Effects of Hazardous Processes**
3. **Linkages Exist Between Natural Hazards**
4. **Hazardous Events That Previously Produced Disasters Are Now Producing Catastrophes**
5. **Consequences of Hazards May Be Minimized**

1) Atmospheric conditions can be forecast, and short-term weather predictions can be made with the use of radar, satellite sensors, and computer modeling. However, because severe weather, such as tornadoes, ice storms, and blizzards, is the consequence of specific conditions, involving numerous atmospheric variables, the degree of accuracy with which predictions about the severity of the weather can be made is often limited. For example, short-term tornado watches may be issued over large areas several hours in advance when large-scale mesocyclones are detected; however, the storm may never produce a tornado. When radar does detect a tornado on the ground, a tornado watch can be issued, but, depending on the proximity to the location, little time may be available to seek shelter.

2) Risk assessment for severe weather is based on the probability that an event will occur and the possible, resulting consequences, which are both fairly straightforward. As Figure 9.21 illustrates, the long-term occurrence of tornadoes around the United States can be estimated, and, therefore, the probability of an event can be assessed. For example, many homeowners in the Midwest have tornado insurance; the cost of the insurance policy is based on the location of the house with respect to tornado occurrences and the value of the house and property should it be damaged or destroyed.

3) Linkages exist between severe weather and both short-term events and long-term changes in global climate. Over the short-term, heavy rainfall resulting from severe weather can cause widespread flooding in lowland areas and flash floods in mountainous areas (see Chapter 6) and trigger mass movement events (see Chapter 7). In addition, lightening ground strikes may ignite small fires, which, if conditions are appropriate, may become a wildfire (see Chapter 13). By definition, long-term changes in global climate (see Chapter 12) will result in the reorganization of Earth's weather system, causing short-term events such as drought, flooding, heatwaves, and even blizzards to occur in parts of the globe where these hazards were uncommon before.

4) Severe weather events that previously produced disasters are more commonly now producing catastrophes. This results because greater population density in regions of severe weather will lead to greater loss of life and property during an event. Additionally, population growth has forced people to migrate to less ideal locations where water supply may be limited or where temperatures are more extreme, and, thus, the severity of such events as drought, heatwaves, and blizzards is greater.

5) The consequences of severe weather may be minimized by long-term changes to infrastructure, warning systems, and education. Buildings can be better designed to withstand the effects of high winds, constructed with better insulation to guard against extreme heat and cold, and built with basements and safe zones where human life can be preserved even if the structure is completely destroyed by severe weather such as a tornado. Advancements in remotely and directly sensed weather conditions should enable greater lead time for warning systems that combined with a well-informed and prepared population should greatly limit the loss of life by severe weather.

Summary

Earth receives energy from the sun, and this energy affects the atmosphere, oceans, land, and all living things before being radiated back into space. Although Earth intercepts only a tiny fraction of the total energy emitted by the sun, this energy sustains life on Earth while it drives many processes at or near Earth's surface, such as the circulation of air masses on a global scale.

Potential, kinetic, and heat energy are three primary forms of energy. In the atmosphere, heat energy occurs as sensible heat that can be measured or latent heat that is stored. Latent heat is either absorbed or released in phase changes. Evaporation absorbs heat and cools the air, whereas condensation, which forms clouds, releases latent heat and warms the air. Energy transfer occurs by convection, conduction, and radiation. Of these, convection is the most significant in producing clouds and severe weather.

The Earth receives primarily short-wavelength, electromagnetic energy from the sun. This energy is reflected, scattered, transmitted, or absorbed on Earth. Dark-colored surfaces reflect less and absorb more solar energy and thus generally heat up. Most absorbed solar energy is radiated back into the atmosphere as long-wavelength, infrared radiation.

Most weather occurs in the troposphere, the lowest of the five major layers of the atmosphere. Clouds in the troposphere are made of water droplets and ice crystals. Changes in atmospheric pressure and temperature are responsible for air movement. Air flows from high-pressure to low-pressure areas. Convergence of air produces low atmospheric pressure and divergence produces high pressure.

Atmospheric stability is the tendency of a parcel of air to remain in place or change its vertical postion. The atmosphere is unstable if air parcels rise until they reach air of similar temperature and density. Severe weather is associated with unstable air.

Winds blowing over long distances curve because the Earth rotates beneath the atmosphere. Called the Coriolis effect, this curvature is to the right in the Northern Hemisphere. Wind patterns are controlled by horizontal changes in atmospheric pressure; the Coriolis effect; and, for surface winds, friction.

Boundaries between cooler and warmer air masses, called fronts, are described as cold, warm, stationary, or occluded. Many thunderstorms, tornadoes, snowstorms, ice storms, and dust storms are associated with fronts.

Rain storms with lightning and thunder, called thunderstorms, form where there is moist air in the lower troposphere, rapid cooling of rising air, and updrafts to create cumulonimbus clouds. Thunderstorm development proceeds through cumulus, mature, and dissapative stages. Most thunderstorms form during maximum daytime heating, either as individual air-mass storms or as lines or clusters of storms associated with fronts.

Severe thunderstorms have winds more than 93 km (58 mi.) per hour, hail diameters greater than 1.9 cm (0.75 in.), or a tornado. Hail forms by ice added in layers onto ice pellets that rise and fall within a storm. Severe thunderstorms form where there is strong vertical wind shear, uplift of air, and dry air above moist air. Downdrafts in severe thunderstorms create gust fronts and microbursts that can cause local damage. Three major types of severe thunderstorms are clusters of storms referred to as multiple convective systems (MCSs), linear squall lines, and large individual storms called supercells. MCSs can produce derechos, straight-line windstorms that are as damaging as tornadoes. Squall lines develop ahead of cold fronts and along drylines, air-mass boundaries similar to fronts. Supercell thunderstorms produce the strongest tornadoes.

Tornadoes are generally funnel-like columns of rotating winds of 65 to more than 450 km (40 to more than 280 mi.) per hour that extend downward from a cloud to the ground. With diameters in tens of meters, these storms usually travel at 50 km (30 mi.) per hour from southwest to northeast. Tornadoes form by wind shear in severe thunderstorms, especially supercells, and are most common in the central United States. Damage by tornadoes is rated on the Enhanced Fujita (EF) Scale, with EF5 the most severe. Waterspouts are generally weaker tornadoes forming under fair-weather conditions along ocean coastlines and over lakes.

Lightning is an underestimated safety hazard associated primarily with thunderstorms. Large differences in electrical charge develop within a thunderstorm cloud or between the cloud and the ground. Flow of electrical charges within the cloud, or between the ground and the cloud, produce lightning. Electrical current from a lightning strike can move through the ground and kill or disable people.

Blizzards are severe winter storms in which large amounts of falling or blowing snow reduce visibilities for extended periods of time. Most blizzards are produced by extratropical cyclones that have crossed the Rocky Mountains or have moved along the East Coast as nor'easters. Safety during cool and cold weather is based on the wind chill temperature, a combination of air temperature and wind speed.

Ice storms occur with prolonged freezing rain along a stationary or warm front, below-freezing surface temperature, and a shallow layer of cold air at the surface. Sleet and snow occur where there is a progressively greater thickness of cold air at the surface.

Weather conditions such as a dust storm or fog can greatly reduce visibility, resulting in deadly accidents. Fog is simply a cloud in contact with the ground. Other hazardous weather conditions include sandstorms, mountain windstorms, drought, and heatwaves. Drought and heatwaves are often linked to high-pressure centers that stagnate over regions for extended periods of time. Heatwaves are intensified in cities because of the urban heat island effect. Safety in hot weather requires knowing the heat index, a combination of air temperature and relative humidity. Severe weather produces the much-feared tornadoes and hurricanes (see Chapter 10), but it is heatwaves and blizzards that continue to cause the majority of human deaths from weather phenomena.

Potential human interactions with weather and its hazards are varied. At the local level, land use such as type of housing, landscaping, and agricultural practices may increase the effect of severe weather. On the global scale, atmospheric warming in response to burning of fossil fuels is changing our planet's weather systems. This warming of both the atmosphere and oceans may feed more energy into storms, potentially increasing the incidence of severe weather events.

Minimizing hazards associated with severe weather such as thunderstorms, tornadoes, heatwaves, droughts, blizzards, and ice storms requires a multifaceted approach. This approach should include the following: (1) more accurate prediction that leads to better forecasting and warnings; (2) mitigation techniques designed to prevent or minimize death and loss of property, such as constructing buildings to better withstand severe weather; (3) hazard preparedness, such as short-term activities that individuals and communities can take once they have been warned of severe weather; and (4) education and insurance programs to reduce risk.

Key Terms

atmosphere (p. 286)
blizzard (p. 303)
Coriolis effect (p. 292)
drought (p. 307)
dust storm (p. 307)
EF Scale (p. 301)
fog (p. 306)
heat index (p. 311)
heatwave (p. 311)
ice storm (p. 304)
lightning (p. 297)
relative humidity (p. 286)
tornado (p. 281)
troposphere (p. 286)
urban heat island effect (p. 312)
warning (p. 313)
watch (p. 313)
wind chill (p. 304)

Review Questions

1. Describe the differences among force, work, and power.
2. What are the three types of energy? How do they differ from one another?
3. What is the difference between sensible heat and latent heat?
4. What are the three types of heat transfer? How do they differ from one another?
5. Describe how the Earth's energy balance works.
6. What is electromagnetic energy? How are the different types of electromagnetic energy distinguished?
7. List the following types of electromagnetic energy in order from shortest wavelength to longest wavelength: radio waves, ultraviolet radiation, gamma radiation, visible light, infrared radiation, X-rays, and microwaves.
8. Explain why the sun radiates 16 times more energy than the Earth.
9. How is color related to energy absorption?
10. Describe the characteristics of the troposphere. How do meteorologists identify the top of the troposphere?
11. What is the tropopause? How high is it above the Earth's surface?
12. Why does atmospheric pressure decrease with increasing altitude?
13. What is the difference between stable and unstable air?
14. Explain the Coriolis effect. How does it influence weather?
15. What conditions are necessary for a thunderstorm to form? A severe thunderstorm?
16. Describe the three stages of thunderstorm development.
17. What are supercells, mesoscale convective complexes, and squall lines? How do they differ? Why are they significant natural hazards?
18. What is hail? How does it form? Where is it most common in the United States?
19. Characterize a tornado in terms of wind speed, size, typical speed of movement, duration, and length of travel.
20. Describe the five stages of tornado development.
21. What is a blizzard? How does a blizzard develop?
22. What is a nor'easter? How is it related to blizzards?
23. Describe the weather conditions that cause an ice storm.
24. How are the heat index and wind chill temperature alike? How are they different? When is each of these indices important?
25. How is global warming expected to affect severe weather?
26. What are some natural service functions of severe weather?
27. What is the difference between a severe weather watch and warning?
28. How do preparedness planning and mitigation differ?
29. Explain how drought, soil moisture, and a heatwave can be interrelated.
30. What causes the urban heat island effect? How can this effect be mitigated?

Critical Thinking Questions

1. What severe weather events are potential hazards in the area where you live? What are some steps you might take to protect yourself from such hazards? Which of these hazards is your community the least prepared for?
2. Lightning is the deadliest weather hazard and the one that is likely to affect many people. Use the Web resources in the next section to
   a. Explain why news reports about lightning "survivors" might be misleading.
   b. Determine when you need to take shelter from lightning and how long you should stay in the shelter.
   c. Determine what behaviors outside and inside your house increase the possibility that you might be struck by lightning.
   d. Explain how people living in Florida are likely to be affected by lightning strikes.
3. Tornadoes can often be spotted on weather radar, whereas many other clouds cannot. What makes tornadoes visible?
4. Study the diagrams of cold fronts and warm fronts, and read the description about the development of ice storms. Explain why sleet (small pellets of ice) is more likely to accompany cold fronts than is freezing rain.

5. Why does hail form in thunderstorms and not in other rainstorms or snowstorms?
6. Has your community ever experienced a heatwave? If so, when did it occur and how were people affected? Does your community have a heat health warning system? If so, what actions do local officials and emergency personnel take when the system is activated? If not, what type of system would you recommend? What actions could you take to mitigate the effects of a heatwave on your living conditions?

Selected Web Resources

National Weather Service Storm Prediction Center
www.spc.noaa.gov/—current severe weather information from NOAA

Welcome to NSSL's Weather Room
ww.nssl.noaa.gov/edu/—questions and answers about tornadoes, thunderstorms, lightning, and hurricanes from the NOAA National Severe Storms Laboratory

National Climatic Data Center
lwf.ncdc.noaa.gov/oa/ncdc.html—weather data from NOAA

Hazards: Informing the Public About Hazards
www.fema.gov/hazards/—fact sheets, booklets, and background information about thunderstorms, tornadoes, heatwaves, and winter storms from the Federal Emergency Management Agency

Weather
www.msc-smc.ec.gc.ca/weather/contents_e.html—fact sheets about lightning, blizzards, hailstorms, tornadoes, waterspouts, humidity, and wind chill from the Meteorological Service of Canada

U.S. Hazards Assessment
www.cpc.noaa.gov/products/expert_assessment/threats.html—current U.S. weather hazards from the National Weather Service Climate Prediction Center

NOAA and American Red Cross Publications on Weather Hazards
weather.gov/om/brochures.shtml—from the National Weather Service Office of Climate, Water, and Weather Services

Weather Safety
weather.gov/safety.php—links to NOAA Web pages on weather safety from the National Weather Service

The Weather World 2010 Project
ww2010.atmos.uiuc.edu—multimedia guide to meteorology linked to current weather conditions; from the University of Illinois

The Tornado Project Online!
www.tornadoproject.com/—a comprehensive Web site on tornadoes from a small company that has been collecting information since 1970

Extreme Weather Sourcebook 2001
sciencepolicy.colorado.edu/sourcebook/—economic and societal aspects of severe weather from the National Center for Atmospheric Research

Heat Island Effect
www.epa.gov/heatislands/—the heightened effects of extreme heat on urban areas and ways to mitigate these effects from the U.S. Environmental Protection Agency

10 Hurricanes and Extratropical Cyclones

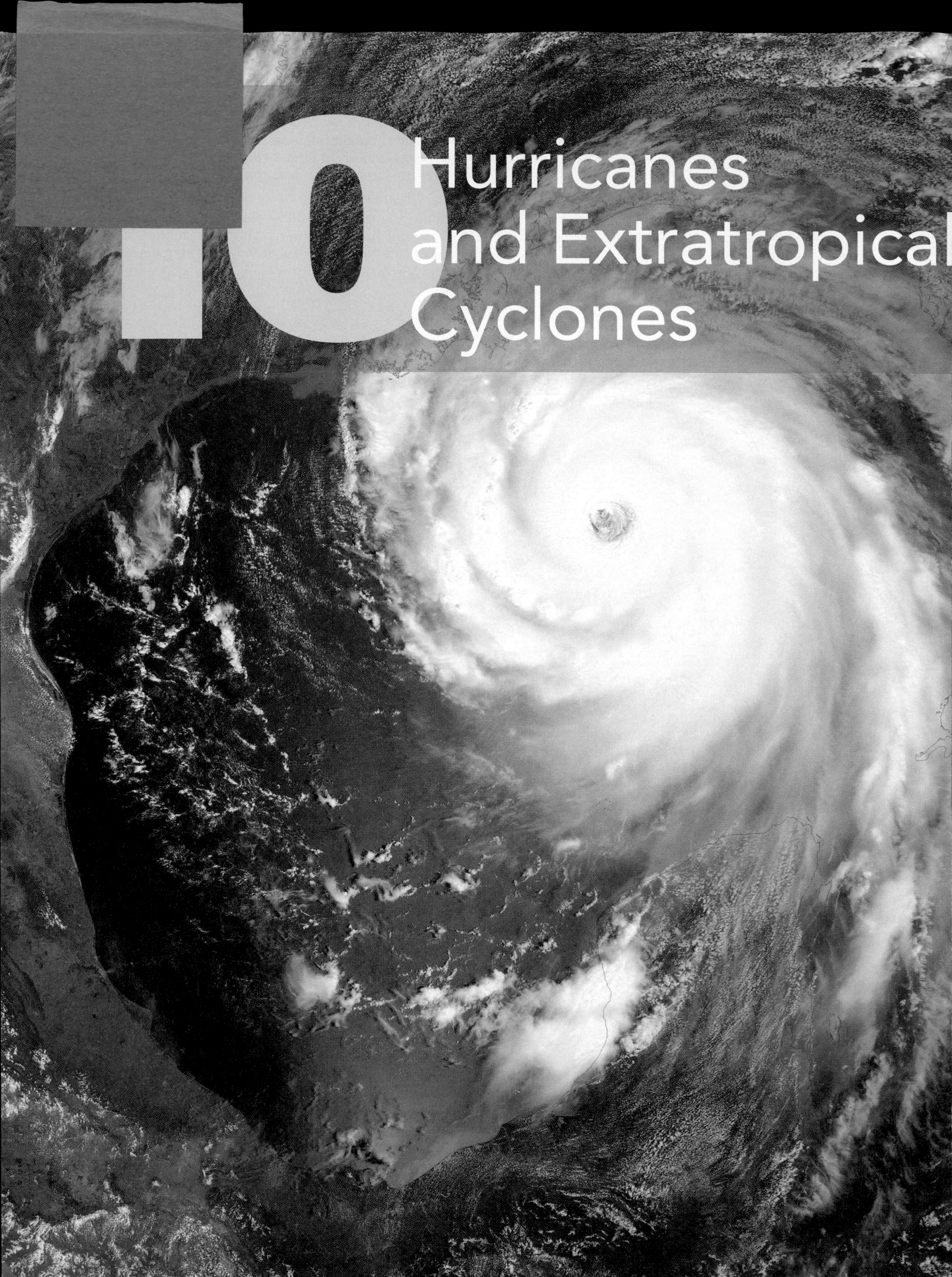

Hurricane Katrina

Unlike the huge lateral blast from the Mount St. Helens volcano in 1980, which took many scientists and public officials by surprise, the flooding of New Orleans and adjacent communities by a major hurricane was widely predicted for decades in scientific publications, including the first edition of this book, and in magazine and newspaper articles. A Pulitzer Prize–winning series of articles about the future hurricane catastrophe was published in New Orleans's major newspaper, the *Times-Picayune,* in 2002. Overall, as David Brooks of the *New York Times* later wrote, "Katrina was the most anticipated natural disaster in American history."[1]

Why was this disaster so anticipated? First, coastlines in the southeastern United States have a high incidence of hurricanes; in the past 50 years, they have flooded New Orleans and devastated coastal Mississippi, Alabama, and Louisiana on more than one occasion. Second, 95 percent of New Orleans has subsided an average of 1.5 m (5 ft.) below sea level and the city relies on a patchwork of levees, floodwalls, and pumps for protection.[2] Parts of the flood protection and drainage system are more than 100 years old. Finally, in the past century, coastal Louisiana has lost wetlands and barrier islands that help mitigate the effects of hurricane storm surges (see Chapter 8).

Katrina was the most anticipated natural disaster in American history.

One effective way of mitigating the effects of hurricanes and extratropical cyclones is to design and conduct realistic disaster exercises. In July 2004, more than a year before Hurricane Katrina flooded New Orleans, scientists and public officials participated in a weeklong "Hurricane Pam" disaster exercise that simulated a slow-moving Category 3 hurricane making landfall just west of New Orleans. This exercise, sponsored by the Federal Emergency Management Agency (FEMA), demonstrated that a Category 3 hurricane would cause catastrophic flooding of New Orleans (Figure 10.1). A hurricane's category indicates its intensity, with Category 3 being the least intense major hurricane (see Table 10.1). Although many lessons from this exercise required funding that

◀ **Hurricane Katrina** Satellite image of the Gulf Coast of the United States on August 29, 2005 as hurricane Katrina makes landfall in Louisiana as a Category 3 hurricane. *(NASA)*

LEARNING OBJECTIVES

Hurricanes, and their mid-latitude relative, extratropical cyclones, are the most powerful storms on Earth and among the most deadly and costly of natural hazards—Hurricane Katrina alone was the most expensive natural disaster in history. Called cyclones, these storms affect parts of the United States and Canada daily and are responsible for most severe weather. Understanding how these storms work helps us appreciate how they pose a threat to our highly technological society. As our climate changes and sea level rises, these storms will pose a greater hazard to coastal regions. Despite this threat, government flood insurance and disaster relief policies continue to encourage development in hurricane-prone regions. As Hurricane Katrina demonstrated, choices made by individuals, businesses and government officials on how to prepare for and respond to hurricanes can turn an "Act of God" into an "Act of Man." Your goals in reading this chapter should be to

- understand the weather conditions that create, maintain, and dissipate cyclones.
- understand the difficulties in forecasting cyclone behavior.
- know what geographic regions are at risk for hurricanes and extratropical cyclones.
- understand the effects of cyclones in coastal and inland areas.
- recognize linkages between cyclones and other natural hazards.
- know the benefits derived from cyclones.
- understand adjustments that can minimize damage and personal injury from coastal cyclones.
- know the prudent actions to take for hurricane or extratropical cyclone watches and warnings.

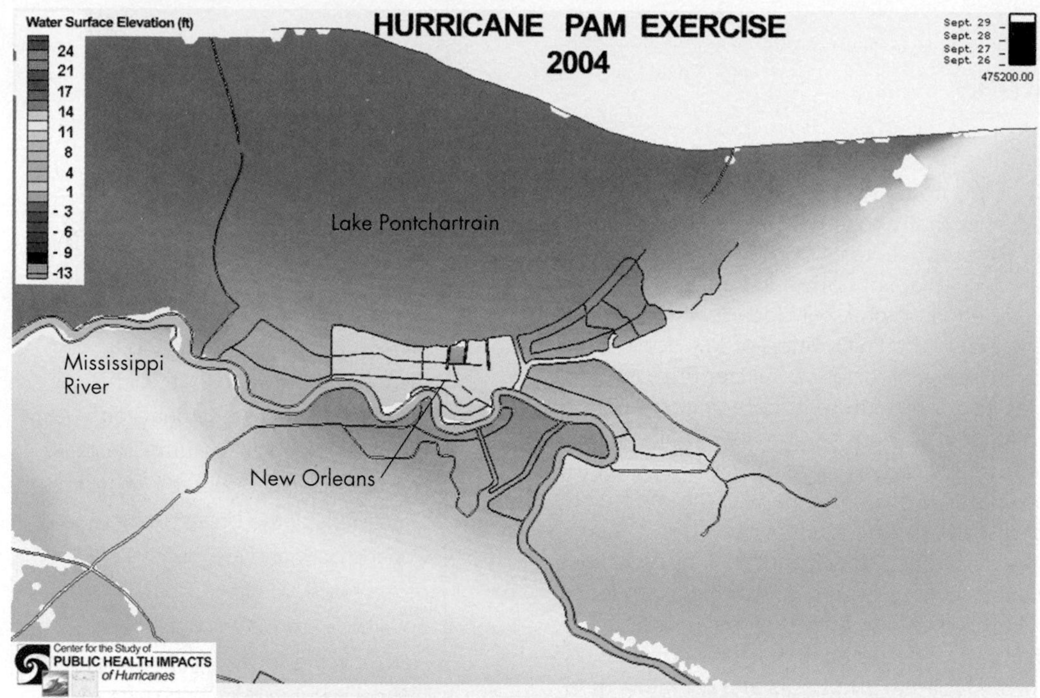

▲ FIGURE 10.1 **COMPUTER STORM-SURGE MODEL FOR FEMA EXERCISE ONE YEAR BEFORE HURRICANE KATRINA** Scientists and engineers at the Louisiana State University Center for the Study of Public Health Impacts of Hurricanes used the ADCIRC computer model to produce this map of storm surge flooding depths for "Hurricane Pam," a week-long FEMA disaster exercise. The exercise was conducted one year before Hurricane Katrina. Red to orange colors indicate a storm surge of 4.3 to 5.2 m (14 to 17 ft.), and light green indicates a surge of around 3.4 m (11 ft.). *(Courtesy LSU Center for the Study of Public Health Impacts of Hurricanes)*

was not forthcoming, the search and rescue coordination established by the participants proved invaluable in saving lives during Hurricane Katrina.[2]

Contrary to initial reports that indicated that Hurricane Katrina was a Category 4 storm when it hit New Orleans, Katrina was a Category 3 hurricane when it made its first landfall on the Mississippi Delta about 80 km (50 mi.) southeast of New Orleans.[3] The storm's intensity then dropped and most of New Orleans experienced the effects of a Category 1 or weak Category 2 hurricane.[3] Both Katrina's drop in intensity and its path were fortunate for New Orleans. Less than 24 hours before landfall, Hurricane Katrina had been a Category 5 storm, one of the most powerful ever observed in the Gulf of Mexico.[3] Also, New Orleans experienced significantly less wind damage than if Katrina's path had been west of the city. In the Northern Hemisphere, hurricanes and extratropical cyclones are most powerful in their right forward quadrant.

From a hazard warning standpoint, the National Weather Service (NWS) did an excellent job of forecasting the landfall of Hurricane Katrina and alerting government officials and the public. NWS forecasters monitored Hurricane Katrina from its inception as a tropical depression over the Bahamas almost a week before it came ashore in Louisiana. Air Force and NOAA reconnaissance aircraft (see Case Study 10.3), satellites, and land-based radars tracked the progress of the storm as it intensified and then doubled in size over the Gulf of Mexico. A day and a half before Katrina made landfall, Dr. Max Mayfield, then director of the NWS National Hurricane Center, made a special telephone call to the governors of Louisiana and Mississippi and the mayor of New Orleans urging evacuations.[2] This was only the second time in Dr. Mayfield's 36-year career that he had believed it necessary to make a call to a governor warning of a hurricane.[4] The next day he briefed the president and other top federal officials in a videoconference that there was "very, very grave concern" that the New Orleans levees would be overtopped.[5] Special NWS advisories for Katrina accurately warned of storm surge heights of up to 8.5 m (28 ft.) and floodwaters overtopping levees.[3] These warnings resulted in a presidential emergency declaration before the storm hit land and the first-ever mandatory evacuation of New Orleans. Katrina's final landfall was within 35 km (22 mi.) of the forecast location.[3]

From an engineering standpoint, Hurricane Katrina breached the New Orleans flood protection system, the single most costly catastrophic failure of an engineered system in history.[6] Flood walls and levees were breached in a least eight locations by water levels lower than design specifications (Figure 10.2).[5,6] On the east side of New Orleans, levees and floodwalls were overtopped and breached by the storm surge because the design and construction of navigation canals acted like a funnel to focus the storm surge

TABLE 10.1

The Saffir-Simpson Hurricane Scale

The Saffir-Simpson Hurricane Scale is a 1 to 5 rating based on the hurricane's present intensity. This rating is used to give an estimate of the potential property damage and flooding expected along the coast from a hurricane landfall. The highest current 1-minute average wind speed is the determining factor in the scale, because storm surge values are highly dependent on the slope of the continental shelf where the hurricane makes landfall.

Category One Hurricane:
Winds 119 to 153 km (74 to 95 mi.) per hour; storm surge generally 1.2 to 1.5 m (4 to 5 ft.) above normal. No real damage to building structures. Damage primarily to unanchored mobile homes, shrubbery, and trees. Some damage to poorly constructed signs. Also, some coastal road flooding and minor pier damage. Hurricanes Irene of 1999 and Katrina of 2005 were Category 1 hurricanes when they made landfall in southern Florida.
Category Two Hurricane:
Winds 154 to 177 km (96 to 110 mi.) per hour; storm surge generally 1.8 to 2.4 m (6 to 8 ft.) above normal. Some roofing material, door, and window damage of buildings. Considerable damage to shrubbery and trees, with some trees blown down. Considerable damage to mobile homes, poorly constructed signs, and piers. Coastal and low-lying escape routes flood 2 to 4 hours before arrival of the hurricane center. Small craft in unprotected anchorages break moorings. Hurricanes Isabel of 2003 and Frances of 2004 were Category 2 hurricanes when they made landfall in North Carolina and Florida, respectively.
Category Three Hurricane:
Winds 178 to 209 km (111 to 130 mi.) per hour; storm surge generally 2.7 to 3.7 m (9 to 12 ft.) above normal. Some structural damage to small residences and utility buildings with a minor amount of wall failures. Damage to shrubbery and trees with foliage blown off trees and large trees blown down. Mobile homes and poorly constructed signs are destroyed. Low-lying escape routes are cut off by rising water 3 to 5 hours before arrival of the hurricane center. Flooding near the coast destroys smaller structures with larger structures damaged by battering of floating debris. Terrain continuously lower than 1.5 m (5 ft.) above mean sea level may be flooded inland 13 km (8 mi.) or more. Evacuation of low-lying residences within several blocks of the shoreline may be required. In 2004, Hurricanes Ivan and Jeanne were Category 3 hurricanes when they made landfall along the Alabama and Florida coasts, respectively.
Category Four Hurricane:
Winds 210 to 249 km (131 to 155 mi.) per hour; storm surge generally 4 to 5.5 m (13 to 18 ft.) above normal. More extensive wall failures with some complete roof structure failures on small residences. Shrubs, trees, and all signs are blown down. Complete destruction of mobile homes. Extensive damage to doors and windows. Low-lying escape routes may be cut off by rising water 3 to 5 hours before arrival of the hurricane center. Major damage to lower floors of structures near the shore. Terrain lower than 3.1 m (10 ft.) above sea level may be flooded, requiring massive evacuation of residential areas as far inland as 10 km (6 mi.). Hurricanes Charley of 2004 and Dennis of 2005 were Category 4 hurricanes when they struck the coasts of Florida and Cuba, respectively.
Category Five Hurricane:
Winds greater than 249 km (155 mi.) per hour; storm surge generally greater than 5.5 m (18 ft.) above normal. Complete roof failure on many residences and industrial buildings. Some complete building failures with small utility buildings blown over or away. All shrubs, trees, and signs blown down. Complete destruction of mobile homes. Severe and extensive window and door damage. Low-lying escape routes are cut off by rising water 3 to 5 hours before arrival of the hurricane center. Major damage to lower floors of all structures located less than 4.6 m (15 ft.) above sea level and within 458 m (500 yd.) of the shoreline. Massive evacuation of residential areas on low ground within 8 to 16 km (5 to 10 mi.) of the shoreline may be required. Three Category 5 hurricanes have made landfall in the United States since records began: the Labor Day Hurricane of 1935, Hurricane Camille of 1969, and Hurricane Andrew of 1992. In 2005, Hurricane Wilma was a Category 5 hurricane at peak intensity in the Gulf of Mexico and is the strongest Atlantic tropical cyclone of record.

Source: Modified after Spindler, T., and Beven, J. 1999. Saffir-Simpson Hurricane Scale, *NOAA, http://www.nhc.noaa.gov/aboutsshs.shtml. Accessed 2/17/07.*

(Figure 10.3).[2] The failure and overtopping of levees and floodwalls resulted in the inundation of 85 percent of the greater New Orleans metropolitan area with 0.3 to 6.1 m (1 to 20 ft.) of water (Figure 10.4).[6] Much of the city remained flooded for two to three weeks before levees could temporarily be repaired and floodwaters pumped out. Other engineering failures included collapse of the I-10 causeway east of New Orleans (Figure 10.5); damage to the Superdome roof (Figure 10.6); loss of cell phone communications and power for residences, businesses, and hospitals; and collapse of several offshore oil and gas production platforms (Figure 10.7).

From a social standpoint, Hurricane Katrina was especially devastating. An estimated 1.2 million people evacuated from the affected area, and more than 400,000 people ended up in public shelters in 18 states.[4] Although evacuation is a difficult, costly, and uncertain process, especially for the poor and elderly, it is the most powerful tool emergency managers have to prevent loss of life in a hurricane.[7] Evacuation was especially problematic in New Orleans where an estimated 250,000 people had no means of private transportation.[8] Those who had no way to evacuate, or chose not to, had few places to go. In the first 10 days of the disaster, Coast Guard boats, ships, and aircraft rescued more than 23,000 people along the Gulf Coast, and other rescuers saved many additional tens of thousands of lives (see Survivor Story 10.1).[9] Hurricane Katrina destroyed more than 300,000 homes in four states.[4]

▲ FIGURE 10.2 **AERIAL VIEW OF TEMPORARY REPAIR OF BREACHED FLOODWALL AND LEVEE ON NEW ORLEANS'S 17TH STREET CANAL** Army helicopters lowered white sand bags to temporarily fill this breach in the floodwall and levee along the 17th Street drainage canal. The failure, caused by the Hurricane Katrina storm surge, occurred at a water level lower than the design limit for the structure. Both faulty design and incomplete geotechnical information may have been responsible for the failure. *(Bob McMillan/FEMA)*

◀ FIGURE 10.3 **NAVIGATION CANAL LEVEE BREACHED BY STORM SURGE** Hurricane Katrina's storm surge flooded through this breach in the Inner Harbor Navigation Canal levee to inundate the New Orleans Lower Ninth Ward. Poor foundation design appears to have been a major cause for the breach. *(Vincent Laforet/AP Images)*

▲ FIGURE 10.4 **NEW ORLEANS INUNDATED AFTER FLOOD PROTECTION SYSTEM FAILS** Levees and flood walls designed to protect New Orleans from a Category 3 storm were breached in at least six places by Hurricane Katrina and were overtopped where the storm surge was funneled down navigational canals. As a result, 85 percent of New Orleans was flooded. This photo, taken the day after landfall, shows cars abandoned on freeway overpasses and smoke rising over parts of the city from fires burning out of control. *(David J. Phillip/AP Images)*

Was Hurricane Katrina an unusual storm for the Gulf Coast? In many ways, the answer to this question is no—hurricanes of equal or greater strength had struck the Louisiana and Mississippi coast a number of times during the previous 50 years. In 1965, Hurricane Betsy, a Category 4 storm, came within 35 miles of New Orleans, breached the levees with a 2.4 to 3.0 m (8 to 10 ft.) storm surge, and flooded parts of the city. Four years later, Category 5 Hurricane Camille barely missed Louisiana and devastated the Mississippi coast with sustained winds of 300 km (190 mi.) per hour. Camille's winds were the strongest hurricane winds ever recorded on the North American continent and were more than three times stronger than Katrina.[10] Although Hurricane Camille's storm surge in Mississippi was nearly as great as Katrina's, the population of the coastal zone in 1969 was 60 percent less than it is today. In 2004, Hurricane Ivan, a Category 4 storm, headed directly toward New Orleans, only to turn to the right before landfall and hit the Alabama coast. Ivan, like Katrina, reached Category 5 strength in the Gulf of Mexico and produced huge waves that damaged oil and gas platforms.[11] Like Katrina in Louisiana and Mississippi, the storm surge from Hurricane Ivan washed away parts of

▲ FIGURE 10.5 **NEW ORLEANS EVACUATION ROUTE DESTROYED BY KATRINA** The storm surge from Hurricane Katrina destroyed these bridges on Interstate-10 crossing Lake Pontchartrain from New Orleans to Slidell, Louisiana. I-10 was one of the major evacuation routes out of the city. *(David Grunfield/Landov Media)*

FIGURE 10.6 HURRICANE WINDS DAMAGE SUPERDOME ROOF The National Weather Service estimates that New Orleans experienced sustained winds of 110 to 125 km (70 to 78 mi.) per hour with gusts of up to 190 km (120 mi.) per hour during Hurricane Katrina. High winds tore two gaping holes in the Superdome roof, letting rain in, but not threatening the safety of thousands of people who had taken refuge there. *(Jocelyn Augustino/FEMA)*

FIGURE 10.7 KATRINA DAMAGES OFFSHORE OIL PLATFORM Storm waves ripped this oil drilling platform, the *Ocean Warwick,* from its anchorage offshore Louisiana and carried it 106 km (66 mi.) to Dauphin Island south of Mobile, Alabama. The accompanying storm surge cut the barrier island in half and destroyed more than half the houses. Overwash pushed most of the sand into Mississippi Sound in the distance. Although Dauphin Island was flooded by storm surges from Hurricanes Lili (2002), Ivan (2004), Dennis (2005), and Katrina (2005), some home owners want to rebuild. *(Adrien Lamarre/US Army Crps of Engineers)*

barrier islands and destroyed highway bridges in Alabama and Florida. Moving inland, Ivan spawned 117 tornadoes and spread heavy rains, causing river flooding as far north as New York State.[11]

Compared to Ivan, the storm surge from Katrina was especially devastating with the water level rising to the height created by a Category 5 hurricane. Although metropolitan New Orleans was paralyzed by a 3.0 to 5.8 m (10 to 19 ft.) storm surge, a more powerful surge in the right forward quadrant of Katrina inundated the Mississippi coastline with 5.5 to 7.3 m (18 to 24 ft.) of water. In places, Katrina's storm surge pushed salt water up to 10 km (6 mi.) inland.[3] Katrina's storm surge completely leveled coastal neighborhoods (Figure 10.8), washed houses onto levees (Figure 10.9), and carried large ships and barges onto land (Figure 10.10). The storm surge was responsible for most of the hurricane-related deaths in Mississippi's coastal counties.[3]

Like many other hurricanes, Katrina produced numerous tornadoes and heavy rainfall as it moved inland. At least 43 tornadoes were reported, including 20 in Georgia, as the storm moved northward.[3] A large area of southeastern Louisiana and southwestern Mississippi received 200 to 250 mm (8 to 10 in.) of rain, and heavy rains extended north into the Tennessee Valley.[3]

If Hurricane Katrina was not extraordinary, and there was ample advance warning, why did more than 1600 people die and damage exceed $125 billion? Katrina was the fourth or fifth most deadly hurricane and the most costly natural disaster in U.S. history.[3] Answers to this question are complex and multifaceted—failure in the design, construction, and maintenance of levees and flood walls (Figure 10.11); an overreliance on technology to protect life and property; social and psychological denial of the hazard; poverty and limited education for many residents of the affected area; diversion of military and government resources overseas; and failures in political leadership, communication, and public policy at all levels. The policy failures include a system of flood insurance and postdisaster aid that encourages people to live in coastal hazard areas and rewards politicians, developers, businesses, and individuals for rebuilding in previously flooded areas.[7]

10.1 Introduction to Cyclones

In meteorological terms Hurricane Katrina was a **cyclone**, an area or center of low atmospheric pressure characterized by rotating winds. Because of the Coriolis effect, the winds in the Northern Hemisphere are deflected to the right and blow in a counterclockwise rotation around the low-pressure center (Figure 10.12),

▲ FIGURE 10.8 **STORM SURGE FLATTENS NEIGHBORHOOD** A 7 to 8.5 m (24 to 28 ft.) storm surge from Hurricane Katrina washed away most of the houses in this Gulfport, Mississippi, neighborhood. *(Tyrone Turner/National Geographic Image Collection)*

SURVIVOR STORY 10.1

Hurricane Katrina

Abdulrahman Zeitoun's Family Evacuated—He Stayed to "Mind the Damage"

After living in New Orleans for more than 30 years, contractor Abdulrahman Zeitoun (Figure 10.A) has seen more than his share of tropical storms. So when forecasters began advising Gulf Coast residents to evacuate for Hurricane Katrina in late August 2005, Zeitoun declined to leave. "I say look, I'm not going. Someone has to stay behind and mind the damage."

On Saturday, August 27, Zeitoun waved goodbye to his wife and four children as they drove to relatives in Baton Rouge and settled in to wait out the storm. The next day, the wind and rain grew stronger and stronger, the power died, and the roof began to leak. But by late Monday, the weather had calmed and the danger seemed over. "I see water on the street a foot and a half deep, a few trees down. I call my wife, I say everything's done, if you want to come back."

In fact, the worst disaster in the history of New Orleans was only just beginning. Nearby, ocean waters surging inland were starting to overtop city levees, bursting through in several places to inundate entire neighborhoods.

Zeitoun heard the floodwaters before he saw them. "There were noises like when you sit by a river. I stick my head out the window and see water in the backyard rising very fast." He moved family photographs, the first aid kit, flashlights and small items upstairs, and then he stacked the first floor furniture in tall piles. "I sacrificed one thing to save another one; anything I could do to minimize the damage."

The water rose for another 24 hours. Zeitoun called his wife and told her not to return, as the city would not be habitable for a long time. Then he climbed in his canoe to check on his tenants elsewhere in town.

What he saw as he paddled the silent streets was beyond his worst imaginings. Water on his block reached the bottom of the stop sign more than six feet high. A neighbor asked for a lift to check on his truck. They found the vehicle in water up to its roof. An older couple waving white flags asked to be rescued. Unable to fit them in the canoe, Zeitoun promised to find help. Further along, they heard a voice from a one-story house. Jumping in the water, Zeitoun managed to open the door. An elderly lady, skirts ballooning around her like a Civil War belle, stood in water up to her shoulders. To rescue her, Zeitoun and his neighbor flagged down one military boat after another, to no avail. Finally, a few civilians with a hunting boat stopped. Zeitoun positioned a ladder beneath her legs, and together the men levered the lady into the vessel. They went on to rescue the older couple and delivered all three seniors to a local hospital.

At his rental, Zeitoun found his tenant healthy and, by some miracle, in possession of a working phone. Neighborhood residents began converging to call friends and relatives.

Zeitoun helped several more people the next day, obtaining a boat to help rescue an older man in a wheelchair. In the evening, Zeitoun, his tenant, a friend, and a man using the phone were gathered at the rental when a military boat floated by. "They jump into the house with machine guns and say, what are you doing here? I say I own the place, it's my house," Zeitoun said. But the soldiers arrested all four and took them to a bus station being used as a temporary jail. They were thrown into a makeshift cage of chicken wire, interrogated for three days, and moved to a nearby prison. After spending a month behind bars, Zeitoun was released on bail. All charges—for suspected looting—have since been dropped.

Despite his ordeal, Zeitoun plans to stay in New Orleans. "My business is here. I have good relations here with my associates, friends, customers. I feel like I have a family in this city," he said.

—KATHLEEN WONG

▲ FIGURE 10.A **HURRICANE KATRINA SURVIVOR** New Orleans resident Abdulrahman Zeitoun evacuated his family but chose to stay behind to face the dangers and devastation of Hurricane Katrina. *(Kathy Zeitoun)*

▲ FIGURE 10.9 **HOUSES WASHED ONTO MISSISSIPPI RIVER LEVEE** A Hurricane Katrina storm surge of more than 4.3 m (14 ft.) washed these houses onto the east bank levee of the Mississippi River at Pointe A La Hache, Louisiana, about 43 km (27 mi.) southeast of New Orleans. *(David Grunfield/Landov Media)*

▲ FIGURE 10.10 **FISHING VESSELS BLOCK STATE HIGHWAY** The Hurricane Katrina storm surge carried these fishing vessels onto Louisiana State Highway 23 at the foot of the Empire Bridge in Plaquemines Parish, southeast of New Orleans. Empire is 8 km (5 mi.) northwest of Buras, the point of initial landfall of Hurricane Katrina. *(George Stringham/US Army Corps of Engineers New Orleans Districts)*

wheras in the Southern Hemisphere, the cyclones rotate clockwise due the Coriolis effect (see Closer Look 9.1). Cyclones rarely develop within 5° of the equator, where the Coriolis effect is weakest. Cyclones are classified as tropical or extratropical based on their place of origin and the temperature of their center or core region. The modifier *extratropical* means "outside of the tropics." **Tropical cyclones** form over warm tropical or subtropical ocean water, typically between 5° and 20° latitude. Unlike extratropical cyclones, they are not associated with fronts and have warm central cores. In contrast, **extratropical cyclones** develop over land or water in temperate regions, typically between 30° and 70° latitude. These mid-latitude cyclones are generally associated with fronts and have cool central cores. Both types of storms are characterized by their **cyclone intensity,** which is indicated by their sustained wind speeds and lowest atmospheric pressure.

Cyclones are associated with most severe weather in North America. Tropical cyclones include tropical depressions, tropical storms, and hurricanes. These three types of tropical weather systems create high winds, heavy rain, surges of rising seawater, and tornadoes. Extratropical cyclones cause strong windstorms, heavy rains, and surges of rising seawater on both the West and East Coasts, and during cooler months, snowstorms and blizzards. These

▲ FIGURE 10.11 **FAILURE OF NEW ORLEANS FLOOD WALL** The storm surge from Hurricane Katrina overtopped and flattened this flood wall constructed with steel pilings and concrete along the Industrial Canal. An American Society of Civil Engineers team inspects the damage to recommend changes in flood wall design. *(Mr. Rune Storesund, P.E.)*

cyclones also produce outbreaks of tornadoes and severe thunderstorms, especially east of the Rocky Mountains in the United States and Canada. Unlike hurricanes, which by definition are severe storms, most extratropical cyclones do not produce severe weather.

Although the destructive effects of tropical and extratropical cyclones can be similar, the two types of storms differ in their source of energy and structure. Tropical cyclones derive energy from warm ocean water and the latent heat that is released as rising air condenses to form clouds. Extratropical cyclones obtain their energy from the horizontal temperature contrast between the air masses on either side of a front. In hurricanes, the most intense of the tropical cyclones, warm air rises to form a spiraling pattern of clouds. The rising and warming air surrounding the center of a hurricane heats the entire core of the storm. In contrast, most extratropical cyclones are fed by cold air at the surface and another flow of cool, dry air aloft. The resulting storm has a cool core from bottom to top. Tropical cyclones that move over land or cooler water lose their original source of heat. These storms either dissipate or become extratropical cyclones moving along a front.

The classification and naming of cyclones are often debated because they are based on a combination of science, custom, and politics. Scientific classification and description of cyclones have their roots in regional names given to these storms. Naming of individual cyclones is generally limited to the more intense tropical cyclones and is a practice that began in the 1940s.[10]

CLASSIFICATION

Although *cyclone* is the general meteorological term applied to a large low-pressure system with winds circulating inward toward its center, a variety of terms are used to describe these systems in various parts of the world. For example, both forecasters and residents use the term *nor'easter* to describe an extratropical cyclone that tracks northward along the east coast of the United States and Canada. Onshore winds from these storms blow from the northeast and can sometimes reach hurricane strength.

The terminology for strong tropical cyclones is especially varied. In the Atlantic and eastern Pacific Oceans, these storms are called **hurricanes** after a Caribbean word for an evil god of winds and destruction.[12] In the Pacific Ocean west of the International Dateline (180° longitude) and north of the equator, hurricanes are referred to as **typhoons** after a Chinese word for "scary wind" or "wind from four directions."[12] Hurricanes in the Pacific Ocean south of the equator and Indian Ocean are referred to

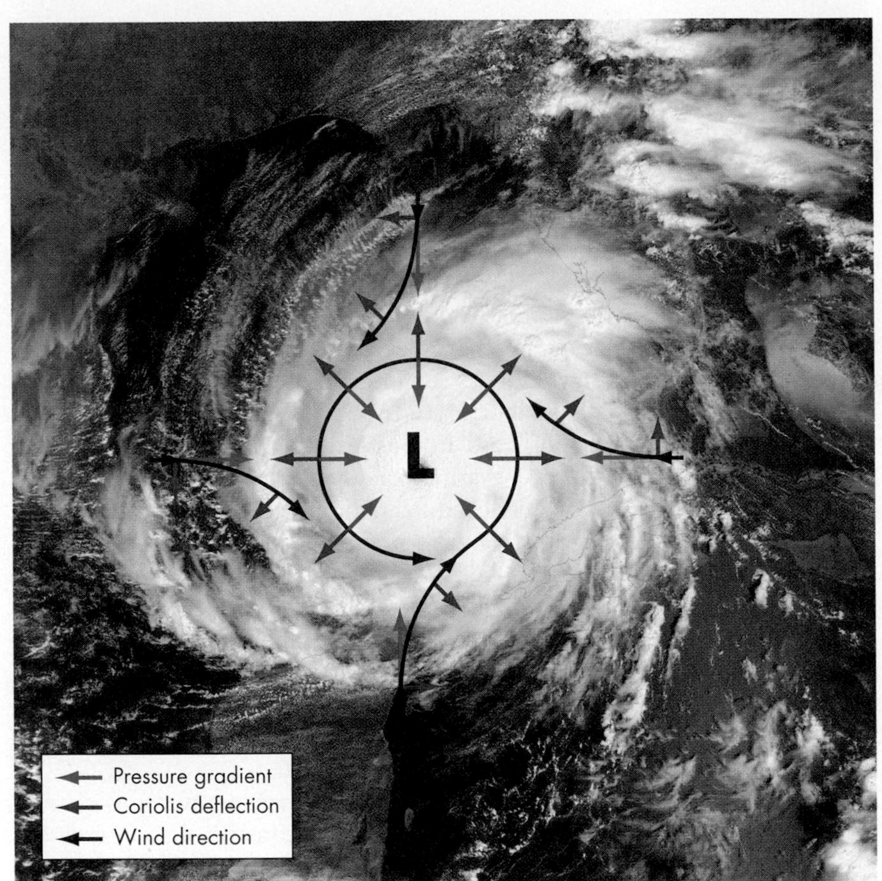

◀ **FIGURE 10.12 HURRICANE ROTATION** High-pressure air masses outside the storm flow along the pressure gradient (blue arrows) toward the central low pressure zone (L) within the storm. The wind created by the moving air is deflected to the right (black arrows) by the coriolis effect (red arrows), which causes the wind to rotate in a counterclockwise direction around the eye of the hurricane. Because the coriolis effect acts at right angles to the wind direction, coriolis deflects wind circling the eye of the storm away from the central low-pressure zone therby apposing the pressure gradient and sustain the storm. *(NASA)*

using some variation of cyclone, a term coined by a curator of the Calcutta Museum in India from the Greek word meaning "coil of a snake."[12] For simplicity, we will refer to all of these strong tropical cyclones as hurricanes.

Hurricanes are classified by their wind speed on a damage-potential scale developed by Herbert Saffir, a consulting engineer, and Robert Simpson, a NWS meteorologist, in the 1970s (see Table 10.1). The Saffir-Simpson Hurricane Scale is divided into five categories based on the highest 1-minute average wind speed in the storm and was first used for NWS public advisories in 1975.[10] A hurricane's category typically changes as it intensifies or weakens, and all but the weakest hurricanes will have more than one category assigned to them during their lifetime. Meteorologists describe Category 3 through 5 hurricanes as major hurricanes. As a hurricane's wind speed, and thus its category, increases, the atmospheric pressure in the storm's center drops. Category 5 hurricanes generally have a central atmospheric pressure of less than 920 millibars (27.17 in.). The record low pressure for an Atlantic hurricane, 882 millibars (26.05 in.), was set by Hurricane Wilma in 2005.[13]

NAMING

A small percentage of all cyclones are given names, either to identify where they form or to track their movement. Extratropical cyclones, especially those that become snowstorms, are sometimes named for the geographic area where they form (e.g., Alberta Clipper) (see Chapter 9). In contrast, all tropical depressions that become tropical storms and hurricanes are given individual names by government forecasting centers. These names are established by international agreement through the World Meteorological Organization (WMO). The standardized naming system adopted by the WMO replaced a patchwork of national practices that once included naming hurricanes in honor of the saint on whose holy day the storm came ashore, the military phonetic alphabet, and the names of girlfriends of weather forecasters.

Today, an official name is assigned once the maximum sustained winds of a tropical depression exceed 63 km (39 mi.) per hour and it becomes a tropical storm. Names are assigned sequentially each year from a previously agreed upon list for the region in which the storm forms. For example, in the Atlantic Ocean the first three names for 2010 are Alex, Bonnie, and Colin; and for 2011 they are Arlene, Bret, and Cindy, and for 2012 Alberto, Beryl, and Chris. Naming tropical storms and hurricanes helps forecasters keep track of multiple storms moving across the ocean at the same time.

In the Atlantic Ocean, the names come from one of six alphabetical lists of alternating men's and women's names. Each list has 21 names derived from English, Spanish, or French. The six lists are used in rotation; therefore, the 2010 list will be used again in 2016 and 2022. Names of especially deadly or damaging hurricanes, such as Katrina,

Rita, Ivan, and Charley, are retired from the rotating lists similar to the retirement of the jersey number of a star basketball player.

In 2005, the number of tropical storms and hurricanes in the Atlantic Ocean exceeded the list for that year. This was the first time since lists were established by the NWS National Hurricane Center in 1953 that more than 21 tropical storms and hurricanes formed in the Atlantic Ocean in a single year. Having run through the entire list of names for 2005, NWS forecasters went to their backup plan of assigning letters of the Greek alphabet to the additional storms (i.e., Tropical Storm Alpha, Hurricane Beta, Tropical Storm Gamma, etc.).

10.2 Cyclone Development

Tropical and extratropical cyclones differ not only in their characteristics, but also in their development. Most tropical and extratropical cyclones form, mature, and dissipate independently. Some tropical cyclones, however, transform into extratropical cyclones if they encounter an upper-level low-pressure trough as they weaken over land or over cooler seawater at higher latitudes.

TROPICAL CYCLONES

A tropical cyclone is a general term for large thunderstorm complexes rotating around an area of low pressure that has formed over warm tropical or subtropical ocean water. These complexes go by a variety of names depending on their intensity and location. Low-intensity tropical cyclones are called tropical depressions and tropical storms. High-intensity tropical cyclones are hurricanes. To be classified as a hurricane, a tropical cyclone must have sustained winds of at least 119 km (74 mi.) per hour somewhere in the storm.[10] Hurricanes require tremendous amounts of heat to develop and generally form only where the sea surface temperature is at least 26°C (80°F).

Most hurricanes start out as a **tropical disturbance,** a large area of unsettled weather that is typically 200 to 600 km (120 to 370 mi.) in diameter and has an organized mass of thunderstorms that persists for more than 24 hours. A tropical disturbance is associated with an elongated area of low pressure called a *trough*. Air in the disturbance has a weak partial rotation caused by the Coriolis effect. Tropical disturbances form in a variety of ways, including lines of convection similar to squall lines, upper-level troughs of low pressure, remnants of cold fronts, and *easterly waves* of converging and diverging winds that develop in the tropics. In the Atlantic Ocean, most tropical cyclones develop from easterly waves that form over western Africa. These waves form where "kinks" occur in the trade winds that blow west over the Atlantic from the West African coast (Figure 10.13).

A tropical disturbance may become a **tropical depression** if winds increase and spiral around the area of disturbed weather to form a low-pressure center. Warm moist air that is drawn into the depression behaves like spinning ice skaters who draw their arms toward the body, thereby increasing their rate of spin. Once maximum sustained wind speeds increase to 63 km (39 mi.) per hour, the depression is upgraded to a **tropical storm** and receives a name. Although the winds are not as strong in tropical storms as in hurricanes, their rainfall amounts can be as intense. If the winds continue to increase in speed, a tropical storm may become a hurricane.

Just as not all tropical depressions develop into tropical storms, even fewer tropical storms develop into hurricanes. Several favorable environmental conditions must be present to allow a hurricane to form. First, warm ocean waters of at least 26°C (80°F) must extend to a depth of 46 m (150 ft.) or more.[14] Warm surface temperatures alone are not enough for hurricane formation; there has to be an ample depth of warm water to provide energy for the storm. Second, the atmosphere must cool fast enough from the surface upward to allow moist air to continue to be unstable and convect. Layers of warm air aloft stop or "cap" hurricane development. Finally, there must be little vertical wind shear, that is, change in wind speed, between the surface and the top of the troposphere (see Figure 9.17). Strong winds aloft prevent hurricane development.

Once developed, a mature hurricane averages around 500 km (310 mi.) in diameter and consists of counterclockwise spiraling clouds that swirl toward the storm's center.[15] This rotation gives hurricanes their characteristic circular appearance when viewed from satellites (see Figure 10.12). Hurricane-strength winds of 119 km (74 mi.) per hour or greater are limited to the interior 160 km (100 mi.) of an average hurricane. The outer area of the storm has gale-force winds, which are winds greater than 50 km (30 mi.) per hour. Thus, much of a hurricane actually has winds that are less than hurricane strength.

The clouds that spiral around a hurricane are referred to as **rain bands** and contain numerous thunderstorms (Figure 10.14). Both these thunderstorms and the surface winds generally increase in intensity toward the center of the hurricane. The most intense winds and rainfall occur in the innermost band of clouds known as the **eyewall.** It is not uncommon for rainfall rates directly beneath the eyewall to reach 250 mm (10 in.) per hour.[15] A hurricane's eyewall is constantly changing as the storm progresses and some intense hurricanes develop double eyewalls. In these storms, the inner eyewall may gradually dissipate and be replaced with a strengthening outer eyewall. This replacement of one eyewall with another often precedes intensification of the hurricane. Trying to predict changes in hurricane intensity is of great interest to meteorologists. Recently, NASA scientists discovered that tall clouds up to 12 km (7 mi.) high, called *hot towers*, can develop within an eyewall 6 hours before a storm intensifies (Figure 10.15).[16]

A hurricane's eyewall surrounds a circular area of nearly calm winds and broken clouds known as the **eye.** Eye diameters range from 5 to more than 60 km (3 to more than 37 mi.).[17] Most hurricane eyes are smaller at the surface

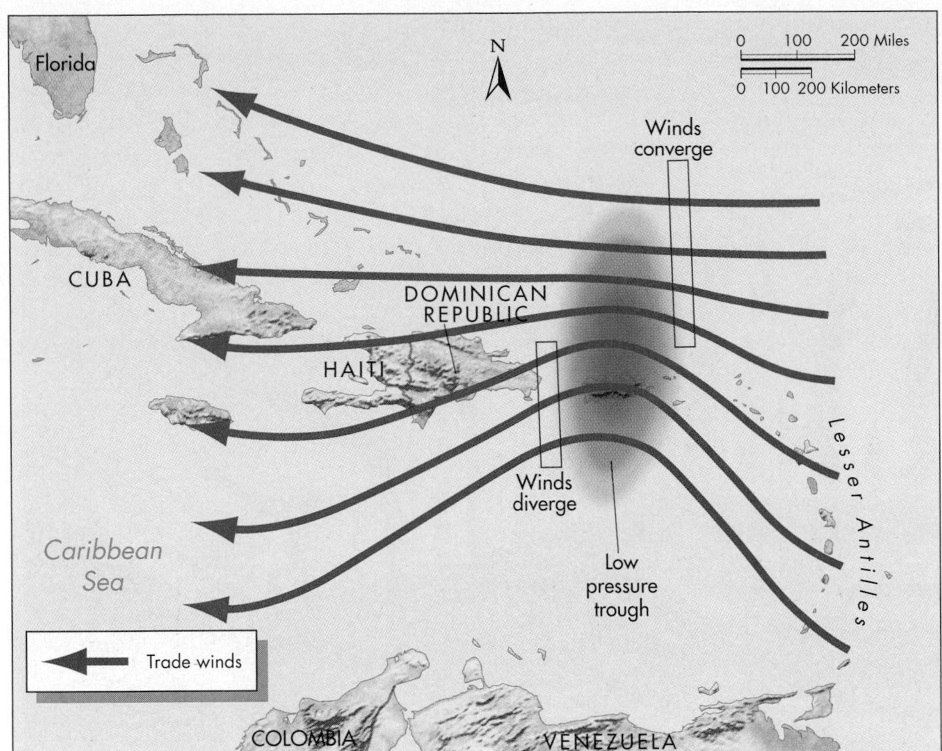

▲ FIGURE 10.13 **TROPICAL DISTURBANCE OVER PUERTO RICO** Tropical disturbances develop as waves or "kinks" in the trade winds as they blow from the east off the African coast. Also called "easterly waves," these disturbances are persistent clusters of thunderstorms associated with a low-pressure trough. Flow lines of the winds shown by the blue arrows converge on the east side of the disturbance and diverge on the west side. *(NASA)*

and widen upward to form a large, amphitheater-like area of nearly cloud-free blue sky surrounded by the white clouds of the eyewall (Figure 10.16). When a hurricane's eye passes directly overhead, there is a "calm before the storm," because the fierce winds of the other side of the eyewall will soon blow from the opposite direction. Birds may become trapped in the eye of a hurricane, and thousands of exhausted birds are sometimes reported roosting as an eye passes overhead.

In a typical hurricane, rising warm and moist air spirals upward around the eyewall. As the air rises, it loses moisture as clouds form and the air becomes drier. This upward spiraling rotation draws air from the eye and causes some of the drier air aloft to sink back downward in the center of the storm. The upward moving air also flows out of the top of the storm and builds a large cloud platform. Outward flowing air at the top of the storm is concentrated in one or more "exhaust jets" (see Figure 10.14).[12] This exhaust system is critical for the continued survival of the hurricane because it allows additional warm, moist air to converge inward in the lower levels of the storm. Outflow of warm air at the top of hurricanes is also a mechanism for the transfer of heat from Earth's tropics to its polar regions.

Movement of a hurricane is controlled by the Coriolis effect, which deflects the storm to the right in the Northern Hemisphere, and by steering winds that are 8 to 11 km (5 to 7 mi.) above the surface. In the Northern Hemisphere, this means that hurricanes commonly track westward in the trade winds across the Atlantic and curve to the north (Figure 10.17). The storms first curve to the northwest, then eventually to the north and northeast. However, deviations from this track are common when steering currents are weak. Hurricanes have even been known to reverse course and make an entire loop in their track. In the North Atlantic, hurricane tracks are also influenced by the location and size of the *Bermuda High*, a persistent high-pressure anticyclone that remains anchored in the North Atlantic during the summer and early fall. Many hurricane tracks tend to curve around the west side of the Bermuda High (Figure 10.17).

A mature hurricane will have a forward speed of 19 to 27 km (12 to 17 mi.) per hour.[18] In most cases, this slow forward speed will mean that a hurricane is a 2-day event for communities directly in the path of the storm. Toward the end of a hurricane's life, its forward speed may suddenly increase to 74 to 93 km (46 to 58 mi.) per hour.[18] This increase in speed happens to many hurricanes that move up the East Coast past Cape Hatteras, toward costal Canada (Figure 10.17). A rapid increase in forward speed can be especially dangerous if it results in an early landfall for the hurricane.

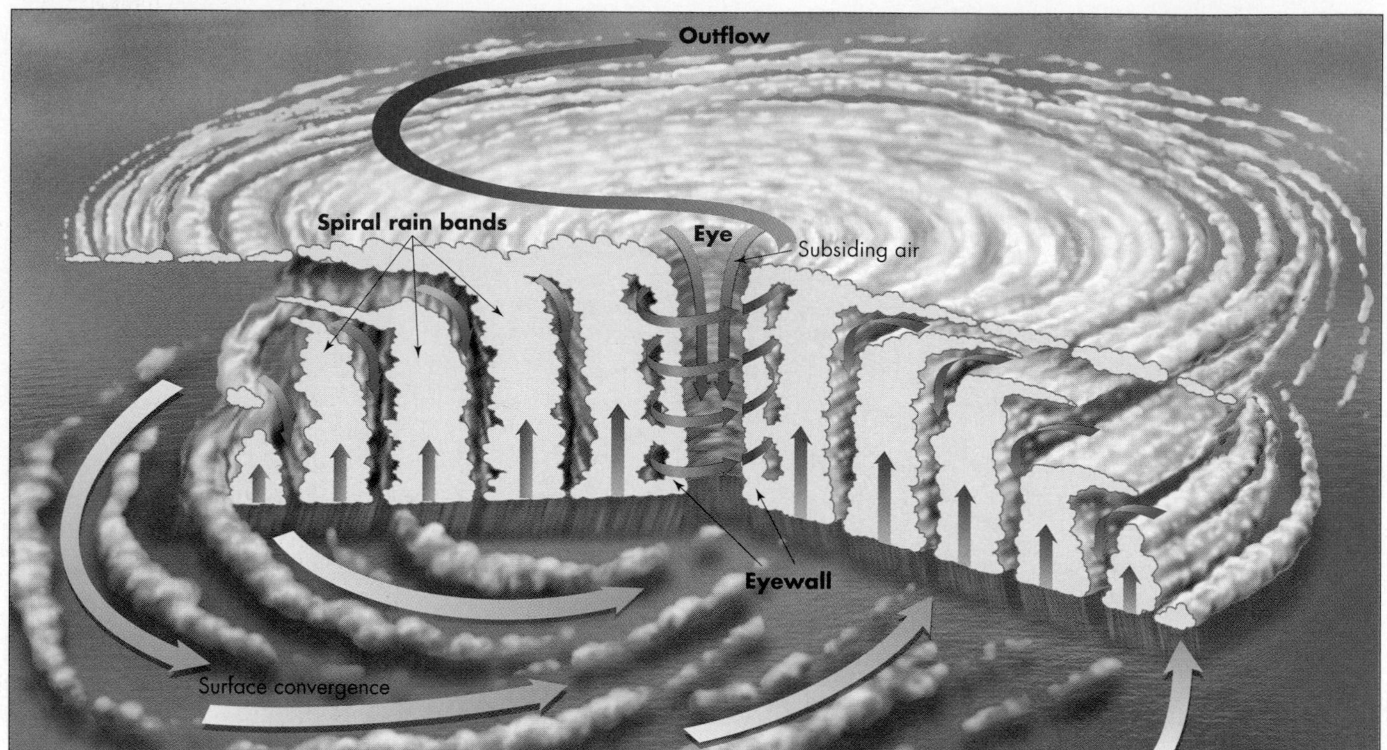

▲ FIGURE 10.14 **CROSS SECTION OF A HURRICANE** A vertically exaggerated cross section of clouds and wind patterns in a hurricane. Warm, moist air at the surface spirals toward the center of the storm and rises to form rain bands. Sinking dry air creates spaces between the bands that have fewer clouds. The innermost rain band is the eyewall cloud that surrounds the eye of the storm. Rising warm air from the eyewall either leaves the storm in one of several outflow jets aloft or loses moisture and sinks back into the eye of the storm. Subsiding dry air in the eye warms by compression, giving the storm its characteristic "warm core," with clearer skies and calmer winds in the eye. *(After NOAA)*

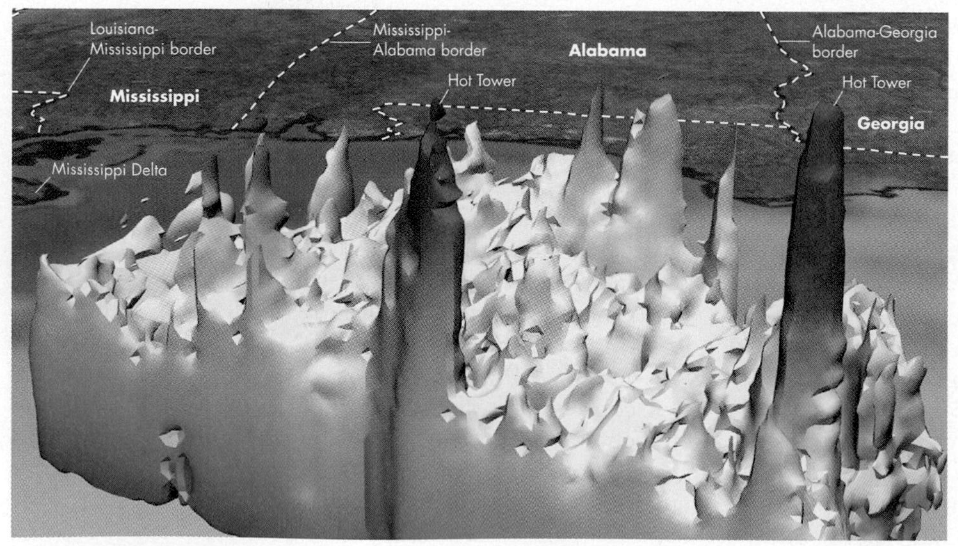

◄ FIGURE 10.15 **"HOT TOWERS" IN HURRICANE KATRINA** NASA's Tropical Rainfall Measuring Mission (TRMM) satellite spotted a pair of extremely tall thunderstorms in the eyewall and outer rain band of Hurricane Katrina shortly before it intensified to Category 5. Referred to as "hot towers," these clouds were more than 15,000 m (50,000 ft.) high. Like cylinders in an engine, these towers release huge amounts of heat and contribute to the rising winds in the storm. Detection of hot towers may help predict changes in hurricane intensity. *(Courtesy NASA/JAXA)*

▲ FIGURE 10.16 **STADIUM-LIKE CLOUDS FORM A HURRICANE'S EYEWALL** High winds and rotating clouds make up this eyewall of Hurricane Katrina. Photo taken from a National Oceanic and Atmospheric Administration (NOAA) Hurricane Hunter aircraft flying through the eye the day before the storm struck the Gulf Coast. Aircraft measurements of a hurricane's wind speed and atmospheric pressure help predict its behavior. *(NOAA)*

In summary, tropical cyclones develop in several stages from tropical disturbances. Each stage represents an increase in sustained winds and a decrease in atmospheric pressure. A tropical disturbance may become a tropical depression, then a tropical storm, and finally a hurricane.

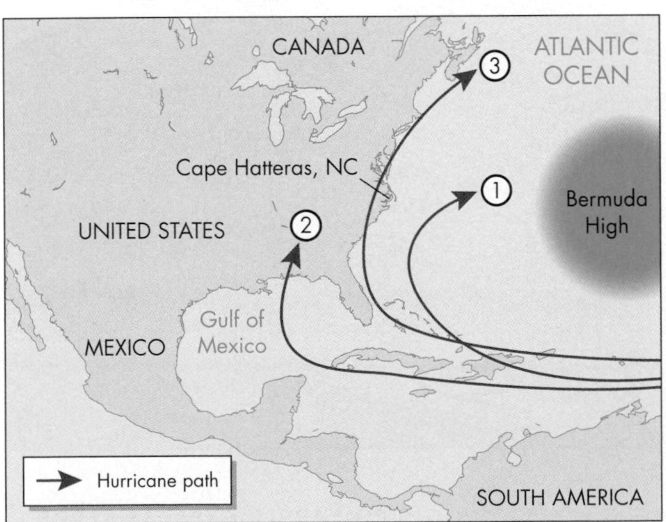

▲ FIGURE 10.17 **HURRICANE PATHS** Three common hurricanes paths showing deflection around the Bermuda High in the Atlantic Ocean. Each path, marked by a curving red arrow showing the direction of storm movement, starts in the central Atlantic Ocean to the east (right) of this map. All three paths threaten Caribbean islands; then path 1 can affect Bermuda, path 2 the southeastern United States, and path 3 the East Coast of the United States and Atlantic Canada.

In the Atlantic Ocean, most hurricanes start out as an easterly wave, a tropical disturbance that forms where trade winds converge and diverge off the West African coast. Hurricane development is favored where (1) there is thick layer warm water at the sea surface; (2) warm, moist air is free to rise upward toward the top of the troposphere; and (3) upper-level winds are relatively weak. Hurricanes have sustained winds of more than 119 km (74 mi.) per hour, average 500 km (310 mi.) in diameter, and have rain bands that spiral inward in a counterclockwise direction in the Northern Hemisphere. The strongest winds and most intense rainfall are in a hurricane's eyewall, a band of clouds that surrounds the calm eye of the storm. Hurricanes obtain their energy from the evaporation and subsequent condensation of warm tropical or subtropical seawater and generally lose strength when they move over land. On land, and over temperate ocean water, most cyclones are extratropical, rather than tropical, in origin.

EXTRATROPICAL CYCLONES

Two key ingredients contribute to the formation of an extratropical cyclone—a strong temperature gradient in the air near the surface and strong winds in the upper troposphere. Surface temperature gradients are generally strongest along a cold, warm, or stationary front (see Chapter 9). Therefore, most extratropical cyclones develop along fronts.

The second ingredient, strong winds in the upper troposphere, occurs in a concentrated flow of air called a **jet stream.** The Northern Hemisphere has two jet streams: one at an average altitude of 10 km (33,000 ft.) called the *polar jet stream*, and the other at an average altitude of 13 km (43,000 ft.) referred to as the *subtropical jet stream* (Figure 10.18 and Figure 9.11).[15] These jet streams vary from less than 100 km (60 mi.) to more than 500 km (310 mi.) in width and are typically several kilometers thick.[19] Jet streams cross North America from west to east at an average speed of 180 km (110 mi.) per hour in the winter and 90 km (55 mi.) per hour in the summer.[16] Peak flow in the jet streams can be twice the average wind speed.

The polar jet stream shifts from a path crossing the conterminous United States in the winter to one crossing southern Canada in the summer. This migration in its path causes a northward shift in the location of severe thunderstorm and tornado activity in the United States during the late winter, spring, and early summer. The subtropical jet stream normally crosses Mexico and Florida and is strongest in the winter.

Large high-pressure ridges and low-pressure troughs in the upper troposphere cause jet streams to bend north or south of their normal path, producing long meanders or waves in their flow. A jet stream may also split in two around isolated high-pressure centers and reunite down flow. Extratropical cyclones often develop in an area where jet stream winds curve cyclonically or diverge, such as on the east side of a low-pressure trough.

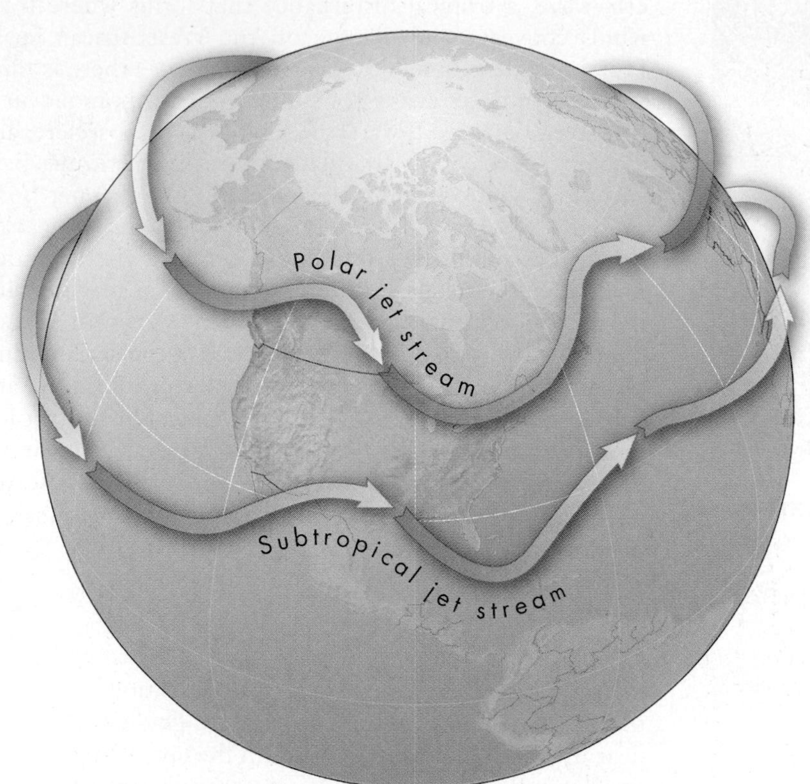

◀ **FIGURE 10.18 POLAR AND SUBTROPICAL JET STREAMS** Strong winds near the boundary between the troposphere and stratosphere are concentrated in jet streams. The strongest winds are in the polar jet stream at around 10 km (33,000 ft.) and generally less intense winds form the subtropical jet stream at about 13 km (43,000 ft.). *(After Lutgens F. K., and Tarbuck, E. J. 2007. The atmosphere: An introduction to meteorology, 10th ed. Upper Saddle River, NJ: Pearson Prentice Hall)*

Bending or splitting can cause the polar jet stream to dip southward and the subtropical jet stream to flow northeastward into the conterminous United States. The southern branch of a split polar jet stream in the Pacific Ocean brings warm moist air out of the tropics and can be recognized on infrared satellite images as a band of clouds extending northeastward from the equatorial Pacific Ocean (Figure 10.19). During some winters, a series of extratropical cyclones will track northeastward along this southern branch. This has led West Coast forecasters to refer to the flow of warm moist air as an "atmospheric river" or as the *Pineapple Express*, because of its origin near Hawai'i. Extratropical cyclones called nor'easters often form when bends of the polar and subtropical jet streams begin to merge off the southeastern coast of the United States.

Extratropical cyclones can intensify when they cross areas with strong low-level temperature gradients. For example, extratropical cyclones that hit the West Coast often weaken as they cross the Rocky Mountains and then deepen and strengthen as they leave the mountains and enter the Great Plains.

Most extratropical cyclones start as a low-pressure center along a frontal boundary, with a cold front developing on the southwest side of the cyclone and a warm front on the east (Figure 10.20). A conveyor-belt–like flow of cold air circulates counterclockwise around the cyclone, wedging

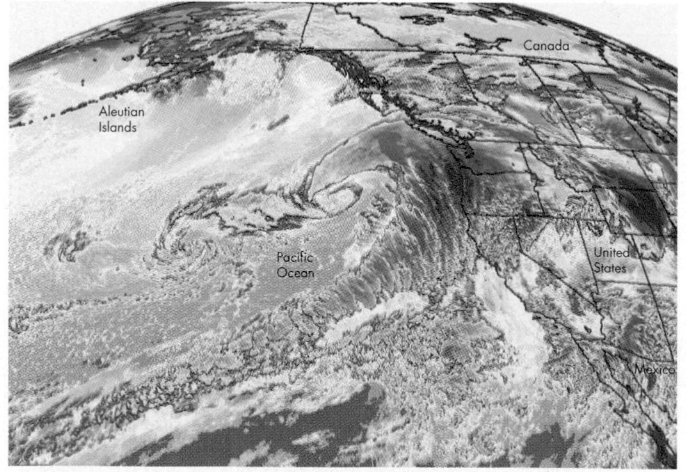

▲ **FIGURE 10.19 THE PINEAPPLE EXPRESS FEEDS MOISTURE TO WEST COAST EXTRATROPICAL CYCLONE** In this image, the Pineapple Express, a stream of warm moist air carried along by a southern branch of the polar jet stream, feeds an extratropical cyclone in the Pacific Ocean. Storms following this stream of tropical air hit the West Coast causing coastal erosion, flooding, and landslides. This color-enhanced infrared image was taken by the NOAA GOES-9 satellite. *(Courtesy of National Oceanic and Atmospheric Administration/National Climatic Data Center)*

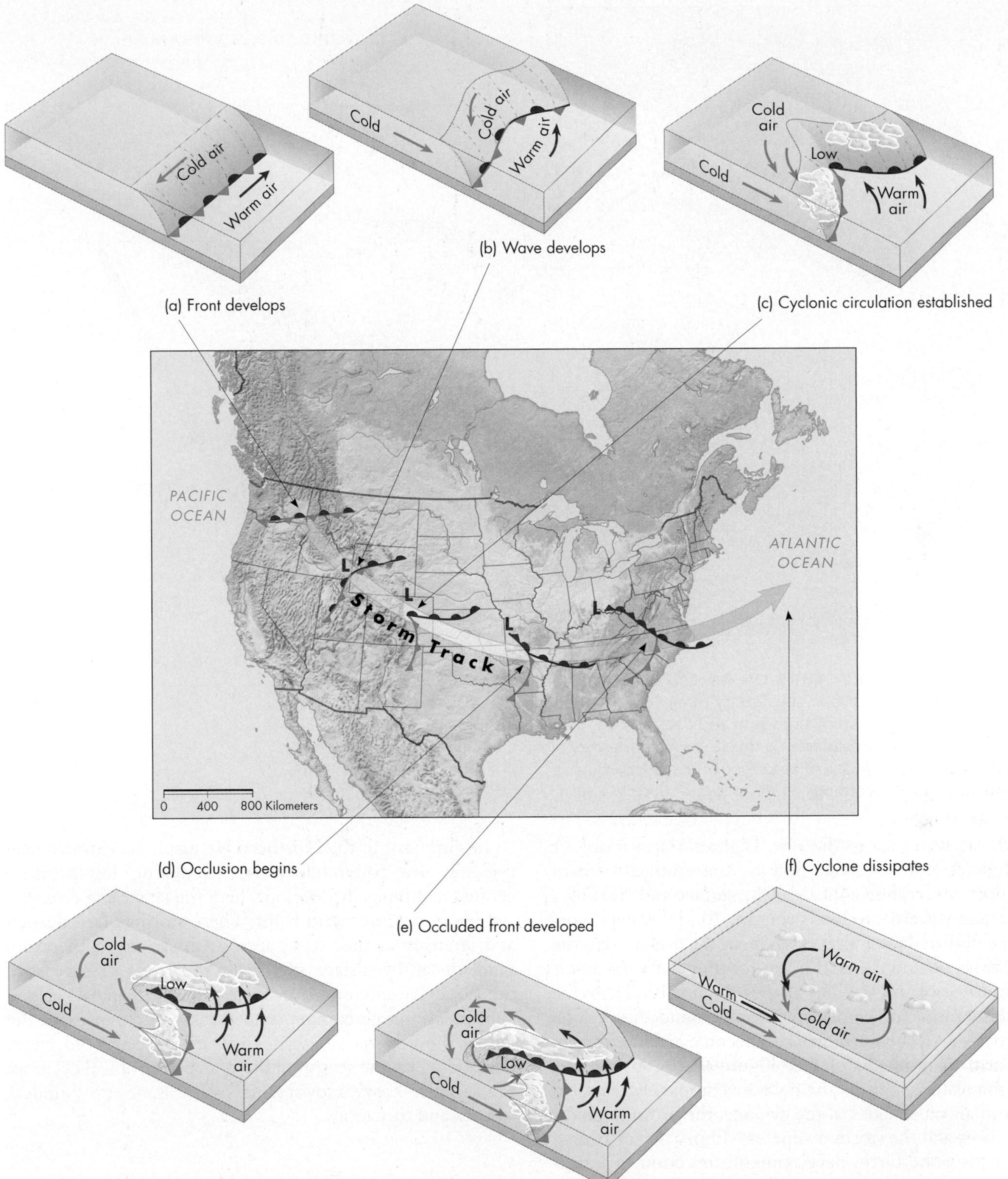

▲ **FIGURE 10.20 DEVELOPMENT OF AN EXTRATROPICAL CYCLONE** Stages in the development of an extratropical cyclone: (a) Air masses, moving in opposite directions along a front, such as the stationary front shown here, provide the temperature contrast for cyclone development; (b) a wave forms along the front at a place where upper-level winds diverge, such as a bend in the jet stream; (c) a surface low-pressure center develops as cold air pushes south on the west side and warm air pushes north on the east side of the cyclone; (d) the cold front advances faster than the warm front, catches up with the warm front, and displaces the warm air upward; (e) an occluded front develops with warm air held aloft by cold and cool air masses at the surface; (f) temperature contrast at the surface disappears, friction slows winds, and atmospheric pressure rises as the cyclone dissipates. *(After Lutgens, F. K., and Tarbuck, E. J. 2007.* The atmosphere: An introduction to meteorology, *10th ed. Upper Saddle River, NJ: Pearson Prentice Hall)*

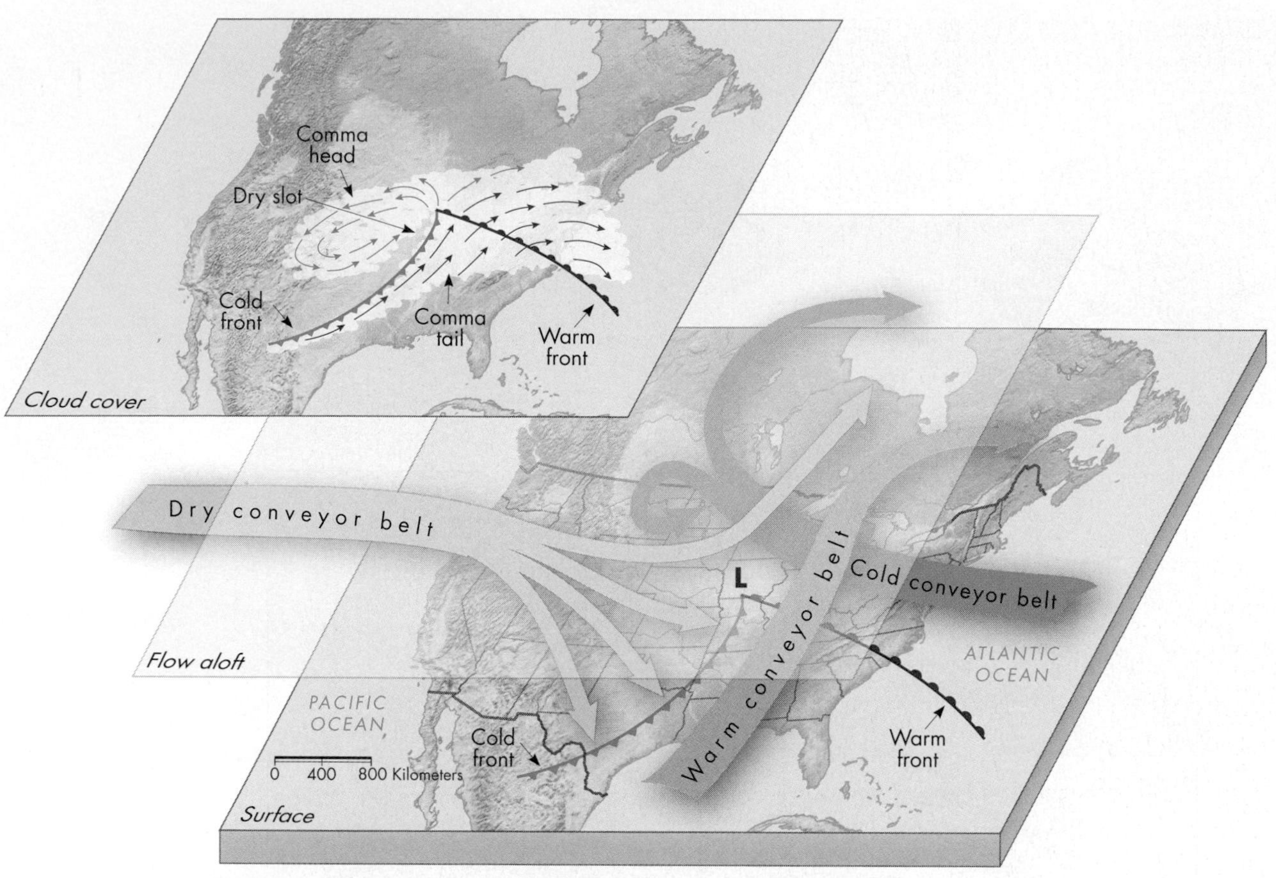

▲ **FIGURE 10.21 STRUCTURE OF AN EXTRATROPICAL CYCLONE** Once an extratropical cyclone develops, it is fed by three "conveyor belts" of air: Warm, generally moist air feeds the storm from the south and rises up over a stream of cold air coming from the east, and dry air aloft feeds the storm from the west sinking behind the advancing cold front at the surface. *(After Lutgens F. K., and Tarbuck, E. J. 2007.* The atmosphere: An introduction to meteorology, *10th ed. Upper Saddle River, NJ: Pearson Prentice Hall)*

beneath the warm air to the east. Lighter, warm moist air rises in a conveyor-belt–like flow on the southeast side of the cyclone, overriding cold air at the surface and creating a comma-like pattern of clouds (Figure 10.21). A conveyor-belt–like flow of dry air aloft often feeds the cyclone and the dry air sinks behind the cold front, forming a *dry slot* that is sometimes visible on satellite images (Figure 10.22). As the cyclone matures, the cold front wraps around the cyclone and merges with the warm front to become an *occluded front* with warm air trapped aloft (see Figure 10.20). At first, the storm intensifies; then, as the cold air completely displaces the warm air on all sides of the cyclone, the pressure gradient weakens and the storm dissipates within a day or two.

Predicting the birth, development, direction of movement, and death of extratropical cyclones is a challenge for forecasters. If all three ingredients described earlier are right, the development and strengthening of an extratropical cyclone can be rapid, occurring within 12 to 24 hours. Once formed, an extratropical cyclone's movement is typically steered by winds in the middle of the troposphere at around the 5600 m (18,000 ft.) level.[15] The forward movement of an extratropical cyclone is generally at half the speed of the steering winds.[17]

In summary, in the Northern Hemisphere, extratropical cyclones are counterclockwise circulating low-pressure centers that typically develop where the jet stream crosses a stationary cold or warm front. These storms often deepen and intensify as they cross areas with strong temperature gradients at the surface, such as on the Great Plains east of the Rocky Mountains. At maturity, an extratropical cyclone appears as a large comma-shaped mass of clouds with the tail of the comma along an eastward-moving cold front. Depending on the season of the year, extratropical cyclones can produce heavy snowstorms, blizzards, severe thunderstorms, and tornadoes.

10.3 Geographic Regions at Risk for Cyclones

In general, tropical and extratropical cyclones pose threats to different parts of the United States and Canada. Tropical cyclones have the greatest impact on coastal areas with warm offshore waters, such as the Gulf of Mexico and the Gulf Stream along the East Coast.

▲ FIGURE 10.22 **FEATURES OF AN EXTRATROPICAL CYCLONE** This NASA Geostationary Observational Environmental Satellite (GOES) image shows the comma-shaped cloud pattern of a mature extratropical cyclone moving across the eastern United States. Clouds in the comma head are formed by counterclockwise circulation of cold air around the cyclone. Sinking dry air behind a cold front forms the dry slot. The comma tail is created by thunderstorms and other clouds ahead of the cold front. *(John Jensenius/National Weather Service)*

Hurricanes pose the most serious threat to the eastern contiguous United States, Puerto Rico, the Virgin Islands, and U.S. territories in the Pacific Ocean. They are a lesser, but real, threat to Hawai'i and Atlantic Canada. On the Pacific coast, hurricanes frequently strike Baja California and the west coast of the Mexican mainland. Moist, unstable air from the remnants of Pacific tropical storms and hurricanes that moves inland over Mexico often contributes to torrential rains in the southwestern United States as far east as Texas.

The East and Gulf Coasts of the United States have the highest risk for tropical storms and hurricanes in North America and will experience, on average, five hurricanes each year. Most hurricanes that threaten the East and Gulf Coasts form off the west coast of Africa and take one of three tracks (see Figure 10.17):

1. Westward toward the east coast of Florida, sometimes passing over Caribbean islands such as Puerto Rico and the Virgin Islands; these storms then move out into the Atlantic Ocean to the northeast without striking the North American continent.

2. Westward over Cuba and into the Gulf of Mexico to strike the Gulf Coast.

3. Westward to the Caribbean and then northeastward skirting the East Coast; these storms may strike the continent from central Florida to New York. A few storms continue north as hurricanes to strike coastal New England or Atlantic Canada.

On occasion, hurricanes and tropical storms also form in the Gulf of Mexico and Caribbean Sea.

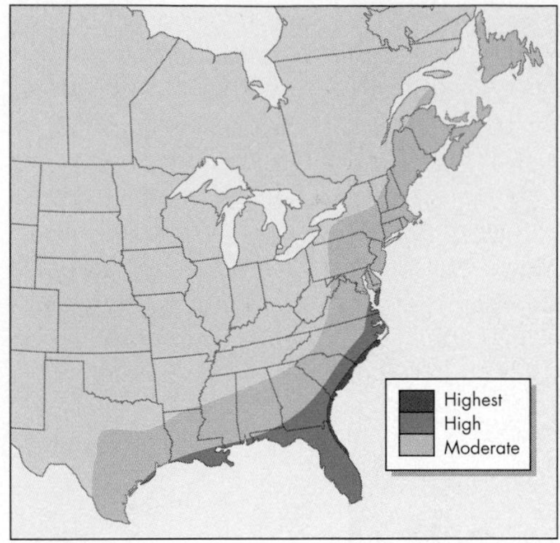

▲ FIGURE 10.23 **NORTH AMERICAN REGIONS AT RISK FOR HURRICANES** Areas shown are based on the likelihood that a hurricane will pass within 160 km (100 mi.). Highest risk area (red area) is likely to experience 60 hurricanes, the high-risk area 40 to 60 hurricanes, and the moderate-risk area fewer than 40 hurricanes in the next 100 years. Map based on observations from 1888 to 2005. *(Modified from U.S. Geological Survey and NOAA Atlantic Oceanographic and Meteorological Laboratory)*

The geographic regions at risk for hurricanes and tropical storms include the entire Gulf Coast from southern Texas to southern Florida, and the Atlantic Coast from southern Florida northward through the coastal provinces of Canada (Figure 10.23). This risk can be assessed with a map showing the probability that a hurricane will strike a particular 80 km (50 mi.) coastal segment in a given year (Figure 10.24). Hurricane-strike probabilities are particularly high along the coasts of southern Florida, Alabama, Mississippi, Louisiana, eastern Texas, and Cape Hatteras.

With the greatest hurricane threat in the southeastern United States, the threat to New England is often underestimated. In 1815, a major hurricane made landfall on Long Island, New York, and then crossed into Massachusetts and New Hampshire. Six years later, a hurricane passed very close to New York City and flooded parts of the city with a 4 m (13 ft.) storm surge.[12] The Great Hurricane of 1938, one of the fastest moving hurricanes on record, had an intensity and path nearly identical to the 1815 hurricane. The hurricane struck Long Island without warning and killed an estimated 680 people in New York, Connecticut, and Massachusetts. An estimated 80 percent of New England lost electrical power from the storm.[12] Winds were clocked at

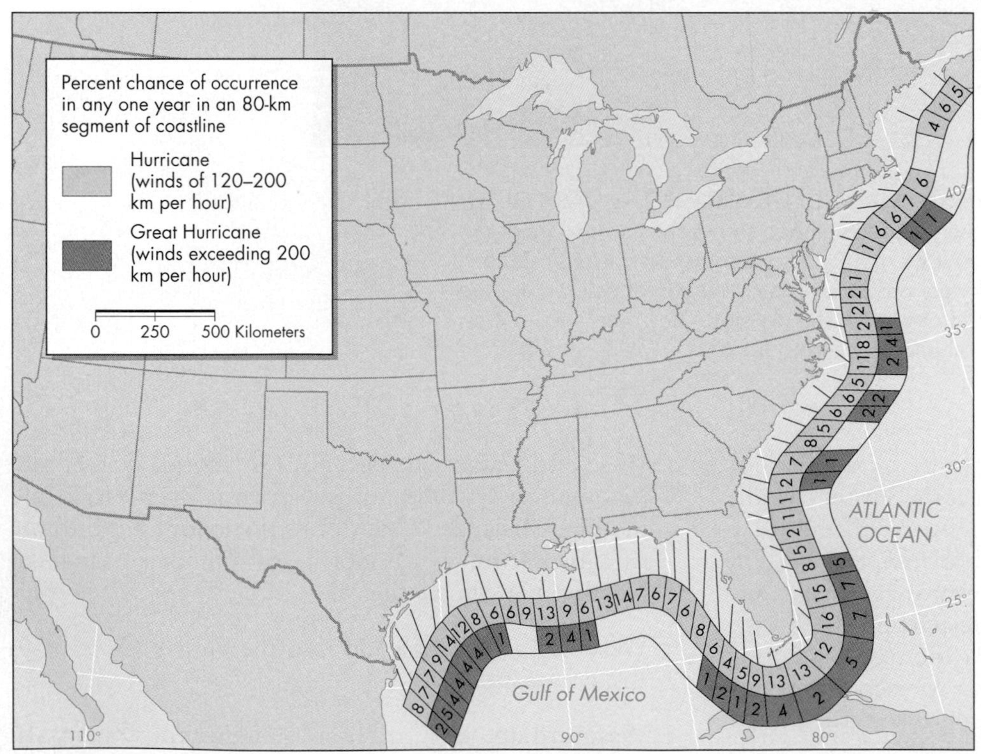

▲ FIGURE 10.24 **MAJOR HURRICANE HAZARD MAP** Probability that a major hurricane will strike a particular 80-km (50-mi.) segment of Gulf and Atlantic Coasts of the United States in a given year is shown in light orange, and the probability that an intense Category 5 storm will strike is shown in darker orange. *(From Council on Environmental Quality, 1981. Environmental trends)*

130 km (80 mi.) per hour in New York City and 160 km (100 mi.) per hour or more from Providence, Rhode Island, to the Boston area.[12] If the 1938 storm hit the same area today, the damage would most likely exceed $18 billion.[12]

Our emphasis on North America might lead one to believe that the North Atlantic Ocean is the most hazardous area for hurricanes and tropical storms. In fact, that is far from the case; the northwest Pacific generally has three times the number of hurricanes as the North Atlantic, and the Pacific and Indian Oceans overall have more hurricanes (Figure 10.25).[10] Two areas, the South Atlantic and southeast Pacific, rarely have hurricanes because of their cold surface waters. Hurricanes do not form close to the equator because of the absence of the Coriolis effect, and once formed, do not cross the equator.

Hurricanes affecting North America generally develop in the summer and early fall with the length of the hurricane season varying from year to year. The official Atlantic hurricane season starts on June 1 and ends November 30. Most Atlantic hurricanes occur in August, September, and October when the sea surface is the warmest. In contrast, the season for tropical cyclones in the Southern Hemisphere is from January to April.

In North America, the geographic region at risk from extratropical cyclones is far larger than from tropical cyclones. Severe weather from extratropical cyclones is greatest in the interior of the continent but may also occur in coastal areas. The greatest threat for this severe weather varies with location and time of year. Extratropical cyclones create strong windstorms in winter months along the Pacific Coast and nor'easters on the Atlantic Coast. In winter months, extratropical cyclones also produce heavy snowstorms and blizzards, especially in the Sierra Nevada and within and east of the Rocky Mountains. During spring and summer months, extratropical cyclones are responsible for the severe thunderstorm and tornado hazard in much of the United States and Canada east of the Rocky Mountains (see Chapter 9).

10.4 Effects of Cyclones

Tropical and extratropical cyclones claim many lives and cause enormous amounts of property damage every year. Both types of cyclones produce flooding, thunderstorms, and tornadoes, and extratropical cycles can create snowstorms and blizzards (see Chapters 6 and 9). Three additional effects of cyclones are especially damaging: storm surge, high winds, and heavy rains. In the case of hurricanes, storm surge by far causes the greatest damage and contributes to 90 percent of all hurricane-related fatalities.[18]

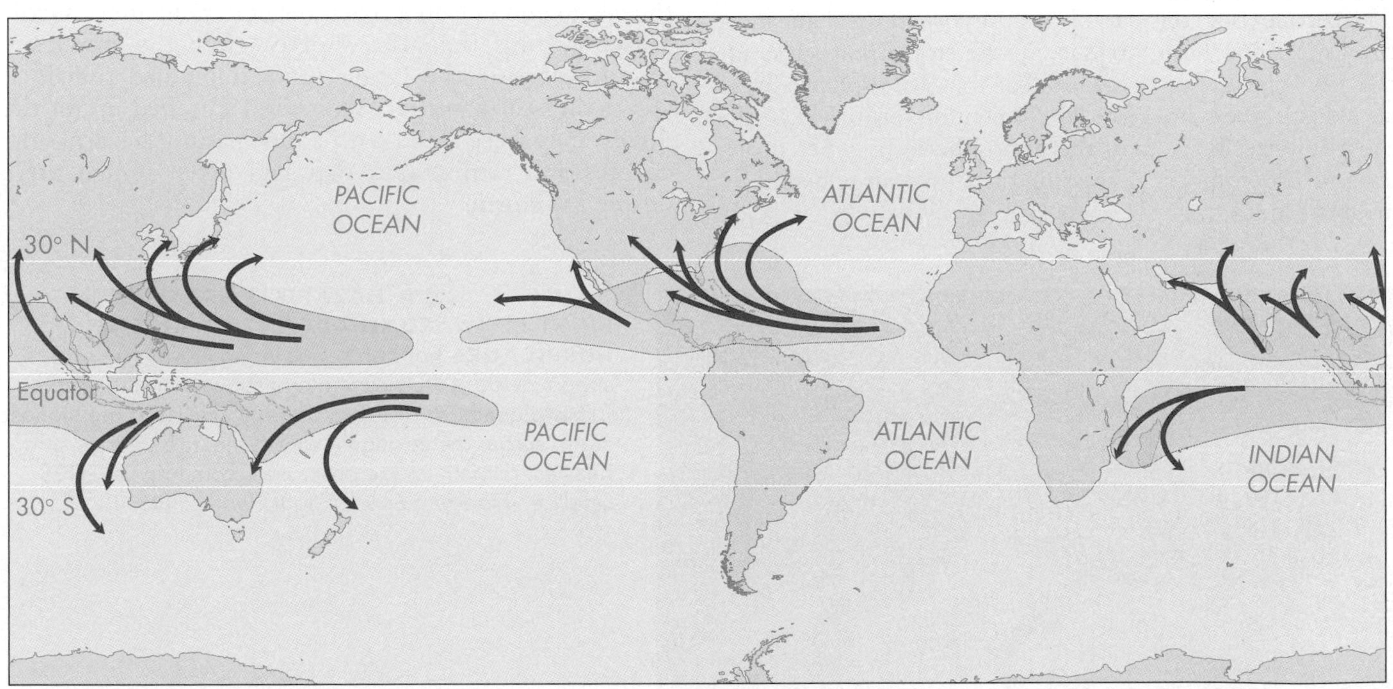

▲ **FIGURE 10.25 TYPICAL TROPICAL CYCLONE PATHS AND REGIONS WHERE THEY FORM** Most tropical cyclones develop between 5° and 20° latitude. Intense tropical cyclones are called hurricanes in the green region, typhoons in the blue region, and severe tropical cyclones or severe cyclonic storms in the tan region. Red arrows show the typical direction and paths of these storms.

STORM SURGE

Storm surge is the local rise of sea level that results primarily from water that is pushed toward the shore by the wind that swirls around a storm (see Figure 10.3). Major hurricanes and intense extratropical cyclones approaching a coastline often generate storm surges of more than 3 m (10 ft.). Storm surges of 12 m (40 ft.) or greater have occurred with hurricanes in Bangladesh and Australia.[19] Larger and more intense cyclones build an elevated dome of water beneath the storm that often creates a higher storm surge on the shore. Higher storm surges also develop when the coastal water depth gradually grows shallower toward shore. On these coasts, wind-driven water piles up until the wind shifts direction or decreases in speed to allow the surge to flow back to the sea. For most hurricanes in the Northern Hemisphere, the storm surge is greatest in the right forward quadrant of the storm as it makes landfall (Figure 10.26). The height of the surge is generally greatest near the time of maximum wind speed and is also greater if landfall takes place at high tide.

Two mechanisms in an intense cyclone cause the storm surge. The first, and by far the most significant one, is stress exerted by wind on the water surface. As winds grow stronger, the water level rises quickly, increasing its elevation by the square of the wind speed. The larger the area over which the wind blows, referred to as the *fetch*, the higher the water will rise. Thus large storms similar to Hurricane Katrina generally produce greater storm surges. The second, but considerably less important mechanism, is the low atmospheric pressure in the storm that sucks up the sea surface. For every millibar that atmospheric pressure drops, the sea surface rises 1 centimeter (0.5 in.).[12] In most intense hurricanes, the atmospheric pressure drops around 100 millibars, increasing the storm surge height by around 1 m (3 ft.).[12]

Storm surge is also affected by the shape of a coastline. In a narrow bay, lagoon, or lake, the height of the storm surge may increase as water sloshes back and forth in the enclosed or partially enclosed body of water. This produces *resonance*, a phenomenon that occurs when a wave reflecting from one shore is superimposed on a wave moving in another direction. The net effect is an amplification of the surge, similar to the amplified shaking from reflected seismic waves in a sedimentary basin (see Chapter 3).

Contrary to depictions in Hollywood movies, a storm surge is not an advancing wall of water; instead, it is a continual increase in sea level as the storm approaches landfall. Much of the storm surge damage actually comes from the large storm waves that are superimposed upon the surge. These storm waves, combined with ocean currents, erode beaches, islands, and roads.

Sand eroded from the beach and coastal sand dunes is carried landward by the storm surge and forms deposits known as **overwash** (see Figure 10.7). Many overwash deposits form as broad fans or deltas at the end of channels that the storm surge cuts through the beach and dunes perpendicular to the shoreline. Called *washover channels*, these features can be problematic where they can cut entirely through islands or peninsulas and isolate one area from another (Figure 10.27). Most washover channels are naturally repaired by coastal currents and wave action in the months following the storm. When Hurricane Allen struck South Padre Island on the Texas coast in 1980, a 4 m (13 ft.) storm surge flooded part of the island and cut numerous washover channels. Geologists visiting the area shortly after the hurricane found that only the larger sand dunes had remained above the storm surge. When they climbed up on the dunes, they were faced with a new natural hazard—the coyotes and rattlesnakes that had taken refuge there during the storm.

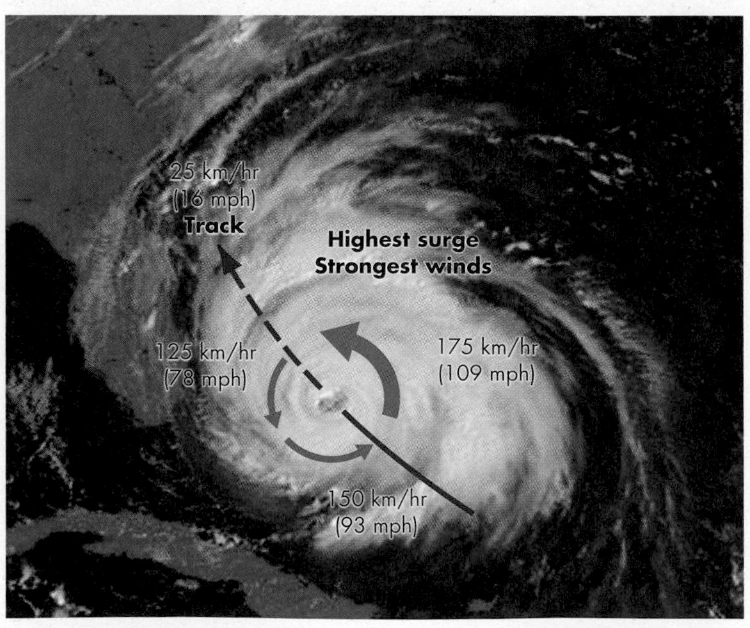

◀ **FIGURE 10.26 HAZARD IS GREATEST IN RIGHT FORWARD QUADRANT OF ATLANTIC HURRICANES** For hurricanes in the Northern Hemisphere, such as Hurricane Floyd shown here, the right forward quadrant of the cyclone typically has the highest storm surge, the strongest winds, most tornadoes, and heaviest rainfall. Image processed from NOAA GOES satellite. *(Goddard Space Flight Center/NASA)*

◀ **FIGURE 10.27 HATTERAS ISLAND BREACHED BY HURRICANE ISABEL STORM SURGE** Three channels, totaling 500 m (550 yd.) in width, were cut through Hatteras Island, North Carolina, by Category 2 Hurricane Isabel in September 2003. The breach, located close to where the eyewall crossed the island, is just west of Cape Hatteras. These washover channels were created by a 2.3 m (7.7 ft.) storm surge in the right forward quadrant of the storm. *(U.S. Geological Survey, Gainesville)*

HIGH WINDS

Although the storm surge from a hurricane, and sometimes an extratropical cyclone, is generally more deadly, the damage caused by high winds is more obvious, in part because the winds affect a much larger area. The Saffir-Simpson Hurricane Scale provides a good summary of the various types of wind damage that commonly occur in a hurricane (see Table 10.1). As mentioned previously, hurricane winds will be the strongest in the eyewall of the storm. Because of this, forecasters try to predict where the eye of the hurricane will make landfall.

The highest sustained wind speed that has been measured in a hurricane is close to 310 km (190 mi.) per hour. This speed was recorded for three Category 5 hurricanes: Camille in the Gulf of Mexico in 1969, Tip in the northwest Pacific in 1979, and Allen in the Caribbean Sea in 1980. Ever higher gusts of 320 km (200 mi.) per hour have been estimated for three hurricanes making landfall in the United States: the Labor Day Hurricane in the Florida Keys in 1935; Camille in Mississippi in 1969; and Andrew near Homestead, Florida, in 1992 (Figure 10.28).[21] The destruction wrought by 320 km (200 mi.) per hour winds is nearly total. In the 1935 Labor Day Hurricane, a train carrying World War I veterans working in the Florida Keys was blown entirely off the tracks (Figure 10.29).[20] Some of the 423 people who died in the Labor Day Hurricane appeared to have been sand blasted to death by sand blown from the beach.[19] The Labor Day Hurricane was the most intense storm to hit the United States in recorded history and served as the inspiration for the classic Humphrey Bogart movie *Key Largo*.[19]

Wind speeds in most hurricanes diminish exponentially once they make landfall; their wind speed is generally reduced in half within about 7 hours after the storm crosses the coastline.[12] However, a few hurricanes do not lose all their strength when they move over land or cold water and instead transition to become extratropical cyclones. During the transition, some of these storms can maintain or even increase their wind speed.[12] In 1954, Hurricane Hazel made landfall in North Carolina, yet it maintained 160 km (100 mi.) per hour winds and caused extensive damage as far north as Toronto, Canada. Transition to an extratropical storm may occur if a dying hurricane merges with an upper-level extratropical cyclone or cold front, or if it rapidly moves from warm to cold water.[12]

To many people's surprise, the strongest recorded wind in the United States has been from an extratropical cyclone, not a hurricane. This record was set by a 372 km (231 mi.) per hour gust at a weather station on Mount Washington in New Hampshire's White Mountains in April 1934. On average, Mount Washington experiences hurricane-strength winds 104 days each year because of the exposed location and mountainous topography.[19]

Strong winds and heavy rains from extratropical cyclones cause most of the severe weather on the West Coast from San Diego to Vancouver. One of the most destructive of these storms was the Big Blow of 1962, which produced wind gusts of 260 km (160 mi.) per hour along the Washington coast and 290 km (179 mi.) per hour on the Oregon coast (Figure 10.30).[12]

In the central United States and Canada, extratropical cyclones are responsible for the high winds in blizzards and tornadoes (see Chapter 9), and for deadly windstorms on the Great Lakes. One 1913 storm sank or grounded 17,300-foot-long ore ships on Lakes Huron and Erie.[19] Another extratropical cyclone in 1975 sank one of the largest ore ships ever to sail on the Great Lakes, the *S.S. Edmund Fitzgerald*, killing the entire crew. Close to the time the *Fitzgerald* sank in Lake Superior, the captain of a

▲ FIGURE 10.28 **AERIALC VIEW OF HURRICANE WIND DAMAGE** This mobile home community near Homestead Air Force Base in southern Florida was devastated by wind gusts of up to 320 km (200 mi.) per hour during Hurricane Andrew in 1992. Injuries were numerous and 250,000 people in Florida were left homeless from the hurricane. *(CORBIS)*

nearby ship estimated wind gusts of more than 160 km (100 mi.) per hour and waves up to 11 m (35 ft.).[17]

HEAVY RAINS

An average hurricane produces about a trillion gallons of rainwater each day, nearly three times the amount of freshwater consumed in the United States each year.[12] The heaviest rainfall in history has occurred where hurricanes have passed over, or close to, mountainous islands. La Reunion Island in the Indian Ocean has five rainfall records from tropical cyclones including 1144 mm (45 in.) in 12 hours, 1825 mm (72 in.) in 24 hours, and 2467 mm (97 in.) in 48 hours.[12] Unlike the storm surge and very high winds that are generally limited to hurricanes and intense extratropical cyclones, heavy rains occur with cyclones of lower intensity. Tropical storms, and even weaker tropical depressions, can cause extensive flooding. In 2001, Tropical Storm Allison, which never became a hurricane, dumped up to 940 mm (37 in.) of rain on Houston, Texas, and more than 690 mm (27 in.) in southern Louisiana.[21] More than 45,000 homes and businesses were flooded in Houston, the fourth largest city in the United States.[23] The storm killed 41 people, caused more than $5 billion in damages, and was the costliest tropical storm in U.S. history.[21] Rainfall from tropical storms and remnants of hurricanes is often most intense when the storms encounter topographic barriers, such as the Appalachian Mountains, and during nighttime hours when the rotational circulation of the storms contracts.

Intense rainfall from cyclones can cause widespread inland flooding (see Chapter 6). Four factors affect the extent of inland flooding: the storm's speed; changes in land elevation over which the storm moves; interaction with other weather systems; and the amount of water in the soil, streams, and lakes before the storm arrives.[22] In 2004, inland flooding from Hurricane Ivan extended from Georgia north through Ohio and Pennsylvania to southern New York State (see Figure 6.19). This flooding was caused, in part, by the slow movement of the storm, the hilly and mountainous terrain of the Appalachians, and saturation of the soil by heavy rains from the remnants of Hurricane Frances less than 10 days before. For some hurricanes, the damage and death toll associated with inland flooding exceeds the storm damage to the coastal zone. Inland flooding along the coast from a hurricane often lasts for many days as floodwaters draining the land surface flow slowly into coastal bays and lagoons.

▲FIGURE 10.29 **HURRICANE BLOWS TRAIN OFF TRACKS IN FLORIDA KEYS** Wind gusts of more than 320 km (200 mi.) per hour from the Labor Day Hurricane of 1935 blew the cars of this railroad train off its tracks in the Florida Keys. Hundreds of World War I veterans working in the Florida Keys were killed on the train as they were being evacuated from the storm. *(Historical Museum of Southern Florida)*

10.5 Linkages between Cyclones and other Natural Hazards

Although cyclones themselves present numerous direct hazards to human life and property, their effects are magnified by linkages to other natural hazards. Cyclones are closely linked to coastal erosion, flooding, mass wasting, and other types of severe weather, such as tornadoes, severe thunderstorms, snowstorms, and blizzards. In coastal areas, and perhaps others, the cyclone hazard is directly linked to climate change.

Some of the fastest rates of coastal erosion occur during the landfall of cyclones. This observation applies to both hurricanes on the Gulf and East Coasts and winter extratropical cyclones on the West Coast. Wind-driven waves superimposed on a storm surge actively erode beaches and coastal dunes, predominantly on the Atlantic Coast, and sea cliffs, primarily on the Pacific Coast (see Chapter 11). Some of the eroded sand in beaches and dunes is replaced during fair-weather conditions, whereas other sand is removed entirely from the coastal zone by waves and currents during storms.

Most cyclones making landfall cause both saltwater flooding from a storm surge and freshwater flooding from heavy rains. Areas where the coast has subsided are more vulnerable to both types of flooding (see Chapter 8). In other cases on land, cyclones cause destructive flooding because weak-steering winds allow the storms to stagnate over the same area for many days. Slow-moving cyclones often cause flash flooding because heavy rains continue to fall on already saturated soil.

In mountainous areas, heavy rains associated with cyclones can cause devastating landslides and debris flows. For example, in 1998, heavy rains from Hurricane Mitch in Guatemala, Honduras, El Salvador, and Nicaragua triggered widespread flooding, landslides, and debris flows, which killed at least 11,000 people.[23] Likewise, many landslides and debris flows in California, the Pacific Northwest, and the Appalachian Mountains are associated with heavy rains produced by extratropical cyclones.

Wind damage from hurricanes is not limited to the winds circulating around the eye of the storm. Hurricanes may generate *downbursts* of precipitation-driven winds with speeds of more than 160 km (100 mi.) per hour and numerous tornadoes.[10] Approximately half the hurricanes that come ashore produce tornadoes. Most of these tornadoes occur within 24 hours after the hurricane makes landfall.[20] Tornado formation is most likely from north to east-southeast of the center of the storm.[20]

Finally, as global sea level rises because of global warming (see Chapter 12), the effects of cyclones making landfall along coastlines will become more severe. This will be especially true in areas of the Gulf and East Coasts that have relatively flat coastal plains. For example, along much of the Texas coast, a 0.3 m (1 ft.) rise in sea level will result in a landward shift of the shoreline by approximately 300 m (1000 ft.). In the coming decades, storm surges from cyclones will be able to penetrate farther inland than ever before in history.

10.6 Natural Service Functions of Cyclones

Cyclones, and the weather fronts associated with them, are the primary source of precipitation in most areas of the United States and Canada. In the eastern and southern United States, hurricanes and tropical storms

◀ **FIGURE 10.30 PACIFIC STORM TOPPLES COLLEGE TOWER** The most destructive extratropical cyclone in Oregon and Washington history came ashore on Columbus Day in 1962 and destroyed the Campbell Hall Tower at Western Oregon State College in Monmouth, Oregon. Winds gusted to 187 km (116 mi.) per hour in Portland, 204 km (127 mi.) per hour in Corvallis, and 222 km (138 mi.) per hour in Newport, Oregon. Forty-six people died in the storm. *(Wes Luchau)*

often provide much needed precipitation to moisture-starved areas. This moisture may come from tropical cyclones that formed in the Atlantic Ocean and Gulf of Mexico, or it may be carried by upper-level winds from hurricanes along the Pacific coast of Mexico. Along the West Coast, extratropical cyclones are the major source of precipitation during the winter rainy season. Extratropical cyclones are also responsible for many heavy snowstorms during winter months.

As mentioned previously, hurricanes also serve an important function in equalizing the temperatures of our planet. Hurricanes elevate warm air from the tropics and distribute it toward polar regions.

Cyclones have important natural service functions in ecosystems. Winds from these storms carry plants, animals, and microorganisms long distances, helping populate volcanic islands with flora and fauna once the islands rise above sea level. Hurricane-generated waves can also stir up deeper, nutrient-rich waters resulting in plankton blooms in the open ocean and in estuaries. Cyclones rejuvenate ecosystems ranging from old-growth forests to tropical reefs. High winds topple weak and diseased trees in forests, and strong waves break apart some types of coral. Overall, these storms contribute to species diversity in many ecosystems.

In summary, cyclones are important sources of precipitation and contribute to the ecological health of many plant and animal communities. When viewed from space, cyclones are awe-inspiring sources of beauty and reminders of the power of natural processes.

10.7 Human Interaction with Cyclones

Human interaction with cyclones, especially hurricanes, has increased markedly in the past four decades. Prior to the advent of global satellite imagery and increases in the population of coastal areas, some hurricanes went undetected for all or part of their existence.

In the past 50 years, population growth in the United States has been the greatest in coastal areas that are especially vulnerable to the effects of cyclones. Today, approximately 53 percent of the people in the United States live in coastal counties.[24] Urban areas continue to expand along coastlines that have been devastated by coastal cyclones in historic times. This includes extensive development in Galveston, Texas; the greater Miami–Ft. Lauderdale and Tampa–St. Petersburg areas of Florida; and the Hamptons on eastern Long Island to name a few. The increase in coastal population has also increased property values in the coastal zone and will continue to raise the cost of damages from hurricanes and coastal extratropical cyclones.

As more people build in coastal counties, the nature of this construction can potentially affect the severity of the cyclone hazard. Destruction of coastal dunes for building sites increases the vulnerability of a shore to storm surge. Seawalls and bulkheads that are built to protect property reflect storm waves and contribute to beach erosion (see Chapter 11). Finally, improperly attached building materials, such as roofing, often become projectiles in hurricane-strength winds.

These projectiles damage other buildings and injure and sometimes kill people.

Many scientists have recently become concerned about the effects that our changing climate could have on tropical cyclones. Global warming, caused by gases that humans are putting into the atmosphere, is raising the temperature of the sea surface and contributing to rising sea level (see Chapter 12). Warm sea-surface temperature is a major factor in tropical cyclone development, and it is possible that warmer ocean water will increase hurricane intensity.[25] Rising sea level will also increase the reach of storm surge and extend the effects of the large waves that ride the surge.

10.8 Minimizing the Effects of Cyclones

Since we cannot prevent the cyclone hazard, the primary way to reduce property damage and avoid loss of life is to accurately forecast these storms and issue advisories to warn people in their path. Other ways to minimize the effects, such as building codes and evacuation procedures, are adjustments to living with hazard and are discussed later (see Section 10.9). Although the following discussion focuses on hurricanes, many of the challenges faced in forecasting their behavior, such as changes in path and intensity, also apply to intense extratropical cyclones.

FORECASTS AND WARNINGS

Minimizing property damage and casualties from hurricanes requires accurate forecasts that predict the behavior of these great storms as they approach landfall. The public must be warned in time to prepare or evacuate. Hurricanes can be difficult to forecast because they encompass many different weather processes, and they generally develop far from shore where there are few, if any, observers (see Case Study 10.2). Once a hurricane has formed, meteorologists must predict if it will reach land, where and when it will strike, how strong the winds will be, how wide an area will be affected, and how much rainfall and storm surge will accompany the storm. Because of the complexity of hurricane prediction, hurricane forecast centers must use a variety of tools.

Hurricane forecast centers are located all across the globe and include the U.S. National Hurricane Center (NHC) on the campus of Florida International University in Miami, Florida, and the Canadian Hurricane Centre (CHC) in Dartmouth, Nova Scotia. The NHC watches the Atlantic Ocean, Caribbean Sea, Gulf of Mexico, and eastern Pacific between May 15 and November 30 each year, and the CHC monitors tropical and posttropical cyclones that are likely to affect Atlantic Canada.[26,27] Both centers issue watches and warnings to the public and support hurricane research. In the United States, a *hurricane watch* is issued when a hurricane is likely to strike within the next 36 hours and a *hurricane warning* is given when the storm is likely to make landfall within the next 24 hours or less.[28] The CHC also issues watches and warnings at appropriate times.[27] Both centers utilize information from weather satellites, hurricane-hunter aircraft, Doppler radar, weather buoys, reports from ships, and computer models to detect, forecast, and predict hurricanes.

Hurricane Forecasting Tools Weather satellites are probably the most valuable tools for hurricane detection, since the great storms form over the open ocean. Satellites can detect early warning signs of hurricanes long before the storm actually begins and alert meteorologists to areas that should be watched closely. Although satellites are excellent for detecting hurricanes over the open ocean, they cannot give accurate detailed information about wind speed and other conditions within the storms. Once a hurricane is detected, other tools are used to develop and refine predictions about the storm's behavior.

Aircraft are an invaluable tool, and special planes are flown directly into a hurricane to gather data, especially regarding intensity. U.S. Air Force hurricane-hunter aircraft perform most of the reconnaissance, although research aircraft of the National Oceanic and Atmospheric Administration (NOAA) also fly missions through hurricanes. These aircraft can begin their data collection when Atlantic tropical cyclones cross 55° W longitude, a north-south line approximately 2600 km (1600 mi.) east of Miami, Florida (Professional Profile 10.3). During winter months, Air Force and NOAA aircraft also collect meteorological data on extratropical cyclones in the Pacific Ocean and along the East Coast.

Doppler radar systems provide another means of collecting data about hurricanes that come within about 320 km (200 mi.) of the United States mainland or some islands, such as Puerto Rico and Key West. Radar can reveal information about rainfall, wind speed, and the direction that the storm is moving (Figure 10.31).

Weather buoys floating along the Atlantic and Gulf Coasts are also used. These buoys are automated weather stations that continuously record weather conditions at their location and transmit information to the U.S. National Weather Service or the Meteorological Service. Some information about offshore weather conditions is also obtained from ships that are in the vicinity of a hurricane.

Meteorologists today use computer models that take into account all available data to make predictions about hurricane tracks (Figure 10.32). Although these models have vastly improved our ability to predict where and when a storm will strike, they are still lacking somewhat in predicting the intensity of the storm; therefore, reconnaissance flights remain an integral part of hurricane prediction.[10] Although such flights are dangerous and extremely expensive, their expense pales in comparison to the cost of evacuating additional miles of coastline with less accurate

CASE STUDY 10.2

Hurricane Intensity and Warm Seas—Opal and Katrina Meet "Eddy"

Determining when and where a hurricane will change intensity is critical to forecasting these destructive tropical cyclones. Scientists must use the forecast intensity to model and predict the extent of the storm surge. Emergency managers rely on storm surge predictions to make decisions on the timing and location of evacuations. If intensity and storm surge are underestimated, people may die if they are not evacuated. Overestimating intensity and storm surge can result in unnecessary evacuation costs. Thus forecasters walk a tightrope prior to hurricane landfall, hoping that they have neither underestimated nor overestimated a storm's intensity. To make things more anxiety producing, the intensity of a hurricane can rapidly increase or decrease in less than 24 hours—literally changing overnight. Although a great deal of research is devoted to improving intensity forecasts, we still do not fully understand how hurricanes rapidly intensify or just as rapidly weaken. Several tantalizing clues to hurricane intensification come from examining the behavior of Hurricanes Opal (1995) and Katrina (2005) as they crossed the Gulf of Mexico.

Both Opal and Katrina encountered warm waters derived from the *Loop Current,* a bend in an ocean current that flows northward through the Yucatan Channel between Cuba and the Mexican Yucatan Peninsula before entering the Gulf of Mexico. Once the Loop Current leaves the Gulf of Mexico, it becomes the Florida Current flowing through the Straits of Florida between the Florida Keys and Cuba. The Florida Current then merges with the Antilles Current along the east coast of Florida to eventually become the Gulf Stream (Figure 9.B). Both Opal and Katrina rapidly intensified as they crossed a large *eddy,* a circular movement of ocean water, which had pinched off a bend in the Loop Current. This eddy, known as the *Eddy Vortex,* is called a warm-core ring by oceanographers because the water temperature in the clockwise circulating water is warmer than the surrounding ocean.[30,31] Both the Loop Current and Eddy Vortex have significant thicknesses of warm water that cause the sea surface to rise 35 to 60 cm (13 to 23 in.) above the surrounding Gulf of Mexico.[32]

In the case of Opal, the hurricane formed from a tropical storm that had crossed the Yucatan Peninsula into the Gulf of Mexico. Opal was predicted to remain a Category 1 or 2 storm (see Table 10.1) as it slowly moved toward Florida. In fact, forecasters weren't even sure that it would maintain hurricane strength until landfall. On the basis of these predictions, the National Hurricane Center (NHC) decided that a hurricane warning was not necessary for the storm's projected landfall in the Florida Panhandle. That decision was made at 5:00 P.M. on the day before the hurricane struck. Later, in the middle of the night, Opal crossed the Eddy Vortex, suddenly intensified to nearly a Category 5 storm, and rapidly headed toward shore (Figure 10.C)![10,30] The NHC issued a hurricane warning, but most people did not receive it until they awoke the next morning with a violent hurricane just offshore. Last-minute evacuations caused massive traffic jams, which further endangered lives. Fortunately, Opal then lost strength and dropped back to Category 3 intensity before landfall, helping to save countless lives.[10] Although the hurricane killed nine people, the death toll would have been much higher if the storm hadn't weakened before making landfall.

For Hurricane Katrina, the story was somewhat similar. Katrina formed as a tropical storm in the Bahamas and became a Category 1 hurricane as it crossed the Florida peninsula. The hurricane then intensified from a low-end Category 3 storm to a Category 5 cyclone in less than 12 hours as it crossed the Loop Current and Eddy Vortex (Figure 10.D).[3,31] As the storm intensified, it also expanded in size; on the day before landfall, its hurricane-force winds extended nearly 170 km (105 mi.) from the center.[3] Fortunately for coastal residents, Katrina then rapidly lost intensity before making landfall on the Mississippi Delta and further weakened before its second landfall along the Mississippi-Louisiana border.[3] Three weeks after Katrina, Hurricane Rita also passed over the Loop Current and Eddy Vortex and intensified to a Category 5 storm (see Figure 10.B).[31,33]

Although the thickness of warm seawater encountered by a hurricane is one of several factors influencing intensity, the behavior of Opal, Katrina, and later Rita suggests that it may be an important one. Scientists can now use a combination of radar altimeter and infrared sensors on multiple NASA, European Space Agency, and military satellites to better understand the link between sea surface temperature and hurricane intensity. You can bet that they will be looking closely each time another hurricane meets "eddy."

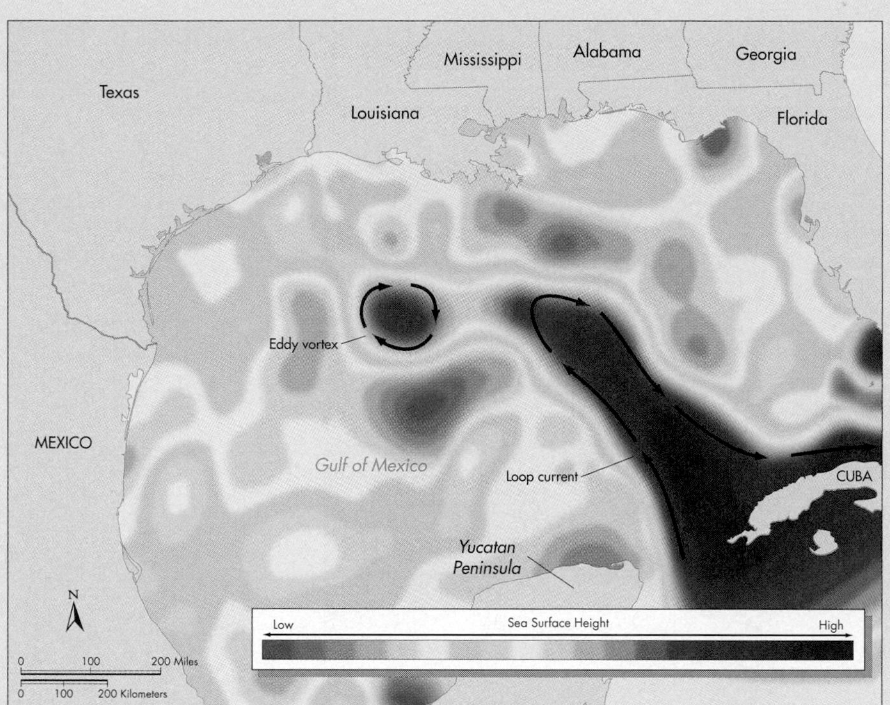

FIGURE 10.B LOOP CURRENT AND EDDY VORTEX IN GULF OF MEXICO Map showing location and direction of the Loop Current and the Eddy Vortex, a warm-core eddy, in September 2004 when Hurricane Rita crossed the Gulf of Mexico. This image is based on radar altimeter measurements of the sea surface made by the NASA Topex/Poseidon and Jason satellites. Red areas show highest sea surface elevation and thus the warmest ocean water. *(Modified from image NASA/JPL/University of Colorado-CCAR)*

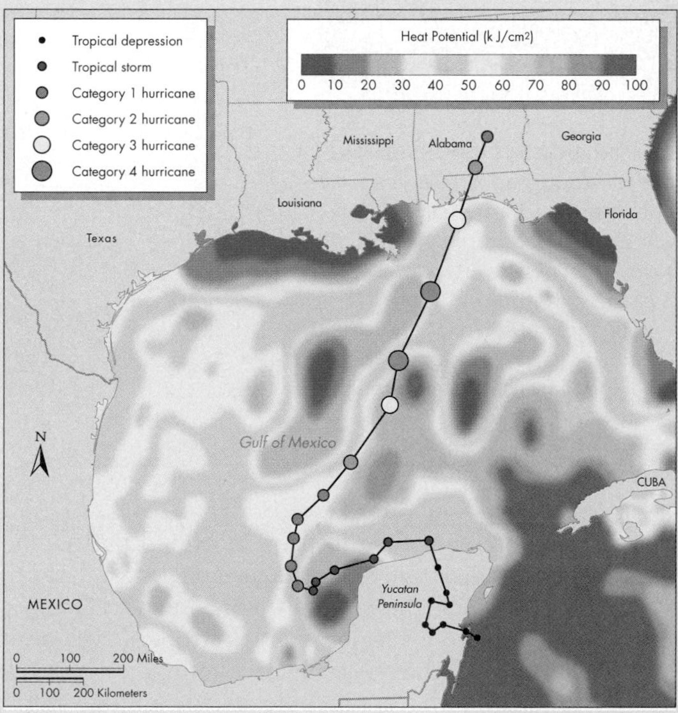

FIGURE 10.C HURRICANE OPAL INCREASES IN INTENSITY WHILE CROSSING EDDY VORTEX This map, based on satellite measurements of the sea surface elevation, shows the heat potential (in kilojoules per square centimeter) of Gulf of Mexico waters for hurricane intensification in October 1995. Red and orange areas have a very high heat potential. Close to the time these measurements were made, Hurricane Opal passed over the Eddy Vortex and increased in intensity from a weak Category 3 storm to nearly a Category 5 hurricane. The upper 100 m (330 ft.) or more of water in the eddy was higher than 26°C (80°F), the temperature needed for hurricane growth. *(After NOAA/Atlantic Oceanographic and Meteorological Laboratory)*

(continued)

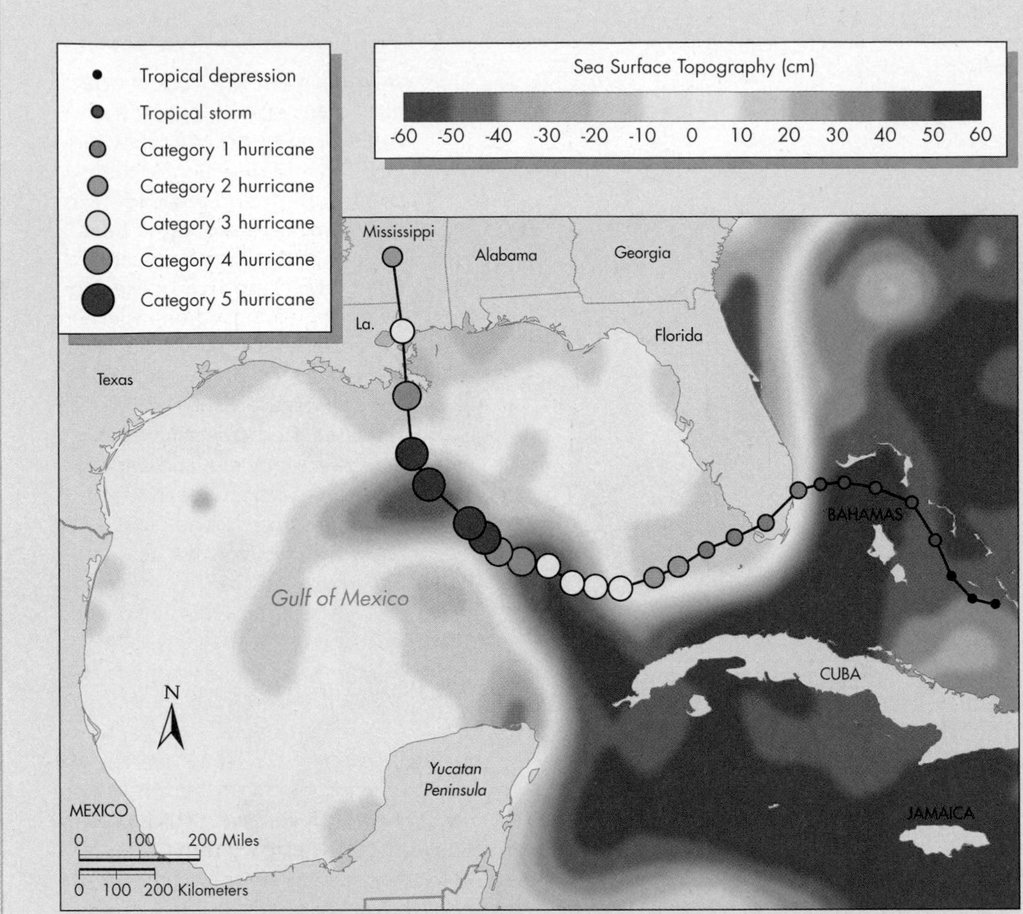

▲ FIGURE 10.D **INTENSIFICATION OF HURRICANE KATRINA AS IT CROSSED THE LOOP CURRENT AND EDDY VORTEX** This map of sea surface topography, prepared from satellite radar altimeters, shows that Hurricane Katrina intensified from Category 3 to Category 5 as it crossed the Loop Current and Eddy Vortex. Oceanographic soundings indicate that sea surface topography correlates with the amount of warm water below the ocean surface. The size and color of the circles marking Katrina's positions show its stage of development from a tropical depression to a tropical storm to the category of hurricane. *(Modified from Scharroo, R., Smith, W. H. F., and Lillibridge, J. L. 2005. Satellite altimetry and the intensification of Hurricane Katrina.* EOS, Transactions American Geophysical Union *86: 366)*

forecasts. Currently only the U.S. government uses such flights, and for hurricanes their use is limited to the western Atlantic Ocean, Caribbean Sea, Gulf of Mexico, and an area around Hawai'i.[10,29]

Storm Surge Predictions Forecasts of the track and the intensity of cyclones making landfall on coastlines are used to help predict the height and extent of the storm surge. As mentioned previously, forecasters can use wind speed, fetch, and average water depth to get a general idea of the elevation of the storm surge. Forecasters must also predict the time of day that the surge will arrive because the additional water piled up by the storm is added to the level established by the astronomical tides. Along the Gulf of Mexico, with its tidal range of only a few feet, this is less important than in parts of New England and Atlantic Canada, where the tidal range is 3 m (15 ft.) or more. Along these parts of the Atlantic Coast, a storm surge from a tropical or extratropical cyclone at high tide would be especially devastating.

Predicting the geographic extent of the storm surge requires detailed information about land elevation along a coastline. An airborne laser surveying technique, known as LIDAR (Light Detection and Ranging), is used to prepare detailed digital elevation models of coastlines. This detailed survey data can be entered into advanced computer programs to predict both the height and extent of the storm surge (see Figure 10.33). Computer programs, such as SLOSH (Sea, Land, and Overland Surges from Hurricanes)

PROFESSIONAL PROFILE 10.3

The Hurricane Hunters

Even though amateur tornado chasers occupy an established, if eccentric, niche in the world of weather enthusiasts, hurricane hunting is still a job for the pros.

In fact, Major Chad "Hoot" Gibson boasted that his crew is the only "weather reconnaissance" squadron in the world. The aptly named "Hurricane Hunters" are a crew of Air Force members who fly planes directly into the center of hurricanes while the storms are still lingering over the ocean (Figure 10.E). During their wild ride, they gather important data about the strength of the storm, which the National Weather Service uses in its prediction of where the storm will strike.

The crew begins 170 km (110 mi.) out from where they estimate the eye of the hurricane will be, and they do their best to fly directly for the center in a straight line. In the course of doing so, Gibson said, they've encountered just about every form of severe weather out there: hail, updrafts and downdrafts, turbulence, and even tornadoes. "We avoid those," he said.

At the center of the hurricane is the furious eyewall, whose winds commonly blow in the neighborhood of 225 km (140 mi.) per hour. And breaching the wall is no smooth ride.

"There are times when the plane vibrates so badly that you can't read your computer screen," Gibson said.

But when they've made it through, the view from the eye, Gibson said, is awe inspiring.

"Amazing. Fantastic," he said. "It's like sitting at the 50-yard line of a football stadium." In a well-formed eye, the sky is blue overhead, although the water below—the plane flies at an elevation of around 3050 m (10,000 ft.)—still "looks like a washing machine."

But the Hurricane Hunters don't dally too long in the eye. One reason is that the winds at the center are so calm, at around 4 to 7 km (2 to 5 mi.) per hour, that the plane itself can throw off the data if it moves too much. Instead, they make a quick pass through the eye as the plane drops a dropsonde, a "weather station in a can," as Gibson described it, to take necessary measurements. Then it's back through the opposite wall, and another bumpy ride.

To frame the sharp contrast in wind speed, Gibson recalled a mission through the center of Hurricane Floyd in September 1999 in which the crew entered the first eyewall, experienced 220 km (137 mi.) per hour winds, then reached the eye 4 minutes later where the winds were blowing at 8 km (5 mi.) per hour, and then 5 minutes later were again confronted with 225 km (140 mi.) per hour gusts at the other side.

Not all eyes are perfectly formed, and some storms will have several smaller centers, so the hurricane hunters use the telltale 180-degree shift in the winds—the trademark of the center of a vortex—as the official core of the storm.

The squadron now flies the WC130J, a specially equipped version of the propeller-driven, Lockheed Martin Hercules military transport plane. Air Force pilots, and their predecessors in the Army Air Corps, have been flying scientific missions through hurricanes since 1944.

—CHRIS WILSON

FIGURE 10.E HURRICANE HUNTER AIRCRAFT This U.S. Air Force Reserve Lockheed Martin WC-130 aircraft based at Kessler Air Force Base, in Mississippi, is specially equipped for taking weather measurements in hurricanes. WC-130 aircraft similar to this one investigate hurricanes at the direction of the National Hurricane Center. *(George Hall/Corbis)*

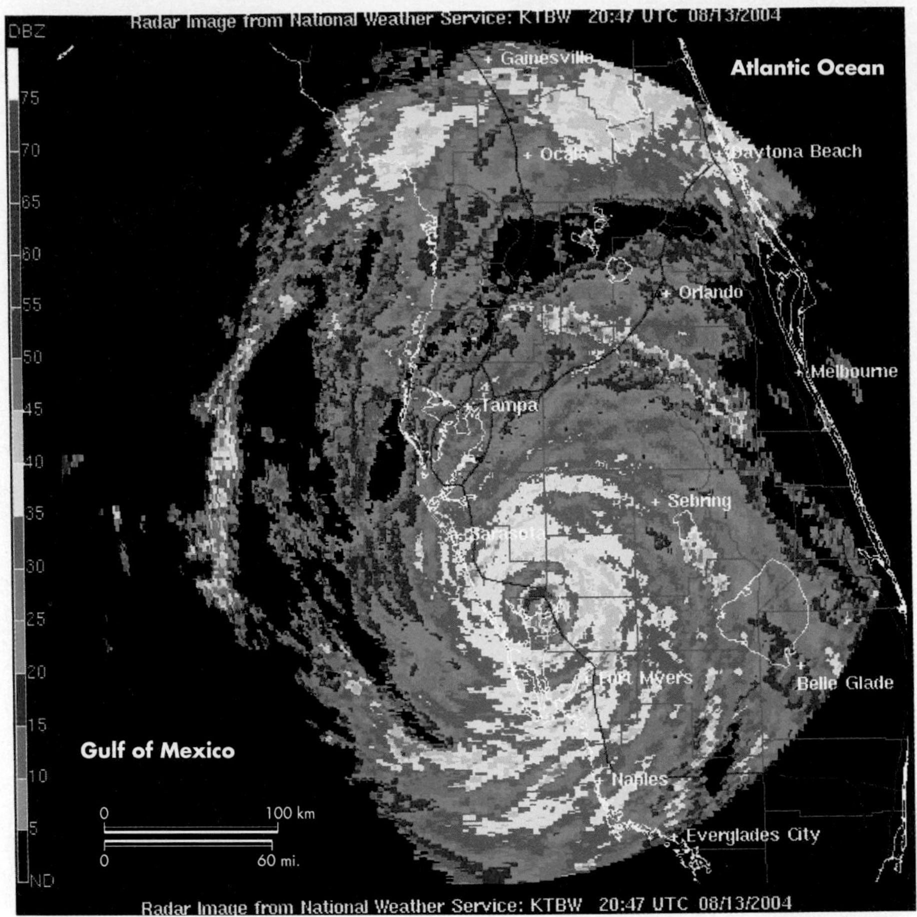

FIGURE 10.31 RADAR IMAGE OF HURRICANE CHARLEY Image from the Tampa Bay weather radar after Hurricane Charley, a strong Category 4 storm, had made landfall between Fort Myers and Sarasota on the west coast of Florida in 2004. Red and orange colors in the spiraling rain bands have the highest winds. *(National Weather Service)*

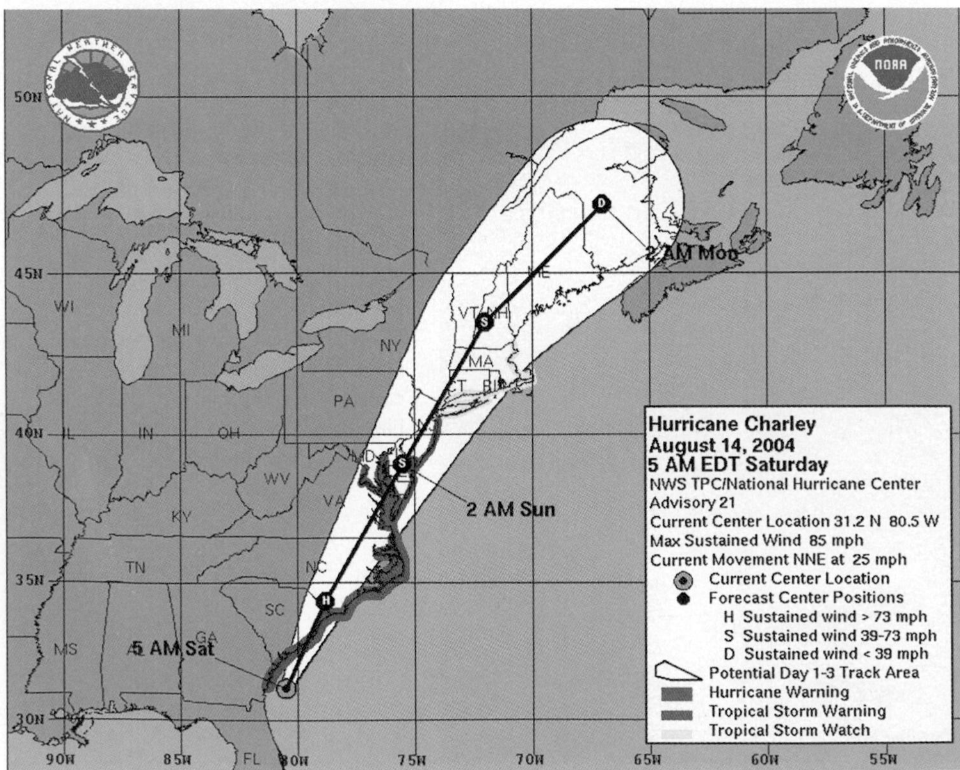

FIGURE 10.32 HURRICANE TRACK MAP This National Hurricane Center map shows the predicted path of Hurricane Charley for the next 3 days. The width of the white area indicates the potential error in the predicted track. Severe weather watches and warnings are shown with red indicating a hurricane warning, blue a tropical storm warning, and yellow a tropical storm watch. *(Courtesy National Weather Service)*

and ADCIRC (ADvanced CIRCulation), take into account the forecast central atmospheric pressure of the storm, its size, forward speed, track, forecast wind speed, and the seafloor topography (i.e., bathymetry). ADCIRC can also take into account the wind and wave stress, tides, and additional water discharged from rivers and was used successfully to predict the storm surge for Hurricane Katrina (see Figure 10.1).[2] Overall these models are accurate to 20 percent or better. This means that if the computer model calculates a 3.0 m (10 ft.) storm surge, you can expect the observed storm surge to be between 2.5 and 3.7 m (8 and 12 ft.).

Hurricane Prediction and the Future Until Hurricane Katrina in 2005, hurricane deaths in the United States had dropped dramatically, largely because of better forecasting, improved evacuation, and greater public awareness.[10] Unfortunately, coastal populations are skyrocketing, and coastal development continues in hazard zones. This trend may result in more hurricane-related deaths if increased traffic slows evacuations, if inexperienced residents are not adequately prepared, or if new residents ignore evacuation procedures.[10]

Whereas increased coastal populations may lead to more hurricane-related deaths, property damage costs have already increased dramatically (Table 10.2). As more and more people build homes and businesses in the coastal zone, we can expect to see these costs continue to rise.

10.9 Perception of and Adjustment to Cyclones

PERCEPTION OF CYCLONES

Residents of regions at risk for hurricanes or coastal extratropical cyclones often have significant experience with them, yet surprisingly, they do not always perceive the danger. During every hurricane, a number of people choose not to evacuate. People's perception of the risk of hurricanes varies according to their experience. Many people living in coastal areas are relatively new residents. Although they may have experienced a few hurricanes, they do not comprehend the threat of a major hurricane, and they may underestimate the hazard. Incorrect predictions of where or when a hurricane will strike may also lower risk perception. For example, if people are repeatedly warned of storms that never arrive, they may become complacent and ignore future warnings. As with any other hazard, accurate risk perception by planners and the public alike is key to reducing the threats associated with coastal hazards.

ADJUSTMENT TO HURRICANES AND EXTRATROPICAL CYCLONES

Warning systems, evacuation plans and shelters, insurance, and building design are key adjustments to hurricanes.

TABLE 10.2

Property Damage Costs Associated with Hurricanes—U.S. Atlantic and Gulf Coasts

Decade	Property Damage[a]
1900–1909	$2.0 billion
1910–1919	$4.9 billion
1920–1929	$3.1 billion
1930–1939	$8.3 billion
1940–1949	$7.6 billion
1950–1959	$18.5 billion
1960–1969	$34.5 billion
1970–1979	$29.6 billion
1980–1989	$29.9 billion
1990–1999	$78.5 billion
2000–2006	$182.7 billion

[a] In 2006 dollars.

Source: Blake, E. S., Jarrell, J. D., Rappaport, E. N., and Landsea, C. W. 2006. The deadliest, costliest, and most intense United States tropical cyclones from 1851 to 2005 (and other frequently requested hurricane facts). NOAA Technical Memorandum NWS TPC-4, Miami, FL: NOAA/National Weather Service, Tropical Prediction Center, National Hurricane Center. From http://www.nhc.noaa.gov/Deadliest_Costliest.shtml, accessed 8/17/10.

▲ FIGURE 10.33 **HURRICANE STORM SURGE FLOODING OF GALVESTON, TEXAS** An estimated 65,500 residents live on Galveston Island, a Gulf of Mexico barrier island south of Houston, Texas. These before (a) and after (b) images of part of the island show flooding that will occur when a 5.8 m (19 ft.) storm surge hits the island from the next Category 5 hurricane. In 1900, catastrophic storm-surge flooding by a Category 4 hurricane drowned more than 6000 people in the most deadly natural disaster in U.S. history. *(University of Austin Bureau of Economic Geology)*

Emergency warning systems are designed to give the public the maximum possible advance notice that a potential hurricane is headed their way. Warning methods include media broadcasts of watches and warnings, and in cases of immediate danger, the local use of sirens. Efficient evacuation plans must be developed and distributed prior to hurricane season to ensure the most well-organized escape possible. At the time of evacuation, public transportation must be provided, shelters opened, and the number of outbound traffic lanes increased on evacuation routes. As was apparent with Hurricanes Katrina and Rita in 2005, it can take days to evacuate heavily populated areas, especially offshore barrier islands that require use of ferries or have few bridges. Last, insurance policies should be available to property holders living in hurricane-prone regions.

Homes and other buildings can be constructed to withstand hurricane-force winds and elevated to allow passage of a storm surge (Figure 10.34). Hurricane and other disaster-resistant building recommendations are available from the Partnership for Advancing Technology

◄ FIGURE 10.34 **HURRICANE-RESISTANT HOUSE** Home in the Florida Keys that has been constructed with strong blocks and space below the living area to allow a hurricane storm surge to pass through the building. *(Edward A. Keller)*

in Housing (PATH), a public-private partnership between homebuilding, product manufacturing, insurance, and financial industries and representatives of U.S. government agencies.

Personal adjustments for shoreline residents of the East and Gulf Coasts include being aware of the hurricane season. Before the season starts, people should prepare their homes and property by trimming dead or dying branches from trees, obtaining flood insurance, and installing heavy-duty shutters that can be closed to protect windows. They should learn the evacuation routes and discuss emergency plans with their loved ones. Disaster preparedness for a hurricane is similar to that for an earthquake and includes having flashlights, spare batteries, a radio, first-aid kit, emergency food and water, can opener, cash and credit cards, essential medicines, and sturdy shoes.[34] As with a tornado watch, once a hurricane watch has been issued, it is imperative to stay tuned to local radio and/or TV stations. If an evacuation order is given, gather emergency supplies and follow the assigned evacuation route.

REVISITING THE FUNDAMENTAL CONCEPTS

Hurricanes and Extratropical Cyclones

The chapter on earthquakes is the first of several in this book that directly discuss a specific natural hazard. For those chapters, we will have, following the summary, a short discussion of how that chapter relates to the fundamental concepts introduced in Chapter 1. Those concepts are

1. **Hazards Are Predictable from a Scientific Evaluation**
2. **Risk Assessment Is an Important Component of Our Understanding of the Effects of Hazardous Processes**
3. **Linkages Exist Between Natural Hazards**
4. **Hazardous Events That Previously Produced Disasters Are Now Producing Catastrophes**
5. **Consequences of Hazards May Be Minimized**

1) Atmospheric conditions and sea surface temperatures can be monitored remotely with the use of radar and satellite sensors and directly with aircraft and ocean sensors to track tropical depressions. Computer modeling of these data may provide short-term predictions about the path the depression will follow. However, like severe weather, cyclone formation and cyclone intensity are the consequence of numerous atmospheric and oceanic variables that continuously change and therefore the accuracy of prediction can be limited. Should a cyclone develop, aircraft (see Professional Profile 10.3) are effective at determining wind strength of the storm that when combined with modeling can accurately predict the category, location, and time that a storm will make landfall or hit a particular location.

2) Risk assessment for cyclones is based on the probability that an event will occur and the possible resulting consequences, which are both fairly straightforward. Specifically, the Gulf Coast and the Eastern seaboard on average are hit by about 6.2 hurricanes per year with 2.3 of them being major hurricanes (greater than Category 3). Their consequences are well known, and, therefore, risk assessment can be easily made.

3) Cyclones are linked to coastal erosion, flooding, mass wasting, tornadoes, and blizzards—some of which may have more serious consequences than the direct effects of the storm. For example, Hurricane Katrina was the most expensive hurricane in the United States, costing nearly $50 billion more than the previous most expensive storm, Hurricane Andrew in 1992. The great cost in property damage and loss of life (more than 1300 deaths) in New Orleans was largely the result of flooding that was caused by levee failure from Katrina's storm surge (see Figure 10.3).

4) In the past few decades, growth in the Unites States has been concentrated in coastal areas, putting more people at risk, while simultaneously driving property values up and increasing the consequences for a particular event. Coastal areas are particularly vulnerable to the effects of tropical cyclones: In addition to the hazard presented by storm surges, coastal erosion, and flooding of low elevations, peak wind strengths are often observed as cyclones make landfall. In the coming decades, population growth and development along the Atlantic coast will likely lead to ever-greater losses in property damage and loss of life as a result of hurricanes and extratropical cyclones.

5) Minimizing the consequences of cyclones requires accurate forecasting and better construction practices. The decrease in loss of life from hurricanes preceding Hurricane Katrina exemplified how advances in remote sensing and computer modeling over the past few decades had aided in minimizing the consequences of cyclones. Yet as the Katrina example above shows (Concept 2), ill constructed levees, which failed as a result of Katrina's storm surge, led to the largest natural disasters in U.S History. Had the levees in New Orleans been designed to withstand a Category 5 hurricane, as they should have been, Katrina, which was a category 4 when it made land fall in Louisiana, certainly would have been far less costly.

Summary

Cyclones, large areas of low atmospheric pressure with winds converging toward the center, are associated with most severe weather. Because of the Coriolis effect, winds blow counterclockwise around cyclones in the Northern Hemisphere.

Cyclones are described as either tropical or extratropical based on their characteristics and origin. Tropical cyclones have warm central cores, are not associated with weather fronts, and form between 5° and 20° latitude over tropical and subtropical oceans. Called typhoons, severe tropical cyclones, or cyclonic storms in most of the Pacific and Indian Oceans, these storms are referred to as hurricanes in the Atlantic and northeast Pacific Oceans and in this book. Extratropical cyclones have cool central cores, develop along weather fronts, and form between 30° and 70° latitude over either the land or ocean. These mid-latitude cyclones produce coastal windstorms, snowstorms, blizzards, and severe thunderstorm outbreaks. Both types of cyclones can have high-velocity straight-line winds, tornadoes, heavy precipitation, and coastal storm surges.

Tropical cyclones are classified by their prevailing wind speed and are referred to as tropical depressions, tropical storms, and hurricanes with increasing wind speed. Tropical storms must have sustained winds of at least 119 km (74 mi.) per hour to become a hurricane. A hurricane is assigned a category on the Saffir-Simpson Hurricane Scale based on measured or inferred sustained winds. Hurricanes reaching Category 3 or above are considered major hurricanes. Tropical storms and hurricanes are given women's or men's names from internationally agreed upon lists for the region where they form. Extratropical cyclones are sometimes named for their geographic area of origin or prevailing wind direction.

Most tropical cyclones start out as a tropical disturbance, a large thunderstorm complex associated with a low-pressure trough. In the Atlantic Ocean, many of these disturbances form off the west coast of Africa as easterly waves in the trade winds. A tropical disturbance that develops a circular wind-circulation pattern becomes a tropical depression and may become a tropical storm if its winds exceed 63 km (39 mi.) per hour. Only a few tropical storms encounter the environmental conditions needed to become a hurricane, that is, waters that are at least 26°C (80°F) to a depth of 46 m (150 ft.), uninhibited convection for moist air evaporation from the sea surface, and weak winds aloft that keep vertical wind shear to a minimum.

Hurricanes develop spiraling rain bands of clouds around a nearly calm central eye. The eye is surrounded by an eyewall cloud that has the storm's strongest winds and most intense rainfall. Warm, moist air condensing in rain bands and the eyewall releases latent heat and provides continual energy for the hurricane. Excess heat is vented out of the storm's top. Most hurricanes move forward at 19 to 27 km (12 to 17 mi.) per hour steered by winds in the middle and upper troposphere and generally lose intensity over land or cooler water.

Extratropical cyclones typically form along fronts where there are strong, diverging winds in the upper troposphere. Strong upper-level winds occur in polar and subtropical jet streams. Extratropical cyclones often intensify as they move east from the Rocky Mountains or along a coastline. In North America, most extratropical cyclones have a cold front trailing to the southwest and a warm front to the east. Cold air circulating around these cyclones collides with warm air rising in the eastern part of the storm. Many mature anticyclones develop an occluded front and a large, comma-shaped cloud formation before they dissipate.

In North America, the Gulf and East Coasts are at highest risk for tropical cyclones. Hurricane-strike probabilities are the greatest in southern Florida, the northern Gulf Coast, and Cape Hatteras. Extatropical cyclones are the primary severe weather hazard on the Pacific Coast and cause storms on the Great Lakes. Nor'easters, named for their strong northeasterly winds, are a hazard for the eastern United States and Atlantic Canada.

Cyclones produce coastal storm surges, high winds, and heavy rains. Major hurricanes and intense extratropical cyclones generate storm surges of more than 3 m (10 ft.) from water piled up by high winds. Higher surges are produced by larger or more intense storms, and on shallow coastlines or narrow bays. In the Northern Hemisphere, the greatest storm surge, highest winds, and most tornadoes are typically in the right forward quadrant of tropical cyclones. Much of the structural damage from coastal storms comes from wind-driven waves on top of the storm surge. Strong currents from storm surges cut channels through islands and peninsulas and deposit sand as overwash. Wind damage from hurricanes is more widespread, but less deadly than the storm surge. Although hurricanes hold most rainfall records, tropical storms also produce intense rains and flooding. Coastal cyclones generally cause both saltwater and freshwater flooding. Inland flooding occurs from hurricane remnants if storms move slowly, encounter hills or mountains, interact with other weather systems, or track over previously saturated ground.

Cyclones are closely linked with other severe weather, flooding, landslide, and debris flow hazards. The hazard posed by storm surges and erosion from coastal cyclones will increase as sea level rises worldwide from global warming. Hurricane intensity is also projected to increase with warmer global temperatures. Both hurricanes and coastal extratropical cyclones will become more destructive to our society as coastal populations and per capita wealth grows. On the positive side, cyclones are major sources of precipitation, help maintain global heat flow, and contribute to long-term ecosystem health.

Accurate forecasts, effective storm warnings, strict building codes, and well-planned evacuations can minimize the effects of coastal cyclones. Hurricane forecasting relies on weather satellites, aircraft flights, Doppler radar, and automated weather buoys. Computer models predict hurricane tracks more accurately than their intensity. Storm surge predictions are based on wind speed, fetch, average water depth, and timing of landfall in relation to astronomical tides.

Perception of the coastal cyclone hazard depends on individual experience and proximity to the hazard. Community adjustments to hurricanes in developed countries involve building protective structures or modifying people's behavior through land-use zoning, evacuation procedures, and warning systems. Individual adjustments to hurricanes include having emergency supplies on hand; preparing once a storm prediction is made; and if required, evacuating before the storm hits. Homes can be constructed to withstand hurricane-force winds and elevated to allow passage of storm surges.

Key Terms

cyclone (p. 331)
cyclone intensity (p. 333)
extratropical cyclone (p. 333)
eye (p. 336)
eyewall (p. 336)
hurricane (p. 334)
jet stream (p. 339)
overwash (p. 346)
rain band (p. 336)
storm surge (p. 346)
tropical cyclone (p. 333)
tropical depression (p. 336)
tropical disturbance (p. 336)
tropical storm (p. 336)
typhoon (p. 334)

Review Questions

1. Describe the characteristics of a cyclone. What distinguishes a tropical cyclone from an extratropical cyclone?
2. How do tropical cyclones, hurricanes, tropical storms, tropical depressions, tropical disturbances, and typhoons differ?
3. Explain how and where hurricanes form and dissipate.
4. What are the conditions necessary for hurricane formation?
5. How does the Coriolis effect influence hurricane winds and movement?
6. What parts of a hurricane have the most intense winds, rainfall, tornadoes, and storm surges in the Northern Hemisphere?
7. How are hurricanes categorized? Which categories are considered major hurricanes? Which category is most intense?
8. Which areas of the United States and Canada have the highest risk for hurricanes?
9. Where are tropical cyclones most common?
10. Describe or sketch the three typical paths taken by Atlantic hurricanes affecting North America.
11. Describe the similarities and differences between the effects of tropical and extratropical cyclones.
12. What are three major causes of hurricane damage? Which is typically the most deadly?
13. Explain the causes and effects of storm surges. What will cause a storm surge to increase?
14. Describe the changes that occur when a hurricane moves inland and becomes an extratropical cyclone.
15. Describe the similarities and differences between tropical storms and hurricanes.
16. How will global warming and population trends interact with cyclones in coastal areas?
17. Explain how cyclones are linked with other natural hazards.
18. What are the natural service functions of cyclones?
19. How do hurricane watches and warnings differ?
20. Describe the tools used in making hurricane forecasts.
21. Describe the adjustments that people need to make to survive hurricanes.

Critical Thinking Questions

1. If you had to evacuate your home and go to a nearby public shelter, where would it be? If you had to evacuate your home and travel at least 160 km (100 mi.) from where you live, where would you go? What would you take with you in either case? What problems might you or your community be faced with in an evacuation (e.g., people in poor health or with disabilities, people without a means of transportation, visitors, pets, domesticated animals)? What would be your concerns?
2. The southern tip of Florida has a very large area of marsh and swampland; the Everglades; and a large lake, Lake Okeechobee. Based on your understanding of how a hurricane works, what effect would these wetlands have on a hurricane that was tracking across southern Florida? What effect would the hurricane have on Lake Okeechobee?
3. What conditions required for hurricane development could explain why hurricanes generally do not form between 5° N and 5° S of the equator, or in the southeastern Pacific and south Atlantic Oceans?
4. Obtain a topographic map (see Appendix C) for an ocean beach you visit, or for an urban area along the shore of the Atlantic or Gulf Coasts. Shade or color the area that would be affected by a 5 m (20 ft.) storm surge. Examine your completed map and assess the damage that would occur, the routes people would take for evacuation, and land-use restrictions you would recommend for future development.

Selected Web Resources

Hurricanes
hurricanes.noaa.gov/—information about hurricanes and hurricane preparedness from the National Oceanic and Atmospheric Administration

www.fema.gov/hazard/hurricane/—Federal Emergency Management Agency information on hurricanes and hurricane preparedness

Canadian Hurricane Centre
www.ns.ec.gc.ca/weather/hurricane/—information about hurricanes from the Meteorological Service of Canada

Hurricane Awareness
weather.gov/om/hurricane/—information about hurricanes and hurricane-tracking charts from the National Weather Service Office of Climate, Water, and Weather Services

NASA's Hurricane Resource Page
www.nasa.gov/mission_pages/hurricanes/main/index.html—information, animations, and videos about hurricanes from the National Aeronautics and Space Administration

Hurricane and Extreme Storm Impact Studies
coastal.er.usgs.gov/hurricanes/—information on the effects of hurricanes from the U.S. Geological Survey

Hurricane Preparedness, Storm Surge
www.nhc.noaa.gov/HAW2/english/storm_surge.shtml—information on storm surge from the National Hurricane Center

Assignments in Applied Geology: Hurricane/Tsunami Hazard Assessment

THE ISSUE

Tsunamis and the storm surge associated with hurricanes are two of the most devastating natural hazards a coastal population can experience. Low-lying, often unprotected populated areas along shorelines can easily fall victim to the powerful force of moving water. Until recently, these hazards have passed quietly under the radar of a fair majority of the public.

YOUR TASK

As a result of recent events such as the December 26, 2004, tsunami and the destruction related to Hurricane Katrina, the village council of Ocean Village, a small town approximately 50 miles east of Hazard City where the Clearwater River enters the ocean, has become concerned about the vulnerability of its fair village. Your job involves learning about the causes and effects of tsunamis and storm surge and then assisting the village council in filing a report. You will have to report to the village council information concerning tsunami travel speed and warning time, how many people need to be housed/evacuated based upon inundation by storm surge or tsunami, and the amount of the village that might need to be rebuilt.

AVERAGE COMPLETION TIME:

2 hours

Coastal Hazards

Folly Island

In many ways, Folly Island, 11 km (7 mi.) south of Charleston, South Carolina, is a typical Atlantic or Gulf Coast **barrier island**. The island is essentially an accumulation of sand that acts as a barrier to ocean waves that would otherwise strike the mainland (Figure 11.1). A little more than 10 km (6 mi.) long and generally less than 1 km (0.6 mi.) wide, most of the island has an elevation between 1.5 and 3 m (5 and 10 ft.) above sea level. A pine forest once covered the island and was the source of its Old English name *folly,* meaning "a tree-crested dune ridge."[1] Folly Island formed 4000 to 5000 years ago from sand dune ridges that accumulated as the rise in sea level from melting of Pleistocene glaciers slowed.[1]

One characteristic that makes the island typical is that its ocean shoreline is eroding landward 0.3 to 1 m (1 to 3 ft.) each year.[1] The island's ocean shoreline has been retreating for at least 160 years and, in the early twentieth century, was receding at an average rate of 2 m (7 ft.) per year.[1] Three factors have caused the shoreline erosion: construction of a nearby jetty, a long wall built perpen-dicular to the coast; destructive beach management practices; and rising sea level produced, in part, by global warming. In 1896, a 3-mile-long rock jetty was constructed 7 km (4.3 mi.) northeast of Folly Island to protect the entrance to Charleston Harbor (Figure 11.1). The jetty partially blocked the delivery of sand to the island by coastal currents and caused severe shoreline erosion on Morris Island immediately to the northeast of Folly Island. Destructive beach management practices on Folly Island began after World War II with the construction of walls both perpendicular and parallel to the shoreline. Between 1949 and 1970, the state highway department built 40 walls of wood pilings perpendicular to the beach (Figure 11.2).[1] Called *groins,* these structures resulted in both beach erosion and deposition. Property owners have also built walls of concrete, boulders, or wood parallel to the beach to protect buildings from flooding (Figure 11.3). Often referred to as *seawalls,* these walls offer some temporary protection for buildings, but, in the long term, they have reflected storm waves and contributed to beach erosion.[1] A nine-story hotel, constructed with a zoning variance in 1985, is located on the beach and is partially protected by a seawall. The hotel's seawall has contributed to beach erosion and failed to protect the building from the 3.7 m (12 ft.) storm surge of Hurricane Hugo in 1989.[1]

The island's ocean shoreline has been retreating for at least 160 years

◀ **Surging Ocean** Coastal erosion may occur slowly over 10s to 100s of years or may happen rapidly during brief periods when combined with Hurricane force winds and storm surges. Kew West, Florida, residents Brian Goss (left), George Wallace and Michael Mooney (right) flee the coast as Hurricane George's 3.6 m (12 ft) storm surge pushes in, washing away houses, cars, roads, and beaches over a 2 day period in September of 1998. *(Dave Martin/AP Wide World Photos)*

LEARNING OBJECTIVES

In this chapter, we focus on one of the most dynamic environments on Earth—the coast, where the sea meets the land. Beaches, composed of sand or pebbles, and rocky coastlines continue to attract tourists and new residents like few other areas, yet most of us have little understanding of how ocean waves form and change coastlines. A major goal of this chapter is to remove the mystery of the processes at work in coastal areas while retaining the wonder. We also seek to explain the hazards resulting from waves, currents, and rising sea level and to discuss how we can learn to live in the ever-changing coastal environment while sustaining its beauty. Your goals in reading this chapter will be to

- understand coastal processes, such as waves, beach forms and processes, and rising sea level.
- understand coastal hazards, such as rip currents and erosion.
- know what geographic regions are at risk for coastal hazards.
- understand the effects of coastal processes, such as rip currents, coastal erosion, and rising sea level.
- recognize the linkages between coastal processes and other natural hazards.
- know the benefits derived from coastal processes.
- understand how human use of the coastal zone affects coastal processes.
- know what we can do to minimize coastal hazards.
- understand the adjustments that can be made to avoid damage from coastal erosion and rising sea level or personal injury from strong coastal currents.

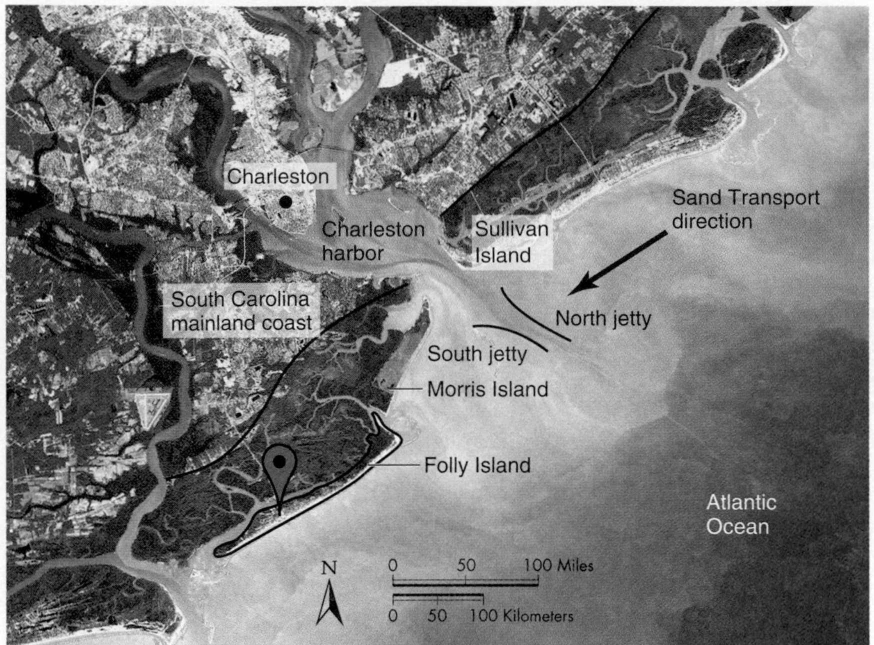

◀ FIGURE 11.1 **COASTAL EROSION** Folly Island is a shrinking barrier island located south of the inlet to Charleston Harbor. One potential cause of the diminishing size of Folly Island is the construction of jetties, built to keep the harbor entrance free of sand, which consequently block sand being transported southward along the coast that would normally replenish the beaches of Folly Island. *(M-Sat Ltd./PhotoResearchers, Inc.)*

Folly Island has not only seen conflict between civilization and nature, but also conflicting approaches to the long-term solution to coastal erosion. Most of the island has walls of various types to protect buildings, but these walls increase beach erosion. To mitigate this erosion, engineers, business interests, and some politicians have favored the periodic addition of sand to the beach, a process called *beach nourishment.* Sand has now been added to the beach three times, in 1979, 1993, and 2005, by contractors for the U.S. Army Corps of Engineers (Figure 11.4). Beach nourishment on Folly Island has cost more than $24 million to date, with more than 75 percent of the nourishment costs paid from federal funds. In fact, in 1996, Congress made promotion of beach nourishment part the U.S. Army Corps of Engineers' mission.[2]

Sand added to a beach often disappears in subsequent storms. The U.S. Army Corps of Engineers estimates that, on average, additional sand will have to be added to Folly Beach every 8 years.[3] This frequency may seem high, but not as high as Jupiter Beach, Florida, and Ocean City, New Jersey, where beach nourishment is projected every 3 years![3] A study of sand added to Folly Beach in 1993 indicates that significant quantities of the sand placed on the beach are being carried out onto the continental shelf and completely removed from the beach system.[4]

Geologists study shoreline behavior over the long term and take into account the natural movement of barrier islands and rising sea level. They have observed that without hard structures, such as seawalls and groins, beaches generally do not disappear; they just shift location. Sea level in Charleston, just north of Folly Island, has been rising on average 3 mm/year since 1920.[5] This rate is likely to increase as the rate of global warming increases.[6]

◀ FIGURE 11.2 **GROINS ON FOLLY ISLAND** Eight of the more than 40 groins on Folly Island, South Carolina, that trap sand moving by coastal currents. Groins create an irregular pattern of shoreline erosion and deposition and, in the end, contribute to coastal erosion downdrift from Folly Island. *(Bob Krist/Corbis)*

▲ FIGURE 11.3 **ATTEMPTS TO MITIGATE COASTAL FLOODING AND EROSION ON FOLLY ISLAND** Property owners use a variety of methods to mitigate the effects of storms and coastal erosion on Folly Island, South Carolina, including this seawall constructed of large boulders, called riprap, and wood and the elevation of this house on wood pilings. Although seawalls may help protect buildings, they generally contribute to beach erosion *(Robert H. Blodgett.)*

(a)

(b)

◄ FIGURE 11.4 **NOURISHMENT OF FOLLY BEACH** (a) A floating hydraulic cutter-head suction dredge in the distance sucked sand from the continental shelf and pumped a sand-and-water slurry to the shore in 2005. The slurry flowed along the beach in the steel pipe that appears in the foreground. (b) Sand and water discharged from the pipe on the beach, and the sand was moved in position by the bulldozer shown in (a). The U.S. Army Corps of Engineers spent more than $12 million to add approximately 1.7 million m^3 (2.3 million $yd.^3$) of sand to the beach in 2005. An estimated 1.9 million m^3 (2.5 million $yd.^3$) was added at a cost of $12.5 million in 1993. *(Robert H. Blodgett)*

Although most ocean beaches in the United States are eroding like Folly Beach, few states and coastal communities are considering sea-level rise in their planning.[7] South Carolina's Beachfront Management Act, implemented in 1990, recognizes the coastal erosion hazard and establishes setback distances for new construction from the shoreline; however, because it has been developed since the 1920s, Folly Beach was at least temporarily exempted from the setback provision.[1] What has yet to be addressed is rising sea level. New houses built on Folly Island are elevated on tall piers (Figure 11.5) at a height that is determined from the base flood elevation shown on FEMA Flood Insurance Rate Maps. This base flood elevation is based on present-day sea level, not the rise in sea level that is likely to occur during the lifetime of the structure.

The challenge for Folly Island, and much of the Atlantic and Gulf coasts, will be adopting coastal management policies and regulations that are forward looking and take into consideration coastal erosion, local subsidence, and rising sea level. In some areas, difficult choices will have to be made between preserving beaches and coastal wetlands and protecting buildings and infrastructure.

▲ FIGURE 11.5 **NEW HOME CONSTRUCTION ON FOLLY ISLAND** After Hurricane Hugo's 3.7 m (12 ft.) storm surge in 1989, more recent construction on Folly Island has been on higher piers. The base flood elevation on the 2004 FEMA Flood Insurance Rate Maps is 4.9 to 7 m (16 to 23 ft.) above sea level for most of the island. With coastal erosion and rising sea level, this elevation may be too low to protect this house from storm surge for the remainder of the century. *(Robert H. Blodgett)*

11.1 Introduction to Coastal Hazards

Coastal areas are dynamic environments that vary in their topography, climate, and organisms. In these areas, continental and oceanic processes converge to produce landscapes that are capable of rapid change.

Overall, coastal topography is greatly influenced by plate tectonics. The east coasts of the United States and Canada, as well as the Canadian Arctic coast, is described as tectonically passive because it is not close to a convergent plate boundary (see Chapter 2). Tectonically passive coasts typically have wide continental shelves with barrier islands and sandy beaches. Less typical are rocky shorelines, which are mostly restricted to New England, Atlantic Canada, and Arctic islands where ancient mountain ranges, such as the Appalachians, meet the Atlantic and Arctic Oceans. Along the Great Lakes, most shorelines are sandy, with rocky shores generally limited to northern areas that have experienced greater uplift following retreat of the Pleistocene glaciers.

In contrast, the west coasts of the United States and Canada are tectonically active because they are close to transform boundaries between the North American and Pacific plates (e.g., the San Andreas fault) and to convergent boundaries between the North American and Juan de Fuca or Pacific plates (e.g., the Cascadia subduction zone). Hawai'i, Puerto Rico, the Virgin Islands, and southern Alaska are also tectonically active. In tectonically active regions, mountain building produces coasts with sea cliffs and rocky shorelines. Although long sandy beaches are present in these regions, they are not as abundant as along the east and Arctic coasts or the southern Great Lakes.

Some American and Canadian coasts are significantly influenced by climate and organisms. For example, shorelines in parts of Alaska, Canada, and the Great Lakes are affected by seasonal ice or by the movement of present-day glaciers. Coastlines in temperate regions are influenced by marsh vegetation, and subtropical and tropical shores are affected by the growth of mangroves and offshore coral reefs.

The impact of hazardous coastal processes is considerable because many populated areas are located near the coast. In the United States, it is expected that most of the population will eventually be concentrated along the nation's 150,000 km (93,000 mi.) of shoreline, including the Great Lakes. Today, the nation's largest cities lie in the coastal zone, and approximately 53 percent of the population lives in coastal counties, even though those counties account for only 17 percent of the land area.[8] The population in coastal counties becomes even larger during peak vacation periods. For example, Ocean City, Maryland, receives an estimated 4 million visitors between Memorial Day and Labor Day.[8] Coastal problems will increase as more people live in coastal areas where the hazards occur.

Once again, our activities continue to conflict with natural processes. Hazards along the coasts, such as coastal erosion, may become compounded by global warming, which is contributing to a worldwide rise in sea level (see Chapter 12).

The most serious coastal hazards are the following:

- Strong coastal currents, including rip currents generated in the surf zone and tidal currents in narrow bays and channels
- Coastal erosion, which continues to produce considerable property damage that requires human adjustment
- Storm surge from tropical and extratropical cyclones, which claims many lives and causes enormous amounts of property damage every year (see Chapter 10)
- Tsunamis, which are particularly hazardous to coastal areas of the Pacific Ocean (see Chapter 4)

11.2 Coastal Processes

WAVES

Waves that batter the coast are generated by offshore winds, sometimes thousands of kilometers from where the waves reach the shoreline. Wind blowing over the water produces friction along the water surface. Because the air is moving much faster than the water, the moving air transfers some of its energy to the water and produces waves. The waves, in turn, eventually expend their energy at the shoreline.

Waves vary in both their size and shape. The size of waves in the ocean or on a lake depends on a combination of the following:

- The velocity or speed of the wind; the stronger the wind speed, the larger the waves.
- The duration of the wind. Winds that last longer, such as during storms, have more time to impart energy to the water, thereby producing larger waves.
- The distance that the wind blows across the water surface. This distance is referred to as the *fetch*. A longer fetch allows larger waves to form. This relationship is one reason that waves are generally larger in the ocean than in a lake.

Waves develop in a variety of sizes and shapes within an area that is experiencing a storm. As waves move away from their place of origin, they become sorted out into groups or sets of waves that have similar sizes and shapes. These sets may travel for long distances across the ocean and arrive at distant shores with little energy loss. If you stand on a beach watching the surf, you may be able to recognize these "sets" of similar-sized waves. Interactions between different sets from different sources produce a regular pattern or "surf beat." Very rarely, a much larger wave will arrive unexpectedly on shore. Called **rogue waves**, these waves can be extremely dangerous (see Case Study 11.1).

Waves moving across deep water have a similar basic shape or wave form (Figure 11.6a). Three parameters describe the size and movement of a wave: *wave height* (*H*), which is the difference in height between the trough and crest of a wave; *wavelength* (*L*), the distance between successive wave crests; and *wave period* (*T*), the time in seconds for successive waves to pass a reference point. Recall that we also used wavelength to describe the electromagnetic radiation and wave height of seismic waves as one indicator of earthquake magnitude. The reference point used to determine wave period could be a pier or similar object, which is anchored in the seafloor or lake bottom.

To understand how waves transmit energy through the water, it is useful to study the motion of an object on the water surface and one below the surface. For example, if you were floating with a life preserver in deep water and could record your motion as waves moved through your area, you would find that you bob up, forward, down, and back in a circular orbit, always returning to the same place (Figure 11.6b). If you were below the water surface with a breathing apparatus, you would still move in circles, but the circle would be smaller. That is, you would move up, forward, down, and back in a circular orbit that would remain in the same place while the waves traveled through (Figure 11.6b).

The shape of the orbital motions changes as waves enter shallow water. At a depth of less than about one-half their wavelength, the waves begin to "feel bottom." This lack of room causes the circular orbits to become ellipses. Motion at the bottom may eventually be a narrow ellipse with a back-and-forth movement that is essentially horizontally (Figure 11.6c). You may have experienced this if you have stood or swam in relatively shallow water on a beach and felt the water repeatedly push you toward the shore, then away from the shore.

Far out at sea, the wave sets generated by storms are called *swells*. As a swell enters shallower and shallower water, transformations take place, such that, eventually, the waves become unstable and break on the shore. In deep-water conditions, mathematical equations can be used to predict wave height, period, and velocity, based on the fetch, wind speed, and duration of time that the wind blows over the water. This information has important environmental consequences: By predicting the velocity and height of the waves generated by a distant storm, we can estimate when storm-generated waves with a particular erosive capability will strike the shoreline and when to expect dangerous coastal currents.

We have said that waves expend their energy when they reach the coastline, but just how much energy are we talking about? The amount is surprisingly large. For example, the energy expended on a 400 km (250 mi.) length of open coastline by waves with height of about 1 m (3 ft.) over a

CASE STUDY 11.1

Rogue Waves

As waves move away from their point of origin, they sort themselves into sets of similarly sized waves. The arrival of alternating sets of small and large waves can be observed on most coastlines. This pattern tends to repeat itself over time, allowing surfers to wait and take advantage of the sets of larger waves that they know will eventually arrive. Occasionally, however, a wave will appear that is much larger than the rest. Known as a *rogue wave,* such a huge wave is created by a number of factors.

Most rogue waves appear to form by *constructive interference.* In this process, multiple, similarly sized waves intersect to create a much larger wave. If they intersect just right, with both their crests and troughs matching, the new wave may be as high as the sum of the intersecting waves. Undersea irregularities, as well as currents, also influence the formation of a rogue wave. Such waves can be extremely dangerous to the unsuspecting beachgoer.

When rogue waves strike the shore, lives may be lost. Such waves may appear out of nowhere, crashing down over a pier or cliff and sweeping unsuspecting people to their deaths. Sadly, each summer, a number of beach visitors are swept into the ocean by rogue waves. Coastal zones are not the only dangerous areas. Rogue waves can also appear out of nowhere in the open ocean—in the middle of hundreds of kilometers of calm water (Figure 11.A). Such waves may be large enough to break in open water and threaten ships. In stormy seas, where almost all waves are large, rogue waves may reach 30 m (100 ft.) in height. An initial study of 3 weeks of global satellite radar data found 10 separate rogue waves in the world's oceans that were taller than 25 m (80 ft.).[10] This is critical information for the shipping industry, since most ships are designed to withstand 15 m (50 ft.) waves.[10] Massive rogue waves were apparently common during the October 1991 "Perfect Storm," the intense extratropical cyclone depicted in both a book and a movie, that was responsible for sinking the fishing boat *Andrea Gail,* killing its entire six-person crew.

FIGURE 11.A ROGUE WAVE This huge wave is approaching the bow of the *JOIDES Resolution,* the scientific drilling ship of the Ocean Drilling Program. Known as rogue waves, these large waves are responsible for sinking supertankers and large cargo-container ships. In April 2005, a 20 m (70 ft.) rogue wave struck the cruise ship *Norwegian Dawn* about 400 km (250 mi.) off the coast of Georgia. *(Ocean Drilling Program, Texas A&M University)*

given period of time is approximately equivalent to the energy produced by an average nuclear power plant over the same time period.[9]

Wave energy is also approximately proportional to the square of the wave height. Thus, if wave height increases from 1 m (3 ft.) to 2 m (7 ft.), the wave energy increases by a factor of 2^2 or 4. If waves continue to grow to the 5 m (16 ft.) height that is typical for large storms, then the energy expended, or wave power, will increase 5^2 or 25 times, as compared with that of waves with a height of 1 m (3 ft.).

When waves enter the coastal zone and shallow water, their shape and movement change, wavelength and velocity decrease, and wave height increases; only the wave

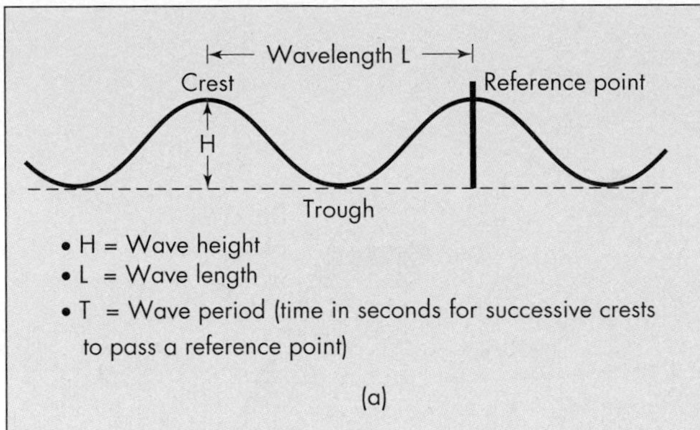

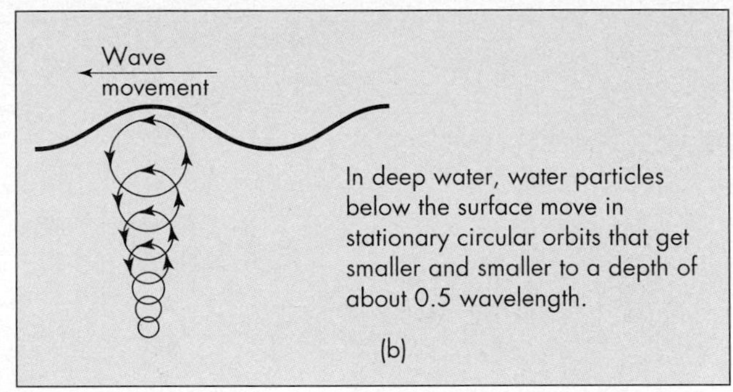

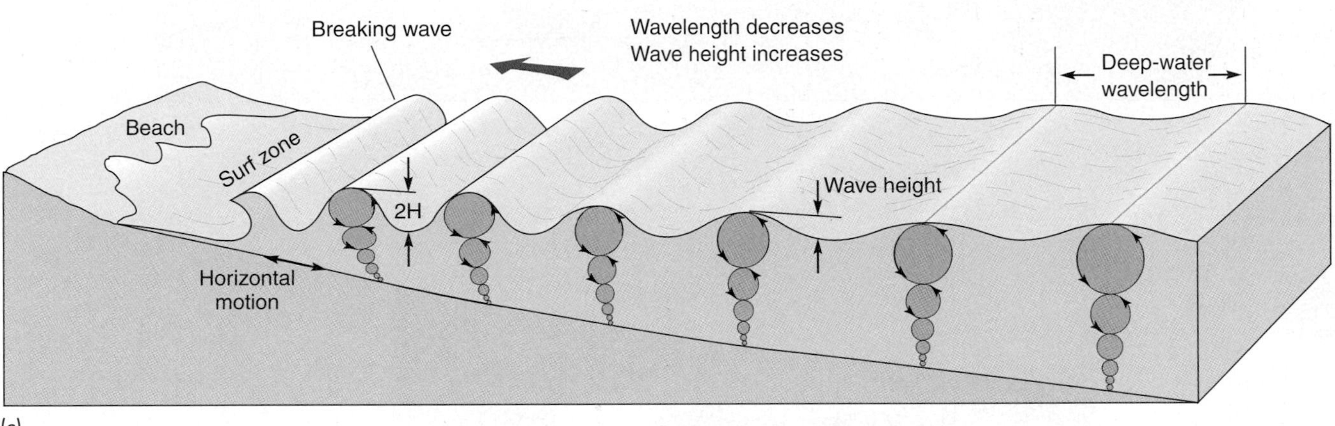

▲ **FIGURE 11.6 WAVES AND BEACHES** (a) Deep-water wave form (water depth is greater than 0.5 L, where L is wavelength). The curving black line is the water surface, and the thick vertical black line is a pier or some other fixed object that can act as a reference point for determining the wave period (T). A black dashed line connects the bottom of the troughs and is the reference line for calculating wave height (H). (b) Motion of water particles associated with wave movement in deep water. The water particles follow the path of the arrows in the black circles. Wave movement is from right to left. (c) Motion of water particles as waves approach the shore. The water particles follow the path of the arrows in the circles. The waves are approaching the shore from right to left.

period remains constant. The waves also change shape from the rounded crests and troughs that are found in deep water to peaked crests with relatively flat troughs in shallow water close to shore. Perhaps the most dramatic feature of waves entering shallow water is their rapid increase in height. As waves approach their breaking point in shallow water, the wave height may increase to twice its deep-water height (Figure 11.6c). The wave crest becomes unstable when the water depth is about 1.3 times the wave height. At that point, the advancing wave no longer encounters enough water ahead of it to complete its symmetrical form, and it collapses or breaks toward the shore. A smaller wave often reforms in deeper water landward of a sand bar and repeats the process, breaking a second or third time before reaching the beach.

Variations along a Coastline Although wave heights offshore from a coastline are relatively constant, the wave height along a coast may increase or decrease as waves approach the shore. These variations are caused by irregularities in the offshore topography and by changes in the shape of the coastline. One way to understand the variations in wave height is to examine the behavior of the long, continuous crest of a single wave, or *wave front,* as it approaches an irregularly shaped coastline (Figure 11.7a).

Irregular coastlines have small rocky peninsulas known as *headlands.* The shoreline between headlands may be relatively straight or somewhat curved. Underwater, the offshore topography surrounding the headland is similar to that of the rest of the coast; that is, the water gets progressively shallower close to shore. This shallowing means that as a wave approaches the coast, it will first slow down in the shallow water off the headland. The slowdown will cause a long wave front to bend when it reaches a headland. This bending, referred to as *refraction,* causes wave fronts to become more nearly parallel to the shoreline.

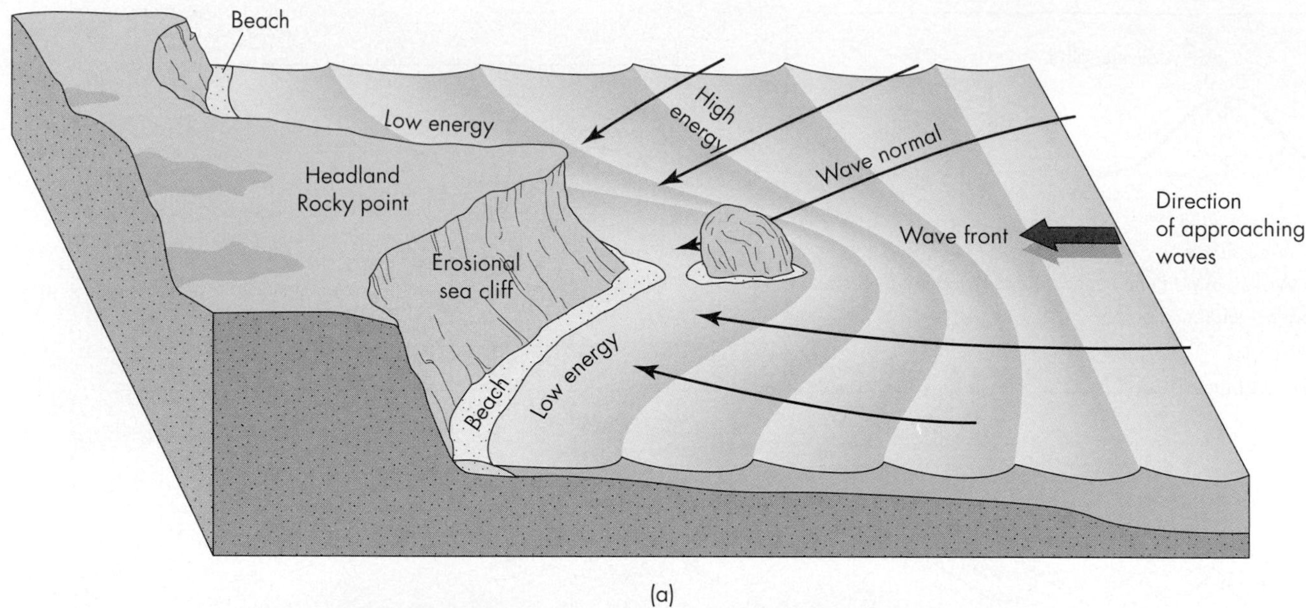

(a)

(b)

▲ **FIGURE 11.7 CONVERGENCE AND DIVERGENCE OF WAVE ENERGY** (a) Idealized diagram of the process of wave refraction and concentration of wave energy at rocky points called *headlands*. Refraction, or bending of wave fronts, causes the convergence of wave normals on the headland and their divergence in low-energy areas along the coast away from the headland. Wave normals are the imaginary long curving black arrows. The red arrow indicates that the waves are approaching the shore from right to left. (b) Large waves striking a rocky headland along the Pacific coast at Pebble Beach in San Mateo County, California. *(Robert H. Blodgett)*

Effects of Wave Refraction To visualize the effects of wave refraction, draw a series of imaginary lines, called *wave normals*, perpendicular to the wave fronts, and add arrows pointing toward the shoreline. The resulting diagram (Figure 11.6a) shows that wave refraction causes a *convergence* of the wave normals at the headland and a *divergence* of the wave normals along the shoreline away from the headland. Where wave normals converge, both wave height and the energy expended by the waves increase. Thus, the largest waves along a shoreline are generally found at the end of a rocky headland (Figure 11.7b). The long-term effect of greater energy expenditure on protruding areas, such as headlands, is that wave erosion tends to straighten the shoreline.

Breaking Waves Waves also vary in how they break along a coastline. Breaking waves may peak up quickly and plunge or surge, or they may gently spill, depending on local conditions. Waves that plunge are called *plunging breakers* (Figure 11.8a). They typically form on steep beaches and tend to be more erosive. Waves that spill are referred to as *spilling breakers* (Figure 11.8b). They commonly develop on

▲ FIGURE 11.8 **TYPES OF BREAKERS** Idealized diagrams and photographs showing (a) plunging breakers on a steep beach and (b) spilling breakers on a gently sloping beach. *([a] Peter Cade/Getty Images [b] Penny Tweedie Photolgraphy)*

wide, nearly flat beaches and are more likely to deposit sand. The type of breaker that occurs along a coastline can change seasonally and with changes in underwater slope and topography. Overall, large plunging breakers that form as the result of storms cause much of the coastal erosion.

BEACH FORM AND PROCESSES

A **beach** is a landform consisting of loose material, such as sand or gravel, which has accumulated by wave action at the shoreline. Beaches may be composed of a variety of loose material. For example, the color and composition of beach sand is directly tied to the source of the sand: Many white Pacific island beaches are made of broken bits of shell and coral, Hawai'i's black-sand beaches are composed of fragments of volcanic rock, and brown Carolina beaches contain grains of the minerals quartz and feldspar.

The Beach Onshore Beaches have a number of features, both onshore and offshore (Figure 11.9). Onshore, the landward extent of the beach is generally either a cliff, called a **sea cliff** along a seashore and a **bluff** along a lakeshore, a line of sand dunes, or a line of permanent vegetation. Sea cliffs and lakeside bluffs develop by the erosion of rock or unconsolidated sediment. In contrast, coastal sand dunes are created by the deposition of wind-blown beach sand.

The onshore portion of most beaches can be divided into two areas: one that slopes landward called a *berm* and another that slopes toward the water called the *beach face*. Berms are flat backshore areas formed by deposition of sediment as waves rush up and expend the last of their energy. They are often places where you'll find people sunbathing. Beaches may have more than one berm or there may be no berm at all. A beach face begins where the beach slope changes direction and steepens toward the water. The part of the beach face that experiences the uprush and backwash of waves is called the *swash zone*. This zone shifts location with changes in water level resulting from storms or, on sea coasts, from tides.

The Beach Offshore Directly offshore from the swash zone are two distinctive zones in the water, the surf zone

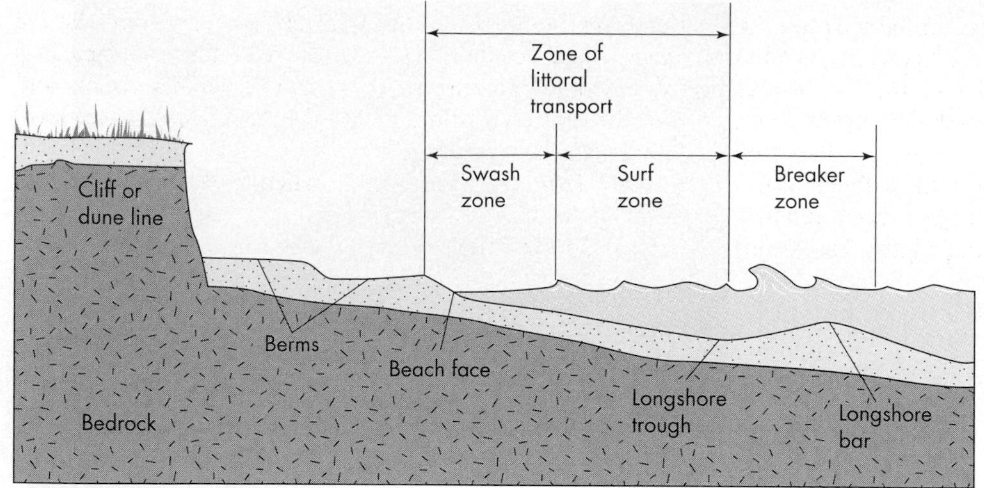

▲ FIGURE 11.9 **BEACH TERMS** Basic terminology for landforms and wave action in the beach and nearshore environment. A sea cliff or line of coastal sand dunes on the left marks the landward extent of the beach. Two berms are shown in the beach sand, each gently sloping toward the cliff or dunes. The beach face, where the land meets the water, is more steeply sloping toward the water. One longshore sand bar and longshore trough are shown underwater; the sand bar forms below the breaker zone. The zone of littoral transport includes both beach drift in the swash zone and longshore drift in the surf zone.

and breaker zone (Figure 11.9). The *surf zone* is that portion of the nearshore environment where turbulent translational waves move toward the shore after the incoming waves break. Beyond the surf zone is the *breaker zone*, the area where incoming waves become unstable, peak, and break. These conditions on the water surface are reflected in the underwater topography. A sand bar, called a *longshore bar*, forms beneath each line of breakers in the breaker zone. Landward from a longshore bar is a *longshore trough*, which is formed by wave and current action. Both the bar and trough are usually elongate, parallel to the breaking waves. Wide and gently sloping beaches may have several lines of breakers and longshore bars.[11]

Sand Transport The sand on beaches is not static; wave action constantly keeps the sand moving in the surf and swash zones. Storms erode sand from the beach and redeposit the sand either offshore or landward from the shoreline. Most of the sand moved offshore during a storm returns to the beach during fair-weather conditions.

A great deal of sand movement occurs parallel to the shoreline by **littoral transport** in the swash and surf zones (Figure 11.10). This movement includes two processes, beach drift and longshore drift. In *beach drift*, the up-and-back movement of beach material in the swash zone causes sediment to move along a beach in a zigzag path (Figure 11.10). *Longshore drift* refers to the transport of

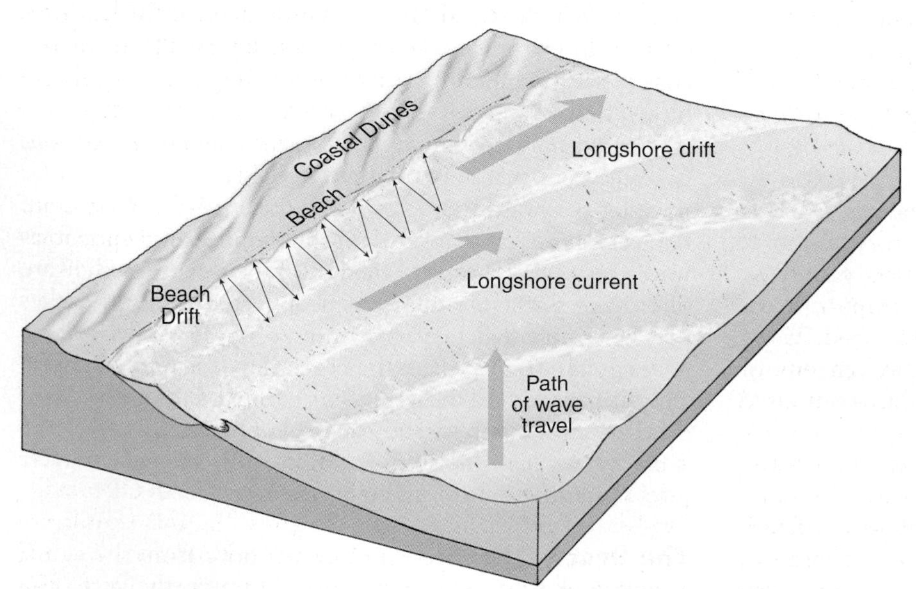

◀ FIGURE 11.10 **TRANSPORT OF SEDIMENT ALONG A COAST** Block diagram illustrating the processes of beach drift and longshore drift. Direction of beach drift is shown by the many arrows in the swash zone. The longshore drift direction is shown by the straight, thick (dark blue) arrow in the surf zone. Collectively, these two types of drift move sand along the coast in a process known as littoral transport.

sediment by ocean currents that flow essentially parallel to the shoreline (Figure 11.10). These currents, called *longshore currents*, are the primary mechanism for littoral transport. Both longshore drift and beach drift occur when waves strike the coast at an angle (Figure 11.10). The terms *updrift* and *downdrift* are often used to indicate the direction in which sediment is moving or accumulating along the shore. For example, updrift in Figure 11.10 is to the lower-left part of the diagram.

The direction of littoral transport along both the East and West Coasts of the United States is most often to the south, although it can be quite variable. Along U.S. seacoasts, most rates of longshore drift are between 150,000 to 300,000 m^3 (200,000 to 400,000 $yd.^3$) of sediment per year. Although rates of longshore drift in the Great Lakes are much less, on the order of 6000 to 69,000 m^3 (8000 to 90,000 $yd.^3$) per year, they are still substantial when you consider that a typical dump truck carries a meager 8 m^3 (10 $yd.^3$) of sand.[12]

11.3 Sea Level Change

Like the beach, the level of the sea at the shore is constantly changing. This change in sea level is caused by a number of processes, some of which operate locally and others that affect all the world's oceans. The position of the sea at the shore, referred to as **relative sea level**, is influenced by both the movement of the land and the movement of the water. These movements can be local, regional, or global in extent. Global sea level, also called **eustatic sea level**, is controlled by processes that affect the overall volume of water in the ocean and the shape of the ocean basins. Changes in eustatic sea level are just one of many factors that cause a change in relative sea level.

Eustatic Sea Level Global sea level rises or falls when the amount of water in the world's ocean increases or decreases or when there is a change in the overall shape of ocean basins. Climate, primarily the average air temperature, is the dominant control on the amount of water in the ocean today. Air temperature influences both the average temperature of the ocean and the amount of water that is stored in ice on land. As the average temperature of the ocean increases, the volume of water expands; and as it cools, the volume of water contracts. Referred to as *thermal expansion* or *contraction*, global warming or cooling of the atmosphere is responsible for this phenomenon. For example, an estimated half of the rise in eustatic sea level in the world's oceans (about 1.6mm per year today) results from thermal expansion caused by the warming of the atmosphere.[6]

Changes in air temperature also cause ice on land to melt or snowfall to increase. Thus, the volume of water frozen in glacial ice, ice caps, ice sheets, and permafrost is closely related to the average air temperature over years or decades. Most recently, global warming of our atmosphere has contributed to melting glaciers, ice caps, and the ice sheets in Greenland and Antarctica. This melting has increased the amount of water in the world's oceans and has contributed to about half of the eustatic rise in sea level (about 1.5 mm per year).[6]

Over longer geologic time spans, ocean basins change shape as the result of plate tectonic processes. These large-scale processes, such as the rate of sea floor spreading at mid-ocean ridges, influence eustatic sea level over long periods of time and are unlikely to contribute significantly to changes in relative sea level observed over decades or centuries.

Relative Sea Level Superimposed on eustatic sea level are local or regional processes that influence the movement of the land and water. The land can rise or fall slowly, as Earth's crust responds to the weight of the now-melted Pleistocene glaciers (see Chapter 12), or rapidly in response to tectonic movements during an earthquake. For example, on Good Friday, 1964, the shoreline of Montague Island in Prince William Sound southeast of Anchorage, Alaska, rose close to 10 m (33 ft.) in just a few minutes during a **M** 9.2 earthquake.[13] Many coastlines that are tectonically active and experience frequent uplift in earthquakes may be less strongly influenced by eustatic sea-level rise caused by global warming.

Movement of the shoreline is also influenced by the rates of deposition, erosion, or subsidence along the coast. Input and output factors in the beach budget described next determine whether a coastline is experiencing net erosion or deposition. Rates of coastal subsidence have a major influence on relative sea level on some shorelines, such as the Louisiana coastline described in Chapter 8.

On a daily basis, astronomical tides and weather conditions primarily control relative sea level. Astronomical tides are produced by the gravitational pull of the moon and, to a lesser extent, the sun. These tides cause relative sea level to fluctuate daily and seasonally as the position of a coast relative to the moon and sun constantly shifts. Although tidal fluctuations are entirely predictable, they can create hazardous currents and affect the height of storm surges. As mentioned previously, storm surge that occurs at high tide is generally more damaging than one that occurs at low tide (see Chapter 10).

Weather conditions also change relative to sea level over a period of hours or days. Changes in wind speed and atmospheric pressure influence the level of the sea. Wind speed has a greater effect than atmospheric pressure on relative sea level. In the open ocean, high winds pile up water and increase wave height, producing the swell described earlier. This swell increases both water level and wave heights when it reaches the shore. Storm surge from tropical storms, hurricanes, and extratropical cyclones crossing the coast are extreme cases of the swell moving landward. As mentioned in Chapter 10, the significant drop in atmospheric pressure in major hurricanes can add a meter or more to the height of the storm surge. For weaker tropical and extratropical cyclones, there is a smaller drop in atmospheric pressure and, thus, less of an effect on the storm surge.

In summary, rapid changes in relative sea level contribute to coastal flooding and hazardous nearshore currents, whereas rising eustatic sea level over decades increases hazards from storm surge and coastal erosion. In the long term, rising sea level threatens coastal cities and the very existence of many islands (see Chapter 12).

11.4 Geographic Regions at Risk for Coastal Hazards

Coastal hazards are present on both sea coasts and lake shores. In particular, large lakes, such as the Great Lakes, Great Bear Lake, Great Slave Lake, and Lake Winnipeg, develop coastal conditions similar to those on the ocean. In North America, coastal processes may have hazardous consequences along the Atlantic, Pacific, Gulf, and Arctic coasts, as well as along the shores of the Great Lakes. Strong nearshore currents, sea-level rise, storm surge from cyclones, and tsunamis are not hazards on all coastlines. For example, tsunamis and hurricanes are generally absent in lakes. In North America, only Lake Okeechobee in Florida and Lake Pontchartrain in Louisiana have experienced significant storm surges from hurricanes in historic times.

Coastal erosion is, however, a more universal hazard. As discussed later, there are various causes for this erosion, but the net effect is the horizontal retreat of the shoreline, sometimes at the rate of meters per year. All 30 states and the Canadian provinces bordering the ocean or the Great Lakes have problems with coastal erosion. Average erosion rates along some barrier islands in the southeastern United States may reach close to 8 m (25 ft.) per year, whereas the Great Lakes shorelines have experienced erosion rates up to 15 m (50 ft.) per year![14] These rates are extreme, however. Generally, rates of coastal erosion are much lower, ranging from a few centimeters per year to about 50 cm (20 in.) per year (Figure 11.11).

Strong nearshore currents, especially the rip currents described later, are common on coastlines, such as those in California and Hawai'i, that have regular, strong surf conditions. These currents can develop on any coast with breaking waves, including the Great Lakes. Strong ocean currents are also a hazard on East and Gulf Coast beaches, where many people are rescued from rip currents each year. Strong coastal currents are especially dangerous around artificial structures, such as groins, jetties, and piers.

Astronomical tides can also produce strong currents in narrow bays and channels. On coastlines with a large range between high and low tide, the incoming flood tide can rush in rapidly. For example, the Bay of Fundy, between Nova Scotia and New Brunswick, Canada, has a tidal range that is 15 m (50 ft.), and the incoming tide can rise 1 m (3.3 ft.) in 23 minutes.[15] Tidal currents rushing through the Golden Gate beneath the Golden Gate Bridge in San Francisco can reach 3 m per second (close to 7 mi. per hour).[15] Although most tidal currents are predictable, the location and intensity of strong coastal currents can take even the strongest swimmers by surprise.

The geographic areas at greatest risk for rising sea level are coastal areas and islands whose elevation is closest to

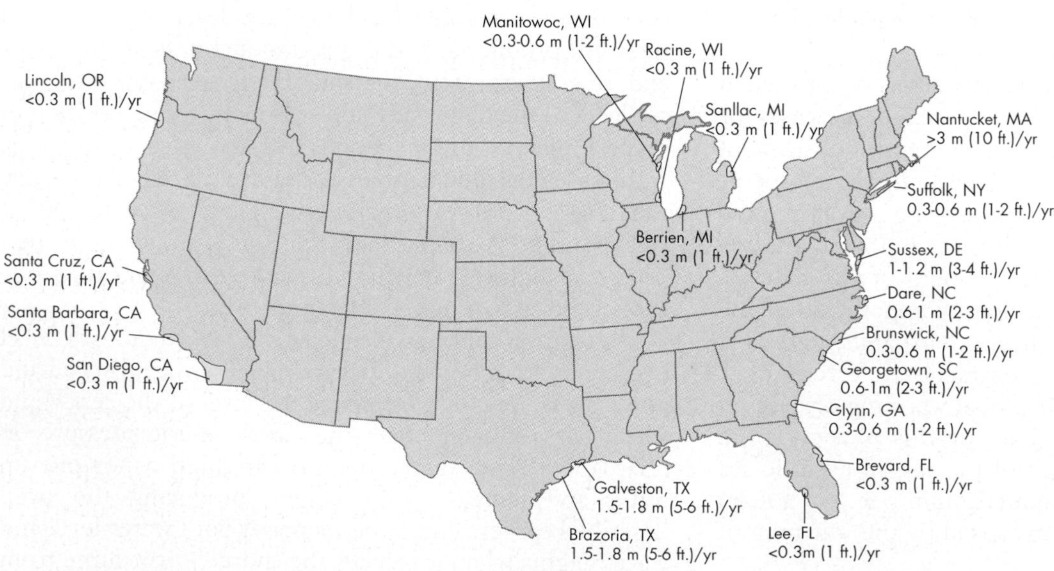

▲ FIGURE 11.11 **COASTAL EROSION RATES** Average annual erosion rates for selected coastal counties in the contiguous United States. *(Modified from* Evaluation of Erosion Hazards Summary: A Collaborative Project of the H. John Heinz III Center for Science, Economics and the Environment. *Prepared for the Federal Emergency Management Agency, 2000;* http://www.fema.gov/pdf/library/erosion.pdf*)*

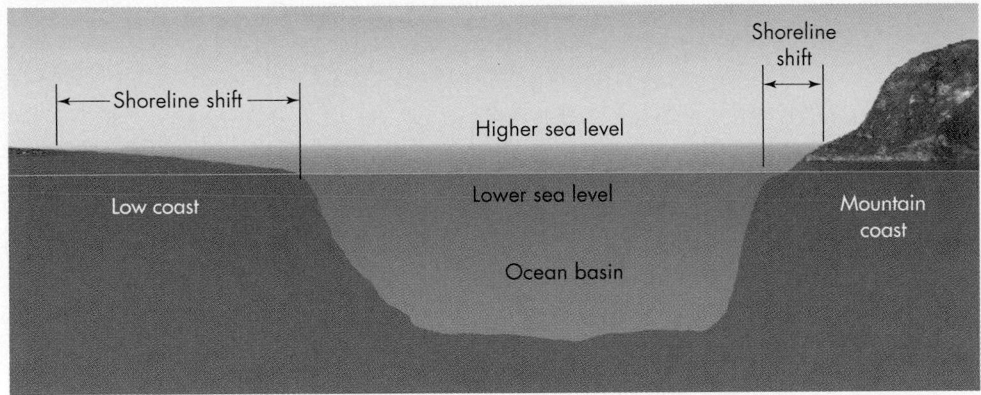

▲ FIGURE 11.12 **RISING SEA LEVEL WILL HAVE THE GREATEST EFFECT ON LOW COASTLINES** Low, tectonically passive coastlines will be most affected by rising sea level. As sea level rises, the shoreline will shift the farthest distance inland on low coastlines.

present-day sea level or whose coastlines are subsiding. Coastal areas in North America that are close to sea level include most of the Gulf Coast, much of the East Coast from Long Island south to Florida, as well as parts of Arctic Canada and the North Slope of Alaska. In these areas, a 0.3 m (1 ft.) rise in sea level can result in a landward shift of the shoreline anywhere from 30 to 300 m (100 to 1000 ft.) (Figure 11.12). Areas that are rapidly subsiding, such as parts of the Louisiana coast, may experience greater inundation from rising sea level (see Chapter 8). Many low islands in the Pacific, called *atolls*, are in danger of becoming inundated with rising sea level. Examples of these islands associated with the United States are found in the Marshall Islands, the Federated States of Micronesia, and the Republic of Palau in the western Pacific Ocean.

11.5 Effects of Coastal Processes

Coastal processes create hazards for both individuals and communities. Individuals face safety hazards from strong coastal currents when they are swimming or wading and hazards to safety and property during storm surges and tsunamis. Communities face long-term hazards related to coastal erosion and rising sea level. Strong coastal currents can be produced by the back flow of water in the surf zone and by the rising and falling of the tide. Coastal erosion affects natural features, such as beaches, dunes, cliffs, and bluffs, and threatens beachfront houses and businesses, lighthouses and docks, as well as roads, streets, and utility lines running close to the shore.

On barrier islands and sandy peninsulas called *spits*, coastal processes constantly shift the location of the shoreline. This can involve movement of an entire island, the inlets between islands, or the separation of a spit from the mainland. The southern end of Hog Island along the Delmarva Peninsula of Virginia has shifted nearly 2.5 km (1.6 mi.) in historic times (Figure 11.13).[16] In Texas, Aransas Pass, the tidal inlet that provides ships access to Corpus Christi Bay, shifted more than 3 km (2 mi.) between 1859 and 1911 before its location was stabilized with the construction of jetties.[17] Bayocean spit near Tillamook, Oregon, and the Nauset Beach spit near Chatham, Massachusetts, were both breached by storms in the twentieth century and separated from the mainland. The continual movement of barrier islands and spits is one of several characteristics that make them poor places for coastal development.

Our discussion of the effects of coastal hazards focuses on the effects of strong nearshore currents, called rip currents, and the effects of beach and cliff erosion. Discussion of the effects of tsunamis (Chapter 4), storm surge (Chapter 10), and rising sea level (Chapter 12) appears elsewhere.

RIP CURRENTS

The longshore currents described previously are not the only currents that develop along a beach. Under specific conditions along a seacoast or lake shore, powerful currents form that carry large amounts of water away from the shore. These currents, called **rip currents**, develop when a series of large waves pile up water between the longshore bar and the swash zone. The water does not return offshore the way it came in but, instead, is concentrated in narrow zones to form rip currents (Figure 11.14a). Beachgoers and lifeguards often incorrectly call this flow "riptide" or "undertow." These currents are not tides and don't pull people under the water, but they can pull people away from shore (see Survivor Story 11.2).

In the United States, around 200 people are killed and 20,000 people are rescued from rip currents each year. Therefore, rip currents constitute a serious coastal hazard, on average killing more people in the United States on an annual basis than do hurricanes or earthquakes; deaths from rip currents are equivalent to those from river flooding. People drown in rip currents because they panic and fight them by trying to swim directly back to shore. This feat is nearly impossible because the current can exceed 6 km

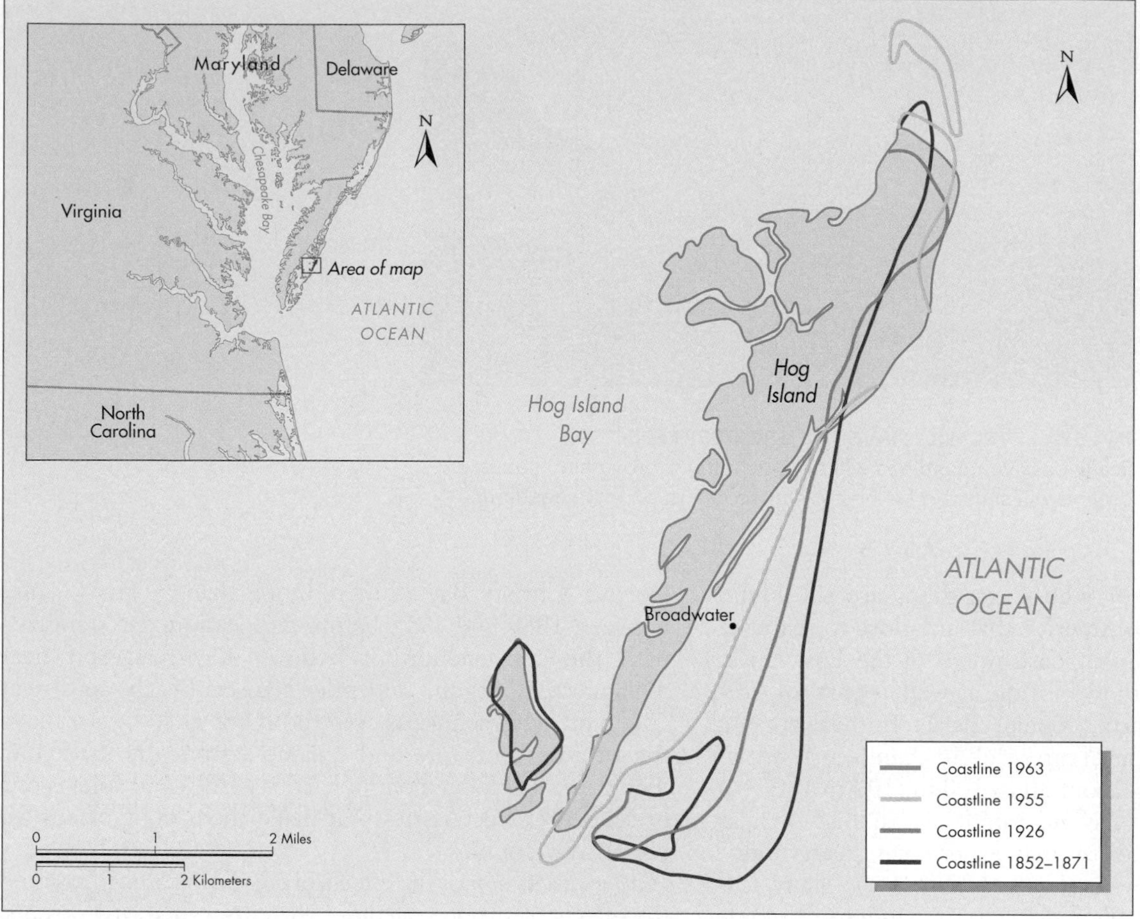

▲ FIGURE 11.13 **BARRIER ISLANDS CONTINUALLY CHANGE SHAPE AND LOCATION** Hog Island in Northampton County, Virginia, shifted location and changed form from the mid-nineteenth to mid-twentieth centuries. The island was once covered with pine forest and was the site of Broadwater, a town of 300 people, 50 houses, a school, church, and lighthouse. A hurricane inundated the entire island in 1933, and, by the early 1940s, the inhabitants had left. The town site is now under several meters of water. *(Modified from Williams, S. J., Dodd, K., and Gohn, K. K. 1991.* Coasts in Crisis. *U.S. Geological Survey Circular 1075)*

▲ FIGURE 11.14 **RIP CURRENTS** (a) The rip current is indicated by the blue milky water spreading outward from this rocky point. The blue milky water has sediment suspended in it. Dark stringers in the water are areas of kelp. (b) The existence of a rip current on this sandy beach is indicated by an area of smooth water where no waves are breaking. *(Edward A. Keller)*

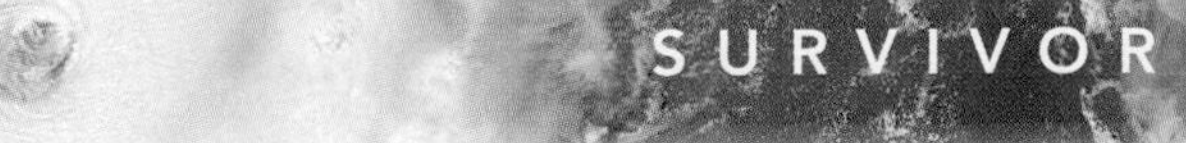

11.2

Rip Current: Two experienced swimmers rescued on Florida beach

It was a hot summer's day in Pensacola, Florida, and Jennifer Kleinbaum and her fiancé Ernie decided to go to the beach. The waters of the Emerald Coast were still as glass, but a red flag warned visitors that it was too dangerous to swim. To be on the safe side, the couple decided to go for a stroll instead (Figure 11.B).

But the sand scorched their feet, and they waded ever farther into the water to stay cool. Before they knew it, they were treading water.

Even so, neither was worried. At 26, both were strong swimmers and certified SCUBA divers. Kleinbaum was certified as a master SCUBA diver, whereas Ernie had grown up in Pensacola, swimming at this very beach.

On the horizon, they spotted a triangular fin. A marine biologist, Kleinbaum dismissed it as a harmless dogfin shark. But Ernie, unnerved, suggested returning to shore.

It took Kleinbaum some time to notice they weren't making any headway. "They say if you're swimming and not getting any closer, don't fight it; swim parallel or float out of the path of the rip current. But I was fighting without realizing it," Kleinbaum said.

They drifted into an area where the waves crashed directly over their heads and found themselves unable to swim away. "I was treading water and trying to keep myself up. I can do this for hours in a pool. But I was struggling trying to avoid the waves, and that made me so much more tired," Kleinbaum said.

Fatigued and frightened, Kleinbaum began to panic. She knew she couldn't hang on to Ernie, for fear of drowning him. "But I wanted to stay close to him, because there was nothing else I could see—he was my mental life raft." Gasping, she told him she couldn't keep her head above water much longer.

Ernie began waving his arm back and forth, yelling for help. Kleinbaum joined him. "Even though I was so tired, I somehow had no problem finding energy for this; it was something I could focus on," Kleinbaum said.

On the beach, they saw people jump up and run away. The next thing Kleinbaum remembers, a man was pushing a red, sausage-shaped lifebuoy toward them, then helping tow the pair toward shore. An off-duty lifeguard, he sat them down on the sand to rest and called an ambulance.

Only then did Kleinbaum realize they had been caught in a rip current. "After he got us out of the water, our overriding emotion wasn't relief or thank you—it was oh, my god, I am so embarrassed. We kept looking at each other saying, how did this happen to us?"

She had thought it couldn't happen to a strong swimmer. Ernie believed it happened only to tourists. "There couldn't have been two people who knew more about the ocean and rip tides, what they were and how to get out of them. Yet we still made all the wrong choices," she said.

—KATHLEEN WONG

FIGURE 11.B A HAPPIER DAY ON THE BEACH Jennifer Kleinbaum plays with her nephew Jake on a Florida beach. An earlier trip to a Pensacola, Florida, beach with her fiancé Ernie nearly ended in tragedy when they were both caught in a rip current while swimming near shore. *(Aaron Kleinbaum)*

(4 mi.) per hour, which even strong swimmers cannot maintain for long. A swimmer trying to fight a rip current soon becomes exhausted and may not have the energy to keep swimming.

Fortunately, rip currents are relatively narrow, only a few meters to a few tens of meters wide. They form in the surf zone and extend out perpendicular to the shoreline, a distance of tens to hundreds of meters offshore (Figure 11.15). Rip currents are fed by longshore currents and make their own channel as they pass through the longshore bar. Fortunately, they widen and dissipate once they have passed the line of breaking waves.

To safely escape a rip current, a swimmer must first recognize the current and then swim parallel to the shore until he or she is outside the current. Only then should the swimmer attempt to swim back to shore. If you cannot reach shore, wave an arm and yell for help. The key to survival is not to panic. When you swim in the ocean, watch the waves for a few minutes before entering the water and note the "surf beat," the regularly arriving sets of small and larger waves. Rip currents can form quickly after the arrival of a set of large waves. They can be recognized as a relatively quiet area in the surf zone where fewer incoming waves break (Figure 10.14b). You may see the current as a mass of water and debris moving out through the surf zone. The water in the current may also be a different color because it carries suspended sediment. Remember, if you do get caught in a rip current, don't panic—swim parallel to shore before heading back to the beach, and you should be able to safely escape.

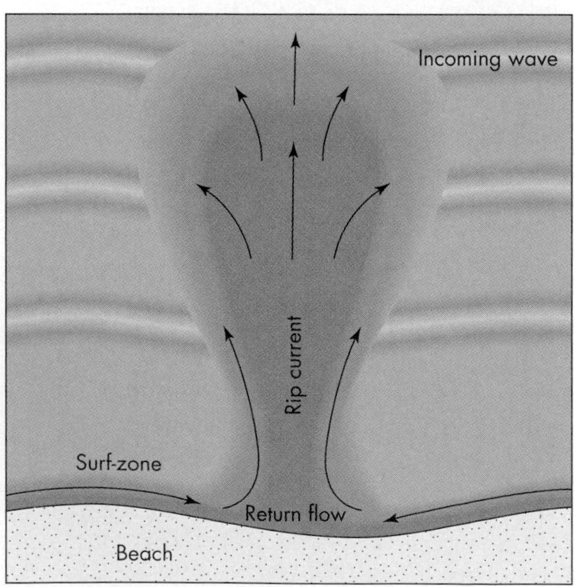

▲ FIGURE 11.15 **FLOW IN A RIP CURRENT** Bird's-eye view of the surf zone showing a rip current, which is the return flow of water that forms as a result of incoming waves. Return flow starts in the surf zone and goes through a low area in the longshore bar; flow expands outward once the current gets beyond the breaker zone.

COASTAL EROSION

As a result of the continuing global rise in sea level and extensive development in the coastal zone, coastal erosion is becoming recognized as a serious national and worldwide problem (see Professional Profile 11.3). Coastal erosion is generally a more continuous, predictable process than other natural hazards, such as earthquakes, tropical cyclones, and floods. Large sums of money are spent in attempts to control coastal erosion, but many of the fixes are only temporary. If extensive development of coastal areas for vacation and recreational living continues, coastal erosion will certainly become a more serious problem.

Beach Erosion An easy way to visualize erosion at a particular beach is to take a beach-budget approach. A **beach budget** is similar to a bank account. In a bank account, deposits made to the account are input; the account balance reflects the storage of money in the account; and withdrawals by cash, check, or debit card are output from the account. Similarly, we can analyze a beach in terms of input, storage, and output of the sand or gravel that makes up the beach (see A Closer Look 11.4). Most input of sediment to a beach is caused by the coastal processes that move the sediment along the shore (see Figure 11.10) or by local wave erosion. Coastal processes, particularly beach drift and longshore drift, bring sediment from updrift of a beach. Local wave erosion of sand dunes or sea cliffs landward of the beach also inputs sediment to the beach. Erosion of sand dunes and cliffs is generally most intense during storms when water floods the entire beach and large waves can reach inland. Sediment moving along a shoreline comes from one or more updrift sources, such as a river delta or erosion of another beach. The sediment that is in storage on a beach is what you see when you visit the site. Output of sediment occurs when coastal processes move sand or gravel away from the beach. These output processes can include littoral drift, a return bottom flow from storm waves that carries sediment into deeper water, and onshore winds eroding sand from the beach and blowing it inland.

If input exceeds output, the beach will grow as more sediment is stored and the beach widens. When input and output are about the same, the beach will stay fairly constant in width. If output of sediment exceeds input, the beach will erode and there will be fewer grains of sand or pieces of gravel on the beach. Thus, we see that the budget represents a balance of sediment on the beach over a period of years. Short-term changes in sediment supply from the attack of storm waves will cause seasonal or storm-related changes to the balance of sediment on a beach. Long-term changes in the beach budget caused by climate change or human impact result in either deposition or erosion of a beach. One long-term change on some East and Gulf Coast beaches in the United States has been a reduction in sand supply from rivers. Construction of dams on rivers reaching the coast, such as the Savannah River in Georgia and the Rio Grande in Texas, has decreased the sand supply to beaches and amplified the coastal erosion problem.

PROFESSIONAL PROFILE 11.3

Rob Thieler, Marine Geologist

Growing up on the New Jersey shore, Rob Thieler (Figure 11.C) learned that beaches are literally as changeable as the weather. As a teenage lifeguard, he watched entire beaches run short on sand, leaving waves to break perilously close to buildings and streets.

Today, Thieler's coastal observations have expanded considerably. As a research geologist with the U.S. Geological Survey, he studies how sandy coastlines shift and change over time.

"We don't have good models to predict where the shoreline's going to be in a hundred years. A beach can lose 30 meters of sand in a single storm, and we certainly can't predict when and where storms will hit for the next century. We have a really incomplete understanding of how sand moves around on beaches and the inner continental shelf," Thieler said. These problems aren't simply academic; the answers have major implications for the estimated 160 million Americans who live near the coast.

In past centuries, Thieler said, residents threatened by disappearing beaches merely moved their houses farther up the dunes. But, for modern homeowners, that's no longer an option—higher ground generally means impinging upon someone else's property.

Most years, Thieler spends about a month at sea, gathering information about coastal geology. The ship tows sidescan sonar, seismic reflection profilers, and swath bathymetric gear in linear swaths that resemble the path of a lawnmower over the shallow waters of the continental shelf. The echoes reveal the texture, topography, and structure of bottom sediments in exquisite detail.

"I love going to sea. There's nothing like exploring a part of the ocean no one has seen before," Thieler said. Much of the technology Thieler uses to study ocean topography was invented only in the past few decades. As a result, the majority of the seafloor remains uncharted territory.

Thieler combines his marine data with land observations, such as the historic locations of shorelines and the elevation of bluffs and beaches. His results can help explain why waves break in certain locations and how currents shape nearshore marine topography.

Thieler has also put this data to use analyzing how vulnerable U.S. coastlines are to sea-level rise. Already, some planners are using his findings to adapt to rising seas. "Before, we've often chosen to hold the line to protect property. As we improve our understanding of how beaches and coasts evolve in a time of rising sea levels and changing climate, the choice is no longer clear," he said.

—KATHLEEN WONG

▲ FIGURE 11.C **DR. ROB THIELER—MARINE GEOLOGIST** Standing on a rock along the Massachusetts shore near his U.S. Geological Survey office in Woods Hole, Dr. Thieler is in one of his favorite places—the coast. Dr. Thieler studies coastal processes as part of a team of scientists assessing the vulnerability of the U.S. coastline to rising sea level. *(Dann Blackwood)*

Cliff Erosion A sea cliff or lakeshore bluff along a coastline may be subject to additional erosion problems. Both features are exposed to a combination of wave action and land erosion by running water and landslides that erode the cliff at a greater rate than either process could do alone. The problem is further compounded when people alter a cliff through poor development.

Erosion by plunging breakers during storms causes many sea cliffs and lakeshore bluffs to retreat. This susceptibility to erosion is not always apparent during low-water periods when a beach is present at the base of the cliff (Figure 11.16). The beach sediment helps protect the cliff from wave erosion during periods between storms. On the West Coast of the United States, most of the sea-cliff erosion takes place during winter storms, and, in the northeastern United States and Atlantic Canada, considerable cliff erosion occurs during nor'easters. Storm waves also help erode lakeshore bluffs on the Great Lakes.

A variety of human activities can increase the erosion of sea cliffs and lakeshore bluffs. These activities create effects that include uncontrolled surface runoff, increased groundwater discharge, and the addition of weight to the top of a cliff. Increased surface runoff, as often occurs with urbanization, can enhance the erosion of coastal cliffs and bluffs unless the drainage is controlled and diverted from the cliff face. Erosion can also intensify if there is an increase in the amount of groundwater discharging from the base of the cliff. Watering lawns and gardens on top of a sea cliff or lakeside bluff can increase the amount of water in the earth material. When this water emerges as seeps or springs from the cliff, it reduces its stability and facilitates both erosion and landslides.[18] Structures such as walls, buildings, swimming pools, and patios may also decrease the stability of a cliff or bluff by increasing the driving forces for mass wasting (Figure 11.17). Strict regulation of development in many areas of the coastal zone now forbids most risky construction, but we continue to live with some of our past mistakes.

The rate of sea cliff and lakeshore bluff erosion is variable, and few measurements have been available. More are on the way because of the increasing use of remote sensing devices, such as an aircraft-mounted laser system called "light detection and ranging" (LIDAR). This system can record several thousand elevation measurements each second with a vertical resolution of 15 cm (6 in.). Once a baseline set of elevations is recorded, subsequent LIDAR flights can detect changes in the coastal zone, such as the shape of the beach, dunes, bluff, or cliff. For example, LIDAR flights along the coast near Pacifica, California, 6 months apart, showed beach erosion of 1 to 2 m (3 to 7 ft.) and sea cliff erosion of about 10 m (30 ft.).

Moderate rates of erosion of 15 to 30 cm (6 to 12 in.) per year occur along other parts of the California coast near Santa Barbara, in comparison to 2 m (7 ft.) per year along parts of the English Norfolk coast and up to 4.6 m (15 ft.) per year along the eastern side of Cape Cod. The rate of erosion depends primarily on the resistance of earth materials and the energy of the waves.[18]

Sea cliff and lakeshore bluff erosion is a natural process that cannot be completely controlled, even if large

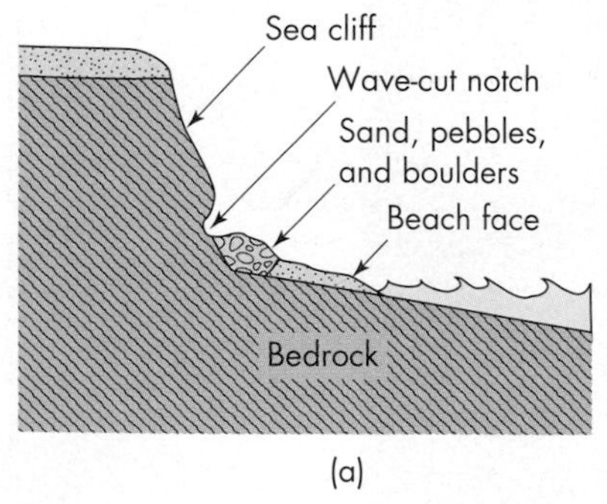

▲ FIGURE 11.16 **SEA CLIFF AND BEACH** (a) Generalized cross section and (b) photograph of sea cliff, beach, and bedrock exposed at low tide at Santa Barbara, California. *(Donald W. Weaver)*

A CLOSER LOOK 11.4

Beach Budget

For a given shoreline segment, the total volume of sand added (gained) to the beach can be balanced (compared) to losses. This produces the "beach budget"[4]:

- If losses are greater than gains, erosion results.
- If losses are less than gains, the beach grows by accretion of sand.
- We can evaluate a budget over a set period of time, such as 1 year or 10 years.

As an example, imagine a simple coast with rivers supplying sand, a sea cliff, beach, and submarine canyon (Figure 11.D). This coast defines a *littoral cell,* which is a segment of coast (a system) that includes sources and transport of sand to and along the beach (in this simple case, rivers deliver sand, as does sea cliff erosion). Sand is transported along the coast, and some is transported from the near shore environment down a *submarine canyon* (an offshore canyon that may head in the surf zone and remove sand from the beach transport system). Here, we determine the budget before and after a dam was constructed, confirming the erosion observed on beaches south of the submarine canyon where homes are threatened.

Sources of Sand for Our Beach Budget Example (+ = gain; − = loss)

Littoral transport (+)	South	200,000 m^3/yr
	North	50,000 m^3/yr
	Net	150,000 m/^{3}yr to south
(Scf) Sea cliff erosion (+)	Erosion rate is 0.5 m/yr; average height of sea cliff is 6 m; total length 3000 m. Assume 50 percent of material eroded remains on beach.	

$$\text{Scf} = (0.5\ \text{m/yr})(6\ \text{m})(3000\ \text{m})(0.5) = 4{,}500\ \text{m}^3/\text{yr}$$

(Sr) River source (+) Assume drainage area of 800 km^2

A dam was constructed 15 years ago, reducing the drainage basin delivering sediment to the coast to 500 km^2.

To estimate the sediment delivered to the coast from the river, we can use equations (numerical models) or regional graphs.[4] The basin before the dam has an area contributing sediment of 800 km^2. From our analysis of sediment production, we estimate the sediment delivered to the beach from the river is about 300 $m^3 km^2$/yr. Following dam construction, the sediment-contributing area is reduced to 500 km^2, with a yield of 400 m^3/km^2/yr. The increase per unit area results because smaller tributaries below the dam have a higher sediment yield.

Assume that 30 percent of sediment delivered from the river will remain on the beach, and that it is sand sized and larger:

Before dam:

$$\text{Sr}(+) = (300\ \text{m}^3/\text{km}^2/\text{yr})(800\ \text{km}^2)(0.3) = 72{,}000\ \text{m}^3/\text{yr}$$

After dam:

$$\text{Sr}(+) = (400\ \text{m}^3/\text{km}^2/\text{yr})(500\ \text{km}^2)(0.3) = 60{,}000\ \text{m}^3/\text{yr}$$

(Scy) Down submarine canyon (−):

Estimated from offshore observations: 220,000 m^3/yr.

(continued)

Budget before dam:		
Longshore drift (Sl)	+	150,000 m^3/yr
Cliff erosion (Scf)	+	4,500 m^3/yr
River (Sr)	+	72,000 m^3/yr
Submarine canyon (Scy)	−	220,000 m^3/yr

Budget +6,500 m^3/y; therefore, because more sand is arriving at point X on Figure 11.D than is leaving, the beach is growing.

Budget after dam:		
Sl	+	150,000 m^3/yr
Scf	+	4,500 m^3/yr
Sr	+	60,000 m^3/yr
Scy	−	220,000 m^3/yr

Budget −5,500 m^3/yr; therefore, more sand is leaving point X than is arriving, and erosion is observed.

Assume there are several beach homes near point X. What advice would you give them? What would be your response if there were high-rise coastal resorts near point X?

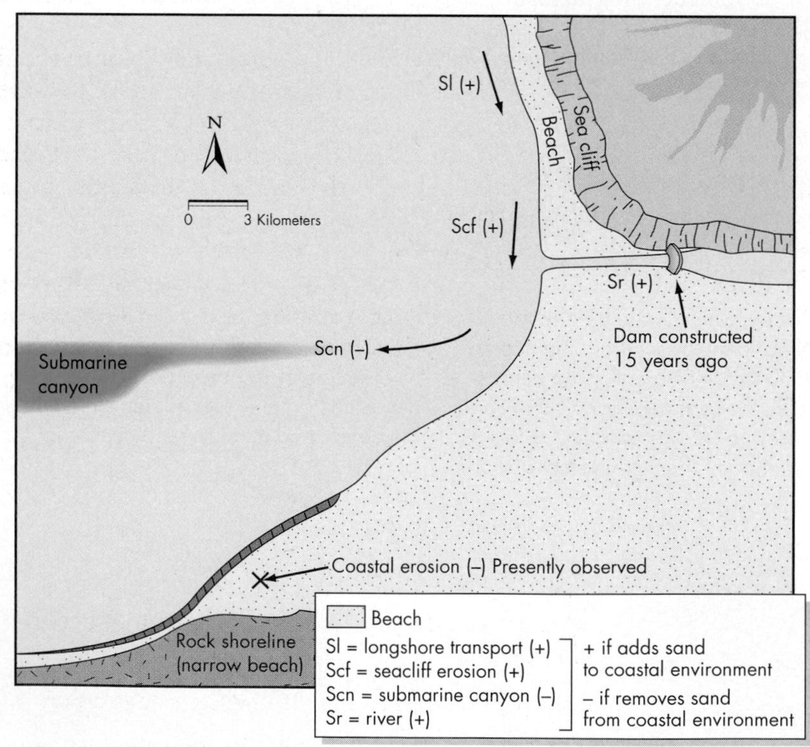

◀ **FIGURE 11.D EXAMPLE OF A BEACH BUDGET** Beach erosion is observed at Point X. See text for calculation of the beach budget before and after construction of the dam.

▲ FIGURE 11.17 **SEA CLIFF EROSION** Apartment buildings on the edge of a sea cliff in the university community of Isla Vista, California. The sign states that the outdoor deck is now open. Unfortunately, the deck is not particularly safe, as it is overhanging the cliff by at least 1 m (3 ft.). Notice the exposed cement pillars in the sea cliff, originally in place to help support the buildings. These decks and apartment buildings are in imminent danger of collapsing into the sea and were condemned in 2004. *(Edward A. Keller)*

amounts of time and money were invested. Therefore, we must learn to live with some erosion. We must modify land-use practices, such as controlling water on and in cliffs and not placing large structures close to the top edge of a cliff.

11.6 Linkages between Coastal Processes and Other Natural Hazards

Coastal processes themselves, such as waves and currents, present numerous hazards to human life and property. These hazards are often linked to other natural hazards, such as earthquakes, volcanic eruptions, tsunamis, cyclones, flooding, landslides, subsidence, and climate change. Earthquakes, such as the **M** 9.1 Sumatran earthquake of 2004, coastal volcanic eruptions, and tsunamis can radically change the shape of the shoreline. Storm waves, storm surge, and coastal flooding from cyclones are closely linked to coastal erosion (Figure 11.18).

Intense precipitation associated with tropical storms, hurricanes, and extratropical cyclones drives many other coastal hazards, such as flooding, erosion, and landslides. As mentioned previously, the storm surge and heavy rainfall inland combine to cause widespread coastal flooding. Areas where the coast has subsided are more vulnerable to both freshwater flooding and storm surge.

Another common hazard caused by coastal processes is landslides. Wave erosion at the base of sea cliffs and lakeside bluffs undercuts the slope and frequently produces landslides. These cliffs may be more susceptible to sliding if structures, such as homes, add weight at the top (Figure 11.19).

Coastal erosion is also linked to climatic conditions that change over decades. For example, an interaction between surface winds and sea surface temperatures in the Pacific Ocean produces a climatic condition known as El Niño (see Case Study 12.2). Strong El Niño conditions can increase the intensity and frequency of extratropical cyclones making landfall on the West Coast. Large waves and storm surge from storms produced by a strong El Niño in 1982 to 1983 significantly increased coastal erosion in California and Oregon.[19]

The giant 2010 oil spill (as much as 170 million gallons by mid-July 2010) from a blowout from an oil well drilled from a platform in deep water off the coast of Louisiana in the Gulf of Mexico emphasizes the fact that coastal processes are linked to environmental problems. The spill is considered the most serious oil spill disaster in U.S. history. As oil moved with wind and Gulf currents to the coastal zone, beaches, and salt marshes, coastal processes became an important factor (Figure 11.20). Oil on beaches is moved with the tides and longshore transport systems driven by waves. Oil contaminates the beaches of barrier islands and moves through inlets to salt marshes. Predicting the movement of the oil along the beaches and in salt marshes requires detailed coastal information on

- Wave height and frequency
- Direction and rate of longshore transport
- Strength of tidal flow into and out of barriers and island inlets
- How far inland tidal flow inundates salt marshes
- How salt marsh vegetation and sediment interact with the oil
- The effects of the oil on life on beaches and salt marshes

Overall, coastal processes are closely linked to other natural hazards that affect the coast: tsunamis; freshwater flooding of coastal plains, bays, and lagoons; and the mass wasting of cliffs along the ocean and lakes.

▲ FIGURE 11.18 **STORMS CONTRIBUTE TO COASTAL EROSION** Two hurricanes, 21 days apart, eroded sand from this coastline in Vero Beach on Florida's east coast. The top picture, taken in August 1997 before the hurricanes, shows a sand dune providing some protection for the homes. In the middle picture, taken on September 8, 2004, after Hurricane Frances, the dune has migrated next to the houses. The third picture, taken on September 29, 2004, after Hurricane Jeanne, shows a house that has been undermined and damaged by storm waves. A red arrow points to the same place in all three photographs. *(U.S. Geological Survey/U.S. Department of the Interior)*

11.7 Natural Service Functions of Coastal Processes

As is true of many of the other hazards we have discussed, it is difficult to imagine the benefits of rip currents, coastal erosion, and sea-level rise; however, some coastal processes do provide benefits. Although erosion is a problem for property owners in coastal areas, the beauty of the coastal zone results in part from wave action and erosion. Many tourists drive the Pacific Coast highway from Washington to California each year to experience the beauty of the coastal zone. The stunning cliffs, rocky headlands, and sea arches found along this scenic road are the direct result of erosion and are important aesthetic resources (Figure 11.21).

For many beaches, coastal erosion is the only significant input of sand. Without erosion of dunes, cliffs, and bluffs inland from the beach, or the erosion of updrift beaches, there would be little sand to form a beach. Such is the case for most beaches on the east shore of Cape Cod in Massachusetts.

Beach processes, such as longshore drift, maintain sandy beaches on all coasts, including lake shores. Disturbance of the coastal zone and coral reefs by storm waves renews these ecosystems and maintains their diversity. Finally, coastal processes provide us with much-loved recreational opportunities, including swimming, surfing, sailing, fishing, and sunbathing. In summary, coastal processes contribute to the ecological health and aesthetic value of the coastal zone and provide a variety of recreational activities.

11.8 Human Interaction with Coastal Processes

Human interference with natural shore processes has caused considerable coastal erosion. Most problems arise in areas that are highly populated and developed. Efforts to stop coastal erosion often involve engineering structures that impede littoral transport. These artificial barriers interrupt the movement of sand, causing beaches to grow in some areas and erode in others, thus damaging valuable beachfront property. This type of human interaction with coastal processes is especially prevalent along the Atlantic, Gulf, and Pacific coasts of the United States, the Great Lakes, and in some parts of Canada.

THE ATLANTIC COAST

The Atlantic coast from northern Florida to New York is characterized by barrier islands, long narrow islands of sand that are separated from the mainland by a lagoon or bay (Figure 11.22). Many barrier islands have been altered to a lesser or greater extent by human use.

The history of barrier islands along the Maryland coast illustrates the interaction of human activity with coastal processes. Demand for the 50 km (30 mi.) of oceanfront beach in Maryland is very high. This limited resource is used seasonally by residents of the Washington and Baltimore metropolitan areas (Figure 11.23). Probably the greatest repercussions from this use have taken place in Ocean City on Fenwick Island and on Assateague Island to the south of Fenwick Island. Since the early 1970s, Ocean City has promoted high-rise condominium and

▲ FIGURE 11.19 **A HAZARDOUS ROOM WITH A VIEW** These homes perched on a sea cliff along Puget Sound west of Port Townsend, Washington, were spared in this 1997 landslide. Additional landslides in the future will cause further retreat of the cliff face. *(Geology and Earth Resources Washington Division)*

hotel development on its waterfront. To make way for new construction, the coastal dunes of this narrow island were removed in many locations. This activity and serious beach erosion have increased the vulnerability of the island to hurricanes. As mentioned previously with respect to South Padre Island in Texas, coastal dunes can survive a hurricane storm surge. They act as natural barriers to storm waves and partially protect structures built behind them. Without coastal dunes, a shoreline has no protection from storm waves or from the formation of washover channels by storm surge.

Ocean City experienced the effects of hurricane storm surge in 1933 when the Ocean City inlet formed to the south of the city. Despite attempts to stabilize the inlet by coastal engineering, there is no guarantee that a new inlet will not destroy part of the city in the future.[20]

Assateague Island is located to the south, across the Ocean City inlet. It encompasses two-thirds of the Maryland coastline and contrasts with the highly urbanized Fenwick Island. Assateague is in a much more natural state and is used for passive recreation, such as sunbathing, swimming, walking, and wildlife observation. However, both islands are in the same littoral cell, meaning they share the same sand supply. At least that was the case until 1935, when jetties were constructed to stabilize the Ocean City inlet. Since construction of the jetties, beaches on the north end of Assateague Island have lost about 11 m (36 ft.) per year, nearly 20 times the long-term rate of shoreline retreat for the Maryland coastline. During the same time, beaches immediately north of the inlet became considerably wider, requiring the lengthening of a recreational pier.[21]

Observed changes on Maryland's Atlantic coast are clearly related to human interference with the longshore drift of sand. Longshore drift is to the south, at an average annual volume of about 150,000 m^3 (196,000 $yd.^3$). Construction of the Ocean City inlet jetties interfered with the natural southward movement of sand. The jetties diverted sand offshore rather than allowing it to continue southward to nourish the beaches on Assateague Island. Starved of sand, the northern portions of the island have experienced serious shoreline erosion during the past 50 years. This example of beach erosion associated with engineering structures has been cited as the most severe in the United States.[21]

THE GULF COAST

Coastal erosion is also a serious problem along the Gulf of Mexico, which, like the Atlantic coast, has numerous barrier islands. One study in the Texas coastal zone suggests that in the past 100 years, human modification of the coastal zone has accelerated coastal erosion by 30 to 40 percent as compared with prehistoric rates.[22] Much of the accelerated erosion appears to be caused by coastal engineering structures, subsidence as a result of groundwater and petroleum withdrawal, and damming of rivers that supply sand to the beaches.

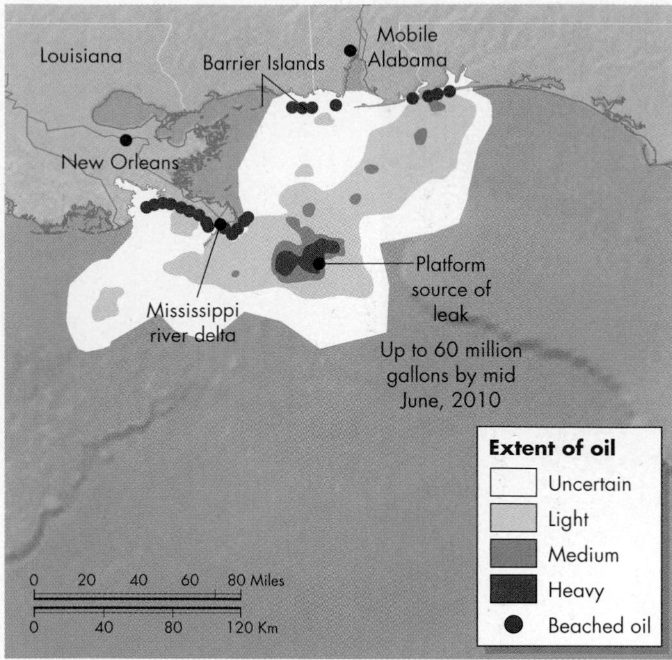

(a)

(b)

▲ **FIGURE 11.20 GULF OF MEXICO OIL SPILL 2010**
(a) Map of the extent of the spill by mid-June *(NOAA with modification)* (b) cleaning oil from a Louisiana beach. *(UPI/Newscom)*

THE GREAT LAKES

Erosion is a periodic problem along the coasts of the Great Lakes and has been particularly troublesome on the Lake Michigan shoreline. Damage is most severe during high lake levels caused by above-normal precipitation. Measurements by the U.S. Army Corps of Engineers since 1860 show that the level of Lake Michigan has fluctuated about 2 m (6.5 ft.). During periods of high water, the considerable wave erosion has destroyed many buildings, roads, retaining walls, and other structures (Figure 11.24).[23] For example, in 1985 high lake levels from fall storms caused an estimated $15 to $20 million in damages.

During periods of below-average lake level, wide beaches develop that dissipate energy from storm waves and protect the shore. However, as lake levels rise, beaches become narrower and storm waves exert considerable energy against coastal areas. Even a small rise in water level on a gently sloping shore will inundate a surprisingly wide section of beach.[23]

Erosion of lakeshore bluffs has also been a problem. Many bluffs along Lake Michigan have eroded back at an average rate of 0.4 m (1.3 ft.) per year.[24] Severity of erosion at a particular site depends on several factors:

- The existence of coastal dunes—dune-protected bluffs erode at a slower rate.
- The orientation of the coastline—sites exposed to high-energy storm winds erode faster.
- Groundwater seepage—seepage at the base of a bluff causes slope instability and increased erosion.
- The existence of protective structures—engineering structures are locally beneficial but often accelerate coastal erosion in adjacent areas.[23, 24]

In recent years, beach nourishment has been attempted for some Great Lakes beaches. For some projects, the added sands have been deliberately chosen to be coarser and heavier than the natural sands, theoretically to reduce the erosion potential.

CANADIAN SEACOASTS

Canada has the longest ocean coastline in the world and, like the United States, it faces a serious erosion problem. As is true elsewhere, Canadian sandy and muddy beaches are eroding more rapidly than rocky beaches. Although most erosion rates are less than 1 m (3 ft.) per year, some areas of Atlantic Canada are eroding at rates of up to 10 m (30 ft.) per year! Even at close to the average rate of erosion, entire islands off Nova Scotia have disappeared into the sea. In the Canadian Arctic along the Beaufort Sea, erosion rates are a bit higher, at 1 to 2 m (3 to 7 ft.) per year. Each year, the Canadian government spends about $1 billion on projects related to the coastline, including erosion control.[25]

11.9 Minimizing the Effects of Coastal Hazards

On the surface, it may appear that coastal hazards would be much easier to control than other natural hazards, such as earthquakes, volcanoes, and hurricanes, that release tremendous amounts of energy. In practice, interactions among coastal processes are complex, and

◀ FIGURE 11.21 **SEA ARCH ERODED FROM ROCKY HEADLAND** Sea arches, like this one along the Pacific coast in Mendocino, California, are formed by the erosion of a rocky headland. Located in a public park, this coastal erosion feature, at least temporarily, adds to the beauty of the coastline. *(Robert H. Blodgett)*

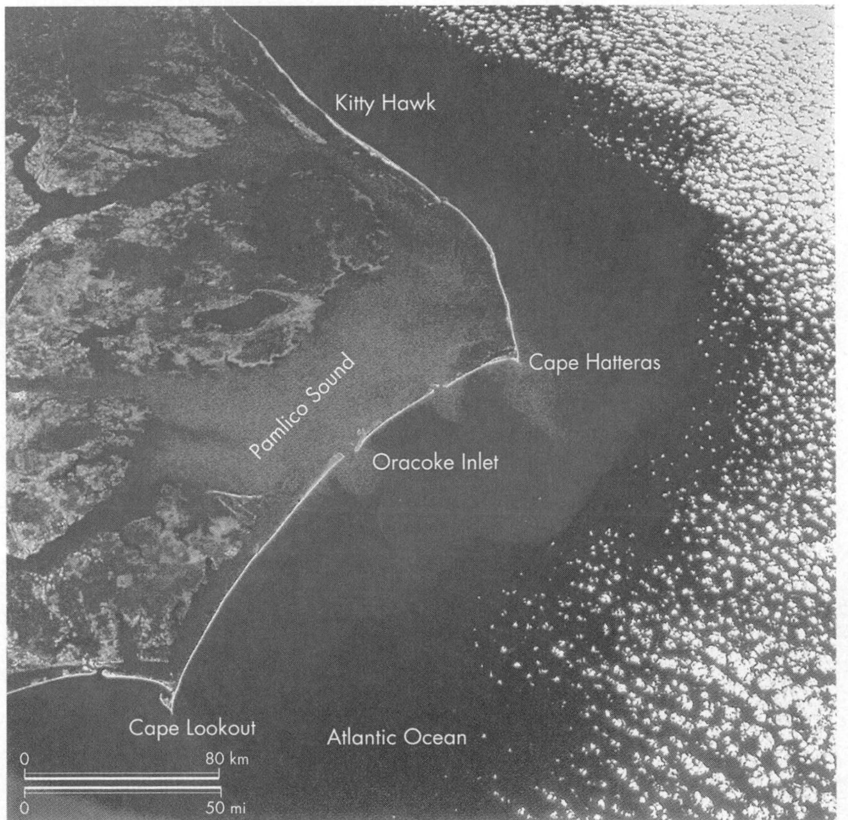

◀ FIGURE 11.22 **THE OUTER BANKS OF NORTH CAROLINA** View looking north from the Apollo-9 mission. The barrier islands appear as thin white ribbons of sand separating the Atlantic Ocean on the right from Pamlico Sound and the mainland on the left. The yellowish brown color in the water is caused by suspended sediment, such as clay, moving within the coastal system. Lobe-shaped fans of suspended sediment are entering the Atlantic Ocean at Ocracoke Inlet and the inlet to the northeast. Kitty Hawk, the site of the Wright Brothers' first powered aircraft flight, is near the top of the image. The distance from Cape Lookout to Cape Hatteras is approximately 100 km (60 mi.). *(NASA)*

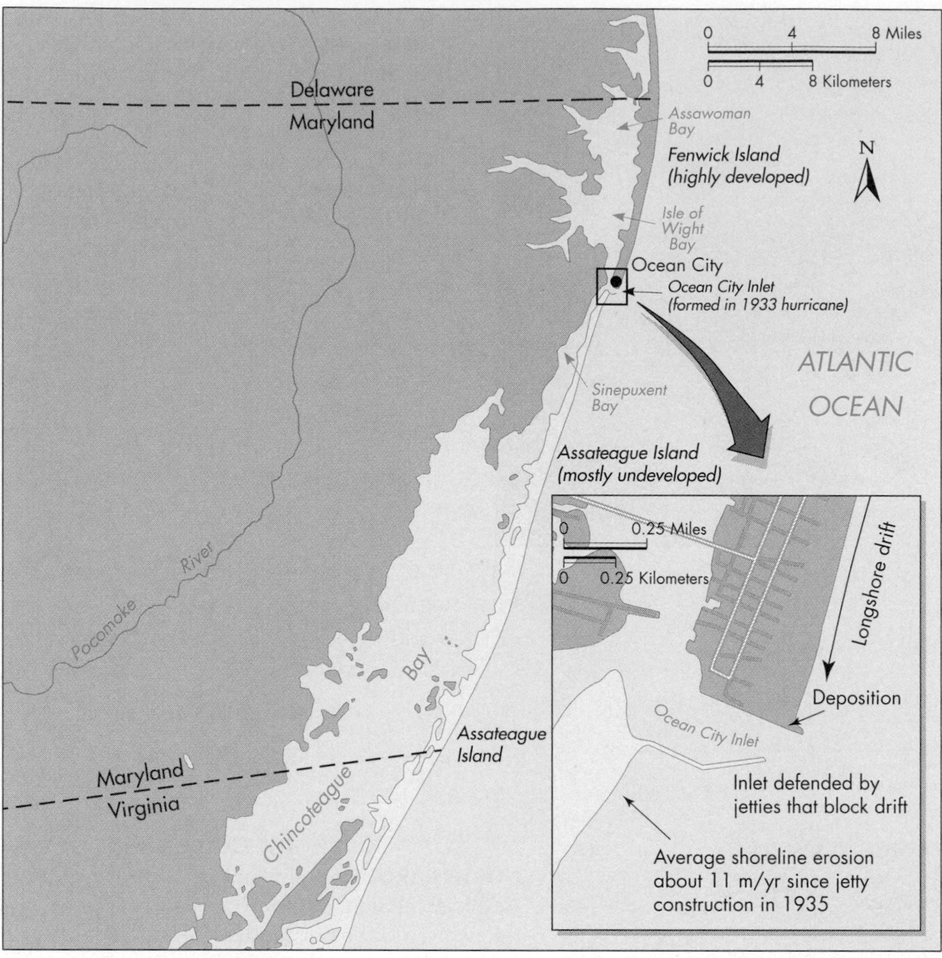

▲ FIGURE 11.23 **URBAN DEVELOPMENT WITH JETTY INCREASES BEACH EROSION** Fenwick Island, a barrier island on the Maryland coast, is experiencing rapid urban development, and potential hurricane damage is a concern. What if a new inlet forms during a hurricane in Ocean City to the north of the Ocean City inlet? Inset shows details of the Ocean City inlet and effects of jetty construction. The northern end of Assateague Island, shown in yellow, is relatively undeveloped and has experienced rapid shoreline erosion since a jetty was constructed along the inlet.

◄ FIGURE 11.24 **LAKES HAVE COASTAL EROSION PROBLEMS** Coastal erosion along the shoreline of Lake Michigan has destroyed this home. *(Steve Leonard)*

efforts to combat coastal erosion and rising sea level have often met with failure or unintended consequences. The following discussion focuses on minimizing beach erosion, one of the most widespread coastal hazards.

Most efforts to minimize beach erosion focus on stabilizing a beach at its present location. These efforts use two approaches:

- Hard stabilization: engineering structures to protect a shoreline from direct wave erosion
- Soft stabilization: adding sand to replace sand that has been eroded from the beach, a process known as beach nourishment.

Both approaches have major drawbacks that are discussed next. Where practical, a third option, sometimes referred to as *managed retreat*, may have better long-term success in preserving beaches. This option recognizes that most shorelines are going to continue to erode for the foreseeable future, and that we need to design buildings, infrastructure, and land-use practices to adapt to this movement.

HARD STABILIZATION

Engineering structures in the coastal environment, such as seawalls, groins, breakwaters, and jetties, are generally constructed to improve navigation or retard erosion. However, because these structures interfere with the littoral transport of sediment along the beach, they all too often cause undesirable deposition and erosion in their vicinity.

Seawalls **Seawalls** are structures built on land parallel to the coastline to help retard erosion and protect buildings from damage. They may be constructed with concrete, large stone blocks called *riprap*, pilings of wood or steel, cemented sand bags, or other materials. The construction of a seawall at the base of a sea cliff or lakeside bluff may not be particularly effective because the cliff is eroded from both the land and the water. The use of seawalls has been criticized because their vertical design reflects incoming storm waves and redirects their energy to the shore. Therefore, seawalls, over a period of decades, promote beach erosion and produce a narrower beach with less sand (Figure 11.25). Unless carefully designed to complement

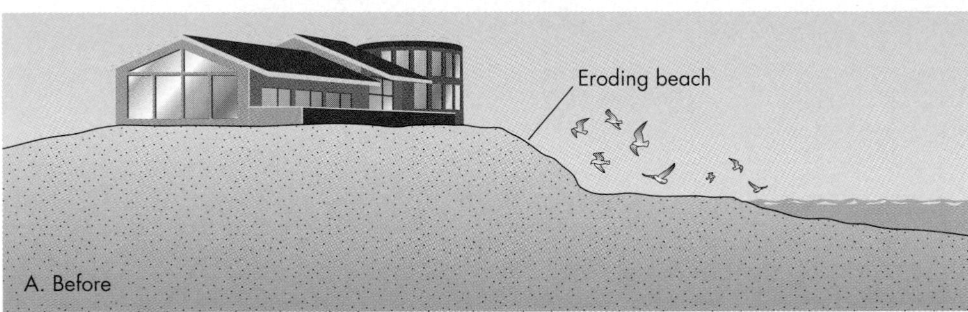

▲ FIGURE 11.25 **SEAWALLS INCREASE BEACH EROSION** Construction of seawalls to mitigate coastal erosion and protect buildings from rising sea level results in beach erosion. A seawall acts as a reflector for storm waves and can result in the permanent loss of a beach.

existing land uses, seawalls generally cause environmental and aesthetic degradation. The design and construction of seawalls must be carefully tailored to specific sites. Because of these difficulties, many geologists believe that seawalls cause more problems than they solve and should be used rarely, if ever, where beach preservation is a goal.

Groins **Groins** are linear structures placed perpendicular to the shore, usually in groups called *groin fields* (see Figure 11.2). Each groin is designed to trap a portion of the sand from the longshore drift (Figure 11.26). A small amount of sand accumulates updrift of each groin, thus building an irregular, but wider beach.

However, there is an inherent problem with groins—although trapped sand is deposited updrift of a groin, erosion occurs downdrift of the structure because the longshore drift has been depleted of sand. Thus, a groin or groin field results in a wider, more protected beach in a desired area but causes erosion of the adjacent shoreline. Once a groin has trapped all the sediment it can hold, sand in the groin is transported around its offshore end to continue its journey along the beach. Therefore, erosion may be minimized by artificially filling each groin. Known as **beach nourishment**, this process requires extracting sand from the ocean floor and placing it onto the beach (see Figure 11.4). When nourished, the groins will draw less sand from the longshore drift, and the downdrift erosion will be reduced.[11] Despite beach nourishment and other precautions, groins often cause undesirable erosion in the downdrift area; therefore, their use should be carefully evaluated.

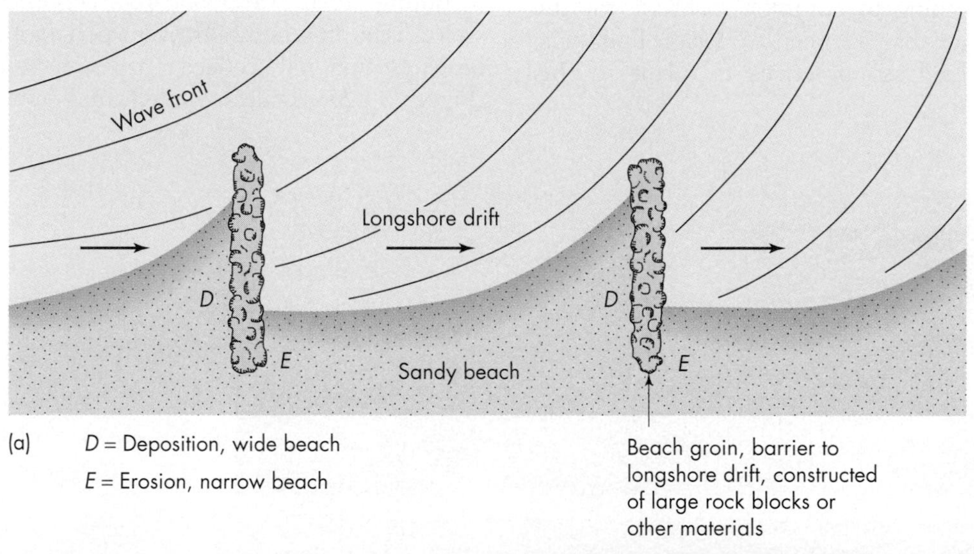

(b)

(c)

▲ FIGURE 11.26 **BEACH GROINS** (a) Diagram of two beach groins built to trap sand; updrift depositional areas (D) are on the left, and the downdrift erosional areas (E) are on the right of the groins. Waves approach from the upper left, and longshore drift is from left to right. (b) Close-up of the eroded area downdrift of a groin. (c) A groin extends seaward in the center of the photograph; contrast the width and height of the beach on the updrift side to the right and the downdrift side to the left. *(Edward A. Keller)*

Breakwaters and Jetties Breakwaters and jetties are linear structures of riprap or concrete that protect limited stretches of the shoreline from waves. A **breakwater** is designed to intercept waves and provide a protected area or harbor for mooring boats or ships. This structure may be attached to, or separated from, the beach (Figure 11.27a, b). In either case, a breakwater blocks littoral transport and alters the shape of the coastline as new areas of deposition and erosion develop. This can cause serious erosion, or it can block a harbor entrance with newly trapped sand. The trapped sand may have to be moved by *artificial bypass*, that is, the dredging and pumping of sand from around the breakwater to redeposit it downdrift.

Jetties are usually built in pairs, perpendicular to the shore at the mouth of a river or at the entrance of an inlet to a bay or lagoon (Figure 11.27c). They are designed to keep a ship or boat channel open and in a fixed location, with minimal dredging to remove sediment. Jetties also shelter a channel from large waves. Like groins, jetties block littoral transport, thus causing the updrift beach adjacent to one jetty to widen whereas downdrift beaches erode.

Unfortunately, it is impossible to build a breakwater or jetty that will not interfere with longshore movement of beach sediment. These structures must, therefore, be carefully planned and incorporate protective measures that eliminate or at least minimize adverse effects. Protective measures may include a dredging and artificial sediment-bypass system, a beach-nourishment program, seawalls, or some combination of these measures.[11]

SOFT STABILIZATION

Beach Nourishment The use of beach nourishment to reduce rates of shoreline retreat is not limited to adding sand to engineering structures. In fact, in most cases it is

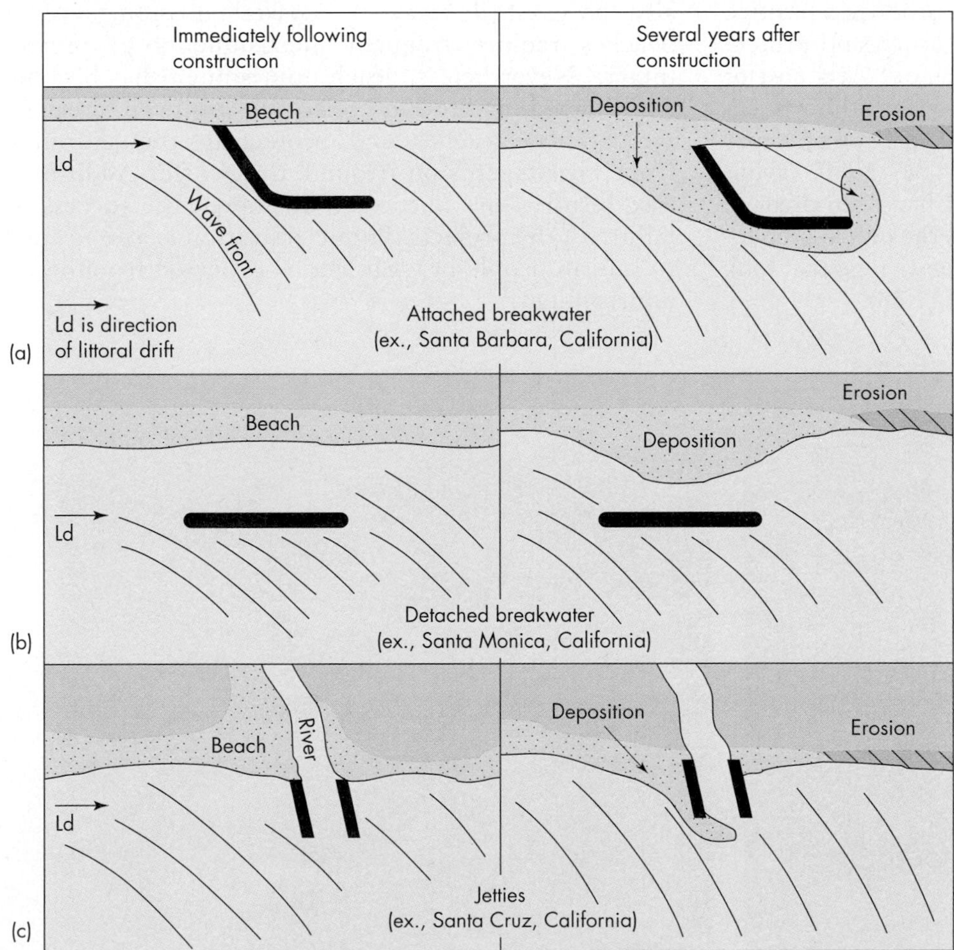

▲ FIGURE 11.27 **ENGINEERING STRUCTURES BUILT IN THE SURF ZONE CAUSE CHANGE** Diagrams illustrating the effects of breakwaters, the thick black lines in (a) and (b), and jetties, the short black lines in (c), on local patterns of deposition (tan) and erosion (brown). The left column of the diagrams shows the structures immediately following their construction, and the right column shows deposition and erosion after the structures were built. Curved, thin black lines show approaching waves and the arrows indicate the direction of littoral transport (Ld).

used as an alternative to engineering structures. In its purest form, beach nourishment consists of artificially placing sand on beaches in the hope of constructing a positive beach budget, where more sand is in storage than leaves in output. Beach nourishment is sometimes referred to as the "soft solution" to coastal erosion, as contrasted with "hard solutions," such as groins or seawalls. Ideally, a nourished beach will protect coastal property from the attack of waves. Beach nourishment is aesthetically preferable to many engineering structures, and it provides a recreational beach as well as some protection from shoreline erosion. Unfortunately, beach nourishment is expensive and must be repeated at regular intervals to stabilize a beach. Considerable care must be taken in selecting the appropriate range of sand sizes for a beach replenishment project, and locating a nearby source for this sand is sometimes difficult.

In the mid-1970s, the city of Miami Beach, Florida, and the U.S. Army Corps of Engineers began an ambitious beach-nourishment program to reverse serious beach erosion that had plagued the area since the 1950s. The program was designed to produce a positive beach budget that would widen the beach and protect coastal resort areas from storm damage.[26] At a cost of approximately $62 million over 10 years, the project nourished about 160,000 m^3 (210,000 $yd.^3$) of sand per year to replenish erosion losses. By 1980, about 18 million m^3 (24 million $yd.^3$) of sand had been dredged and pumped from an offshore site onto the beach, producing a 200-m (660-ft.) wide beach.[27] The change that took place over a decade is dramatic (Figure 11.28).

Part of the Miami project included building vegetated coastal dunes to function as a buffer to wave erosion and storm surge (Figure 11.29). Special wooden walkways allow public access through the dunes, but other areas of the dunes are protected. The successful Miami Beach nourishment project has functioned for more than 20 years and survived major hurricanes in 1979 and 1992; it is certainly preferable to the fragmented erosion-control methods that preceded it.[11]

More than 600 km (370 mi.) of U.S. coastline has received some sort of beach nourishment. Not all this nourishment has had the positive effects reported for Miami Beach. For example, in 1982, Ocean City, New Jersey, nourished a stretch of beach at a cost of just over $5 million. A series of storms that subsequently struck the beach eroded the sand in just 2 1/2 months. The beach sands of Miami may yet be eroded and require replacement at a greater cost. Beach nourishment remains controversial, and some consider it nothing more than "sacrificial sand" that will eventually be washed away by coastal erosion.[27] Most beaches require frequent nourishments to remain intact.[3] Nevertheless, beach nourishment has become a preferred method of restoring or even creating recreational beaches and protecting the shoreline from coastal erosion around the world. Additional case histories are needed to document the success or failure of the projects. Public education is also needed to inform people of what can be expected from beach nourishment.[11]

(a)

(b)

▲ FIGURE 11.28 **BEACH NOURISHMENT** Miami Beach (a) before and (b) after beach nourishment. *(U.S. Army Corps of Engineers, Headquarters)*

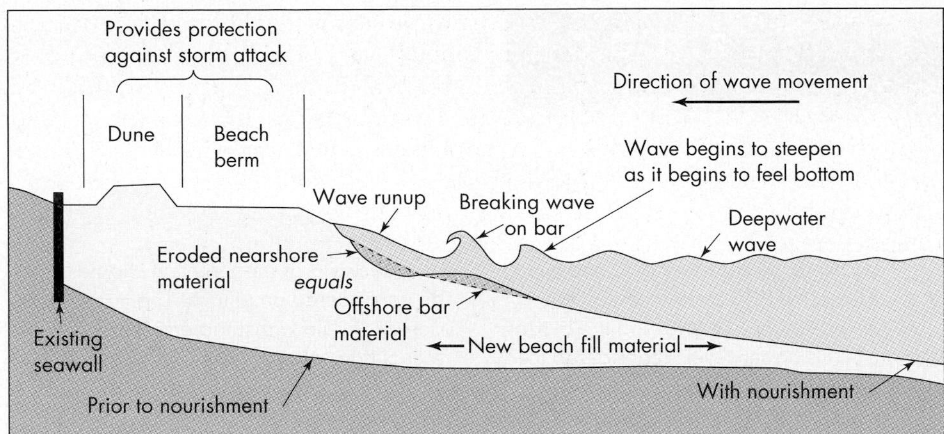

◀ **FIGURE 11.29 MIAMI BEACH SHAPE AFTER NOURISHMENT** Diagrammatic cross section of the Miami Beach nourishment project. The dune and beach berm system provides protection against storm attack. An existing seawall is the thick black line on the left. New beach fill material is in yellow. Orange areas are beach sand that may be eroded and deposited during a storm. A dashed line shows a pre-storm profile of the beach. *(Courtesy of U.S. Army Corps of Engineers)*

11.10 Perception of and Adjustment to Coastal Hazards

PERCEPTION OF COASTAL HAZARDS

People generally perceive land as being stable, and, in many cultures, it is viewed as a permanent asset that can be bought and sold.[16] This perception of stability is appropriate for most land where geologic changes are relatively slow and take centuries or millennia to alter the land. Coastlines, however, are dynamic, and shorelines erode, accumulate, and move in time frames of months and years (see Figure 11.18).

Even if individuals understand that erosion can be rapid, their past experience, proximity to the coastline, and the probability of suffering property damage influence their perception of coastal erosion as a natural hazard. One study of sea-cliff erosion near Bolinas, California, 24 km (15 mi.) north of the entrance to San Francisco Bay, found that coastal residents in an area who were likely to experience damage in the near future were generally well informed and saw the erosion as a direct and serious threat.[28] Other people living a few hundred meters from a possible hazard, although aware of the hazard, knew little about its frequency, severity, and predictability. Still further inland, people were aware that coastal erosion exists but had little perception of it as a hazard.

Many people do not perceive waves and currents as hazards. Often, their first visits to beaches are as children, where their parents closely limit their activities. In fact, waves that have large heights and wavelengths and plunging waves can be dangerous for wading and swimming. These waves are considerably more powerful than spilling waves that generally provide safe conditions for waders and swimmers. Rip currents, which are commonly produced by plunging waves with large wavelengths, are responsible for about 80 percent of all lifeguard rescues on the seashore.[29] Many beachgoers do not recognize a rip current and try to enter the water where there are few breaking waves (see Figure 11.14b), only to be caught in a rip current.[29]

ADJUSTMENT TO COASTAL HAZARDS

As with most other natural hazards, there are a variety of ways that we can adjust to shorelines with strong currents, coastal erosion, and rising sea level. Since strong currents cannot be prevented—they exist naturally, as well as around engineering structures—the best adjustments are education about and awareness of the conditions that create this hazard. For many ocean beaches, and some beaches on large lakes, it may be best to restrict swimming to supervised locations when life guards are on duty. On these beaches, warnings about rip currents and tidal currents are often posted. In the United States, general beach and surf forecasts are issued by the National Weather Service. Information about coastal wave and weather conditions are also available through NOAA's nowCOAST Web site (*nowcoast.noaa.gov*).

For the coastal erosion hazard, adjustments generally fall into one of three categories:[30]

- Beach nourishment that tends to imitate natural processes—the "soft solution."
- Shoreline stabilization through structures, such as groins and seawalls—the "hard solution." For example, a sea wall may be constructed along an entire beach or a coastal point of particular interest may be defended by coastal engineering (Case Study 11.7).
- Land-use change that attempts to avoid the problem by not building in hazardous areas or by relocation of threatened buildings—the "managed retreat solution."

One of the first tasks in managing coastal erosion is determining the rates of erosion. Estimates of future erosion rates are based on historic shoreline change or on statistical analysis of the oceanographic environment, such as the waves, wind, and sediment supply that affect coastal erosion. Recommendations are then made concerning setbacks, which are considered to be minimum standards for state or local coastal erosion management programs. A setback is the distance from the shoreline beyond which development, such as homes, is allowed. Only a few states, such as California, Florida, New Jersey, New York, North

A CLOSER LOOK 11.5

E-Lines and E-Zones

A special committee of the National Research Council (NRC), at the request of the Federal Emergency Management Agency (FEMA), recently developed coastal zone management recommendations, some of which are listed next:[30]

- Estimates of future erosion rates should be based on historic shoreline change or statistical analysis of local wave, wind, and sediment supply conditions.
- Once local erosion rates have been determined, maps should be made showing erosion lines and zones, referred to as E-lines and E-zones (Figure 11.E). An *E-line* is the location of expected erosion in a given number of years; for example, the E-10 line is the location in 10 years. *E-zones* are similar to hazard zones in floodplains; that is, the E-10 zone would be the area between sea level and the E-10 line. Rising sea level is considered to be an imminent hazard in the E-10 zone, and no new habitable structures should be allowed. The system of E-lines and E-zones can be used to establish setback distances. For example, if the erosion rate is 1 m (3.3 ft.) per year, the E-10 setback is 10 m (33 ft.).
- Movable structures are allowed in the intermediate- and long-term hazard zones (E-10 to E-60).
- Permanent large structures are allowed at setbacks greater than the E-60 line.
- With the exception of those on high bluffs or sea cliffs, all new structures built seaward of the E-60 line should be constructed on pilings. Their design should withstand erosion associated with a high-magnitude storm with a recurrence interval of 100 years.

NRC recommendations concerning setbacks are considered to be minimum standards for state or local coastal erosion management programs. At present, only a small number of states, such as Florida, New Jersey, New York, North Carolina, and South Carolina use a setback based on the rate of erosion. Nevertheless, the concept of E-lines and E-zones based on erosion-designated setbacks and allowable construction has real merit in coastal zone management.

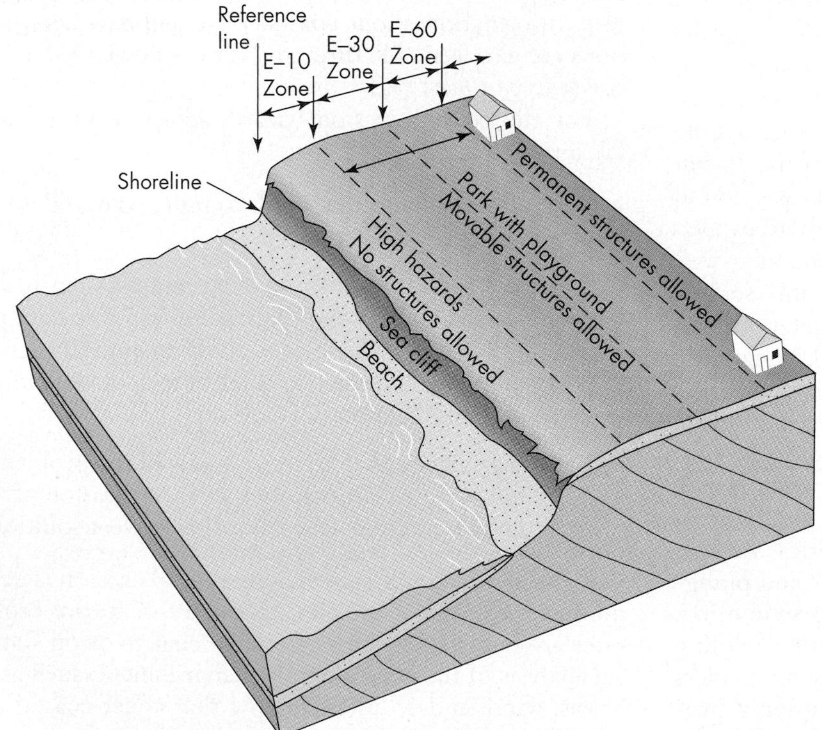

FIGURE 11.E E-LINES AND E-ZONES Conceptual diagram of E-lines and E-zones based on the rate of coastal erosion from a reference point, such as the sea cliff or dune line. Width of the zones depends on the rate of erosion and defines setback distances. Of course, after 60 years, the structures will be much closer to the shoreline and vulnerable to erosion. It is a form of planned obsolescence. *(Modified from National Research Council. 1990. Managing coastal erosion. Washington, DC: National Academy Press)*

Carolina, and South Carolina, use a setback distance for buildings based upon the rate of erosion (see A Closer Look 11.5).[31] The concept of setback has real merit in coastal zone management and is at the heart of land-use planning that minimizes damage from coastal erosion.

We are at a crossroads today with respect to adjustment to coastal erosion. One road leads to ever-increasing coastal defenses in an attempt to control erosion. The second path involves learning to live with coastal erosion through flexible environmental planning and wise land use in the coastal zone.[3,30] The first path follows history in our attempt to control erosion through the construction of engineering structures, such as seawalls. In the second path, most structures in the coastal zone are considered temporary and expendable; only a few critical facilities may be considered to be permanent (see Case Studies 11.6 and 11.7). Development in the coastal zone must be in the best interests of the general public rather than for a few who profit from it. This concept is at odds with the attitude of developers who consider the coastal zone "too valuable not to develop." In fact, development in the coastal zone is not the problem; rather, the problem lies in building in hazardous areas and areas better suited for other uses. In other words, beaches belong to all people to enjoy, not only those fortunate enough to purchase beachfront property. The states of Hawai'i and Texas have taken this philosophy to heart. In Hawai'i, all beaches are public property, and local property owners cannot deny access to others. Likewise, in Texas, virtually all of the ocean beaches are public property, and coastal zoning now requires avenues for public access.

Accepting the philosophy that, with minor exceptions, coastal zone development is temporary and expendable and that consideration should first be given to the general public, requires an appreciation of the following five principles:[3]

1. *Coastal erosion is a natural process rather than a natural hazard.*
2. *Any shoreline construction causes change.* This change interferes with natural processes and produces a variety of adverse secondary and tertiary changes.
3. *Stabilization of the coastal zone through engineering structures protects property, not the beach itself.* Most protected property belongs to relatively few people at a larger expense to the general public
4. *Engineering structures designed to protect a beach may eventually destroy it.*
5. *Once constructed, shoreline engineering structures produce a costly trend in coastal development that is difficult, if not impossible, to reverse.*

If you consider purchasing land in the coastal zone, remember these guidelines: (1) allow for a good setback from the beach, sea cliff, or lakeshore bluff; (2) be high enough above the water level to avoid flooding, and take into account rising sea level in making this determination; (3) construct buildings to withstand adverse weather, especially high winds; and (4) if hurricanes are a possibility, be sure there are adequate evacuation routes.[3] Remember, it is always risky to buy property where land meets water.

CASE STUDY 11.6

Moving the Cape Hatteras Lighthouse

Managed retreat is one method for dealing with coastal erosion and rising sea level. That was the option chosen by the National Park Service for protecting the historic Cape Hatteras Lighthouse. When the lighthouse was built in the late nineteenth century, it was approximately 0.5 km (0.3 mi.) from the Atlantic Ocean. By the early 1990s, it was only 100 m (328 ft.) from the sea and in danger of being destroyed by a major storm. As you may recall from Chapter 9, Cape Hatteras is frequently a target for both hurricanes and nor'easters. What follows is a first-hand description of moving this national monument.

One probably would not expect the task of moving a 4400-ton lighthouse to be "mostly nuts and bolts and hydraulics." But this is how Rick Lohr, president of the International Chimney Corporation (ICC), described the task of moving the Cape Hatteras Lighthouse—the tallest brick lighthouse in the world—more than a half mile when coastal erosion threatened to claim the historic structure for the fishes (Figure 11.F).

The ICC was contracted to move the lighthouse to a safe location away from the shoreline in 1998, and it immediately set about making the necessary calculations to ensure that the project would go smoothly.

Among the preparatory tasks was testing the 880-m (2900-foot) path to the structure's new location for any hidden regions that could collapse during the move.

"Sand compresses well, better than a lot of soils," Lohr said of the terrain. "What you're concerned about is, are there any sinkholes?"

In order to find out, the ICC conducted what is known as "proof rolling," in which it "hauled a similar amount of weight across the move route" to test for any such sinkholes, Lohr said. This did not require an extraordinary amount of weight; the total surface area of the structure at ground level, including the hydraulic jacks and

(continued)

rollers the ICC installed, was about 4900 ft.^{2} (70 ft. × 70 ft.), making the average pressure at any given point roughly 1 ton per square foot.

As for the move itself, the company used a system known as the "Unified Jack System," invented in the 1950s by ICC employee Pete Freisen, who, in his mid-80s, still served as a technical director for the Hatteras project. The system, in this project, used 120 individual jacks, which allowed for every point to rise equally, no matter what the pressure.

After cutting the lighthouse from its foundation, the 120 jacks were placed under the structure and divided into three groups of 40 jacks each. The practice of dividing the temporary support into three "centers of effort" is standard for all large moves, Lohr said.

"We move in three zones, no matter what the structure looks like," he said, adding that they used the same method for moving an entire terminal at Newark International Airport in 2000. "We develop three points of effort, and they form a triangle."

The theory behind this strategy, he said, is that one of the three central points can theoretically sink without stressing the structure itself. To clarify, Lohr compared their process to a tricycle: If a tricycle gets a flat tire, the whole vehicle tips in that direction, but its structure is not stressed. By contrast, in a wagon, if one of the four wheels goes, the structure suffers stress in its base near the missing support.

According to the ICC's method, Lohr said, "If you hit a soft spot, you just tip a little," and the jacks can be readjusted to compensate.

Even though computers now aid companies such as ICC greatly in their initial calculations, such as determining the center of gravity of a structure, the science of moving large objects has existed since the turn of the century.

"Strangely enough," Lohr said, "everything is pretty basic."

Thanks to Rick Lohr and his engineers, the lighthouse is expected to be safe in its new location until the middle or end of the twenty-first century.

—CHRIS WILSON

(a)

(b)

▲ FIGURE 11.F **LIGHTHOUSE MOVED** The historic Cape Hatteras Lighthouse was moved in summer 1999 to keep it from being destroyed by coastal erosion. (a) Cape Hatteras Lighthouse before it was moved. Only around 100 m (330 ft.) from the water, the lighthouse could have been destroyed in a major storm. *(Don Smetzer/Photo Edit)* (b) Cape Hatteras lighthouse being moved; at its new location, the lighthouse should be safe for the next 50 to 100 years. *(Reuters)*

CASE STUDY 11.7

Coastal Erosion at Pointe du Hoc, France

Pointe du Hoc is located on the Normandy coast of France. It is located on a promontory point several hundred kilometers northwest of Paris, on the shore of the English Channel (Figure 11.G). The point is a historical site located on a coastal plain of low relief. The ocean side is a sea cliff, approximately 30 meters (100 ft) high (Figure 11.H). The site is visited by many thousands of people annually to view the location where the invasion to free Europe in World War II began on June 6, 1944. The site is considered by many to be so important that expending a large amount of money to preserve the most vulnerable parts of the point is worthwhile. Of course, this is a value judgment. The main problem is that the point has experienced as much as 10 m (33 ft.) of erosion in the past 60 years. As a result, it was feared that the observation post (OP) might actually fall into the sea. Predictions were that it probably could not last more than a few decades at most.

Those who want to preserve the last bit of rock on the point, rather than, say, move some of the structures, obtained funding to evaluate if the point might be preserved for another few decades by reinforcing it. This is an interesting approach to coastal erosion because it is a site-specific solution to a particular problem that involves coastal stabilization rather than, say, adjustment, such as managed retreat. The case is also interesting because it involves detailed geologic evaluation and study of processes of sea cliff retreat, linked to a specific strategy to reduce the rate of coastal erosion, which has been estimated at 0.1–0.2 m per year.[32] Scientists and engineers studying the sea cliff realize that any attempt to absolutely stop sea cliff erosion is futile. Rather, their objective is a temporary solution to buy as much as 50 years. Total erosion in the past 60 years has been approximately 10 m (33 ft.), and, in future decades, that rate might be significantly reduced.

The sea cliff was carefully evaluated and the geology studied in detail. The idealized profile of the sea cliff and the geology is shown in Figure 11.I. Three geologic layers are present:[32]

- The top layer, which is up to about 9.0 m (30 ft) thick, is a silty, gravelly, sandy, clay soil. This layer has relatively low strength and is prone to small landslides and grain-by-grain erosion resulting from precipitation.
- The middle layer is a fractured limestone that, with depth, becomes sandy and enriched with calcium carbonate muds. The layer is about 15 m (50 ft) thick. The limestone beds are nearly flat lying but have numerous nearly vertical fractures that are filled, to a lesser or greater extent, with weathered products of the limestone that are rich in clay. Precipitation from the surface can remove some of the clay, resulting in the fractures being more open. The fractured limestone is vulnerable to larger slope failures, particularly if the

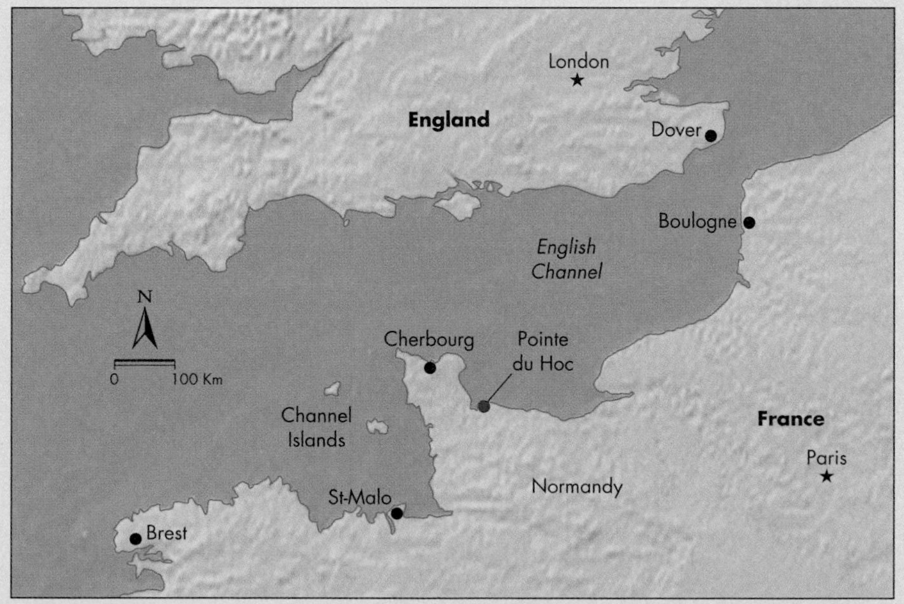

FIGURE 11.G **LOCATION OF POINTE DU HOC.**

(continued)

rock is underlain by areas of coastal erosion that have produced wave-cut notches and sea caves. In fact, investigators studying the site have observed that this is the dominant failure mechanism that results in the most rock being lost from the sea cliff. When wave erosion undermines the rocks at the base, they may fail along the fractures, falling onto the beach or into the sea.

- The lowest layer present is a black marl, which is a calcium carbonate–rich mudstone. Both the fractured limestone and the black marl have relatively high rock strength, compared to that of the overlying soil.

A study of the sea cliff suggests that the attack of waves at the rock face near the beach dislodges blocks of rock that creates caverns or sea caves. These are clearly visible along the base of the sea cliff and can be quite large, as illustrated in Figure 11.H. The caves grow larger and deeper with time and, eventually, leave the overlying rock unsupported, and a failure may occur along fractures. It was concluded that the primary reason for collapse of the sea cliff is the presence of the caverns at the base of the cliff. As a result, the primary recommendation for reducing the rate of erosion is to fill the caverns and stabilize the base of the sea cliff. Work on this project began in 2010, and the objective is to fill the caverns with several hundred cubic meters of concrete. Rock bolts are also used to secure the concrete and stabilize the rocks. Following filling of the caverns, an outermost layer of cement, which is specially designed to look like native rock will be added. This is obviously a temporary solution because new sea caves can form and the cement itself may be undermined and eroded. As a result, the study recommended a maintenance program based on inspections and recognized that future work will undoubtedly be necessary. It is also important to recognize that the development of sea caves and large landslides is not the only process affecting the sea cliff. The top unit of the sea cliff is the weak soil, and there are numerous, small landslides occurring on a regular basis. In fact, the top of the sea cliff has a wavy form, resulting from the many small landslides. Therefore, stabilizing the base of the sea cliff will not eliminate sea cliff retreat. It will probably reduce it by about one-half. Therefore, the work on the lower part of the sea cliff, which was completed by September 2010, is be a temporary solution at best. In the long run, the environmentally preferable solution will, by necessity, mean moving structures of historical importance back from the edge of the cliff. This is essentially the same solution reached in the case history that discussed the Cape Hatteras Lighthouse, which was moved inland.

In summary, the decision to use a "hard solution" to the sea cliff at Pointe du Hoc is an interesting one and based on the value judgment that the historical significance of the site is so great that measures need to be taken to preserve it. The cost of the work is modest to date, at about $6 million; however, future maintenance costs will be incurred, and, often, structures that are built in the sea cliff environment are fortified even more in subsequent years, so the total cost into the future is unknown. The solution to coastal erosion and the decision to minimize it at Pointe du Hoc are obviously special circumstances, in which a small length of sea cliff is stabilized. Should such an approach be applied to a long section of sea cliff, the cost would be far too great to maintain. For now, the point is better protected, and the erosion rate will probably decrease, but, in the future, if the site is to be maintained, important structures will need to be moved inland.

◀ **FIGURE 11.H POINTE DU HOC** This historical site is on a coastal plain underlain by Cretaceous limestone. Pits are bomb craters from World War II.*(Patrick Forget/Sagaphoto/Alamy)*

FIGURE 11.I IDEALIZED GEOLOGY AT POINTE DU HOC (OBSERVATION POINT). *(Data from Briand, J-L., Nouri, H. R., and Darby, C. 2008.* Pointe du Hoc stabilization study. *Texas A&M University)*

REVISITING THE FUNDAMENTAL CONCEPTS

Coastal Erosion

1. **Hazards Are Predictable from a Scientific Evaluation**
2. **Risk Assessment Is an Important Component of Our Understanding of the Effects of Hazardous Processes**
3. **Linkages Exist Between Natural Hazards**
4. **Hazardous Events That Previously Produced Disasters Are Now Producing Catastrophes**
5. **Consequences of Hazards May Be Minimized**

1) The scientific understanding of coastal processes is an advanced field. We have instruments that record weather, wave activity, tides, and land use, and land forms have been mapped in the coastal environment. A lot is known about beach processes and form, as well as about processes that operate in other coastal environments, such as sea cliffs, salt marshes, and barrier beaches. As a result, good predictions on the nature and extent of the coastal erosion problem are possible. One potential difficulty is trying to predict sea level rise in the future, which will impact rates of coastal erosion.

2) Establishing the risk resulting from coastal hazards is fairly straightforward. The probability of erosion events can be predicted and potential losses can be estimated. The risk is the product of these two variables. Establishing risk for various segments of our coastline is important because it provides valuable information concerning where to place our efforts in an attempt to control or live with coastal erosion.

3) There are a number of linkages between coastal erosion and other natural hazards, including tsunamis, severe storms, and flooding. In addition, human hazards, such as oil spills, are also linked to coastal processes because it is those processes that are responsible for transporting the oil in the coastal zone. Some of the most serious coastal erosion occurs during hurricanes and other high-magnitude storms that produce flooding. When a hurricane strikes a coastal region, the entire coastline may be changed as beaches in some areas are eroded, and, in other areas, new inlets form and barrier islands erode while other inlets are filled with sediment.

4) More and more people want to live in the coastal zone. As a result, coastal cities and communities have grown considerably in recent decades. As a result, coastal erosion events resulting from storms and other activities are becoming more damaging. What used to be a rural coastline may now be developed with oceanfront homes. As a result, an erosion event, rather than simply changing a rural undeveloped beach, may erode through urban development, damaging homes and supporting structures.

(continued)

5) Consequences of coastal erosion and other coastal hazards may be minimized. The environmentally preferred solution is land use and avoiding areas with a high hazard. In some cases, this may result in managed retreat of the coastline. In other cases, where considerable coastal development has occurred and where cities have been built, large amounts of money will continue to be spent in efforts to control erosion. It will be more difficult to minimize consequences of coastal erosion in the future as sea level continues to rise as a result of global warming. At this point in the history and development of coastal processes, we are at a crossroads. Many people want to protect the coastline at all costs. Others would rather maintain beaches for future generations and choose a softer path of minimizing coastal erosion, such as managed retreat or beach nourishment.

QUESTIONS TO THINK ABOUT

1. **What are the specific factors in the coastal zone that must be known or estimated to predict the future consequences of coastal erosion?**
2. **Assume you are living on a barrier island in the Gulf of Mexico that is experiencing rapid urbanization. Nearly everyone wants to live near the edge of the ocean, in other words, as close to the water as possible. It is expected that, in the future, the number of homes and people will increase dramatically from a few hundred now to a few thousand in 10 years. You are hired by the community to help evaluate land use and potential effects of coastal erosion. What advice would you give to the communities on the barrier island, and what sorts of data would you need to gather to make conclusions concerning future land use? In addition, how would you handle the people problems? That is, how would you talk to people about erosion and what the consequences are likely to be and how the process might be minimized?**

Summary

Most waves are generated by windstorms at sea or on large lakes and expend their energy on the shoreline. The size of these waves in open water depends on a combination of wind speed, duration of the wind, and fetch. Waves approaching the shore are slowed by friction and can double in height. The height of waves grows larger in storms, and wave energy is proportional to the square of this height. Irregularities in the shoreline focus wave energy and account for local differences in wave erosion; these irregularities are largely responsible for determining the shape of the coast.

Beaches consist of sand or gravel deposited at the coast and may have other loose material, such as broken shells, coral, or volcanic rock that was eroded locally. Beach sediment comes from erosion of cliffs, bluffs, or dunes landward of the beach; from the shoreline updrift from the beach; or from rivers bringing sediment to the sea. Wave action by spilling or plunging waves shapes the beach. Most waves strike a beach at an angle, producing a longshore current parallel to the shore. This current, combined with beach drift, produces littoral transport of sediment parallel to the coast.

Any coastal area, including the shorelines of large inland lakes, is at risk from coastal hazards. Coastal hazards include tidal currents, rip currents generated in the surf zone, coastal erosion, rogue waves, rising sea level, tsunamis, and cyclones. Tidal currents can be hazardous in narrow bays and channels, especially on coasts with large tidal ranges. Beaches experiencing waves from storm swells or plunging breakers commonly have hazardous rip currents that can carry even the strongest swimmers offshore. Storm waves, enhanced by rising sea level and deficits in sand supply, contribute to coastal erosion. Hazardous rogue waves can form in the open ocean by the constructive interference of waves of similar size and wavelength.

Relative sea level, the place where the sea meets the land, is influenced by eustatic sea level and other local or regional changes in land or water position caused by tectonic uplift, coastal subsidence, erosion and deposition, and tides. Eustatic sea level is the global position of the ocean surface. Over decades, this level is influened by climate change. Today, rising eustatic sea level is being caused by global warming, which expands ocean volume and melts glaciers, ice caps, and ice sheets. Tsunamis and cyclones can produce deadly surges of water flooding coastal areas. Storm surge, most commonly from hurricanes, is a function of the wind speed of the storm, atmospheric pressure, water depth, and the configuration of the coastline.

Rip currents occur on most seacoasts and sometimes in large lakes. Tidal currents are especially hazardous in areas with large tidal ranges, such as Alaska, the Pacific Northwest, and parts of New England and Atlantic Canada. Although coastal erosion causes a relatively small amount of damage compared to river flooding, earthquakes, and tropical cyclones, it is a serious problem along all the coasts of the United States, Great Lakes shorelines, and many Canadian seacoasts. Factors contributing to coastal erosion include river damming, high-magnitude storms, and the worldwide rise in sea level.

Human interference with natural coastal processes by hard stabilization of the shoreline is occasionally successful, but, in many cases, it causes considerable coastal erosion. Sand tends to accumulate on the updrift side of groins and jetties and then erode on the downdrift side. Seawalls reflect storm waves and cause beach erosion. Most problems with coastal erosion occur in areas with high population density, but sparsely populated areas along the Outer Banks in North

Carolina are also experiencing trouble. Soft stabilization, such as beach nourishment, has had limited success in restoring or widening beaches, but even in "successful" cases, it remains to be seen whether it will be effective in the long term.

Perception of coastal hazards depends mainly on the individual's experience with and proximity to the hazard. Community adjustments in developed countries involve either building protective structures to lessen potential damage or modifying people's behavior through land-use zoning, evacuation procedures, and warning systems.

Adjustment to coastal erosion in developed areas has often been the "technological fix" of building seawalls, groins, and other structures, and more recently, beach nourishment. These approaches to beach stabilization have had mixed success and often cause additional problems in adjacent areas. Engineering structures are expensive; require maintenance; and, once in place, are difficult to remove. The cost of engineering structures may eventually exceed the value of the properties they protect; such structures may even destroy the beaches they were intended to save. Establishing setback distances and managed retreat can be effective adjustments to rising sea level.

Key Terms

barrier island (p. 365)
beach (p. 373)
beach budget (p. 380)
beach nourishment (p. 392)
bluff (p. 373)
breakwater (p. 393)
eustatic sea level (p. 375)
groin (p. 392)
jetty (p. 393)
littoral transport (p. 371)
relative sea level (p. 375)
rip current (p. 377)
rogue wave (p. 369)
sea cliff (p. 373)
seawall (p. 391)

Review Questions

1. How do waves on a lake or ocean form? What affects their height?
2. How is wave height related to the energy of the wave?
3. What conditions favor the development of large wind-driven waves?
4. Describe how water particles behave as a wind-driven wave passes. How does this behavior change in shallow water?
5. Describe what happens to waves as they enter shallow water.
6. Explain wave convergence and divergence at a headland.
7. Describe how breakers form and how the two most common types occur and behave.
8. Explain how littoral transport takes place and how longshore drift is related to direction of wave approach.
9. What causes tides and strong tidal currents?
10. Explain the difference between eustatic and relative sea level and the factors that control each.
11. What are the most serious hazards affecting the coasts?
12. Describe how you recognize rip currents, how they form, and what you should do if you are caught in one.
13. What are the primary causes of coastal erosion? Why is coastal erosion becoming more common and also more of a problem?
14. Describe the processes that cause erosion of sea cliffs and lakeshore bluffs.
15. How are coastal processes related to flooding, landslides, subsidence, and climate change?
16. Describe the natural service functions of coastal processes.
17. Explain the purpose and effects of seawalls, groins, breakwaters, and jetties.
18. List arguments for and against beach nourishment.
19. What affects people's perception of coastal hazards?
20. What are the five general principles that should be accepted if we choose to live with, rather than control, coastal erosion?
21. Describe the adjustments that people need to make in order to protect themselves from strong ocean currents and to reduce coastal erosion.

Critical Thinking Questions

1. Do you think that human activity has increased coastal erosion? Outline a research program that could test this question.
2. Do you agree or disagree with the following statements: (1) All structures in the coastal zone, with the exception of critical facilities, should be considered temporary and expendable. (2) Any development in the coastal zone must be in the best interest of the general public rather than the few who developed the oceanfront. Explain your position on both statements.
3. You have been asked by a coastal community to evaluate the feasibility of a beach-nourishment project. Describe the types of information that you would require for your evaluation and how you would determine how often nourishment will be needed in the future.
4. Compare and contrast the shoreline features of tectonically active and passive coasts of the United States, and describe how coastal hazards differ in the two coasts that you have evaluated.

Assignments in Applied Geology: Shoreline Property Assessment

THE ISSUE

Shorelines are among the most dynamic places on our planet. Shorelines are places of constant change, and, in most locations, either erosion or deposition is taking place through the actions of winds, waves, and currents. The rate of change is generally slow; however, during severe storms, the change can be dramatic, with tons of sediment being removed or deposited in just a few hours.

Shorelines are also locations that attract people. Their scenic beauty and recreational opportunities make them highly desirable for both year-round and vacation housing. However, the dynamic and sometimes violent nature of shorelines can make some of them risky locations for development.

Shoreline erosion is a significant problem, and each year it destroys or damages millions of dollars worth of property. However, those who are well informed about how erosion occurs and willing to research the history of a property can generally determine if it is at risk. These types of studies enable citizens and communities to avoid development or investment in risky areas, saving millions of dollars and, possibly, lives.

YOUR TASK

Hazard City is about 50 miles from the coast, and many Hazard City residents enjoy visiting the beach. Many have purchased second homes, beachfront cottages, or time-shares. The most popular locations for these purchases are along the ocean beachfront and along the banks of the Clearwater River just upstream from where it enters the ocean. Care must be taken when purchasing properties in these areas because shoreline erosion can be a significant problem. Prudent buyers seek professional advice before they commit to such an important purchase. You plan to visit the beach next weekend, and a Hazard City real estate office has asked you to assess a few properties and report on how they might be affected by shoreline erosion.

AVERAGE COMPLETION TIME:

1 hour

12 mate and Climate Change

What Does Our Recent History Tell Us about Potential Consequences of Global Warming?

The famous philosopher George Santayana stated in 1905 that "those that cannot remember the past are condemned to repeat it." It is a matter of debate whether cycles in human history actually repeat themselves, but the repetitive nature of many natural processes and hazards, even if they result from different reasons, is indisputable. So, what can we learn by examining the past with reference to global climate change and, in particular, global warming? It turns out that over an approximate 300-year period from A.D. 950 to 1250, the Earth was considerably warmer than normal (normal meaning the average surface temperature during the past century or some shorter interval, such as between 1961 and 1990). This period is known as the Medieval Warming Period (MWP), and we can possibly learn some lessons from that event. Although it occurred only a few hundred years ago, we don't know much about the MWP except that it was, in fact, warm and perhaps nearly as warm as it is today. Most scientists who study the indicators of past climate state that it probably wasn't quite as warm, but, certainly, parts of the world, in particular, Western Europe and the North Atlantic, may have been warmer some of the time than they were in the last decade of the 20th century. During the MWP, there were winners and losers. In Western Europe, there was a flourishing of culture and activity, as well as an expansion of the population, as harvests were plentiful and people generally prospered. During that period, many of the Europe's grand cathedrals were constructed.[1,2] Finally, during the MWP, the first global trade routes opened from Europe to China, helped by favorable climate and connected by camel caravans.

Those that cannot remember the past are condemned to repeat it.

During the MWP, sea temperatures in the North Atlantic evidently were warmer, and there was less sea ice. The famous Viking explorer, Erik the Red, embarked on a voyage of exploration near the end of the 10th century A.D. When he arrived at Greenland with his ships and people, they set up settlements that flourished for several hundred years, and they were able to grow a variety of crops, including corn, that had never before been cultivated in Greenland. They were also able to raise

LEARNING OBJECTIVES

Many natural hazards, such as drought, heat waves, floods, hurricanes, blizzards, and wildfire, are related to climate. Climate change is also a potential cause of extinction of plants and animals. A basic understanding of climate science is needed to comprehend the mechanisms of these hazards. Your goals in reading this chapter are to

- understand the difference between climate and weather and how their variability is related to natural hazards.
- know the basic concepts of atmospheric science, such as structure, composition, and dynamics of the atmosphere.
- understand how climate has changed during the past million years through glacial and interglacial conditions and how human activity is altering our current climate.
- understand the potential causes of climate change.
- know how climate change is related to natural hazards.
- know the ways we may mitigate climate change and associated hazards.

◀ **River Thames, England** During the winter of 1683–1684, the Thames River in London was frozen solid. Depicted here is the Frost Fair, which was held on the frozen river with the Old London Bridge shown in the distance. c.1685 (oil on canvas' painting formerly attributed to Jan Wyck (1Bridgeman Art Library))

their animals and, in general, enjoy a prosperous life. During the same warm period, Polynesian people in the Pacific, taking advantage of winds flowing throughout the ocean, were able to sail to and colonize islands over vast areas of the Pacific, including Hawai'i.[2]

Although some prospered in Western Europe and the Pacific during the MWP, other cultures were not so fortunate. Associated with the warming period were long, persistent droughts (think human generational long) that appear to have been partially responsible for the collapse of sophisticated cultures in North and Central America. The collapses were not sudden but occurred over a period of many decades, and, in some cases, the people just moved away. These included the people living near Mono Lake on the eastern side of the Sierra Nevada in California, the Chacoan people in what is today Chaco Canyon in New Mexico, as well as the Maya civilization in the Yucatan of southern Mexico and Central America.

Today, we are worried about the present warming trends. Should they continue and accelerate or increase, we may face a prolonged drought in the southwestern United States again, as well as in other parts of the world, which may wreak havoc on the ability to produce the food that the nearly 7 billion people upon the Earth require.[2]

When the MWP ended, it was followed by the Little Ice Age (LIA) that lasted several hundred years from the mid-1400 to 1700.[3] The cooling of the LIA made it more difficult for people in Southeast Asia and Western Europe. Angkor Watt, with its palace and temple in Cambodia, was constructed in the year 1181. The region flourished for several hundred years as canals and reservoirs irrigated vast rice fields. The water supply from summer monsoon rains was reliable, and people prospered. With climate change and the transition to the LIA, monsoon rains weakened, and water became scarce. The Ankorian civilization was in decline by the 15th century and collapsed by the end of the sixteenth century. The collapse was gradual, and people left the city, moving to smaller communities. Why the collapse occurred is not fully understood, but it was in part due to the onset of the LIA and a more volatile climate with reduced monsoon rains and drought. In short, the food supply diminished as water resources became scarce.[2] During the LIA, there were storms, wet periods, extremes of heat and cold, and climate change that caused a number of problems. Some winters were very cold, and, during one such winter in 1683–1684, the River Thames in London was frozen solid for 2 months and sea ice extended for kilometers into the North Sea, closing harbors and significantly restricting shipping. Crop failures occurred in Western Europe during the LIA, and the population was devastated by the Black Plague that reached out into the Atlantic and Iceland by about 1400. With the cooling times, increasing sea ice, and more environmental stress, trade was restricted with Greenland. Eventually, the colonies in North America and Greenland were mostly abandoned. Part of the reason for the abandonment in North America and, particularly, in Newfoundland, was that the Vikings may not have been able to adapt to the changing conditions, as did the Inuit peoples living there. As times became tough, the two cultures collided, and the Vikings, despite their fierce reputation, were not as able to adapt to the changing, cooler times as were the Inuit.

We do not know what caused the MWP, and the details about it are somewhat obscured by the lack of sufficient climate data during that period with which to estimate temperatures. We do know that it was warm in Western Europe and the North Atlantic, and we can't blame that on the burning of fossil fuels. What this may suggest is that more than one factor may cause warming; the present warming, for the, most part, is clearly the result of the emission of carbon dioxide into the atmosphere from burning fossil fuels and land-use changes. Perhaps a lesson for us is that changes in the most recent past warming (MWP) resulted in winners and losers. Today, the world population is approaching 7 billion, and half of us live in cities. The world population during the MWP was about 500 million, and agriculture was blooming. However, persistent droughts during the MWP apparently caused pain and suffering over much of the semiarid areas of the world, from coastal California to the southwestern United States, as well as in Central and South America. This is a red flag! The potential effects of prolonged, persistent drought and loss of food production for the world, should present warming continue, is a serious threat, and we will explore this, along with other aspects of present warming trends, throughout this chapter.[4–6]

12.1 Global Change and Earth System Science: An Overview

Preston Cloud, a famous Earth scientist interested in the history of life on Earth, human impact on the environment, and the use of resources, proposed two central goals for the Earth sciences:[4]

1. Understand how Earth works and how it has evolved from a landscape of barren rock to the complex landscape dominated by the life we see today.
2. Apply that understanding to better manage our environment.

Until recently, it was thought that human activity caused only local or, at most, regional environmental change. It is now generally recognized that the effects of human activity on Earth are so extensive that we are actually involved in an unplanned planetary experiment. To recognize and perhaps modify the changes we have initiated, we need to understand how the entire Earth works as a system. The discipline, called **Earth system science,** seeks to further this understanding by learning how the various components of the system—the atmosphere, oceans, land, and biosphere—are linked on a global scale and interact to affect life on Earth.[5]

12.2 Climate and Weather

Many people who hear about global warming form opinions based on day-to-day or week-to-week changes in the weather. They do not appreciate the important distinction between climate and weather. **Climate** refers to the characteristic atmospheric conditions of a given region over long periods of time, such as years or decades. **Weather** refers to the atmospheric conditions of a given region for short periods of time such as days or weeks. For example, when we think of the coast of the Pacific Northwest, we think of mild temperatures, high humidity, and lots of rain. Thus, when planning a trip to Vancouver, British Columbia, or Seattle, Washington, we might use our knowledge of the climate to help us remember to bring an umbrella! It is possible, however, that during a weeklong stay in Vancouver or Seattle, we enjoy only bright, sunny, dry *weather*, because day-to-day weather conditions can vary greatly.

Although we generally think of temperature and precipitation as climatic factors, climate is more than average high and low temperatures and/or amount of rainfall at a given location. In fact, it is possible for two locales to have the same average annual temperature but be in different climate zones. For example, San Diego, California, has an average annual temperature of around 18°C (64°F), and small temperature changes from month to month. In contrast, El Paso, Texas, has close to the same average annual temperature but experiences large changes in temperature, ranging from a January average of 7°C (45°F) to a July average of 28°C (83°F). Although the two areas have the same average annual temperature, it is clear that they have different climates.

CLIMATE ZONES

Climate may be classified many different ways. The simplest classification is by temperature and precipitation, but climate at a particular place or region may be complex. For example, it may depend on infrequent or extreme seasonal patterns, such as rain in the monsoon season in parts of India. The major climatic zones on Earth are shown in Figure 12.1a. Selected processes and changes that produce and maintain the climate system are shown in Figure 12.1b.

EARTH CLIMATE SYSTEM AND NATURAL PROCESSES

Climate exerts a major influence on natural processes. Flooding is, in part, dependent on rainfall amount and intensity, landslides may be more common in areas with rainy climates, and wildfires are more likely to occur in dry areas. A familiarity with Earth's climate zones is a first step toward recognizing the threat from natural hazards. Applying classification provides not only a map of world climates, but also information about the relationship between climate and vegetation.[6] Not all dry areas, for example, are prone to wildfires, but some dry areas are at significantly more risk than others.

12.3 The Atmosphere

ATMOSPHERIC COMPOSITION

As mentioned in Chapter 9, our atmosphere is composed mainly of nitrogen and oxygen with smaller amounts of other gases. Compounds, such as nitrogen, that form a constant proportion of the mass of the atmosphere are called *permanent gases*, whereas gases whose proportions vary in time and space, such as carbon dioxide, are considered *variable gases*. The atmosphere also contains microscopic particles called *aerosols*, whose proportions also vary in time and space.

PERMANENT AND VARIABLE GASES

The major permanent gases, which are those that remain essentially constant by percentage, include nitrogen, oxygen, and argon. Together, these gases comprise about 99 percent by volume of all atmospheric gases (Table 12.1). Nitrogen generally occurs as molecules of paired nitrogen atoms (N_2) and composes about 78 percent of the volume of all permanent gases. Although N_2 molecules make up the greatest volume of the atmosphere, they are relatively unimportant in atmospheric dynamics. However, when nitrogen forms compounds with other gases, such as oxygen, it can play an important role in Earth's climate dynamics.

The second largest component is oxygen, composing 21 percent of the atmosphere by volume. Like nitrogen, oxygen molecules consist mainly of paired atoms of oxygen (O_2). As we humans well know, oxygen is critical to the existence of most life forms on Earth. Like nitrogen, O_2 is relatively unimportant to atmosphere dynamics. However, as discussed later, other oxygen compounds such as ozone (O_3) play an extremely important role in the atmospheric-climate system.

Argon makes up most of the remaining 1 percent of the permanent gases, along with lesser amounts of neon, helium, krypton, xenon, and hydrogen. With the exception of hydrogen, the remaining 1 percent of the permanent gases are not chemically reactive and have little or no effect on climate.

Although the variable gases (Table 12.1) account for only a small percentage of the total mass of the atmosphere, some of them play important roles in atmospheric dynamics. These gases include carbon dioxide, water vapor, ozone, methane, nitrous oxide, and halocarbons. The atmosphere also contains a variable amount of aerosols. An **aerosol** is not a gas but a microscopic liquid

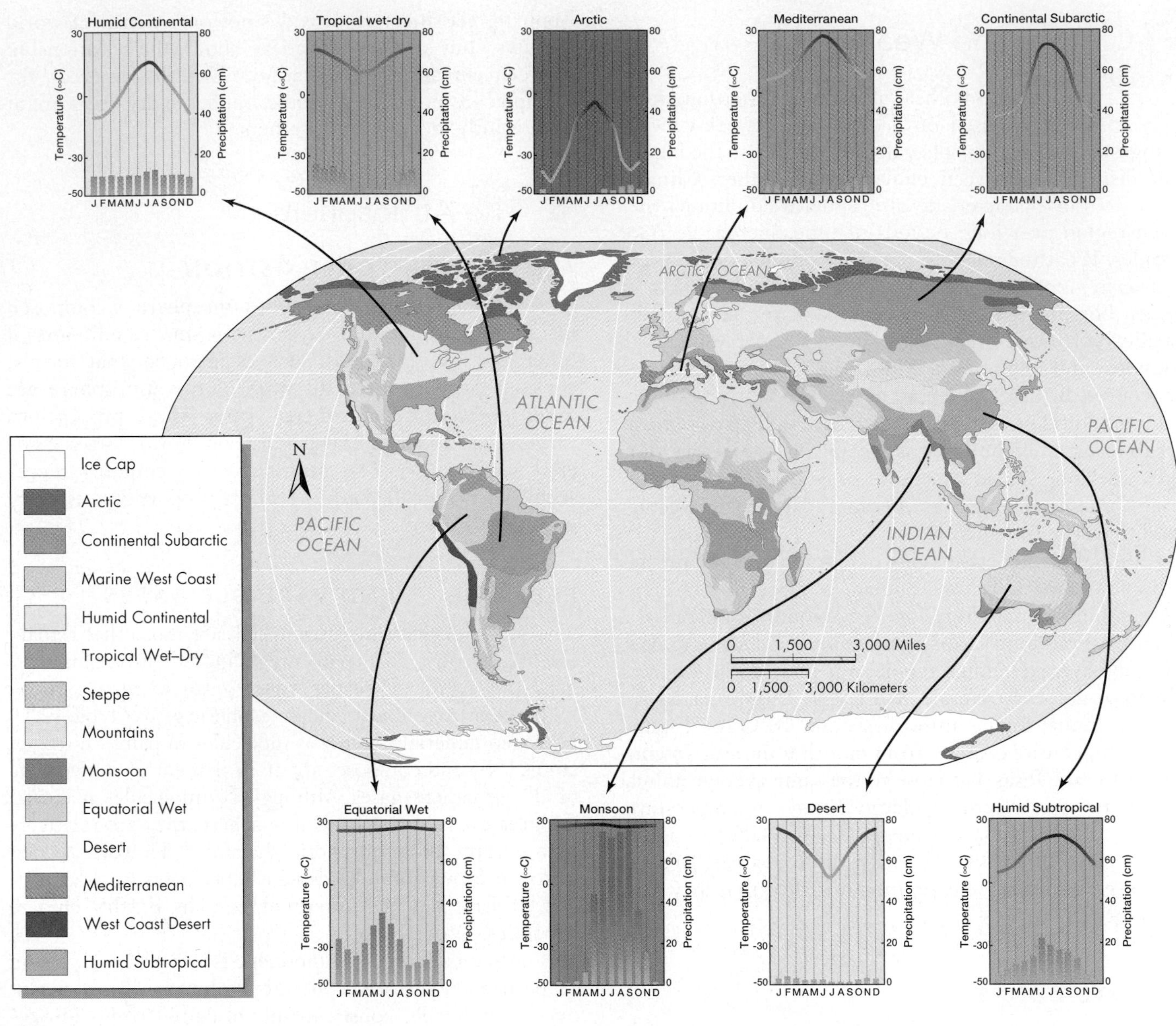

▲ FIGURE 12.1 **CLIMATES OF THE WORLD** (a) Characteristic temperature and precipitation conditions. Temperature is represented by the red line, and precipitation is shown as bars. *(Modified after Marsh, W. M., and Dozier, J.* Landscapes: An introduction to physical geography. *(Copyright © 1981. John Wiley & Sons)* (b) Idealized diagram showing the complex, linked components and changes of the climate system that produce, maintain, and change the climates of the world. *(IPCC. 2007.* The Physical Science Basis: *Working Group I.* Contribution to the Fourth Assessment Report. IPCC. *New York: Cambridge University Press. 989p.)*

or solid particle. In the atmosphere, these particles act as nuclei around which water droplets condense to form clouds. Aerosols have a wide variety of natural and anthropogenic sources. Natural sources include desert dust, wildfires, sea spray, and major periodic contributions from volcanoes.[7] Anthropogenic sources include the burning of forests in land clearing and sulphate particles and soot from the burning of fossil fuels.

GLACIATIONS

A study of climate change is, to a great extent, the study of changes in the atmosphere and linkages between the atmosphere and the lithosphere, hydrosphere, cryosphere, and biosphere. The cryosphere is the part of the hydrosphere in which most of the water stays frozen year-round. Major components of the cryosphere include permafrost, sea ice, ice caps, glaciers, and ice sheets. The ice in permafrost,

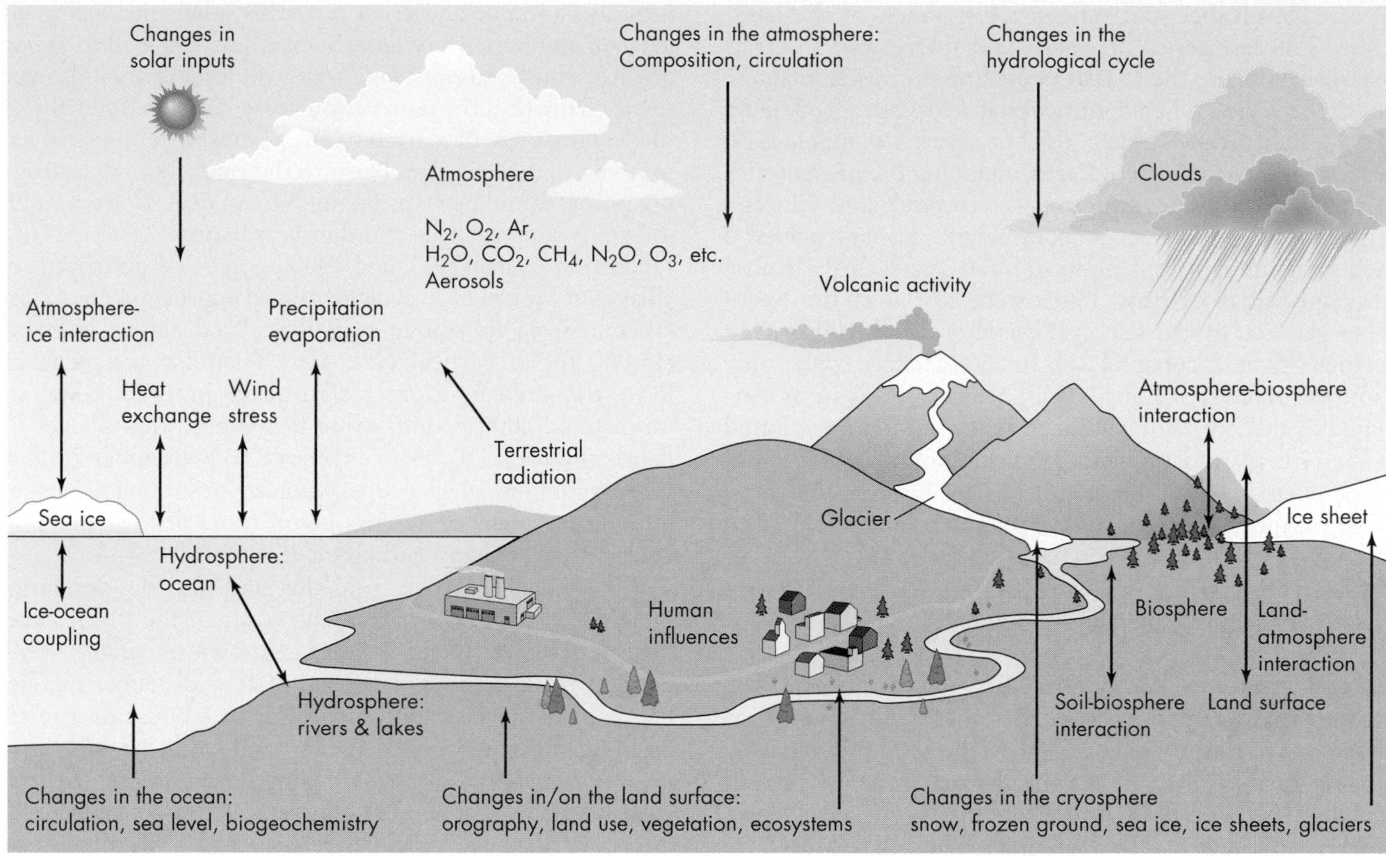

(b)

▲ FIGURE 12.1 (CONTINUED)

sea ice, and ice caps is either fixed in place or floating. In contrast, an individual **glacier,** or cluster of glaciers that forms an ice sheet, flows from high areas to low areas under the weight of accumulated ice. Like beaches, glaciers and ice sheets have budgets with inputs and outputs. Inputs consist of new snow recrystallizing to form ice at higher elevations, and outputs occur at lower elevations where ice is lost by melting, evaporation, and blocks breaking off the front of the glacier or ice sheet. Glaciers and ice sheets advance when their budget is positive and retreat when they have a net loss of ice.

Over the past 3.2 million years, Earth's climate system has fluctuated greatly and alternated between periods of major continental glaciation, referred to as *glacial intervals*, and times of warmer climate with significantly less glaciation, referred to as *interglacial intervals.* During glacial intervals, large ice sheets covered northern North America and Europe, and glaciers expanded in mountainous regions. In interglacial intervals, individual glaciers and large ice sheets retreated. Today, we live in interglacial conditions, with warm temperatures not experienced since the last interglacial interval 125,000 years ago. We will discuss why climate changes from glacial to interglacial later in this chapter with a more specific discussion of why climate changes. At this point, we will discuss glaciers and the hazards they present in more detail.

Ice sheet *advances* and *retreats,* or growth and melting back of glacial ice, have been relatively rare during the 4.6 billion years of Earth history. However, during the past 1 billion years, several major glacial events have occurred. We are now living during one of these events—one that began approximately 2.5 million years ago when glaciers began to

TABLE 12.1

Composition of the Atmosphere

Permanent Gases		Variable Gases	
Nitrogen	78.08%	Water Vapor	0.2–4%
Oxygen	20.95%	Carbon Dioxide	0.038%
Argon	0.93%	Methane	0.00017%
Neon	0.0018%	Nitrous Oxides	0.000032%
Helium	0.00052%	Ozone	0.000004%
Krypton	0.00011%	Halocarbons	0.00000002%
Xenon	0.00009%		
Hydrogen	0.00005%		

Percentage by volume. Data in part from Bryant, E. 1997. Climate process & change. *New York: Cambridge University Press.*

repeatedly advance and retreat across much of the landscape. The last series of glacial and interglacial intervals took place during the **Pleistocene Epoch,** which included multiple ice ages when continental glaciers advanced. During the Pleistocene Epoch, glaciers covered as much as 30 percent of the land area of Earth, including the present sites of major cities such as New York, Toronto, and Chicago (Figure 12.2). Pleistocene continental glaciers achieved their maximum extent about 21,000 years ago. Because large quantities of fresh water were stored in the ice of these glaciers, global sea level was more than 100 m (330 ft.) lower than its current level. Today glacial ice covers only about 10 percent of Earth's land. Nearly all this ice is contained in the Antarctic ice sheet; lesser amounts are found in the Greenland ice sheet and in the mountain glaciers of Alaska, British Columbia, southern Norway, the Alps, New Zealand, and elsewhere. We are presently in an interglacial interval, so the abundance of ice is relatively low. Nevertheless, we are probably still living in a glacial event, and ice sheets may again advance in the future.

Glacial Hazards Although at first glance glaciers may not seem to present a particularly threatening hazard to people, that is not the case. Glaciers are huge, actively flowing masses of ice and rock debris whose movement and melting have been responsible for property damage, injuries, and deaths. Their irregular, cracked surfaces have always presented hazards to people crossing them for scientific exploration or mountaineering; surface snow may hide deep cracks in the ice, known as *crevasses,* into which a traveler may fall. On some steep slopes, glacial ice falling from above is an additional hazard.

Changes in the flow of glaciers present hazards that have been well documented, particularly in the Alps of southern Switzerland.[8] Glaciers can expand or surge and overrun villages, fields, roads, or other structures. They may also advance and cross a stream valley, producing an ice dam and temporary lake. Such a dam can lead to major downstream flooding if lake water undercuts or spills over the ice dam, or if the dam is otherwise destroyed. Finally, if the leading edge of a glacier retreats upslope, blocks of ice may fall off in ice avalanches into the valley below, destroying property and perhaps taking human lives. For example, in 1962 a combined ice and debris avalanche, known as the Huascaran Avalanche, killed 4000 people and destroyed six villages in Peru. The avalanche moved approximately 16 km (10 mi.) from its source as it descended almost 4000 m (13,000 ft.) of vertical elevation. A similar avalanche in Peru, triggered by a great earthquake in 1970, killed an estimated 20,000 people while burying several villages in debris (Figure 12.3).[8] More recently, in September 2002, a huge mountain glacier broke loose in southern Russia, sending a torrent of ice, mud, and rocks downslope. The debris buried villages and killed at least 100 people.

When glaciers move from the land into the sea, large blocks of ice break off from the front of the glacier and drop into the water in a process known as *calving.* This process is spectacular to observe, at a distance, that is, because the blocks are often as large as a house or a small building. Calving produces blocks of floating ice, known as *icebergs,* which can float into shipping lanes and create a hazard to navigation. This danger was dramatically illustrated on April 14, 1912, when the *Titanic,* a luxury ocean liner on its maiden voyage from England to the United States, struck an iceberg and sank in the North Atlantic. That tragic accident claimed 1501 lives. It is interesting that large icebergs may also be beneficial to us. They are being looked at as potential sources of fresh water if they could be towed economically to areas where water is scarce.

Our discussion of ice avalanches in Peru and other areas shows that glaciers may trigger other hazardous events.

(a)

(b)

▲ FIGURE 12.2 **EXTENT OF ICE SHEETS** (a) Ice sheets about 21,000 years ago. (b) Ice sheet today. Accessed 9/25/10 at www.ncdc.noaa.gov

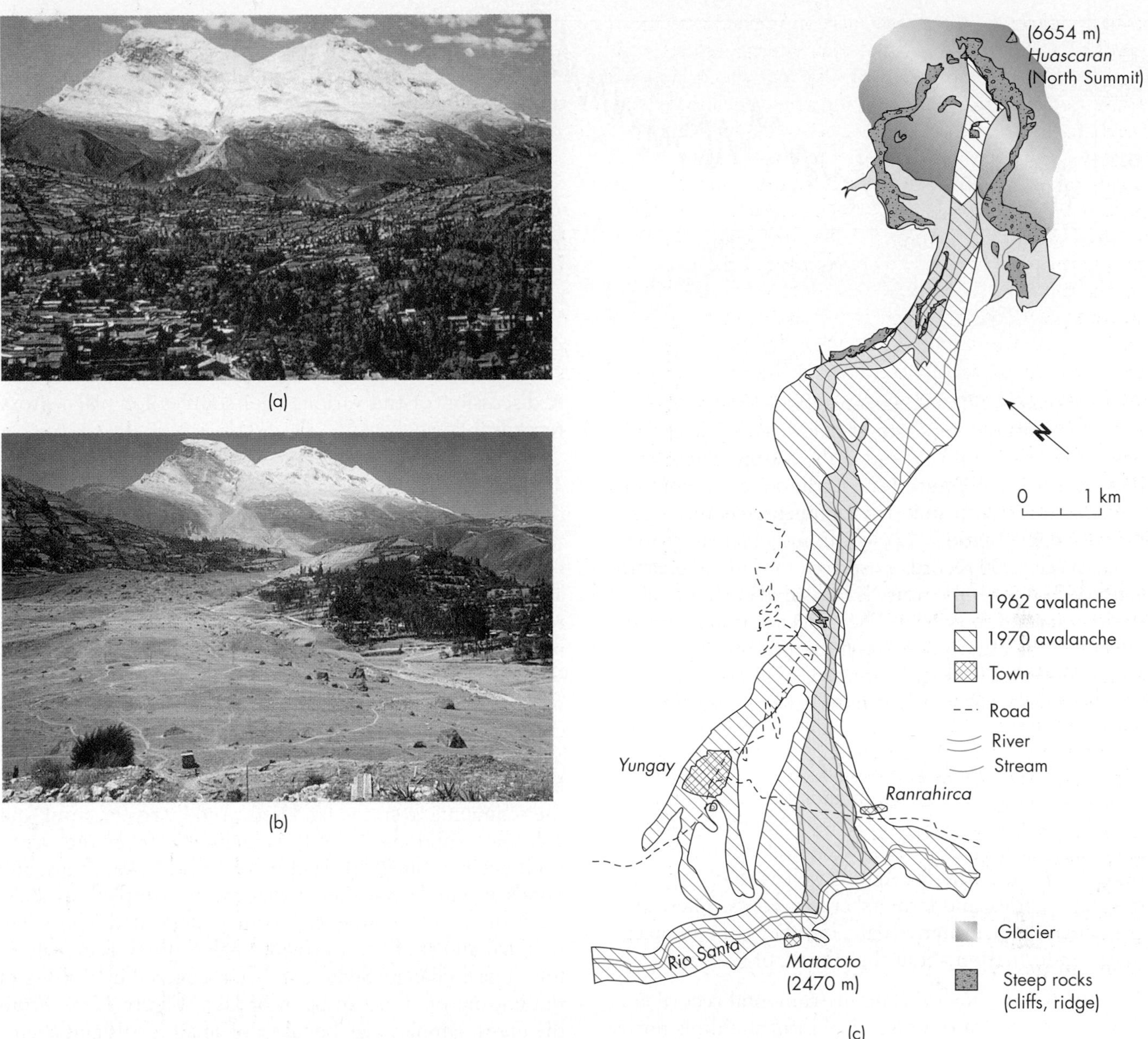

▲ FIGURE 12.3 **GLACIAL ICE CONTRIBUTES TO DEBRIS AVALANCHES** (a) Yungay, Peru, prior to the earthquake and debris avalanche. Snow-covered Mount Huascaran is on the skyline. *(Lloyd S. Cluff)* (b) The town site after the 1970 earthquake and avalanche. *(Lloyd S. Cluff)* (c) Tracks of the 1962 (yellow) and 1970 (area shaded with diagonal lines) avalanches down Mount Huascaran. *(From Welsh, W., and Kinzl, H. 1970. Der Gletschersturz vom Huascaran [Peru] am 31. Mai 1970, Die grosste Gletscherkatastrophe der Geschichte.* Zeitschrift für Gletscherkunde und Glazialgeologie *6: 181–92)*

This tendency is particularly true for active volcanoes that have glaciers on their slopes. A volcano that erupts hot volcanic ash and lava onto a glacier can rapidly melt the ice and trigger downslope movement of glacial ice and volcanic debris. This sequence of events has produced some of the largest and most hazardous volcanic mudflows or *lahars* known from both the historical and recent geologic record. We will now move our discussion to how we study past climate change and predict future change.

12.4 How We Study Past Climate Change and Make Predictions

The data gathered to document and better understand climate change are from a variety of time scales and variable regions, from continents, to the oceans, to hemispheres, and to the entire planet. Data are available for three main time periods:[9]

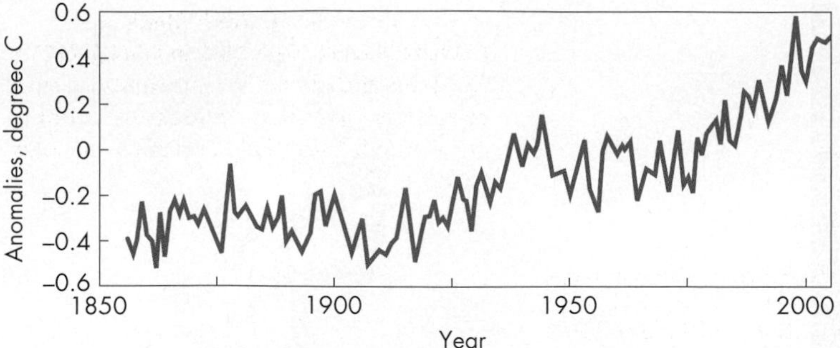

◀ FIGURE 12.4 **CHANGE IN TEMPERATURE** Diagram showing temperature changes from about 1850 to present with departure from the average temperature from 1960 to 1990. In recent years, the anomaly has been about 0.5 to 0.6°C from the 1960 to 1990 average (shown as "0" on the graph). Accessed 11/29/10 at www.ncdc.noaa.gov

- The *Instrumental Record.* Starting about 1860, measurements of temperatures have been made at various locations on land and in the oceans. These are the data shown in Figure 12.4. Earliest records are from the late seventeenth and early eighteenth centuries, and the network of stations has significantly increased over time. About 1000 records exist from the late nineteenth century. Today, temperature is measured at about 7000 stations around the world. The concentration of carbon dioxide in the atmosphere has been measured since about 1960. Accurate measurement of the production of solar energy has been taken over the past several decades.
- The *Historical Record.* A variety of historical records go back several hundred years. Included are people's written recollections (books, newspapers, journal articles, personal journals, etc.) of the Medieval Warming Period and the Little Ice Age, as well as ship's logs, traveler's diaries, and farmers' crop records. These are not generally quantitative data, but they contain quantitative information about the climate of the past.
- The *Paleo-Proxy Record.* The instrumental record is short, and most of the historical information is not quantitative. As a result, there has been a need for extending the record back further. Paleoclimatology (study of past climate) is part of Earth science. It is clear that the paleo record of Earth's climate has provided some of the strongest data to support and test recent climate change. The term **proxy data** refers to data that is not strictly climatic but can be correlated with climate, such as temperature of the land or sea. Some of the information gathered as proxy data includes natural records of climate variability, as indicated by tree rings, ocean sediments, ice cores, fossil pollen, corals, and carbon-14 (^{14}C). The disadvantages of paleoclimate proxy data are obvious; the data are not a direct measurement of temperature, and therefore temperature must be inferred from the data. In spite of this, the paleoclimate proxy data preserved in the geologic record provide the best evidence of change that predates the historical and instrumental records.

A discussion of the various data sources for paleo-proxy records follows.

Tree Rings The growth of trees is influenced by climatic conditions, such as the amount of rainfall and the variability in temperature. Most trees put on one growth ring per year, and the width, density, and isotopic composition of annual rings provide information about past climate—referred to as *dendroclimatology.* By dating dead trees radiometrically or by counting rings within living trees, scientists have developed a dendroclimatology proxy record that extends back more than 10,000 years (Figure 12.5).

Sediments The oceans of the world are a repository for the sediments from the land delivered by rivers, wind, and volcanic eruptions, as well as sediment from the ocean itself (such as shells of dead organisms). Lakes, bogs, and ponds accumulate sediment that can be sampled for a climate signal. Over time, sediment is deposited, and when sampled and studied, it provides paleo-proxy data sources for climate change. Sediments are recovered by drilling in the bottom of an ocean basin or lake (Figure 12.6). From the cores, samples may be taken of small fossils and chemicals that are contained within the sediments that may be interpreted to better understand past climate change. Some of the strongest evidence for past climate change comes from these proxy records.

Ice Cores Glaciers contain an accumulation record of snow that has been transformed into glacial ice over hundreds of thousands of years. Ice cores are obtained by drilling into the ice and obtaining a core, which may be studied in detail in order to learn about past conditions (Figure 12.7). Ice cores from glaciers often contain small bubbles of air deposited at the time of the snow. The composition and ratio of past atmospheric gases preserved in the ice may be studied and used to infer a number of paleoclimatic variables. Ice cores also contain a variety of chemicals and other materials, such as volcanic ash and dust, which can provide proxy data to assist in evaluating climate change. The ice itself may be studied to determine the paleo-isotopic composition of the water, which provides

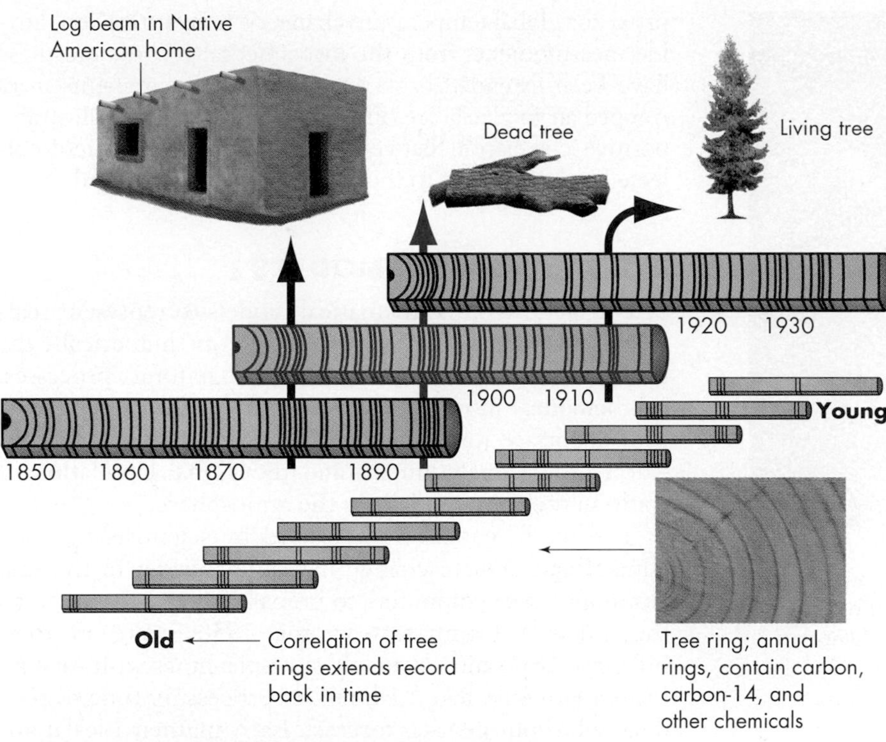

◀ FIGURE 12.5 **TREE RING CHRONOLOGY (DENDROCHRONOLOGY)** Contains proxy climate information, such as relative rainfall or periods of drought and carbon-14 activity (indicator of solar variability.)

▲ FIGURE 12.6 **EXAMINING CORES** Marine core from seafloor under the Antarctic Ice Shelf is being examined. *(Science Source/Photoresearchers)*

information about the volume of ice on the land, as well as processes occurring in the paleo-oceans.

Pollen Pollen (from flowers, trees, and other plants), along with other sediment, accumulates in a variety of environments, including oceans, bogs, and lakes. Scientists study the abundance and types of pollen in order to investigate past climate. For example, if the climate cools, there will be a change in the assemblage of pollen in sediments that reflects the change in climate. Pollen, if found in sufficient quantity, may be dated. Since the grains are preserved in sedimentary layers that also might be dated, a chronology can be developed. Based on the types of plants found at different times, the climatic history can be reconstructed.

Corals Coral reefs consist of corals (and other organisms) that have hard skeletons composed of calcium carbonate ($CaCO_3$) extracted by the corals from the seawater. The calcium carbonate contains isotopes of oxygen, as well as a variety of trace metals, that can be used to estimate the temperature of the water in which the coral grew. Thus, corals are a source of paleo-proxy data that can help us interpret climate change. Corals may be dated by several dating techniques, and a chronology of change over time may be constructed.

Carbon-14 The production of ^{14}C produced in the upper atmosphere is caused by collisions between neutrons and nitrogen-14 (^{14}N). The nitrogen is part of what are called *cosmic rays* that come from outer space and are a product of the energy from the sun. Solar

▲ FIGURE 12.7 **ICE CORE** Collecting an ice core in 2005 on Greenland Ice Cap for examination of the climate record it contains. *(Alfred-Wegener Institute fur Polar-und Meeresforschung)*

activity can be observed from the frequency of sunspots, which are dark areas on the sun surrounded by a lighter area. As sunspot activity increases, the energy from the sun that reaches Earth also increases. The frequency of sunspots has been accurately measured for decades and observed by people for about 1000 years. When sunspot activity is high, an associated solar wind, which produces ionized particles consisting predominately of protons and electrons, emanates from the sun. The solar wind reflects cosmic rays (including nitrogen-12, ^{12}N). As a result, the amount of ^{14}C is reduced. The record of ^{14}C in the atmosphere is correlated to tree ring chronology (dendrachronology, as discussed earlier). Each ring of wood of known age contains carbon, and the amount of ^{14}C can be measured. If the climatic record for a period of time is known, it may be correlated with ^{14}C. Thus, we can examine solar energy output, extending back thousands of years, by studying tree rings and the carbon-14 they contain. Based upon the ^{14}C record and its link to solar energy, it appears that the production of solar energy was slightly higher around A.D. 1000, during the Medieval Warming Period, and was slightly lower during the Little Ice Age from about A.D. 1400 to 1800. The effect of solar radiation on recent climate change (during the instrumental record) can account for a small percentage of observed changes in climate. The variability of solar energy is not sufficient to explain the warming since about 1960.

Carbon dioxide The concentration of global carbon dioxide in the atmosphere is arguably the most important proxy for global temperature change. There are carbon dioxide measurements from the instrumental record, and these have been extended back through the measurements from trapped air in glacial ice (Figure 12.8). Figure 12.9 illustrates positive correlation between CO_2 in the atmosphere collected from ice cores to the paleo-temperature record.

GLOBAL CLIMATE MODELS

Scientists develop mathematical models to represent real-world phenomena. These models describe numerically the linkages and interactions between natural processes. Mathematical models have been developed to predict the flow of surface water and groundwater, erosion and deposition of stream sediment, and the global circulation of water in the ocean and air in the atmosphere.

The mathematical models used by scientists to study climate and climate change have their origin in the first attempts to use computers to prepare weather forecasts in the 1950s.[10] Computers in the 1950s were in their infancy—large and extremely slow machines with vacuum tubes that could take 24 hours of processing time to produce a 24-hour weather forecast. Early mathematical models run on computers were, by necessity, regional in extent and used primarily to describe the general circulation of the atmosphere. Called *General Circulation Models*, these mathematical models were the foundation for the first truly *Global Climate Models*. From the 1980s through the early 2000s, atmospheric General Circulation Models were coupled with mathematical models of other Earth subsystems, such as the land surface, ocean and sea ice, aerosols, and the carbon cycle.[11]

The framework for the General Circulation Model is that of a large stack of boxes (Figure 12.10a). Each box is a three-dimensional cell that is several degrees of latitude and longitude on each side. A typical cell is generally rectangular in shape and has an area about the size of the state of Oregon. Each cell varies in height, depending on how many cells are used to subdivide the lower atmosphere. Most models use 6 to 20 layers of cells to represent the atmosphere up to an altitude of around 30 km (nearly 19 mi.).[6] Data are arranged into each of the cells, and boundary conditions are established for the overall model. Mathematical equations based on the principles of physics are used to describe the major atmospheric processes that interact between the cells. Models are run backward to see if they accurately describe historic, and even prehistoric, changes in climate. Like General Circulation Models for forecasting weather, Global Climate Models are constantly being adjusted and evaluated to better make predictions.[12] Global Climate Models are reasonably consistent with global temperature change from 1900 to the present (Figure 12.10b).

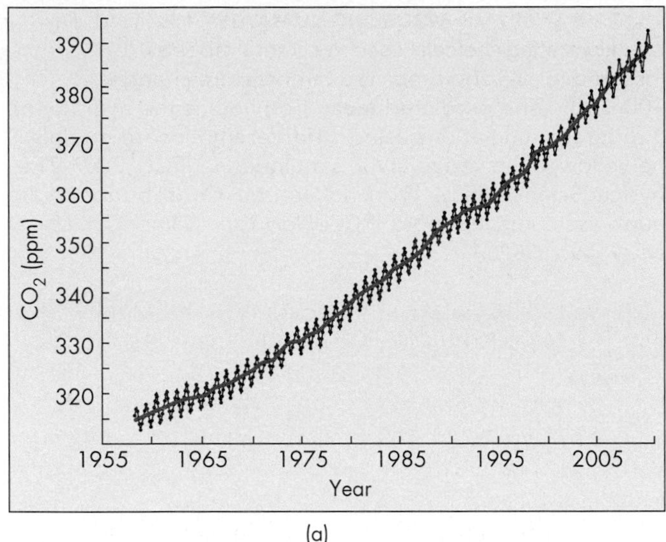

(a)

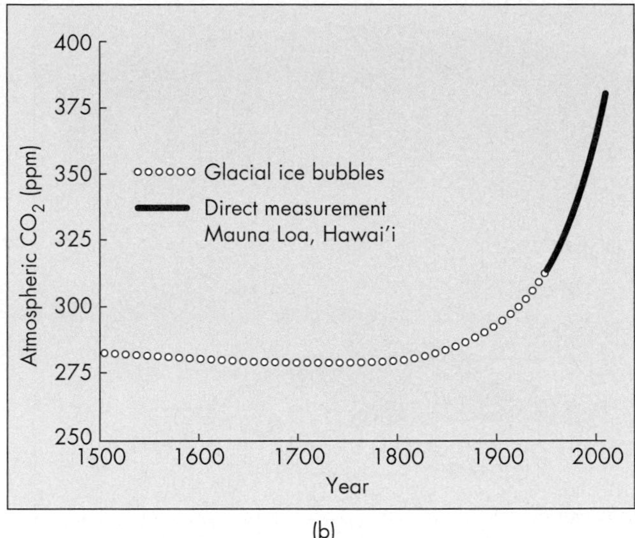

(b)

▲ FIGURE 12.8 **CARBON DIOXIDE IN THE ATMOSPHERE** (a) Atmospheric concentration of carbon dioxide measured at Maun a Loa, Hawai'i. *(Data from Scripps Institute of Oceanography, NOAA, and C. D. Keeling, at http://www.esrl.noaa.gov/gmd/ccgg/trends/co2_data_mlo.html)* (b) Average concentration of atmospheric carbon dioxide from 1500 to 2000, based on measurements of air bubbles trapped in glacial ice and on direct measurements from the NOAA Mauna Loa Observatory in Hawai'i. *(Data in part from Post, W.M., et al. 1990.* American Scientist *78[4]:210–26)*

Global Climate Models are a valuable first approach to solving the complex problems of climate change. These models provide information for evaluating Earth as a system. The models also identify additional data that are needed to produce better, more-detailed models in the future. One of the greatest difficulties in using Global Climate Models to predict future climate change is our inability to anticipate human behavior and the paths that our civilization will take in dealing with climate change. Today, Global Climate Models provide useful guidance on regions that are likely to be relatively wetter or drier and hotter or colder in the future. Current models are sophisticated, and predictions

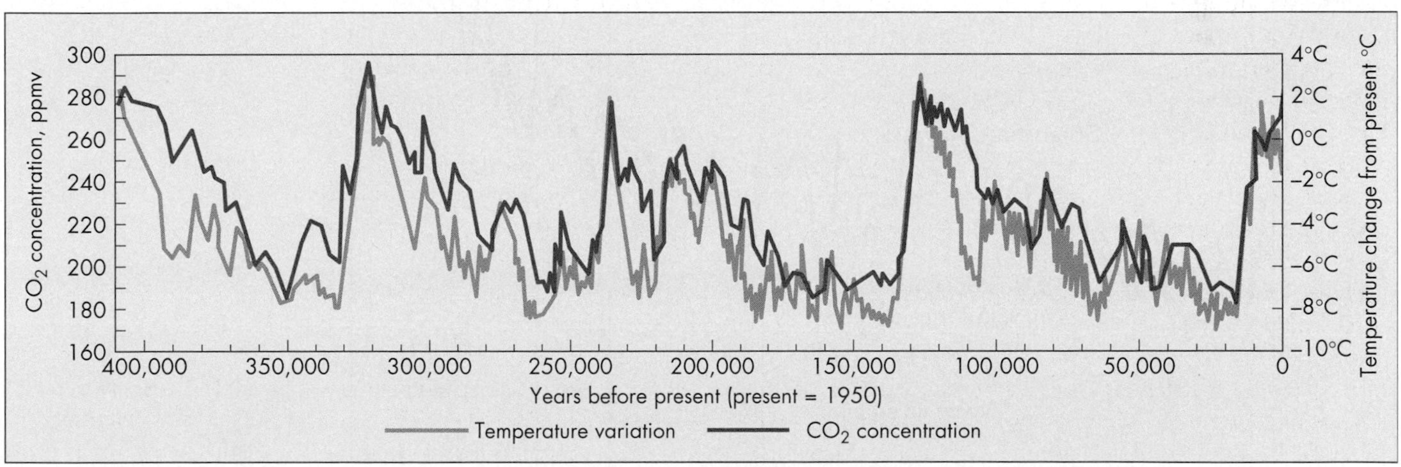

▲ FIGURE 12.9 **AIR TEMPERATURE CHANGES CORRESPOND CLOSELY TO ATMOSPHERIC CO_2** Measurement of the carbon dioxide content in air bubbles and isotopic ratios in glacial ice from cores taken at Vostok, Antarctica, show that atmospheric CO_2 levels have corresponded closely to air temperature for more than 410,000 years. Records from other ice cores show a similar pattern extending back more than 800,000 years. *(Modified from Petit, J. R., Jouzel, J., et al. 1999. Climate and atmospheric history of the past 420,000 years from the Vostok ice core in Antarctica.* Nature *399:429–36)*

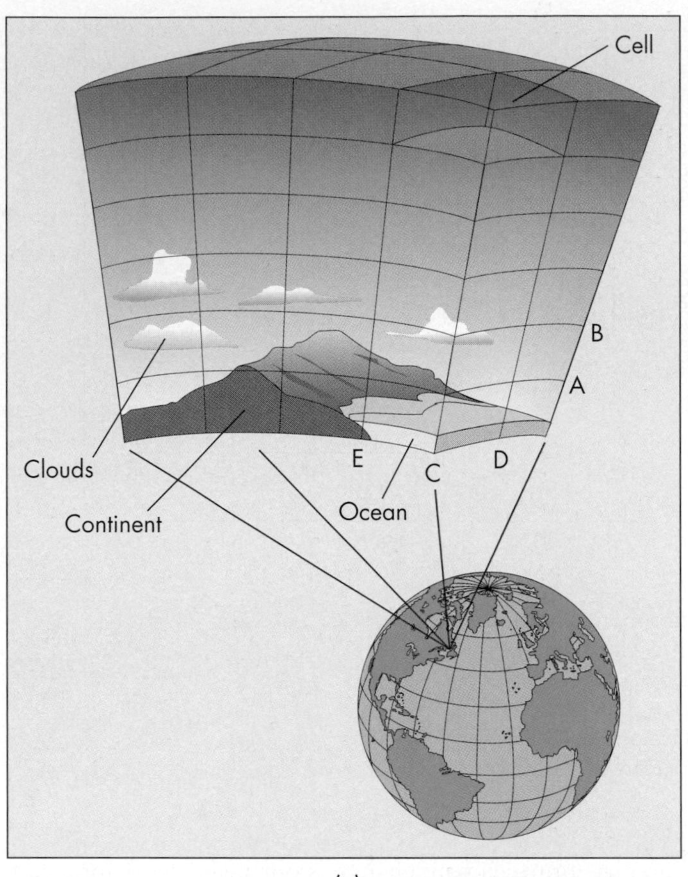

(a)

◀ FIGURE 12.10 **MODELING CLIMATE** (a) Idealized diagram illustrating the cells used in climate models . (b) Observed (black) and predicted temperature changes 1900–2005. The predicted mean (red line) is the average of 58 different simulations using 14 different climate models. The yellow is the variability in simulations. *(IPCC. 2007.* The Physical Science Basis*: Working Group I.* Contribution to the Fourth Assessment Report. IPCC. *New York: Cambridge University Press. 989p.)*

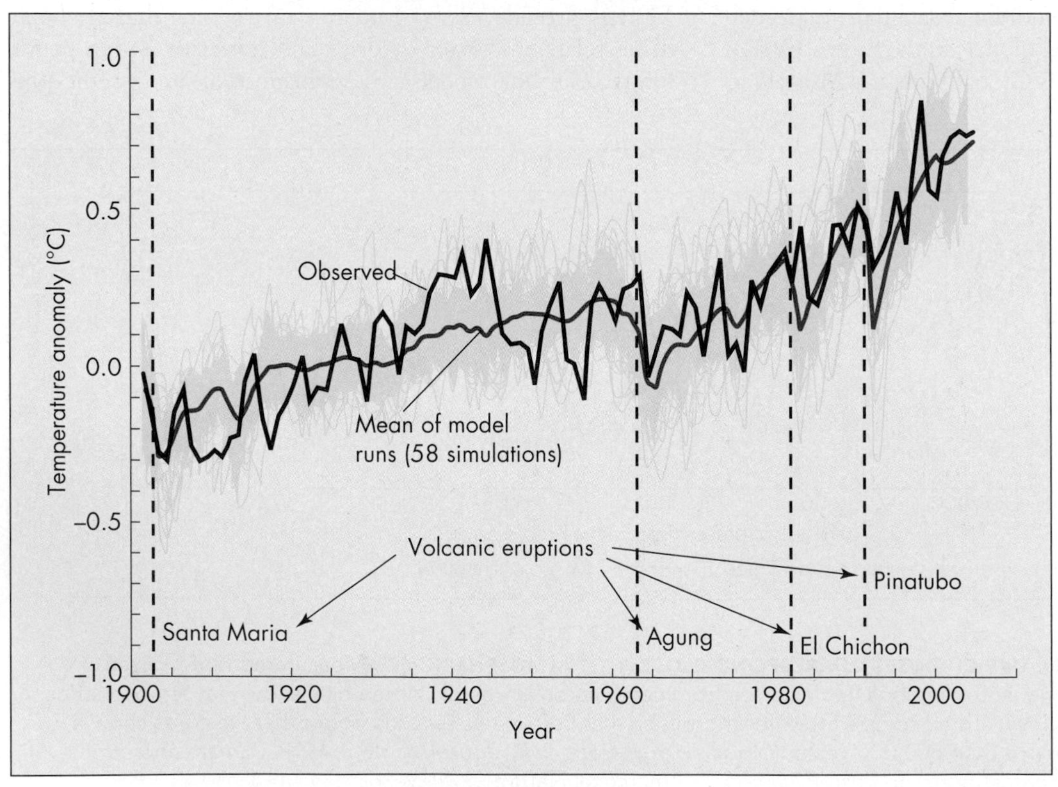

(b)

from modeling are generally consistent. Modeling the climate generates hypotheses to test with data. It is important to acknowledge that models do not produce data. Models use data-linked mathematical calculations with the purpose of better understanding the global climate systems and making predictions. The most important evidence (and only data) for global warming is from the measurements in the past 150 years, from proxy-evidence (data from the geologic record such as ocean sediments, and tree rings over the past few thousand to few hundred thousand years).

12.5 Global Warming

Global warming is defined as the observed increase in the average temperature of the near-surface land and ocean environments of Earth during the past 50 years. A growing volume of evidence suggests that we are now in a period of global warming, resulting from burning vast amounts of fossil fuels. Does this mean we are experiencing human-induced global warming? Many scientists now believe that human processes, as well as natural ones, are contributing significantly to global warming.[13–15]

THE GREENHOUSE EFFECT

For the most part, the temperature of Earth is determined by three factors: the amount of sunlight Earth receives, the amount of sunlight Earth reflects (and, therefore, does not absorb), and atmospheric retention of reradiated heat.[14] See Figure 9.4, which shows the basics of Earth's energy balance between incoming and outgoing energy.

Earth's energy balance today is slightly out of equilibrium, with about 1 Watt per m^2 (W/m^2) more energy coming from the sun than is lost to space. The energy we are talking about, W/m^2, is energy per unit time (joules/sec) per unit area (m^2). The units W/m^2 are power per unit area, but we speak of them as solar energy. The units W/m^2 are widely used in global warming and climate change research.[13–15]

Earth receives energy from the sun in the form of electromagnetic radiation (Figure 12.11). Radiation from the

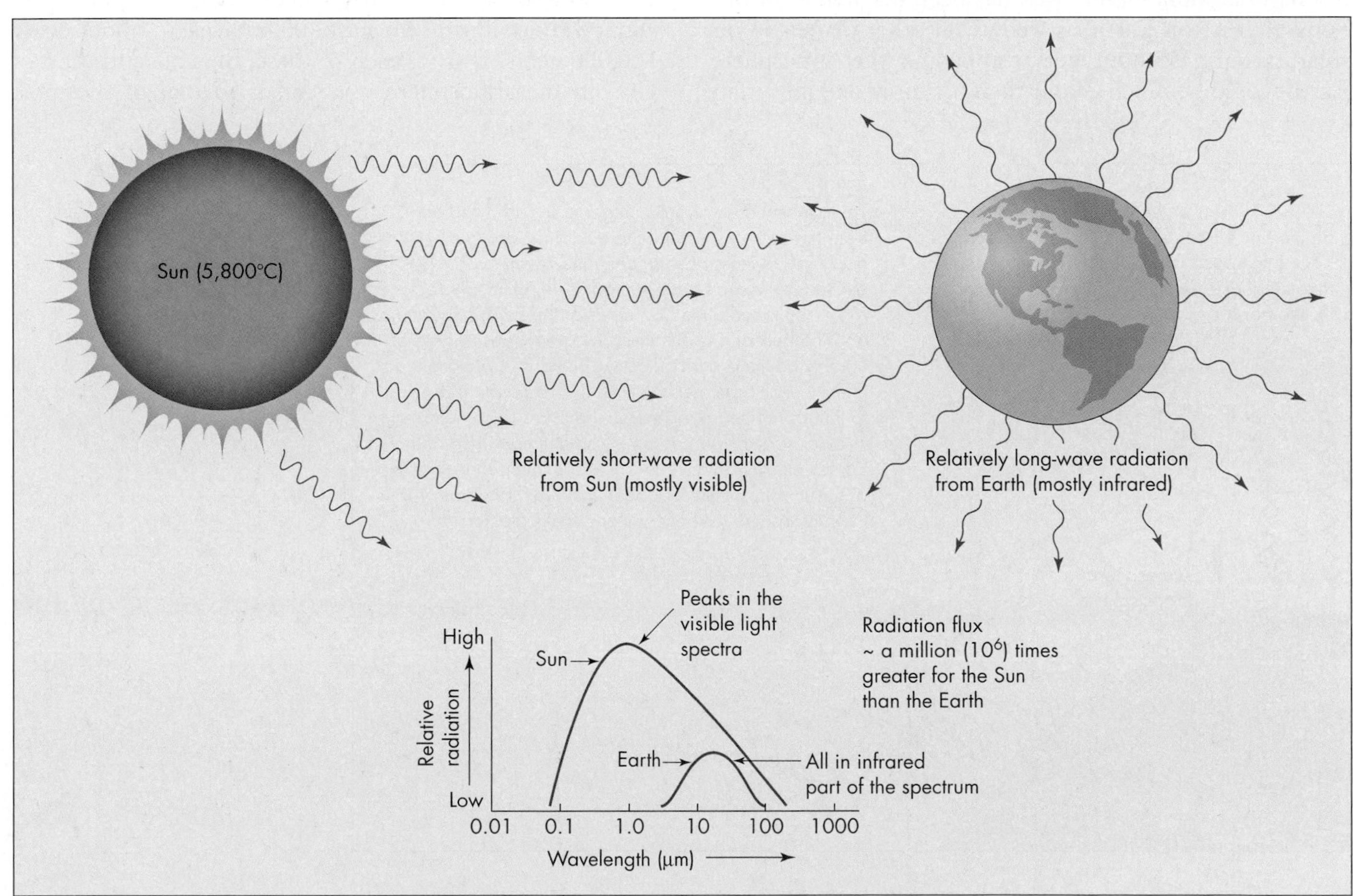

▲ FIGURE 12.11 **EARTH-SUN** Idealized diagram comparing the emission of energy from the sun with that from Earth. Notice that the solar emissions have a relatively short wavelength, whereas those from Earth are relatively long wavelengths. *(Modified after Marsh, W. M., and Dazier, J. Landscapes:* An introduction to physical geography, Copyright © 1981, New York, NY: John Wiley and Sons; *developed by M. S. Manalis and E. A. Keller, 1990)*

sun is relatively short wave and mostly visible, whereas Earth radiates relatively long-wave infrared radiation. The hotter an object, whether it is the sun, the Earth, a rock, or a lake, the more electromagnetic energy it emits. The sun, with a surface temperature of 5800°C (10,472°F), radiates much more energy per unit area than does Earth, which has an average surface temperature of 15°C (59°F).

Absorbed solar energy warms Earth's atmosphere and surface, which then reradiate the energy back into space as infrared radiation.[13] Water vapor and several other atmospheric gases—including carbon dioxide (CO_2), methane (CH_4), and chlorofluorocarbons (CFCs, human-made chemicals used in air conditioners and refrigerators) tend to trap heat. That is, they absorb some of the energy radiating from Earth's surface and are thereby warmed. As a result, the lower atmosphere of Earth is much warmer than it would be if all of the infrared radiation escaped into space without this intermediate absorption and warming. This effect is somewhat analogous to the trapping of heat by a greenhouse and is therefore referred to as the **greenhouse effect** (Figure 12.12).

It is important to understand that the greenhouse effect is a natural phenomenon that has been occurring for millions of years on Earth, as well as on other planets in our solar system. Without heat trapped in the atmosphere, Earth would be much colder than it is now, and all surface water would be frozen. Most of the natural "greenhouse warming" is due to water vapor and small particles of water in the atmosphere. However, potential global warming resulting from human activity is related to carbon dioxide, methane, nitrogen oxides, and chlorofluorocarbons. In recent years, the atmospheric concentrations of these gases and others have been increasing because of human activities. These gases tend to absorb infrared radiation from Earth, and it has been hypothesized that Earth is warming because of the increases in the amounts of these so-called greenhouse gases. Table 12.2 shows the rate of increase of these atmospheric gases resulting from human-induced emissions and their relative contribution to the *anthropogenic*, or human-caused, component of the greenhouse effect. Notice that carbon dioxide produces 60 percent of the relative contribution.

Measurements of carbon dioxide trapped in air bubbles of the Antarctic ice sheet suggest that during most of the past 160,000 years, the atmospheric concentration of carbon dioxide has varied from a little less than 200 ppm to about 300 ppm.[14] The highest levels are recorded during major interglacial periods that occurred approximately 125,000 years ago and today. Major interglacials occurred about 4 times during the past 400,000 years—about every 100,000 years. During each of these, the concentration of CO_2 in the atmosphere was similar to that of the most

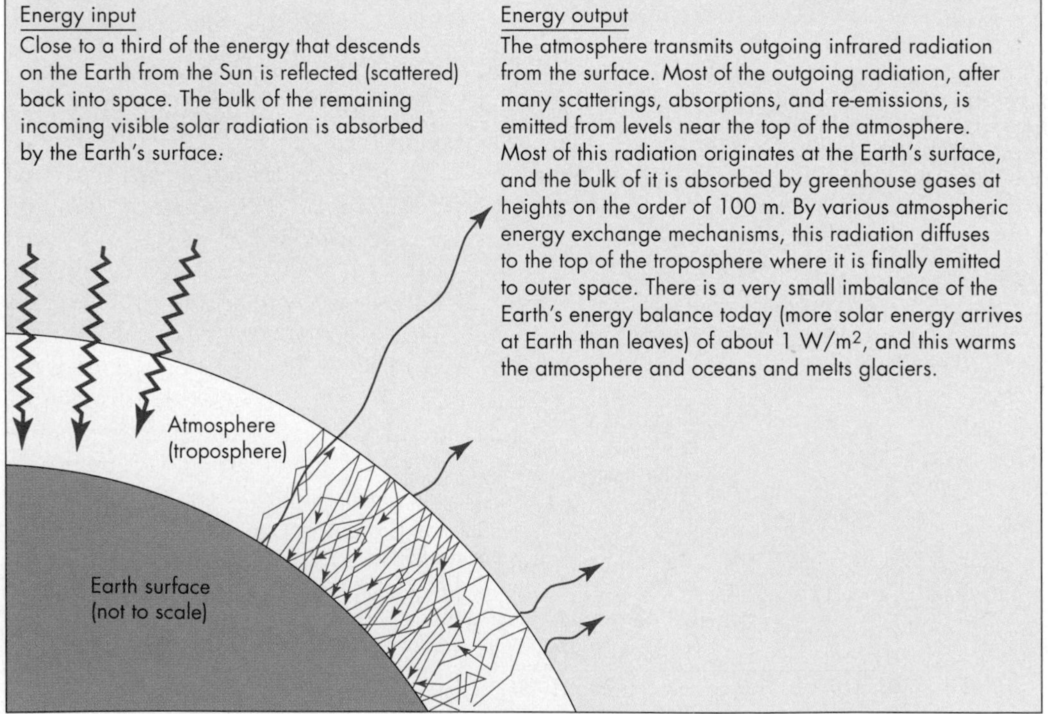

▲ FIGURE 12.12 **GREENHOUSE EFFECT** Idealized diagram showing the greenhouse effect. Incoming visible solar radiation (sunshine) is absorbed by Earth's surface, warming it. Infrared radiation is then emitted at the surface of Earth as earthshine to the atmosphere and outer space. Most of the infrared radiation emitted from Earth is absorbed by the atmosphere, heating it and maintaining the greenhouse effect. *(Developed by M. S. Manalis and E. A. Keller, 1990)*

TABLE 12.2

Rate of Increase and Relative Contribution of Several Gases to the Anthropogenic Greenhouse Effect

	Rate of Increase (% per year)	Relative Contribution (%)
CO_2	0.5	60
CH_4	<1	15
N_2O	0.2	5
O_3^a	0.5	8
CFC-11[b]	4	4
CFC-12[b]	4	8

[a] In the troposphere.

[b] CFC—chlorofluorocarbon (human made chemical)

Source: Data from Rodhe, H. 1990. A comparison of the contribution of various gases to the greenhouse effect. Science *248:1218, table 2. Copyright 1990 by the AAAS.*

recent interglacial event about 125,000 years ago.[16] At the beginning of the Industrial Revolution, the atmospheric concentration of carbon dioxide was approximately 280 ppm. Since 1860, fossil fuel burning has contributed to the exponential growth of the concentration of carbon dioxide in the atmosphere. The change from approximately 1500 to 2000 is shown in Figure 12.8b. Data for before the mid-twentieth century are from measurements made from air bubbles trapped in glacial ice. The concentration of carbon dioxide in the atmosphere today is approximately 390 ppm, and it is predicted to reach at least 450 ppm—more than 1.5 times the preindustrial level—by the year 2050. Changes in atmospheric concentration of carbon dioxide at Mauna Loa, Hawai'i, from the mid-twentieth century to today, are shown in Figure 12.8a. The annual cycles with a lower concentration of CO_2 during the summer are due to the summer growing season in the Northern Hemisphere, when plants extract more carbon dioxide from the atmosphere by photosynthesis.

GLOBAL TEMPERATURE CHANGE

The Pleistocene ice ages began approximately 2 million years ago, and, since then, there have been numerous changes in Earth's mean annual temperature. Figure 12.13 shows the changes of approximately the past million years on several time scales. The first scale shows the entire million years (Figure 12.13a), during which major climatic changes involved swings of several degrees Celsius in mean temperature. Low temperatures have coincided with major glacial events that have greatly altered the landscape; high temperatures are associated with interglacial conditions. Interglacial and glacial events become increasingly prominent in the scales, showing changes over 150,000 and 18,000 years. The last major interglacial warm period, even warmer than today, was the Eemian (Figure 12.13b). During the Eemian, sea level was a few meters higher than it is today. The cold period that occurred about 11,500 years ago is known as the Younger Dryas; it was followed by rapid warming to the Holocene maximum, which preceded the Little Ice Age (Figure 12.13c).

A scale of 1000 years shows several warming and cooling trends that have affected people (Figure 12.13d). For example, a major warming trend from approximately A.D. 1100–1300 allowed the Vikings to colonize Iceland, Greenland, and northern North America. When glaciers made a minor advance around A.D. 1400, during a cold period known as the Little Ice Age, the Viking settlements in North America and parts of Greenland were abandoned.

In approximately 1750, an apparent warming trend began that lasted until approximately the 1940s, when temperatures cooled slightly. Over the past 140 years, more changes are apparent, and the 1940s event is clearer (Figure 12.13e). It is evident from the record that in the past 100 years, global mean annual temperature increased by approximately 0.8°C (1.4°F). Most of the increase has been since the 1970s, and the 1990s, and the first 8 years of the twenty-first century had the warmest temperatures since global temperatures have been monitored.[5] This period may become known as the early twenty-first century increase in global temperature. Table 12.3 lists evidence of the recent warming.

WHY DOES CLIMATE CHANGE?

The question that begs to be answered is this: Why does climate change? Examination of Figure 12.13 suggests that there are cycles of change lasting 100,000 years, separated by shorter cycles of 20,000 to 40,000 years in duration. These cycles were first identified by Milutin Milankovitch in the 1920s as a hypothesis to explain climate change. Milankovitch realized that the spinning Earth is like a wobbling top, unable to keep a constant position in relationship to the sun; this instability partially determines the amount of sunlight reaching and warming Earth. He discovered that variability in Earth's orbit around the sun follows a 100,000-year cycle that is correlated with the major glacial and interglacial periods shown on Figure 12.13a. Earth's orbit varies from a nearly circular ellipse to a more elongated ellipse. Over the 100,000-year cycle of Earth, when the orbit is most elliptical, solar radiation reaching Earth is greater than during a more circular orbit. Cycles of approximately 40,000 and 20,000 years are the result of changes in the tilt and wobble of Earth's axis respectively. *Milankovitch cycles* reproduce most of the long-term cycles observed in the climate, and they do have a significant effect on climate. However, the cycles are not sufficient to produce all of the observed large-scale global climatic changes. Therefore, these cycles, along with other processes, must be invoked to explain global climatic change. Thus, the Milankovitch cycles that force (push)

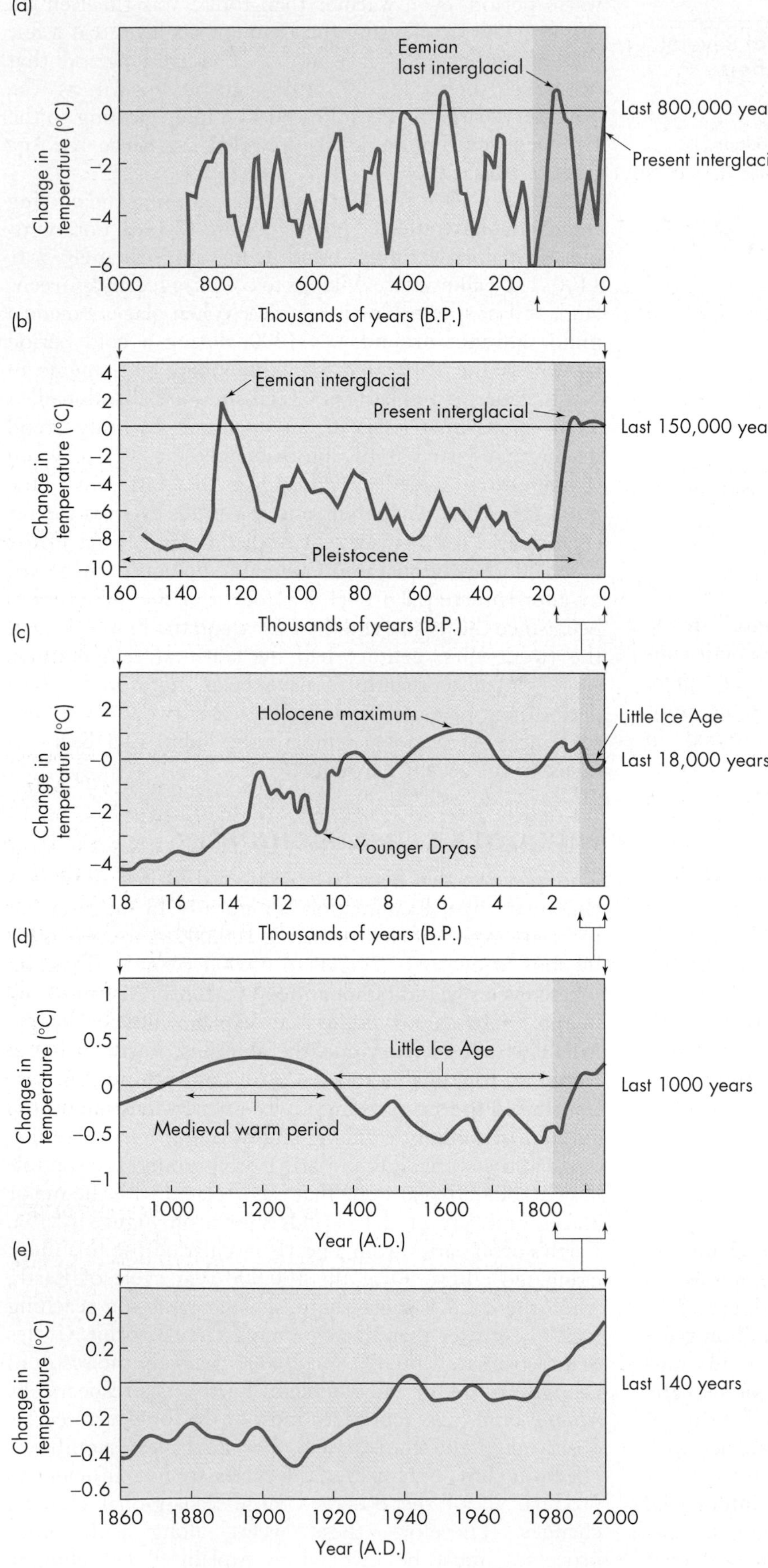

◀ **FIGURE 12.13 CHANGES IN GLOBAL TEMPERATURE** Change in temperature over different periods of time during the past million years. Graphs (a) to (d) are over time periods of 100,000 to 1000 years. The rapid rise from about 1900 to 2008 of nearly 0.9°C is shown on (e). Note the very rapid rise, since about 1970, of about 0.2°C per decade. See text for further explanation. *(Modified after University Corporation for Atmospheric Research, Office for Interdisciplinary Studies. 1991. Science capsule, Changes in time in the temperature of the earth.* EarthQuest *5[1]; and the UK Meteorological Office. 1997.* Climate change and its impacts: A global perspective)

TABLE 12.3

Evidence Supporting the Early Twenty-first Century Rise in Global Temperature: Global Temperature Data from the United States (NOAA) and Europe (WMO)

Warming, since the mid-1970s, has been about 3 times as rapid as in the preceding century.
The 2000s was the warmest decade in the past 142 years, according to historical records, and the past 1000 years, according to geologic data.
The 15 warmest years have all occurred since 1990.
The warmest decade on record was 2000–2009; 2005 was the warmest year and 2007 and 2009 tied for the second warmest years; 2002, 2003, and 2006 were very close to 2007 and 2009; 2008 was the coolest year of the decade, because of strong cooling in tropical Pacific oceans.
In 2003, the United States was cooler and wetter than average in much of the eastern part of the country, and warmer and drier in much of the West. Ten western states were much warmer than average; New Mexico had its warmest year. Alaska was warmer in all four seasons and had one of its 5 warmest years since Alaska began taking measurements in 1918.
Europe, in 2003, experienced summer heat waves, with the warmest seasonal temperatures ever recorded in Spain, France, Switzerland, and Germany. Approximately 15,000 people died in heat waves in Paris during the summer.
Warm conditions, along with drought, in 2003 contributed to severe wildfires in Australia; Southern California; and British Columbia, Canada.
Since 1970, the average temperature of Earth at the surface has increased about 0.2°C per decade.

Note: A few years of high temperatures with drought, heat waves, and wildfires are not by themselves an indication of longer-term global warming. The persistent trend of increasing temperatures over several decades is more compelling evidence that global warming is real and happening.

the climate in one direction or another can be looked at as natural processes (forcing) that when linked to other processes (forcing), produce climatic change.[17,18] We now will consider this concept of climate *forcing* in more detail.

Climate forcing is defined as an imposed change of Earth's energy balance. The units for the forcing are W/m^2, and they can be positive if a particular forcing increases global mean temperature or negative if temperature is decreased. For example, if the energy from the sun increases, then Earth will warm (this is positive climate forcing). If CO_2 were to decrease, causing Earth to cool, that would be an example of negative climate forcing.[19] *Climate sensitivity* refers to the response of climate to a specific climate forcing after a new equilibrium has been established, and the time required for the response to a forcing to occur is the *climate response time*.[20] A significant implication of climate forcing is that if you maintain small climate forcing for a long enough time, large climate change can occur.[13] Figure 12.14 shows the climate forcing that produced the last ice age 22,000 years ago (last glacial maximum).[13,20] Notice that 1 W/m^2 produces a temperature change of about 0.75°C (1.3°F). Climate forcing in the industrial age is shown in Figure 12.15. Total positive forcing is about 1.6 W/m^2, most of which is due to greenhouse gas forcings (CO_2, CH_4, N_2O). Rates of emissions of these gases are shown in Figure 12.16. They have increased dramatically in the past 100 years.[14]

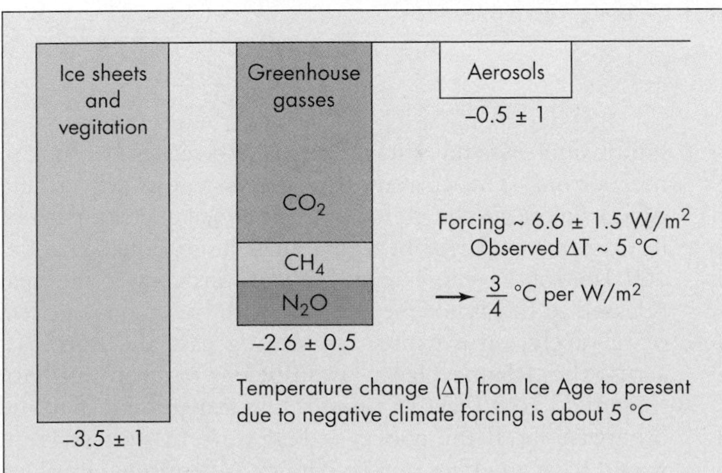

FIGURE 12.14 **CLIMATE FORCING (W/m^2) DURING LAST ICE AGE** *(Modified after Hansen, J. 2003. Can we defuse the global warming time bomb? Edited version of presentation to the Council on Environmental Quality. June 12. Washington DC., also Natural Science. http://www.naturalscience.com)*

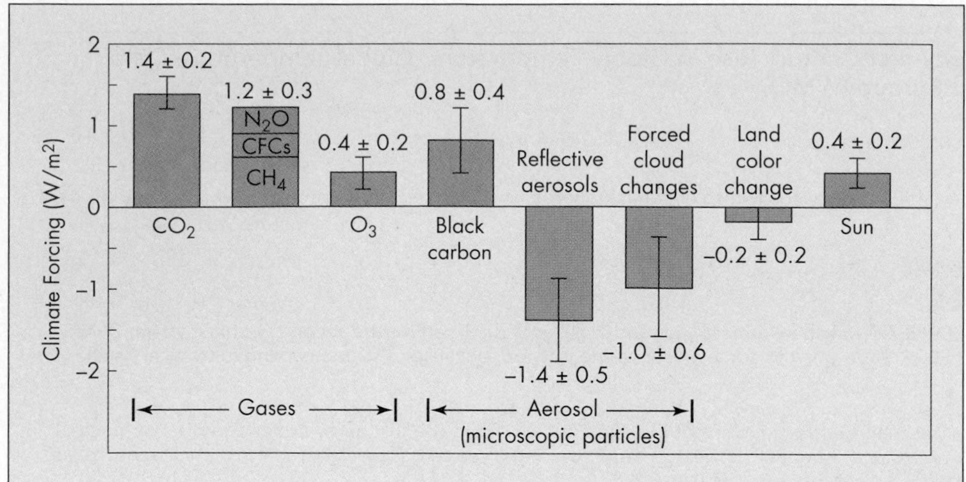

- Increases of greenhouse gases (except O_3) are known from observations and bubbles of air trapped in ice sheets. The increase of CO_2 from 285 parts per million (ppm) in 1850 to 368 ppm in 2000 is accurate to about 5 ppm. The conversion of this gas change to a climate forcing (1.4 W/m^2), from calculation of the infrared opacity, adds about 10% to the uncertainty.

- Increase of CH_4 since 1850, including its effect on stratospheric H_2O and tropospheric O_3, causes a climate forcing about half as large as that by CO_2. Main sources of CH_4 include landfills, coal mining, leaky natural gas lines, increasing ruminant (cow) population, rice cultivation, and waste management. Growth rate of CH_4 has slowed in recent years.

- Tropospheric O_3 is increasing. The U.S. and Europe have reduced O_3 precursor emissions (hydrocarbons) in recent years, but increased emissions are occurring in the developing world.

- Black carbon ("soot"), a product of incomplete combustion, is visible in the exhaust of diesel-fueled trucks. It is also produced by biofuels and outdoor biomass burning. Black carbon aerosols are not well measured, and their climate forcing is estimated from measurements of total aerosol absorption. The forcing includes the effect of soot in reducing the reflectance of snow and ice.

- Human-made reflective aerosols include sulfates, nitrates, organic carbon, and soil dust. Sources include burning fossil fuel and agricultural activities. Uncertainty in the forcing by reflective aerosols is at least 35%.

- Indirect effects of aerosols on cloud properties are difficult to compute, but satellite measurements of the correlation of aerosol and cloud properties are consistent with the estimated net forcing of −1 W/m^2, with uncertainty of at least 50%.

▲ FIGURE 12.15 **CLIMATE FORCINGS IN THE INDUSTRIAL AGE THAT STARTED IN 1750** Positive forcings warm and negative forcings cool. Human-caused forcings in recent years dominate over natural forcings. Total forcing is about 1.6 ± 0.1 W/m^2, consistent with observed rise in air surface temperature over the past few decades. *(Modified after Hansen, J. 2003. NASSA Goddard Institute for Space Studies and Columbia University Earth Institute.)*

We now believe that our climate system may be inherently unstable and capable of changing quickly from one state to another in as short a time as a few decades.[21] However, very short or abrupt climate change is unlikely. Part of what may drive the climate system and its potential to change is the *ocean conveyor belt*, a global-scale circulation of ocean waters, characterized by strong northward movement of 12° to 13°C (53° to 55°F) near-surface waters in the Atlantic Ocean that are cooled to 2° to 4°C (35° to 39°F) when they arrive near Greenland (Figure 12.17).[21] As the water cools, it becomes saltier; the salinity increases the water's density and causes it to sink to the bottom. The current then flows southward around Africa, adjoining the global pattern of ocean currents. The flow in this conveyor belt current is huge, equal to about 100 Amazon Rivers. The amount of warm water and heat released to the atmosphere, along with the stronger effect of relatively warm winter air moving east and northeast across the Atlantic Ocean, is sufficient to keep northern Europe 5° to 10°C (8.5° to 17°F) warmer than it would otherwise be. If the conveyor belt were to shut down, it would have an effect on the climate of Europe. However,

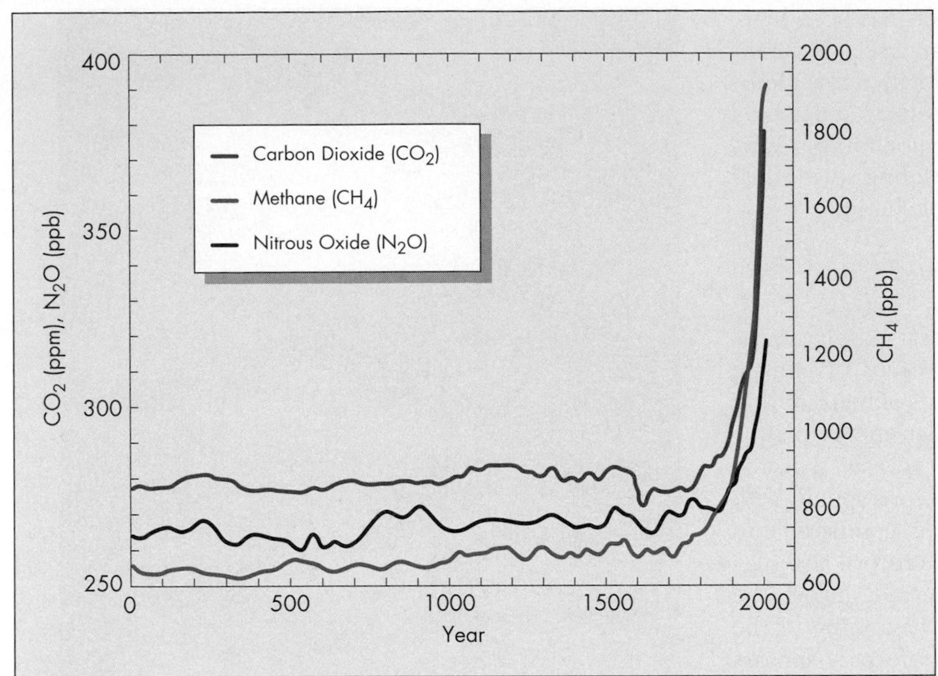

◀ FIGURE 12.16 **CONCENTRATIONS OF GREENHOUSE GASES DURING THE PAST 2,000 YEARS.** Atmospheric concentrations of carbon dioxide, methane, and nitrous oxide have increased rapidly in recent decades. *(Modified after IPCC. 2007. The physical science basis: Working Group I.* Contribution to the Fourth Assessment Report. IPCC. *New York: Cambridge University Press. 989p.)*

the effect would not be catastrophic to England and France in terms of producing extreme cold and icebound conditions.[22]

Although scientific uncertainties exist, sufficient evidence exists to state that (1) there is a discernible human influence on global climate, (2) warming is now occurring, and (3) the mean surface temperature of Earth will likely increase by between 1.5° to 4.5°C (2.6° to 7.8°F) during the twenty-first century.[16] The human-induced component of global warming results from increased emissions of gases that tend to trap heat in the atmosphere. There is good reason to argue that increases in carbon dioxide and other greenhouse gases are related to an increase in the mean global temperature of Earth. Over the past few hundred thousand years, there has been a strong correlation between the concentration of atmospheric CO_2 and global

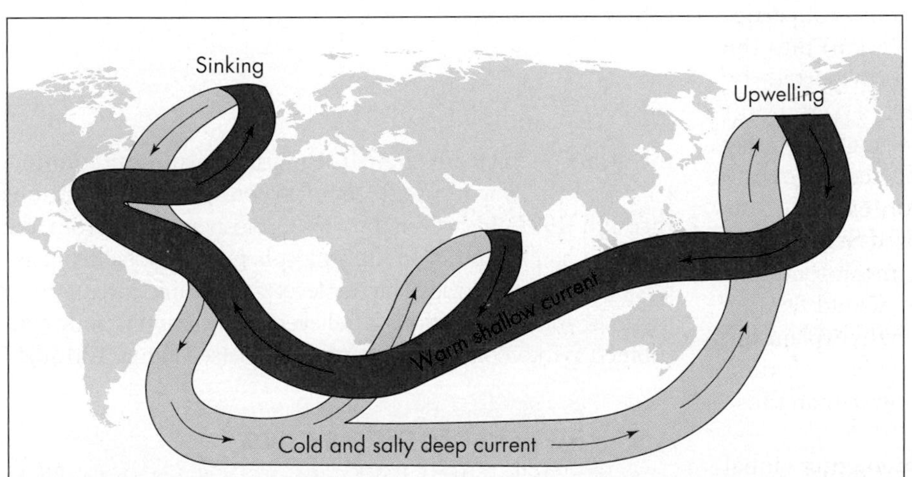

▲ FIGURE 12.17 **OCEAN CONVEYOR BELT** Idealized diagram of the ocean conveyor belt. The actual system is more complex, but, in general, warm surface water (red) is transported westward and northward (increasing in salinity owing to evaporation) to near Greenland, where it cools from contact with cold Canadian air. As the water increases in density, it sinks to the bottom and flows south, then east to the Pacific, where upwelling occurs. The masses of sinking and upwelling waters balance, and the total flow rate is about 20 million m^3 (700 million $ft.^3$) per second. The heat released to the atmosphere from the warm water helps keep northern Europe warmer than it would be if the oceanic conveyor belt were not present. *(Modified after Broecker, W. 1997. Will our ride into the greenhouse future be a smooth one?* Geology Today *7[5]:2–6)*

temperature (see Figure 12.9). When CO_2 has been high, temperature has also been high, and, conversely, low concentrations of CO_2 have been correlated with a low global temperature. However, in order to better understand global warming, we need to consider major forcing variables that influence global warming, including solar emission, volcanic eruption, and anthropogenic input.

SOLAR FORCING

Since the sun is responsible for heating Earth, solar variation should be evaluated as a possible cause of climate change. When we examine the history of climate during the past 1000 years, the variability of solar energy plays a role. Examination of the solar record reveals that the Medieval Warm Period (A.D. 1000–1300) corresponds with a time of slightly increased solar radiation, comparable to that which we see today. Evaluation of the record also suggests that minimum solar activity occurred during the fourteenth century, coincident with the beginning of the Little Ice Age (see Figure 12.13d). Therefore, it appears that variability of the input of solar energy to Earth can partially explain climatic variability during the past 1000 years. However, the effect is relatively small, only 0.25 percent; that is, the difference between the solar forcing from the Medieval Warm Period to the Little Ice Age is only a fraction of 1 percent.[23,24] Brightening of the sun is unlikely to have had a significant effect on global warming since the beginning of the Industrial Revolution.[25]

▲ FIGURE 12.18 **VOLCANIC ERUPTION TEMPORARILY COOLS CLIMATE** This eruption of Mount Pinatubo in the Philippines in 1991 injected vast amounts of volcanic ash and sulfur dioxide aerosols up to about 30 km (19 mi.) into the stratosphere *(US Geological Survey/Denver)*

VOLCANIC FORCING

Upon eruption, volcanoes can hurl vast amounts of particulate matter, known as aerosols, high (15–25 km) into the atmosphere. The aerosol particles are transported by strong winds around Earth. They reflect a significant amount of sunlight and produce a net cooling that may offset much of the global warming expected from the anthropogenic greenhouse effect.[14,26] For example, increased atmospheric aerosols over the United States (from air pollution) are probably responsible for mean temperatures roughly 1°C (1.7°F) cooler than they would be otherwise. Aerosol particle cooling may, thus, help explain the disparity between model simulations of global warming and actual recorded temperatures that are lower than those predicted by models.[27]

Volcanic eruptions add uncertainty in predicting global temperatures. For instance, consider the cooling effect of the 1991 Mt. Pinatubo eruption in the Philippines (Figure 12.18). Tremendous explosions sent volcanic ash to elevations of 30 km (19 mi.) into the stratosphere, and, as with similar past events, the aerosol cloud of ash and sulfur dioxide remained in the atmosphere, circling Earth, for several years. The particles of ash and sulfur dioxide scattered incoming solar radiation, resulting a climatic forcing of about 3 W/m^2, cooling Earth about 2.3°C (4°F) in 1991 and 1992. Calculations suggest that aerosol additions to the atmosphere from the Mt. Pinatubo eruption counterbalanced the warming effects of greenhouse gas additions through 1992. However, by 1994, most aerosols from the eruption had fallen out of the atmosphere, and global temperatures returned to previous higher levels.[28] Volcanic forcing from pulses of volcanic eruptions is believed to have significantly contributed to the cooling of the Little Ice Age (see Figure 12.13d).[24]

ANTHROPOGENIC FORCING

- Evidence of anthropogenic climate forcing, resulting in a warmer world, is based, in part, on the following:
- Recent warming over the past few decades of 0.2°C (0.4°F) per decade cannot be explained by natural variability of the climate over recent geologic history.
- Industrial age forcing of 1.6 W/m^2 (see Figure 12.15) is mostly due to emissions of carbon dioxide that, with other greenhouse gases, have greatly increased in concentration in the past few decades (see Figure 12.16).

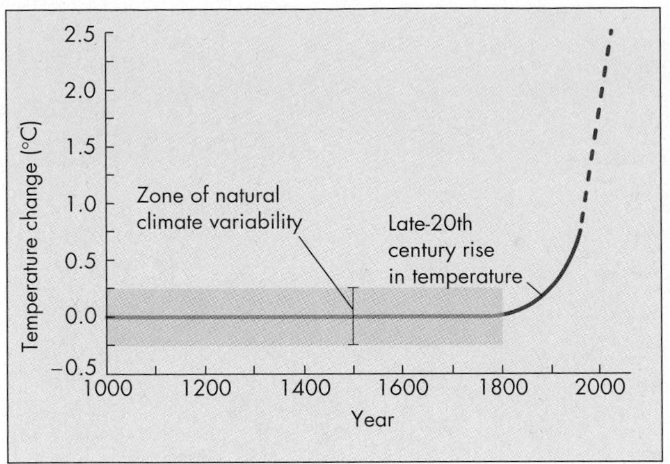

▲ FIGURE 12.19 **LATE-TWENTIETH-CENTURY RISE IN TEMPERATURE** Global temperature change during the past 1000 years. This graph was generated by a computer model that removed the effects of solar and volcanic forcing so you could visualize the impact of anthropogenic forcing from greenhouse gases. *(Modified after Crowley, T. J. 2000. Causes of climate change over the past 1000 years.* Science *289:270–277)*

- Climate models suggest that natural forcings in the past 100 years cannot be responsible for what we know to be a nearly 1°C (1.8°F) rise in global land temperature (see Figure 12.4). When natural and anthropogenic forcing are combined, the observed changes can be explained (Figure 12.15).

Figure 12.19 suggests that present warming greatly exceeds the natural variability (climate forcing) and closely agrees with the response predicted from models of greenhouse gas forcing.[14]

Human processes are also causing a slight cooling. Reflection from air pollution particles (aerosols) has reduced incoming solar energy by as much as 10 percent. This is termed *global dimming*. Negative forcing from aerosols in the industrial age is -1.4 W/m^2 (Figure 12.15) and may be offsetting up to 50 percent of the expected warming resulting from greenhouse gases.

12.6 Potential Effects of Global Climate Change

Our discussion of global warming can be summarized as follows: (1) Human activity is increasing the concentration of greenhouse gases in the atmosphere, (2) the mean temperature of Earth increased by about 0.8°C (1.4°F) in the past 100 years, and (3) a significant portion of the observed increase in the mean temperature of Earth results from human activity.

All climate models are consistent in predicting that warming will continue, as a result of the greenhouse gases now in the atmosphere, and possibly accelerate in the coming decades. That holds true even if greenhouse gas emissions by people are nearly eliminated—warming of about 0.5 to 1.0 °C (0.9 to 1.8°F) would still occur in coming decades. Therefore, we need to carefully examine the potential effects of such warming.

If carbon dioxide in the atmosphere doubles from preindustrial levels, as expected, it is estimated that the average global temperature will rise about 1.5° to 4.5°C (2.6° to 7.8°F), with significantly greater warming at the polar regions.[14] Specific effects of this temperature rise in a specific region are difficult to predict, but two of the likely possibilities are change in the global climate pattern and a rise in sea level (Figure 12.20), resulting from the expansion of seawater as it warms and the partial melting of glacial ice.

GLACIERS AND SEA ICE

There is concern that global warming is resulting in accelerated melting of glacial ice of the Greenland ice sheet (Figure 12.21 and Figure 12.22) and mountain glaciers. The latter are of particular importance in Europe and South America as water resources for people and ecosystems farther down the mountain. In Bolivia, a 70-year-old ski resort on a glacier is all but shut down because the small glacier on which the ski runs are built has just about disappeared.

Many glaciers in the Pacific Northwest of the United States are currently retreating (Figure 12.23); those in Europe, China, and Chile are retreating or decreasing more, in terms of thickness of ice (typically 10 to 30 percent of ice thickness since 1977), than are advancing or increasing in thickness.[29,30] The increase in the number of retreating glaciers is accelerating in the Cascades, Switzerland, and Italy. Evidently, this acceleration is in response to a mean global temperature that has averaged 0.4°C (0.68°F) above the long-term mean temperature during the years from 1977 to 1994. On Mt. Baker in the Northern Cascades, for example, all eight glaciers were advancing in 1976. By 1990, all eight were retreating. In addition, 4 of 47 alpine glaciers observed in the Northern Cascades have disappeared since 1984. Most of the glaciers in Glacier National Park may be gone by 2030, those of the European Alps by the end of the century.[29, 31]

Melting of the ice of the Greenland ice sheet has doubled since about 1998. Melting produces surface water that flows through openings and fractures (crevasses) and down to the base of the glacier, where it lubricates the bottom of the ice, resulting in an acceleration of glacial movement. Most glaciers in the southern half of the ice sheet are accelerating and losing ice more rapidly than previously. In 2005 alone, about 200 km^3 (50 mi^3) of glacial ice was lost.[29]

When glacial ice melts and bare ground is exposed, there is a positive feedback because white ice reflects sunlight (absorbing less sunlight), whereas darker rock reflects less sunlight while absorbing more. The more ice that

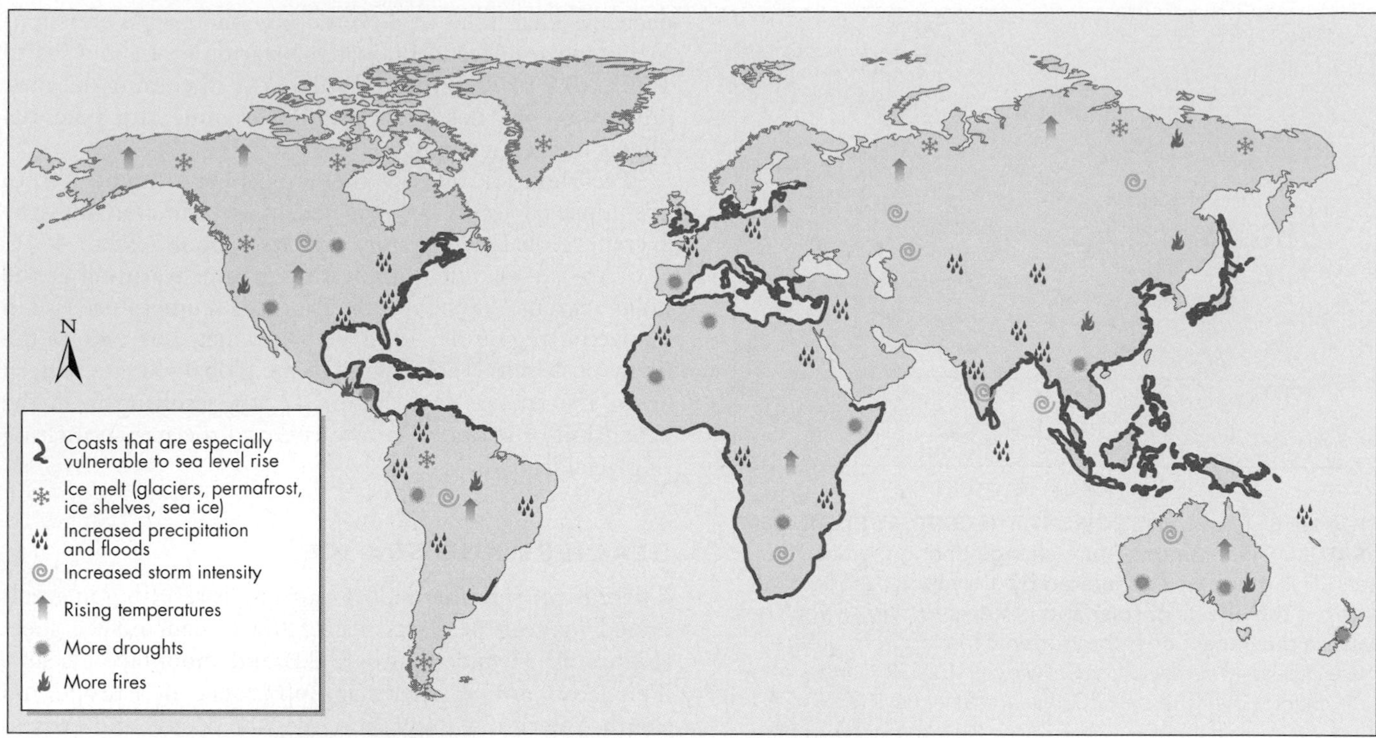

▲ FIGURE 12.20 **PROJECTED DISASTER-RELATED IMPACTS OF CLIMATE CHANGE** General overview of hazards that may be associated with climate change. Events will be spread across the globe and include sea-level rise, flooding, ice and snow melt, increased precipitation and storm intensity, rising temperatures, and more droughts and fires. *(Modified from Abramowitz, J. 2001.* Unnatural disasters. *Worldwatch Paper 158, Washington, DC: Worldwatch Institute, based on data from R. T. Watson, et al. 1998.* The regional impacts of climate change: *An assessment of vulnerability. Special Report of IPCC Working Group II, ed. J. J. McCarthy et al., Cambridge, England: Cambridge University Press;* Climate change 2001: Impacts, adaptation, and vulnerability: Contribution of Working Group II to the Third Assessment of the Intergovernmental Panel on Climate Change. *Cambridge, England: Cambridge University Press; and Revenga, C., et al. 1998.* Watersheds of the world. *Washington, DC: World Resources Institute and Worldwatch Institute)*

◀ FIGURE 12.21 **GREENLAND ICE SHEET IS SHRINKING** This summer meltwater stream on the Greenland ice sheet illustrates that surface melting is one of several ways that ice is being lost from the large Greenland ice sheet. The ice sheet is also losing large amounts of ice through outflow glaciers shedding icebergs into the ocean. *(H.Jay Zwally, Ph D./Science Magazine)*

▲ FIGURE 12.22 **INCREASED MELTING OF GREENLAND ICE SHEET** Between 1992 (salmon colored) and 2005 (salmon and red color combined) the portion of the Greenland ice sheet that experienced surface melting during the summer increased considerably. The area of greatest increase was in the southern third of Greenland where the entire ice sheet experienced surface melting during summer months. *(Courtesy Russell Huff and Konrad Steffan/ University of Colorado CIRES)*

(a)

(b)

▲ FIGURE 12.23 **MOST GLACIERS ARE RETREATING** Global warming is the primary cause for retreat of most of the world's glaciers, like Muir Glacier in Glacier Bay National Park, Alaska, shown here. Photograph (a) was taken on August 13, 1941, *(US Geological Survey, Denver)* and (b) was taken from the same vantage on August 31, 2004. Between 1941 and 2004, this glacier retreated more than 12 km (7 mi.) and thinned by more than 800 m (875 yd.). Vegetation in the foreground of (b) is now growing where there was only bare rock in 1941. *(US Geological Survey, Denver)*

melts, the faster are the warming and increased melting. This is a classic *positive feedback cycle*—warming leads to more warming. This explains, in part, why warming at higher latitudes and elevations may exceed that in other areas.[14]

Sea ice in the Arctic Ocean is declining in area. Since the 1970s, when satellite remote sensing became possible, ice coverage in September (when sea ice is at a minimum) has declined about 11 percent per decade. The lowest extent of sea ice was in 2007 (Figure 12.24). The rapid decline in 2007 was partly in response to atmospheric circulation that favored melting. There is concern that the Arctic is in a transitional phase toward a seasonal ice-free condition by 2030.[32]

The Antarctic Peninsula (the relatively narrow land that points toward South America) is known to be one of the most rapidly warming regions on Earth. New studies suggest that the warming extends well beyond the peninsula to include most of West Antarctica. Although the trend is partly offset by East Antarctic's cooling in the autumn, the trend for Antarctica is positive climate forcing, resulting in warming. The causes of warming are complex, but they require changes in atmospheric circulation with temperature changes (warmer sea surface water and sea ice). The most likely direct cause is anthropogenic greenhouse gas forcing, resulting from increasing concentrations of greenhouse gases in the atmosphere.[33]

As Earth warms, more snowfall on Antarctica is predicted. Satellite measurement from 1992 to 2003 suggested that the East Antarctic ice cap on Antarctica increased in mass during the period of measurement by about 50 billion tons per year [34] (a very small amount, given the enormous size of Antarctica—less than 0.5 percent of Antarctica, covered to a depth of 1 m). Another study, using computer simulation and ice core records, reported that there has been no statistically significant increase in snowfall over Antarctica since the 1950s, suggesting that there is no reduction of sea level rise by increased Antarctic snowfall, nor an increase in mass of glacial ice that would store water.[35] What is clear is that Antarctica is a complex place where more basic data research is needed to better understand the consequences of global warming.

A series of particularly warm years or an apparent increase in melting or breakup of glacial ice is not proof of global warming resulting from humans' burning of fossil fuels. Nevertheless, evidence to that effect is consistent with global warming and is a source of concern.

CLIMATE PATTERNS

Global warming might change the frequency and intensity of violent storms as warming oceans feed more energy into the atmosphere. More or larger coastal storms will increase the hazard of living in low-lying coastal areas, many of which are experiencing rapid growth of human populations. Global and regional climatic changes also have an effect on the incidence of hazardous natural events, such as storm damage, landslides, drought, and fires, as is dramatically illustrated by El Niño. **El Niño** is a natural climatic event that occurs on an average of once every few years, most recently in 1997–1998. El Niño is both an oceanic

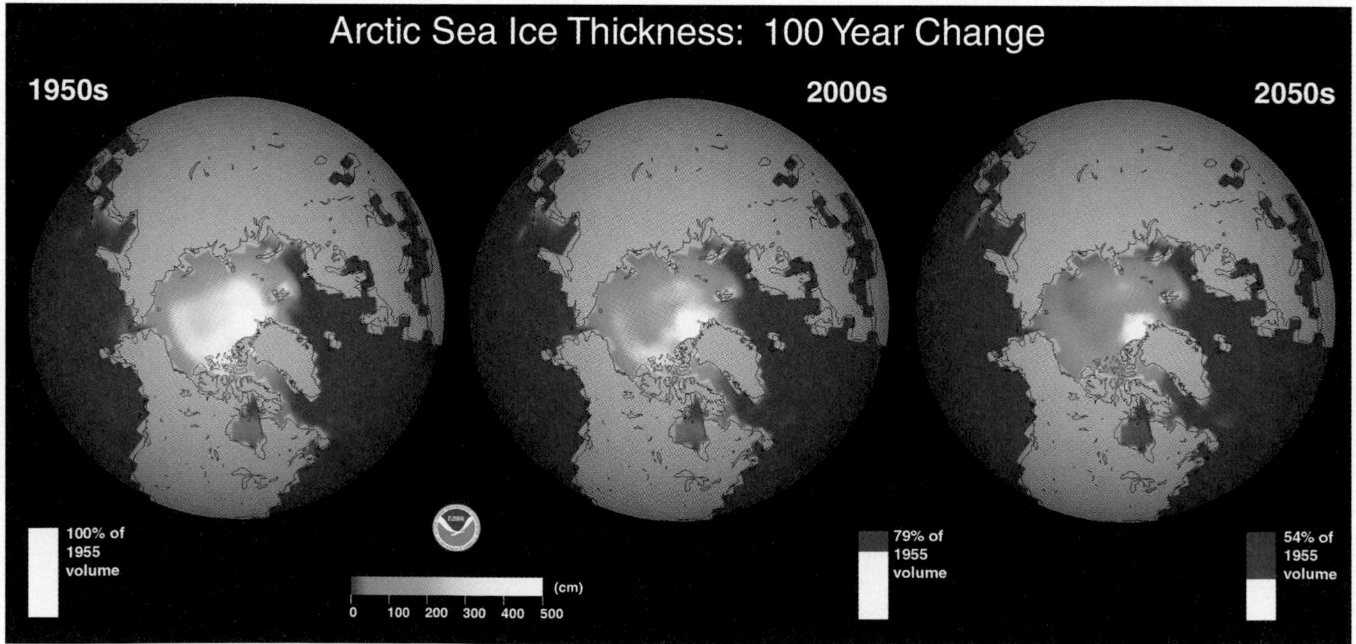

▲ FIGURE 12.24 **ARCTIC SEA ICE (MINIMUM) 1953–2007** Rapid decline in 2007 was partly due to atmospheric circulation that favored melting. *(National Geophysical Data Center.)*

A CLOSER LOOK 12.1

El Niño

El Niño events probably start from random, slight reductions in the trade winds that, in turn, cause warm water in the western equatorial Pacific Ocean to flow eastward (Figure 12.A). This change further reduces the trade winds, causing more warm water to move eastward, until an El Niño event is established.[36] El Niños are thought to bring about an increase in some natural hazards on a nearly global scale by putting a greater amount of heat energy into the atmosphere. The heat energy in the atmosphere increases as more water evaporates from the ocean to the atmosphere. The increase of heat and water in the atmosphere produces more violent storms, such as hurricanes.

Figure 12.B shows the extent of natural disasters that have been attributed, in part, to the El Niño event of 1997–1998, when worldwide hurricanes, floods, landslides, droughts, and fires killed people and caused billions of dollars in damages to crops, ecosystems, and human structures. Australia, Indonesia, the Americas, and Africa were particularly hard hit. There is some disagreement about how much damage and loss of life is directly attributable to El Niño, but few disagree that it is significant.[37,38] We do not yet completely understand the cause of El Niño events. They occur every few years, to a lesser or greater extent, including the large El Niño events of 1982–1983 and 1997–1998.

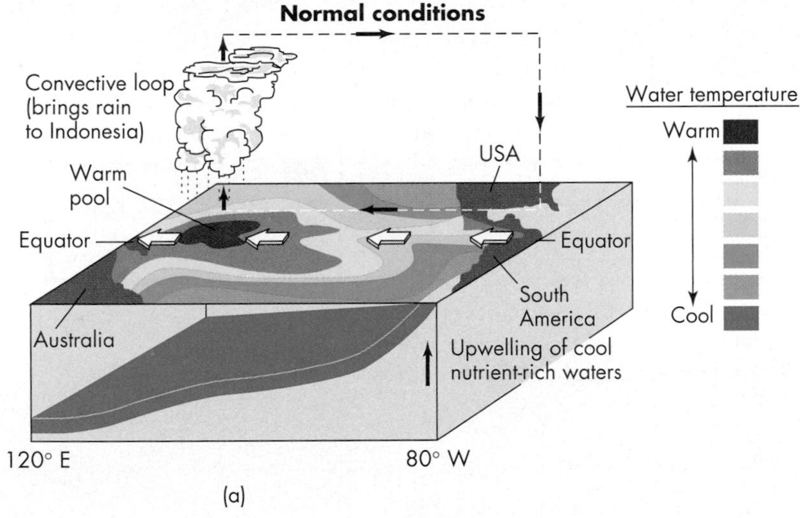

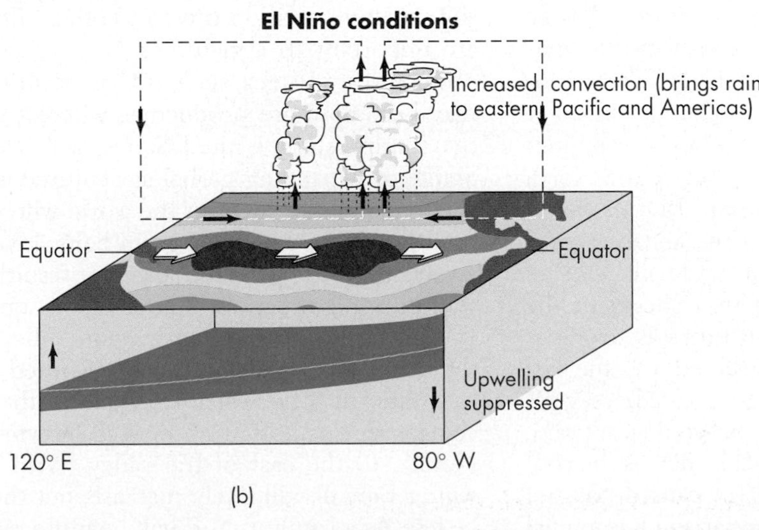

FIGURE 12.A EL NIÑO Idealized diagram contrasting (a) normal conditions and processes with (b) those of El Niño. *(Modified after National Oceanic and Atmospheric Administration.* www.elnino.noaa.gov/. *Accessed 3/3/99)*

(continued)

El Niño events can cause havoc by increasing the occurrence of hazardous natural events, and there is concern that human-induced climatic change may, through global warming, produce more and stronger El Niño events in the future. This effect would result, in part, because, as we burn more fossil fuels and emit more greenhouse gases into the atmosphere, the oceans will continue to warm, and differential warming in various parts of the ocean may increase the frequency and intensity of El Niño events in the Pacific Ocean.

The opposite of El Niño is La Niña, in which eastern Pacific waters are cool, and droughts, rather than floods, may result in Southern California. The alternation from El Niño to La Niña is a natural Earth cycle that has only recently been recognized.[39]

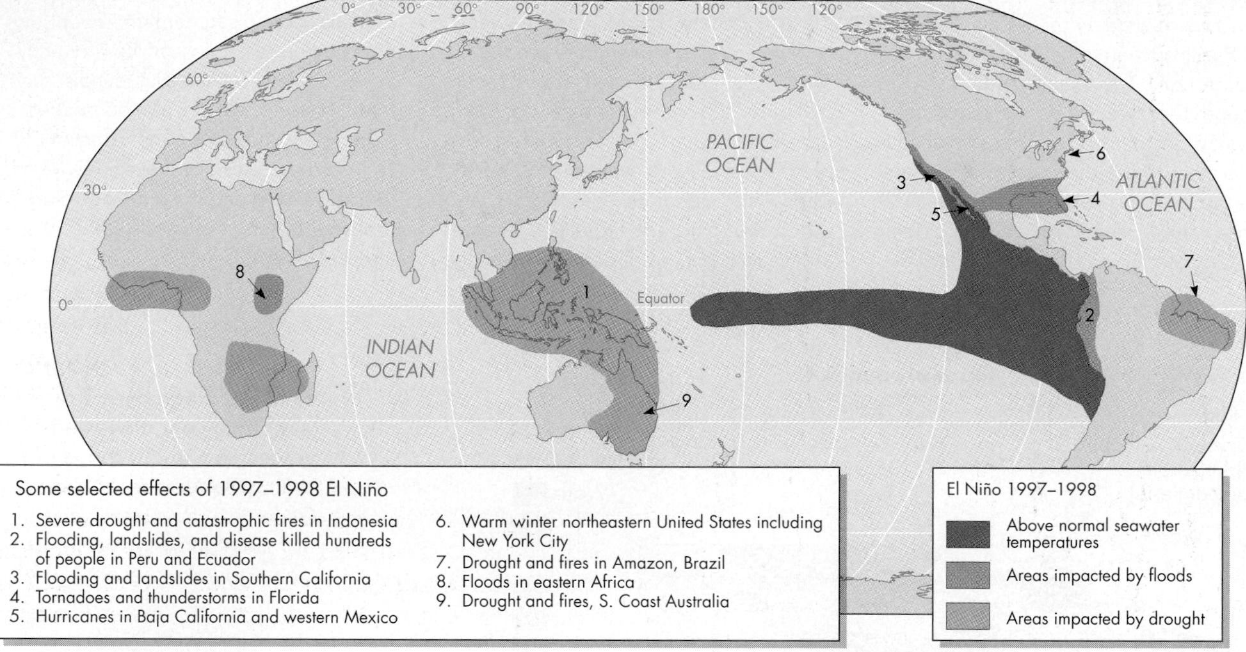

▲ FIGURE 12.B **THE 1997–1998 EL NIÑO EVENT** Map shows the general extent of the El Niño effects and the regions damaged by floods, fires, or drought. *(Data from NationalOceanic and Atmospheric Administration, 1998)*

and an atmospheric phenomenon, involving unusually high surface temperatures in the eastern equatorial Pacific Ocean and droughts and high-intensity rainstorms in various places on Earth (see A Closer Look 12.1). The oceans of the world have oscillations related to changes in water temperature, air pressure, storms, and weather over periods of a few years to decades. Oscillations occur in the North Pacific (the Pacific Decadal Oscillation, PDO), South Pacific, Indian, and North Atlantic oceans and can (over short periods of a few years to a decade) influence climate more than human-caused global change. Effects of the oscillations in any given year can be 10 times as strong as the long-term warming that we have observed over the past century. Yearly variability in temperature can be several degrees, producing a series of warm or cool years and, sometimes, a much warmer or cooler year, such as the cool winter of 2009–2010. A cool year in 1911 caused Niagara Falls in New York to freeze over. By comparison, the annual increase of warming attributed mostly to human activity is about two hundredths of a degree Celsius per year.

Global rise in temperature will change rainfall patterns, soil moisture relationships, and other climatic factors that are important to agriculture. It has been predicted that some northern areas, such as Canada and Eastern Europe, may become more productive, whereas others, including the southwestern United States, will become more arid. Stable or expanding global agricultural activities are crucial to people throughout the world who depend upon the food grown in the major grain belts. Hydrologic changes associated with climatic change resulting from global warming might seriously affect food supplies worldwide. In California and the San Joaquin Valley, where much of the fruits and vegetables for the United States are raised less water may be available. Much of the irrigation water comes from runoff from spring snowmelt in the Sierra Nevada to the east of the valley. With global warming, winter rainfall will likely increase, but the snow pack will be less. As a result, runoff will be more rapid, filling reservoirs before it is needed to irrigate crops. Water may have to be released from filled reservoirs and lost to the sea.

A lesson learned from the Medieval Warming Period 1000 years ago is that warming and reduced rainfall in the southwestern United States, Mexico, Central America, and parts of Africa, China, India, and Southeast Asia may result in prolonged (intergenerational) drought that will threaten water supply and the ability to grow crops. As a result, if long droughts occur with global warming, the most serious potential threat coming at a time in world history when human population is the greatest ever is the threat that we will not have an adequate supply of food for the expected 9 to 10 billion people by 2050.

SEA-LEVEL RISE

A rise of sea level is a potentially serious problem related to global warming (see Case Study 12.2). Global ocean temperature to a depth of at least 3 km has increased since 1961. The oceans of the world, in that time, have absorbed about four-fifths of the heat that has been added to Earth's climate system.

Warming of the oceans causes the water to expand (thermal expansion), producing a rise in sea level. The rate of sea-level rise resulting from thermal expansion from 1961 to 2003 is estimated to be about 0.42 ± 0.2 mm/year, increasing to about 1.6 ± 0.5 mm/year from 1993 to 2003. Thus, it appears that sea level rise, resulting from warming of ocean water, is accelerating.[14]

Melting of glacial ice is thought to be contributing to sea-level rise. Melting of sea ice (floating ice) does not cause a rise in sea level (you can verify this by placing ice in a glass of water and observing the water level in the glass while waiting for the ice to melt). The estimated increase in sea level from melting glaciers is estimated to be about 1.38 ± 0.4 mm/year from 1961 to 2003, increasing to about 1.5 ± 0.6 mm/year from 1993 to 2003.[14]

Several tentative conclusions may be put forth concerning rise in sea level:

- Both thermal expansion and melting glacial ice contribute significantly to the observed sea-level rise since 1961.
- The difference between the observed and estimated rise in sea level is considerable, suggesting that additional research is needed to better understand sea-level rise.
- The rates of both thermal expansion and melting glacial ice are accelerating.
- The Greenland ice sheet's contribution to sea-level rise has increased about 4 times in recent decades, consistent with surface observations of melting glacial ice.

Estimates of the rise expected in the next century vary widely, from approximately 18 to 59 cm (7 to 23 in.), and precise estimates are not possible at this time. However, a 40-cm (16-in.) rise in sea level would have significant environmental impacts. Such a rise could cause increased coastal erosion on open beaches of up to 80 m (260 ft.), rendering buildings and other structures more vulnerable to waves generated from high-magnitude storms. In areas with coastal estuaries, a 40-cm (16-in.) sea-level rise would cause a landward migration of existing estuaries, again putting pressure on human-built structures in the coastal zone. Communities would have to choose between making substantial investments in controlling coastal erosion or allowing beaches and estuaries to migrate landward over wide areas.[14,40,41]

Rise in sea level is already threatening some small islands in the tropical Pacific Ocean. The island nation of Tuvalu consists of about 9 atolls (Figure 12.25) that developed as rising sea level since the last glacial maximum about 20,000 years ago overtopped a degraded carbonate platform. The highest place on the island is only about 4.5 m (15 ft.) above sea level. People first inhabited the island of Funafuti and other atolls of Tuvalu about 2000 years ago when stable inlets formed.[42] Recent sea-level rise is now threatening Funafuti, especially during high tides, which enhance wave attack and flooding. During high tide flood events (Figure 12.26), people report water bubbling up from the ground, contributing to the flooding of low-lying areas. With continued sea-level rise, the future ability of Funafuti and other atolls to support people is uncertain. The 12,000 people on Funafuti could become the first to be displaced by rising sea level.

The coastline of northwest and north Alaska is experiencing rapid erosion from the rise in sea level, melting, permafrost soils, and loss of summer sea ice that can protect a coastline from erosion.[43] Some small islands may disappear, and some villages have already been moved inland.

WILDFIRES

There is a complex relationship between climate change and fires, and the two phenomena interact on a variety of levels. Global warming is predicted to lead to an increase in both droughts and El Niño events, which will set the

▲ FIGURE 12.25 **FUNAFUTI ATOL** Part of the island nation of Tuvala in the tropical Pacific Ocean, 3000 km (1850 mi.) northeast of New Guinea. *(Global Warming Images/Alamy)*

SURVIVOR STORY 12.2

Residents of the Maldive Islands

The Maldives, a nation of nearly 1200 small coral islands in the Indian Ocean, cannot afford to see ocean levels rise.

The vast majority of these islands, part of atolls that are tropical paradises famous for their beaches of white sand, prismatic coral gardens, and crystalline lagoons, lie at an extremely low elevation—the average elevation is less than a meter above sea level. As global warming continues to cause sea levels to rise, much of this nation could face a watery burial beneath the waves.

This low elevation is already proving problematic. Because they lie so close to the sea, the Maldive Islands, about 200 of which are inhabited for a total population of more than 300,000 people, are particularly susceptible to tsunamis and storm waves. In addition to the physical damage, such hazards can spell disaster for the country's economy—nearly 20 percent of the labor force is in the fishing industry, and tourism accounts for a staggering one-third of the country's gross national product (GNP).

Long-time president of the Maldives, Maumoon Abdul Gayoom, has for more than two decades now been a vocal advocate for an international curb on emissions of carbon dioxide and other gases that contribute to global warming, particularly among major industrial nations such as the United States. The nation even has a permanent mission to the United Nations headquartered in New York.

But the Maldives are not about to wait out the results of diplomatic efforts while the islands slowly submerge. After the capital city, Malé, which currently has about 80,000 residents, suffered large-scale damage from storm waves in the late 1980s, the Japanese government provided funding for a massive seawall surrounding the island (Figure 12.C). The single seawall, measuring 3 m (10 ft.) tall, costs millions of dollars, a cost that is prohibitively high for the many other inhabited islands.

Should sea levels rise significantly in the next decades—even a rise of 1 m (3 ft.) would spell disaster for much of this archipelago—the only option for many of the residents will be evacuation.

Although experts do not entirely agree on the exact threat that climate change poses to this island nation, Gayoom and others from this small country continue to pressure the United Nations to address the problem now, before the Maldives become a modern-day Atlantis.

—CHRIS WILSON

◀ FIGURE 12.C **AERIAL VIEW OF MALÉ IN THE MALDIVE ISLANDS** A seawall surrounds most of Malé, capital of the Maldives. Constructed to protect the island from waves up to 2 m (7 ft.) high, the seawall cost about $4000 per square meter. Seawalls have been built on only a few of the 1192 islands that make up this country. Approximately 80 percent of the land area of the Maldives is less than a meter above mean sea level. *(Peter Essick/Aurora Photos, Inc.)*

▲FIGURE 12.26 **FLOODING OF FUNAFUTI, TUVALA, During High Tides of February 2005.** *(Global Warming Images/Alamy)*

stage for wildfires. For example, scientists in Spain studied a strip of Mediterranean coastline between 1968 and 1994, comparing the number of forest fires to the temperature and aridity of the area. They found that as the coastal area warmed and became more arid, both the number of fires and the size of the area burned increased.[44] Another investigation, a computer study, used four different global circulation models (GCMs) for the atmosphere. Historical weather data were input into the models to project forest fire danger levels in Canada and Russia under a warmer climate. These models suggested that there will be more frequent and severe forest fires, and that the number of years between successive fires in a given location will decrease.[44,45]

CHANGES IN THE BIOSPHERE

A growing body of evidence indicates that global warming is initiating a number of changes in the biosphere, threatening both ecological systems and people. An *ecosystem* is a community of species and their nonliving environment in which energy flows through the system and chemicals cycle. Sustained life is a function of ecosystems, not individual species. Changes to ecosystems resulting from global warming include risk of regional extinction of species in the system as land and hydrology changes (for example, forested land changes to grassland). Other changes include shifts in the range of plants and animals, with a variety of potential consequences: For example, mosquitoes carrying diseases, including malaria and dengue fever in Africa, South America, Central America, and Mexico, are migrating to higher elevations; butterfly species are moving northward in Europe; some bird species are moving northward in the United Kingdom; subalpine forests in the Cascade Mountains in Washington State are migrating to higher meadows; alpine plants in Austria are shifting to higher elevations; sea ice melting in the Arctic is placing stress on seabirds, walruses, and polar bears; and warming and increasing acidity of shallow water in the Florida Keys, Bermuda, Australia's Great Barrier Reef, and many other tropical ocean areas is believed to be contributing to the bleaching of coral reefs.[14,45,46] (see A Closer Look 12.3).

Global warming is likely to impact many ecosystems in North America over the next 100 years for the following reasons:[45]

- Climate change may be accelerating.
- Warming is expected to be 2° to 4°C (3.6° to 7.2°F).
- Precipitation in some regions is projected to be less frequent but more intense.
- The temperature of streams and rivers will likely increase.
- Wildfires will be more frequent.
- Growing seasons will be lengthened, with earlier spring and greater primary productivity, especially at higher latitudes.
- Rainfall and wind speed from hurricanes and other violent storms are likely to increase as warming continues
- The oceans, especially at water depths less than about 1 km (0.6 mi.) are warming and becoming more acidic.

As a result,

- Some species will experience stress. Most vulnerable will be those that are not mobile, such as some vegetation on land and shellfish in the ocean.
- Many species will migrate toward higher altitudes in an attempt to adapt to warming.

ADAPTATION OF SPECIES TO GLOBAL WARMING

With global warming, temperature and precipitation change in different regions, and this stresses ecosystems. During the past 25 years or so, plants and animals have shifted their ranges by about 6 km (3.8 mi.) per decade toward the polar areas. In addition, spring is arriving earlier, and plants, as a consequence, are blooming earlier, frogs are breeding earlier, and migrating birds are arriving earlier in various places. The rate of change has been about 2.3 days per decade. In addition, it has been reported that tropical pathogens have moved up in latitude and elevation, affecting species that may not be adapted to them. For example, in Costa Rica, more than 60 species of frogs may have gone extinct because of warmer temperatures that have affected their immune system response to a lethal fungus that has taken advantage of the warmer temperature. Climate change may have caused the first extinction of a mammal, a species of opossum that has apparently disappeared from Queensland, Australia. It was reported that the opossum lived only at an elevation of 1000 m (3280 ft.) and was susceptible to being exposed to higher temperatures, even for a short period of time.[50]

Marine Ecosystems and Climate Change

Approximately one-third of carbon dioxide emissions from human activities enter the oceans of the world. This decreases the atmospheric concentration of carbon dioxide, reducing global warming, but at a cost to the oceans. Dissolved carbon dioxide and water form carbonic acid ($CO_2 + H_2O \rightarrow H_2CO_3$). The acid dissociates to bicarbonate ion HCO_3^- and hydrogen ion H^+. As H^+ increases, the water becomes more acidic (Figure 12.D) and lowers the pH; Over the past 15 years, ocean acidity has increased about 30 percent. The increased acidity of the ocean and ecosystem is fundamentally changing ocean chemistry structure and function by stressing marine organisms from plankton to coral and shellfish as they attempt to adapt to increasing acidity. The stress for some species may be manifested by reduced growth rates and reproduction. These changes could occur if organisms need to expend increased energy to maintain their internal pH, resulting in less energy for growth and reproduction. How individual species and ecosystems will respond to future increased acidification is a complex problem that is the subject of increasing research.[47,48] Increased acidity of sea water may make it more difficult for some marine organisms that produce carbonate shells, such as oysters, coral, and some plankton, to draw carbonate from the seawater to make their shells.

In response to the warming atmosphere over the past 100 years, heat has been transferred to the upper water of the oceans. As a result, the average temperature of the upper 700 m (2300 ft.) of the sea has increased by about 0.6°C (1°F) over the past century. Temperature increase has a fundamental effect on biological processes of marine organisms through increased metabolic rates that impact ecosystem processes of energy expenditure and chemical cycling. Marine organisms have the ability to adapt to temperature change within a particular range of temperatures. If the range is exceeded, the ability to adapt is reduced and populations may decline or be driven to local extinction.[49] Mobile species such as fish can swim to other areas or deeper water where the temperature is closer to what they prefer. Those species tied to the bottom such as shellfish, may slowly migrate if change is sufficiently slow. The change of temperature and acidity of seawater appears to be accelerating, and that may make adaptation more difficult for some species.

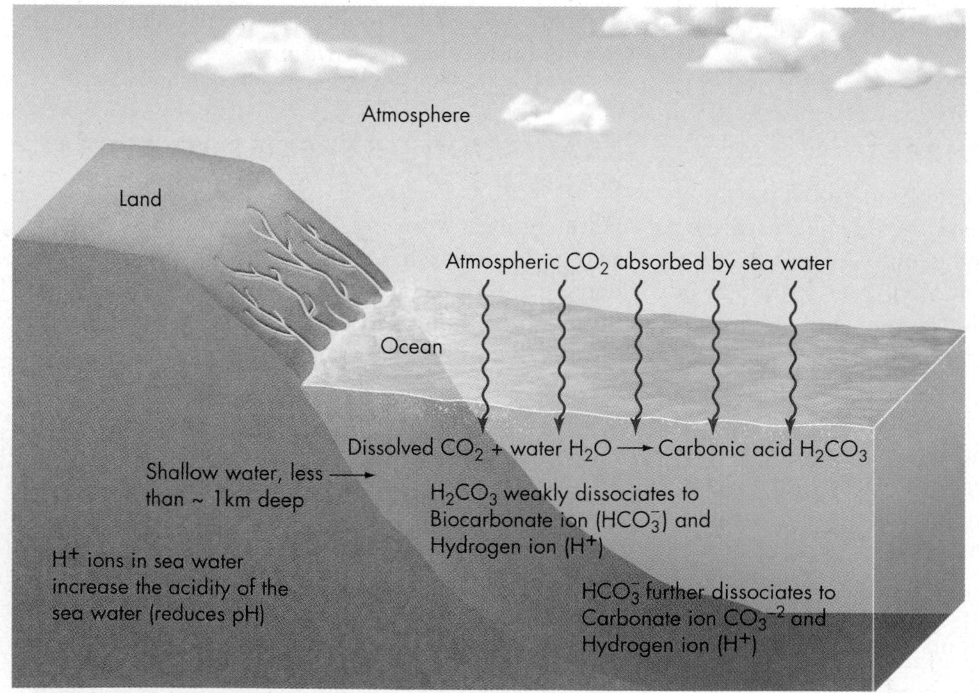

◀ **FIGURE 12.D**
CHANGES IN SEA WATER
Idealized diagram showing processes that changes the acidity of seawater.

A problem with past evaluations of the potential threats to species or their possible regional extinction was that they were based on simple models that primarily considered temperature and precipitation, perhaps along with soil type and hydrology, that determine, in part, where a particular species lives. This information would then be put into a standard climate model to predict where a particular species might migrate in order to avoid adverse effects of climate change. We now know that these simple envelope models that use a few climate and topographic variables are not sufficient, and newer models that incorporate additional biological elements, including competition among species and genetics, are being used to evaluate evolutionary response to climate change.[50,51]

A controversial suggestion is that we might assist migration of some species that are unable to migrate with climate change. To some, this is a drastic, controversial step that may be unacceptable because of the risk of, essentially, creating an invasive species. Assisted migration worries ecologists and conservation biologists because they have spent a lot of time and effort working against invasive species, some of which cause ecosystem problems and the extinction of other species. Before seriously considering assisted migration, much more research is needed to better understand the habitats of threatened and endangered species, including where they can do well, what might threaten them, and what they might threaten.[50]

12.7 Predicting the Future Climate

Predicting the future climate is problematic because predicting the future has always been difficult. People who make predictions have often been wrong. For climate and its effects on living things, we can attempt to apply the geological concept of *uniformitarianism*, which assumes that processes that occurred in the past occur today and will occur in the future. This approach has led to investigations of atmospheric concentrations of greenhouse gases in the geologic record, which we have explored in this chapter. However, the time that is of particular interest to us is the one that is the most difficult to collect climate data from. That time is the past few centuries to 1000 years ago, during the scientific and technological revolution.

The problem we face is that temperature records have been kept, at best, for only about 150 years, and the early records are from only a few places. Until the advent of satellite remote sensing, measurements of air temperature over the oceans were taken only where ships traveled, and this did not provide the kind of sampling that satisfies statisticians. Many places on Earth have never had good, long-term, ground-based temperature measurements. So, when we want to know what the temperatures were like in the 19 centuries before carbon dioxide concentrations began to rise from the burning of fossil fuels, experts have to find ways to extrapolate, interpolate, and estimate. The Hadley Meteorological Center in Great Britain is an outstanding example of a group of scientists who are attempting to reconstruct such temperature records from the mid-nineteenth century to the present. The situation has improved greatly in recent years, with the establishment of ocean platforms with automatic weather-monitoring equipment, coordinated by the World Meteorological Organization. Thus, we have records that are particularly good since about 1960. What is emerging from the climate data is that the warming we have experienced over the past few decades exceeds that of anytime in the past 400 years. This is supported by a large variety of reliable data. There is less confidence in temperature reconstructions from about A.D. 950 to A.D. 1250 because the data are sparse and not yet quantified. This period of sparse data includes the Medieval Warming Period (MWP) from about A.D. 950 to A.D. 1250. Limited data suggest that some specific locations during the MWP may have been as warm as or even a bit warmer than today. Data available for most specific locations suggest that the latter part of the twentieth century was warmer than the MWP. Prior to A.D. 900, there is little confidence in global mean surface or water temperatures because uncertainties associated with the proxy evidence for past temperatures are large over both space and time.[13]

12.8 Strategies for Reducing the Impact of Global Warming

An important unsolved question about climate change being vigorously researched is the following: What is the origin of rapid climate change over decades to about 100 years that we believe has happened during geologic time (past 1,000,000 years)? Two other important questions concerning the Earth climate system and people are (1) What changes have occurred? and (2) What changes could occur in the future? Answering these questions requires geologic evaluation of prehistoric changes, and the prediction, through modeling and simulation, of future change (Figure 12.27). Because we now know global warming is due, in part, to the increased concentration of greenhouse gases, reduction of these gases in the atmosphere is a primary management strategy. This was the subject of the 1997 United Nations Framework Convention on Climate Change in Kyoto, Japan. The objective of the convention was to produce an international agreement to reduce emissions of greenhouse gases, especially carbon dioxide. The United States originally agreed to reductions but, in 2001, refused to honor the agreement, much to the disappointment of other nations, especially its European allies. As a result, the leadership in controlling global warming has shifted from the United States to the European Union. The Kyoto Protocol was signed by 166 nations and became a formal international treaty in February 2005. To date the United States has not

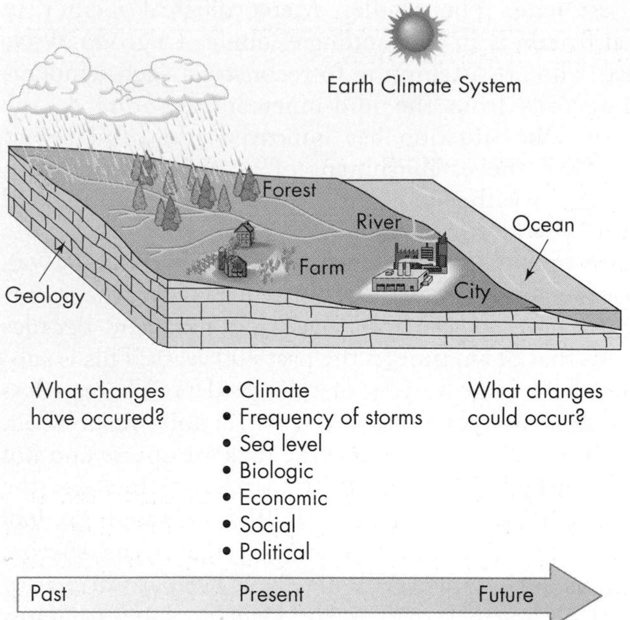

▲ FIGURE 12.27 **TWO BIG QUESTIONS CONCERNING EARTH'S CLIMATE SYSTEM LINKED TO PEOPLE AND ENVIRONMENT** What changes have occurred? What changes could occur? *(Modified after International Panel on Climate Change 2001, at www.ipcc)*

signed any international agreements to address climate change. A comprehensive energy plan that would have included climate change was killed in the Senate in the summer of 2010. Powerful lobbies from the fossil fuel industry combined with some senators from states with abundant fossil fuels or those who fear changing energy policy or do not believe global warming from burning fossil fuels is occurring used a filibuster to stop the legislation that would have initiated a more rapid change to alternative energy sources.

We are faced with a dilemma with respect to burning fossil fuels. On one hand, fossil fuels are vital to our society and necessary for continued economic development and growth, as well as for human well-being. On the other hand, scientific evidence suggests that burning fossil fuels is contributing significantly to global warming. Burning fossil fuels is linked to environmental problems that include a rise in sea level, increased surface temperatures, air pollution, and increased frequency and intensity of storms, such as hurricanes. Impacts associated with fossil fuels depend upon how much the global temperature will actually rise. Estimates of temperature rise from climate models range from about 2° to 4°C. (3.6° to 8.2° F)[14]. If the increase is close to 2°C, we can probably adapt with minimal disturbance. If the increase in temperature is near the high end, then significant impacts are likely. One way to estimate potential increase in temperature that is independent of the model predictions is to examine the geologic record for past change. A study of the geologic record from ocean sediments deposited during the past several hundred thousand years suggests that warming in the next century will be about 5°C, consistent with models that predict warming as great as 4.5°C[52] If global temperature rises 4°C, then we will need a strategy to reduce emissions of carbon (about 50 percent over several decades) to avoid serious environmental disruptions that would result because, at this level of temperature increase, significant environmental impacts are much more likely. Also, it is important to recognize that following the peak of emissions of carbon dioxide, it will probably take several hundred years or longer before temperature and sea level stabilize (Figure 12.28). Therefore, the sooner we act to reduce emissions, the lower the impacts will be, and the sooner we will be on a path to stabilizing climate change. Even if carbon emissions were reduced to zero, warming will continue during this century. There is 0.5° to 1.0°C warming in the Earth climate system now.

Assuming that the reduction of carbon emissions is necessary to reduce impacts of global warming, we need to take action. The reduction must occur at a time when human population and energy consumption are increasing. There are several ways that we might reduce emissions of carbon dioxide into the environment: improved engineering of fossil fuel–burning power plants (develop and use coal-burning power plants that sequester the carbon and place it in safe storage); use those fossil fuels that release less carbon into the atmosphere, such as natural gas, which, on burning, releases less carbon dioxide than does coal or oil (Table 12.4); conserve energy to reduce our dependence on fossil fuels; use more alternative energy sources; and store carbon in Earth's systems, such as forests, soils, and rocks below the surface of Earth.[53–56]

Storage of carbon in plants, soils, and the ocean has received considerable study. The option of sequestration of carbon in the geological or rock environment has received less attention. Sequestration is attractive because the residence time of carbon in the geologic environment is

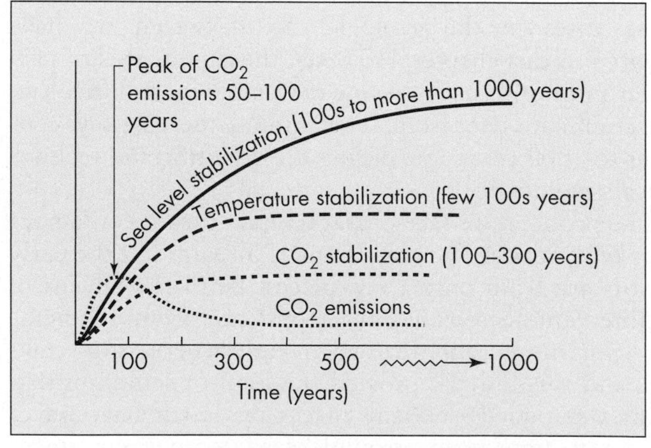

▲ FIGURE 12.28 **LAG TIMES TO STABILIZATION** Diagram indicates lag times to stabilization of CO_2, temperature, and sea level, following peak in emission. *(Modified after International Panel on Climate Change 2001, at www.ipcc)*

TABLE 12.4

Comparison of Common Fossil Fuels in Terms of Percentage of Total Carbon Released

Fossil Fuel	Carbon Emitted as % of Total Carbon from Fossil Fuels	Comparisons
Coal	36	Releases much more carbon per unit of energy than oil or gas.
Oil	43	Releases 23% less carbon per unit of energy than coal.
Natural Gas	21	Releases 28% less carbon per unit of energy than oil or gas.

Source: Data from Dunn, J. 2001. Decarbonizing the energy economy. World Watch Institute State of the World 2001. *New York: W. W. Norton and Company. Chapters IV, V.*

generally long, potentially thousands to hundreds of thousands of years.

The general principle of geologic sequestration of carbon is fairly straightforward. The idea is to capture carbon dioxide from power plants and industry and inject the carbon into the subsurface geologic environment. Two geologic environments have received considerable attention in this regard. The first is sedimentary rocks that contain salty water. These rocks, known as saline aquifers, are fairly widespread at numerous locations on Earth and have large reservoir capacity with the potential to sequester many years of human-produced carbon dioxide emissions. They could provide the necessary time to transition to an energy economy not dependent on fossil fuels.[53] The second (but related) geologic environment for sequestration of carbon is depleted oil and gas fields. The process of placing carbon dioxide in the geologic environment involves compressing the gas to a mixture of liquid and gases and then injecting it underground, using wells. Injecting carbon dioxide into depleted oil and gas fields has an added advantage, in that the carbon dioxide is not only stored but also serves as a way to enhance recovery of remaining oil and gas in the reservoir. The injected carbon dioxide helps move the oil toward production wells.

A demonstration project is now ongoing in Saskatchewan, Canada. The Weyburn oil field started production in the 1950s and was considered to be depleted. However, with enhanced recovery and storage of carbon dioxide, production is likely to last for several more decades. The source of carbon dioxide for the Weyburn field comes from a coal-burning plant in North Dakota, by way of a pipeline that delivers several thousand tons of carbon per day.[54]

Another carbon sequestration project is located beneath the North Sea. Carbon dioxide is injected into a salt-water aquifer located below a natural gas field. The project, which started in 1996, injects about 1 million tons of carbon dioxide into the subsurface environment each year. It is estimated that the entire facility can hold about the amount of carbon dioxide that is projected to be produced from all of Europe's fossil fuel plants in the next few hundred years.[53] The cost of sequestering the carbon beneath the North Sea is high. However, it saves the company from having to pay taxes for emitting carbon dioxide into the atmosphere. Finally, pilot projects to demonstrate the potential and usefulness of storing carbon in the United States have been initiated in Texas, beneath depleted oil fields. The good news is that immense salt aquifers are common beneath many areas of the United States, including the Gulf Coast, Texas, and Louisiana, and the potential to store carbon is immense.[55]

Recent studies suggest that global warming is probably not an immediate emergency, and we will have a decade or so to develop alternatives to continued intensive burning of fossil fuels. If we decide that we must stabilize the concentration of atmospheric carbon dioxide in the future, it will be necessary to go through a transition from fossil fuel energy sources to alternative sources that produce much less carbon dioxide. However, based on studies of glacial ice cores from Greenland, there are indications that significant climatic change may occur quickly, perhaps in as short a time as a few years. A quick natural or human-induced warming or cooling is probably unlikely, but, if one does occur, the potential impacts could be fast and serious (see A Closer Look 12.4). Human civilization began and has developed into our present highly industrialized society in only about 7000 years. That period has been characterized by a relatively stable, warm climate that is probably not characteristic of longer periods of Earth history. It is difficult to imagine the human suffering that might result late in the twenty-first century from a quick climate change to harsher conditions, when there are as many as 9 to 10 billion or more people on Earth to feed.[21]

12.4

Abrupt Climate Change

Abrupt climate change is defined as a large-scale change in the global climate system that takes place over a few decades or less. Rapid changes are thought to have occurred in the past, but little is known about them or why they happened. Such change is anticipated to persist for at least a few decades and will cause substantial disruptions to both human and natural systems.[57] Several types of abrupt climate change that could cause a serious risk to humans and the natural environment in terms of our ability to adapt include

- A rapid change of sea level as a result of changes in the glaciers and ice sheets.
- Droughts and floods, resulting from widespread rapid changes to the hydrologic cycle.
- Abrupt change in the pattern of circulation of water in the Atlantic Ocean that is characterized by northward flow of warm, salty water in the upper layers of the ocean.
- A rapid release of methane (a strong greenhouse gas) to the atmosphere from both melting permafrost and the ocean's sediment.

The U.S. Climate Change Science Program has addressed these four potential abrupt changes.[57]

One of the major questions is whether there will be an abrupt change in sea level. We know that even small changes in sea-level rise may have significant repercussions for society, with serious economic impacts, including coastal erosion as well as an increase in coastal flooding and a loss of coastal wetlands. The present climate models do not adequately capture all aspects of sea-level changes resulting from melting glacial ice. However, given this, a concern exists that projections for sea-level rise in the future, as presented by the Intergovernmental Panel on Climate Change,[14] probably underestimate the amount of sea-level rise during the twenty-first century.

The second question involves the potential for abrupt changes of the hydrologic cycle. Of particular concern are those changes that affect water supply, especially through protracted, long droughts. Of particular significance is the recognition that droughts can develop faster than people and society are able to adapt to them. Thus, droughts that last from several years to a decade or more have serious consequences to society. It has been pointed out that long droughts have occurred in the past and are still likely to reoccur in the future, even in the absence of global warming, as a result of increased greenhouse-gas forcing.[57]

A third question concerning abrupt change involves the Atlantic Ocean circulation system, which carries warm water to the North Atlantic, where it becomes saltier and sinks to become a cold water bottom current that moves south. The warm water is partly responsible for keeping Western Europe more comfortable and hospitable. However, it is primarily the large expanse of the ocean itself that accomplishes this. It is expected that the strength of current may decrease approximately 25 to 30 percent as a result of the global warming in the twenty-first century. However, it appears unlikely that the ocean current system in the Atlantic will undergo a collapse or an abrupt transition to a weakened state in the next 100 years. [57]

A final question concerning abrupt change is whether or not there will be a rapid change in atmospheric methane. This is a significant question, because methane is a strong greenhouse gas that, if greatly increased in concentration in the atmosphere, would accelerate global warming. It is generally concluded that a very rapid change in the release of methane during the next 100 years or so is very unlikely. However, ongoing warming will increase emissions of methane from both ocean sediments and wetlands. Wetlands in the northern high latitudes are particularly more susceptible to releasing additional methane because there is accelerated warming, along with enhanced precipitation, in permafrost areas that contain a lot of stored methane. Thus, it appears that, in future decades, methane levels will likely increase and cause additional warming.[57]

In conclusion, it appears that abrupt climate change during the next century is unlikely, and, thus, we will probably have time to respond to potential adverse consequences of global warming. However, time is growing short (because it takes time to initiate change in policy), and a serious response to global warming from all countries is necessary in the near future.

A better understanding of past abrupt changes in climate, through the collection and analysis of geologic data, is necessary to fully understand what may cause future change. Geologic data provide the most direct evidence of past change. Geologic data from sediments and glacial ice, along with monitoring, are assisting in understanding the causes of long-term changes in ocean and glacial conditions and how these are linked to atmospheric response. With this, we may be better able to forecast both long- and short-term droughts that have serious consequences to humans and the natural environment.

PROFESSIONAL PROFILE 12.5

Sally Benson—Climate and Energy Scientist

Sally Benson (Figure 12.E) has spent her career researching solutions to the most pressing environmental problems of our time. In the mid-1970s, as a young scientist with Lawrence Berkeley National Laboratory, she tackled the first oil shortage by investigating ways to harness the power of geothermal energy.

Ten years later, she was elbow-deep in the mud of Kesterson Reservoir in California's Central Valley. Irrigation runoff had caused selenium from local soil to accumulate in the water, causing local birds to hatch chicks with horrifying deformities. Benson's studies made her realize that microbes could be environmentally friendly tools to clean up other sites with toxic metal contamination.

"The experience was so positive because I was doing this cutting edge science at the same time regulators had to make a decision about how to clean up the site. I got an idea of the impact that science can make, and how research could get results," Benson said.

By the mid-1990s, Benson was directing all earth science research at the laboratory. "I talked with many people, read a lot, and came to the conclusion that climate change was the most significant issue facing the world," she said. Benson organized research programs that developed regional models of climate change to help residents plan for droughts and temperature shifts and that studied the carbon cycle of the oceans, among other projects.

In 2007 Benson was appointed director of the Global Climate and Energy Project at Stanford University, which seeks to develop energy sources that release fewer greenhouse gases. "Energy efficiency in lighting, heating, and cooling systems makes sense in the world. But at the end of the day, we need to do a lot more than that. If our current understanding is correct, we need to cut overall emissions by 80 percent of today's levels," Benson said.

Benson sees many promising ways to achieve that goal. One is renewable energy. "Today's biofuels don't provide much advantage in terms of carbon dioxide emissions. But alternatives such as cellulosic ethanol (producing ethanol with plant fibers), where there are low emissions in the process of growing and making them, are incredibly important," Benson said.

Another is to continue using some fossil fuels to run power plants and ships, but to capture the greenhouse gases they emit. "More than half of the electricity worldwide is produced by burning coal, a plentiful resource in many places. We need to find a way to make it carbon neutral," Benson said. One way to remove such emissions from the atmosphere altogether would be to inject them in aquifers deep within the earth. Benson has studied ways to use technology developed by the oil and gas industries to select sites where the rock offers a good seal, inject the gas, and monitor the area for leaks.

"I get to do important work solving critical problems, get to be outdoors, do experiments at scale, and have fun while I do it. Interacting with the huge amount of people impacted by these issues makes the earth sciences very rewarding," Benson said.

—KATHLEEN WONG

▲ FIGURE 12.E **DR. SALLY BENSON** Professor Benson is executive director of Stanford University's Global Climate and Energy Project, a $225 million effort to develop energy resources that are not harmful to the environment, especially energy sources that release fewer greenhouse gases. Like many successful scientists, Dr. Benson has a multidisciplinary background with degrees in geology, materials science, and minerals engineering, and with applied research in hydrology and reservoir engineering. *(Sally M. Benson)*

REVISITING THE FUNDAMENTAL CONCEPTS

Climate Change

1. **Hazards Are Predictable from a Scientific Evaluation**
2. **Risk Assessment Is an Important Component of Our Understanding of the Effects of Hazardous Processes**
3. **Linkages Exist between Natural Hazards**
4. **Hazardous Events That Previously Produced Disasters Are Now Producing Catastrophes**
5. **Consequences of Hazards May Be Minimized**

1) The study of climate change is an area of intense research. Most of our data comes from studying past climate, but climate models are being used to predict what may happen in the future. Assuming we will be able to accurately predict global change and, in particular, global warming in the future, then that information can assist us in gaining a better understand of hazards associated with climate change.

2) Risk assessment associated with climate change is fairly straightforward once we know how much change we're actually talking about. For example, predictions have been made concerning rise in sea level that can be compared to maps of existing low-lying areas and islands, and, from that, the risk to these areas can be evaluated. The risk is the probability of an event occurring times the consequences. For example, if we're fairly confident that sea level will rise a few tens of centimeters in the coming century, then we can assess the risk in terms of which areas would be inundated in the future. Risk associated with processes such as prolonged drought is more difficult. Prolonged droughts are an aspect of global change we are very concerned about because widespread droughts in areas where our food is grown could lead to major famine in a number of areas around the planet. We know that during the Medieval Warming Period, when areas of the southwestern United States and South America experienced droughts, some civilizations collapsed. The collapse was not specifically from the drought, but the loss of water resources with growing population results in less availability of food, and that certainly played a role in the collapse of several civilizations.

3) Global warming is linked to a number of natural hazards. For example, as the oceans warm and sea levels rise, there will be greater pressure on the shorelines and coastal erosion hazards will increase. Discussion and ongoing research has tried to evaluate linkages between sea-level rise and temperature of seawater and intensity of storms, from thunderstorms to hurricanes. Linkages also exist between water processes and global change. As the Earth warms, some places will become drier and others wetter, and this will change the nature of the runoff, as well as the magnitude and frequency of flooding.

4) All of the hazard events associated with global change are likely to have greater consequences in the future because human population continues to increase. As a result, more of us are vulnerable to change in water resources that are linked to food supply. Droughts of, say, 50 years ago would have an effect on far fewer people than if a prolonged drought occurs again now or in the future.

5) Consequences of global change may be minimized. However, most scientists think that we need to work on minimizing the hazard sooner rather than later. Consequences are not likely to be rapid, and most changes will be gradual. However, gradual changes can lead to serious consequences. Probably the best way to minimize global climate change consequences is to attack the process itself and control emissions of greenhouse gasses. Beyond this, some coastal areas may need to be defended, particularly on shores where large urban centers exist. Minimizing the effects of prolonged drought will be difficult, given the huge increase in human population. However, it is prudent to consider worst case scenarios and what the implications of prolonged drought in various parts of the world are likely to be. Then we might plan for the amount of grain and other foodstuffs necessary to increase our surplus of food supplies to better withstand the effects of warming.

Summary

The main goal of the emerging integrated field of study known as Earth system science is to obtain a basic understanding of how our planet works and how its various components, such as the atmosphere, oceans, and solid Earth, interact. Another important goal is to predict global changes that are likely to occur within the next several decades. The global changes involved include temperature, or climate, changes and the resulting changes in seawater and on land. Because these are short-term predictions, Earth system science is relevant to people everywhere.

Methods of studying global change include examination of the geologic record from lake sediments, glacial ice, and other Earth materials; gathering of real-time data from monitoring stations; and development of mathematical models to predict change.

Climate refers to characteristic atmospheric conditions, such as precipitation and temperature over seasons, years, and decades. The atmosphere is a dynamic, complex environment in which many important chemical reactions occur.

Human activity is contributing significantly to global warming. The trapping of heat by the atmosphere is generally referred to as the greenhouse effect. Water vapor and several other gases, including carbon dioxide, methane, and chlorofluorocarbons, tend to trap heat and warm Earth because they absorb some of the heat energy radiating from Earth. The effects of a global rise in temperature include a rise in global sea level and changes in rainfall patterns, high-magnitude storms, soil moisture, agriculture, and the biosphere.

Natural climate-forcing mechanisms that may cause climatic change include Milankovitch cycles, solar variability, and volcanic activity. Anthropogenic causes include air pollution and increase in greenhouse gases, especially carbon dioxide. We now understand that global climate can change rapidly over a time period of a few decades to a few hundred years. Two important questions are (1) What is the nature and extent of past climate change? and (2) What climate changes will occur in the future?

The science of global warming is well understood. Human-induced global warming is occurring. There is no reason for gloom and doom, but we need to take appropriate action soon to slow or stop global warming and associated environmental consequences.

Adjustments to global warming will range from adapting to change to reducing emissions of carbon dioxide and sequestration of carbon. Several different, simultaneous adjustments are likely.

Key Terms

aerosol (p. 409)
climate (p. 409)
climate forcing (p. 423)
Earth climate system (p. 410)
Earth system science (p. 408)
El Niño (p. 430)
glacier (p. 411)
global warming (p. 419)
greenhouse effect (p. 420)
Pleistocene Epoch (p. 412)
proxy data (p. 414)
weather (p. 409)

Did You Learn ?

1. What is the distinction between climate and weather?
2. What is the basis for climate classification?
3. List the different forms of water in the atmosphere.
4. What is climate proxy data and why is it important?
5. How do aerosols influence climate? Where do they come from?
6. Explain or diagram the general global pattern of atmospheric circulation.
7. What is an interglacial?
8. What are the potential causes of continental glaciation?
9. How can glaciers be hazardous?
10. What is climate forcing and how does it work?
11. Explain or diagram the greenhouse effect. Describe the types of radiation that are involved and the wavelength of the radiation.
12. List the major greenhouse gases and their natural and anthropogenic sources. Which gases are enhancing the greenhouse effect the most?
13. What is the Milankovitch effect? What are the three changes in the Earth's orbit that cause this effect?
14. Explain solar, volcanic, and anthropogenic forcing.
15. Describe the natural hazards associated with climate change.
16. How is climate change likely to affect the weather?
17. Which processes are contributing the most to rising sea level?
18. How will climate change affect the biosphere?
19. How is drought related to global warming?
20. Which parts of our planet are experiencing the greatest effects of climate change?
21. What are the projected changes to our climate for the remainder of this century?
22. How have international agreements dealt with ozone depletion and climate change?
23. What are the proposed methods for carbon sequestration? Which method holds the greatest promise and why?
24. Explain El Niño and its effects.
25. What changes should individuals, the private sector, and government make to reduce greenhouse gases?

Critical Thinking Questions

1. In this chapter, we discussed some possible effects of continued global warming. Which types of hazards would you expect to increase in your area as the climate changes? Would you expect to see any decrease? Think about the ways that climate change might alter the lives and lifestyles of people living in your area. Do you see any evidence of climate change where you live?
2. Assessing the rate and cause of change is important in many disciplines. Have a discussion with a parent or someone of similar age and write down the major changes that have occurred in his or her lifetime and in your lifetime. Characterize these changes as gradual, abrupt, surprising, chaotic, or another descriptive word of your choice. Analyze these changes and discuss which ones were most important to you personally. Which of these changes affected your environment at the local, regional, or global level?
3. How do you think climate change is likely to affect you in the future? What adjustments will you have to make? What can you do to mitigate the effects?
4. Some people, for cultural, political, religious, or other reasons, do not accept the conclusions that climate change is primarily caused by human activities. What do you see as the basis for their opinions or beliefs? How do you think that they might be convinced otherwise? If you share their opinions or beliefs, indicate why you do and what it would take to convince you that climate change is the result of human activities.

Assignments in Applied Geology: Snowpack Monitoring

THE ISSUE

At high elevations, the winter's snowfall accumulates into "snowpack." A frozen reservoir of sorts, the snowpack stores water until temperatures rise to trigger melting. Geologists continually monitor snowpack levels to determine how much water they contain for a number of important reasons. First, in many areas, the snowpack runoff collects in reservoirs and supplies much of the nearby residents' yearly water supply. (If you live near the Sierra Nevada mountain range, for instance, you are probably familiar with the snowpack level updates on the nightly news.) However, geologists are also concerned with the rate at which the snowpack will melt; if the melting is rapid, the system of streams, rivers, and reservoirs may not be able to handle the sudden inrush of water, resulting in flooding. As the climate changes, snowpack monitoring will take on added importance in planning for potential hazards and resource availability.

YOUR TASK

Hazard City has been dealing with two surface water problems: (1) spring flooding caused by rapid melting of the Lava Mountain snowpack and (2) water supply shortages during late summer and fall. You have been asked to devise a plan that will more efficiently use Johnson Reservoir to control flooding and to maintain a year-round water supply. This simple goal consists of two elements: (1) entering the spring snowmelt season with a low water level in Johnson Reservoir, which will allow the full reservoir capacity to be used for capturing flood waters and (2) leaving the spring snowmelt season with the water level of Johnson Reservoir as high as is safely possible, which will maximize the amount of water in storage for the long, dry summer.

Although the goal is simple, there is one complexity: The volume of water produced by the spring snowmelt exceeds the capacity of Johnson Reservoir. This means that if the reservoir is filled too early in the snowmelt season, it will not be able to contain any remaining floodwaters. The best way to solve this problem is to allow the initial floodwaters to pass through the reservoir and then, when there is just enough snow on Lava Mountain to "top off" the reservoir, capture all remaining snowmelt water. To do this successfully, reservoir managers need a way to estimate the amount of water tied up in the Lava Mountain snowpack at any given time. Your job is to learn about snowpack monitoring, review the data that have been collected on Lava Mountain, and determine the general pattern of snowpack accumulation and melting.

AVERAGE COMPLETION TIME:

1 hour

13 Wildfires

Wildfires in 2002—Colorado and Arizona

During the past several decades, thousands of people seeking a more relaxed lifestyle than that found in our large cities have moved into rural areas on the fringes of cities—close enough to commute to their jobs, but far enough to be comfortable in their rural community. This population shift has affected the Front Range of the Rocky Mountains southwest of Denver, Colorado, as well as the mountains of Arizona, California, and most other western states. The lives of tens of thousands of residents of these areas were changed abruptly in June 2002, when wildfires ignited by an illegal campfire and by a lost person signaling for help quickly grew out of control. These wildfires became some of the largest and worst fires ever to occur in the region.

The Hayman fire southwest of Denver, Colorado, burned approximately 560 km^2 (138,000 acres or 215 $mi.^2$) of forest. By the third week of June, several hundred structures, including homes and other buildings, had been destroyed, and many thousands of people were evacuated.

What is clear is that our past fire policies have not reduced the wildfire hazard

Two factors promoted rapid growth of the fire: (1) A severe drought, one that was estimated to occur once per 100 years, left the forests very dry and (2) overly dense forests with abundant fuel—the result of nearly a century of suppressing fires.

What is clear is that our past fire policies have not reduced the wildfire hazard. A fire with temperatures that can reach as high as 1100°C (2000°F) moving through the forest cannot be stopped. By the end of June, more than 8100 km^2 (2 million acres or 3100 $mi.^2$) in the western United States had burned, an area significantly larger than the state of Delaware. The Hayman fire (Figure 13.1) was only 1 of about 16 large fires that summer, some of which were even larger.

The story of wildfires in the southwestern United States in 2002 emphasizes some of the fundamental principles involved in living with a natural hazard: First, large wildfires depend on the distribution of fuel; second, the natural process of wildfires is causing disasters and catastrophes today as more people move into brush and forest lands at the wildland/urban interface; and third, our fire management policies need to be reevaluated if we are to minimize destruction by fires.

LEARNING OBJECTIVES

As Earth's population grows, more people are moving into and near brushlands, forest, and other wildlands where wildfires naturally occur. This trend increases the risk of property damage and loss of life from wildfires. Your goals in reading this chapter should be to

- understand wildfire as a natural process that becomes a hazard when people live in or near wildlands.
- understand the effects of fires.
- know how wildfires are linked to other natural hazards.
- know potential benefits provided by wildfires.
- know the methods employed to minimize the fire hazard.
- know the potential adjustments to the wildfire hazard.

◀ **The Mount Calvary Monastery** This monastery, above Santa Barbara, burned during a wildfire in 2008. A grove of eucalyptus trees just downslope may have contributed to the intense fire that claimed the structure. *Edward A. Keller*

◀ FIGURE 13.1 **SMOKE FROM FOREST FIRE CREATES ITS OWN CLOUDS** Type One helicopter water tanker takes off with 3800 liters (1000 gallons) to fight the Hayman fire in Colorado. Large pyrocumulus clouds formed over the fire and grew to an estimated height of 6400 m (21,000 ft.). *FEMA)*

13.1 Introduction to Wildfire

Wildfires are one of nature's oldest phenomena, dating back more than 350 million years ago when trees evolved and spread across the land. Wildfire behavior changed markedly about 20 million years ago when grasses evolved. Grasses provided a new type of fuel that grew quickly and could sustain more frequent fires. Then about 11,500 years ago (beginning of the Holocene Epoch), when the climate warmed after the last major glacial advance, wildfire behavior intensified again. The geologic record shows a significant increase in the amount of charcoal found in sediment from this time, indicating an increase in fires. Part of this increase probably reflects early humans who used fire to clear land and assist in hunting.[1,2]

Before humans evolved, fires that were ignited by lightning strike or volcanic eruption would burn until they ran out of fuel or were extinguished naturally. After a fire, plant regrowth begins from roots, spores, and seeds, and the cycle starts again. This cycle is so ancient that some plants evolved to rely on and use fire to their advantage. For example, oak and redwood trees have bark that resists fire damage, and some species of pine trees have seed cones that open only after a fire. In the Mediterranean climate of Southern California, chaparral (brush) species have adapted to fire and sprout new roots after being burned.

Natural fires started by lightning strikes and volcanic eruptions allowed early humans to harness fire for heat, light, and cooking; these benefits of fire, in turn, helped human populations spread across continents.[1] The ability to use fire for cooking and warmth allowed humans to expand their diet and settle in colder areas. Fires were also set to encourage new growth and attract game. Native Americans used fires as a tool in hunting, warfare, and agriculture. Early Europeans, including members of the famous Lewis and Clark expedition that explored America from coast to coast, commented on the many fires set by people for a variety of purposes. This practice continues today, sometimes with serious consequences (see Case Study 13.1)

13.2 Wildfire as a Process

Wildfire is a self-sustaining, rapid, high-temperature biochemical oxidation reaction that releases heat, light, and other products.[1,2] This reaction requires three things—fuel, oxygen, and heat—that are sometimes visualized as a fire triangle (Figure 13.2). For the chemical reaction to proceed, a fire must have appropriate fuel, adequate oxygen, and external heat that is high enough to start the reaction. If any of the three requirements of this triangle is removed, the fire goes out.

During a wildfire, plant tissue and other organic material are rapidly oxidized and broken down by combustion, or burning. Grass, brush, and forestlands burn because, over long periods of time, these systems establish a balance between plant productivity and decomposition. Microbes alone do not decompose plants fast enough to balance plant growth, so wildfire helps in this decomposition. Therefore, in a sense, the primary cause of fire is vegetation, which has removed carbon dioxide from the atmosphere to produce organic matter by photosynthesis.[4,5] Fire, like the rotting or decomposition of dead plants by bacteria, essentially reverses the process of photosynthesis. In this simplified view of wildfire, carbon dioxide, water vapor, and heat are released when plants burn. Actual combustion, however, is complex and releases numerous chemical compounds in solid, liquid, and gas form. Of the gases

CASE STUDY 13.1

Indonesian Fires of 1997–1998

During the winter of 1997–1998, Indonesia received international attention for severe wildfires that were burning out of control on the islands of Borneo and Sumatra (Figure 13.A). The fires, many deliberately set to clear land, were intensified by a long drought caused by a strong El Niño event. Before they stopped, the fires burned nearly 100,000 km^2 (39,000 mi.^{2}) of land, an area slightly larger than state of Indiana.[3] The dry season began in late May 1997 and by September the fires were virtually beyond the control of firefighters. Although some rain fell in November, the drought intensified in late 1997 and caused fires to rage again in early 1998. These fires were not extinguished until April when heavy rains fell at last.

The fires released so much smoke that Indonesia, Malaysia, and Singapore were blanketed in smog, what governments and news media in the region euphemistically called "haze." This haze even reached as far north as southern Thailand and the Philippines (Figure 13.A), causing health and economic problems throughout the region. Extreme levels of pollution reduced visibility in some places to less than 30 m (100 ft.)[3] and forced many residents in all three countries to wear protective facemasks. In the most severely affected areas, many people developed chronic respiratory, eye, and skin ailments; schools and businesses were forced to close for days to weeks; and many premature deaths were blamed on the haze.[3] Perhaps most tragic, low visibility was thought to be partly responsible for the crash of a Garuda Airlines jet into a mountainside in northern Sumatra, killing all 234 people on board.

The fires also had an adverse effect on the orangutan populations in the area. Some Bornean orangutans, whose populations were already on the decline, were forced out of the forest by the fires and subsequently killed by villagers as they raided fields for food. Furthermore, large areas of the orangutan's preferred habitat were burned, forcing them to adapt to new environmental conditions.[3]

Economic concerns aside, the impact of the wildfires on tropical forest ecosystems, biodiversity, and the atmosphere was so severe that some observers called them a global natural disaster. So what caused these drastic fires? Rainforests, which cover most of Indonesia's 13,500 islands, are generally damp, and fire is not common in their natural regime. However, throughout history, small-scale farmers have used slash-and-burn methods to clear land for cultivation, by cutting down and burning the existing forest or woodlands. More recently, large agribusinesses have been employing the same methods; in addition, oil, lumber, and mining companies have been using fires to clear land for development. These fires are set with the intention of burning only the land designated for clearing. However, in the 1997–1998 fires, a severe drought caused by El Niño provided the perfect setting for the fires to rage out of control. In fact, this instance was the fifth time that wildfires had become unmanageable since El Niño–related fires of 1983.

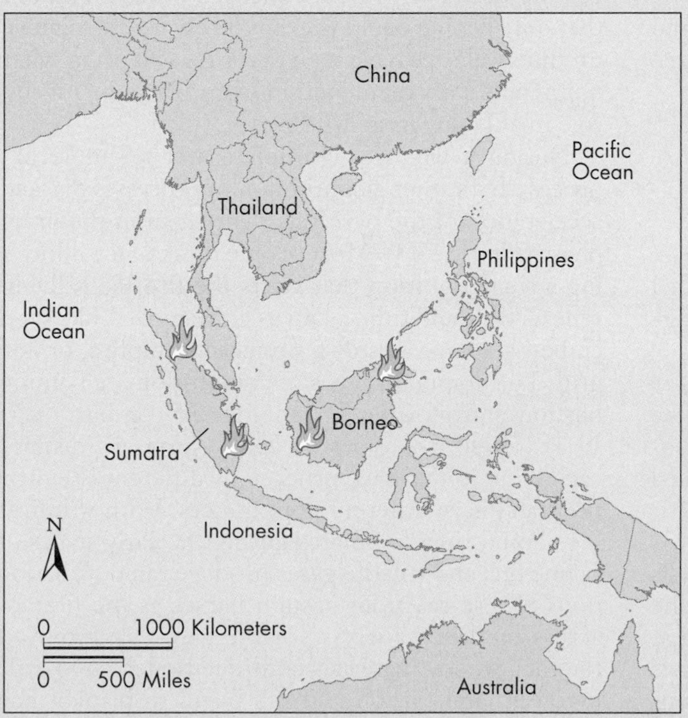

◀ FIGURE 13.A **LOCATION OF INDONESIAN FIRES** Many of the 1997–1998 fires in Indonesia were on the islands of Borneo and Sumatra.

▲ FIGURE 13.2 **THE FIRE TRIANGLE** If any part of this triangle is missing, a fire cannot start; likewise, if any part of the triangle is removed, a fire goes out. *(After Cottrell, W. H. Jr. 2004.* The book of fire, *2nd ed. Missoula, MT: Mountain Press.)*

released, carbon dioxide and water vapor are the most abundant; other gases occur in trace quantities.[1] Common trace gases released are nitrogen oxides, carbonyl sulfide, carbon monoxide, methyl chloride, and hydrocarbons such as methane.[4] These gases, along with solid particles of ash and soot, compose part of the smoke observed during a wildfire. Both ash and soot are powdery residues that accumulate after burning. Ash consists primarily of mineral compounds, and soot is made of unburned carbon.

The image of flames and smoke is accurate as a mental picture of wildfire, but a closer look can identify three phases in a wildfire:

- preignition
- combustion
- extinction

In the first phase, **preignition,** fuel is brought to both a temperature and water content that favors ignition. Preignition involves two processes, preheating and pyrolysis.

In preheating, the fuel loses a great deal of water and other volatile chemical compounds. A *volatile compound* is one that is easily vaporized to a gas. For example, because gasoline contains many volatile chemical compounds, spilled gasoline evaporates quickly to a vapor that you can smell.

The other important process of preignition is **pyrolysis,** which literally means "heat divided." Pyrolysis is actually a group of processes that chemically degrade the fuel. Degradation takes place as heat divides, or splits, large fuel molecules into smaller ones. The products of pyrolysis include volatile gases, mineral ash, tars, and carbonaceous char.[5] You have probably caused pyrolysis when you didn't want to. Pyrolysis takes place when you scorch a piece of toast and turn it black. The burnt toast is covered with char, and smoke coming out of the toaster contains small black droplets of tar. A similar result occurs when you scorch cotton fabric with a hot iron. The two processes of preignition, preheating and pyrolysis, operate continuously in a wildfire. Heat, radiating from flames, causes both preheating and pyrolysis in advance of the fire. These processes produce the first fuel gases, which can ignite in the next phase of a fire.

The second phase, **combustion,** begins with ignition. Combustion marks the start of a set of processes completely different from those related to preignition. Preignition processes absorb energy. In the combustion phase, external reactions involving flaming or glowing liberate energy in the form of heat and light.[2] Although there are many sources of ignition, such as lightning, volcanic activity, and human action, ignition does not automatically lead to a wildfire. In fact, many more ignitions occur than do full-blown wildfires, because wildfires will not become established unless sufficient fuel is present. Wildfires develop when vegetation is mature and has accumulated in sufficient quantities to carry fire across the land.[5] At a time scale that is relevant for grass, brush, and forestlands, ignitions are common; in a period of 50 to 100 years, nearly every acre of land is struck by lightning. On this time scale, wildfires set by humans are not nearly as relevant. If a person starts a large wildfire, he or she is simply preceding a lightning strike that would soon cause the fire anyway. Once wildfire has crossed an area and most of the vegetation has been destroyed, the low fuel supply reduces future ignitions. Fire will not threaten again until there is sufficient new fuel. This argument is contrary to twentieth-century fire management concepts, which commonly held that only people could prevent forest fires. On an ecological time scale, we have seen that fires will occur whether or not people start them, and humans play a relatively minor role in affecting large wildfires.

During a wildfire, ignition is not a simple or single process, but rather a continuum of processes marked by an acceleration of the pyrolysis that began in the preignition phase (Figure 13.3). Most people think of a wildfire as having a single ignition that starts the fire. Actually, ignition repeats time and time again as a fire moves, like sparks and embers popping out of a fireplace, campfire, or barbeque grill. The dominant types of combustion are flaming combustion and glowing or smoldering combustion (Figure 13.3). These two types of combustion are distinct from each other in that they proceed by different chemical reactions and have different appearances. With wildfire, flaming combustion dominates during the early stages of a fire as fine fuel and volatile gases produce rapid oxidation reactions. These reactions sustain flames as the fire advances across the landscape. As volatile gases are removed from the fuel, woody materials continue to decompose through pyrolysis, and carbon and ash begin to blanket new fuel. This blanket of noncombustible material can hinder flaming combustion and lead to glowing or smoldering

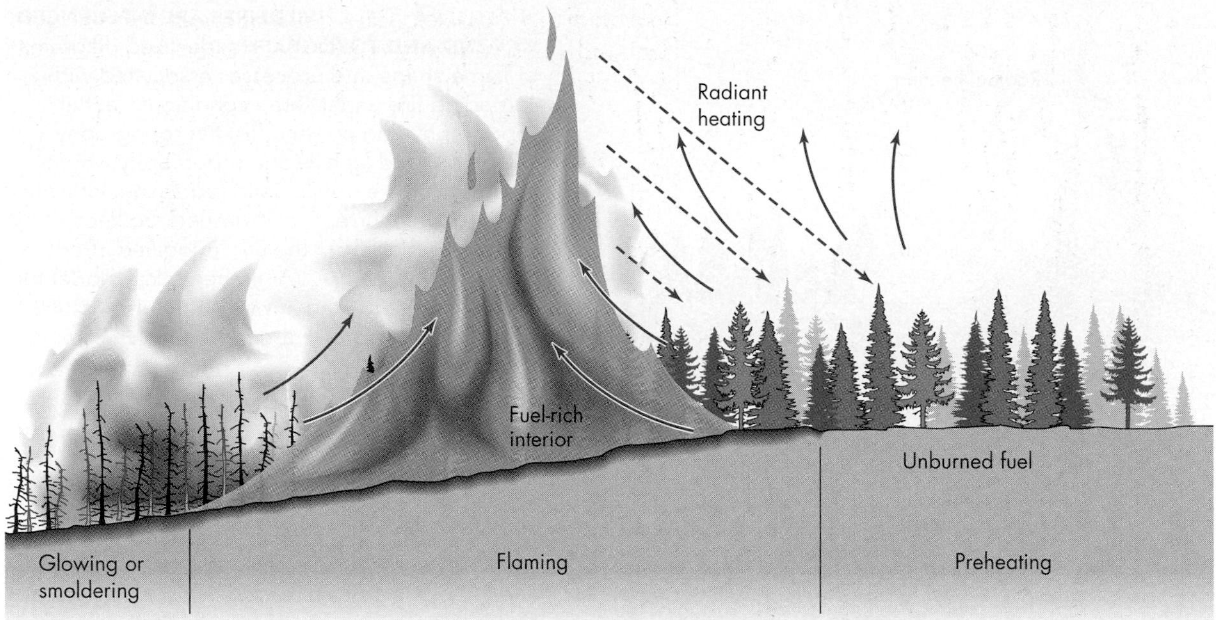

▲ FIGURE 13.3 **PARTS OF A WILDFIRE** Idealized diagram of an advancing wildfire showing phases of a wildfire: (1) an area of pyrolysis where unburned vegetation is preheating, (2) an area of flaming combustion with rising air currents, and (3) an area of glowing or smoldering combustion. Fire is advancing from left to right. Solid arrows indicate motion of air currents and dashed lines show radiant heating from the fire. *(Modified after Ward, D. 2001. Combustion chemistry and smoke. In* Forest fires: Behavior and ecological effects*, ed. E. A. Johnson and K. Miyanishi, pp. 55–77. San Diego, CA: Academic Press)*

combustion instead. The ultimate product of glowing or smoldering combustion can be charcoal. In summary, the two types of combustion are flaming combustion and glowing or smoldering combustion. Flaming combustion is the rapid, high-temperature conversion of fuel to thermal energy and is characterized by a large amount of residual unburned material. Glowing or smoldering combustion can take place at lower temperatures and does not require rapid pyrolysis for its growth.

As wildfire moves across the landscape, three processes control the transfer of heat and as topography and wind direction change, the zone of flaming combustion can take on several forms (Figure 13.4). The three primary ways by which heat is transferred are *conduction*, which is the transmission of heat through molecular contact; *radiation*, heat transfer though electromagnetic waves; and *convection*, heat transfer by the movement of heated gases driven by temperature differences in a fluid, such as air. Wildfires transfer heat dominantly by convection, although radiation, which generates radiant heat, also plays a role (Figure 13.4). Transfer of both convective heat and radiant heat increases the surface temperature of the fuel. As heat is released, the air and other gases become less dense and rise. In wildfires, the oxidation chemical reaction in the fire releases heat, and this heat is transferred by convection of the air. The rising air and other hot gases remove both heat and combustion products from the zone of flaming. This process shapes the fire as it pulls in fresh air that is needed to sustain combustion.[2] Convection thus aids in continually providing the fire with oxygen.

Finally, **extinction** is the point at which combustion, including smoldering, ceases. When there is no longer sufficient heat and fuel to sustain combustion, the fire is considered extinct.

FIRE ENVIRONMENT

The behavior of a large wildfire can be explained by three factors in its environment: fuel, topography, and weather. With sufficient information concerning these three factors, we can better understand and predict wildfire behavior.[5,6]

Fuel Wildfire fuels are complex and differ in type, size, quantity, arrangement, and moisture content. Types of fuel include leaves, twigs, and decaying material on the forest floor; grass and shrubs of various size; small trees; large woody debris; and large living trees. Where forests, swamps, bogs, or marshes are underlain by organic soils, the fuel for a wildfire can also include underground accumulations of **peat,** an unconsolidated deposit of partially decayed wood, leaves, or moss.

The size of the fuel can influence ignition and how rapidly a fire moves. Finer fuels, such as grass and pine needles, burn more readily than large woody material. For example, if disease or a storm downs a large number of trees, after 15 to 30 years or so, the larger pieces of wood

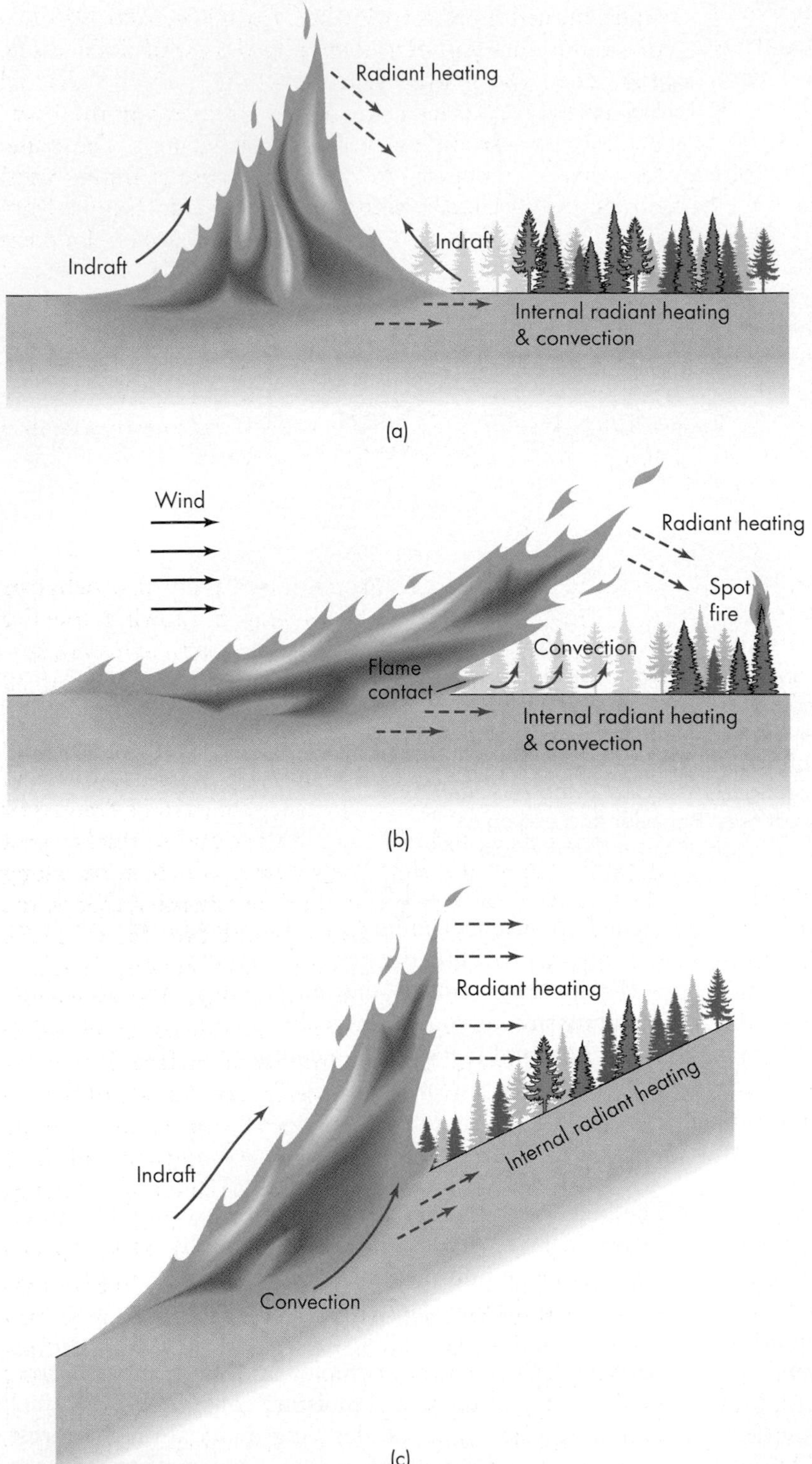

◀ **FIGURE 13.4 WILDFIRES ARE INFLUENCED BY WIND AND TOPOGRAPHY** Idealized diagrams of flame shape and processes associated with a spreading fire under three conditions: (a) flat topography and no wind, (b) flat topography with wind, and (c) hillslope topography where slope drives fire uphill. Solid red arrows indicate motion of air currents and dashed red lines show radiant heating from the fire. *(Modified after Rothermel, R. C. 1972.* A mathematical model for predicting fire spread in wildland fuels. *U.S. Forest Service Research Paper INT-115)*

may dry out and become partly decayed, allowing them to sustain combustion during a wildfire. Fuel arrangement can also be important. Landslides, tornadoes, and hurricanes arrange downed woody debris in clumps or accumulations in various parts of the landscape.[6] Finally, during droughts, even organic-rich swamp and marsh soils, which are usually saturated with water, dry out and become fuel for wildfires. These fires, commonly referred to as *peat fires*, burn slowly and can last for many months.

Topography Topography can have a profound effect on fire for several reasons. First, the moisture content of fuel is influenced by its location on the landscape. In the Northern Hemisphere, south-facing slopes are relatively warm and dry because they face the sun for a longer period of time. Thus, fuel on a south-facing slope is typically drier and burns more easily. Topography also influences air circulation. Slopes exposed to prevailing winds tend to have drier vegetation than do slopes sheltered from the wind and thus are more prone to combustion. Also, in mountainous areas, winds tend to circulate up canyons during the daytime, providing an easy path for wildfires. Finally, once a fire starts, topography can strongly influence its movement. Wildfires burning on steep slopes tend to preheat fuel upslope from the flames (Figure 13.4c). This makes it easier for a fire to move upslope and often causes it to spread more rapidly in that direction. Fires may also advance downslope, especially if they are driven by the wind.[6]

Weather Weather, especially temperature, precipitation, relative humidity, and winds, has a dominant influence on wildfire. Large wildfires are particularly common following droughts that reduce fuel moisture content. Hot, dry conditions associated with drought, as well as summers in the southwestern United States, lead to the formation of "dry thunderstorms." During these storms, rain evaporates before it reaches the ground and is not available to extinguish fires started by lightning. Changes in relative humidity, even during a single day, influence the moisture content of fuel and how fast a fire burns. Wildfires burn more rigorously when relative humidity is at its lowest point, generally during the midafternoon.[2] Finally, winds greatly influence the spread, intensity, and form of a wildfire. Strong winds and changing wind conditions help a fire preheat unburned fuel in the surrounding area. Winds can also carry burning embers away from a wildfire to ignite *spot fires* well ahead of the flaming front (Figure 13.4b).[6]

The influence of weather on fires was dramatically illustrated during the 1998 to 2002 drought in Florida, one of the worst in the state's history (Figure 13.5).[7] Hot, dry conditions during the drought reduced fuel moisture in swamps and wetlands. This resulted in wildfires in normally wet bald cypress swamps and grassy marshes, in addition to the more typical fires in upland pine/oak forests and scrublands. These wetland fires included the Mallory Swamp fire in 2001, which burned more than 250 km^2 (61,000 acres or 95 mi.^{2}) of swamp and commercial woodland west of Tallahassee, and the Deceiving Fire in 1999, which charred 700 km^2 (173,000 acres or 270 mi.^{2}) of marsh in the Everglades and surrounding upland near Miami.[8] The name "Deceiving" was given to the fire because changing wind directions continually shifted its movement.[7] Smoke from the Deceiving Fire closed I-75 (Alligator Alley), the main east-west highway across southern Florida, for several weeks, and drifted into Miami, where cars had to drive with headlights on in the middle of the day.[7] Wind-blown smoke was also a major problem in 2001, when it smothered theme parks in central Florida, including Walt Disney World, closed I-4, the main east-west route across central Florida, and caused a 20-vehicle accident.[9]

▲ FIGURE 13.5 **DROUGHT CONTRIBUTES TO FLORIDA WILDFIRES** During the 1998–2002 drought, more than 25,000 wildfires burned more than 6100 km^2 (1.5 million acres or 2300 mi.^{2}) in Florida. Fires burned parts of the Everglades in south Florida and Mallory Swamp in north-central Florida. Roads such as this one had to be closed and people evacuated. *(Charlotte Sun)*

TYPES OF FIRES

Wildfire scientists and firefighters classify fire behavior according to the layer of fuel that is allowing the fire spread. Although there are three fire types—ground, surface, and crown—most fires are complex blends of all three. *Ground fires* creep along slowly just under the ground surface, with little flaming and more smoldering combustion (Figure 13.6). In forests, these fires burn in *duff*, decaying organic matter in the soil, and in drained or temporarily dry swamps and marshes, and in thicker peat deposits below the soil. *Surface fires*, which move along the ground, may vary greatly in their intensity, that is, the amount of heat energy released by the fire (Figure 13.7). Low-intensity surface fires burn grass, shrubs, dead and downed limbs, leaf litter, and other debris. They burn relatively slowly with glowing or smoldering combustion and limited flaming combustion. Some surface fires, however, such as those in the Mediterranean climate of Southern California that burn chaparral brushlands, can be extremely intense and release large amounts of heat energy as they move swiftly across a landscape. Finally, *crown fires* are those in which flaming combustion is carried through the canopies of trees (Figure 13.8). Crown fires may begin when a surface fire moves up trunks into limbs of trees through various layers of fuel, or it may spread independently of surface fires. Large crown fires are generally driven by strong winds and are often aided by steep slopes. Such fires will grow and expand as long as conditions for combustion are favorable. Large wind-driven crown fires are nearly impossible to stop; humans and other animals need to evacuate from their path.

13.3 Geographic Regions at Risk from Wildfires

Most areas of the United States and Canada that are in, or in close proximity to, grasslands, shrublands, woodlands, or tundra are at risk for wildfires during dry weather or drought conditions. Even the sparsely vegetated

◀ FIGURE 13.6 **GROUND FIRE** A ground fire burns roots and some surface vegetation. Commonly there is more smoldering combustion than flame. *(Dr. Florian Siegert/Remote Sensing Solutions Gmbh)*

◀ FIGURE 13.7 **SURFACE FIRE** A surface fire, such as this one in a ponderosa pine forest, burns low vegetation rather than the upper part of tall trees. *(Kent & Donna Dennen/PhotoResearchers, Inc.)*

◀ FIGURE 13.8 **CROWN FIRE** Crown fires burn the upper part of tall trees. They sometimes spread ahead of the fire on the surface. In 1995, high winds turned this crown fire, the Green Mountain fire on Mount Lemmon outside of Tucson, Arizona, into a firestorm. *(A.T. Willett/Alamy)*

Sonoran desert in Arizona, rainforests in the Pacific Northwest, Hawai'i, Puerto Rico, and normally wet marsh and swamplands in the southeastern United States, Alaska, and Canada can experience wildfire. Large wildfires, however, are most common in Alaska and the western contiguous states in the United States (Figure 13.9), and in Canada, in the Canadian Rockies, and in a belt that extends from the Yukon Territory southeast to Lake Superior then east to Labrador.[10,11] The geographic region at greatest risk for wildfires shifts from year to year. For example, in 2006 the greatest number of acres burned in the United States were in Texas, in 2004 and 2005 in Alaska, in 2003 in the Northern Rockies and California, and in 2002 Alaska and Arizona took the lead.[12] In 2007 the spotlight shifted to Georgia, where a 2340 km^2 (903 mi.^{2}) fire in the Okefenokee Swamp was the largest in southeastern United States in more than a century, and Utah, where a 1470 km^2 (567 mi.^{2}) brushfire was the largest fire in state history.

13.4 Effects of Wildfires and Linkages with Other Natural Hazards

Wildfires affect many aspects of the local environment: They burn vegetation, release smoke into the atmosphere, char soil, create favorable conditions for landslides, increase erosion and runoff, and harm wildlife. Although we may primarily be concerned with the effects a fire has on the local biota of a region, it is important to consider the changes that may also occur in the geologic and atmospheric environments.

EFFECTS ON THE GEOLOGIC ENVIRONMENT

Wildfires have differing effects on soils depending on the type and moisture content of the soil and the duration and intensity of the fire. The amount and intensity of

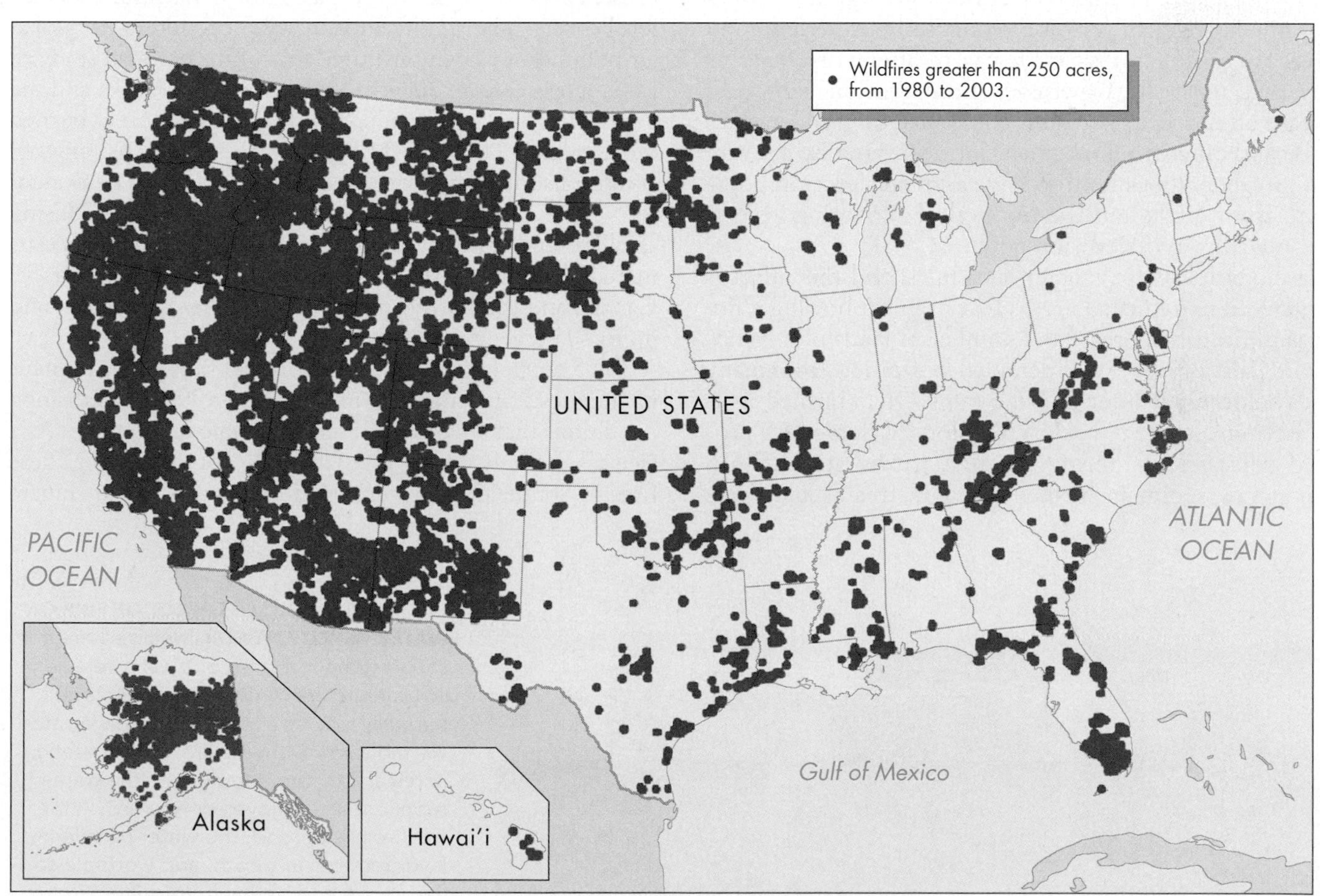

▲ FIGURE 13.9 **ALASKA AND WESTERN CONTIGUOUS STATES HAVE GREATEST WILDFIRE HAZARD** Map depicts areas of the United States that experienced wildfires greater than 1 km^2 (250 acres) from 1980 to 2003. There were no fires of this size in Puerto Rico during this time period. A similar map for Canada produced for a 40-year period by the Canadian Forest Service can be found at http://fire.nofc.cfs.nrcan.gc.ca/maps/Canada/lfdb/lfdb_59-99_lrg.png. *(U.S. Geological Survey)*

precipitation after a fire also play a role in determining how a wildfire affects the soil.[13]

Extremely hot fires that scorch dry, coarse soil may leave a non-wettable, or water-repellent, layer—called a *hydrophobic layer*—in the soil. Water repellency is caused by an accumulation of water-repellent chemicals in the lower layers of debris on the forest or brushland floor. These chemicals are derived from the burning vegetation and penetrate the soil as a gas that when cool forms a "waxy" substance, which coats soil particles (Figure 13.10).[2] This water-repellent layer in the soil increases surface runoff and thus erosion because the burned surface now lacks vegetation to hold the loose soil above the layer. Water flows rapidly over a water-repellent layer, like water down a water-repellent raincoat. Loose soil at the surface is easily saturated and then moved by the water.[14] The presence of hydrophobic layers in soils may increase the flood hazard in burned areas for several years following a fire until the layer decays.

Soil erosion and landslides, including debris flows (see Chapter 6), are common following wildfires. However, the overall effects of wildfire on erosion are variable. Areas naturally subject to erosion commonly experience an increase in erosion rates for a few years after a fire, whereas areas that normally experience little erosion show less increase after a fire.[2] Erosion and landslide problems are significantly increased on steep slopes charred by a severe burn. Steeper slopes tend to have a greater erosion problem to begin with, and the removal of anchoring vegetation only adds to the erosion potential.

Precipitation helps trigger landslides, and this effect is exaggerated in a burned area. Heavy rains following a fire may significantly increase the number of landslides. Wildfires in California in 1997 denuded vegetation from many slopes before the winter El Niño rains. Of 25 burned areas that were mapped by the U.S. Geological Survey, 10 produced debris flows during the first winter storm.[15] On other slopes, sediment flushed through streams and rivers because the rate of sediment erosion on burned slopes was greatly increased (Case Study 13.2).

EFFECTS ON THE ATMOSPHERIC ENVIRONMENT

In the short term, large wildfires create their own clouds (see Figure 13.1) and release smoke, soot, and invisible gases that contribute to air pollution (see Case Study 13.1). Smoke and soot significantly increase the concentration of very fine particles, referred to as *particulates*, in the atmosphere. This increase can be observed thousands of miles downwind of large, long-lasting fires. Several times during the past decade, fires burning in Mexico, Guatemala, and other countries in Central America have increased particulates in Texas and other parts of the southern United States, causing some cities to violate federal Clean Air standards.

Wildfires can also contribute to the formation of smog through the release of large quantities of carbon monoxide, volatile organic compounds, and nitrogen oxides. In the presence of sunlight, these gases form harmful ground-level ozone. The significance of wildfires adding to global air pollution was demonstrated in an international study of gases released from 2004 wildfires in central Alaska and the Yukon Territory of Canada. Wildfires in these areas burned more than 45,000 km^2 (11 million acres or 17,000 mi.^{2}) of coniferous forest and organic soil, a land surface equivalent to New Hampshire and Massachusetts combined.[16] Using satellites and computer models, atmospheric scientists estimated that these fires increased ozone levels in parts of Canada and the northern United States by 25 percent and up to 10 percent in Europe.[16,17]

On a much larger scale, wildfires and fires set by humans in Indonesia and Malaysia in 1997 and 1998 created smog conditions that affected millions of people in Southeast Asia (see Case Study 13.1). The favorable conditions for these fires in Southeast Asia were created by an El Niño climate

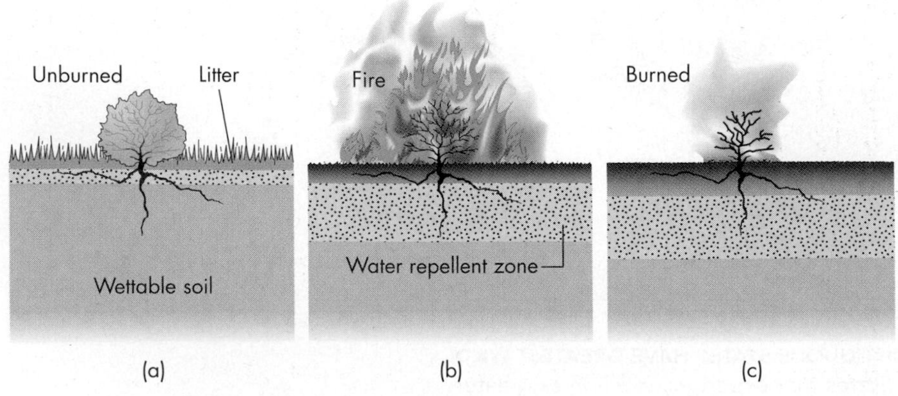

◀ **FIGURE 13.10 EFFECT OF FIRE ON WATER-REPELLENT SOIL** Idealized diagram showing the response of a water-repellent soil layer to fire. A water-repellent layer normally resides under the litter layer, atop wettable soil. During a fire, the vegetation and leaf litter are burned, and the water-repellent layer moves downward. After the fire has passed, the water-repellent layer remains at depth, and wettable soil lies above and below it. *(Modified from DeBano, L. F. 1981. Water repellent soils: A state-of-the-art review. U.S. Forest General Technical Report PSW-46. In Pyne, S. J., Andrews, P. L., and Laven, R. D. 1996. Introduction to wildland fire, 2nd ed. New York: John Wiley & Sons, Inc.)*

CASE STUDY 13.2

Wildfire in Southern California

The last two days of December 1933 delivered approximately 25 cm (10 in.) of rainfall to an already saturated landscape approximately 25 km (15 mi.) north of the city of Los Angeles. Three then-emerging communities of Montrose, La Cañada, and La Cresenta experienced catastrophic flooding and debris flows that destroyed several hundred homes and killed more than 30 people. These events were recorded in newspapers around the country on New Year's Day, 1934. The floods and debris flows resulted from high-intensity rainfall on a landscape that had been stripped of vegetation by a fire the previous November. The fire burned about 20 km^2 (5000 acres) in several watersheds in steep mountainous terrain of the San Gabriel Mountains. Runoff from the burned areas was 10 to 40 times greater than for unburned areas. Floodwaters, sand, and large boulders roared out of the mountain canyons causing great damage.

The New Year's Day floods in the Los Angeles area clearly illustrated that catastrophic flooding and debris flows were likely to follow a wildfire. As a result, some river channels were lined with concrete, and debris basins were constructed to catch sediment following future fires. How likely are catastrophic floods and debris flows following wildfire in Southern California? We can't answer this question precisely because the historic record is short and the time between wildfires, their *return interval,* is about 40 years. However, if catastrophic flooding and debris flows generally are responses to intense rainfall after a wildfire, then the return period is probably several hundred to several thousand years. This estimate assumes that the probability of intense precipitation in the first year following a fire is low.

Although catastrophic flooding and debris flows certainly do not follow every fire, flushing of sediment from stream channels is common. Following both the Wheeler Fire in 1985 and the Painted Cave fire in 1990, scientific studies documented the flushing of sediment from small burned watersheds.[18,19]

In Southern California, and probably in most similar settings, sediment moves downslope in dry weather following a wildfire. This gravity-driven process, known as *dry ravel,* moves a large volume of sand, gravel, and organic material that was stored upslope of brush vegetation before the fire (Figure 13.B). Much of this material accumulates at the base of the slope or in adjacent stream channels. Other material that remains on the slope

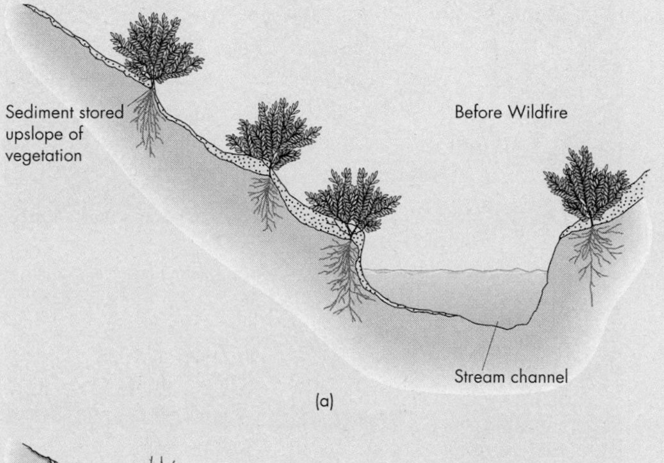

FIGURE 13.B SEDIMENTATION AFTER WILDFIRES (a) Sediment stored on hillslopes between fires is trapped upslope of chaparral vegetation. (b) After fire, sediment moves downslope by the process of dry ravel and accumulates in irregularities on hillslopes or channels. Moderate stream flows later flush the sediment from the channel. *(Modified from Florsheim, J. L., Keller, E. A., and Best, D. W. 1991. Fluvial sediment transport in response to moderate storm flows following chaparral wild fire, Ventura County, southern California. Geological Society of America Bulletin, 103:504–11)*

(continued)

is often easily washed downslope in the first rainfall event after a fire, because the water-repellent layer is now deeper below the surface (see Figure 13.10). The voluminous amount of material that moves downslope from dry ravel or surface runoff subsequently clogs stream channels with sediment (Figure 13.Cb). In a second moderate rainstorm and streamflow event, this sediment is flushed from the channel, leaving it much as it was before the fire (Figure 13.Cc). Less sediment was available on the hillslopes in the small watershed, and the stream flow was sufficient to transport the available sediment and also scour the channel. Less often, large catastrophic debris flows may be produced when intense precipitation of sufficient duration falls on burned slopes with abundant coarse debris.

Engineering solutions are often used to mitigate the high rates of sedimentation that can occur following a wildfire. Sediment control basins are constructed at the mouth of canyons to catch sediment flushed from stream channels as well as debris flows. These basins, sometimes known as debris basins, have been constructed on many streams in Southern California (Figure 13.D). For example, a debris basin was constructed following the Painted Cave fire of 1990 (Figure 13.Db). Sediment was flushed from stream channels in the winter following the fire, and the debris basin trapped approximately 23,000 m^3 (30,000 $yd.^3$) of it that otherwise would have been transported downstream, where it would have reduced the capacity of the stream channel and increased the flood hazard.

Studies of wildfire in Southern California brushland have shown that fire is only the first half of an event that alters the environment. Following a fire, runoff and erosion often increase, even in response to rather modest rainfall. In Southern California, the frequency of the combined fire/sedimentation events is directly correlated with population density in the wildland/urban interface. Large catastrophic wildfires occur during periods of warm weather and high winds when suppression of fire is nearly impossible. Therefore, prescribed burns to reduce the fuel load of Southern California brushlands are not likely to be successful in reducing or stopping a catastrophic wildfire.[20]

Also important to understanding wildfire in Southern California is

(a)

(b)

(c)

▲ FIGURE 13.C **INCREASED SEDIMENT IN STREAMS** (a) A small stream channel in Southern California shortly after a wildfire that burned vegetation in the drainage basin. Some trees near the channel survived the fire. *(Edward A. Keller)* (b) The scene after the first winter rainstorm. Note the voluminous amount of sediment deposited. *(Edward A. Keller)* (c) The same stream channel after a second winter rainstorm scoured the channel, removed much of the deposited sediment, and returned the channel to a shape that is similar to its shape before the fire. *(Edward A. Keller)*

reconstructing fire history for grassland, brush, and forest ecosystems. European settlers living in Los Angeles prior to the twentieth century were not, evidently, very concerned about wildfires that periodically burned for months in the mountains. This attitude changed as agriculture and urbanization transformed the land. As early as 1892, it was recognized that fire suppression was a potential management tool to preserve native vegetation to control erosion and flooding. This recognition led to the idea of setting small controlled fires to reduce excessive accumulation of fuel and thereby reduce the risk of large catastrophic wildfires.[21,22]

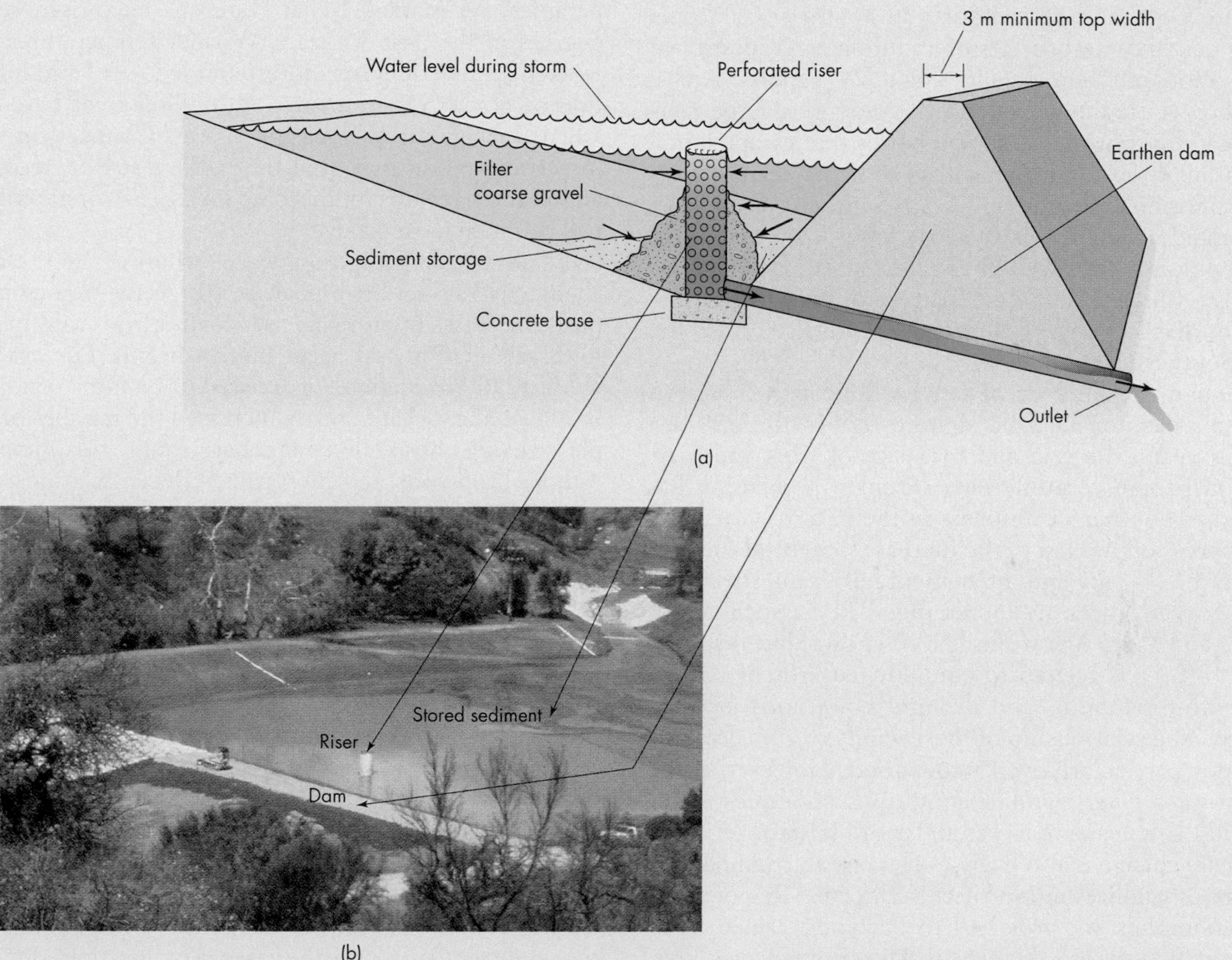

▲ FIGURE 13.D **SEDIMENT CONTROL BASIN** A sediment control basin can mitigate the effects of increased erosion and sedimentation after a wildfire. (a) Cross section of a sediment control basin. Sediment carried by storm water settles out in the basin, while the water filters through loose gravel into a riser, where it then drains through a pipe to an outlet. Accumulated sediment is periodically removed mechanically. Black arrows indicate direction of water flow. *(After Soil Conservation Service. 1974. Erosion and sediment control)* (b) A sediment basin constructed to trap sediment that eroded after a wildfire in Southern California. *(Edward A. Keller)*

event in the Pacific Ocean (see Case Study 13.2). Thus, wildfires have both short-term interactions with weather and long-term interactions with climate.

LINKAGES WITH CLIMATE CHANGE

One projected effect of climate change in this century is an increase in the intensity and frequency of wildfires. This increase may be brought about by changes in temperature, precipitation, and the frequency and intensity of severe storms. It may also be linked to biological changes that have taken place in the type and quantity of fuel available for wildfires. Some of these changes may already be underway in the western United States and Canada.[23,24]

In late October 2003, wildfires in Southern California burned nearly three-quarters of a million acres, destroyed several thousand homes, and killed 24 people. Twelve major fires started in a single week. Most were large catastrophic fires known as "runaway" fires that escape initial firefighting efforts and then spread to damage large areas (see Professional Profile 13.3). In Southern California, autumn and winter are particularly hazardous seasons for catastrophic wildfires. Wildfires often accompany hot, dry Santa Ana winds, a large-scale pattern of atmospheric circulation that produces dangerous conditions over the entire Southern California region.[25,26]

Although the number of large wildfires in Southern California has not changed significantly in the past few hundred years, the size and intensity of fires and their effect on people certainly have. Human population has doubled in Southern California in the past 50 years, and fire suppression has led to the increase in natural fuels; as a result, the severity and intensity of burns and their ecological consequences are increasing.[25] The association of wildfire and Santa Ana winds is well known, but the exact timing of fires is related to complex patterns of rainfall during winter months and drought conditions prior to the fires. The relationship of these winds to global warming is not particularly well understood. However, if the trend toward drier, windier conditions continues, wildfires will likely become larger and more intense.[26,27]

Climate change is also likely to increase the number and intensity of wildfires in Northern California. Regional climatic conditions are projected to become warmer, drier, and windier—three factors that promote wildfires. Warmer, drier climatic conditions will result in changes to plant communities with grasslands and chaparral (scrub forest) replacing the larger coniferous trees. These changes in plant types will increase fire frequency and intensity because grasslands and brush burn more frequently than do forests. As a result of these changes, more property will be destroyed, more lives lost, more floods and mudslides will follow fires, and ash and soot will reduce air and water quality. In addition, the cost of firefighting and insurance against fire will increase as damages increase.[26]

Increases in the size or intensity of severe storms may also promote wildfires by increasing the incidence of lightning, one method of igniting wildfires. In addition, warmer and drier conditions will make trees more susceptible to beetle infestations, weakening them and rendering them more vulnerable to fire.

Changes in vegetation resulting from global warming may also affect wildfire frequencies and intensities. Precipitation changes in parts of the western United States are likely to change some landscape dominated by sagebrush to widespread grasslands. As a whole, grasslands tend to burn more frequently than the sagebrush that grows there today.[26,27]

These trends in fire frequency and intensity are not limited to California and the Southwest. Similar changes are taking place in Alaska and Canada.[23,24] For example, temperatures on Alaska's Kenai Peninsula have risen several degrees in the past 30 years. Warmer temperatures have weakened the coniferous forest and led to a large infestation of beetles. An estimated 40 million trees have been killed throughout an area the size of Connecticut. This insect infestation may well be the largest ever recorded, and as the trees die and become fuel, catastrophic wildfire will be much more likely.[26]

Finally, the linkage between wildfires and climate change will also affect human health. As mentioned previously, increases in airborne particulates from wildfires can lower air quality and harm human health. For example, wildfires in Florida in 1998, related to La Niña conditions in the Pacific Ocean, greatly increased the number of people seeking hospital emergency room treatment for asthma, bronchitis, and chest pain.[26] Recently, atmospheric scientists and geologists have become concerned with increases in peatland fires in both the boreal forests of Canada and Russia and the tropical forests of Malaysia and Indonesia. Burning peat releases smog-producing gases, higher levels of particulates than the combustion of wood or grass, and significant quantities of toxic mercury into the environment (Figure 13.11).[3,28]

EFFECTS ON THE BIOLOGICAL ENVIRONMENT

Fires have many direct and indirect effects on the biological environment, which includes plants, animals, and human beings. The effects may vary from moderate to severe depending upon the type, size, location, duration, and intensity of the fire.

Vegetation Fire's effects on vegetation are numerous, varied, and complex. Some plants, such as scrub oak, juniper, and mesquite, are susceptible to fire whereas others, such as Douglas fir and Ponderosa pine, are not.[29] Douglas fir and Ponderosa pine are adapted to frequent fires with thick bark that insulates them from fire damage.[29] Even if a plant does not die in a wildfire, a fire can make it vulnerable to later destruction by disease or drought. For some plants, such as a number of native prairie wildflowers, fires are required for the continuation

PROFESSIONAL PROFILE 13.3

Wildfires

Bob Krans, Fire Division Chief in Poway, California, had been in the firefighting profession for 30 years and had never seen anything like this. For one, the smoke, usually a shade of black or gray, was orange.

The Cedar Fire began on October 25, 2003, when a lost hunter lit a signal fire somewhere in the Cleveland National Forest, and it would become, before it was extinguished, the worst wildfire in California state history.

"I saw conditions that I'd never seen before," Krans said. "It was extremely frustrating to deal with."

The smoke appeared orange because of the sheer volume of the blaze, which was lighting the smoke from within, Krans said. He described the fire as sounding like a "freight train" because of the burning fuel and the surrounding wind, some of which was being generated by the fire itself because of the extremely large amount of oxygen it was consuming.

Fighting a wildfire comes with its own set of hazards beyond the intense heat and quickly changing conditions. Krans said that firefighters near enough to the blaze can actually suffer "steam burns," in which the heat turns the sweat inside their protective suits to a vapor, causing up to third-degree burns on a firefighter's skin.

At times during the Cedar Fire, Krans said officials estimated that the blaze was consuming 6000 to 12,000 acres per hour, urged on by 65-mph winds—a combination of the existing weather and the self-generated wind from the burn.

At that rate, Krans said one could easily observe the movement of the fire as it drew closer (Figure 13.E).

Although departments have extensive safety checks in place, one firefighter did perish and several others were badly injured while battling the Cedar Fire when a second blaze closed off their escape route.

For larger blazes such as the one that tore through Southern California in October and early November of 2003, the smoke can be so thick that vehicles require headlights.

"Daytime literally turns to nighttime," Krans said.

—CHRIS WILSON

▲ FIGURE 13.E **FIRE DIVISION CHIEF BOB KRANS, POWAY, CALIFORNIA** Chief Krans examines the Cedar Fire burning steadily toward him through a field of thick cedars. The fire reached the spot where he is standing in 3 minutes, consuming the home that stood directly behind Krans at the time. *(Bob Krans)*

of their species. Fire also helps the propagation of many woody plants. For example, seeds of sumac and bearberry remain dormant in the soil until opened by the heat from a fire.[2] Overall, fire is an important long-term control on the types of plants that grow in many areas of North America. The type, intensity, and frequency of fires selectively favor some plant species and eliminate others.[2]

Wildfire at the urban interfaces in Southern California is perhaps the most serious of the frequently occurring natural hazards. It has been observed that, in some cases, dense oak trees provided a partial barrier to the fire's advance in. In some cases, fire burns some of the oaks only halfway through, stopping with a knife edge line (a vertical line between burned and unburned tree). Oaks can release visible water vapor from heating as the fire approaches. Large oaks live several hundred years and usually survive a number of fires. Oak trees, even those that appear dead after a fire, will usually sprout quickly and recover. Of course, extremely hot, intense, and lingering fire may kill oak trees and set homes on fire, but the homes with oak trees around them have a better chance of surviving. It is much more difficult for fire to advance through oak trees than through chaparral. The possible natural service function of oak trees to retard fire moving across the land needs to be studied in more detail.

Eucalyptus trees (imported from Australia) are also well adapted to wildfire, but, in an entirely different way from oak trees. Eucalyptus trees have loose bark that hangs down, inviting fire to move up into the canopy. Sometimes the trees literally explode sending burning embers through the air for hundreds of meters. With the high winds that often accompany fire, the embers can travel even greater distances, spreading spot fires.

▲ FIGURE 13.11 **BURNING PEATLANDS CONTRIBUTE TO SMOG AND GLOBAL WARMING** Fires in organic soils, such as this one in a bog in the Poca das Antas Biological Reserve northeast of Rio de Janeiro, Brazil, release gases to the atmosphere that cause smog and contribute to global warming. Carbon, originally stored for many hundreds or thousands of years in organic soil or underlying peat deposits, is released to the atmosphere as carbon dioxide, a greenhouse gas contributing to global warming. *(Douglas Engle/AP Images)*

Fire is important in enabling eucalyptus seedpods to open. During fire, wind and flames may disperse the pods. On hot days, a faint blue flammable gas from the rich oil the trees produce has been reported. Burning eucalyptus trees can significantly increase the temperature of the fire, greatly increasing the hazard in the immediate area. During the Tea Fire (November 14, 2008), the Mount Calvary Monastery burned (see photograph opening this chapter). Just down slope of the structure is a group of eucalyptus trees that the fire moved through just before encountering the monastery. The trees are still there, having sprouted following the fire. There is discussion that eucalyptus trees should be removed in areas with a wildfire hazard. Some investigators have stated that eucalyptus trees are the most dangerous trees in a wildfire. Certainly that is the experience in Australia.

Animals Although extremely large, intense wildfires may kill even the fastest running animals, in general, fire has few direct effects on wildlife. Most animals are able to escape or hide from the danger posed by a fire. Mobile species, such as birds, deer, cougars, and bears, are generally able to move away from a fire as it advances. Rodents that nest beneath the ground also have a good chance of survival because soil is a good insulator. However, wildfires may have indirect effects on wildlife. In fire-denuded areas, loss of plant cover exposes the soil to solar heating, thereby increasing ground temperatures and altering habitats. In intensely burned areas, the pioneering plant species that initially colonize the land will determine both the type and number of vertebrates that can thrive in the area. In streams, fish and other aquatic species may suffer from increased sedimentation from wildfires, and water temperatures may increase because plants along the stream banks have been destroyed.[13]

Human Beings When a fire removes much or all of the vegetation in a watershed, subsequent rains will have much greater erosive potential. Erosion produces large quantities of sediment and plant debris that affect the water quality of streams and lakes. For example, in May 1996, a fire burned two watersheds that contribute runoff into the drinking water reservoir for Denver and Aurora, Colorado. Two months after the fire, a storm produced floods in the watersheds and sent large amounts of floating debris and high dissolved concentrations of manganese into the reservoir. Even two years after the fire, the water quality was still not as good as it had been before the fire.[15]

Smoke and haze produced by fires can harm human health (see Case Study 13.1). Exposure to smoke and haze commonly produces eye, respiratory, and skin problems. After prolonged exposure to smoke, people, especially firefighters, may experience respiratory problems that become permanent, or even fatal.

Wildfire, of course, has the potential to destroy personal property. As human population continues to increase, more and more people are living on the fringes of wild lands, in an area called the **wildland/urban interface.** When a wildfire occurs in brushland or forest areas near a city, such as in the Santa Monica Mountains of Los Angeles or the Front Range near Denver, Colorado, hundreds to thousands of homes may be at risk.

13.5 Natural Service Functions of Wildfires

Although fires threaten human lives and property, they are also beneficial to ecosystems. Fire is a part of the natural cycle of the landscape and, as mentioned earlier, has been so since terrestrial plants evolved.

BENEFITS TO SOIL

Fires tend to increase the nutrient content of a soil and leave an accumulation of carbon on the surface in the form of ash. Under the right conditions, when erosion does not remove the ash from the environment, a nutrient reservoir may form that is beneficial to local plants. For plants that are well adapted to fire, such as those that release seeds only after a fire, these nutrients provide a valuable resource for new seedlings. Fires also tend to reduce populations of soil microorganisms. Decline in microorganisms may benefit those plants with which they compete for nutrients. In some cases, the destruction of microorganisms may also be beneficial since some microorganisms are parasites or carry diseases.[2]

BENEFITS TO PLANTS AND ANIMALS

When a wildfire reduces the number of individuals of a species of plant in a given area, the result may be beneficial to the plant community as a whole. The removal of some species temporarily reduces competition for moisture, nutrients, and light, allowing the surviving species, and new species, to thrive.

In species that depend on fire for reproduction, a wildfire may trigger the release of seeds or stimulate flowering. Two species of trees, Lodgepole pine and aspen, are strongly dependent on fire for reproduction and growth. In aspens, fire stimulates the release of a growth enzyme that can cause an aspen tree to produce as many as a million sprouts per acre.[30] These sprouts are important food for elk and moose. In Lodgepole pines, seeds remain sealed in pinecones for decades until fire melts a resin surface coating on the cones to release the seeds.[2] The seeds then fall to a nutrient-rich, ash-covered forest floor that is devoid of competition for the growth of pine seedlings. Species with these reproductive adaptations tend to exist mainly in environments where fire is common and frequent. If wildfires were eliminated from their environment, some species would eventually become extinct.

Fire is also important to ecosystems as a whole. The North American prairie is an example of an ecosystem that is well adapted to wildfires. Grasses, the dominant class of prairie plant, are especially suited for growth after a fire because they grow from their base rather than leaf tips and 90 percent of their biomass is underground and generally undamaged by fire.[31] For grasses, fire removes surface litter and allows nutrients and moisture to reach their roots. Native Americans long ago understood the importance of fire, wind, drought, and grazing animals for the prairie ecosystem.[31] Following a fire, colonization and replacement of plant species follow a regular pattern known as *secondary succession.*

Burning of plant material recycles nutrients in the system by quickly decomposing organic matter and allowing new plants to grow. Fire also reduces competition among species for sunlight, water, and nutrients, and crown fires allow more sunlight to reach the understory.

Many species of birds, insects, reptiles, and mammals may benefit from wildfire. For example, new grassland and forage for deer and bison emerge following removal of trees by fire. Also, burned, decaying logs provide homes for insects and supply food for mice, coyotes, birds, and bears.

YELLOWSTONE FIRES OF 1988

Because wildfires potentially benefit ecosystems, many scientists believe that natural fires in woodland areas should not be suppressed. In 1976 Yellowstone National Park, which covers 9000 km^2 (2.2 million acres or 3500 mi.^{2}) of Wyoming and Montana, began a policy of allowing natural burns in all areas of the park managed as wilderness. This meant that any fire that started naturally was allowed to burn, provided it did not endanger human life, threaten visitor areas such as Old Faithful, or risk spreading to areas around the park. Any human-caused fire would be extinguished immediately. Before the summer of 1988, the worst fire in Yellowstone's history had burned about 100 km^2 (25,000 acres or 40 mi.^{2}). By the end of the 1988 fire season, more than 3200 km^2 (800,000 acres or 1250 mi.^{2}) had burned, causing a major controversy over the natural-burn policy.

Lightning strikes during the summer of 1988 ignited 50 fires in the park. Of those, 28 were allowed to burn, according to the natural-burn policy. Initially small, the fires quickly expanded in the hot, dry conditions, fueled further by high winds in mid-July. On July 17, bowing to political pressure, Yellowstone officials sent fire crews in to fight a natural fire. By July 21, multiple natural fires in and around Yellowstone were being fought; by July 22, however, it was clear that the fires were beyond the control of the crews. The fires did not slow until September 11, when rains fell throughout the area. The snows of November finally extinguished the fires.

The large area burned at Yellowstone in 1988 caused criticism of Yellowstone's natural-burn policy. Although most scientists agree that fire is good for the natural environment, it was difficult for some people to sit by and watch the park burn. Eventually 9500 firefighters were deployed at a cost of $120 million of taxpayers' money. Nonetheless, the fires continued until they were naturally extinguished. Critics stated that the fires would not have been so big if park officials had fought them from the beginning. Others argued that the fires would not have been so severe if years of fire suppression prior to the mid-1970s had not allowed fuel to accumulate in the area. Post-fire studies have shown that the effects of the 1988 Yellowstone wildfires were beneficial to the environment, and the park still adheres to a natural-burn policy. This policy is correct because Yellowstone's ecosystems have, through geologic time, adapted to and become dependent upon wildfire. The fires of 1988 did not destroy the park; rather, they revitalized ecosystems through natural transformations that cycle energy and nutrients through solids, plants, and animals.

13.6 Minimizing the Wildfire Hazard

Clearly it is desirable to minimize the potential of wildfires to destroy human life and property. There are several components of fire management.

FIRE MANAGEMENT

The task of *fire management* is a difficult one because large wildfires cannot be prevented. Furthermore, not all fires should be suppressed; remember the benefits of fire to the natural environment. However, suppressing fires when they threaten human lives and property is desirable. The primary approaches to fire management include science, education, data collection, and use of prescribed burns.

Science Scientific understanding of wildfire and its role in the structure and function of ecosystems is critical to fire management; we cannot manage what we do not understand. Of particular importance in fire ecology is an understanding of the **fire regime** of an ecosystem. The fire regime is broadly defined to include (1) the types of fuel that are found in plant communities; (2) typical fire behavior as described by fire size, intensity, and amount of biomass removed; and (3) the overall fire history of the area, including fire frequency and recurrence interval. Reconstruction of fire histories is difficult in many areas because fires have been suppressed for nearly 100 years.[19,20] Nevertheless, fire management is more likely to be successful if fire regimes for specific ecosystems can be defined.

Education Public education is an integral part of fire management. People can be made aware of wildfire as a hazard and what they can do to minimize their personal risk.

Data Collection Remote sensing has become an important tool for fire management, especially the use of satellite imagery to map vegetation and determine fire potential. Since the early 1990s, the U.S. Geological Survey (USGS), in cooperation with the National Oceanographic and Atmospheric Administration (NOAA), has prepared weekly and biweekly maps for the contiguous United States and Alaska illustrating plant growth and vigor, vegetation cover, and biomass production. These maps, combined with determinations of the moisture content of vegetation in a given area, are an invaluable resource for fire managers. A management tool called the Fire Potential Index (FPI) was developed by the USGS and the U.S. Forest Service to characterize the relative fire potential of forests, rangelands, and grasslands. The FPI takes into account the total amount of burnable plant material or fuel load, the water content of dead vegetation, and the percentage of the fuel load that is living vegetation. Regional and local FPI maps are prepared daily in a geographic information system (GIS) to help land managers develop plans for minimizing the threat from fires.

Prescribed Burns For the past century, fire management in the United States has been guided by a policy of wildfire suppression, mainly to protect human interests. In forests, this policy has led to a buildup of fuel and the potential for larger, high-intensity fires. One way to counter this dangerous buildup of fuel is to ignite **prescribed burns.** The use of prescribed or controlled burns for forest management is not new; such fires have been used for years as an alternative to absolute fire suppression (Figure 13.12).[20]

The main goal of prescribed burns is to reduce the amount of fuel and thus the likelihood of a catastrophic wildfire.[15] Fire ecologists have found that more frequent, smaller fires will lead to fewer large fires.

Each prescribed burn has a written plan that outlines the objectives, where and how it is to be carried out, under whose authority, and how the burn will be evaluated. Those in charge of a prescribed burn take on a great deal of responsibility—they have to predict the natural behavior of the fire and successfully keep it under control. Planners face the difficult task of predicting the acceptable fuel and weather conditions under which they can safely control the fire. Factors such as temperature, humidity, and wind must be taken into account.

Unexpected changes in wind direction during a prescribed burn can have catastrophic results. The unexpected happened in spring 2000 when fire managers lost control of a prescribed burn at Bandelier National Monument in New Mexico. A prescribed burn ignited on May 4 quickly raged out of control and sliced through 80 km^2 (20,000 acres or 30 mi.^{2}). Within 24 hours, the burn was declared a wildfire as a result of sporadic, unexpected changes in wind direction. After 2 weeks, the fire had engulfed 190 km^2 (46,000 acres or 72 mi.^{2}) and was still on the move. Nearby drought-parched pine forest and

▲ FIGURE 13.12 **FOREST MANAGEMENT WITH PRESCRIBED BURNS** This old-growth longleaf pine forest near Thomasville, Georgia, has been managed with annual prescribed burns during lightning season by the Tall Timbers Research Station for more than 25 years. The result is a healthy, diverse ecosystem similar to what existed when Europeans first explored the region in the sixteenth century. *(Jim Cox/Tall Timbers Research Station)*

grassland supplied ample fuel for the fire that destroyed 280 homes and forced 25,000 people to evacuate the town of Los Alamos, just 110 km (70 mi.) north of Albuquerque.

After the flames were extinguished and the damage assessed, the town and the nation were left only with questions. How could a controlled burn, ordered by the National Park Service itself, end up nearly destroying a town? A formal review following the fire called for changes in policy. The changes included a more careful analysis and review of burn plans, better coordination and cooperation among federal agencies in developing burn plans, and use of a standard checklist before setting a prescribed fire. These and other changes have been instituted. Furthermore, because of the near-disaster at Los Alamos, the existing 1995 Federal Fire Management Policy was reviewed and then updated in 2001. Specific changes include emphasizing the role of science in developing and implementing fire management programs.

In 2003 the U.S. Congress passed the Healthy Forests Restoration Act, a forest management plan to reduce large damaging wildfires by thinning (logging) trees on federal lands. The act reduces or limits the environmental analysis that is normally required to initiate logging projects. Those in support of the act state that the new management procedures will reduce the risk of catastrophic fires to towns in and around national forests, will save lives of forest residents and firefighters, and will protect wildlife including threatened and endangered species. Those opposing the act argue that large-scale logging will be promoted far from communities at risk from wildfire and overall will damage our forests. They further argue that the best approach to minimizing risk from wildfire is through the selective removal of vegetation around communities and homes and through education and planning.

13.7 Perception of and Adjustment to the Wildfire Hazard

PERCEPTION OF THE WILDFIRE HAZARD

In general, people who live or work in the wildland/urban interface do not adequately perceive a risk from wildfires. California provides a good example of this apparent lack of concern. Residents of California, where fires burn almost every year, should be familiar with the risks associated with wildfires, yet development continues on brush-covered hillslopes. The demand for hillside property has increased property prices in these areas, which means that the people whose property is most at risk of fire have paid a premium for that "privilege." Fire insurance and disaster assistance may actually worsen the situation; if people believe that the government will reimburse their fire losses, they may not see any reason not to live where they choose, regardless of the risk. In the past hundred years, dozens of fires have struck brushlands of California, while population and property values continue to increase. The result of a lack of perception of a real risk from fires was tragically illustrated in October 1991, when a wildfire destroyed almost 6.5 km^2 (1600 acres) and about 3800 houses and apartments in the cities of Oakland and Berkeley (Figure 13.13). The fire killed 25 people and caused more than $1.68 billion in damages. When it was over, it was labeled one of the worst urban disasters in U.S. history.

The fire started on October 19 from cooking fires in a camp of homeless people. Firefighters thought it was extinguished and left the site around midnight. The next day was hot, with high winds, and embers left from the night before reignited the fire. Urbanization had added additional fuel to the area, which was previously composed of grass-covered slopes with a small number of oak and redwood trees. The additional fuel included numerous wood structures and imported trees such as eucalyptus. For decades prior to the fire, the density of trees and structures had increased. Open land was reduced from 47 percent of the area in 1939 to 20 percent in 1988, and the number of plant species increased from a few to a few hundred. When the fire reignited, it quickly became uncontrollable. All that firefighters could do was evacuate residents. The fire moved through the urban landscape very quickly; during the first hour, it consumed a home every 5 seconds! Although the fire was started by people, other factors contributed: an ample fuel supply of buildings, brush, and trees and hot, windy weather. Furthermore, from a fire-hazards perspective, land-use planning in the area was insufficient. Structures were not required to be fire resistant, the density of buildings and placement of utilities as related to fire hazard were not restricted, and no rules required removal of excess vegetation around buildings.[2]

ADJUSTMENTS TO THE WILDFIRE HAZARD

Adjustment to the fire hazard may be accomplished through fire danger alerts and warnings, education, codes and regulation, insurance, and evacuation.

Fire Danger Alerts and Warnings Both U.S. and Canadian federal agencies have developed fire danger rating systems to alert land managers, residents, and visitors to daily changes in conditions affecting the development of wildfires. Fire danger ratings combine information about fuel conditions, topography, weather, and risk of ignition to assess the wildfire hazard. National Weather Service forecast offices also issue fire weather watches 12 to 72 hours in advance of extreme fire conditions and **red flag warnings** when extreme fire conditions either are occurring or will take place is less than 24 hours. These watches and warnings put citizens, public officials, and firefighters on alert.

Fire Education Community awareness programs and presentations in schools about fire safety may help reduce the fire hazard. Unfortunately, for many people who have never experienced a fire, the risk may not seem "real," even with the help of fire education programs.

FIGURE 13.13 **KILLER WILDFIRE** A wildfire in the hills of Oakland, California, in 1991 devastated an entire neighborhood, killing 25 people and destroying more than 2500 homes. Damages were more than $1.68 billion. *(Tom Benoit Photography)*

Codes and Regulations One way to reduce risk in fire-prone areas is to enforce building codes that require structures be built with fire-resistant materials. For example, using stone or brick for building and roofing materials, rather than wooden shingles, helps fireproof homes. Unfortunately, in California, stone roofing tiles are not always safe during earthquakes. Making appropriate fire-resistant materials mandatory in new structures could significantly reduce the amount of damage caused by fires. Many local governments and state/province and federal land management agencies also exercise their regulatory authority by issuing bans on outdoor burning and fireworks when conditions are favorable for grass, brush, and forest fires. These bans help reduce the number of fires ignited by people.

Fire Insurance Fire insurance is another adjustment to the threat of fires. Such insurance ensures that people whose property has been destroyed by a fire will be reimbursed. However, insurance may provide a false sense of security and prompt more people to live in fire-prone areas.

Evacuation Evacuation of people from danger zones is probably the most common adjustment to the fire hazard. Evacuation of people from homes helps ensure their safety, but it does not protect their homes or personal belongings (see Survivor Story 13.4).

PERSONAL ADJUSTMENT TO THE FIRE HAZARD

There are many things you can do to protect yourself from fire (Table 13.1). If you live or work in the wildland/urban interface, you may want to obtain a book that describes how to prepare for the inevitable wildfire.[32]

TABLE 13.1

Reducing your fire hazard at home

Maintain Home Heating Systems
■ Have your chimney regularly inspected and cleaned.
■ Remove branches hanging above and around the chimney.
Have a Fire Safety and Evacuation Plan
■ Install smoke alarms on every level of your home.
■ Test smoke alarms monthly and change the batteries at least once a year.
■ Practice fire escape and evacuation plans.
■ Mark the entrance to your property with signs that are clearly visible.
■ Know which local emergency services are available and have those numbers posted.
■ Provide emergency vehicle access through roads and driveways at least 3.7 m (12 ft.) wide with adequate turnaround space.
Make Your Home Fire Resistant
■ Use fire-resistant and protective roofing and materials such as stone, brick, and metal to protect your home. Avoid using wood materials, which offer the least fire protection.
■ Keep roofs and eaves clear of debris.
■ Cover all exterior vents, attics, and eaves with metal mesh screens having openings no larger than 6 millimeters.
■ Install multi-pane windows, tempered safety glass, or fireproof shutters to protect large windows from radiant heat.
■ Use fire-resistant draperies for added window protection.
■ Keep tools for fire protection nearby: 30 m (100 ft.) of garden hose, shovel, rake, ladder, and buckets.
■ Make sure that water sources, such as hydrants and ponds, are accessible to the fire department.
Let Your Landscape Defend Your Property
■ Trim grass on a regular basis up to 30 m (100 ft.) surrounding your home.
■ Create defensible space by thinning trees and brush within 10 m (30 ft.) around your home.
■ Beyond 10 m (30 ft.), remove dead wood, debris, and low tree branches.
■ Landscape your property with fire-resistant grasses and shrubs to prevent fire from spreading quickly.
■ Stack firewood at least 10 m (30 ft.) away from your home and other structures.
■ Store flammable materials, liquids, and solvents in metal containers outside the home, at least 10 m (30 ft.) away from structures and wooden fences.

Source: Modified after Federal Emergency Management Agency. 1993. Wildfire: Are you prepared? *FEMA Publication L-203.* http://www.usfa.fema.gov/downloads/pdf/publications/wildfire.pdf.

SURVIVOR STORY 

Two Wildfires in the Hills above Santa Barbara, California

Wildfire is probably the most serious, common, recurring hazard in the hills of Southern California. The return period at a particular location is often several decades, and, with the rapidly growing population of Southern California, these fires may be catastrophic. The hills behind Santa Barbara are the south flank of the Santa Ynez Mountains that extend for several tens of kilometers from, roughly, the east to the west. The entire flank of the mountain range has burned in the past 75 years or so. During that period, many people have built homes in the wildland-urban interface where there are several different ecosystems, including oak woodlands and chaparral (a small, dense brush that is fire adaptive and burns quickly). The oak trees are also adapted to fire, and they often live several hundred years, which means they have survived several fires.

My wife Valery and I (Ed Keller, one of your authors) returned to the University of California, Santa Barbara following a sabbatical in Europe in 2000. On a Sunday after church, we were visiting Rattlesnake Creek, where I had worked for 20 years on the geology, hydrology, and, more recently, endangered steelhead trout habitat. On the way back to the city, we noticed a home for sale. For a number of years, my wife had mentioned that she would like to have a place with one large oak tree. To our surprise, the property for sale was several hectares and included hundreds of oak trees. The home on the site was a prefabricated structure built by one of the first families in the canyon, and, after a number of weeks and talking to the owner, we purchased the property. I was aware of the potential fire hazard and believed that the location was defendable for several reasons. First, there was little chaparral, being mostly oak woodlands along Rattlesnake Creek; and, second, there were several fire hydrants along the road in front of the property, and so a house could be designed that could be defended from fire. There were also several large eucalyptus trees, which are an invasive species from Australia. Eucalyptus trees are fire adaptive but burn fiercely during fires. In fact, the trees encourage fire with their loose bark that burns readily and lets the fire move up the tree. In Australia, it is well known that it is dangerous to live in close proximity to eucalyptus trees. As a result, on our property, we removed all of the eucalyptus trees, even though they are often stately, beautiful trees.

Valery took several years off of work as an administrator with the Santa Barbara School District and designed and served as the general contractor for our new home, which is a one-story Spanish Colonial style home. It was designed with fire in mind, with thick walls, recessed windows, no eaves for smoke to get in, and a fireproof tile roof. We also built a circular drive that was large enough for fire trucks to move easily to and from and around the front of the house. Finally, we trimmed all the trees so that no forking branches and leaves were below the roof or close to the home. We created a defensible space clear of most fuel for a fire.

Everything went well, and we finished the construction and moved in at the end of 2003. In November 2008, the first wildfire came into the area from the north and the east. Several hundred homes burned, and Westmont College in nearby Montecito was heavily damaged. The well-known and loved Mount Calvary Monastery on a nearby ridge burned (see opening photograph of this chapter). We had planned on evacuating up the hill to a location known as St. Mary's Seminary and Retreat House, which is on a hilltop that has been cleared of vegetation and, I believed, would be a good place to evacuate to during a fast-moving fire. We gathered up our dogs in the car and were surprised that the location had been evacuated! Furthermore, flames about 60 m (200 ft.) high were racing down the canyon, and our home is in that canyon. We evacuated to the coastal area and sat out the fire. We could not return for almost a week. I called the home the next morning and was relieved to hear our answering machine was still working. When we did return, we found that the flames had come within about 15 m (50 ft.) of the house. During the fire, a number of firemen from all over the state moved into the Santa Barbara area and, in many instances, camped out at homes to defend them. This happened at our house, and, when we returned, there were fire hoses and fire lines cut to stop creeping fire through the oak trees. It was clear that the oak trees did not burn completely. As the flames came from the chaparral into the oak trees, water vapor is exuded from the oak trees, and, sometimes, only the first line of trees burns and sometimes only half of an individual tree before the fire stopped and moved around the tree, keeping to the chaparral. Nevertheless, firebrands swirl through the air with the wind, and these can land on decks and roofs and other places, so spot fires can occur up to several hundred meters and more in front of the fire. Thus, it is important to have fire personnel present after the fire has moved through to put out the creeping fires and spot fires. Sometimes, homes will catch on fire hours after the fire has moved through, if these spot fires are not discovered.

We moved back into the house, and everything seemed okay again. We thought we'd survived our fire in the canyon, although we knew that the canyon to our west, Mission Canyon, had still not burned in 40 years. Then, in May 2009, I received a phone call from my wife saying that she had noticed from her school in lower Santa Barbara smoke in the foothills just west of our canyon home. I turned on the television and learned that a small fire had broken out on what is known as Jesusita Trail—it turned out later that the fire was caused by brush-clearing equipment. At any rate, that fire flared up rapidly. I gathered up the dogs, closed the windows and the damper to the chimney, and evacuated to the north into the area that had burned during the first fire.

The second fire was fierce, moving through Mission Canyon and Rattlesnake Canyon, and 11 homes were burned within about a 10-minute walk from our house. Firemen, again, from all over the state, converged on the wildfire, and a truck or two was parked in our driveway, according to a couple of our neighbors who stayed to fight the wildfire. Evidently, during the worst of the firestorm, which was very intense, with high winds and flames probably several hundred feet high and firebrands falling everywhere, they retreated to St. Mary's Seminary and Retreat House and, after the fire went through, came back and put out creeping fires. This time, the fire came from the west, and it burned, again, within just a few meters of our home.

One of our neighbors who lost their home had a line of several large, tall eucalyptus trees along the road in front of their home (Figure 13.F). I talked to a neighbor who was there during the fire who lived in a home in oak woodlands very close to homes that burned. He said that the eucalyptus trees went up like torches, practically exploding into fire that caught the home near his on fire. There was no chance for the structure to survive the intense heat from the burning eucalyptus trees. We watched on TV as commentators reported about the terrible fires in Rattlesnake Canyon. We believed that our home might have been lost—the answering machine was not responding. As it turns out, we were lucky compared to some of our neighbors. I do believe that we made the right moves in terms of where we purchased our property, its accessibility to firefighting, and the existence of the oak woodlands that surround the house. These factors played a big role, along with the firefighters who assisted during the two fires. During the second fire, a fire whirlwind went up the street about 30 meters (100 ft.) from our home, taking out parts of two trees. These fire whirlwinds are like tornadoes from hell, and they can move quickly. They often have flames and smoke hundreds of meters high and can be extremely dangerous. If that fire whirlwind had gone through our property, nothing could have saved our house.

At this point, you might say, "What is a geology professor who writes a book on hazards doing living in an area with a fire hazard?" We moved there knowing that there was a hazard, but I had carefully calculated the probability of the house being destroyed during wildfire as low (because of the reasons discussed earlier). It is about a 5-minute drive from the downtown area within the city limits. It is a beautiful location in the oak woodlands that is incredibly quiet and peaceful. Our survival story is that we survived two fires in a 6-month period, and the major oak woodlands surrounding our home and in the valley remain intact. Those oak trees that were burned and singed are now sprouting and recovering. However, the fires were a tragedy in Santa Barbara, as several hundred homes were lost.

◄ FIGURE 13.F **HOME LOST DURING THE SANTA BARBARA JESUSITA FIRE OF 2009.** For The fire raced up Rattlesnake Canyon claiming more than 10 homes. This home had large eucalyptus trees just upslope that burned extremely hot and most likely contributed to the loss of the home. *(Edward A. Keller)*

REVISITING THE FUNDAMENTAL CONCEPTS

Wildfire

1. **Hazards Are Predictable from a Scientific Evaluation**
2. **Risk Assessment Is an Important Component of Our Understanding of the Effects of Hazardous Processes**
3. **Linkages Exist between Natural Hazards**
4. **Hazardous Events That Previously Produced Disasters Are Now Producing Catastrophes**
5. **Consequences of Hazards May Be Minimized**

1) The frequency and return periods of wildfire are well known for many regions around the world. We know, for example, that the return period of fire in Southern California is 30 to 50 years, compared to return periods of 100 years or more for conifer forests. What is becoming more difficult is the linkage to global warming that seems to be occurring. As the southwestern United States warms, droughts will become more common and the fire season extended. Fires are also expected to be more intense and frequent. Of course, with more drought conditions, vegetation may grow more slowly, and that will have an effect on the recurrence of fire as well. Basically, we are entering uncharted areas of understanding concerning wildfire frequency; nevertheless, our general understanding of the science of fires and their occurrence is a well-studied and mature area of research.

2) The risk from wildfire may be defined as the product of the probability of a fire occurring times the consequences. The probability of fire has remained fairly constant, but the consequences have increased dramatically in recent years in many parts of the country, as a result of more and more people moving into the wildland-urban interface; that is, people are moving farther from cities into more rural areas where wildfire is more likely. Sometimes, people move into areas with a high fire risk. As a result, the risk of fires is increasing.

3) Wildfires are linked to several other natural hazards. For example, following wildfire, runoff and soil erosion may greatly increase. As a result, the flood hazard increases. Following wildfire, certain types of landslides also become more frequent. Dry ravel on slopes increases, and there is a threat of debris flows in the first few years following wildfire. Most of the debris flows will be small, but, occasionally, with the return period probably of several hundred years, larger debris flows to catastrophic debris flows are possible.

4) In recent decades, there has been a move to suppress wildfire. As a result, natural fires that occur may become larger and more intense. From that perspective, events that were previously disasters may now be catastrophes. The same change from disaster to catastrophe can occur because more and more people are living in areas with a high fire hazard. With more people living in the wildland-urban interface, the impacts of wildfire are going to increase in the coming decades.

5) Consequences of wildfire may be minimized by a variety of strategies. Most important in fire management is for homeowners to develop a defendable space around their homes and be prepared to evacuate. Emergency planning at the community level is also important. In some areas, reverse 911 calls warn people that a fire has started and update them with advised or required evacuations. Evacuation of a large number of people is often a difficult process, and community-planning groups work on this issue, so that when evacuation is necessary, it can be done in an orderly manner.

QUESTIONS TO THINK ABOUT

1. **Do you think development in areas with a high fire hazard should be restricted? Why or why not?**
2. **How might you educate people living in an area with a fire hazard that has not had a fire in two generations?**
3. **What are reactive policies following a wildfire and how do they compare to proactive steps taken before a fire? Which is preferable? Why?**

Summary

Wildfire, one of nature's oldest natural processes, is a self-sustaining, rapid, high-temperature, biochemical oxidation reaction that releases heat and light. Most wildfires in natural ecosystems maintain a rough balance between plant productivity and decomposition.

For a fire to burn, it must have fuel, oxygen, and heat. The two main processes that generate wildfire are preignition and combustion. Preignition involves heating and pyrolysis of fuel to drive off moisture and break down large carbon molecules into smaller ones. The smaller molecules create a cloud of flammable gas directly above the fuel, which then ignites. Ignition often starts with a lightning strike and then continues with windblown embers from the existing fire. Combustion typically occurs first by flaming, followed later by glowing and smoldering.

Wildfires can be classified based on what part of the landscape burns. Surface fires are those that burn along the forest floor or across the surface of brush and grasslands. Ground fires burn beneath the forest floor by smoldering. In forests, swamps, and marshes with organic soils, ground fires can smolder in peat deposits for many months. Fast-moving crown fires begin when surface fires ignite treetops. Spot fires are ignited ahead of the main fire by embers carried on the winds.

Wildfire behavior is influenced by fuel, weather, topography, and the fire itself. Fuel varies in size, shape, arrangement, and moisture content and ranges from fast-burning grasses and conifer needles to slow-burning logs and organic soil matter. Weather conditions favoring wildfires include high winds and temperatures, low humidity, and dry thunderstorms. Longer-term drought conditions are especially favorable for wildfires. Fires spread rapidly up steep slopes and are driven by winds up canyons. Predictions of fire behavior rely on an understanding of the interaction between a fire's environment and the fire regime.

Fire can increase runoff, erosion, flooding, and landslides. Climate extremes, such as El Niño and La Niña, and global warming can increase wildfire activity in many regions. Natural service functions of fires include increasing the nutrient content of soils, initiating regeneration of plant communities, creating new habitat for animals by altering landscapes, and potentially reducing the risk of large fires in the future.

Fire management includes education, data collection and mapping, and prescribed burns. Large wildfires are difficult to prevent and generally cannot be suppressed. Evacuation remains a primary adjustment for the wildfire hazard.

Key Terms

combustion (p. 450)
extinction (p. 451)
fire regime (p. 464)
peat (p. 451)
preignition (p. 450)
prescribed burn (p. 464)
pyrolysis (p. 450)
red flag warning (p. 466)
wildland/urban interface (p. 463)

Review Questions

1. How has the nature of wildfires and human interaction changed over geologic and historic time?
2. How are wildfires related to plant photosynthesis and decomposition?
3. What are the major gases and solid particles produced by a wildfire?
4. What are the three requirements for fire to start and for combustion to continue? What happens when one of these requirements is removed?
5. What are the three phases of a wildfire?
6. Explain how the two processes in the preignition phase prepare plant material for combustion.
7. What are sources for the initial ignition of wildfires? How often does ignition occur in a wildfire?
8. How do the processes of combustion differ from those of ignition?
9. What happens during pyrolysis?
10. Explain how the two types of combustion differ.
11. What are the three processes of heat transfer in a wildfire? Rank them in order of their importance.
12. What three factors control the behavior of a wildfire?
13. How do topography and weather influence a wildfire?
14. Describe the weather conditions that are most favorable for wildfires.
15. Describe the three types of fire.
16. Explain how wildfires affect erosion of the land.
17. What effects do wildfires have on the atmosphere?
18. How are some plants specially adapted to fire?
19. Describe how climate change is likely to affect the frequency and intensity of wildfires.
20. Explain how wildfires in peatlands are especially hazardous.
21. How do wildfires affect vegetation, animals, and humans?
22. What are the natural service functions of wildfires?
23. What are the four primary approaches to fire management?
24. What are the difficulties associated with prescribed burns?
25. Explain how people can adjust to the wildfire hazard.

Critical Thinking Questions

1. You live in an area with a significant wildfire hazard. What can you to do protect your home and belongings from fires? Make a list of actions you can take to protect yourself.
2. The staff of a large national park is reviewing its fire policy. You, a wildfire expert, have been asked for advice. The park's current policy is to suppress all fires as soon as they begin. It does not use prescribed burns and is considering switching to a policy of allowing natural burns. What would you suggest? List pros and cons of each policy before making your decision.
3. Describe the features in and surrounding your home that might make it vulnerable to a wildfire.
4. Most discussion of the wildfire hazard focuses on the potential destruction, injury, or death that can take place from the flames. Discuss the hazards to humans and the environment that come from the smoke produced by wildfires.

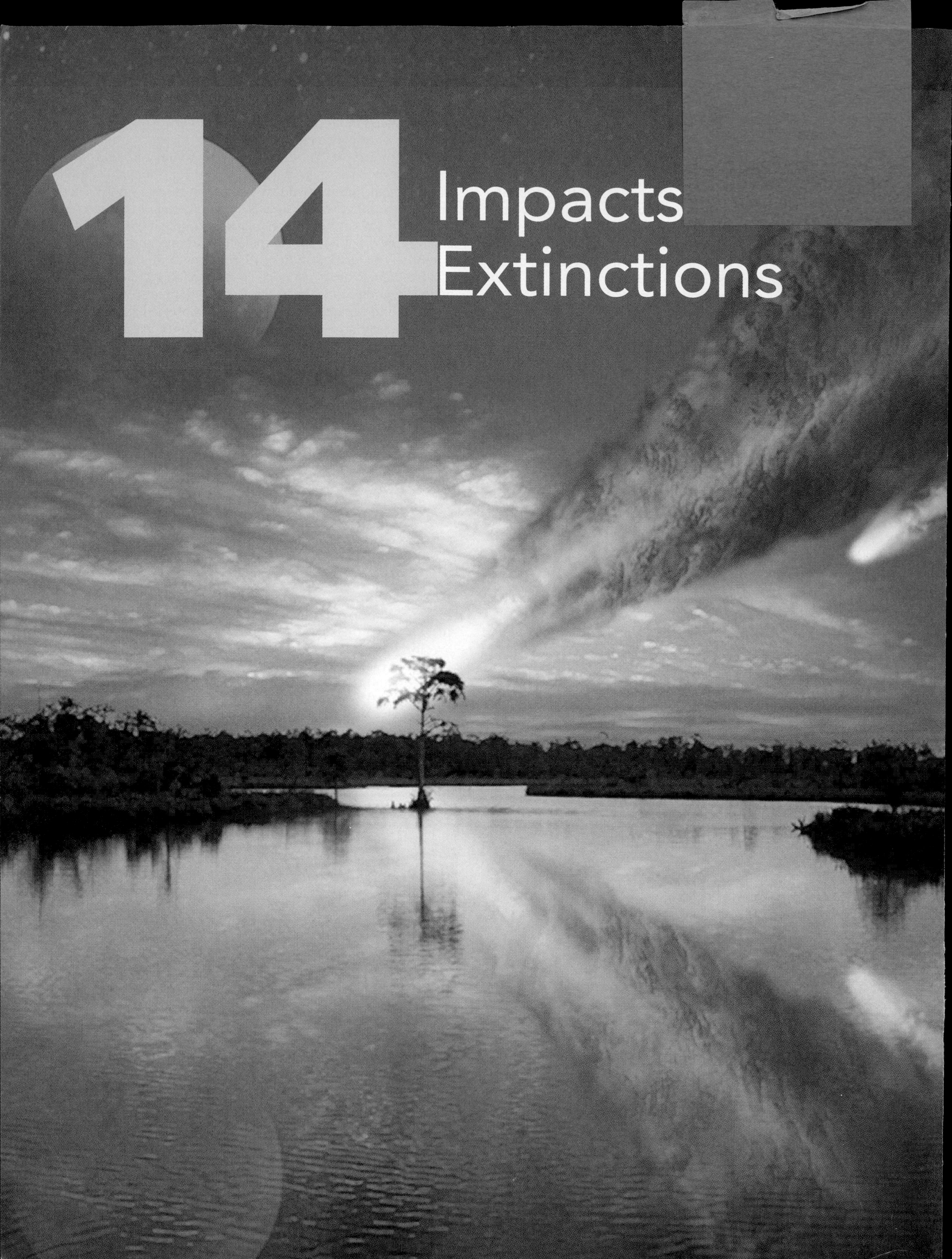

14 Impacts Extinctions

The Tunguska Event

On June 30, 1908, shortly after 7 A.M., witnesses in Siberia reported observing a blue-white fireball with a glowing tail descending from the sky. The fireball exploded above the Tunguska River Valley in a heavily forested, sparsely populated area. Later, calculations would show that the explosion had the force of 10 megatons of TNT—equivalent to 10 hydrogen bombs. Although there were few witnesses close to the event, the sounds from the explosion were heard hundreds of kilometers away, and the blast wave was recorded at meteorological recording sites throughout Europe. A tremendous air blast caused more than 2000 km^2 (770 mi.^{2}) of forest to be flattened and burned (Figure 14.1). The devastated area was more than twice the size of New York City.

The devastated area was more than twice the size of New York City.

A herdsman in the vicinity of the blast was one of the few people who witnessed the devastation on the ground. His hut was completely flattened by the blast, and its roof was blown away. Other witnesses a few tens of kilometers from the explosion reported that they were physically blown into the air and knocked unconscious; they awoke to find a transformed landscape of smoke and burning trees that had been blasted to the ground.

At the time of the explosion, Russia was in the midst of political upheaval. As a result, there was no quick response to or investigation of the Tunguska event. Finally, in 1924, geologists who were working in the region interviewed surviving witnesses and determined that the blast from the explosion was probably heard throughout an area of at least 1,000,000 km^2 (386,000 mi.^{2}), an area the size of Texas and New Mexico combined! They also found that the fireball had been witnessed by hundreds of people. Russian scientists went into the area in 1927, expecting to find an impact crater produced by the asteroid that had apparently struck the area. Surprisingly, they found no crater, leading them to conclude that the devastation had been caused by an aerial explosion, probably at an elevation of about 7 km (23,000 ft.). Later calculations estimated the size of the asteroid responsible for the explosion to be about 25 to 50 m (80 to 160 ft.) in diameter. It was most likely composed of relatively friable (easy to crumble) stony material.[1,2] The people of Earth were lucky that the Tunguska event occurred in a sparsely populated forested region. If the fireball had exploded over a city such as London, Paris, or Tokyo, the lives of millions of people would have been lost. Tunguska-type events are thought to occur on the order of every 1000 years.[3]

LEARNING OBJECTIVES

Earth has been bombarded by objects from space since its birth. Such impacts have been linked to the extinction of many species, including the large dinosaurs. The risk of impact from asteroids, comets, and meteoroids continues today. Your goals in reading this chapter are to

- know the difference between asteroids, meteoroids, and comets.
- understand the physical processes associated with airbursts and impact craters.
- understand the possible causes of mass extinction.
- know the evidence for the impact hypothesis that produced the mass extinction at the end of the Cretaceous Period.
- know the likely physical, chemical, and biological consequences of impact from a large asteroid or comet.
- understand the risk of impact or airburst of extraterrestrial objects and how that risk might be minimized.

◀ **Deep Impact**
Artist's conceptual drawing of two large extraterrestrial meteors entering Earth's atmosphere.
(The Virginian Pilot)

◀ FIGURE 14.1 **TUNGUSKA FOREST, SIBERIA, 1908** An aerial blast downed trees over an area of about 2000 km^2 (770 mi.^{2}). *(Sovfoto/Eastfoto)*

14.1 Earth's Place in Space

Preston Cloud, a famous geologist, wrote in 1978:

> Born from the wreckage of stars, compressed to a solid state by the force of its own gravity, mobilized by heat of gravity and radioactivity, clothed in its filmy garments of air and water by the hot breath of volcanoes, shaped and mineralized by 4.6 billion years of crustal evolution, warmed and peopled by the Sun, this resilient but finite globe is all our species has to sustain it forever.[4]

Cloud was referring to the original evolution of our planet that started about 4.6 billion years ago. One can go back even further and consider the start of the universe—some 14 billion years ago. The starting point is thought to have been an explosion, known as the "Big Bang." This explosion produced the atomic particles that later formed galaxies, stars, and planets (Figure 14.2). The first stars formed within the first 1 billion years after the Big Bang, and stars continue to form today. A star's life span depends on its mass; large stars have higher internal pressure and burn up more quickly than small stars. Objects with a mass equal to about that of our sun last around 10 billion years, and large stars with masses of 100 times that of our sun last only about 100,000 years. Thus, our sun is now a middle-aged star, about halfway through its life span.

When one of the massive, short-lived stars dies, it does so in a particularly spectacular way—as a supernova. A *supernova* occurs when the star is no longer capable of sustaining its mass and collapses inward, resulting in a high-energy explosion that scatters its mass into the void of space creating a vast *nebula.* Ultimately the force of gravity wins out and the matter within the nebula begins to collapse back inward on itself and new stars are born in what is called a *solar nebula* (Figure 14.3). It is from such a nebula that our sun formed 5 billion years ago. The sun grew by accretion of matter from a flattened pancake-like rotating disk that was dominated by hydrogen and helium molecules and trace amounts of dust composed of the other elements of the periodic table. As the solar nebula condensed under gravitational forces, our sun, which contains approximately 99.8 percent of the mass of the solar system formed at the center; the remaining mass became trapped in solar orbits as rings, similar to the rings around the planet Saturn today. Gravitational forces from the largest, densest particles attracted other particles in the rings until they collapsed into the planetary system we have today (Figure 14.4). During their

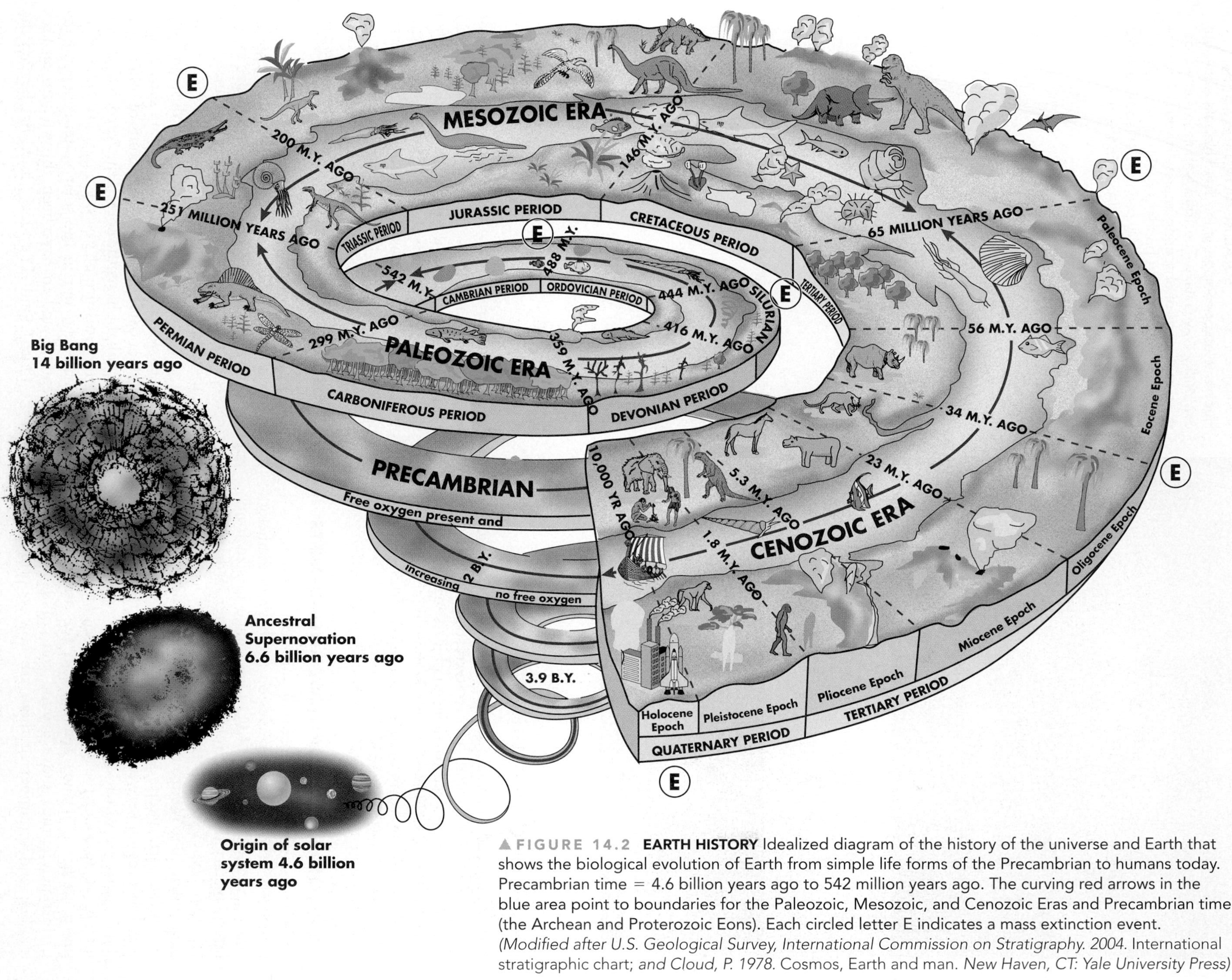

▲ FIGURE 14.2 **EARTH HISTORY** Idealized diagram of the history of the universe and Earth that shows the biological evolution of Earth from simple life forms of the Precambrian to humans today. Precambrian time = 4.6 billion years ago to 542 million years ago. The curving red arrows in the blue area point to boundaries for the Paleozoic, Mesozoic, and Cenozoic Eras and Precambrian time (the Archean and Proterozoic Eons). Each circled letter E indicates a mass extinction event. *(Modified after U.S. Geological Survey, International Commission on Stratigraphy. 2004.* International stratigraphic chart; *and Cloud, P. 1978.* Cosmos, Earth and man. *New Haven, CT: Yale University Press)*

FIGURE 14.3 **SOLAR NEBULA** Hubble telescope image of the Trifid Nebula within our own Milky Way Galaxy. The Trifid Nebula is named for the three bands of obscuring interstellar dust. Near the intersection of the dust bands, a group of recently formed, bright stars is easily seen. The Nebula is about 9000 light-years from Earth. *(NASA/Photoresearchers)*

FIGURE 14.4 **MASS OF THE SOLAR SYSTEM** Artist depiction of objects in our solar system, showing the relative sizes of the sun, planets, and Pluto (Plantesmal). *(NASA)*

early history, all the planets in our solar system, including Earth, were bombarded by extraterrestrial objects that ranged from dust-sized particles to objects many kilometers in diameter. However, the bombardment did not cease 4.6 billion years ago; it continues, although at a lesser rate, to the present.

ASTEROIDS, METEOROIDS, AND COMETS

Trillions of particles remain in our solar system. Astronomers group these particles on the basis of diameter and composition (Table 14.1). They range in size from interplanetary dust a fraction of a millimeter in diameter to larger bodies such as **asteroids** that range in diameter from about 10 m (30 ft.) to 1000 kilometers (620 mi.). For the most part, asteroids are found in the *asteroid belt,* which is a region between Mars and Jupiter (Figure 14.5). The asteroids, which are composed of rock material, metallic material, or rocky-metal mixtures, would pose no threat to Earth if they remained in the asteroid belt. Unfortunately, they move around and collide with one another, and a number of them are now in orbits that intersect Earth's orbit. Asteroids sometimes break into smaller particles known as **meteoroids** (Table 14.1), which range in size from dust particles to objects a few meters in diameter. When a meteoroid enters Earth's atmosphere, it is known as a **meteor.** As they streak through the atmosphere, frictional heating of meteors produces lightsometimes called *shooting stars.* Meteors occurring in large numbers produce the familiar *meteor showers.*

The last type of particle is a **comet.** Comets are distinguished from meteoroids and asteroids by their glowing tail of gas and dust (Figure 14.6). They range in size from a few meters to a few hundred kilometers in diameter, and they are thought to be composed of a rocky core surrounded by ice and covered in carbon-rich dust. In addition to frozen water, the ice contains carbon dioxide (dry ice), carbon monoxide, and smaller amounts of other compounds. As sunlight heats the comet, the "dirty ice" evaporates to form a mixture of gases. Comets are believed to have originated far out in the solar system, beyond the planet Neptune, and to have been thrown into an area called the Oort Cloud. The Oort Cloud lies beyond the

TABLE 14.1

Meteorites and Related Objects

Type	Diameter	Composition	Comments
Asteroid	10 m to 1000 km	Stony or metallic	Strong and hard if made of metal or some rock types; hard types may hit Earth, and weak types will explode in Earth's atmosphere at heights of several to hundreds of kilometers. Most originate in asteroid belt between Mars and Jupiter.
Comet	A few meters to a few hundred kilometers	Frozen water and/or carbon dioxide form an icy core that is surrounded by rock fragments and dust; like a "dirty snowball"	Weak, porous, will often explode in Earth's atmosphere at heights of kilometers to hundreds of kilometers. Most originate outside solar system in the Oort Cloud, 50,000 AU[a] from the sun, or within the solar system in the Kuiper Belt of comets; a tail, pointed away from the sun, forms as the icy core vaporizes and dust particles are shed from the object.
Meteoroid	Less than 10 m to larger than dust size	Stony, metallic, or carbonaceous (containing carbon)	Most originate from collisions of asteroids or comets. May be made of strong or weak rock.
Meteor	Centimeters to dust size	Stony, metallic, carbonaceous, or icy	Destroyed in Earth's atmosphere as shooting stars; their light is produced by frictional heating in the atmosphere.
Meteorite	Centimeters to asteroid size	Stony, metallic, or carbonaceous	Object that has actually hit Earth's surface. Stony variety, called a *chondrite*, is most abundant.[b]

[a]AU is the distance from Earth to the sun, about 240,000,000 km (150,000,000 mi.).

[b]There are many types of chondrites. They contain chondrules, which are small (less than 1 mm) spheroidal inclusions that are glassy or crystalline. Planets are constructed from chondrite meteorites (asteroids).

Source: Data from Rubin, A. F. 2002. Disturbing the solar system. *Princeton, NJ: Princeton University Press.*

Kuiper Belt and extends out as far as 50,000 times the distance from the Earth to the sun.[2] Early in the history of the evolution of Earth, bombardment by asteroids and comets contributed the building blocks of our planet, which was built up from the collision of innumerable smaller bodies.

14.2 Airbursts and Impacts

When entering Earth's atmosphere, asteroids, comets, and meteoroids travel at velocities that range from about 12 to 72 km per second (27,000 to

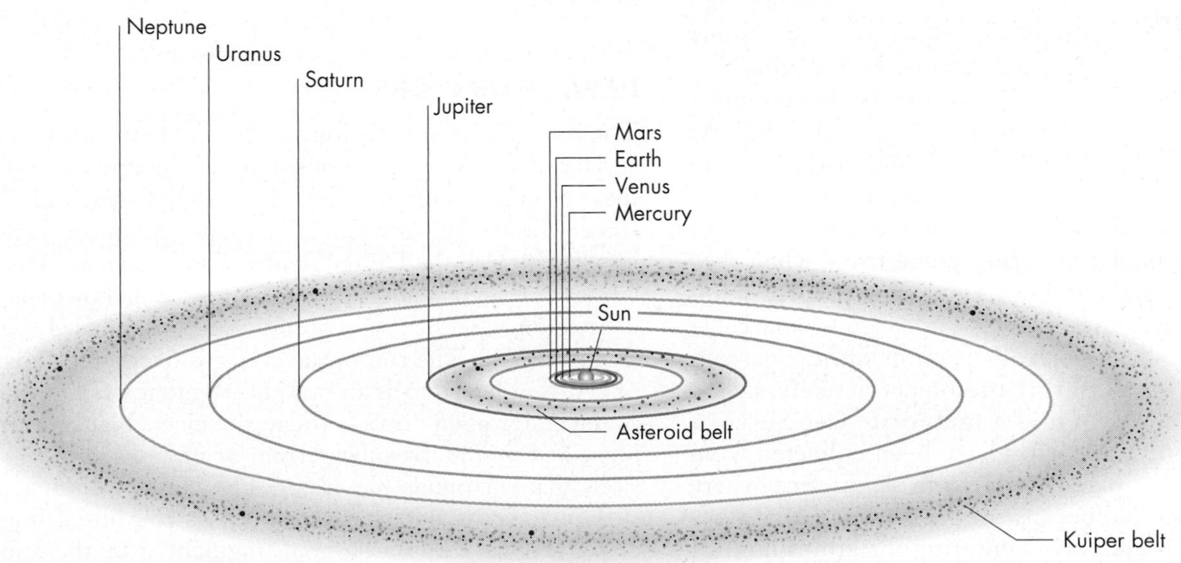

▲ FIGURE 14.5 **SOLAR SYSTEM** Diagram of our planetary system showing the asteroid belt and Kuiper belt. The Oort Cloud is too far away from the sun to be seen in this perspective. Orbits of the planets and belts are not to scale.

FIGURE 14.6 COMET HALE-BOPP Comet Hale-Bopp as it appeared in the night sky in 1997 and remained visible with the naked-eye for 18 months. Comets have a nucleus of ice and dust and a tail of evolved gases and dust grains. *(Aaron Horowitz/Corbis)*

161,000 mi. per hour).[1] The composition of asteroids and meteoroids varies (see Table 14.1). Some contain carbonaceous material, whereas others are composed of native metals, such as iron and nickel. Others are stony, consisting of silicate minerals, such as olivine and pyroxene—common minerals in igneous rocks. Stony meteoroids and asteroids are *differentiated*, meaning that they have undergone igneous, and sometimes metamorphic, processes as part of their geologic histories. As previously mentioned, meteoroids and asteroids are derived from the asteroid belt, between the orbits of Mars and Jupiter, whereas comets come from the Oort Cloud. Regardless of where they come from, when they intersect Earth's orbit and enter our atmosphere, meteoroids, asteroids, and comets undergo remarkable changes as they heat up from friction during descent and produce bright light. If the object actually strikes Earth, then we speak of it as a **meteorite** (see Survivor Story 14.1). Many meteorites have been collected from around the world, particularly from Antarctica where they are concentrated because of particular meteorological conditions. A meteoroid entering the atmosphere at about 85 km (53 mi.) above Earth's surface will become a meteor and emit light (Figure 14.7). It will then either explode in an **airburst** in the atmosphere at an altitude between 12 and 50 km (7 and 31 mi.), or collide with Earth. The Tunguska event that opened this chapter was a giant airburst, but there are a significant number of meteorite craters on the surface of Earth, and more than 175 individual craters or small crater fields have been identified.[6]

IMPACT CRATERS

The most direct and obvious evidence for impacts on the surface of Earth comes from studies of the **impact craters** they produce (Figure 14.8). The 49,000-year-old Barringer Crater in Arizona is perhaps the most famous impact crater in the United States. Also known as "Meteor Crater," it is an extremely well-preserved, bowl-shaped depression with a pronounced, upraised rim (Figure 14.9a). The rim of the crater is overlain by a layer of debris, referred to as an *ejecta blanket* that was blown out of the crater upon impact. Today, the ejecta blanket can be identified by the irregular terrain of mounds and depressions that surrounds the crater. The crater we see at the surface today is not nearly as deep as the initial impact crater (Figure 14.9b). Material fragmented by the impact has fallen back into the crater, and some of the walls have collapsed. Rocks that the asteroid hit were shattered and deformed, and the angular, broken pieces have been naturally cemented or fused together to form a rock type

SURVIVOR STORY 14.1

Meteorites in Chicagoland

When a meteorite shower lit up the sky, Pauline Zeilenga assumed the worst

When the thunder wouldn't stop, Pauline Zeilenga's first thought was that it must be nuclear war.

Zeilenga was in her living room with her husband, Chris, in their home in Park Forest, Illinois, when, shortly after midnight, the night sky was emblazoned with light.

"It wasn't like lightning," she said. "It was pitch black outside, and then the night literally turned to day. I said, 'What the heck was that?"'

About a minute later came "the loudest thunder you've ever heard," she said. "But it kept going and going and going. And, at a certain point, we realized it wasn't thunder, but an explosion."

It occurred to her that perhaps the Sears Tower had been hit—a thought not altogether far-fetched; the date was March 27, 2003, shortly after the United States and its allies began an extended bombing campaign in Iraq.

Then, when the thunderous rumbling finally subsided, the couple heard a sound like that of hail as small objects struck the exterior of the house.

Going outside to investigate, Chris, an avid collector of artifacts, Pauline said, quickly identified the real culprit behind the disturbance: a meteorite, or, by this point, many meteorites, falling across the Park Forest region.

But Pauline wasn't convinced: "I was telling my husband, 'Find out what it was.'"

The area was teaming with scientists and amateur meteorite enthusiasts in the days after the impact. Zeilenga said they went "meteorite hunting" with some others she had met over the telephone. Larger meteorites, including the ones that punched holes in the ceilings of residences in other areas of the region, can sell for thousands of dollars (Figure 14.A). But by the time Pauline and company had scoured the area, others had already cleared away most of the debris.

All in all, Zeilenga said the experience was one-of-a-kind once the shock had subsided. "It was the coolest," she said.

—CHRIS WILSON

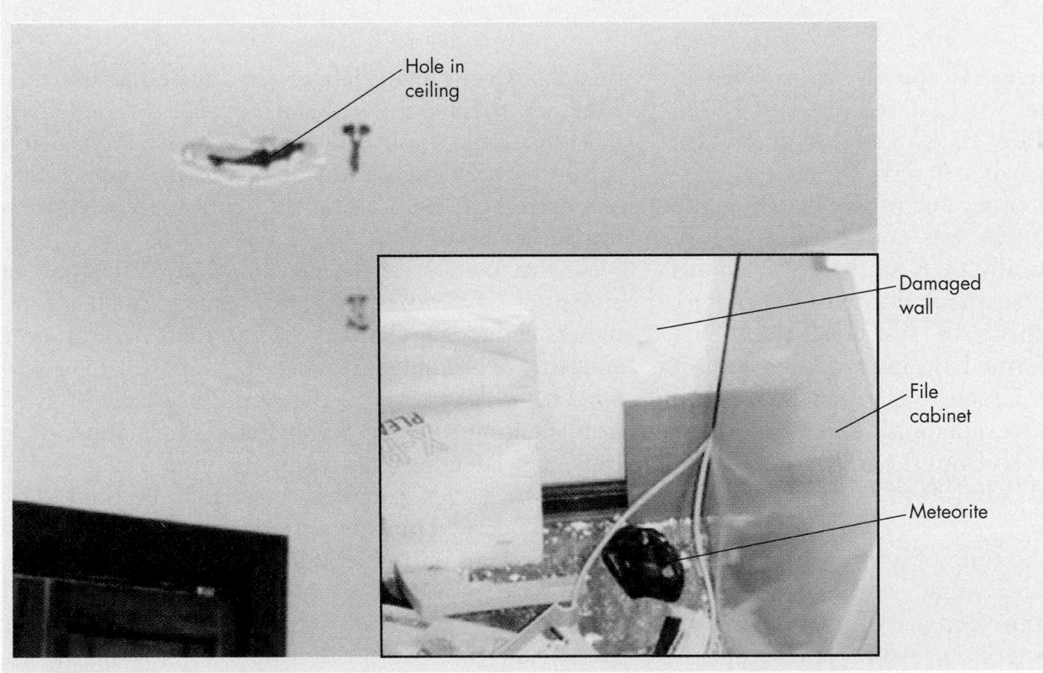

▲ FIGURE 14.A **PARK FOREST METEORITES** Pauline and Chris Zeilenga's experience was one of many in the Chicago suburbs of Park Forest, Steger, and Olympia Fields, Illinois. These pictures of Ivan and Colby Navarro's home show a hole through which a meteorite crashed into the house. The meteorite is sitting on the floor in the inset picture. Colby was in the room working at a computer when the meteorite crashed through the roof, struck the printer, banged off the wall, and came to rest next to the filing cabinet. Fortunately no injuries were reported; however, a number of roofs, windows, walls, and cars were damaged. This area was the most populated region to be hit by a meteorite shower in modern times. *(Ivan & Colby Navarro)*

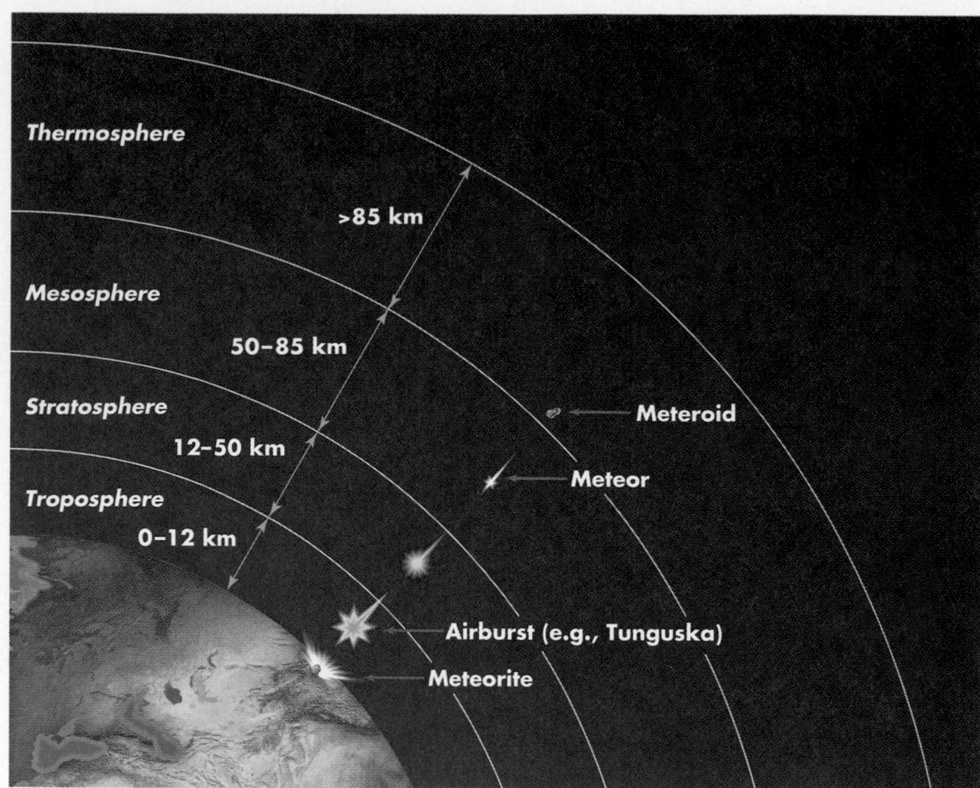

◀ FIGURE 14.7 **METEOROID ENTERING EARTH'S ATMOSPHERE** Idealized diagram showing a meteoroid entering Earth's atmosphere. A light phenomenon that results from a meteoroid's entry into the atmosphere is called a meteor. Most meteors are extremely small dust-to sand-size meteoroids. A large meteoroid may break apart in an aerial burst or crash into Earth's surface. *(Modified after R. Baldini: http://www.th.bo.infn.it/tunguska/impact/fig1_2.jpg)*

known as *breccia*. Barringer Crater is approximately 1.2 km (0.7 mi.) in diameter with a depth of about 180 m (590 ft.). The rim of the crater rises about 260 m (850 ft.) above the surrounding Arizona desert. When the existence of the Barringer Crater became widely known in the late nineteenth century, there was considerable debate concerning its origin. Ironically, G. K. Gilbert, the famous geologist who postulated that the majority of the moon's craters were formed by impacts, did not believe that Barringer Crater had been formed by impact. The impact origin for the Barringer Crater was later established through careful study and evaluation. This study concluded that the crater formed from the impact of a small asteroid, probably about 25 to 100 m (80 to 330 ft.) in diameter.[7]

Features that form at the time of impact differentiate impact craters from craters that result from other processes, such as volcanic activity. Impacts involve extremely high velocity, energy, pressure, and temperatures, which are normally not experienced or produced by other geologic processes. Most of the energy of the impact is in the form of kinetic energy, or energy of movement. This energy is transferred to Earth's surface through a shock wave that propagates into Earth. The shock wave compresses, heats, melts, and excavates earth materials. It is this transfer of kinetic energy that produces the crater.[8] The shock can metamorphose some rocks in the impact area, while others are melted and mixed with the materials of the impacting object itself. Most of the metamorphism consists of high-pressure modifications of minerals, such as quartz. These modifications are characteristic of meteorite impact, so they are extremely helpful in confirming the origin of an impact crater.

Impact craters can be grouped into two types, simple and complex. *Simple craters* are typically small, a few kilometers in diameter. Barringer Crater has the features characteristic of a simple impact crater (Figure 14.9b). *Complex impact craters* experience the same processes of vaporization, melting, ejection of material, formation of ejecta rims, and later infilling typical of simple craters, but the shape of a larger, complex crater may be quite different (Figures 14.10). During a period of seconds to several minutes following impact, complex craters may grow to sizes of tens of kilometers to more than 100 km (60 mi.) in diameter. In these craters, the rim collapses more completely, and the central crater uplifts following the impact. Typically, impact craters on Earth larger than about 6 km (4 mi.) are complex, whereas smaller craters tend to be of the simpler variety.

Geologically, ancient impact craters are difficult to identify because they are commonly either eroded or filled with sedimentary deposits that are younger than the impact. For example, subsurface imaging and drilling below the present Chesapeake Bay has identified a crater about 85 km (53 mi.) in diameter, now

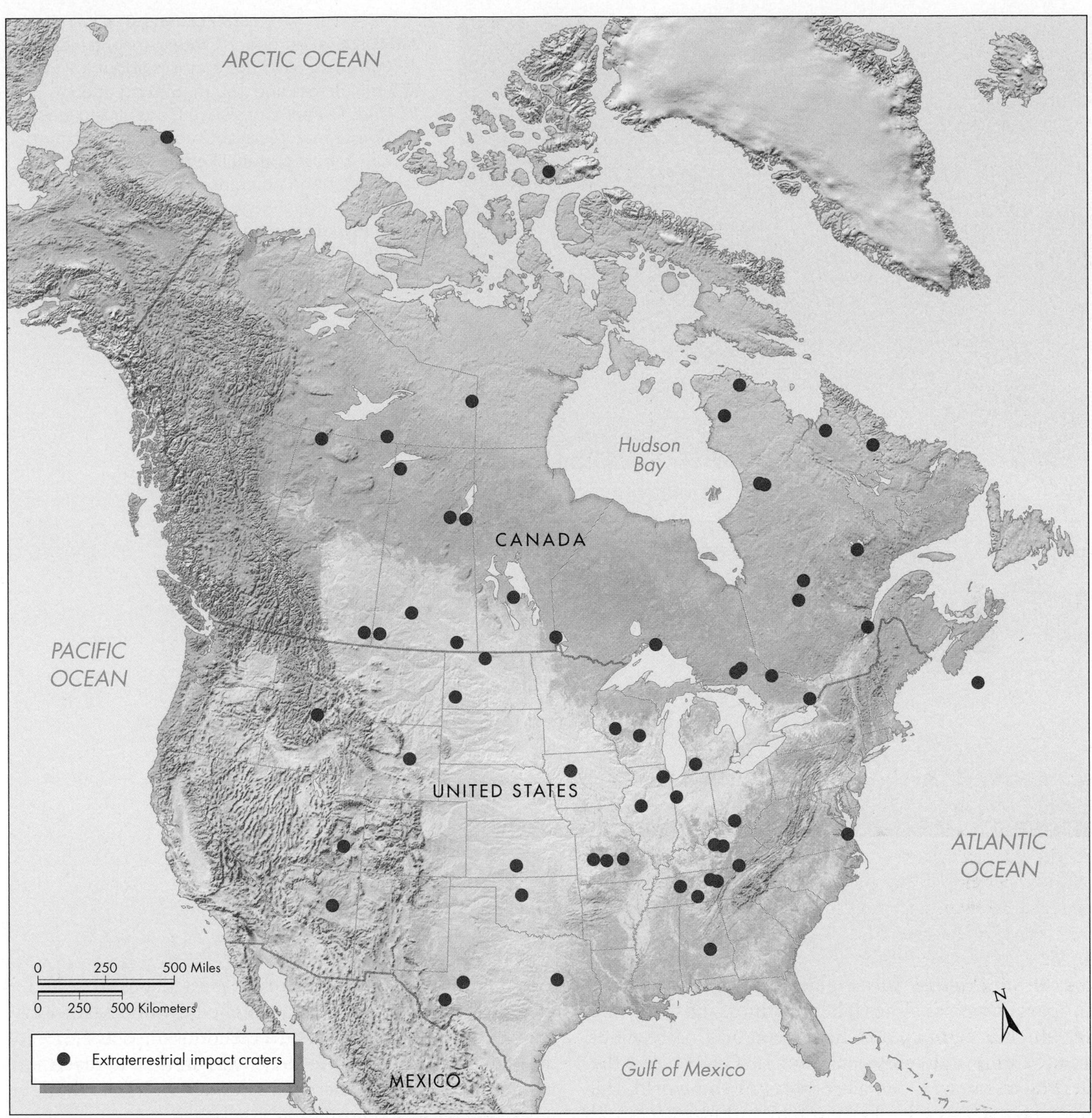

▲ FIGURE 14.8 **EXTRATERRESTRIAL IMPACT CRATERS OF THE UNITED STATES AND CANADA** Map showing locations of individual craters that have been identified at the surface and in the subsurface. Not shown are more than a half million Carolina Bay craters that cover the coastal plain from Florida to New Jersey. *(Modified after Grieve, R. A. F., 2006.* Impact structures in Canada. *St. Johns, Newfoundland and Labrador: Geological Association of Canada GEOtext 5; and Evans, K. R., Horton, Jr., J. W., Thompson, M. F., and Warme, J. E. 2005. The sedimentary record of meteorite impacts: An SEPM research conference.* The Sedimentary Record *March 2005, pp. 4–69)*

buried by about 1 km (0.6 mi.) of sedimentary deposits (Figure 14.10). The crater was produced by the impact of a comet or asteroid about 3 to 5 km (2 to 3 mi.) in diameter about 35.5 million years ago.[9] Compaction and faulting of earth materials above the buried crater may be, in part, responsible for the location of Chesapeake Bay.

A good example of an eroded impact crater, the Manicouagan Crater, can be found northeast of Quebec, Canada (Figure 14.11). At this crater, a ring-shaped reservoir about

(a)

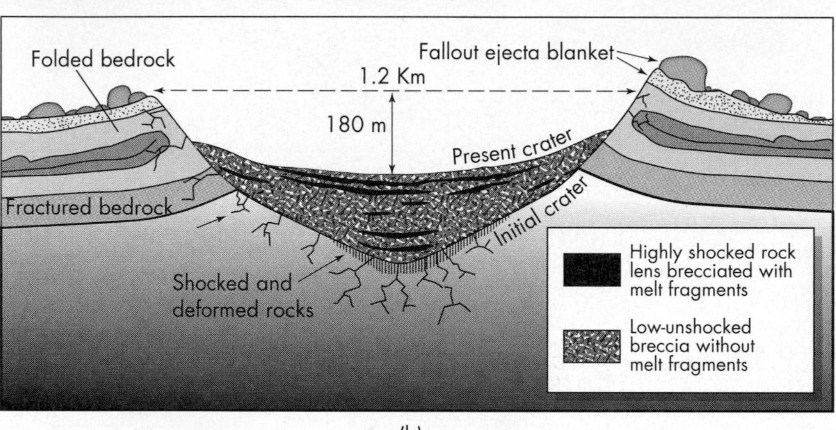

(b)

FIGURE 14.9 SIMPLE IMPACT CRATER IN ARIZONA (a) Barringer Crater, Arizona (about 49,000 years old). The crater is about 1.2 km (0.7 mi.) across and 180 m (590 ft.) deep. *(Charles O'Rear/Corbis)* (b) Generalized cross section of features associated with the crater. Simple impact craters like this typically have raised rims and no central uplift or peak. *(Modified after Grieve, R., and Cintala, M. 1999. Planetary impacts. In P. R. Weissman, L. McFadden, and T. V. Johnson, eds.,* Encyclopedia of the solar system. *San Diego, CA: Academic Press)*

65 km (40 mi.) across fills a glacially eroded valley in the impact breccia. The lake outlines most of the crater, which is estimated to have been originally about 100 km (62 mi.) in diameter (Figure 14.11).[6] One of the five largest terrestrial impact craters in the world, the Manicouagan Crater is close to 215 million years old (Table 14.2).[6]

Most detailed remote sensing studies of craters are of those on the moon and Mars. Impact craters are much more common on the moon than on Earth for three reasons: (1) Most impact sites on Earth are in the ocean where craters are subsequently buried by marine sediment or destroyed by plate tectonic processes; (2) impact craters on land are now generally subtle features that are eroded, buried, or covered with vegetation; and (3) smaller meteoroids and comets tend to burn up and disintegrate in Earth's atmosphere before striking the surface. In 1993, Gene and Carolyn Shoemaker and David Levy discovered a comet, later known as Shoemaker-Levy 9, from photographs they had taken through a telescope on Palomar Mountain in Southern California. Less than a year and a half later, astronomers watched Shoemaker-Levy 9 explode in one of the most tremendous impacts ever witnessed. From the start, the comet was unusual in that it was composed of several discrete fragments with several bright tails, and it was reported as a misshapen comet. Shoemaker-Levy 9 was one of about 50 comets in the Jupiter family that circle between Jupiter and the sun. The comet's orbit was tied to that of Jupiter, and, after years of orbiting the planet, it separated into 21 fragments known as a "string of pearls."

From telescopes on Earth and the Hubble Space Telescope in Earth orbit, astronomers watched as fragments of the comet entered Jupiter's atmosphere at speeds of 60 km (37 mi.) per second. The fragments then exploded, releasing between 10,000 and 100,000 megatons of energy, depending on their size. More energy was released than if all of Earth's nuclear weapons were detonated at the same time. Hot, compressed gases expanded violently upward from the lower part of Jupiter's atmosphere at speeds as

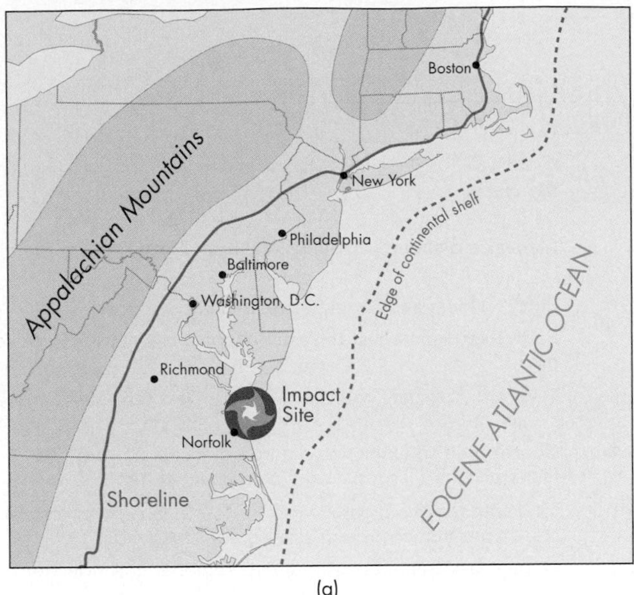

(a)

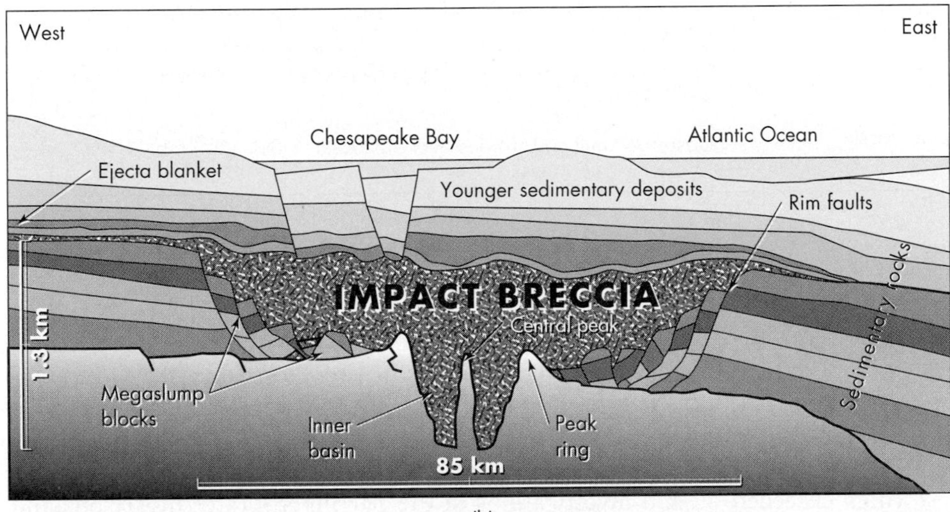

(b)

◀ **FIGURE 14.10 CHESAPEAKE BAY IMPACT CRATER** (a) Map showing 35.5-million year-old impact crater at the mouth of Chesapeake Bay. Crater formed during the Eocene Epoch of the Tertiary (Paleogene) Period. Approximate shoreline of the Atlantic Ocean during the Eocene is shown as a solid blue line between the Appalachian Mountains and labeled U.S. cities. Blue dashed line offshore is the seaward edge of the continental shelf. (b) Note that there is about a 13X vertical exaggeration in the cross section. The crater is approximately 85 km (53 mi.) in diameter and 1.3 km (4300 ft.) deep. Most of the crater is filled with angular rock fragments that have been cemented together to form a rock called breccia. This buried crater was discovered by the U.S. Geological Survey during a groundwater study. *(Williams, S., Barnes, P., and Prager, E. J. 2000. U.S. Geological Survey Circular 1199).*

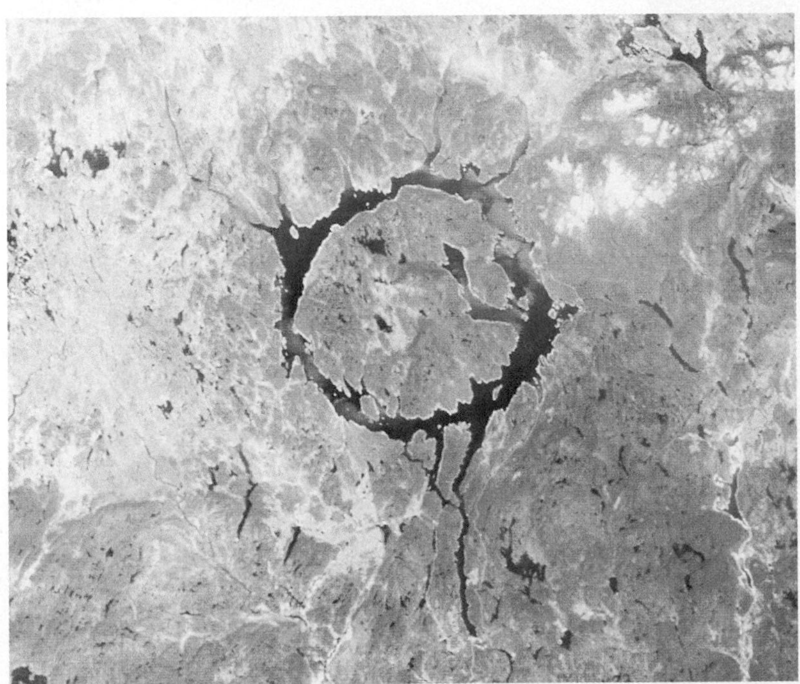

◀ **FIGURE 14.11 COMPLEX IMPACT CRATER IN QUEBEC** Satellite image of the Manicouagan impact structure northeast of the city of Quebec in Quebec, Canada. Complex craters typically have a faulted rim and central peak. This type of crater develops when the impact is so large that the Earth's crust rebounds from the impact and forms a central peak. The ring-shaped lake is about 65 km (40 mi.) in diameter. *(NASA)*

TABLE 14.2

Notable Impacts and Airbursts of Extraterrestrial Objects

Age[a]	Feature/Event	Location	Significance
4,450,000,000	Birth of the moon	Earth	Mars-size planetary body collides with Earth to form the moon.
2,023,000,000	Vredefort Dome	South Africa	Earth's oldest and largest terrestrial impact crater.
1,850,000,000	Sudbury Crater	Ontario, Canada	Earth's second largest terrestrial impact crater; rich in nickel and copper ores.
214,000,000	Manicouagan Crater	Quebec, Canada	Tied with Popigai Crater for fourth largest terrestrial impact crater.
64,980,000	Chicxulub Crater	Below Yucatán Peninsula and Gulf of Mexico	Contributed to extinction of large dinosaurs; produced enormous tsunamis; third largest terrestrial impact crater.
35,700,000	Popigai Crater	Siberia, Russia	Tied with Manicouagan Crater for fourth largest terrestrial impact crater; has numerous industrial-grade diamonds.
35,500,000	Chesapeake Bay Crater	Buried below Virginia, United States	Largest U.S. impact crater; affects regional groundwater flow.
49,000	Barringer Crater (aka Meteor Crater)	Arizona, United States	First impact crater identified on Earth; training site for Apollo astronauts.
12,900	Younger Dryas Airburst	Southern Canada	Contributed to extinction of large Pleistocene mammals and the Clovis civilization.
42	Tunguska Airburst	Siberia, Russia	Destroyed millions of trees over a 2200-km^2 (800-$mi.^2$) area.

[a]Years before 1950.

Source: Modified from Grieve, R. A. F. 2006. Impact structures in Canada. *St. Johns, Newfoundland and Labrador: Geological Association of Canada, GEOtext 5.*

high as 10 km (6 mi.) per second. Gas plumes from the larger impacts reached elevations of more than 3000 km (1900 mi.), a height that is around 340 times as high as Mt. Everest, the tallest mountain on Earth. Tremendously large rings developed in Jupiter's atmosphere around the sites of the impact (Figure 14.12). These rings exceeded the diameter of the Earth! It was truly a remarkable show for astronomers and a sobering event for those who consider that impacts such as this might one day occur on Earth.[2]

After the impact of the fragmented comet on Jupiter and investigations of the Barringer Crater in Arizona and

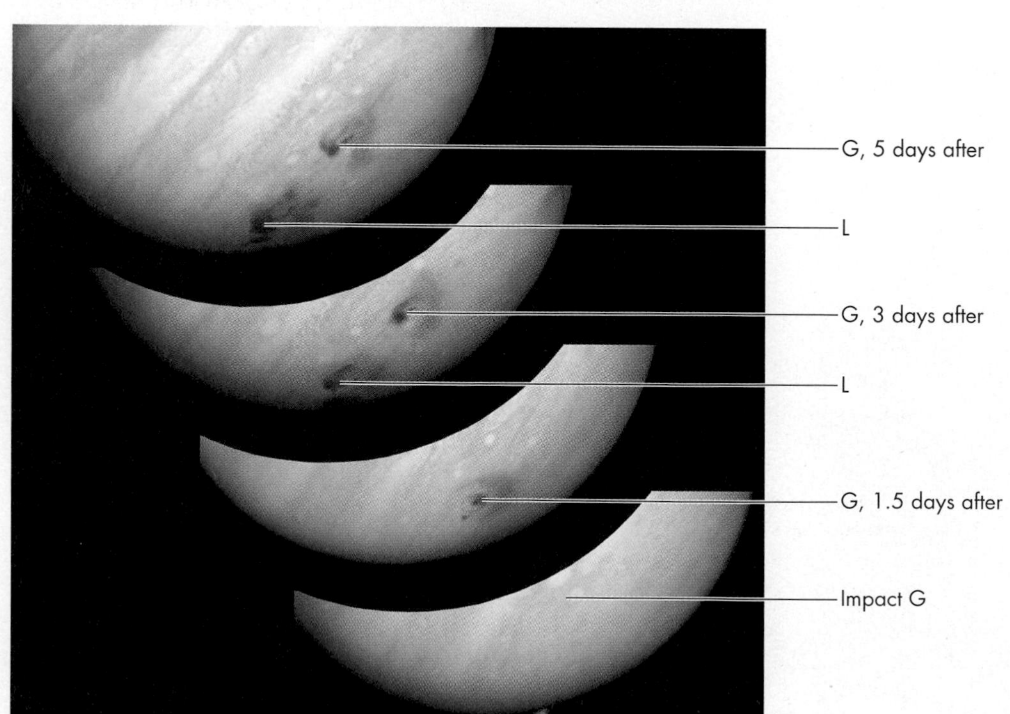

FIGURE 14.12 COMET HITS JUPITER Spectacular impact of the comet Shoemaker-Levy 9G on the planet Jupiter in 1994. The comet was composed of 21 fragments known as a "string of pearls." One fragment after another entered Jupiter's atmosphere and exploded. From bottom to top, these four images were taken several days apart. The reddish-brown dots and rings mark the impact sites of fragments G and L. The diameter of these rings exceeded the diameter of Earth. *(NASA)*

several other craters on Earth, the idea that there could be a catastrophic impact on Earth was finally accepted. At the time of the Tunguska event, described at the beginning of this chapter, several bizarre ideas were suggested to explain it, such as nuclear explosions and the explosion of an alien spaceship! As we shall see, the idea that impacts by asteroids and comets might cause catastrophes on Earth, and even mass extinction of life, was greatly resisted by scientists until very recently (see Case Study 14.2).

14.3 Mass Extinctions

A **mass extinction** is characterized by the sudden loss of large numbers of plants and animals relative to the number of new species being added.[10] Because the geologic time scale was originally based upon the appearance and disappearance of various fossil species, mass extinctions generally coincide with the boundaries of geologic periods or epochs of the time scale (Figure 14.13). Most hypotheses for mass extinctions involve relatively rapid climate change. This climate change can be triggered by plate tectonics, volcanic activity, or extraterrestrial impact. Plate tectonics is generally a relatively slow process that moves the position of continents and, thus, habitats to different locations. On occasion, plate movements create new patterns of ocean circulation, which have a major effect on climate. Extremely large volcanic eruptions can also cause significant climate change. Large basaltic eruptions producing huge volumes of *flood basalts* can release large quantities of carbon dioxide into the atmosphere and cause global warming. In contrast, large explosive volcanic eruptions of more silica-rich lava can inject tremendous quantities of volcanic ash into the upper atmosphere and cause global cooling. Finally, climate change is one of several effects of extraterrestrial impacts or airbursts that can contribute to extinctions.

During the past 550 million years of Earth's history, at least six major mass-extinction events occurred. The earliest mass extinction occurred approximately 446 million years ago, near the end of the Ordovician Period (Figure 14.13). One of the largest mass extinctions in Earth's history, with around 100 families of animals becoming extinct, this event coincided with a major continental glaciation in the Southern Hemisphere.[11] In fact, it appears that this event was actually two extinctions: one when the climate cooled and the other when the climate warmed following the glacial interval.[11] The largest recorded mass extinction occurred near the end of the Permian Period, about 250 million years ago, when 80 to 85 percent of all species alive died out (Figure 14.13).[11] Although there is now evidence for an impact at the Permian-Triassic boundary, it is also believed that this mass extinction may not have been caused by a single catastrophe but may have spanned a period of about 7 million years. The extinction also corresponds to eruption of the massive Siberian flood basalts, which likely emitted voluminous greenhouse gasses to the atmosphere and may have dramatically changed the climate. In addition, sea-level fall and assembly of the supercontinent of Pangaea at this time would have dramatically reduced the area of shallow marine environment where many of the species lived.

The third mass extinction at the Triassic-Jurassic boundary (Figure 14.13), 202 million years ago, also appears to

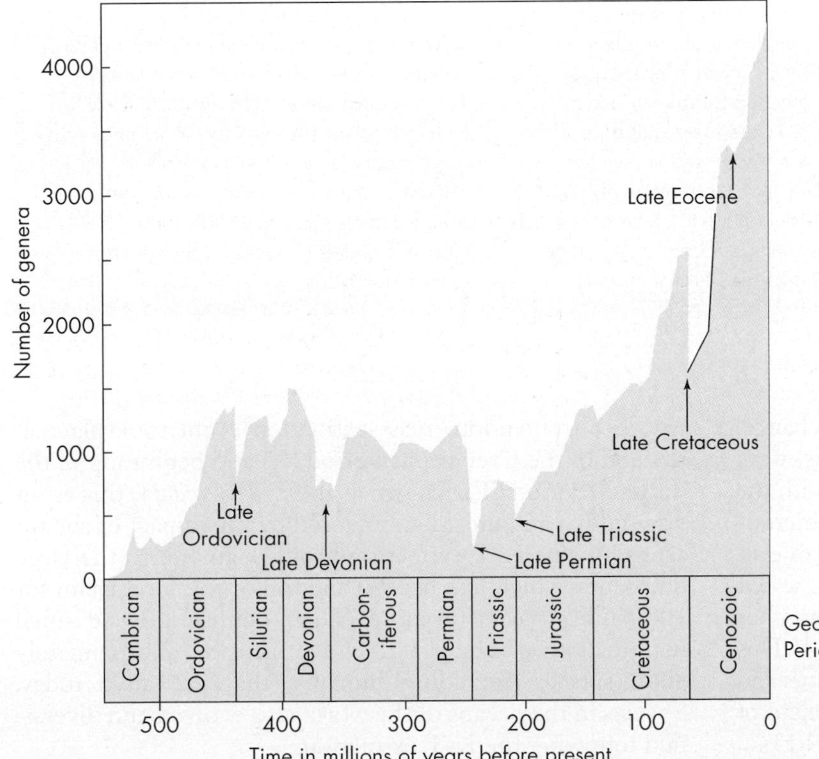

FIGURE 14.13 MASS EXTINCTION EVENTS Biodiversity diagram illustrating the increasing diversification of life through the Phanerozoic Period, punctuated by mass extinction events.

CASE STUDY 14.2

Uniformitarianism, Gradualism, and Catastrophe

The idea that Earth could be hit by large objects from outer space was not accepted for many hundreds of years, in spite of the numerous eyewitness accounts in various parts of the world. These accounts may have started with a lethal meteorite that landed in Israel in the year 1420 B.C.[1] Up until the fifteenth century, when Galileo invented the telescope and clearly demonstrated that the planets, including Earth, revolve around the sun, the religious establishment did not accept the idea; in fact, Galileo was jailed for his beliefs. In the year 1654, Irish Archbishop Ussher proclaimed that the Earth was created in the year 4004 B.C., making it approximately 6000 years old. This was the dogma of the time and not to be disputed. However, people studying the formation of mountains, such as the Alps, large river valleys, and other features, had a hard time understanding how these features could be formed in only 6000 years. Finally, in 1785, the Scottish doctor James Hutton wrote an influential book that introduced the concept of *gradualism*, or *uniformitarianism*. This concept states that present geological processes may be studied to learn the history of the past—"the present is the key to the past." He also argued that the Earth must be much older than 6000 years to allow the gradual processes of erosion, deposition, and uplift to form mountain ranges and other features on Earth's surface. In 1830, Charles Lyell wrote another influential book on geology, popularizing the role of gradual processes and casting aside the dogma of a young Earth. He proclaimed that the Earth had a long history that could be understood by studying present-day processes and the rock record. When Ussher's young Earth became the accepted belief, Earth scientists were forced to conclude that most of the processes that formed our planet were catastrophic in nature. This perspective explained biblical events, such as Noah's Flood, but not much else. After scientists overturned the young Earth idea, they could then explain the history of Earth in terms of processes that could be carefully observed, such as uplift and erosion. Charles Darwin was impressed with Charles Lyell's book and applied the ideas of an ancient Earth and uniformitarianism to his concept of biological evolution. The concept of gradualism lasted into the twentieth century and culminated with the discovery of plate tectonics, perhaps the crowning glory of uniformitarianism. The theory of plate tectonics explains the present position and origin of continents by the relatively slow processes of seafloor spreading and uplift.

However, occasionally, scientists also discovered evidence of the role of catastrophic events. Some scientists pointed to large craters on Earth's surface, which they believed were the result of asteroid impacts. Also, it was well known that there were a number of relatively rapid extinctions during Earth's history in which a large percentage (commonly half or more) of the species of plants and animals perished relatively suddenly. Five of these extinctions are in the relatively distant geologic past; the most recent mass extinction is ongoing today as a result of human activity (see Figure 14.2). The mass extinction that occurred 65 million years ago at the K-T (Cretaceous-Tertiary) boundary appears to have been caused, to a great extent, by the impact of an asteroid. The asteroid that struck the Yucatán Peninsula in Mexico was approximately 10 km (6 mi.) in diameter. Even faced with direct evidence for the impact, many scientists were skeptical that a mass extinction might be triggered by the impact of an asteroid. Also, it wasn't until 1947 that Barringer Crater (aka Meteor Crater) in Arizona was finally listed as a probable impact event. Up to that time, it was still often referred to as a *cryptovolcanic* event. This means that the volcanic activity was "hidden," yet it produced a crater or disturbed area at the surface. Eventually Barringer Crater was shown to be the result of an asteroid impact, probably with airbursts; since then, more than 175 other large impact features have been identified on all the continents and on the ocean floor.

These observations have led to a new concept, *punctuated uniformitarianism*. This concept states that although uniformitarianism explains the long geologic record of gradual mountain building, canyon erosion, and landscape construction, periodic catastrophic events do occur and can cause mass extinctions.

have been related to volcanic activity and climate change. Large amounts of carbon dioxide, a greenhouse gas, were released from basaltic volcanic eruptions associated with the breaking apart of the supercontinent Pangaea. These eruptions may have been the largest volume of basaltic lava ever released by volcanoes on land.[11] Earth's temperatures, which were already 3°C (5°F) warmer than today, increased another 3° to 4°C (5° to 7°F) at the end of the Triassic Period.[11] This rapid climate change resulted in the extinction of plants and animals on land and in the ocean. Half of all the genera of marine animals became extinct in this event (Figure 14.13).[11]

Another tremendous mass-extinction event took place at the end of the Cretaceous Period (K) and beginning of the Tertiary Period (T). Known as the *K-T boundary*, this event was sudden and most likely caused by the impact of a giant asteroid. The K-T extinction brought an end to the large dinosaurs, which had been at the top of the food chain for 100 million years or more. Their demise allowed small mammals to diversify and evolve into the approximately 4000 species, including humans, that are alive today. Species in the oceans of the world also evolved and diversified following the K-T extinction.

A fifth mass extinction took place near the end of the Eocene Epoch about 34 million years ago (Figure 14.13). Although there is some evidence for an asteroid or comet impact at that time, most scientists link this extinction to climate change brought on by plate tectonics. The movement of tectonic plates at this time allowed a cold ocean current to develop around Antarctica; the resulting cooling produced the Antarctic ice cap, which is still present today. This resulted in a global cooling and drying of the climate.[11]

The final mass extinction occurred near the end of the Pleistocene Epoch. This extinction of mammals, reptiles, amphibians, birds, fish, and plants continues today. The initial cause of this event was first thought to be some combination of overhunting by Stone Age man and climate change. However, new evidence indicates that it was initiated by an airburst of a comet (see Case Study 14.3). In many ways, this mass extinction continues today because of human activities, including loss of habitat from land-use changes and global warming, widespread deforestation, and the application of pesticides and other chemicals.[10]

Although most of these extinction events are believed to have been related to climate changes, the case for impact-related mass extinction 65 million years ago, at the K-T boundary, is well documented. This event produced the third largest crater on Earth and provides a good example of the effects of a catastrophic impact—a blast-produced shock wave, fallout of ejecta, a large tsunami, global wildfires, dust loading in the atmosphere, and finally a collapse of ecological food chains in the ocean and on land.

K-T BOUNDARY MASS EXTINCTION

One of the great geologic detective stories of the past 50 years is the investigation of the K-T mass extinction. We now have evidence that 64,980,000 years ago, an asteroid with a diameter of about 10 km (6 mi.) hit Earth along the northern shore of what is now the Yucatán Peninsula. That event altered Earth's history forever. Although much of the physical landscape of Earth remained unchanged after the impact, the planet's inhabitants were largely changed. The large dinosaurs disappeared, as did many species of plants and other animals in both the oceans and on land. Approximately 70 percent of all genera and their associated species died off (Figure 14.13). Some reptiles, such as turtles, alligators, and crocodiles, and some birds, plants, and smaller mammals survived, either because of their habitat requirements or their location at the time of the impact. Not all areas experienced widespread wildfires following the impact, and some plants and animals were better adapted to subsequent changes in climate.[12] The impact contributed to the demise of the large dinosaurs on land and swimming reptiles in the oceans. Extinction of these animals set the stage for the evolution of mammals. This evolution eventually produced primates and humans. What would the world look like today if the K-T extinction had never happened? There's a good chance that humans never would have evolved!

Since this extinction has been so important for our evolution, we will look more closely at how scientists developed the hypothesis that the K-T mass extinction was caused by the impact of a large asteroid. The story is full of intrigue, suspense, rivalries, and cooperation—typical of many of the great scientific discoveries.[13] Scientists with backgrounds in geology, physics, chemistry, biology, geophysics, and astronomy worked together to develop and test the hypothesis that the K-T mass extinction was triggered by an impact. Walter Alvarez, a professor at the University of California, asked the question that started it all: What is the nature of the boundary between rocks of the Cretaceous and Tertiary Periods?

Alvarez was interested in Earth's history and, particularly, in reading that history as recorded by rocks. Early in his studies of the K-T boundary, Alvarez teamed up with his physicist father, Luis, and they decided to measure the concentration of a platinum-group metal called *iridium* in the thin clay layer that marks the K-T boundary in Italy. Walter Alvarez and colleagues had initially gone to the site in Italy to study the magnetic history of Earth. What they found at the K-T boundary was a very thin layer of clay (Figure 14.14), and it looked like the extinction of many species occurred abruptly at the clay layer—fossils found in rocks below the clay were simply not there in the rocks above the clay.[13]

They then asked the question: How much time was involved in the deposition of the clay layer? Was it a few years, a few thousand years, or millions of years? The approach they took was to measure the amount of iridium in the clay. They chose iridium because it is found in small concentrations in meteorites, and because the global rate of accumulation of meteoritic dust on Earth is constant. The deep ocean floor accumulates meteoritic dust like a tabletop or picture frame in your home—the longer ocean-floor sediment sits undisturbed, the more iridium it will accumulate. Clay washing into the ocean from the land can cover previously deposited meteorite dust. The faster the rate of sedimentation, the more diluted the meteorite dust becomes. Slow rates of sedimentation allow time for more meteorite dust to accumulate in ocean deposits.

What Walter Alvarez and his colleagues found was entirely unexpected. The team had anticipated measuring approximately 0.1 parts per billion of iridium in the clay layer, which they thought would represent slow accumulation through time. If the clay layer were deposited rapidly, then the amount of iridium would be even less. What they actually found was about 3 parts per billion—30 times as much as expected. As you recall from our discussion of CO_2 concentration in the atmosphere, 3 parts per million (ppm) is a very small quantity, and 3 parts per billion (ppb) is 1000 times smaller than 3 ppm! Although 3 ppb is an extremely small amount of iridium, it was much more than could be explained by their previous hypothesis of slow deposition over time. They reevaluated the data, this time including samples that had been removed for treatment before measurement, and got a final value of about 9 ppb,

◀ **FIGURE 14.14 EVIDENCE FOR IMPACT** The Cretaceous/Tertiary (K-T) boundary in Italy lies in a thin clay layer. The scientist on the left is pointing to the layer below his left knee. This layer has an anomalously high concentration of the metal iridium. High concentrations of this metal are found in extraterrestrial objects such as meteorites. Its presence here is consistent with an asteroid impact. *(Walter Alvarez)*

which is nearly 100 times as great as expected. This discovery led them to a new hypothesis: The iridium might be the remains of a single asteroid impact. The team's iridium discovery, along with its hypothesis of an extraterrestrial cause for the extinction at the K-T boundary, was published in 1980.[14] In that paper, the team also reported elevated concentrations of iridium in deep-sea sediments in Denmark and in New Zealand, all at the K-T boundary. Although the discovery of the iridium anomaly at several places around the world made the team more confident of its impact hypothesis, the scientists still had no crater. This absence prompted other scientists to search for potential craters that formed 65 million years ago.[13]

Would it be possible to find the crater? The K-T crater might have been completely filled with sediment and no longer recognizable, or the crater might have been on the ocean floor and subsequently destroyed by plate subduction. Fortunately, the site of the crater was identified in 1991.[15] Geologists in Mexico studying the structural geology of the Yucatán Peninsula discovered what they determined to be a buried impact crater with a diameter of at least 180 km (112 mi.). The crater, named Chicxulub for a nearby village, is nearly circular and has a clear boundary between unfractured rocks within the crater and fractured rocks outside the crater. About half of the crater lies beneath the seafloor of the Gulf of Mexico and half beneath sedimentary rocks and vegetation on the northern end of the Yucatán Peninsula (Figure 14.15). On land, the researchers found a semicircular pattern of sinkholes, known to the Mayan people as *cenotes*, which corresponded directly to the edge of the proposed impact crater. The cenotes range from about 50 to 500 m (160 to 1640 ft.) in diameter. They were presumably formed by the chemical weathering of fractured limestone on the outside of the crater boundary. Thus, both the distribution of fractured rock and the cenotes must be related to the circular structure because there are no other geologic features in the area that could explain the curved pattern. The crater, which is now filled with other rock, is believed to have been as deep as 30 to 40 km (18 to 25 mi.) at the time of impact. However, subsequent slumping and sliding of materials from the sides soon filled in much of the crater, and sedimentation over the past 65 million years completely buried the structure. In drilling within the crater, geologists have found glassy melt rock, interpreted to be impact breccia, below a massive layer. The force of the impact excavated the crater, fractured the rock on the outside, and produced the breccia. The glassy nature of the breccia suggests that sufficient heat was present to melt rocks immediately after the impact.[16] Another study of the crater found glass mixed in with and overlain by the breccia, as well as evidence of shock metamorphism commonly associated with impact features.[17] Most scientists now believe that the asteroid impact that struck the area nearly 65 million years ago did, in fact, contribute significantly to the K-T mass extinction.

After identifying the site of the crater and the evidence to support its existence, questions naturally arose regarding how such an event could cause a global mass extinction.[13] The asteroid that formed the Chicxulub crater was huge; its diameter is estimated at about 10 km (6 mi.). By comparison, many jet aircraft fly at cruising altitudes of around 10 km above Earth's surface. The summit of Mt. Everest is not as high as 10 km, but it is close. Consider, too, that the asteroid struck the atmosphere of Earth at a speed of about 30 km (19 mi.) per second, which is about 150 times faster than a jet airliner travels. The amount of energy that was released is estimated to have been about 100 million megatons, roughly 10,000 times as great as the entire nuclear arsenal of the world.

▲ **FIGURE 14.15 LARGE IMPACT CRATER IN MEXICO** Map showing location of the Chicxulub impact crater on the north shore of the Yucatán Peninsula in Mexico. This crater is buried below sedimentary rocks on the peninsula and the seafloor of the Gulf of Mexico. Formed by an asteroid impact about 65 million years ago, the age of this crater coincides with the Cretaceous/Tertiary (K-T) mass extinction. *(D.Van Ravenswaay/PhotoResearchers, Inc.)*

Walter Alvarez and numerous other geologists, physicists, paleontologists, and astronomers have reconstructed a likely sequence of events for the impact and its aftermath (Figure 14.16). The hole blasted in Earth's crust was nearly 200 km (125 mi.) across and 40 km (25 mi.) deep. At an altitude of 10 km (6 mi.) in the upper atmosphere, and moving at 30 km (19 mi.) per second, the asteroid would have taken less than half a second to reach Earth and almost instantaneously produce the large crater. When it contacted Earth, shock waves quickly crushed the rocks beneath, filling in all the cracks, partially melting the rocks, producing the breccia, and blasting bits and pieces high into the atmosphere. All this activity probably took about 2 seconds as the shock wave and heat vaporized rocks on the outer fringes of the impact. The tremendous amount of debris built a huge ejecta blanket around the crater. A gigantic cloud of vaporized rocks and gases would have produced an equally large fireball that rose up and formed a huge mushroom cloud. The explosion itself and the rising materials would likely have been sufficient to accelerate and eject material far beyond Earth's surface. Particles of rock were blasted into ballistic trajectories before they fell back to the ground. The fireball would have produced sufficient heat to set fires around the globe. Vaporization of gypsum salts in the bedrock and ocean water produced sulfuric acid in the atmosphere. Additional acids were added as a result of burning nitrogen in the atmosphere. Thus, following the impact, acid rain probably fell for a long period. The dust in the atmosphere circled Earth, and, for months, essentially no sunlight reached the lower atmosphere. The lack of sunlight stopped photosynthesis on land and in the ocean. Acid rain was toxic to many living things, particularly terrestrial and shallow marine plants and animals. As a result, the food chain virtually stopped functioning because the base of the chain had been greatly damaged. Part of the impact occurred in the ocean and would have significantly disturbed the seafloor, generating tsunamis that could have reached heights of more than 300 m (1000 ft.). These waves raced across the Gulf of Mexico and inundated parts of North America.[13,18] Wildfires ravaged southern North America, all of Central America, and parts of South America, Africa, Asia, and Australia.[12] The climate first cooled from lack of sunlight and then significantly warmed as aerosols and carbon dioxide from the pulverized limestone of the Yucatán Peninsula enhanced the greenhouse effect.[11] Finally, large numbers of ferns restored plant cover on the burned landscape.[11]

In summary, the impact of the asteroid caused a global catastrophic killing, which we refer to as a mass extinction.

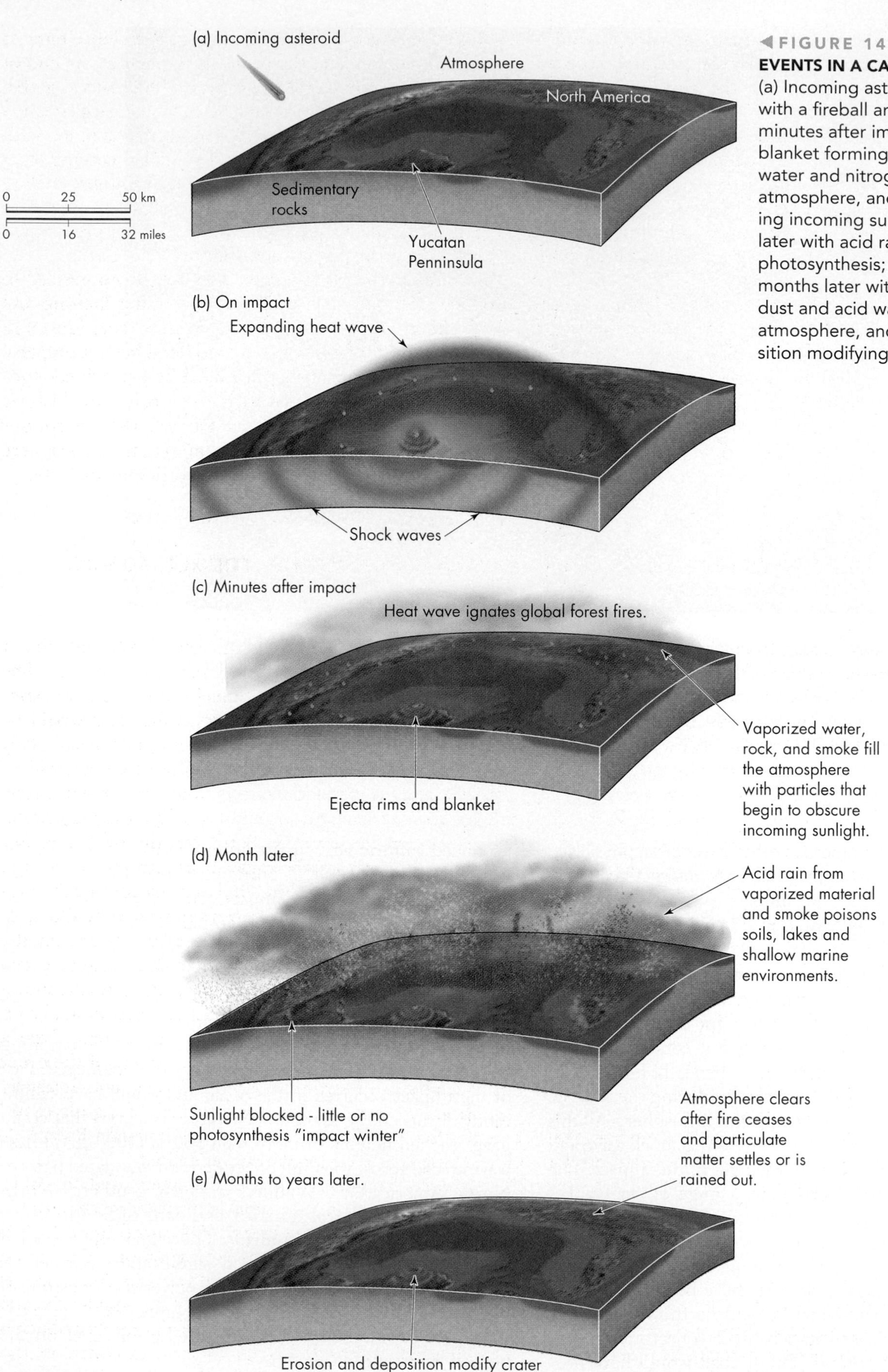

◀ **FIGURE 14.16 SEQUENCE OF EVENTS IN A CATASTROPHIC IMPACT** (a) Incoming asteroid; (b) on impact, with a fireball and shockwaves; (c) minutes after impact with an ejecta blanket forming, wildfires starting, water and nitrogen vaporizing in atmosphere, and a dust cloud blocking incoming sunlight; (d) a month later with acid rain and little or no photosynthesis; and (e) several months later with sunlight returning, dust and acid washed out of the atmosphere, and erosion and deposition modifying the crater.

Although evidence indicates that some species of dinosaurs were dying off before the impact, the event certainly seems to be responsible for wiping out the remaining large dinosaurs. So many other species of terrestrial and marine animals and plants died that there is little doubt that a massive impact was the likely cause of their extinction. Should such an event occur again, the loss of species would again be immense. It might well mean the extinction of humans and many of the larger mammals and birds. These extinctions would lead the way to yet another period of rapid evolution.

What we have learned from the K-T extinction is sobering. On the other hand, we know that impacts from objects as large as 10 km (6 mi.) in diameter are rare and occur only every 40 to 100 million years or so. However, large impacts are not the only hazard from comets, asteroids, and meteoroids. Smaller impacts are much more probable, and they can wreak havoc on a region, causing great damage and loss of life.

14.4 Linkages with Other Natural Hazards

The impact or airburst of an asteroid or comet is a direct cause for a number of other natural hazards including tsunamis, wildfires, earthquakes, mass wasting, climate change, and, possibly, volcanic eruptions. Most impacting objects land in the world's oceans, and large objects will cause tsunamis. The asteroid that hit the Yucatán Peninsula close to 65 million years ago created large, complex tsunamis that spread across the Gulf of Mexico and are recorded in sedimentary deposits in Mexico, Texas, Alabama, and Cuba.[18] Researchers have estimated that 200- to 300-m (660- to 1000-ft.) high waves originated from the impact blast and subsequent landslides in the Gulf of Mexico.[18]

Wildfires of regional or global extent have resulted from three well-documented airbursts or impact events: the K-T boundary, the Pleistocene Younger Dryas boundary (see Case Study 14.3), and the Tunguska events. Superheated clouds of gas and debris apparently reach temperatures needed to dry out and then ignite living vegetation.[12] Computer simulations suggest that wildfire patterns following a large impact are complex and involve rock blasted into space that then returns to Earth on the other side of our planet.[12] If it is any consolation, the simulations suggest that the wildfires did not burn the entire land surface of Earth.[12]

Seismic waves from a large impact most likely activate numerous landslides both on the land and under water. For example, the asteroid that formed the Chicxulub impact on the Yucatán Peninsula appears to have produced a **M** 10 earthquake and caused the mass wasting of the continental slope of North America as far north as the Grand Banks of Newfoundland.[21]

Both the asteroid that created the Chicxulub Crater at the K-T boundary and the probable airburst at the end of the Pleistocene Epoch (see Case Study 14.3) caused global changes in climate. Impacts on land can inject large quantities of dust into the atmosphere. This dust, combined with smoke from wildfires, would result in global cooling for a number of years after any major impact. The global cooling would then be followed by prolonged global warming from the large amounts of carbon dioxide and other greenhouse gases produced by postimpact wildfires.[12]

Extraterrestrial impacts have also been hypothesized to produce large volcanic eruptions by causing melting and instability in Earth's mantle.[29] This melting could result in huge eruptions of lava, referred to as flood basalts, and create large igneous provinces on land. These provinces contain 100 times more lava than that from any historic volcanic eruptions.[30] In the geologic past, these eruptions appear to have produced global changes to the atmosphere and oceans and contributed to major extinctions of life.

14.5 Minimizing the Impact Hazard

RISK RELATED TO IMPACTS

The risk of an event is related both to the probability that it will occur and to the consequences should it take place. The consequences of an airburst or direct impact from an extraterrestrial object several kilometers in diameter would be catastrophic. Although 7 out of 10 such events would likely occur in the oceans, the enormous size of an asteroid or comet would mean that the effects would be felt worldwide. Certainly, there would be significant differences, depending on the site of impact or airburst, but the overall consequences would constitute a global catastrophe with high potential for mass extinction. On Earth, events of this magnitude probably have return periods of tens to hundreds of millions of years (Figure 14.17). Smaller objects, on the order of a few tens of meters, would cause a regional catastrophe if they produced an airburst or surface impact near a populated area. The size of the devastated area would be on the order of several thousand square kilometers. Even a small impact could result in millions of deaths if the event occurred over or in an urban region. Smaller, but regionally significant, events are sometimes called "Tunguska-type" events after the 1908 airburst over the Tunguska River Valley in Siberia (see Chapter 14 Introduction).

Airbursts from asteroids with diameters of about 50 to 100 m (165 to 330 ft.) are capable of causing catastrophic damages to a region about every 1000 years (Figure 14.17). Using the Tunguska event as typical of one that could occur on our planet every thousand years, an urban area is likely to be destroyed every few tens of thousands of years. The basis for this assessment comes from the distribution of 300 small asteroids known to have exploded in the atmosphere. Extrapolation of the data allows estimates to be made for larger, more damaging events.[3]

CASE STUDY 14.3

Possible Extraterrestrial Impact 12,900 years ago

Picture a North American landscape with grasses, low shrubs, and clusters of coniferous spruce trees, along with elephant-like mammoths; short-faced bears bigger than grizzly bears; camels larger than today's two-humped Bactrian camels; dire wolves; and large, lumbering ground sloths. Add to your image a small hunting party of Clovis people—men and women dressed in fur clothing with large spears waiting near a watering hole for an unsuspecting mammoth (Figures 14.B and 14.C). This would have been a typical day toward the end of the Ice Age, the Pleistocene Epoch, around 12,900 years ago. Along the horizon to the north, you would have seen the massive white and blue Laurentide ice sheet, a vast continental glacier about 100 m (330 ft.) tall along its edges. Returning to the same area only 200 years later, you would find that everything has changed—the large animals and people are gone, and the ice sheet has advanced a considerable distance; the glacial interval known as the *Younger Dryas* (named after tundra fossil pollen of the dryas plant)[19,20] has started. What happened? This is a question that has perplexed geologists, paleontologists, and archeologists for more than two centuries.

Until the 1790s, most scientists and scholars didn't seriously entertain the idea that large Ice Age mammals, such as the mammoths and mastodons, were extinct. Even President Thomas Jefferson, who had written about the buried remains of a ground sloth found in West Virginia, thought that the animals might be alive somewhere.[21] It wasn't until 1796 that the French naturalist Georges Cuvier demonstrated that the ground sloth and elephant relatives, the mammoths and mastodons, were indeed extinct.[21]

The possible cause of the extinction of the megafauna and the termination of the Clovis culture has been a long-standing and controversial subject. The two major competing hypotheses to explain the extinction of the megafauna are (1) overkill by humans and (2) abrupt cooling (i.e., rapid climate change). Both of these hypotheses have shortcomings. Human overkill seems unlikely in light of the large numbers of animals that went extinct, including many that the Paleo Americans evidently did not regularly hunt. Abrupt cooling definitely happened; the cause is what is being debated. Was the cooling due to climate shifts that have occurred frequently during Earth's history, or was it something much different?[19]

The magnitude of the abrupt cooling at the onset of the Younger Dryas was little different from what had often occurred over the previous 80,000 years, and none of the earlier events were associated with major extinctions. The conclusion is that the extinction of Pleistocene megafauna is unique during the latest Pleistocene period and was too abrupt and widely distributed to have resulted from human overkill or climatic cooling.[19]

At many Clovis sites, archeologists have identified what is known as a black mat, which is a thin layer of carbonaceous, dark, organic-rich clays (Figure 14.D). The base of the black mat coincides with the beginning of Younger Dryas cooling, after which there is no evidence for "in place" remains of the megafauna or artifacts from the Clovis culture. Murray Springs, in Arizona, is a well-known and well-studied Clovis site, where the youngest mammoth bones

FIGURE 14.B MASTODONS DIE IN MASS EXTINCTION Large mammals, such as these mastodons in an Indiana bog, roamed North America at the end of the Pleistocene Epoch. They, and their elephant relatives, the mammoths, disappeared in a mass extinction at the end of the Pleistocene. A new hypothesis proposes that an airburst of a comet over the Laurentide ice sheet contributed to their extinction. *(Karen Carr/ Indiana State Museum)*

▲ FIGURE 14.C **CLOVIS PROJECTILE POINTS** The Clovis people fashioned distinctive fluted projectile points from chert and used them as spear tips and butchering knives to hunt mammoths, mastodons, and other now-extinct animals. For several centuries, this technology was widespread in North America until it disappeared around 12,900 years ago at the start of the Younger Dryas glaciation. *(Warren Morgan/Corbis)*

and Clovis tools are directly in contact with the base of the black layer. At that site, it is apparent that the termination of Pleistocene megafauna and Clovis culture was sudden, coinciding with the deposition of the black mat.[19,22,23]

Two other geologic mysteries (other than the mass extinction) about the time of the Younger Dryas are

1. The formation of an estimated half-million elliptical wetlands in the southeastern United States known as "Carolina Bays." These features are not actually coastal bays but are, instead, elliptically shaped marshes, swamps, and lakes named after the Loblolly Bay tree (Figure 14.E).[24] Found mostly in Atlantic coastal plain uplands from Florida to New Jersey, these wetlands fill depressions that range up to 11 km (7 mi.) long and 15 m (50 ft.) deep.[25] The Carolina Bays are most abundant in eastern South and North Carolina. Geologists have proposed dozens of hypotheses to explain their unusual shape, generally consistent northwest-southeast orientation, and sand rims. It turns out that one of the earliest hypotheses, that they were produced by a meteorite shower, may be the closest to the truth.[26]
2. The 1000-year-long Younger Dryas cooling episode that brought the Pleistocene Epoch to a close. As the Younger Dryas began, several events occurred that suggest that the Laurentide and Greenland ice sheets became unstable: (a) Meltwater drainage from a large Canadian glacial lake, Lake Agassiz, shifted abruptly from the Mississippi to the St. Lawrence River and began to drain into the Atlantic Ocean; (b) the edges of the Greenland ice sheet began to melt rapidly; (c) a large armada of icebergs calved off the ice sheets and floated south into the North Atlantic; and (d) there was a major reduction in the Atlantic Ocean conveyor belt current, the ocean current that moderates temperatures in southern Greenland and Europe (see Chapter 12).[19] What caused this unusual combination of abrupt glacial melting followed by climatic cooling?

A hypothesis being tested to explain the mass extinction of the Pleistocene megafauna, disappearance of the Clovis culture, destabilization of the ice sheet, and, possibly, the formation of the Carolina Bays is that an extraterrestrial (cosmic) impact 12,900 years ago contributed to these events. Evaluation of numerous Clovis sites, selected because they were well documented and dated, supports the hypotheses that there may have been a major extraterrestrial

◄ FIGURE 14.D **BLACK LAYER AT THE YOUNGER DRYAS BOUNDARY** Called the "black mat," this carbon-rich layer at the Murray Springs, Arizona, mammoth-kill site drapes mammoth bones that were butchered by Clovis hunters. A black layer of similar composition marks the Younger Dryas boundary in many places in North America and Europe. *(GeoScience Consulting)*

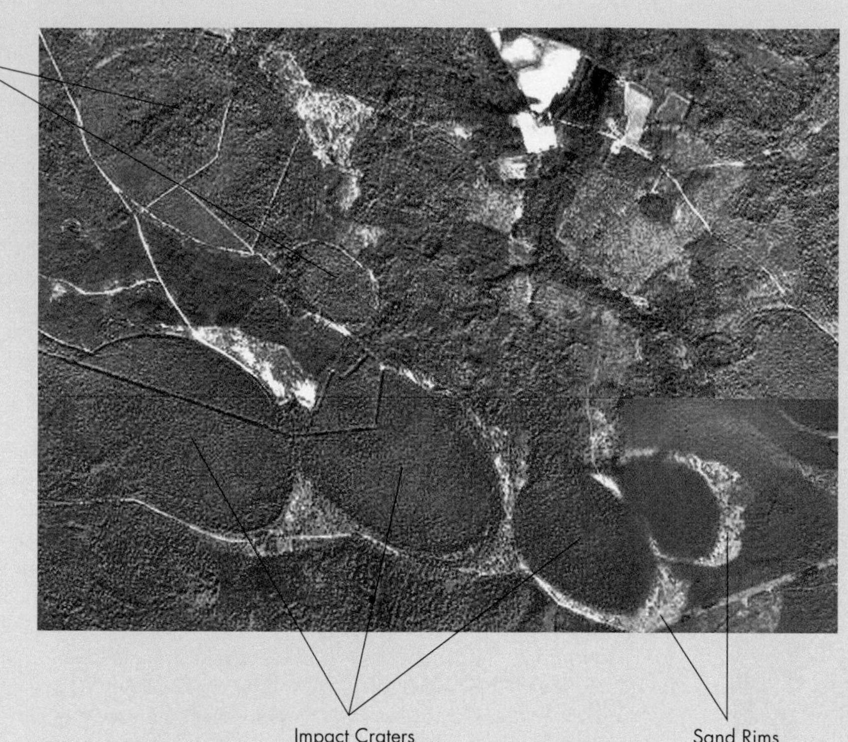

◀ **FIGURE 14.E CAROLINA BAY CRATERS** This color infrared aerial photograph shows numerous elliptical, northwest-southeast–oriented Carolina Bay craters just north of Myrtle Beach, South Carolina. The bright red color indicates living vegetation in swamps that fill the craters. Rims of white sand surround each of the craters. *(U.S. Geological Survey/U.S. Department of the Interior)*

collision over North America 12,900 years ago at the beginning of the Younger Dryas event. Hypothetically, the event consisted of one or more (perhaps many) low-density extraterrestrial objects (probably comets) that exploded in the atmosphere over much of North America. At that time, much of Canada was covered by the Laurentide ice sheet, which was partially destabilized by the proposed event. It is speculated that many objects exploded over the continent, including the ice sheet, setting off heat flashes and shock waves and generating intense winds (hundreds of kilometers per hour) over North America. The event also would have spawned fireballs that destroyed forests, grasslands, and animals, while producing an abundance of charcoal, ash, soot, and toxic fumes.[19] Evidence of this is found as far west as Santa Rosa Island in the Santa Barbara Channel.[27]

Physical evidence for the Younger Dryas event includes the discovery of a variety of possible extraterrestrial markers, including magnetic grains with iridium, magnetic micro spherules, carbon spherules, charcoal and soot, and evidence for intense wildfires. An abundance of diamonds, including nanodiamonds (including forms normally thought to be formed only by cosmic impact), micro diamonds, and shock diamonds (also thought to form from cosmic impact), have been discovered within the black mat, but much less frequently above and below the mat. As the hypothesis of the Younger Dryas event continues to be tested, more data will be forthcoming, and we will learn more about the catastrophic event that brought about the demise of the Pleistocene megafauna, the end of the Clovis culture, and the climatic change from warm to colder glacial conditions that lasted for about 1000 years.[19,28] Although no craters of Younger Dryas age have yet been found in North America near the suspected impact region, as the hypothesis is thoroughly tested over the next decade, it is likely that more direct physical evidence of the cosmic impact will be discovered.

Predictions of the likelihood and type of future impacts contain enormous statistical uncertainty. Deaths caused by an impact during a typical century may range from zero to as high as several hundred thousand. Computer-run simulations suggest that during a given century, approximately 450 deaths per year may be attributed to an impact. A truly catastrophic event could kill millions of people; averaging this number over thousands of years produces a relatively high average annual death toll.

Overall, the risk from impacts is relatively high. For example, the probability that you will be killed by an impact-driven catastrophe at the global level is approximately 0.01 percent to 0.1 percent. By comparison, the probability that you will be killed in a car accident is approximately 0.008 percent and by drowning is 0.001 percent. According to these probabilities, it appears that the risk of dying from a large impact or airburst from a comet or asteroid is considerably greater than other risks we normally face in life. However, we emphasize again that this risk is spread out over thousands of years. Although the average death toll in any one year may appear high, it is just that—an *average*. Remember that such events (and

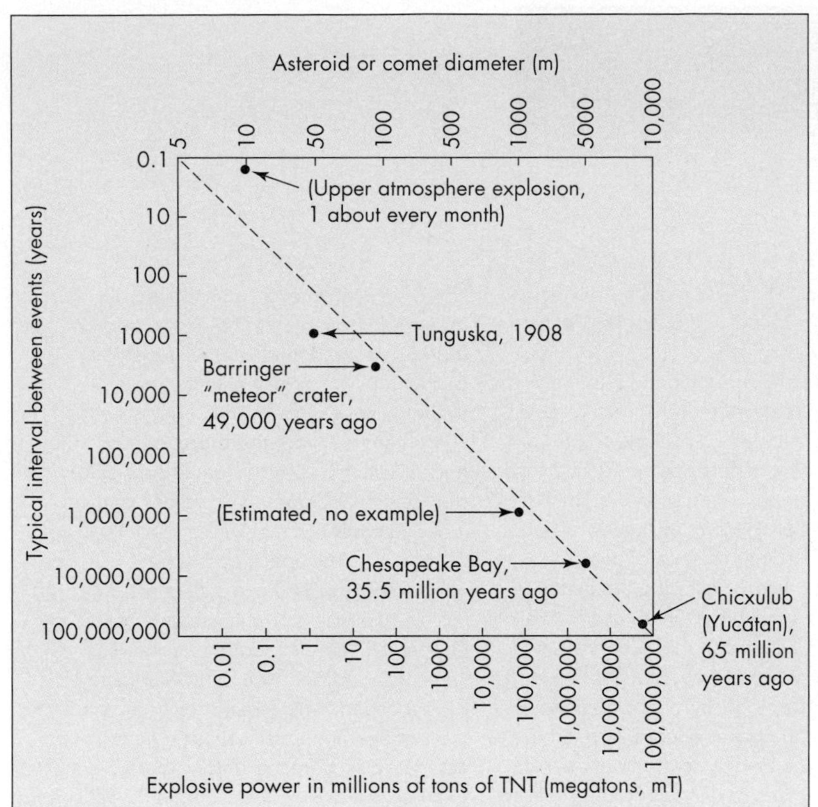

◀ **FIGURE 14.17 ENERGY FROM IMPACT** Estimate of relationship between energy of Earth impact and time interval between events for various sizes of asteroids or comets. Also shown is the estimated diameter for the impacting comet or asteroid. *(Weissman, P. R., McFadden, L., and Johnson, T. V., eds. 1999.* Encyclopedia of the solar system. *San Diego, CA: Academic Press; and Brown, P., Spalding, R. E., ReVelle, D. O., Tagliaferri, E., and Worden, S. P. 2002. The flux of near-Earth objects colliding with the Earth.* Nature *420:294–96)*

related deaths) occur very infrequently. Certainly there is a risk, but the time period between events is so long that we shouldn't lose any sleep worrying about a global catastrophic event caused by an extraterrestrial object.

MINIMIZING THE IMPACT HAZARD

It is only recently that scientists and policy makers have appreciated the risk of an impact of an asteroid or comet. Are we helpless, or is there something that can be done to minimize the hazard? The first and foremost need is to identify nearby objects in our solar system that might threaten Earth (see Case Study 14.4). Identification and categorization of comets and asteroids that cross Earth's orbit are already in progress and could be scaled up to include objects in smaller size classes. The additional size classes would include objects with diameters less than 50 m (160 ft.), those 50 meters to several hundred meters, and those with a diameter of several kilometers. A program known as *Spacewatch*, which has been operating since 1981, is attempting to inventory the region surrounding Earth with expansion to the entire solar system. Based upon the inventory to date, scientists believe that there are around 85,000 objects with a diameter of 100 m (330 ft.) or greater that are Earth-crossing asteroids. Another program, known as the *Near-Earth Asteroid Tracking Project (NEAT)*, was started in 1996. The objective of this program, which is supported by the National Aeronautical and Space Administration (NASA), is to study the size distribution and dynamic processes associated with *near-Earth objects* (NEOs) and, specifically, to identify those objects with a diameter greater than 1 km (3300 ft.). Both observation programs utilize telescopes with digital imaging devices. Using these devices, astronomers take three to five images of a given area of the night sky at intervals of 10 to 60 minutes. Computer software is then used to compare the images to identify fast-moving objects.[32] The programs and systems to identify NEOs are expected to intensify in the future, and increasing numbers of objects will be cataloged. This is a first step toward evaluating the potential hazard from NEOs. However, evaluation will take a long time because many of the objects have orbits that may not bring them close to Earth for decades, and the average amount of time between potentially catastrophic impacts is at least thousands of years for the smallest objects. The good news is that most of the objects identified as being potentially hazardous to Earth will likely not collide with our planet until several thousand years after they have been discovered. Therefore, we will have an extended period of time to learn about a particular extraterrestrial object and attempt to develop appropriate technology to minimize the hazard.[1]

By one estimate, about 20 million extraterrestrial bodies in near-Earth orbits have the potential for a significant impact.[1] Only about 4 percent of these bodies are likely to penetrate Earth's atmosphere and excavate a crater. More than half are structurally weak and prone to explode at altitudes of about 30 km (20 mi.) above Earth's surface. Although these explosions are spectacular, they are not a significant hazard at the surface of Earth. The remaining

CASE STUDY 

Near-Earth Objects

Near-Earth objects (NEOs) either reside and orbit between the Earth and the sun or have orbits that intersect with Earth's orbit. Although these objects include numerous meteoroids, asteroids, and comets, it is only the asteroids and comets that are tens of meters to a few kilometers in diameter that are a significant hazard (Figure 14.F). If a NEO with a diameter of a few tens of meters were to burst in the atmosphere or impact Earth's surface, we would experience a Tunguska-type event of regional destruction. A NEO of several kilometers in diameter would cause a global catastrophe.

As mentioned earlier, most asteroids originate in the asteroid belt, located between Mars and Jupiter (see Figure 14.5). If an asteroid remains in the belt, it does not become a NEO; however, if an asteroid is disturbed by a collision or by the gravitational pull of Jupiter its orbit may become more elliptical, cross Earth's orbit, or enter the space between the sun and Earth. The current estimate of Earth-crossing asteroids with diameters larger than 100 m (330 ft.) is close to 85,000. Larger Earth-crossing asteroids are scarcer. An estimated 1100 have a diameter greater than 1 km (3300 ft.) and 30 are greater than 5 km (3 mi.).[32]

Comets, the other type of NEO, are generally a few kilometers in diameter and are giant masses of dirty ice with a covering of rock particles and dust. Most originate far out in the solar system in a spherical cloud known as the Oort Cloud. This cloud is at a distance of about 50,000 AU, with 1 AU being the distance from the Earth to the sun, or about 150 million km (93 million mi.). Comets are best known for the beautiful light they create in the night sky, a glow that results from dust and gases released from the ice as it is heated by solar radiation. An expanding stream of gas and dust produces a spherical cloud around the comet. A tail of gas and dust is "blown" out from the cloud by the force of solar radiation, sometimes called the solar wind; thus, the tail is always directed away from the sun. Many of the dust and gas particles burn up in the atmosphere when the Earth's orbit passes through a comet's tail. This creates streams of meteors, referred to as a meteor shower. One of the most famous comets, Halley's Comet, is perhaps the best studied of all comets because of a 1986 spacecraft mission to observe it. The expedition found that the comet is a fluffy, porous body with little strength. In fact, the entire nucleus of Halley's Comet is about 20 percent ice, and, therefore, it must have 80 percent empty space, consisting of a network of cracks and voids between loosely cemented materials.[1] Halley's Comet visits the space above Earth approximately once every 76 years, giving every generation a chance to view it. Of the 4357 NEOs identified as of December 2006, only 63 of the objects have a gaseous tail when they approach the sun and are classified as comets.[33] An estimated 15 percent of the other objects are thought to be extinct comet nuclei that no longer generate a tail when heated by solar radiation.[32]

NEOs, both asteroids and comets, apparently have a relatively short life span. Consequently, new NEOs must continuously be ejected from the asteroid belt to explain the large number thus far identified. Likewise, the orbits of some comets from the outer solar system are perturbed by planets or other objects to eventually become NEOs as well.

FIGURE 14.F ASTEROID 243 IDA Numerous impact craters on the surface of asteroid 243 Ida show that collisions between objects in the asteroid belt are common. Ida is approximately 52 km (32 mi.) long and orbits the sun at a distance of 441 million km (274 million mi.). It takes nearly 5 years for Ida to complete one orbit of the sun. Photograph taken by the NASA Galileo spacecraft in 1993. *(JPL/NASA)*

objects, which constitute about 40 percent of the total population of objects, are moderately strong stony asteroids (see Table 14.1). These relatively slow-moving asteroids can penetrate the atmosphere and produce a serious threat at the surface of Earth. Such bodies produce Tunguska-type events. Identifying all these objects will be extremely difficult; there are believed to be about 10 million of them. With a diameter of only 25 m (80 ft.), they are difficult to identify and track.[1] It is currently impossible for us to identify and catalog 10 million small asteroids or comets. We are much better prepared to identify and track objects of a few hundred meters to a kilometer or so in diameter. In 2007, NASA proposed to Congress that existing surveys be extended to identify 90 percent of potentially hazardous NEOs greater than 140 m (460 ft.) in size by 2020.[32]

As survey efforts to identify smaller objects increase, more NEOs that are potentially hazardous to Earth will be detected. Communication of this hazard to the public and public officials must be both timely and authoritative. As with earthquake predictions, pending impacts must be evaluated by committees of scientists before predictions should be acted upon. The International Astronomical Union has established a technical review process for predictions that has had mixed results.[33] NASA's NEO Program Office maintains a Web site (http://neo.jpl.nasa.gov) that classifies potentially hazardous NEOs by the Torino Impact Hazard Scale (Table 14.3).

Once a large NEO is recognized to be on a collision path with our planet, options available to avoid or minimize the hazard from an airburst or crater-forming event are somewhat limited. The largest events are the real killers, and there would be no safe place on the planet. Living things, including people, within the blast area would be killed immediately; those farther away are likely to be killed in the ensuing months from the cold and the destruction of the food chain. Even if we could identify and intercept the object, blowing it apart into smaller pieces would likely cause more damage than would one impact from the larger body, because then each of the smaller pieces, some of which would become radioactive in the explosion, would rain down upon Earth. A more thoughtful approach would be to try to gently divert the object so that it misses Earth. Let's assume we identify a 400-m (1300-ft.) diameter asteroid that we believe will strike Earth in approximately 100 years. In all likelihood, the body has been crossing Earth's orbit for millions of years without an impact, and if it were possible to slightly nudge it and change its orbit, it would miss, rather than strike, Earth. This scenario is not unlikely because the probability that we would identify the object at least 100 years before impact is about 99 percent. We have the technology to change the orbit of a threatening asteroid with small nuclear explosions that are close enough to the asteroid to nudge it, but far enough away to avoid breaking it up. NASA has determined that such a "standoff" explosion would be 10 to 100 times more effective than nonnuclear alternatives, such as pulsed lasers or focused solar radiation, in changing the orbit of a potentially hazardous object.[33] To do so would require a coordinated mission between the world's militaries and space agencies, and the cost of such an expedition would likely exceed $1 billion. However, this seems a small price to pay considering the potential damages from a Tunguska-type event in an urban area.

Another option for smaller events might be evacuation. If, months in advance, we could precisely predict the point of impact, evacuation is theoretically possible. However,

TABLE 14.3

The Torino Impact Hazard Scale

No Hazard	0	Low collision hazard or object will burn up in atmosphere
Normal	1	Object will pass near Earth with collision extremely unlikely
	2	Somewhat close encounter; collision unlikely and does not merit public attention
Merits Attention by Astronomers	3	Close encounter with localized destruction possible; merits public attention if collision less than a decade away
	4	Close encounter with regional destruction possible; merits public attention if collision less than a decade away
	5	Close encounter with serious, but uncertain, regional destruction threat; merits contingency planning if less than a decade away
Threatening	6	Close encounter with serious, but uncertain, global catastrophe threat; merits contingency planning if less than three decades away
	7	Very close encounter with unprecedented, but still uncertain, global catastrophe threat; merits international contingency planning if less than a century away
	8	Collision will occur with object capable of localized destruction on land or tsunami if offshore; once every 50 to 100 years
Certain Collisions	9	Collision will occur with object capable of regional devastation or major tsunami; once every 10,000 to 100,000 years
	10	Collision will occur with object capable of global climatic catastrophe that threatens civilization; once every 100,000 years or more

Source: Modified after Morrison, D., Chapman, C. R., Steel, D., and Binzel, R. P. 2004. Impacts and the public: Communicating the nature of the impact hazard. In Mitigation of hazardous comets and asteroids, *ed. M. J. S. Belton, T. H. Morgan, N. H. Samarasinha, and D. K. Yeomans, pp. 353–90. New York: Cambridge University Press.*

evacuating an area of several thousand square kilometers would be a tremendous, and likely impossible, undertaking.[1]

In summary, we continue to catalog extraterrestrial objects that intersect Earth's orbit. We are beginning to think about our options to minimize the hazard (Professional Profile 14.5). Given the potential long-term warning before an object would actually strike the surface of Earth, we may be able to devise methods to intercept and minimize the hazard by nudging the object into a different orbit so that it misses Earth.

PROFESSIONAL PROFILE 14.5

Dr. Michael J. S. Belton: Is Earth Headed for Another Deep Impact?

According to Dr. Michael J. S. Belton, even the smallest asteroid capable of penetrating the Earth's atmosphere could cause widespread destruction equaling 1000 Hiroshima bombs or two simultaneous eruptions of Mount St. Helens.

And it gets better: There's a 20 percent chance it will happen in your lifetime.

Belton (Figure 14.G), a retired astronomer with the National Optical Astronomy Observatories, a consortium of 31 U.S. institutions and three international affiliates, now directs his own company, Belton Space Exploration Initiatives, LLC. His company is currently evaluating the technology that would be required to avert such an extraterrestrial rocket, the next time one comes our way.

But before we start sending manned crews equipped with nuclear detonators à la the movies *Deep Impact* and *Armageddon,* we need a lot of information about the nature of the asteroids, about which surprisingly little is currently known.

In February 2004, a conference organized by the American Institute of Aeronautics and Astronautics (AIAA) drew together top experts on asteroidal and cometary impacts to discuss the steps that need to be taken to eventually protect Earth from such a disaster.

"Basically, the answer is a lot needs to be done," said Belton, who attended the conference. "We're going to have to learn a lot more."

Among the necessary pieces of information that need to be gathered are data on the interior and the basic structure of asteroids, Belton said. The surface material and physical parameters are also vital to the discussion on how an asteroid could be deflected or destroyed.

In order to fill the gaps in our understanding of asteroids, some scientists are proposing manned missions to asteroids to gather information. But such missions, some of which are in the planning stages, require a great deal of funding to complete.

"Right now, there's serious discussion in the community about this, but there's not much discussion in the administration, in the Congress," Belton said.

A paper Belton prepared for the AIAA conference estimated that it would take at least 20 years for an adequate program to be developed that could successfully avert such a catastrophe.

In the meantime, several organizations, both within NASA and internationally, are working to catalog every NEO larger than 1 kilometer (0.6 mi.) in width. A NASA project called the Spaceguard Survey was officially kicked off in 1998; its objective is to identify 90 percent of all NEOs greater than 1 km in diameter by 2008.

Additionally, the report from the AIAA conference in February recommended that the Spaceguard Survey extend its findings to all NEOs of 100 meters (330 ft.) or wider (Earth's atmosphere usually protects the planet from asteroids smaller than about 50 m (165 ft.). in width, although even an object this size could cause thousands of fatalities). When this project is completed, scientists could have a great deal more warning in the event that an asteroid did have Earth's name on it.

In the meantime, we may want to keep our fingers crossed that this doesn't happen.

—CHRIS WILSON

FIGURE 14.G **MICHAEL J. S. BELTON** Dr. Belton was deputy principal investigator of NASA's Deep Impact Project, which successfully launched an SUV-sized satellite to rendezvous with Comet Tempel 1 on July 3, 2005. The "flyby" spacecraft shot a blunt "impactor" into the comet at a velocity 10 times that of a rifle bullet. Observations of the impact found that the comet is made of microscopic dust, water and carbon dioxide ice, and hydrocarbons. *(National Optical Astronomy Observatories)*

REVISITING THE FUNDAMENTAL CONCEPTS

Impacts and Extinctions

1. **Hazards Are Predictable from a Scientific Evaluation**
2. **Risk Assessment Is an Important Component of Our Understanding of the Effects of Hazardous Processes**
3. **Linkages Exist between Natural Hazards**
4. **Hazardous Events That Previously Produced Disasters Are Now Producing Catastrophes**
5. **Consequences of Hazards May Be Minimized**

1) Sufficient studies of comets and asteroids and their interaction with planets enable us to make fairly straightforward predictions of the return time that an object of a particular size would impact Earth. For example, an event the size of Tunguska in 1908, which could wipe out a large urban area, occurs on average every 1000 years. Of course, 70 percent of these events will strike the ocean. On the other hand, the return period for an impact of an asteroid the size of the one that struck the Yucatán 65 million years ago reoccurs approximately every 100 million years. This information is not particularly useful for, say, urban planning, because return periods are simply way too long. Because the consequences of being struck by a large asteroid or comet are so great, there is an effort to identify near-Earth objects, and an impact hazard scale has been developed.

2) Risk assessment is difficult for events with long return periods. Nevertheless, the potential consequences from an impact of a comet or asteroid are so large that the product of the risk, which is very small, with the consequences, which are very large, produces a relatively large risk, relative to other hazards that we normally think about. For example, the risk states that an urban area is likely to be destroyed every few tens of thousands of years. Given the size of our urban areas and the millions of people who live in them, the consequences are huge. However, the probability of such an impact is so low in any one year that it is beyond our ability to comprehend the event ever happening.

3) Many linkages exist between impacts of comets and asteroids with Earth and other natural hazards. For example, a hit in the ocean can cause a tsunami. An impact on land would cause large-scale fires across the landscape. Linkages to atmospheric hazards would include cooling of the atmosphere, acid rain, and fallout of particulates that would be harmful to life. A truly large comet or asteroid could be what is sometimes termed a "planet cleansing" event that would cause mass extinction of much of the life on Earth.

4) The human population of Earth today is about 7 billion, compared to about 1 billion in the year 1830. Because human population has grown dramatically, natural hazards have greater consequences because of the crowding of people into large urban centers. Thus, an event in 1830 that might have killed a million people would more likely kill more than 10 million people today. Much smaller events could still have serious ramifications. However, any impact with a medium-to-large asteroid or comet will be a truly catastrophic event.

5) It is possible to minimize the hazard of impact of a comet or asteroid. The most encouraging note is that a large comet or asteroid would probably be seen a hundred years or more before it is likely to strike Earth. Therefore, we might have time to try to minimize the hazard through a variety of physical means. Smaller events that might catastrophically damage an urban area are more difficult to identify because they are not very large. If we assume that we might see such an object a year or so before impact, there might be time to evacuate if we can accurately predict where it will strike. However, evacuation of large urban centers with millions of people is difficult. The plan is to try and identify any near-Earth objects that are larger than 1 km (0.6 mi.) in diameter. It has been recommended that smaller objects, those with diameters of 100 m (330 ft.) or less, also be identified. Even an object about 50 m (165 ft.) in diameter could cause many thousands of fatalities.

Summary

Asteroids, meteoroids, and comets are extraterrestrial objects likely to intercept Earth's orbit. Small objects may burn up in the atmosphere and be visible as meteors at night. Depending on their size, velocity, and composition, large objects from a few meters to 1000 kilometers in size may disintegrate in the atmosphere in an airburst or hit the surface of Earth. Objects that impact Earth's surface produce both simple and complex craters. Many craters have yet to be identified because they are severely eroded, on the seafloor, or buried by younger deposits.

Large objects can cause local to global catastrophic damage and mass extinction of life. The best-documented impact occurred 65 million years ago at the end of the Cretaceous Period (K-T boundary) and likely produced the mass extinction of many species, including the large dinosaurs. Recent studies suggest that an airburst was responsible for the extinction of many large Pleistocene mammals and the disappearance of the Clovis people at the start of the Younger Dryas glaciation. In addition to mass extinctions, impacts and airbursts of large asteroids and comets are likely to cause tsunamis, wildfires, earthquakes, mass wasting, climate change, and possibly volcanic eruptions.

The risk from an airburst or direct impact of an extraterrestrial object is a function of its probability and the consequences should it occur. Relatively small events such as the 1908 Tunguska explosion are likely somewhere on Earth about every 1000 years. The oceans will receive 70 percent of the impacts and airbursts. A significantly larger event would be capable of causing catastrophic damage to an urban area. Such events can be expected every few tens of thousands of years. Programs such as Spacewatch and NEAT (Near-Earth Asteroid Tracking project) will, with high certainty, identify NEOs of diameters greater than a few hundred meters at least 100 years before possible impact. With this warning, sufficient time should then be available to intercept and divert the object by use of nuclear explosions or other techniques. There are about 10 million smaller objects (potential Tunguska-type objects) that could produce catastrophic damage to urban areas. Identifying all these objects will be extremely difficult. Thus, we are particularly vulnerable to these smaller objects.

Key Terms

airburst (p. 478)
asteroid (p. 476)
comet (p. 476)
impact crater (p. 478)
mass extinction (p. 485)
meteor (p. 476)
meteorite (p. 478)
meteoroid (p. 476)

Review Questions

1. Where is Tunguska? What happened there? Why is it important to our discussion of natural hazards? How often do these events occur?
2. What is the difference between an asteroid, meteor, comet, meteoroid, and meteorite?
3. What are meteorites made of? Comets?
4. Where do comets and asteroids originate?
5. Describe the general characteristics of an impact crater. How can it be distinguished from other types of craters?
6. How do simple and complex impact craters differ?
7. Why does Earth apparently have so few impact craters?
8. Explain the significance of Comet Shoemaker-Levy 9.
9. What is the significance of Barringer Crater?
10. What are the hypotheses for the cause of mass extinctions?
11. When was the greatest mass extinction in Earth's history?
12. Summarize the causes for the six major mass extinctions in Earth's history.
13. What is the K-T boundary? Why is it significant in the study of natural hazards?
14. How are other natural hazards linked to extraterrestrial impacts and airbursts?
15. Explain how the risk of dying from an asteroid impact is greater than the risk of a car accident.
16. What can be done to minimize the extraterrestrial impact hazard?

Critical Thinking Questions

1. Describe the likely results if a Tunguska-type event were to occur over or in central North America. If the event were predicted with 100 years warning, what could be done to mitigate the effects, if changing the object's orbit were not possible? Outline a plan to minimize death and destruction.
2. How would the effects of an asteroid impact in water differ from those of an asteroid impact on land? Consider what would happen physically and chemically with water and how the impact craters might differ.
3. Compare the velocity of an asteroid or comet, seismic waves, and sound waves. Why do they differ?
4. How did the effects of the impact at the K-T boundary differ from the airburst at the Younger Dryas boundary? How were the two events similar?

APPENDIX A: Minerals

Characteristic Properties of Minerals

Minerals are naturally occurring substances with defined physical and chemical properties. Physical properties are commonly used for identification and chemical composition is the basis for mineral classification. Mineral properties can also have an aesthetic or utilitarian value. For example, native copper, when discovered in the Great Lakes region by Native Americans, was first valued for its metallic properties of luster and malleability and was used for jewelry and tools. Many ancient cultures, in addition to our own, have valued gemstones such as rubies and sapphires for their beauty. Historically, one of the most valued minerals has been halite (NaCl), or common table salt, because all animals, including people, need it to live. Halite has been mined and obtained from coastal evaporation ponds for thousands of years. Some clay minerals have been highly valued because they can be molded into useful pots and decorative statues and painted with black, red, and orange paint made from still other minerals. Today, specific physical properties of a wide variety of minerals are used for everything from ceramics to electronics, metallurgy, agriculture, water treatment, and personal jewelry. An understanding of the physical properties will help us identify common minerals.

Identifying Minerals

Identification of mineral specimens that are large enough to fit in the palm of your hand, referred to as hand specimens, is a combination of pattern recognition and testing for particular properties or characteristics. These properties include hardness, specific gravity, fracture, cleavage, crystal form, color, luster, as well as other properties that are diagnostic of a particular mineral or group of minerals.

HARDNESS

Most physical properties of minerals reflect characteristics of their internal atomic structure. For example, a mineral's *hardness*, that is, its resistance to scratching, is related to the size, spacing, and strength of bonding of the atoms within the mineral. Absolute hardness is not easy to measure, so in 1822 Austrian mineralogist Friedrich Mohs came up with an ingenious 1-through-10 scale of *relative hardness* (H) that is still widely used. On the Mohs scale the softest mineral is talc ($H = 1$) and the hardest mineral is diamond ($H = 10$) (Table A.1). The change in absolute hardness from one number to the next is not the same; any mineral or substance will scratch a mineral with a lower hardness on the Mohs scale. You can use common objects, such as a fingernail (H about 2.2), a copper penny (a U.S. penny

TABLE A.1

Mohs Hardness Scale

	Relative Hardness[1]	Mineral	Comment	Hardness of Common Materials
Softest	1	Talc	Softest mineral known; used to make powder	Graphite, "lead" in pencils (1–2)
↓	2	Gypsum	Used to make wall board	Fingernail (~2.2)
	3	Calcite	Main mineral in marble and limestone	Copper penny[2] (~3.2)
	4	Fluorite	Mined for fluorine, used in glass and enamel	
	5	Apatite	Enamel of your teeth	Pocketknife (~5.1), Glass (~5.5)
	6	Orthoclase	Common rock-forming mineral	Hard steel file or needle (~6.5)
	7	Quartz	When purple is gemstone amethyst, birthstone for February	
	8	Topaz	When transparent is gemstone, birthstone for November	
	9	Corundum	When red is ruby, gemstone, birthstone for July; when blue is sapphire, gemstone, birthstone for September	
Hardest	10	Diamond	Hardest known mineral; gemstones are extremely brilliant	

[1] Note: The scale is relative, and unit increases in hardness do not represent an equal increase in true hardness.

[2] U.S. penny minted before 1982 or Canadian penny minted before 1997.

minted before 1982, or a Canadian penny minted before 1997, H about 3.2), and a glass plate (H = 5.5) to approximate the relative hardness of minerals.

For example, you can determine if a mineral is harder or softer than 5.5 by taking a fresh, unweathered surface of the mineral and attempting to scratch a glass plate. You must be certain that the mineral is actually making a scratch and not just leaving a line of powdered mineral on the surface. Do not attempt to use a glass test plate if it becomes fractured or broken; be sure to follow proper procedures in disposing of broken or damaged glass. If the mineral is softer than glass, then you can determine if it is harder than a copper penny. By this process you can bracket the relative hardness (e.g., >3.2 and <5.5) of the unknown mineral sample. Hardness tests can sometimes be misleading, so never rely on hardness alone to identify a mineral.

SPECIFIC GRAVITY

Some minerals have unusually light or heavy specific gravity. The *specific gravity* of a mineral is the density of a mineral compared with the density of water. Water has a specific gravity of 1. The specific gravity of minerals varies from about 2.2 for halite to 19.3 for gold. Most minerals have specific gravities of about 2.5 to 4.5. In practice, we take a sample of a mineral in hand, heft it, and estimate if it has a light, medium, or heavy specific gravity.

CLEAVAGE

Many minerals have a distinctive way in which they break (Table A.2). Those that break along smooth, even planar surfaces are said to have *cleavage*. Cleavage develops because the forces that bond atoms together in a crystalline structure are rarely equal in all directions. Most cleavage will appear as a series of small, reflective steps with each parallel step breaking along the same direction of weak bonding within the mineral. One technique for recognizing cleavage is to rotate the specimen in front of you with a light source, such as the Sun or a ceiling light, over your shoulders. Each time you encounter a direction of cleavage, the small, parallel cleavage steps will reflect light all at once.

Two characteristics of cleavage are important in identifying minerals: the number of cleavage directions and the angle between the cleavage directions. Minerals may have one, two, three, four, or six cleavage directions (Figure A.1). When you count the number of cleavage directions, be sure not to count the same direction twice because cleavage surfaces commonly repeat themselves on opposite sides of a specimen. In determining the angle between two cleavages, it is generally sufficient to decide whether or not the angle is close to 90°. In some specimens there is the potential to confuse a cleavage surface with a crystal face that developed as the crystal grew. Unlike cleavage surfaces, crystal faces are generally not repetitive, nor do they form imperfect steps.

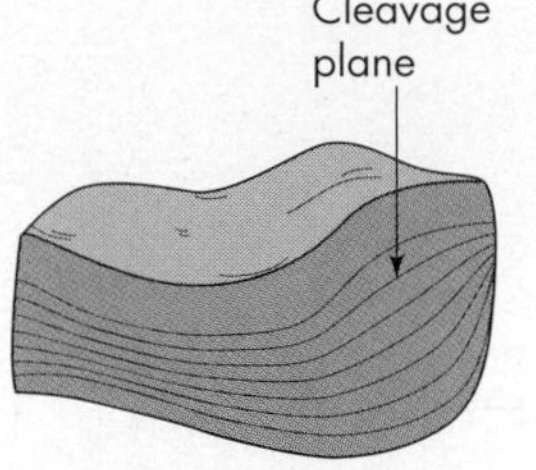

(a) One direction of cleavage. Mineral examples: muscovite and biotite micas

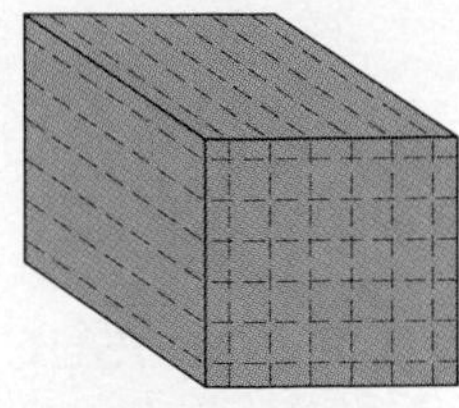

(b) Two directions of cleavage at right angles or nearly right angles. Mineral examples: feldspars, pyroxene

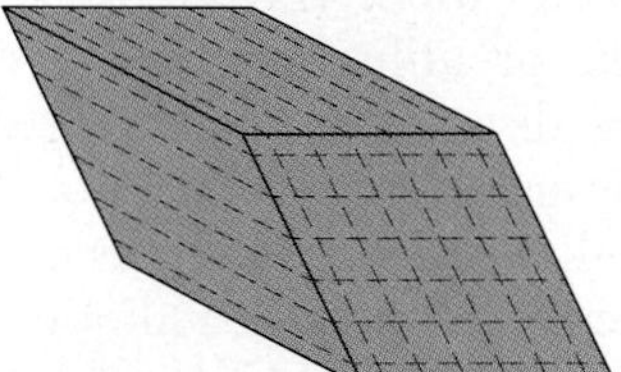

(c) Two directions of cleavage not at right angles. Mineral example: amphibole

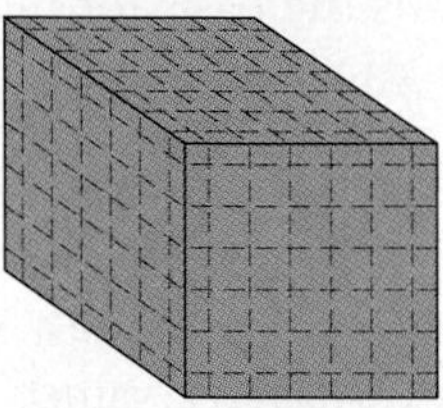

(d) Three directions of cleavage at right angles. Mineral examples: halite, galena

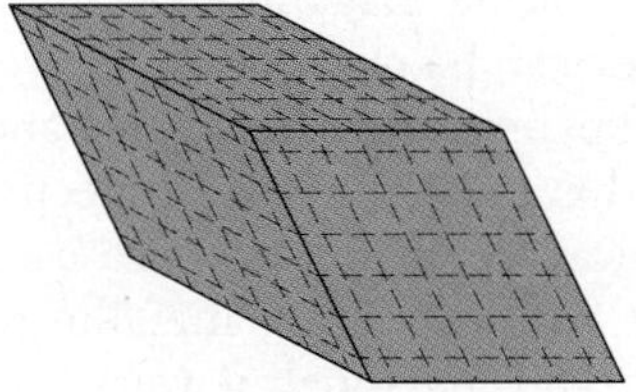

(e) Three directions of cleavage not at right angles. Mineral example: calcite

▲ FIGURE A.1 **Common Types of Cleavage** Except for mica, where cleavage is indicated with solid lines, cleavage directions in this diagram are shown with dashed lines. (a) Mica has one cleavage direction and breaks into sheets. (b) Feldspar and pyroxene have two cleavage directions at nearly right angles. (c) Amphibole has two directions of cleavage that do not meet at right angles. (d) Halite and galena have three cleavage directions that meet at right angles. This relationship causes both minerals to break into cube-shaped fragments. (e) Calcite also has three cleavage directions, but they do not meet at right angles. This relationship causes calcite to break into rhombohedrons.

CRYSTAL FORM

Crystal faces are important, however, in recognizing *crystal form*, which is the geometric shape taken on by the mineral as it crystallized. Unfortunately most mineral specimens do not show crystal form; however, when it is present, crystal form can be a useful diagnostic property for identifying some minerals. For example, the pointed and elongate hexagonal crystal form of quartz is diagnostic, as are the cube-shaped crystals of pyrite.

TABLE A.2

Properties and Environmental Significance of Selected Common Minerals

Mineral Class	Mineral	Chemical Formula	Color	Hardness	Other Characteristics	Comment
Silicates	Plagioclase feldspar	$(Na, Ca)Al(Si, Al)Si_2O_8$	Usually white or gray, but may be others	6	Two good cleavages at approximately 90ˆ and may have fine striations on one of the cleavage surfaces	One of the most common of the rock-forming minerals; is a group of feldspars ranging from sodium to calcium rich; important industrial minerals
	Alkali feldspar	$(Na, K)AlSi_3O_8$	Gray, white to pink or salmon color	6	Two good cleavages at 90° may be translucent to opaque with glassy luster	One of the most common of the rock-forming minerals; used in porcelain and a wide variety of industrial processes
	Quartz	SiO_2	Varies from colorless to white, gray, pink, purple, and several others, depending on impurities	7	Often has good crystal shape with six sides; conchoidal fracture; coarse crystalline varieties have glassy luster and microcrystalline varieties have dull to waxy luster	Very common rock-forming mineral; resistant to most chemical weathering; basic constituent of glasses and fluxes; commonly used as an abrasive; colored varieties, such as amethyst and agate, are semiprecious gemstones
	Pyroxene	$(Ca, Mg, Fe)_2Si_2O_6$	Usually greenish to black	5–6	Crystals are commonly short and stout; two cleavages at about 90°;	Important group of rock-forming minerals; particularly common in igneous rocks; weathers rather quickly
	Amphibole	$(Na, Ca)_2(Mg, Al, Fe)_5Si_8O_{22}(OH)_2$	Generally black to green	5–6	Distinguished from pyroxene by the cleavage angle being 120°; rather than 90°; amphibole also generally has better cleavage with higher luster than pyroxene	Important group of rock-forming minerals particularly for igneous and metamorphic rocks; relatively nonresistant to weathering
	Olivine	$(Mg, Fe)_2SiO_4$	Generally green, but also may be yellowish	6.5–7	Conchoidal fracture; commonly occurs in aggregates of small glassy grains	Important rock-forming mineral, particularly for igneous and metamorphic rocks; relatively nonresistant to chemical weathering
	Clay	Various hydrous aluminum silicates with elements such as Ca, Na, Fe, Mg, and K	Generally white, but may vary due to impurities	1–2	Generally found as soft, earthy masses composed of very fine grains; may have an earthy odor when moist; often difficult to identify the particular clay mineral from hand specimen	Clay minerals are very important from an environmental viewpoint; many uses in society today; clay-rich soils often have many engineering geology problems
	Chlorite	Hydrous Mg, Fe, Al silicate	Green to dark green	2–2.5	Commonly in layered or scaly masses with one direction of cleavage; thin sheets are flexible and nonelastic; glassy to pearly luster	Important group of rock-forming minerals in metamorphic rocks

(Continued)

TABLE A.2 *(Continued)*

Mineral Class	Mineral	Chemical Formula	Color	Hardness	Other Characteristics	Comment
Silicates (*Cont.*)	Talc	$Mg_3Si_4O_{10}(OH)_2$	Pale green to white or gray	1	In layered or compact masses with one direction of cleavage; thin sheets flexible and nonelastic; soapy feel with pearly to greasy luster	Used to make ceramics, paint, paper roofing, cosmetics, and ornamental stone
	Biotite (black mica)	$K(Mg, Fe)_3AlSi_3O_{10}(OH)_2$	Black to dark brown or dark green	2.5–3	Breaks apart in parallel sheets as a result of excellent cleavage in one direction	Important rock-forming mineral common in igneous and metamorphic rocks
	Muscovite (white mica)	$KAl_2(AlSi_3O_{10})(OH)_2$	White, light yellow, brown, pink, or green; colorless in cleavage sheets	2–3	Breaks apart into thin sheets due to excellent cleavage in one direction	Important rock-forming mineral, particularly for igneous and metamorphic rocks; used for several purposes, including industrial roofing materials, paint, and rubber
Carbonates	Calcite	$CaCO_3$	Colorless to white, but may have a variety of colors resulting from impurities	3	Effervesces strongly in dilute hydrochloric acid; often breaks apart to characteristic rhombohedral pieces as a result of two good cleavages at 78° transparent varieties display double refraction, where a single dot on white paper looks like two dots if viewed through the mineral	Main constituent of the important sedimentary rock limestone and the metamorphic rock marble; associated with a variety of environmental problems with these rocks, including the development of sinkholes; chemically weathers rapidly; used in a variety of industrial processes including asphalts, fertilizers, insecticides, cement, and plastics
	Dolomite	$Ca, Mg(CO_3)_2$	Generally white, but may be a variety of other colors, including light brown and pink	3.5–4	When powdered, will slowly effervesce in dilute hydrochloric acid; two cleavages at 78ˆ; may be transparent to translucent and have a vitreous to pearly luster	Common mineral in the sedimentary rock dolostone and dolomitic limestone
	Malachite	$Cu_2CO_3(OH)_2$	Bright to emerald green, but may be dark green	3.5–4	Effervesces slightly in dilute hydrochloric acid and turns the acid solution green; often displays a swirling structure of light and darker green bands	Valued as a decorative stone; used in jewelry and as a copper ore

Oxides and Hydroxides	Hematite	Fe_2O_3	Usually various shades of reddish brown to red to dark gray	1–6.5	Streak is dark red	A common mineral found in small amounts in many igneous rocks, particularly basaltic rocks
	Magnetite	Fe_3O_4	Black	6	Magnetic	The most important ore of iron
	Goethite	FeO(OH)	Generally yellow to yellow-brown or black	1–5.5	Often present as earthy masses or in the form of a crust; streak is yellow-brown	Formerly referred to as limonite, which is now known to be relatively rare. Often forms from the chemical weathering of iron minerals and is present as "rust"
Sulfides	Pyrite	FeS_2	Generally a pale brassy yellow tarnishing to brown	6–6.5	Often present as well-formed cubic crystals with striations on crystal faces	Has been used in the production of sulfuric acid, but mostly known for adding sulfur content to coal and contributing to the formation of acid-rich waters that result from the weathering of the mineral
	Chalcopyrite	$CuFeS_2$	Dark brassy to golden yellow, tarnishing to iridescent films that are reddish to blue-purple	3.5–4	Easily disintegrates; greenish black streak lacks cleavage	An important ore of copper; often associated with gold and silver ores
	Galena	PbS	Silver gray	2.5	Gray to black streak; high specific gravity; general metallic luster; cubic cleavage	Primary ore of lead
Sulfates	Gypsum	$CaSO_4 \cdot 2H_2O$	Generally colorless to white, but may have a variety of colors due to impurities	2	Often is transparent to opaque with one perfect cleavage; may form fibrous crystals, but often is an earthy mass	Several industrial uses in making of plaster of Paris for construction material, fertilizer, and flux for pottery
	Anhydrite	$CaSO_4$	Commonly white or gray, but may be colorless	3–3.5	Commonly observed as massive fine aggregates; may be translucent to transparent	Used for the production of sulfuric acid, as a filler material in paper, and occasionally as an ornamental stone

(Continued)

TABLE A.2 *(Continued)*

Mineral Class	Mineral	Chemical Formula	Color	Hardness	Other Characteristics	Comment
Halides	Fluorite	CaF_2	Variable colors, often purple, yellow, white, or green	4	Often observed as cubic crystals but may also be massive; four perfect cleavages	A number of industrial uses, including a flux in the metal industry, optical lenses, and in the production of hydrofluoric acid
	Halite	NaCl	Generally colorless to white, but may be a variety of colors due to impurities	2.5	Salty taste; often has a cubic form due to three perfect cleavages	Common table salt; makes up rock salt, which is used for ice removal and as a host for buried nuclear waste
Native Elements	Gold	Au	Gold yellow	2.5–3	Crystals are rare; extremely high specific gravity; very ductile and malleable	Primary use as a monetary standard; used in jewelry and in the manufacture of computer chips and other electronics
	Diamond	C	Generally colorless, but also occurs in various shades of yellow, brown, blue, pink, green, and orange	10	Commonly cut to display brilliant luster	Used in industrial abrasives and, of course, jewelry
	Graphite	C	Black to gray	1–2	Often occurs as foliated masses; black streak and marks paper as well; greasy feel	A variety of industrial uses including lubricants, dyes, and in pencils as the "lead"
	Native Sulfur	S	Usually yellow when pure, but may be brown to black from impurities	1.5–2.5	Characteristic sulfurous smell, similar to the smell of fireworks or gunpowder	Often a by-product of the oil industry; may be used to manufacture sulfuric acid

Modified after Davidson, J. P., Reed, W. E. and Davis, P. M. 1997. Exploring Earth. Upper Saddle River, NJ: Prentice Hall; and Birchfield, B. C., Foster, R. J., Keller, E. A., Melhom, W. N., Brookins, D. G., Mintt, L. W., and Thurman, H. V. 1982. Physical geology. Columbus, OH: Charles E. Merrill.

FRACTURE

Broken surfaces of a mineral that do not show cleavage are described as *fracture*. The type of fracture that is present is generally important only if the specimen has no cleavage. Although most fracture is uneven or irregular, there are two types of fracture that are distinctive: *conchoidal fracture* and *fibrous fracture*. In conchoidal fracture, the broken surface is curved and can be at least partially smooth. A broken glass bottle will exhibit conchoidal fracture. In many cases a conchoidal fracture surface will reflect light continually as a specimen is rotated.

COLOR

The *color* of a mineral can be misleading because a given mineral may have several different colors, or several different minerals may have the same color. For example, depending on the impurities present, normally clear quartz may be white, pink, purple, red, brown, yellow, green, or black.

A more reliable way of using color to identify some minerals is to powder part of a specimen on an unglazed porcelain plate. The resulting *streak* made by the powder is especially useful in identifying metallic minerals. For example, the hand specimen of the mineral hematite (Fe_2O_3) may be dull black or a shiny, dark metallic silver color, but its streak is always red. Streak tests can be done only for minerals that are softer than the porcelain "streak plate," which typically has a hardness of about 6.5.

LUSTER

Luster refers to the way light is reflected from a mineral. Most minerals can be described as having either a metallic or nonmetallic luster. Minerals with metallic luster appear silver, brass, or a shiny, but not an enamel-like black and are opaque, even on thin edges. Nonmetallic luster can be quite variable with descriptive terms for luster such as glassy, pearly, greasy, earthy, and resinous (like pine sap or dried turpentine). Light will also pass through thin edges of a nonmetallic mineral.

Some minerals have more than one luster. For example, some specimens of graphite have metallic luster and others are nonmetallic.

STEPS IN IDENTIFYING A MINERAL

1. Decide if the mineral has a metallic or a nonmetallic luster and refer to Table A.3. If the mineral is metallic, determine the color of its streak.
2. Determine the relative hardness of the specimen by first testing to see if it will scratch glass or alternatively, if it can be scratched by a knife (Table A.1). Note that if the mineral has a hardness close to that of glass (H about 5.5) or a pocket knife (H about 5.1), determining hardness may be difficult and the mineral may appear in Table A.3 in either the "harder than glass" or "softer than glass" sections.
3. Decide whether the mineral has cleavage. If cleavage is present, determine the number of directions and whether the angle between the directions is close to 90°. If cleavage is not present, determine the type of fracture.
4. Once you think you have identified the mineral, test for other physical or chemical properties (e.g., taste, smell, reaction with dilute hydrochloric acid) listed for that mineral in Tables A.2 and A.3 to confirm your identification.

The identification of minerals in the field or in the laboratory from small specimens, using simple tests and perhaps a hand lens to determine cleavage and other properties, is basically a pattern recognition exercise. After you have looked at a number of minerals over a period of time and have learned how particular minerals vary, you will become more proficient at mineral identification. When positive identification is necessary, mineralogists use a variety of sophisticated analytical equipment to determine the chemical composition and internal structure of a sample.

TABLE A.3

Key to Assist Mineral Identification

To use this table; (1) decide whether the luster of the mineral to be identified is nonmetallic or metallic. If it is nonmetallic, decide whether it is a light-colored or a dark-colored mineral; (2) use a glass plate to determine if the mineral specimen is harder or softer than the glass plate; (3) look for evidence of cleavage; (4) compare your samples with the other properties shown here and use Table A.2 for final identification.

Luster	Hardness	Cleavage	Properties	Mineral
Light-colored nonmetallic luster	Harder than glass	Shows cleavage	White, pink or salmon-colored; 2 cleavage planes at nearly right angles; hardness, 6.	Orthoclase (Alkali feldspar)
			White, gray, or green-gray; 2 cleavage planes at nearly right angles; hardness, 6; striations on one cleavage.	Plagioclase
		No cleavage	White, clear, or any color; glassy luster; transparent to translucent; hexagonal (six-sided) crystals; hardness, 7; conchoidal fracture.	Quartz
			Various shades of green and yellowish green; glassy luster; granular masses and crystals in rocks; Olivine hardness, 6.5–7 (apparent hardness may be much less in granular masses).	Olivine
	Softer than glass	Shows cleavage	Colorless to white; salty taste; 3 perfect cleavages forming cubic fragments; hardness, 2.5.	Halite
			White, yellow to colorless; 3 perfect cleavages forming rhombohedral fragments; hardness, 3; effervesces with dilute hydrochloric acid.	Calcite
			Pink, colorless, white, or brown; rhombohedral cleavage; hardness, 3.5–4; effervesces with dilute hydrochloric acid only if powdered.	Dolomite
			White to transparent; 1 perfect cleavage; hardness, 2.	Gypsum
			Pale green to white; feels soapy; 1 cleavage; hardness, 1.	Talc
			Colorless to light yellow, brown, or green; transparent in thin elastic sheets; 1 perfect cleavage; hardness, 2–3.	Muscovite
			Colorless to yellow to light blue, green, or purple; 4 cleavages not at 90°; hardness, 4.	Fluorite
		No cleavage	Pale green to white; feels soapy; hardness, 1.	Talc
			White to transparent; hardness, 2.	Gypsum
			Yellow; resinous luster; smells like fireworks; hardness, 1.5–2.5.	Sulfur
Dark-colored nonmetallic luster	Harder than glass	Shows cleavage	Black to dark green; 2 cleavage planes at nearly 90°; hardness, 5–6.	Pyroxene
			Black to dark green; 2 cleavage planes about 120°; hardness, 5–6.	Amphibole
		No cleavage	Various shades of green and yellow; glassy to dull or waxy luster; granular masses and crystals in rocks; hardness, 6.5–7 (apparent hardness may be much less).	Olivine
			White, clear, or any color; glassy luster; transparent to opaque; hexagonal (six-sided) crystals; hardness, 7; conchoidal fracture.	Quartz
			Steel gray to black; reddish brown streak; earthy appearance; hardness, 5.5–6.5.	Hematite
	Softer than glass	Shows cleavage	Black to brownish or greenish black; cleavage, 1 direction; hardness, 2.5–3.	Biotite
			Violet, yellow, green, pink, white, and colorless; 4 cleavages other than 90ˆ; hardness, 4.	Fluorite
			Various shades of green; cleavage, 1 direction; in layered or scaly masses; hardness, 2–2.5.	Chlorite
		No cleavage	Dull to bright red; reddish brown streak; earthy appearance; hardness, 1–6.5.	Hematite
			Lead-pencil black; smudges fingers; hardness, 1–2; 1 cleavage that is apparent only in large crystals.	Graphite
			Yellow-brown to dark brown; may be almost black; streak yellow-brown; earthy; hardness, (usually soft). 1–5.5	Goethite
Metallic luster			Black; strongly magnetic; hardness, 6.	Magnetite
			Lead-pencil black; smudges fingers; hardness, 1–2; 1 cleavage that is apparent only in large crystals.	Graphite
			Light brass yellow; black streak; cubic crystals, commonly with striations; hardness, 6–6.5.	Pyrite
			Dark brass yellow; may be tarnished; greenish black streak; hardness, 3.5–4; massive.	Chalcopyrite
			Shiny gray; black streak; very heavy; cubic cleavage; hardness, 2.5.	Galena

Modified after Birchfield, B. C., Foster, R. J., Keller, E. A., Melhom, W. N., Brookins, D. G., Mintt, K. W., and Thurman, H. V. 1982. Physical geology. Columbus, OH: Charles E. Merrill; and Hamblin, W. K. and Howard, J. D. 2005. Exercises in physical geology, 12th ed. Upper Saddle River, NJ: Pearson Prentice Hall.

APPENDIX B: Rocks

In Chapter 1 we stated that a *rock* is an aggregate of one or more minerals. In traditional geologic investigations, this is the most commonly used definition. Another definition is used by geologists and geological engineers who are concerned with the properties of earth materials that affect engineering design and environmental problems associated with natural hazards. In environmental and engineering geology, the term rock is reserved for earth materials that cannot be removed without blasting using explosives, such as dynamite; earth materials that can be excavated with normal earth-moving equipment, for example a shovel or a bulldozer, are called *soil*. Thus, loosely compacted, poorly cemented sandstone may be considered a soil, whereas well-compacted clay may be called a rock.

This pragmatic definition of rock described above also applies to the use of the term *soil* in engineering and environmental geologic reports. To an engineer, soil is earth material that can be excavated without blasting, regardless of whether it is on land or under-water. In traditional geologic investigations the term soil is restricted to surficial earth material that has developed on land by in-place weathering. Other loose earth material not formed by in-place weathering is generally considered *sediment*, distinguished from sedimentary rock which requires a hammer to break it apart. These differences in definitions are more than academic; they can affect communication between engineers, scientists, architects, planners, and emergency responders.

Identifying Rocks

Rocks are composed of minerals, and in some cases other substances such as natural glass and pieces of organisms. Whereas a few rocks contain essentially only one mineral, most contain several. As with minerals, the identification of hand specimens of rocks is primarily through the recognition of patterns. However, there are some useful hints that will assist you. The first task is to decide if the rock is igneous, sedimentary, or metamorphic. Sometimes this decision is not as easy as you might think. It is particularly difficult to identify rocks that are fine grained. A specimen is considered fine grained if you cannot see individual mineral grains with your naked eye. Even if a specimen is not fine grained, it is advisable to use a magnifying glass or hand lens to examine a mineral or a rock.

Some general rules of thumb will assist you in getting started: (1) If the sample is composed of bits and pieces of other rocks, it is most likely sedimentary, especially if the pieces have rounded corners; (2) if the rock specimen has minerals that are all oriented in the same direction, such as parallel sheets of mica, then it is most likely metamorphic; (3) if individual mineral crystals are relatively coarse grained, meaning you can see them with the naked eye or easily with a hand lens, and the crystals are interlocking and composed of minerals such as quartz and feldspar, it is probably an igneous rock that cooled underground, such as granite; (4) if the rock is mostly fine grained but contains some larger crystals, it is probably an igneous rock that cooled on or just below Earth's surface; (5) if the rock is relatively soft and effervesces (fizzes) when dilute hydrochloric acid is applied, it is probably a carbonate sedimentary rock, such as limestone, or a metamorphic rock, such as marble; (6) if it is a really grungy, soft, highly weathered, or altered rock, as many in nature seem to be, you are going to have some problems identifying it!

Physical Properties of Rocks

Physical properties of rocks include color, specific gravity, relative hardness, porosity, permeability, texture, and strength.

COLOR

The color of a rock varies depending upon the minerals present and the amount of weathering that has occurred. Rocks encountered in their natural environment are various shades of light gray to brown to black. Chemical weathering can produce surface stains of black iron or manganese oxides, and brown to orange stains of iron oxides and hydroxides.

SPECIFIC GRAVITY AND RELATIVE HARDNESS

The same definition of specific gravity we gave for minerals applies to rocks and refers to the weight of the rock relative to the weight of water. Some rocks are lighter than water and will float. For example, pumice, a light-colored volcanic rock with many holes produced by gas bubbles, is lighter than water. Rocks become heavier as iron- and magnesium-bearing minerals become more abundant. There is no scale for relative rock hardness similar to the Mohs scale for minerals. Soft rocks can be broken with your fingers, and hard rocks require a sledgehammer to break apart.

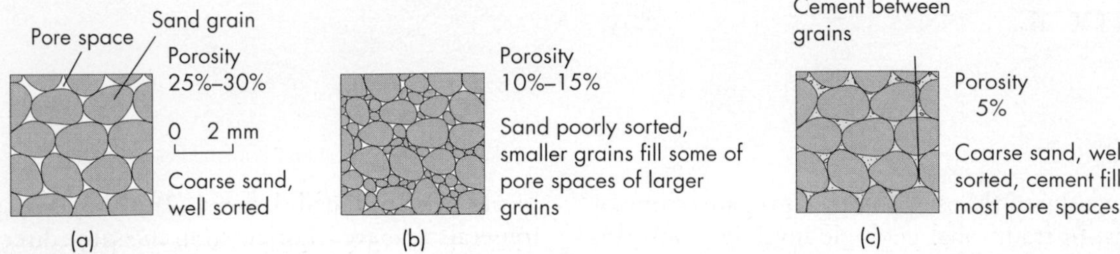

▲ **FIGURE B.1 Sorting and Cementation Affect Porosity** (a) Porosity of a well-sorted coarse sand. (b) Porosity of a poorly sorted sand. (c) Porosity of a cemented, well-sorted coarse sand; examples include calcium carbonate ($CaCO_3$) and silica (SiO_2).

POROSITY AND PERMEABILITY

Porosity is the percentage of a rock's volume that is empty space, that is void space between grains, or open space in fractures. *Permeability* is the capacity of a porous rock to transmit a fluid, such as oil or water. The properties of porosity and permeability are important in understanding many geologic hazards, such as landsliding and earthquakes.

TEXTURE

The *texture* of a rock refers to the size, shape, and arrangement of the crystals or grains within it. Crystal or grain size can be measured directly or described qualitatively as fine or coarse grained. A rock with crystals or grains that are visible with the naked eye or easily with a magnifying hand lens is considered coarse grained; one that is not is fine grained. The degree to which grains in a rock are similar in size is referred to as *sorting*. For example, a sandstone in which most sand grains are similar in size is described as well sorted (Figure B.1a, Figure c), whereas one that contains a variety of grain sizes is poorly sorted (Figure B.1b). Crystal or grain shape can vary from spherical with rounded corners to irregular with sharp, angular corners. Likewise the arrangement of crystals or grains can be variable. Crystals can be interlocking, such as in granite or marble; aligned, such as mica grains in schist; or small grains may occupy spaces between large grains, such as in poorly sorted sandstone (Figure B.1b). Other rock textures are characterized by vesicles, or cavities, formed by gas expansion during their formation. For example, vesicular rocks commonly occur near the surface of a lava flow where gas is escaping. The texture of a rock can also influence its porosity. Rocks that are poorly sorted or well cemented generally will have lower porosity than those that are well sorted and only partially cemented. For example, in sandstone, porosity decreases from about 30 percent to 5 percent as pore spaces are filled with smaller grains or cementing material such as calcite ($CaCO_3$) or quartz (SiO_2) (Figure B.1).

STRENGTH

The *strength* of a rock is commonly reported in terms of its compressive strength, which is the mechanical compression necessary to break or fracture the specimen. As measured in the laboratory, the strength of a rock is given as a force per unit area, such as Newtons per meter squared (N/m^2). Ranges of compressive strength for some of the common rock types are shown on B.1.

TABLE B.1

Strength of Common Rock Types

	Rock Type	Range of Compressive Strength (10^6 N/m^2)	Comments
Igneous	Granite	100 to 280	Finer-grained granites with few fractures are the strongest; generallysuitable for most engineering purposes.
	Basalt	50 to greater than 280	Zones of broken rock holes or fractures reduce the strength.
Metamorphic	Marble	100 to 125	Solution openings and fractures weaken the rock.
	Gneiss	160 to 190	Generally suitable for most engineering purposes.
	Quartzite	150 to 600	Very strong rock.
Sedimentary	Shale	Less than 2 to 215	May be a very weak rock for engineering purposes; careful evaluation is necessary.
	Limestone	50 to 60	May have clay layers, solution openings, or fractures thatweaken the rock.
	Sandstone	40 to 110	Strength varies with nature and extent of fractures, composition of grains, degree and type of mineral cement.

Data primarily from Bolz, R. E., and Tuve, G. L. eds. 1973. Handbook of tables for applied engineering science. Cleveland, OH: CRC Press.

APPENDIX C: Maps and Related Topics

Topographic Maps

A *topographic map* depicts the shape and elevation of the land surface as well as the location of human-made features, such as roads, schools, and political boundaries. Elevations and some other natural features are shown by *contour lines*, which on topographic maps are lines of equal elevation above sea level. Individual contour lines on a topographic map are a fixed vertical distance apart, known as the *contour interval* (CI). Common contour intervals are 5, 10, 20, 40, 80, or 100 meters or feet. The contour interval selected for a particular map depends upon the topography being represented as well as the scale of the map. If there is relatively little difference in elevation, or relief, between the highest and lowest points on a map, then the contour interval selected may be relatively small; however, if there is higher relief, the map maker will choose a larger contour interval so that the reader can resolve individual contour lines on steep slopes. This brings up an important point: Where contour lines are close together, the inclination or slope of the land is relatively steep, compared to flatter areas where contour lines are farther apart.

The *scale* of a topographic map, or of any other map, may be communicated in several ways. First, a scale may be given as a ratio, such as 1 to 24,000 (1:24,000), which means that 1 inch on a map is equal to 24,000 inches (2000 ft.) on the ground. Second, a topographic map may have several graphic or bar scales that are subdivided in feet, kilometers (kilometres), and miles. These scales are generally found in the lower margin of the map and are useful for measuring distances. Finally, the scale of some maps is stated in terms of specific units of length on the map, such as 1 inch equals 200 feet. This means that 1 inch on the map is equivalent to 200 feet (2400 in.) on the ground. In this example we could also state that the scale is 1 to 2400. The most common scale for U.S. Geological Survey topographic maps is 1:24,000 and for Natural Resources Canada maps is 1:50,000, but scales of 1:125,000 or smaller are also used. Remember 1 ÷ 24,000 is a larger number than 1 ÷ 125,000, so 1:125,000 is the smaller scale. Generally, the smaller the map scale, the more area is shown. In addition to contour lines, topographic maps also show a number of cultural features, such as roads, houses, and other buildings. Features such as streams and rivers are often shown in blue. In fact, a whole series of symbols are commonly used on topographic maps. These symbols are shown in Figure C.1 for United States maps, and similar symbols are printed on the back of most Canadian maps.

READING TOPOGRAPHIC MAPS

Reading or interpreting topographic maps is as much an art as it is a science. After you have looked at many topographic maps that represent the variety of landforms and features found at the surface of Earth, you begin to recognize these forms by the shapes of the contours. This is a process that takes a fair amount of time and experience in looking at a variety of maps. However, there are some general rules for reading topographic maps:

- Valleys containing streams of any size have contours that form Vs, in which the apex of the V points in the up-stream direction. This is sometimes known as the *rule of Vs*. Thus, if you are trying to draw the drainage pattern that shows all the streams, you should continue the stream in the upstream direction as long as the contours are still forming a V-pattern. The Vs will no longer be noticeable near the drainage divide, that is, the high point between two drainage basins.
- Where contour lines are spaced close together, the slope or inclination of the land surface is relatively steep, and where contour lines are spaced relatively far apart, the slope is relatively low and the land is flatter. As you travel up an incline or slope, you may notice that the contour lines are spaced relatively far apart and that at some point the angle of inclination or slope changes and the contour lines become closer together; this point is known as "break in slope." A break in slope is commonly observed at the foot of a mountain or where a valley side meets the surface of a floodplain.
- Contours near the upper parts of hills or mountains may show closure, that is, the contour line meets itself rather than extending to the edge of the map. These closed contours may be relatively oval or round for a conical peak, or longer and narrower for a ridge. Remember, the elevation of the top of a mountain or hill is higher than the last contour shown and may be estimated by taking half the contour interval and adding that value to the elevation of the highest contour line. That is, if the highest contour on a peak were 1000 m with a contour interval of 50 m, then the approximate elevation of the mountain top is 1025 m.
- Most topographic depressions are also shown with closed contours; the difference is that a depression contour has small *hachure* or tic marks on the contour lines that point toward the center of the depression.
- Sometimes the topography on a slope is unusually *hummocky*, that is, undulatory or uneven, when

Control data and monuments

Vertical control

Third order or better, with tablet	BM × 16.3
Third order or better, recoverable mark	× 120.0
Bench mark at found section corner	BM 18.6
Spot elevation	× 5.3

Contours

Topographic

- Intermediate
- Index
- Supplementary
- Depression
- Cut; fill

Bathymetric

- Intermediate
- Index
- Primary
- Index primary
- Supplementary

Boundaries

- National
- State or territorial
- County or equivalent
- Civil township or equivalent
- Incorporated city or equivalent
- Park, reservation, or monument

Surface features

Levee	Levee
Sand or mud area, dunes, or shifting sand	Sand
Intricate surface area	Strip mine
Gravel beach or glacial moraine	Gravel
Tailings pond	Tailings pond

Mines and caves

Quarry or open pit mine	
Gravel, sand, clay, or borrow pit	
Mine dump	Mine dump
Tailings	Tailings

Vegetation

Woods	
Scrub	
Orchard	
Vineyard	
Mangrove	Mangrove

Glaciers and permanent snowfields

- Contours and limits
- Form lines

Marine shoreline

Topographic maps

- Approximate mean high water
- Indefinite or unsurveyed

Topographic-bathymetric maps

- Mean high water
- Apparent (edge of vegetation)

Coastal features

Foreshore flat	Mud
Rock or coral reef	Reef
Rock bare or awash	
Group of rocks bare or awash	
Exposed wreck	
Depth curve; sounding	3
Breakwater, pier, jetty, or wharf	
Seawall	

Rivers, lakes, and canals

Intermittent stream	
Intermittent river	
Disappearing stream	
Perennial stream	
Perennial river	
Small falls; small rapids	
Large falls; large rapids	
Masonry dam	
Dam with lock	
Dam carrying road	
Perennial lake; Intermittent lake or pond	
Dry lake	Dry lake
Narrow wash	
Wide wash	Wide wash
Canal, flume, or aquaduct with lock	
Well or spring; spring or seep	

Submerged areas and bogs

Marsh or swamp	
Submerged marsh or swamp	
Wooded marsh or swamp	
Submerged wooded marsh or swamp	
Rice field	Rice
Land subject to inundation	Max pool 431

Buildings and related features

- Building
- School; church
- Built-up area
- Racetrack
- Airport
- Landing strip
- Well (other than water); windmill
- Tanks
- Covered reservoir
- Gaging station
- Landmark object (feature as labeled)
- Campground; picnic area
- Cemetery: small; large (Cem)

Roads and related features

Roads on Provisional edition maps are not classified as primary, secondary, or light duty. They are all symbolized as light duty roads.

- Primary highway
- Secondary highway
- Light duty road
- Unimproved road
- Trail
- Dual highway
- Dual highway with median strip

Railroads and related features

- Standard gauge single track; station
- Standard gauge multiple track
- Abandoned

Transmission lines and pipelines

Power transmission line; pole; tower	
Telephone line	Telephone
Aboveground oil or gas pipeline	
Underground oil or gas pipeline	Pipeline

▲ FIGURE C.1 **TOPOGRAPHIC MAP SYMBOLS** This chart shows some of the common symbols that the U.S. Geological Survey uses on its topographic maps. Older maps may have slightly different symbols. Symbols on Canadian topographic maps are printed on the reverse side of the map. *(From U.S. Geological Survey)*

compared with the general topography of the area. Such hummocky topography may indicate mass wasting processes and landslide deposits.

In summary, after observing a variety of topographic maps and working with them for some time, you will begin to see the pattern of contours as an actual landscape consisting of hills, valleys and other features.

LOCATING YOURSELF ON A MAP

The first time you take a topographic map outdoors you may have some difficulty locating where you are. Determining where you are is crucial to mapping floodplains, landslides, or other features. One way to locate yourself on a map is to identify two or three distinctive features that you can see—such as a mountain peak, road intersection, or a prominent bend in a road or river—that are also shown on the map. You can then use a compass to determine your location. This is done by taking a compass direction, that is, a bearing, to each of the prominent features and drawing these bearings on the map; your location is where they intersect.

Today we can also use orbiting *global positioning system (GPS)* satellites for locating ourselves and working with maps. Hand-held GPS satellite receivers are readily available at a very modest price and can generally locate your position on the ground with an accuracy of 5 to 10 m (16 to 33 ft.). GPS receivers work by receiving signals from at least four satellites and by calculating the distance from each satellite to your location. This is done with extremely accurate clocks that determine the time delay in receiving an expected radio signal from each satellite; the longer the delay, the further the satellite is from the receiver. With signals from at least four satellites, the computer in your GPS receiver can calculate a three-dimensional position on Earth's surface. Using correction signals transmitted from the Federal Aviation Administration *Wide Area Augmentation System* (WAAS), the accuracy of defining your position can be reduced to less than 3 m (10 ft.) (Figure C.2). Another way to increase positional accuracy is to average multiple measurements taken at the same location over a period of 10 minutes or more.

Your GPS receiver can be linked directly to *geographic information system (GIS)* computer software, so when a position is known, it may be plotted directly on a map viewed on a computer display in the receiver, your vehicle, or a laptop computer. GPS and GIS technology have revolutionized the way we do mapping in the field. A note here concerning the expression "*in the field*"; this expression means "outside on the surface on Earth," not the "field of geology." For example, a geologist about to go out and study landslides triggered by heavy rains in the Appalachians would say to her colleagues, "I am going into the field."

AN EXAMPLE FROM A COASTAL LANDSCAPE

In this example we will look at topographic and cultural features in a coastal landscape (Figure C.3). If we were flying along the coast in a plane we would have a sideways or oblique view of this landscape consisting of two hills with an intervening valley (Figure C.3a). The coast along the eastern hill to the right is a sea cliff, and in the center is a peninsula that was formed as a sand spit by longshore coastal processes. A hook on the end of the peninsula suggests that the direction of littoral sand transport in the surf zone and beach is from the east (right) to the west (left).

The topographic map for the area has a contour interval (CI) of 20 ft. (Figure C.3b). This map indicates that the elevation of the highest hill on the east side of the map is approximately 290 ft. above sea level; the estimated elevation is half way between 280 ft., the highest contour on the hill, and 300 ft., which would be the next contour if the hill were higher. Three streams drain the hill, two into the ocean and one into a river; notice that in all three streams the topographic contours create a V in the upstream direction toward the top of the hill.

Other information that may be "read" from the topographic map includes the following:

- The landform on the western portion of the map (left side) is a hill with elevation of about 275 ft, a gentle slope to the west, and a steep slope to the east (right) toward the ocean and the river valley. The eastern slope is particularly steep near the top of the mountain where the contours are very close together.
- A river in the center part of the map is flowing into a bay protected by a hooked sand bar. The relatively flat land next to the river is a narrow floodplain. Along the west side of the river is a light-duty road that parallels the river and then extends along the coastline. A second, unimproved road crosses the river and extends out to the head of the sand spit, providing access to a church and two other buildings. The floodplain is delineated on both sides of the river by the 20-ft. contour, above which to the 40-ft. contour is a break in slope at the edge of the valley.
- The eastern and southern slopes of the 275-foot hill on the west (left) side of the map have a number of small streams flowing mostly southward into the ocean. These streams appear to be relatively steep gullies that are eroding into the hillslope. In particular, the stream directly south of the 275-foot elevation mark appears to be cutting headward into the steep hillside.

To continue a study of this area, we might construct an east-west *topographic profile* across the area along line A-A'. In constructing a profile we must decide how much to exaggerate the vertical scale of the view (Figure C.3c).

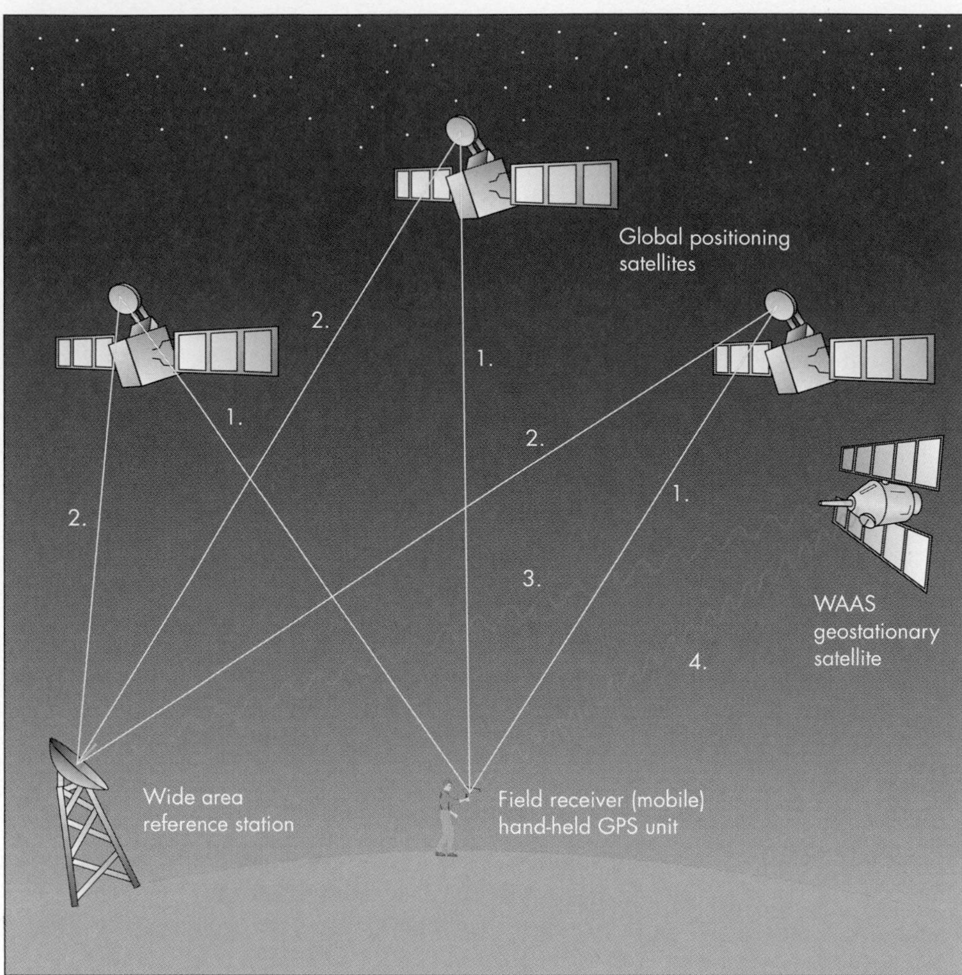

▲ FIGURE C.2 **USING THE GLOBAL POSITIONING SYSTEM** U.S. Air Force NAVSTAR Global Positioning System (GPS) satellites orbit approximately 22,200 km (13,800 mi.) above Earth. Each satellite has its own distinct signal, shown with #1 yellow lines, which can be recognized by a handheld GPS receiver. Using extremely accurate timekeeping, the receiver calculates the distance to a satellite by determining how long its signal took to arrive. A receiver must process signals from at least four satellites to calculate a three-dimensional position on Earth's surface. As long as the satellite signal is not altered, most GPS receivers will achieve a positional accuracy of within 5 to 10 m (16 to 33 ft.) of their true position on Earth's surface. Increased accuracy can be obtained with a correction signal obtained a Wide Area Reference Station that is part of the Federal Aviation Administration's Wide Area Augmentation System (WAAS). The WAAS reference station receives signals from the same GPS satellites, shown as #2 yellow lines, at the same time as the mobile GPS receiver. The WAAS then compares the position calculated from the GPS satellites with its true position to establish a correction factor. The correction factor is then transmitted to a WAAS geostationary satellite, shown as the #3 wavy line, and then broadcast back to the mobile GPS receiver, shown as the #4 wavy line. A GPS receiver with WAAS correction typically has a positional accuracy of within 3 m (10 ft.) or less of the true position.

The resulting *vertical exaggeration (VE)* is determined by comparing the same unit on both the vertical and horizontal scale; for example, the number of feet represented by 1 inch on both scales. After constructing topographic profiles, your next task might be to obtain aerial photographs and geologic maps to draw more conclusions concerning the topography and geology.

Geologic Maps

Geologists are interested in the types of earth materials found in a particular location and in their spatial distribution. Producing a geologic map is a very basic step in understanding the geology of an area. In constructing a *geologic map* a geologist must make interpretations that are consistent with the

(a)

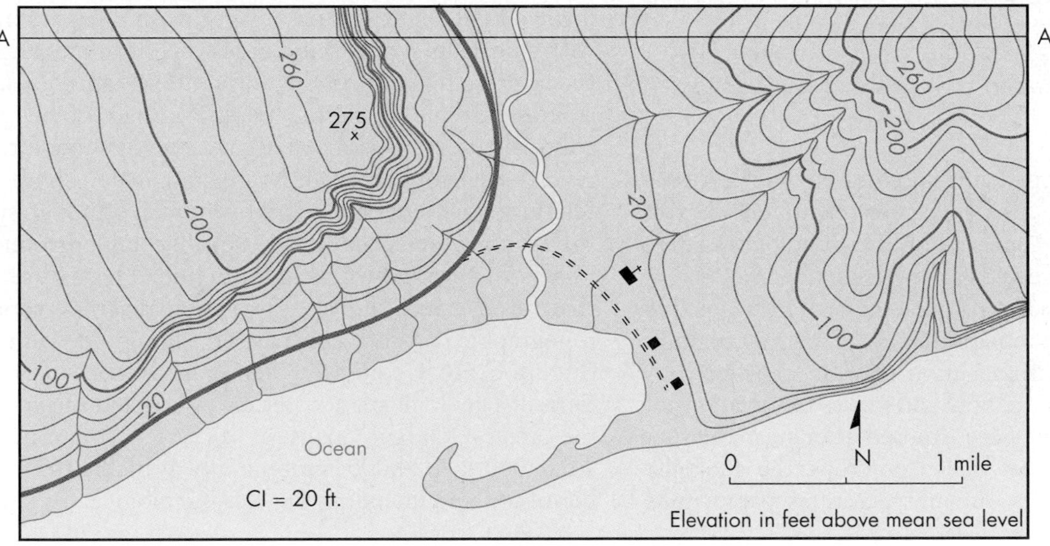

(b)

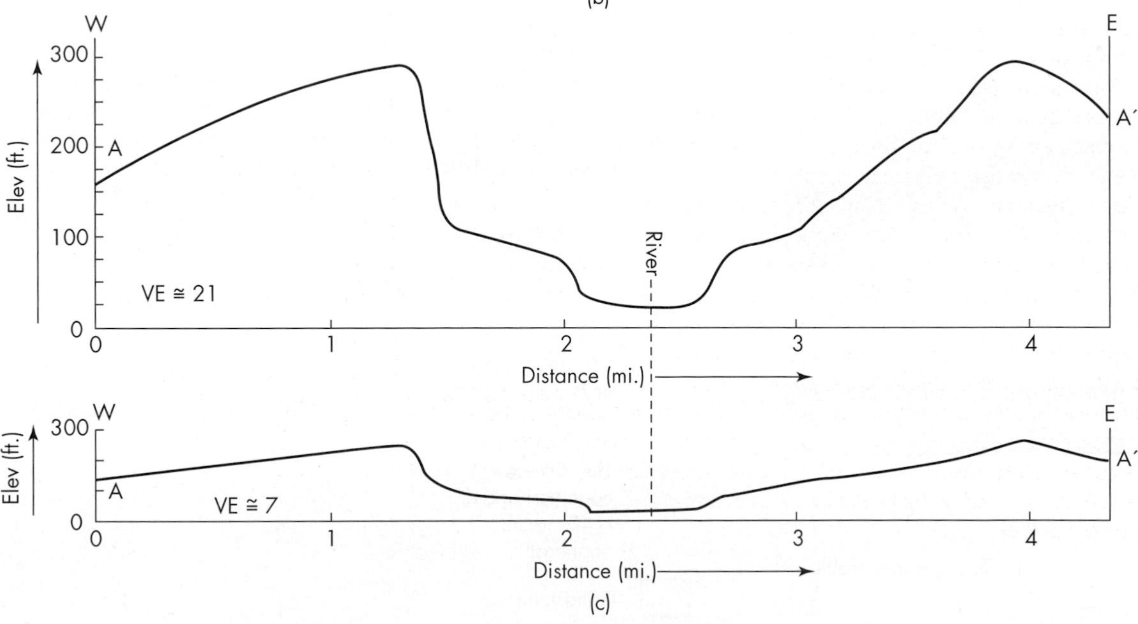

(c)

▲ FIGURE C.3 **TOPOGRAPHY OF A COASTAL LANDSCAPE** (a) An oblique view of the coastline mapped in part (b) of this figure. (b) Topographic map for the same area with a contour interval of 20 ft. (c) Topographic profiles along line A-A′ of the topographic map shown with vertical exaggeration of approximately 21 times and 7 times. Vertical exaggeration is the ratio of the vertical to horizontal scale for the topographic profile. The ratio for the upper profile is approximately 21, so it has a vertical exaggeration of approximately 21 times. For the lower profile, the vertical exaggeration is about 7 times. In the real world, of course, there is no vertical exaggeration (the vertical and horizontal scales are the same). As an experiment, you might try to make a topographic profile along line A-A′ with no vertical exaggeration. What do you conclude? *(From U.S. Geological Survey)*

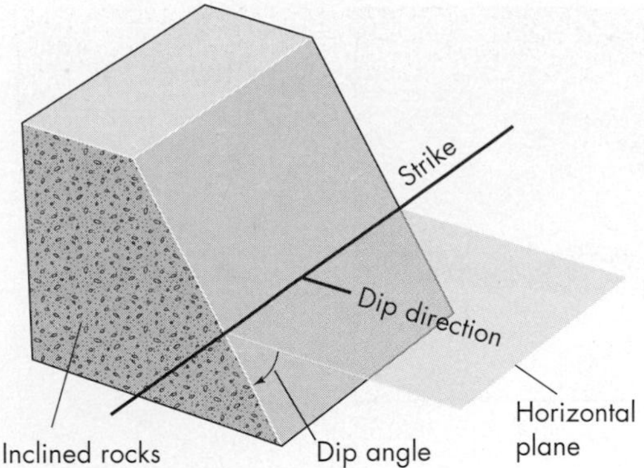

▲ FIGURE C.4 **STRIKE AND DIP** Idealized block diagram showing the strike and dip of inclined sedimentary rocks. Strike is the long direction of a T-shaped symbol and the dip direction is shown with the short line perpendicular to strike.

Earth history of the mapped area and the geologic processes that have been at work. Like any other scientific endeavor, each new observation becomes a test of the geologist's interpretations, sometimes referred to as working hypotheses.

The first step in preparing a geologic map is to obtain a good base map (usually topographic) or aerial photograph on which the geologic information may be transferred. A geologist then goes into the field and makes observations of earth materials where they are exposed at or near the surface. From these exposures, called outcrops, the geologist groups distinctive rock or sediment types into separate units called *formations*. Each formation must be thick enough to be depicted on the scale that is being mapped. On the geologic map, each formation is separated from adjacent formations by boundaries known as *contacts*, which appear on the map as thin lines between swaths of different colors.

As the geologist maps formations, he or she also takes measurements of the three-dimensional orientation, or attitude, of the mapped units. These measurements, referred to as *strike and dip*, are generally made with a compass and a device for measuring vertical angles known as a clinometer. Each strike and dip measurement is commonly recorded directly on the map with a T-shaped strike and dip symbol. Strike is the compass direction of the line formed by the intersection of layering in the earth material with a horizontal plane, and dip is the maximum angle that the layering makes with the horizontal (Figure C.4).

For example, a simple geologic map of an area of approximately 1350 km^2 (520 mi.^{2}) might show three rock types—sandstone, conglomerate, and shale (Figure C.5a). On this map the arrangement of the strike and dip symbols suggests the presence of an arch-like underground geologic structure known as an anticline. The nature of this structure becomes apparent in a geologic cross section and topographic profile constructed along the line E–E' (Figure C.5b). Geologists often make a series of cross sections of a geologic map to better understand the geology of the area. Geologic maps at a variety of scales from 1:250,000 to 1:24,000 are generally available from a number of sources, including the U.S. Geological Survey.

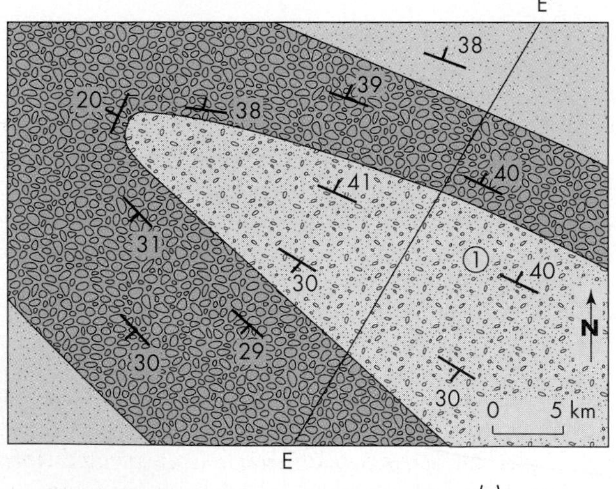

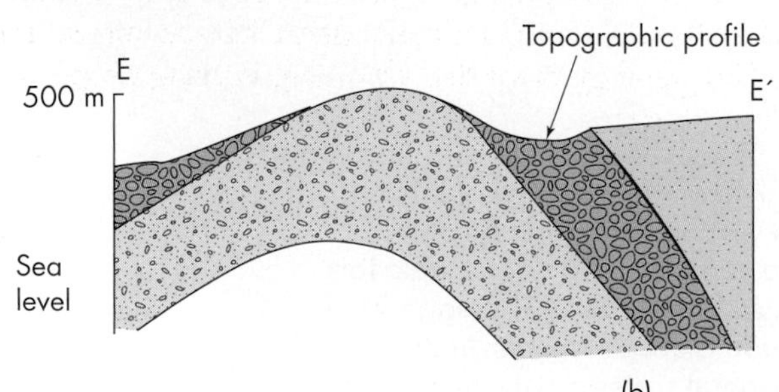

40 Strike and dip of sedimentary rock units (showing angle of dip); for example, at ① on the map. The strike is northwest and dip is 40° to the northeast.

Sandstone

Conglomerate

Shale

◀ FIGURE C.5 **GEOLOGIC MAP** (a) A very simple, idealized geologic map showing three formations, each consisting of a different type of sedimentary rock. (b) Geologic cross section and topographic profile along line E-E′. The cross section shows that the pattern of the formations on the geologic map is one made by an arch-like geologic structure known as an anticline.

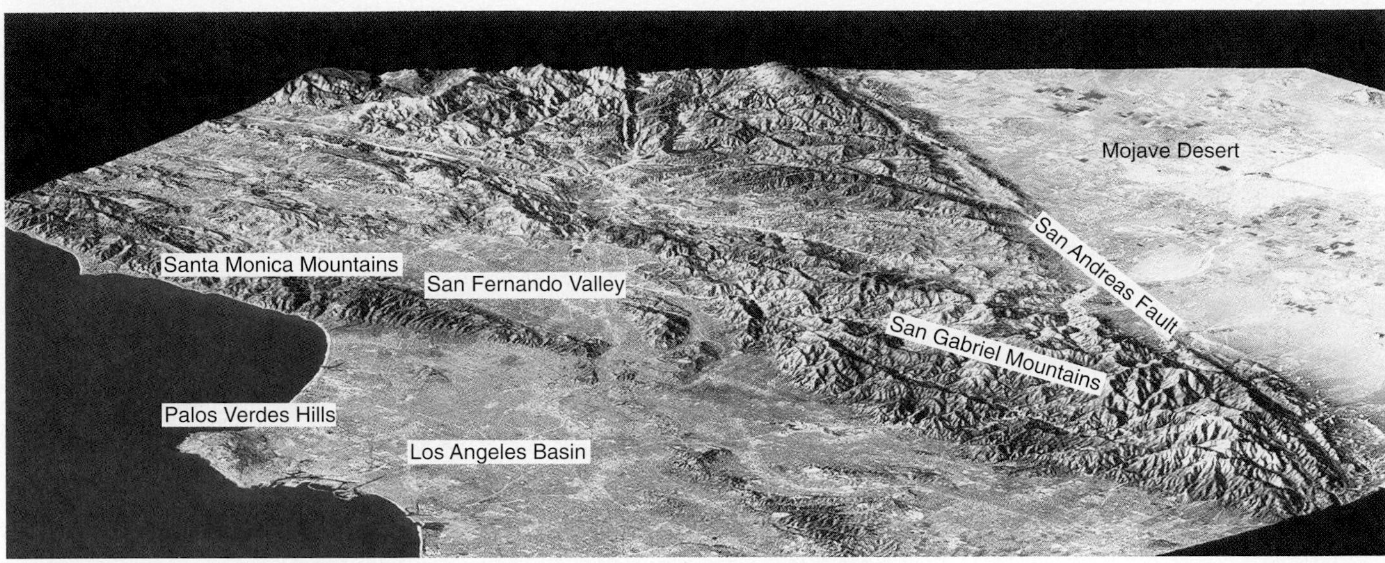

▲ FIGURE C.6 **DIGITAL ELEVATION MODEL** This digital elevation model (DEM) of the Los Angeles area is constructed as an oblique view from the southwest with a fair amount of vertical exaggeration. The flat, light blue area in the lower half of the image is the Los Angeles Basin. The smaller basin above the Los Angeles Basin is the San Fernando Valley, site of the epicenter of the 1994 Northridge earthquake. It is separated from the Los Angeles Basin by the Santa Monica Mountains. To the left of the Los Angeles Basin are the Palos Verdes Hills on the Palos Verdes Peninsula, the site of the Portuguese landslide discussed in Chapter 6. Finally on the right side of the map is the San Andreas fault, a strike-slip fault that is one of the major geologic hazards in southern California. *(Courtesy of Robert Crippin, NASA, Jet Propulsion Laboratory)*

Digital Elevation Models

Topographic data for the United States and many other parts of the world are now available as digital files. These data sets contain arrays of elevation values at a specific spacing, for example, the surface ground elevations on a 30 m grid for a 900 m^2 (9,700 $ft.^2$) area. Computer programs are then used to synthesize and view the data; color shading may be added to show the topography. The resulting graphic representation of Earth's surface is known as a *digital elevation model (DEM)*. DEMs can be examined from a variety of angles using software such as Google Earth, and you may view the topography obliquely from any compass direction. The vertical dimension may also be exaggerated so that minor topographic differences may become more apparent. A DEM for the Los Angeles Basin clearly illustrates that Los Angeles is nearly surrounded by mountains and hills, which have been uplifted by recent tectonic activity (Figure C.6). DEMs are becoming important research tools in evaluating the topography of an area. They may be used to delineate and map features such as fault scarps, landslides, karst topography, floodplains, and volcanic and impact craters.

Summary

Several types of maps and images are useful in evaluating geology and Earth processes. Of particular importance are topographic maps and profiles, and geologic maps and cross sections. Digital elevation models may be constructed from topographic data, and a variety of other special purpose maps are also available. Examples include maps of recent landslides, floodplain and coastal flood hazard maps, and engineering geology maps that show engineering properties of earth materials.

Hazard City: Assignments in Applied Geology

Map Reading

THE ISSUE

The citizens of Hazard County have become aware of contamination in Mill Stream, a tributary of the Clearwater River. They are concerned that this change in the quality of the stream will affect the river's ecosystem.

YOUR TASK

You have been hired to investigate this situation. The investigation will involve sampling water from Mill Stream and three of its tributaries to determine if any of them are the source of the contamination. To complete your study, you will have to use a topographic map to determine the distance that you will travel and the easiest way to complete your sampling. This will involve using a scale on a topographic map, interpreting topographic contours, and determining the relief of the area—the difference in elevation between high and low points on the land surface.

AVERAGE COMPLETION TIME:

1 hour

APPENDIX D: How Geologists Determine Geologic Time

In order to understand the history of Earth, we must determine the actual age of earth materials with the science of *geochronology*. In geochronology, scientists combine *relative and absolute dating* techniques to reconstruct Earth history. Relative dating establishes a chronological order of events using fossil evidence and geologic relationships, the use of which is governed by geologic principles and laws. The oldest of these laws, the law of superposition, was proposed by Nicholas Steno in 1668, and has been applied and tested by scientists for over 300 years. This law states that in any succession of sediment or sedimentary rock that has not been deformed, the oldest layer lies at the bottom, with successively younger layers above. Although a relative geochronology can be established based upon superposition and other laws and principles of relative dating, it is absolute dating that provides the information necessary to establish rates of geologic processes and the numeric ages of earth materials.

From a natural hazards perspective, it is important to establish rates of geologic processes and the timing of past geologic events such as volcanic eruptions, earthquakes, tsunamis, floods, and landslides. The chronology of these events is critical in estimating their return period or recurrence interval. Knowing how frequently an event takes place helps us better understand the hazard and predict when a similar event is likely to occur in the future.

Geologic time is much different, in a way, than our normal time framework. Although geologic time and "normal" time use the same units of measure—years—they differ vastly in duration and in the instruments we use to measure duration.[1] Normal time is counted in hours, days, seasons, or decades, and the instrument used to measure time is a clock. In contrast, geologic time—sometimes called "deep time"—is measured in tens of thousands to hundreds of millions to several billion years. To measure this vast amount of time, geologists take advantage of naturally occurring forms of uranium (U), potassium (K), carbon (C), and other chemical elements to date the earth materials in which they are found.

The natural forms of these chemical elements are called *isotopes* and are distinguished from each other by their weight. The weight of an isotope is the sum of the weight of smaller, subatomic particles called protons and neutrons that are found in the center of the atom, the nucleus. Each proton and neutron has a weight of 1, so the number of protons plus the number of neutrons determines the atomic weight of an isotope. All isotopes of a chemical element have the same number of protons, because it is the number of protons that defines a chemical element. Where isotopes vary is in their number of neutrons. For example, carbon has three common isotopes, carbon-12, carbon-13, and *carbon-14*. All three isotopes have 6 protons because they are carbon, but the number of neutrons varies from 6 to 7 and 8, respectively.

Isotopes also differ in their stability. Some isotopes, such as carbon-12 and carbon-13, have a stable number of neutrons and protons in their nucleus and do not split apart. Most isotopes, however, are unstable and will undergo spontaneous splitting, sometimes called decay, at some point in the future. Unstable isotopes are commonly called *radioactive isotopes* because radiation is released when they decay. Although it is impossible to predict when an individual unstable atom will decay, laboratory studies can determine the rate at which a large number of atoms of a radioactive isotope will decay. Each radioactive isotope has a constant rate of decay that is unaffected by physical forces.

Absolute dating is possible if a measurable quantity of a radioactive isotope is incorporated into a mineral or substance when it forms. The decay of the unstable radioactive isotope is irreversible and becomes a clock that is running at a constant rate. In absolute dating, the unstable radioactive isotope is known as the *parent* and the *stable isotope* that eventually forms from the decay of the parent is known as the *daughter product*. Radioactive isotopes, particularly those of very heavy elements, undergo a series of radioactive decay steps that finally ends when a stable, nonradioactive isotope is produced. For example, the isotope uranium-238 (U-238) undergoes 14 nuclear transformations in different steps to finally decay to stable lead-206 (Pb-206), which is not radioactive.

An important characteristic of a radioactive isotope such as U-238 is its *half-life*, which is the time required for one-half of a given amount of the isotope to decay to another form. Every radioactive isotope has a unique and characteristic half-life (Table D.1). As decay proceeds over time, the amount of parent radioactive isotope in a substance decreases and the amount of the stable daughter isotope proportionally increases (Figure D.1). Without knowing the actual quantity of radioactive isotope that was initially present, it is often possible to obtain a numeric date by comparing the relative proportions of parent and daughter isotope in a substance. In most cases, two important conditions must be met to obtain an accurate numeric date. First, no new atoms of the parent isotope have been added to the substance since it formed; this is true for many minerals in igneous rocks where the radioactive isotope becomes part of the atomic structure when a mineral crystallizes from magma. Second, all atoms of the daughter isotope produced by decay remain trapped in the substance; again this occurs for minerals in igneous rocks where atoms are strongly bonded in the crystal structure. Should

TABLE D.1

Half-Lives of Radioactive Isotopes Commonly Used in Absolute Dating

Parent Radioactive Isotope	Daughter Stable Isotope	Half-Life
Uranium-238	Lead-206	4.5 billion yr
Uranium-235	Lead-207	700 million yr
Potassium-40	Argon-40	1.3 billion yr
Carbon-14	Nitrogen-14	5730 yr

any daughter product "leak" from the sample, it would give a numeric date that was too young. Because radioactive isotopes such as U-238, U-235, and K-40 have relatively long half-lives (Table D.1), their decay is useful in dating rocks on the order of millions to billions of years. For example, the ratio of Pb-207/ Pb-206, the daughter products of U-235 and U-238, has been used to date the oldest piece of continental crust, a crystal of the mineral zircon from sandstone in Western Australia, at 4.4 billion years before the present. The actual methods for measuring amounts of parent and daughter isotopes and calculating numeric dates are complex and tedious, although the concept is easy to grasp. These methods have been successfully used to assign numeric dates to the geologic time table and to delineate important Earth history events such as mountain building, ice ages, and the appearance of life forms.

Physical anthropologists and archeologists concerned with human history are interested in developing a geochronology for the last 8 million years of Earth history; that is the period from which human and ancestral human fossils are found. From a natural hazards perspective, we are often interested in establishing the geochronology of the past few hundred to few hundred thousand years of Earth's history, and for this we have several potential methods. For example, the radioactive isotope U-234 undergoes decay to thorium (Th)-230 at a known rate, and the ratio of these two isotopes is useful in numerically dating a variety of materials, such as coral, back to several hundred thousand years.

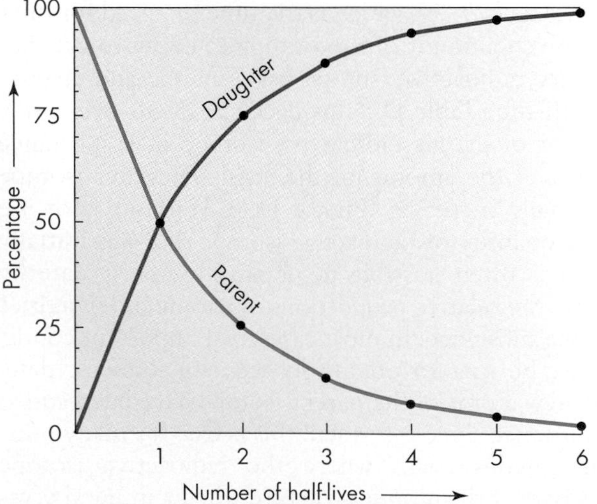

▲ **FIGURE D.1 RELATIVE PROPORTION OF PARENT AND DAUGHTER ISOTOPES OVER TIME** This graph shows the reduction of a parent radioactive isotope to a daughter stable isotope with each additional half-life.

For sediment younger than about 40,000 years, C-14 is used extensively for numeric dating. The most common form of carbon is stable C-12, but C-14, which is radioactive, occurs in small quantities and undergoes radioactive decay to the stable daughter isotope nitrogen-14 (N-14). The half-life of the decay of C-14 to N-14 is 5730 years. The C-14 method works because C-14 is only incorporated into organic matter while an organism is alive; when the organism dies the C-14 in its tissue undergoes radioactive decay without being replaced with new C-14. Common materials dated with C-14 include wood, bone, charcoal, and other types of buried organic material. C-14 has a relatively short half-life, and, as a result, after about 40,000 years the amount of C-14 remaining is very small and difficult to measure. Consequently, use of this technique is limited.

The field of geochronology is expanding rapidly, and new techniques are constantly being developed. For example, it is now possible to directly date the amount of time that a landform has been exposed at Earth's surface. This is known as *exposure dating*. The basic idea is that certain isotopes, such as beryllium-10, aluminum-26, and chlorine-36, are produced when cosmic rays interact with Earth's atmosphere and accumulate in measurable quantities in surface materials such as soil, alluvium, and exposed rock surfaces. Thus, the amount of these isotopes that has accumulated is a measure of the minimum time of exposure to the surface environment.

Another innovative technique, known as *lichenometry*, uses the slow, but constant growth of *lichens* on rock surfaces to determine a minimum time that a rock is exposed. Lichens are plant-like growths of photosynthetic algae or bacteria with a fungus. These organisms tend to grow in circular patches with a rate of growth that is known for a particular species. Careful measurements of the size of lichen patches, coupled with known growth rates, provide minimum numeric ages for exposure of the rocks on which they are found. This method has been successfully used in California and New Zealand to date regional occurrences of rockfalls generated by large earthquakes, thus dating past seismic activity.[2,3] Lichenometry can provide dates to about 1000 years before present.

Sediment may also be accurately dated through the use of *dendrochronology*, which is the analysis of annual growth rings of wood. Because of variations in annual precipitation, the width of tree rings differs from year to year. This produces a distinctive sequence of wide and thin rings similar to a bar code on merchandise at a store. Once a regional chronology is developed using growth rings in both living and dead trees, it then becomes possible to determine the age of a growth ring pattern from a piece of buried wood. This method has been used extensively by archeologists to date prehistoric sites of human habitation and also by climatologists as a method of reconstructing prehistoric patterns of precipitation. Growth rings record climate because during dry years, tree rings are narrow relative to wet years.

Very accurate numeric geochronology may also be obtained from counting varves. A *varve* is a layer of sediment representing one year of deposition, usually in a lake or an ocean. Careful counting of varves may extend the chronology back several thousand years.

Although the most accurate chronology is the historical record, its length varies from only a few hundred years in the United States and Canada to several thousand years in China. Given the brevity of human history compared with the great length of geologic history, we resort to numeric dating methods to establish geochronology. There are more than 20 methods useful in establishing geochronology that yield numeric dates. These methods are crucial to understanding rates of geologic processes and the recurrence interval of natural hazards.

REFERENCES

1. **Ausich, W. I.**, and **Lane, G. N.** 1999. *Life of the past*, 4th ed. Upper Saddle River, NJ: Prentice Hall.
2. **Bull, W. B.**, and **Brandon, M. T.** 1998. Lichen dating of earthquake-generated regional rock fall events. Southern Alps, New Zealand. *Geological Society of America Bulletin* 110: 608–84.
3. **Bull, W. B.** 1996. Dating San Andreas fault earthquakes with lichenometry. *Geology* 24: 111–14.

References

CHAPTER 1

1. **USGS.** 2010. Magnitude 7.0 Haiti region. http://earthquake.usgs.gov.
2. **USGS.** 2009. October 17,1989 Loma Prieta earthquake. http://earthquake.usgs.gov.
3. **Eberhard, M. O.,** and **four others.** 2010. The Mw 7.0 of January 12,2010: USGS/EERI Advance reconnaissance team report. U.S. Geological Survey Open-File Report 2010-1048. Executive Summary. Washington D.C.
4. **Hoyvis, P., Below, R., Scheuren, J.-M.,** and **Guha-Sapir, D.** 2007. *Annual disasters statistical review: Numbers and trends 2006.* Center for Research on the Epidemiology of Disasters (CRED). Brussels, Belgium: University of Louvain.
5. **Renner, M.,** and **Chafe, Z.** 2007. *Beyond disasters.* Washington, DC: World Watch Institute.
6. **Abramovitz, J. N.** 2001. Averting unnatural disasters. In *State of the world 2001,* ed. L. R. Brown et al., Worldwatch Institute, pp. 123–42. New York: W.W. Norton & Co.,
7. **Guha-Sapir, D., Hargitt, D.,** and **Hoyois, P.** 2004. *Thirty years of natural disasters 1974–2003: The numbers.* Brussels, Belgium: University of Lonvain, Center for Research on the Epidemiology of Disasters (CRED).
8. **Advisory Committee on the International Decade for Natural Hazard Reduction.** 1989. *Reducing disaster's toll.* National Research Council. Washington, DC: National Academy Press.
9. **Crossett, K. M., Culliton, T. J., Wiley, P. C.,** and **Goodspeed, T. R.** 2004. *Population trends along the coastal United States: 1980– 2008.* National Ocean Service, National Oceanic and Atmospheric Administration.
10. **White, G. F.,** and **Haas, J. E.** 1975. *Assessment of research on natural hazards.* Cambridge, MA: The MIT Press.
11. **Crowe, B. W.** 1986. Volcanic hazard assessment for disposal of high-level radioactive waste. In *Active tectonics,* ed. Geophysics Study Committee, pp. 247–60. National Research Council. Washington, DC: National Academy Press.
12. **Jones, R. A.** 1986. New lessons from quake in Mexico. *Los Angeles Times,* September 26.
13. **Brown, L. R., Flavin, C.,** and **Postel, S.** 1991. *Saving the planet.* New York: W.W. Norton & Co.
14. **Population Reference Bureau.** 2000. *World population data sheet.* Washington, DC: Population Reference Bureau.
15. **Smil, V.** 1999. How many billions to go? *Nature* 401:429.
16. **Abramovitz, J. N.,** and **Dunn, S.** 1998. *Record year for weather-related disasters.* Vital Signs Brief 98-5. Washington, DC: World Watch Institute.
17. **Kates, R. W.,** and **Pijawka, D.** 1977. From rubble to monument: The pace of reconstruction. In *Disaster and reconstruction,* ed. J. E. Haas, R. W. Kates, and M. J. Bowden, pp. 1–23. Cambridge, MA: The MIT Press.
18. **Costa, J. E.,** and **Baker, V. R.** 1981. *Surficial geology: Building with the Earth.* New York: John Wiley.
19. **Trenberth, K. E., Jones, P. D., Ambenje, P., Bojariu, R., Easterling, D., Klein Tank, A., Parker, D., Rahimzadeh, F., Renwick, J. A., Rusticucci, M., Soden, B.,** and **Zhai, P.** 2007. Observations: Surface and atmospheric climate change. In *Climate change 2007: The physical science basis; contribution of Working Group I to the fourth assessment report of the Intergovernmental Panel on Climate Change,* ed. S. Solomon, D. Qin, M. Manning, Z. Chen, M. Marquis, K. B. Averyt, M. Tignor, and H. L. Miller. New York: Cambridge University Press.

CHAPTER 2

1. **Wysession, M.** 1995. The inner workings of Earth. *American Scientist* 83:134–47.
2. **Glatzmaier, G. A.** 2001. *The geodynamo.* www.es.ucsc.edu/glatz/geodynamo.html. Accessed 2/21/01.
3. **Fowler, C. M. R.** 1990. *The solid Earth.* Cambridge, England: Cambridge University Press.
4. **Le Pichon, X.** 1968. Sea-floor spreading and continental drift. *Journal of Geophysical Research* 73:3661–97.
5. **Isacks, B. L., Oliver, J.,** and **Sykes, L. R.** 1968. Seismology and the new global tectonics. *Journal of Geophysical Research* 73:5855–99.
6. **Cox, A.,** and **Hart, R. B.** 1986. *Plate tectonics.* Boston: Blackwell Scientific Publications.
7. **Keller, E. A.,** and **Pinter, N.** 1996. *Active tectonics.* Upper Saddle River, NJ: Prentice Hall.
8. **Pinter, N.,** and **Brandon, N. T.** 1997. How erosion builds mountains. *Scientific American* 276(4):60–65.
9. **Dewey, J. F.** 1972. Plate tectonics. *Scientific American* 22: 56–68.
10. **Heirtzler, J. R., Le Pichon, X.,** and **Baron, J. G.** 1966. Magnetic anomalies over the Reykjanes Ridge. *Deep Sea Research* 13:427–43.
11. **Cox, A., Dalrymple, G. B.,** and **Doell, R. R.** 1967. Reversals of Earth's magnetic field. *Scientific American* 216(2):44–54.
12. **Claque, D. A., Dalrymple, G. B.,** and **Moberly, R.** 1975. Petrography and K-Ar ages of dredged volcanic rocks from the western Hawaiian Ridge and southern Emperor Seamount chain. *Geological Society of America Bulletin* 86:991–98.

CHAPTER 3

1. **U.S. Geological Survey.** 2006. *Magnitude 7.6—Pakistan.* U.S. Geological Survey Earthquake Hazards Program. http://earthquake. usgs.gov/eqcenter/eqinthenews/2005/usdyae/. Accessed 6/9/07.
2. **Achenbach, J.** 2006. The next big one. *National Geographic.* 209(4):120–147.
3. **U.S. Geological Survey.** 1996. *USGS response to an urban earthquake, Northridge '94.* U.S. Geological Survey Open File Report 96–263.
4. **USGS.** 2010. Magnitude 7.0 Haiti region. http://earthquake.usgs.gov. Accessed (3/15/10)
5. **Eberhard, M. O., Baldridge, S., Marshall, J., Mooney, W., Rix, G. J.,** 2010. The M_w 7.0 of January 12, 2010: USGS/EERI Advance reconnaissance team report. U.S. Geological Survey Open-File Report 2010-1048. Executive Summary. Washington, D.C.
6. **Wald, D., Wald, L., Worden, B.,** and **Goltz, J.** 2003. *Shake Map—A tool for earthquake response.* U.S. Geological Survey Fact Sheet FS-087-03.
7. **Cervelli, P.** 2004. The threat of silent earthquakes. *Scientific American* 290:86–91.
8. **Radbruch, D. H.,** and **Lennert, B. J.** 1966. Damage to culvert under Memorial Stadium University of California. In *Tectonic creep in the Hayward fault zone California,* pp. 3–6. U.S. Geological Survey Circular 525.
9. **Steinbrugge, K. V.,** and **Zacher, E. G.** 1960. Creep on the San Andreas fault. In *Focus on environmental geology,* ed. R. W. Tank, pp. 132–37. New York: Oxford University Press.
10. **Melbourne, T. I.,** and **Webb, F. H.** 2003. Slow but not quite silent. *Science* 300:1886.
11. **Bolt, B. A.** 2006. *Earthquakes: 2006 Centennial update,* 5th ed. San Francisco: W.H. Freeman.
12. **Fisher, R.** 2009. Seismic boom: Breaking the quake barrier. *New Scientist,* issue 2719.
13. **Hough, S. E., Friberg, P. A., Busby, R., Field, E. F., Jacob, K. H.,** and **Borcherdt, R. D.** 1989. Did mud cause freeway collapse? *EOS, Transactions, American Geophysical Union* 70(47):1497, 1504.
14. **Yeats, R. S.** 2001. *Living with earthquakes in California: A survivor's guide.* Corvallis: Oregon State University Press.
15. **Jones, R. A.** 1986. New lessons from quake in Mexico. *Los Angeles Times,* September 26.
16. **Hanks, T. C.** 1985. *The National Earthquake Hazards Reduction Program: Scientific status.* U.S. Geological Survey Bulletin 1659.
17. **U.S. Geological Survey.** 2010. *Magnitude 8.8—offshore Maule Chile.* http://earthquake.usgs.gov/earthquakes/eqinthenews/2010/us2010tfan/ (Accessed 8/11/10)
18. **Advisory Committee on the International Decade for Natural Hazard Reduction.** 1989. *Reducing disaster's toll.* National Research Council. Washington, DC: National Academy Press.
19. **Hough, S.** 2002. *Earthshaking science: What we know (and don't know) about earthquakes.* Princeton, NJ: Princeton University Press.
20. **Sieh, K.,** and **LeVay, S.** 1998. *The Earth in turmoil: Earthquakes, volcanoes and their impact on humankind.* New York: W.H. Freeman.
21. **Hamilton, R. M.** 1980. Quakes along the Mississippi. *Natural History* 89:70–75.
22. **Mueller, K., Champion, J., Guccione, E. M.,** and **Kelson, K.** 1999. Fault slip rates in the modern New Madrid Seismic Zone. *Science* 286:1135–38.
23. **Yeats, R. S., Sieh, K.,** and **Allen, C. R.** 1997. *The geology of earthquakes.* New York: Oxford University Press.
24. **Youd, T. L., Nichols, D. R., Helley, E. J.,** and **Lajoie, K. R.** 1975. Liquefaction potential. In *Studies for seismic zonation of the San Francisco Bay region,* ed. R. D. Borcherdt, pp. 68–74. U.S. Geological Survey Professional Paper 941A.
25. **Hansen, W. R.** 1965. *The Alaskan earthquake, March 27, 1964: Effects on communities.* U.S. Geological Survey Professional Paper 542A.
26. **Liu, J. G.,** and **Kusky, T.** 2008. After the quake. *Earth* 53(10):48–51.
27. **U.S. Geological Survey.** 1996. *USGS response to an urban earthquake; Northridge '94.* U.S. Geological Survey Open-File Report 96–263.
28. **Pakiser, L. C., Eaton, J. P., Healy, J. H.,** and **Raleigh, C. B.** 1969. Earthquake prediction and control. *Science* 166:1467–74.
29. **Evans, D. M.** 1966. Man-made earthquakes in Denver. *Geotimes* 10(9):11–18.
30. **Reed, C.** 2002. Triggering quakes with waste. *Geotimes* 47(3):7.
31. **Frohlich, C.,** and **Davis, S. D.** 2002. *Texas earthquakes.* Austin: University of Texas Press.
32. **Page, R. A., Boore, D. M., Bucknam, R. C.,** and **Thatcher, W. R.** 1992. *Goals, opportunities, and priorities for the USGS Earthquake Hazards Reduction Program.* U.S. Geological Survey Circular 1079.

33. **State of California Uniform Building Code.** 1997. Chapter 16.
34. **State of California.** 2008. California Geological Survey—Probabilistic seismic hazards assessment—Peak ground acceleration. http://www.conservation.ca .gov. Accessed 11/30/08.
35. **U. S. Geological Survey.** 2008. Deterministic and scenario ground-motion maps. http://www.earthquake.usgs.gov/research/hazmaps/scenario. (Accessed 8/11/10)
36. **Wells, D. L.**, and **Coppersmith, K. J.** 1994. New empirical relationships among magnitude, rupture length, rupture width, rupture area and surface displacement. *Bulletin of the Seismological Society of America* 84:974–1002.
37. **Eberhart-Phillips, D., Haeussler, P. J., Freymueller, J. T., Frankel, A. D., Rubin, C. M., Craw, P., Ratchkovski, N. A., Anderson, G., Carver, G. A., Crone, A. J., Dawson, T. E., Fletcher, H., Hansen, R., Harp, E. L., Harris, R. A., Hill, D. P., Hreinsdóttir, S., Jibson, R. W., Jones, L. M., Kayen, R., Keffer, D. K., Larsen, C. F., Moran, S. C., Personius, S. F., Plafker, G., Sherrod, B., Sieh, K., Sitar, N.,** and **Wallace, W. K.** 2003. The 2002 Denali fault earthquake, Alaska: A large magnitude, slip-partitioned event. *Science* 300:1113–18.
38. **Fuis, G. S.**, and **Wald, L. A.** 2003. *Rupture in South-Central Alaska—the Denali earthquake of 2002.* U.S. Geological Survey Fact Sheet 014–03.
39. **Field, E. H., Milner, K. R.**, and the **2007 Working Group on California Earthquake Probabilities.** 2008. Forecasting California's earthquakes—What can we expect in the next 30 years? *U.S. Geological Survey Fact Sheet* 2008-3027.
40. **Committee on the Science of Earthquakes.** 2003. *Living on an active Earth: Perspectives on earthquake science.* National Research Council. Washington, DC: National Academies Press.
41. **Scholz, C.** 1997. Whatever happened to earthquake prediction? *Geotimes* 42(3):16–19.
42. **Press, F.** 1975. Earthquake prediction. *Scientific American* 232:14–23.
43. **Scholz, C. H.** 2002. *The mechanics of earthquakes and faulting*, 2nd ed. New York: Cambridge University Press.
44. **Rikitakr, T.** 1983. *Earthquake forecasting and warning.* London: D. Reidel.
45. **Silver, P. G.**, and **Wakita, H.** 1996. A search for earthquake precursors. *Science* 273:77–78.
46. **Allen, C. R.** 1983. Earthquake prediction. *Geology* 11:682.
47. **Hait, M. H.** 1978. Holocene faulting, Lost River Range, Idaho. *Geological Society of America Abstracts with Programs* 10(5):217.
48. **Gori, P. L.** 1993. The social dynamics of a false earthquake prediction and the response by the public sector. *Bulletin of the Seismological Society of America* 83:963–80.
49. **Yeats, R. S.**, 2004. *Living with earthquakes in the Pacific Northwest*, 2nd ed. Corvallis: Oregon State University Press.
50. **Holden, R., Lee, R.**, and **Reichle, M.** 1989. *Technical and economic feasibility of an earthquake warning system in California.* California Division of Mines and Geology Special Publication 101.
51. **Whitman, R. V., Anagon, T., Kircher, C. A., Lagurio, H. J., Lawson, R. S.**, and **Schneider, P.** 1997. Development of a national earthquake loss estimation methodology. *Earthquake Spectra* 13(4):643–61.
52. **Southern California Earthquake Center.** 2007. *Putting down roots in earthquake country.* Los Angeles: University of Southern California.
53. **Coburn, A.**, and **Spence, R.** 1992. *Earthquake protection.* New York: John Wiley & Sons.
54. **Stein, W.**, and **Wysession, M.** 2003. *An introduction to seismology, earthquakes, and earth structure.* Malden, MA: Blackwell Publishing.

CHAPTER 4

1. **Subarya, C., Chlieh, M., Prawirodirdjo, L., Avouac, J.-P., Bock, Y., Sieh, K., Meltzner, A. J., Natawidjaja, D. H.**, and **McCaffrey, R.** 2006. Plate-boundary deformation associated with the great Sumatra-Andaman earthquake. *Nature* 440:46–51.
2. **Kerr, R. A.** 2005. Failure to gauge the quake crippled the warning effort. *Science* 307:201.
3. **U.S. Geological Survey.** 2005. *Magnitude 9.1—off the west coast of northern Sumatra.* U.S. Geological Survey Earthquake Hazards Program. http://earthquake.usgs.gov/eqcenter/eqinthenews/2004/ usslav/. Accessed 6/03/07.
4. **Chapman, C.** 2005. The Asian tsunami in Sri Lanka: A personal experience. *EOS, Transactions, American Geophysical Union* 86(2):13–14.
5. **Bendeich, M.** 2005. *Elephants saved tourists from tsunami.* Reuters. http:// savetheelephants.org. Accessed 5/25/07.
6. **Achenbach, J.** 2006. The next big one. *National Geographic* 209(4):120–147.
7. **Sieh, K.** 2006. Sumatran megathrust earthquakes: From science to saving lives. *Philosophical Transactions Royal Society* 364:1947–1963.
8. **Jaffe, B. E, Gelfenbaum, G., Buckley, M. L., Watt, S., Apostos, A., Stevens, A. W., Richmond, B. M..** 2010.*The limit of inundation of the September 29, 2009 tsunami in Tutuila, American Samoa.* U.S. Geological Survey Open File Report 2010-1018.
9. **Bryant, E.** 2001. *Tsunami: The underrated hazard.* New York: Cambridge University Press.
10. **Bolt, B. A.** 2006. *Earthquakes*, 5th ed.; 2006 Centennial update. New York: W.H. Freeman.
11. **U.S. Geological Survey.** 2005. *Life of a tsunami.* Western Coastal and Marine Geology Program. http://walrus.wr. usgs.gov/tsunami/ basics.html. Accessed 5/25/07.
12. **Hokkaido Tsunami Research Group.** 1993. Tsunami devastates Japanese coastal region. *EOS, Transactions American Geophysical Union* 74(37):417–432.
13. **Tappin, D. R., Watts, P., McMurtry, G. M., Lafoy, Y.**, and **Matsumoto, T.** 2001. The Sissano, Papau New Guinea tsunami of July 1998—offshore evidence of the source mechanism. *Marine Geology* 175:1–23.
14. **Stover, C. W.**, and **Coffman, J. L.** 1993. *Seismicity of the United States, 1958–1989* (revised). U.S. Geological Survey Professional Paper 1527.
15. **Risk Management Solutions.** 2006. *2004 Indian Ocean Tsunami Report.* Newark, CA: Risk Management Solutions, Inc. http://www.rms.com/Publications/IndianOceanTsunamiReport. pdf. Accessed 5/27/07.
16. **Satake, K., Wang, K.**, and **Atwater, B. F.** 2003. Fault slip and seismic moment of the 1700 Cascadia earthquake inferred from Japanese tsunami descriptions. *Journal of Geophysical Research* 108(B11):148–227, doi:10.1029/2003JB002521.
17. **Nelson, A. R., Atwater, B. F., Bobrowsky, P. T., Bradley, L.-A., Clague, J. J., Carver, G. A., Darienzo, M. E., Grant, W. C., Krueger, H. W., Sparks, R., Stafford Jr., T. W.**, and **Stuiver, M.** 1995. Radiocarbon evidence for extensive plate-boundary rupture about 300 years ago at the Cascadia subduction zone. *Nature* 378:371–374.
18. **Atwater, B. F.** 1992. Geologic evidence for earthquakes during the past 2000 years along the Copalis River, southern coastal Washington. *Journal of Geophysical Research* 97(B2):1901–19.
19. **Potera, C.** 2005. In disasters wake: Tsunami lung. *Environmental Health Perspectives* 113(11):A734.
20. **California Seismic Safety Commission.** 2005. *The tsunami threat to California; Findings and recommendations on tsunami hazards and risks.* Report CSSC 05-03.
21. **Danielsen, F., Serensen, M. K., Olwig, M. F., Selvam, V., Parish, F., Burgess, N. D., Hiraishi, T., Karunagaran, V. M., Rasmussen, M. S., Hansen, L. B., Quarto, A.**, and **Suryadiputra, N.** 2005. The Asian tsunami: A protective role for coastal vegetation. *Science* 310:643.
22. **Geist, E. L.** and **Parsons, T.** 2006. Probabilistic analysis of tsunami hazards. *Natural Hazards* 37:277–314.

CHAPTER 5

1. **Wright, T. L.**, and **Pierson, T. C.** 1992. *Living with volcanoes.* U.S. Geological Survey Circular 1073.
2. **Pendick, D.** 1994. Under the volcano. *Earth* 3(3):34–39.
3. **Decker, R.**, and **Decker, B.** 2006. *Volcanoes*, 4th ed. New York: W.H. Freeman.
4. **Smith, G. A.**, and **Pun, A.** 2010. *How does Earth work*, 2nd ed. Upper Saddle River, NJ: Pearson Prentice Hall.
5. **Fisher, R. V., Heiken, G.**, and **Hulen, J. B.** 1997. *Volcanoes.* Princeton, NJ: Princeton University Press.
6. **Christiansen, R. L., Lowenstern, J. B., Smith, R. B., Heasler, H., Morgan, L. A., Nathenson, M., Mastin, L. G., Muffler, L. J. P.**, and **Robinson, J. E.** 2007. *Preliminary assessment of volcanic and hydrothermal hazards in Yellowstone National Park and vicinity.* U.S. Geological Survey Open-file Report 2007–1071.
7. **Francis, P.** 1983. Giant volcanic calderas. *Scientific American* 248(6):60–70.
8. **Kious, W. J.**, and **Tilling, R. I.** 1996. *This dynamic Earth: The story of plate tectonics.* Washington, DC: U.S. Geological Survey.
9. **Schmincke, H.** 2004. *Volcanism.* New York: Springer-Verlag.
10. **IAVCEE Subcommittee on Decade Volcanoes.** 1994. Research at decade volcanoes aimed at disaster prevention. *EOS, Transactions, American Geophysical Union* 75(30):340, 350.
11. **Crandell, D. R.**, and **Waldron, H. H.** 1969. Volcanic hazards in the Cascade Range. In *Geologic hazards and public problems*, conference proceedings, ed. R. Olsen and M. Wallace, pp. 5–18. Office of Emergency Preparedness Region 7.
12. **Nakada, S.** 2000. Hazards from pyroclastic flows and surges. In *Encyclopedia of volcanoes*, ed. H. R. Sigurdsson, B. F. Houghton, S. R. McNutt, H. Rymer, and J. Stix, pp. 945–55. San Diego, CA: Academic Press.
13. **Neal, C. A., Casadevall, T. J., Miller, T. P., Hendley, J. W. II**, and **Stauffer, P. H.** 1998. *Volcanic ash—Danger to aircraft in the North Pacific.* U.S. Geological Survey Fact Sheet 030–97.
14. **Mazzocchi, M., Hansstein, F., Ragona, M.**, 2010. The 2010 volcanic ash cloud and its financial impact on the European airline industry. CESifo forum / a joint initiative of Ludwig-maximilians-Universitat and the I fo Institue for Economic Research, vol. 1, no. 1, p. 92–100.
15. **Simkin, T., Siebert, L.**, and **Blong, R.** 2001. Volcano fatalities—lesson from the historical record. *Science* 291:255.
16. **Holloway, M.** 2000. The killing lakes. *Scientific American* 286(3): 90–99.
17. **Scarth, A.** 1999. *Vulcan's fury: Man against the volcano.* New Haven, CT: Yale University Press.
18. **Sutton, J., Elias, T., Hendley II, J. W.**, and **Stauffer, P. H.** 2000. *Volcanic air pollution—A hazard in Hawai'i.* U.S. Geological Survey Fact Sheet 169–97, version 1.1.
19. **Johnson, J., Branley, S. R., Swanson, D. A., Stauffer, P. H.**, and **Hendley II, J. W.** 2000. *Viewing Hawai'i's lava safely—Common sense is not enough.* U.S. Geological Survey Fact Sheet 152-00, version 1.1.
20. **U.S. Geological Survey.** 1999. *Pilot project, Mount Rainier volcano lahar warning system.* http://volcanoes.usgs.gov/About/ Highlights/RainierPilot/Pilot_highlight.html. Accessed 6/17/07.
21. **Watts, A. B.**, and **Masson, D. G.** 1995. A giant landslide on the north flank of Tenerife, Canary Islands. *Journal of Geophysical Research* 100:24487–98.
22. **Hammond, P. E.** 1980. Mt. St. Helens blasts 400 meters off its peak. *Geotimes* 25(8):14–15.
23. **U.S. Geological Survey.** 2007. *Thoughts on the 27th anniversary of catastrophic eruption of Mount St. Helens and on ongoing activity.* Vancouver, WA: USGS Cascade Volcano Observatory, http:// vulcan.wr.usgs.gov/Volcanoes/MSH/Eruption04/MediaInfo/ May07/talking_points_may2007.pdf. Accessed 6/18/07.
24. **Gardner, C.** 2005. Monitoring a restless volcano: The 2004 eruption of Mount St. Helens. *Geotimes* 50(3):24–29.
25. **Pendick, D.** 1995. Return to Mount St. Helens. *Earth* 4(2):24–33.
26. **Self, S.** 2005. Effects of volcanic eruptions on the atmosphere and climate. In *Volcanoes and the environment*, ed. J. Martí, and G. G. J. Ernst, pp. 152–74. New York: Cambridge University Press.

27. **Oppenheimer, C.** 2003. Climatic, environmental and human consequences of the largest known historic eruption: Tambora volcano (Indonesia) 1815. *Progress in Physical Geography* 27(2): 230–59.
28. **Anonymous.** 1991. Pinatubo cloud measured. *EOS, Transactions, American Geophysical Union* 72(29):305–306.
29. **Duffield, W. A.** 2005. Volcanoes, geothermal energy, and the environment. In *Volcanoes and the environment*, ed. J. Martí, and G. G. J. Ernst, pp. 304–332. New York: Cambridge University Press.
30. **Heiken, G.** 2005. Industrial uses of volcanic materials. In *Volcanoes and the environment*, ed. J. Martí, and G. G. J. Ernst, pp. 387–403. New York: Cambridge University Press.
31. **Kilburn, C. R. J.**, and **Sammonds, P. R.** 2005. Maximum warning times for imminent volcanic eruptions. *Geophysical Research Letters* 32: L24313, doi 10.1029/ 2005GL024184.
32. **McGuire, B.** 2006. *Hazard and risk review 2006*. London: Benfield Hazard Research Centre, University College London.
33. **Francis, P.** 1976. *Volcanoes*. New York: Pelican Books.
34. **Richter, D. H., Eaton, J. P., Murata, K. J., Ault, W. U.**, and **Krivoy, H. L.** 1970. *Chronological narrative of the 1959–60 eruption of Kilauea Volcano, Hawaii*. U.S. Geological Survey Professional Paper 537E.
35. **Tilling, R. I.** 2000. Volcano notes. *Geotimes* 45(4):19.
36. **Gardner, C. A.**, and **Guffanti, M. C.** 2006. *U.S. Geological Survey's alert notification system for volcanic activity*. U.S. Geological Survey Fact Sheet 2006-3139.
37. **Murton, B. J.**, and **Shimabukuro, S.** 1974. Human adjustment to volcanic hazard in Puna District, Hawaii. In *Natural hazards; Local, national, global*, ed. G. F. White, pp. 151–159. New York: Oxford University Press.
38. **Geotimes** 2005. Vesuvius' Next Eruption, April 50 (4).
39. **Williams Jr., R. S.**, and **Moore, J. G.** 1973. Iceland chills a lava flow. *Geotimes* 18(8):14–17.

CHAPTER 6

1. **Rahn, P. H.** 1984. Flood-plain management program in Rapid City, South Dakota. *Geological Society of America Bulletin* 95:838–43.
2. **U.S. Department of Commerce.** 1973. *Climatological data, national summary* 24(13).
3. **Anonymous.** 1993. The flood of '93. *Earth Observation Magazine*, September:22–23.
4. **Mairson, A.** 1994. The great flood of 1993. *National Geographic* 185(1):42–81.
5. **Bell, G. D.** 1993. The great Midwestern flood of 1993. *EOS, Transactions, American Geophysical Union* 74(43):60–61.
6. **Noaa.** 2008. *Climate of 2008. Midwestern U.S. flood overview*. www.ncdc.noaa.gov. Accessed 4/22/10.
7. **Anonymous.** 1993. Flood rebuilding prompts new wetlands debate. *U.S. Water News*, November:10.
8. **Pinter, N., Thomas, R.**, and **Wollsinski, J. H.** 2001. Assessing flood hazard on dynamic rivers. *EOS, Transactions, American Geophysical Union* 82:333–39.
9. **Pinter N.** 2009. Non-stationary flood occurrence on the upper Mississippi River–lower Missouri River system: Review and current status. In *Finding a balance between floods, flood protection and river navigation*, eds. R. E. Criss and T. M. Kusky. P. 34–40. St. Louis, MO: Saint Louis University Center for Environmental Science. Available at http://www.ces.slu.edu.
10. **Hey, D. Kostel, J.** and **Montgomery, D.** 2009. An ecological solution to the flood damage problem. In *Finding a balance between floods, flood protection and river navigation*, eds. R. E. Criss and T. M. Kusky. P. 72–79. St. Louis, MO: Saint Louis University Center for Environmental Science. Available at http://www.ces.slu.edu.
11. **Saulny, S.** 2007. Development rises on St. Louis area flood plains. *New York Times*, May 15.
12. **Pinter, N.** 2005. One step forward, two steps back on U.S. floodplains. *Science* 308:207–08.
13. **Committee on Alluvial Fan Flooding.** 1996. *Alluvial fan flooding*. National Research Council. Washington, DC: National Academy Press.
14. **Edelen Jr., G. W.** 1981. Hazards from floods. In *Facing geological and hydrologic hazards, earth-science considerations*, ed. W. W. Hays, pp. 39–52. U.S. Geological Survey Professional Paper 1240–B.
15. **Keller, E. A.**, and **Capelli, M. H.** 1992. Ventura River flood of February, 1992: A lesson ignored? *Water Resources Bulletin* 28(5):813–31.
16. **Keller, E. A.**, and **Florsheim, J. L.** 1993. Velocity reversal hypothesis: A model approach. *Earth Surface Processes and Landforms* 18:733–48.
17. **Beyer, J. L.** 1974. Global response to natural hazards: Floods. In *Natural hazards: Local, national, global*, ed. G. F. White, pp. 265–74. New York: Oxford University Press.
18. **McCain, J. F., Hoxit, L. R., Maddox, R. A., Chappell, C. F.**, and **Caracena, F.** 1979. Meteorology and hydrology in Big Thompson River and Cache la Poudre River Basins. In *Storm and flood of July 31–August 1, 1976, in the Big Thompson River and Cache la Poudre River Basins, Larimer and Weld Counties, Colorado*. U.S. Geological Survey Professional Paper 1115A.
19. **Shroba, R. R., Schmidt, P. W., Crosby, E. J.**, and **Hansen, W. R.** 1979. Geologic and geomorphic effects in the Big Thompson Canyon area, Larimer County. In *Storm and flood of July 31–August 1, 1976, in the Big Thompson River and Cache la Poudre River Basins, Larimer and Weld Counties, Colorado*. U.S. Geological Survey Professional Paper 1115B.
20. **Bradley, W. C.**, and **Mears, A. I.** 1980. Calculations of flows needed to transport coarse fraction of Boulder Creek alluvium at Boulder, Colorado. *Geological Society of America Bulletin*, Part II, 91:1057–90.
21. **Linsley Jr., R. K., Kohler, M. A.**, and **Paulhus, J. L.** 1958. *Hydrology for engineers*. New York: McGraw-Hill.
22. **Leopold, L. B.** 1968. *Hydrology for urban land planning*. U.S. Geological Survey Circular 554.
23. **Seaburn, G. E.** 1969. *Effects of urban development on direct runoff to East Meadow Brook, Nassau County, Long Island, New York*. U.S. Geological Survey Professional Paper 627B.
24. **Agricultural Research Service.** 1969. *Water intake by soils*. Miscellaneous Publication No. 925. U.S. Department of Agriculture.
25. **Strahler, A. N.**, and **Strahler, A. H.** 1973. *Environmental geoscience*. Santa Barbara, CA: Hamilton Publishing.
26. **Office of Emergency Preparedness.** 1972. *Report to Congress*, Vol. 3: *Disaster preparedness*. Washington, DC: U.S. Government Printing Office.
27. **Slade Jr., R. M.**, and **Patton, J.** 2003. *Major and catastrophic storms and floods in Texas*. U.S. Geological Survey Open-File Report 03–0193.
28. **Mackin, J. H.** 1948. Concept of the graded river. *Geological Society of America Bulletin* 59:463–512.
29. **Dolan, R., Howard, A.**, and **Gallenson, A.** 1974. Man's impact on the Colorado River and the Grand Canyon. *American Scientist* 62:392–401.
30. **Lavender, D.** 1984. Great news from the Grand Canyon. *Arizona Highways Magazine*, January:33–38.
31. **Hecht, J.** 1996. Grand Canyon flood a roaring success. *New Scientist* 151:8.
32. **Lucchitta, I.**, and **Leopold, L. B.** 1999. Floods and sandbars in the Grand Canyon. *Geology Today* 9:1–7.
33. **Terstriep, M. L., Voorhees, M. L.**, and **Bender, G. M.** 1976. *Conventional urbanization and its effect on storm runoff*. Illinois State Water Survey Publication, ISWS CR-177.
34. **Lagmay, A. M. F., Rodolfo, R. S.**, and **Bato, M. G.** 2110. The perfect storm: Floods devastate Manila. *Earth* 55(4):50–55.
35. **Mount, J. F.** 1997. *California rivers and streams*. Berkeley: University of California Press.
36. **Baker, V. R.** 1984. Questions raised by the Tucson flood of 1983. In *Proceedings of the 1984 meetings of the American Water Resources Association and the Hydrology Section of the Arizona–Nevada Academy of Science*, pp. 211–19.
37. **Baker, V. R.** 1994. Geologic understanding and the changing environment. *Transactions of the Gulf Coast Association of Geological Societies* 44:1–8.
38. **U.S. Congress.** 1973. *Stream channelization: What federally financed draglines and bulldozers do to our nation's streams*. House Report No. 93–530. Washington, DC: U.S. Government Printing Office.
39. **Rosgen, D.** 1996. *Applied river morphology*. Lakewood, CO: Wildland Hydrology.
40. **Pilkey, O. H.**, and **Dixon, K. L.** 1996. *The Corps and the shore*. Washington, DC: Island Press.
41. **Baker, V. R.** 1976. Hydrogeomorphic methods for the regional evaluation of flood hazards. *Environmental Geology* 1:261–81.
42. **Smith, K.**, and **Ward, R.** 1998. *Floods*. New York: John Wiley and Sons.
43. **Bue, C. D.** 1967. *Flood information for floodplain planning*. U.S. Geological Survey Circular 539.
44. **Schaeffer, J. R., Ellis, D. W.**, and **Spieker, A. M.** 1970. *Flood-hazard mapping in metropolitan Chicago*. U.S. Geological Survey Circular 601C.

CHAPTER 7

1. **Gurolla, L. D., DeVecchio, D. E., Keller, E. A.**, *2010*, Late Pleistocene landslides in the modern landscape and hazards they present, Journal of Geomorphology.
2. **Pestrong, R.** 1974. *Slope stability*. American Geological Institute. New York: McGraw-Hill.
3. **Nilsen, T. H., Taylor, F. A.**, and **Dean, R. M.** 1976. *Natural conditions that control landsliding in the San Francisco Bay Region*. U.S. Geological Survey Bulletin 1424.
4. **Burroughs, E. R. Jr.**, and **Thomas, B. R.** 1977. *Declining root strength in Douglas fir after felling as a factor in slope stability*. USDA Forest Service Research Paper INT-190.
5. **Campbell, R. H.** 1975. *Soil slips, debris flows, and rainstorms in the Santa Monica Mountains and vicinity, southern California*. U.S. Geological Survey Professional Paper 851.
6. **Terzaghi, K.** 1950. Mechanism of landslides. In *Application of geology to engineering practice*, ed. S. Paige, pp. 83–123. Geological Society of America Berkey Volume.
7. **Leggett, R. F.** 1973. *Cities and geology*. New York: McGraw-Hill.
8. **WestWide Avalanche Network.** 2007. *US & Canada, avalanche fatalities & close calls*. Alta, UT: Center for Snow Science. www.avalanche.org. Accessed 6/30/07.
9. **Abromeit, D., Deveraux, A. M.**, and **Overby, B.** 2004. *Avalanche basics*. U.S. Forest Service National Avalanche Center. www.fsavalanche.org/basics/basic_index.html. Accessed 6/29/07.
10. **Perla, R. I.**, and **Martinelli, M. Jr.** 1975. *Avalanche handbook*. U.S. Department of Agriculture, Agricultural Handbook 489.
11. **Schuster, R. L.** 1996. Socioeconomic significance of landslides. In *Landslides; investigation and mitigation*, ed. A. K. Turner and R. L. Schuster, pp. 12–35. Transportation Research Board Special Report 247, National Research Council. Washington, DC: National Academy Press.
12. **Flemming, R. W.**, and **Taylor, F. A.** 1980. *Estimating the cost of landslide damage in the United States*. U.S. Geological Survey Circular 832.
13. **University of California, SNEP Science Team and Special Consultants.** 1996. Sierra Nevada ecosystems. In Status of the Sierra Nevada; *Sierra Nevada Ecosystem Project, Final report to Congress:* Berkeley, CA, v.1. http://ceres.ca.gov/snep/pubs/web/ PDF/v1_ch01.pdf. Accessed 7/14/07.
14. **Kattleman, R.** 1996, Impacts of floods and avalanches. In Status of the Sierra Nevada; *Sierra Nevada Ecosystem Project, Final report to Congress:* Berkeley, CA, v.2. http://ceres.ca.gov/snep/pubs/web/PDF/vII_C49.PDF. Accessed 7/14/07.

15. **Swanson, F. J.**, and **Dryness, C. T.** 1975. Impact of clear-cutting and road construction on soil erosion by landslides in the Western Cascade Range, Oregon. *Geology* 7:393–96.
16. **Jones, F. O.** 1973. *Landslides of Rio de Janeiro and the Sierra das Araras Escarpment, Brazil.* U.S. Geological Survey Professional Paper 697.
17. **Ehley, P. L.** 1986. The Portuguese Bend landslide: Its mechanics and a plan for its stabilization. In *Landslides and landslide mitigation in southern California*, ed. P. L. Ehley, pp. 181–90. Guidebook for fieldtrip, Cordilleran Section of the Geological Society of America meeting, Los Angeles, California.
18. **Leighton, F. B.** 1966. Landslides and urban development. In *Engineering geology in southern California*, ed. R. Lung and R. Proctor, pp. 149–97, Special Publication, Los Angeles Section, Association of Engineering Geologists.
19. **Committee on the Review of the National Landslide Hazards Mitigation Strategy.** 2004. *Partnerships for reducing landslide risk.* National Research Council. Washington, DC: The National Academies Press.
20. **Briggs, R. P.**, **Pomeroy, J. S.**, and **Davies, W. E.** 1975. *Landsliding in Allegheny County, Pennsylvania.* U.S. Geological Survey Circular 728.
21. **Jones, D. K. C.** 1992. Landslide hazard assessment in the context of development. In *Geohazards*, ed. G. J. McCall, D. J. Laming, and S. C. Scott, pp. 117–41. New York: Chapman & Hall.
22. **Slosson, J. E.**, **Yoakum, D. E.**, and **Shuiran, G.** 1986. Thistle, Utah, landslide: Could it have been prevented? In *Proceedings of the 22nd Symposium on Engineering Geology and Soils Engineering*, ed. S. H. Wood, pp. 281–303. Boise, ID: 22nd Symposium on Engineering Geology and Soils Engineering.
23. **Spiker, E. C.**, and **Gori, P. L.** 2003. *National landslides mitigation strategy—A framework for loss reduction.* U.S. Geological Survey Circular 1244.
24. **Piteau, D. R.**, and **Peckover, F. L.** 1978. Engineering of rock slopes. In *Landslides*, ed. R. Schuster and R. J. Krizek, pp. 192–228. Transportation Research Board Special Report 176, National Research Council. Washington, DC: National Academy Press.
25. **U.S. Geological Survey** and **Pierce County, Washington Department of Emergency Management.** 1999. *Mount Rainier Volcano Lahar Warning System Pilot Project.* http://volcanoes.usgs.gov/ About/Highlights/RainierPilot/ Pilot_highlight.html. Accessed 6/29/07.

CHAPTER 8

1. **Gamma Remote Sensing.** 2000. *Land subsidence in Venezia.* www.gamma-rs.ch/ findit.php?search_term=&content=venezia. Accessed 7/22/06.
2. **Waltham, T.** 2002. Sinking cities. *Geology Today* 18(3):95–100.
3. **Fletcher, C.**, and **Da Mosto, J.** 2004. *The science of saving Venice.* New York: Umberto Allemandi & Co.
4. **Nosengo, N.** 2003. Save our city! *Nature* 424:608–09.
5. **Birkland, P. W.** 1984. *Soils and geomorphology.* New York: Oxford University Press.
6. **Brady, N. C.**, and **Weil, R. R.** 1996. *The nature and properties of soils*, 11th ed. Upper Saddle River, NJ: Prentice Hall.
7. **Keller, E. A.**, **Bonkowski, M. S.**, **Korsch, R. J.**, and **Shlemon, R. J.** 1982. Tectonic geomorphology of the San Andreas fault zone in the southern Indio hills, Coachella Valley, California. *Geological Society of America Bulletin* 93:46–56.
8. **Van der Woerd, J.**, **Klinger, Y.**, **Sieh, K.**, **Tapponnier, P.**, **Ryerson, F. J.**, and **Meriaux, A. S.** 2006. Long-term slip rate of the southern San Andreas fault from 10Be-26Al surface exposure dating of an offset alluvial fan. *J. Geophys. Res.*, 11 B04407: 10.1029/2004 JB00359.
9. **Krynine, D. P.**, and **Judd, W. R.** 1957. *Principles of engineering geology and geotechnics.* New York: McGraw-Hill.
10. **Singer, M. J.**, and **Munns, D. N.** 1996. *Soils*, 3rd ed. Upper Saddle River, NJ: Prentice Hall.
11. **Veni, G.**, **DuChene, H.**, **Crawford, N. C.**, **Groves, C. G.**, **Huppert, G. N.**, **Kastning, E. H.**, **Olson, R.**, and **Wheeler, B. J.** 2001. *Living with karst; A fragile foundation.* Alexandria, VA: American Geological Institute.
12. **Galloway, D.**, **Jones, D. R.**, and **Ingebritsen, S. E.** 1999. Introduction. In *Land subsidence in the United States*, ed. D. Galloway, D. R. Jones, and S. E. Ingebritsen, pp. 1–6. U.S. Geological Survey Circular 1182.
13. **Waltham, T.**, **Bell, F.**, and **Culshaw, M.** 2005. *Sinkholes and subsidence: Karst and cavernous rocks in engineering and construction.* New York: Springer-Verlag.
14. **Bloom, A. L.** 1991. *Geomorphology: A systematic analysis of late Cenozoic landforms*, 3rd ed. Upper Saddle River, NJ: Prentice Hall.
15. **Nelson, F. E.**, **Anisimov, O. A.**, and **Shiklomanov, N. I.** 2001. Subsidence risk from thawing permafrost. *Nature* 410:889–90.
16. **Goldman, E.** 2002. Even in the high Arctic nothing is permanent. *Science* 297:1493–94.
17. **Stanley, J.-D.**, **Goddio, F.**, **Jorstad, T. F.**, and **Schnepp, G.** 2004. Submergence of ancient Greek cities off Egypt's Nile Delta—a cautionary tale. *GSA Today* 14(1):4–10.
18. **Penvenne, L. J.** 1996. The disappearing delta. *Earth* 5(4):16–17.
19. **Scheffe, K. F.** 2005. Collapsible soils in the Rio Grande Valley of central New Mexico. *Geological Society of America Abstracts with Programs* 37(7):327.
20. **Scheffe, K. F.**, and **Lacy, S. L.** 2004. Hydro-compactible soils. In *Understanding soil risks and hazards: Using soil survey to identify areas with risks and hazards to human life and property*, ed. G. B. Muckel, pp. 60–64. Lincoln, NE: U.S. Department of Agriculture, Natural Resources Conservation Service, National Soil Survey Center.
21. **Fischetti, M.** 2001. Drowning New Orleans. *Scientific American* 285(10):76–85.
22. **Ingebritsen, S. E.**, **McVoy, C.**, **Glaz, B.**, and **Park, W.** 1999. Florida Everglades: Subsidence threatens agriculture and complicates ecosystem restoration. In *Land subsidence in the United States*, ed. D. Galloway, D. R. Jones, and S. E. Ingebritsen, pp. 95–106. U.S. Geological Survey Circular 1182.
23. **Grunwald, M.** 2006. *The swamp: The Everglades, Florida, and the politics of paradise.* New York: Simon and Schuster.
24. **Roberts, H. H.** 1997. Dynamic changes of the Holocene Mississippi River delta plain: The delta cycle. *Journal of Coastal Research* 13(3):605–27.
25. **Morton, R. A.**, **Bernier, J. C.**, and **Barras, J. A.** 2006. Evidence of regional subsidence and associated interior wetland loss induced by hydrocarbon production, Gulf Coast region, USA. *Environmental Geology* 50:261–74.
26. **Dokka, R. K.** 2006. Modern-day tectonic subsidence in coastal Louisiana. *Geology* 34:281–84.
27. **Torbjörn, T. E.**, **Bick, S. J.**, **van de Borg, K.**, and **De Jong, A. F. M.** 2006. How stable is the Mississippi Delta? *Geology* 34:697–700.
28. **Louisiana Coastal Wetlands Conservation and Restoration Task Force.** 1998. *Coast 2050: Toward a sustainable coastal Louisiana.* Baton Rouge, LA: Louisiana Department of Natural Resources.
29. **Penland, S.**, **Wayne, L. D.**, **Britsch, L. D.**, **Williams, S. J.**, **Beall, A. D.**, and **Butterworth, V. C.** 2000. *Process classification of coastal land loss between 1932 and 1990 in the Mississippi River delta plain, southeastern Louisiana.* U.S. Geological Survey Open-File Report 00-0418.
30. **Wilding, L. P.**, and **Tessier, D.** 1988. Genesis of vertisols: Shrink-swell phenomena. In *Vertisols: Their distribution, properties, classification and management*, ed. L. P. Wilding and R. Puentes, pp. 55–81. College Station, TX: Texas A&M University Soil Management Support Services Technical Monograph No. 18.
31. **Noe, D. C.**, **Jochim, C. L.**, and **Rogers, W. P.** 1999. *A guide to swelling soils for Colorado homebuyers and homeowners.* Colorado Geological Survey Special Publication 43.
32. **McCarthy, D. F.** 2002. *Essentials of soil mechanics and foundations: Basic geotechnics*, 6th ed. Upper Saddle River, NJ: Prentice Hall.
33. **Johnson, K. S.** 2005. Subsidence hazards due to evaporite dissolution in the United States. *Environmental Geology* 48:395–409.
34. **Olive, W. W.**, **Chleborad, A. F.**, **Frahme, C. W.**, **Schlocker, J.**, **Schneider, R. R.**, and **Schuster, R. L.** 1989. *Swelling clays map of the conterminous United States.* U.S. Geological Survey Miscellaneous Investigations Series Map I-1940.
35. **Schmidt, W.** 2001. Sinkholes in Florida. *Geotimes* 46(5):18.
36. **Comiso, J. C.**, and **Parkinson, C. L.** 2004. Satellite-observed changes in the Arctic. *Physics Today* 57(8):38–44.
37. **Dixon, T. H.**, **Amelung, F.**, **Ferretti, A.**, **Novali, F.**, **Rocca, F.**, **Dokka, R.**, **Sella, G.**, **Kim, S.-W.**, **Wdowinski, S.**, and **Whitman, D.** 2006. Subsidence and flooding in New Orleans: A subsidence map of the city offers insight into the failure of the levees during Hurricane Katrina. *Nature* 441:587–88.
38. **Wray, W. K.**, and **Meyer, K. T.** 2004. Expansive clay soil . . . a widespread and costly geohazard. *Geo-Strata* 5(4):24–25, 27–28.
39. **DiMillo, A.** 2005. *A quarter-century of geotechnical research.* U.S. Department of Transportation Federal Highway Administration. www.fhwa.dot. gov/engineering/geotech/pubs/century/. Accessed 6/25/06.
40. **Jones, D. E.**, and **Holtz, W. C.** 1973. Expansive soils— the hidden disaster. *Civil Engineering—ASCE* 43(6):49–51.
41. **Ritter, D. F.**, **Kochel, R. C.**, and **Miller, J. R.** 2002. *Process geomorphology.* New York: McGraw-Hill.
42. **National Research Council (U.S.) Joint Academies Committee on the Mexico City Water Supply.** 1995. *Mexico City's water supply: Improving the outlook for sustainability.* Washington, DC: National Academies Press.
43. **Harris, R. C.** 2004. Giant desiccation cracks in Arizona. *Arizona Geology* 34(2):1–4.
44. **Slaff, S.** 1993. *Land subsidence and earth fissures in Arizona.* Arizona Geological Survey Down-to-Earth Series 3.
45. **Karst Waters Institute.** 2002. *What is karst (and why is it important)?* www .karstwaters.org/kwitour/whatiskarst.html/. Accessed 7/23/06.
46. **Holzer, T. L.**, and **Galloway, D. L.** 2005. Impacts of land subsidence caused by withdrawal of underground fluids in the United States. In *Humans as geologic agents*, ed. J. Ehlen, W. C. Haneberg, and R. A. Larson, pp. 87–99. Geological Society of America, Reviews in Engineering Geology XVI.
47. **Bull, W. B.** 1973. Geologic factors affecting compaction of deposits in a land subsidence area. *Geological Society of America Bulletin* 84:3783–3802.
48. **Kenny, R.** 1992. Fissures: Legacy of a drought. *Earth* 1(3):34–41.
49. **Craig, J. R.**, **Vaughan, D. J.**, and **Skinner, B. J.** 1996. *Resources of the Earth: Origin, use, and environmental impact*, 2nd ed. Upper Saddle River, NJ: Prentice Hall.
50. **Rahn, P. H.** 1996. *Engineering geology: An environmental approach*, 2nd ed. Upper Saddle River, NJ: Prentice Hall.
51. **Kappel, W. M.**, **Yager, R. M.**, and **Miller, T. S.** 1999. The Retsof Salt Mine collapse. In *Land subsidence in the United States*, ed. D. Galloway, D. R. Jones, and S. E. Ingebritsen, pp. 111–20. U.S. Geological Survey Circular 1182.
52. **Péwé, T. L.** 1982. *Geologic hazards of the Fairbanks area, Alaska.* Alaska Division of Geological & Geophysical Surveys Special Report 15.
53. **Ingebritsen, S. E.**, **Ikehara, M. E.**, **Galloway, D. L.**, and **Jones, D. R.** 2000. *Delta subsidence in California: The sinking heart of the state.* U.S. Geological Survey Fact Sheet FS-005-00.
54. **Coplin, L. S.**, and **Galloway, D.** 1999. Houston-Galveston, Texas: Managing coastal subsidence. In *Land subsidence in the United States*, ed. D. Galloway, D. R. Jones, and S. E. Ingebritsen, pp. 35–48. U.S. Geological Survey Circular 1182.
55. **Galloway, D.**, **Jones, D. R.**, and **Ingebritsen, S. E.** 1999. Mining ground water—Introduction. In *Land subsidence in the United States*, ed. D. Galloway, D. R. Jones, and S. E. Ingebritsen, pp. 7–13. U.S. Geological Survey Circular 1182.

CHAPTER 9

1. **Grazulis, T. P.** 2001. *The tornado: Nature's ultimate windstorm.* Norman: University of Oklahoma Press.
2. **Aguado, E.**, and **Burt, J. E.** 2007. *Understanding weather and climate*, 4th ed. Upper Saddle River, NJ: Pearson Prentice Hall.

3. **Wilson, J. W.**, and **Changnon, S. A. Jr.** 1971. *Illinois tornadoes.* Urbana: Illinois State Water Survey Circular 103.
4. **Smith, G. A.**, and **Pun, A.** 2006. *How does Earth work? Physical geology and the process of science.* Upper Saddle River, NJ: Pearson Prentice Hall.
5. **Smith, J., ed.** 2001. *The facts on file dictionary of weather and climate.* New York: Facts on File, Inc.
6. **Lutgens, F. K.**, and **Tarbuck, E. J.** 2007. *The atmosphere; An introduction to meteorology,* 10th ed. Upper Saddle River, NJ: Pearson Prentice Hall.
7. **Christopherson, R. W.** 2006. *Geosystems; An introduction to physical geography.* 6th ed. Upper Saddle River, NJ: Pearson Prentice Hall.
8. **Ackerman, S. A.**, and **Knox, J. A.** 2003. *Meteorology; Understanding the atmosphere.* Pacific Grove, CA: Thomson Learning.
9. **American Meteorological Society.** 2002. *Updated recommendations for lightning safety—2002.* http://www.ametsoc.org/POLICY/ Lightning_Safety_Article.pdf. Accessed 7/26/06.
10. **Rauber, R. M., Walsh, J. E.**, and **Charlevoix, D. J.** 2005. *Severe & hazardous weather; An introduction to high impact meteorology,* 2nd ed. Dubuque, IA: Kendall/Hunt.
11. **National Weather Service.** 2004. *Lightning—the underrated killer.* http://www.lightningsafety.noaa.gov/overview.htm. Accessed 7/26/06.
12. **National Weather Service.** 2004. *Lightning risk reduction outdoors.* http://www.lightningsafety.noaa.gov/outdoors.htm. Accessed 7/26/06.
13. **Burt, Christopher C.** 2004. *Extreme weather: A guide & record book.* New York: W.W. Norton.
14. **Ashley, W. S.**, and **Mote, T. L.** 2005. Derecho hazards in the United States. *Bulletin of the American Meteorological Society* 86:1577–92.
15. **National Weather Service.** 1992. *Advanced spotters' field guide.* National Oceanic and Atmospheric Administration PA\# 92055. http://www.nws.noaa.gov/om/brochures/adv_spotters.pdf. Accessed 9/16/06.
16. **National Weather Service.** 2008. *Natural hazard statistics.* http://www.nws.noaa.gov/om/hazstats.shtml. Accessed 1/11/09.
17. **National Weather Service.** 2006. *The Enhanced Fujita Scale (EF Scale).* Norman, OK: NOAA/National Weather Service Storm Prediction Center. http://www.spc.noaa.gov/efscale/. Accessed 9/17/06.
18. Environmental Canada. 2004. Tornados. www.pnr-rpn.ec.gc.ca/air/summerservere/ae00s02.en.html. Accessed 7/26/06.
19. **National Weather Service.** 2001. *Winter storms; The deceptive killers.* http://www.nws.noaa.gov/om/brochures/ winterstorm.pdf. Accessed 7/27/06.
20. Environmental Canada. 2002. Blizzards. www.pnr-rpn.ec.gc.ca/air/winterservere/blizzards.en.html. Accessed 7/27/06.
21. **National Snow and Ice Data Center.** 2004. *The blizzards of 1996.* Boulder: University of Colorado. http://nsidc.org/snow/blizzard/ plains.html. Accessed 7/27/06.
22. **Environment Canada.** 2004. *Blizzards and winter weather hazards.* http://www.msc-smc.ec.gc.ca/cd/brochures/blizzard_e.cfm. Accessed 7/27/06.
23. **Lecomte, E. L., Pang, A. W.**, and **Russell, J. W.** 1998. *Ice storm '98.* Toronto: Institute for Catastrophic Loss Reduction Research Paper Series No. 1.
24. **Environment Canada.** 2002. *Ice Storm 1998; The worst ice storm in Canadian history?* http://www.msc-smc.ec.gc.ca/media/icestorm98/ icestorm98_the_worst_e.cfm. Accessed 1/17/07.
25. **Changnon, S. A.**, and **Changnon, J. M.** 2002. Major ice storms in the United States, 1949–2000. *Environmental Hazards* 4:105–11.
26. **DeGaetano, A. T.** 2000. Climatic perspective and impacts of the 1998 northern New York and New England ice storm. *Bulletin of the American Meteorological Society* 81:237–54.
27. **National Climate Data Center.** 2006. *Eastern U.S. flooding and ice storm.* National Oceanic and Atmospheric Administration. http://www.ncdc.noaa.gov/oa/reports/janstorm/janstorm.html. Accessed 1/20/07.
28. **Jones, K. F.**, and **Mulherin, N. D.** 1998. *An evaluation of the severity of the January 1998 ice storm in northern New England; Report for FEMA Region 1.* Hanover, NH: U.S. Army Corps of Engineers Cold Regions Research and Engineering Laboratory. http://www.crrel.usace.army.mil/techpub/CRREL_Reports/reports/IceStorm98.pdf. Accessed 10/9/06.
29. **Coppola, S.**, and **Wear, B.** 2007. Lack of de-icer backs up airport. *Austin American-Statesman.* January 18, 136(174):A4.
30. **Federal Emergency Management Agency.** 1998. *Disaster connections kids to kids; Kids from Maine talk about surviving the Ice Storm of '98.* http://www.fema.gov/kids/me98_04.htm. Accessed 1/20/07.
31. **Whittow, J.** 1980. *Disasters; The anatomy of environmental hazards.* London: Penguin Books Ltd.
32. **Jones, R. L.** 2005. *Canadian disasters—An historical survey.* http://www.ott.igs.net/jonesb/DisasterPaper/ disasterpaper.html. Accessed 10/7/06.
33. **Subcommittee on Disaster Reduction.** 2003. *Reducing disaster vulnerability through science and technology: An interim report of the Subcommittee on Disaster Reduction.* Washington, DC: Committee on the Environment and Natural Resources, National Science and Technology Council, Executive Office of the President of the United States.
34. **Bagnold, R. A.** 1941. *The physics of blown sand and desert dunes.* New York: Chapman Hall.
35. **Saunders, G.** 2004. The silent killer. *Weatherwise* 57(2):27.
36. **De Bono, A., Peduzzi, P., Giuliani, G.**, and **Kluser, S.** 2004. *Impacts of summer 2003 heat wave in Europe.* United Nations Environment Programme, Division of Early Warning and Assessment—Europe, Early Warning on Emerging Environmental Threats. http://www.grid.unep.ch/product/publication/EABs.php. Accessed 12/23/06.
37. **British Broadcasting Corporation News.** 2003. *Sizzling temperatures break UK record.* http://news.bbc.co.uk/1/hi/uk/3138865.stm. Accessed 12/24/06.
38. **Luterbacher, J., Dietrich, D., Xoplaki, E., Grosjean, M.**, and **Wanner, H.** 2004. European seasonal and annual temperature variability, trends, and extremes since 1500. *Science* 303:1499–1503.
39. **U.S. Environmental Protection Agency.** 2006. *Heat island effect.* http://www.epa.gov/heatisland/about. Accessed 9/17/06.
40. **Black, E., Blackburn, M., Harrison, G., Hoskins, B.**, and **Methven, J.** 2004. Factors contributing to the summer 2003 European heatwave. *Weather* 59:217–23.
41. **Stott, P. A., Stone, D. A.**, and **Allen, M. R.** 2004. Human contribution to the European heatwave of 2003. *Nature* 432:610–14.
42. **World Health Organization.** 2003. *The health impacts of 2003 summer heat-waves; Briefing note for the delegations of the fifty-third session of the WHO Regional Committee for Europe.* http://www.euro.who.int/document/Gch/HEAT-WAVES%20RC3.pdf. Accessed 1/7/07.
43. **Bouchama, A.** 2004. The 2003 European heat wave. *Intensive Care Medicine* 30:1–3.
44. **Houghton, J.** 2004. *Global warming: The complete briefing,* 3rd ed. Cambridge: Cambridge University Press.
45. **Trenberth, K. E., Jones, P. D., Ambenje, P., Bojariu, R., Easterling, D., Klein Tank, A., Parker, D., Rahimzadeh, F., Renwick, J.A., Rusticucci, M., Soden, B., and Zhai, P.** 2007. Observations: surface and atmospheric climate change. In: *Climate Change 2007: The Physical Science Basis; Contribution of Working Group I to the Fourth Assessment Report of the Intergovernmental Panel on Climate Change.* ed. S. Solomon, D. Qin, M. Manning, Z. Chen, M. Marquis, K. B. Averyt, M. Tignor, and H. L. Miller. New York: Cambridge University Press.
46. **Pyne, S. J., Andrews. P. L.**, and **Laven, R. D.** 1996. *Introduction to wildland fire.* New York: John Wiley & Sons.
47. **National Oceanic and Atmospheric Administration National Weather Service Storm Prediction Center.** 2002. *Storm Prediction Center.* http://www.spc.noaa.gov/misc/aboutus.html. Accessed 7/27/06.
48. **Golden, J. H.**, and **Adams, C. R.** 2000. The tornado problem: Forecast, warning, and response. *Natural Hazards Review* 1(2):107–18.
49. **Roberts, R. D., Burgess, D.**, and **Meister, M.** 2006. Developing tools for nowcasting storm severity. *Weather and Forecasting* 21:540–58.
50. **Godschalk, D. R., Brower, D. J.**, and **Beatly, T.** 1989. *Catastrophic coastal storms: Hazard mitigation and development management.* Durham, NC: Duke University Press.
51. **Federal Emergency Management Agency. 2004.** *Saferooms; Know your risk and have a safe place to go . . . with time to get there.* http//www.fema.gov/plan/prevent/ saferoom/. Accessed 1/10/07.
52. **Auerbach, P. S.** 2003. *Medicine for the outdoors.* Guilford, CT: The Lyons Press.
53. **Gill, P. G. Jr.** 2002. *Wilderness first aid; A pocket guide.* Camden, ME: Ragged Mountain Press.
54. **Federal Highway Administration Road Weather Management Program.** 2006. *Low visibility.* http://www.ops.fhwa.dot.gov/weather/weather_events/low_visibility.htm. Accessed 7/29/06.

CHAPTER 10

1. **Brooks, D.** 2005. The best-laid plan: Too bad it flopped. *The New York Times,* September 11, 2005.
2. **van Heerden, I.**, and **Bryan, M.** 2006. *The storm; What went wrong and why during Hurricane Katrina—the inside story from one Louisiana scientist.* New York: Viking Penguin.
3. **Knabb, R. D., Rhome, J. R.**, and **Brown, D. P.** 2006. *Tropical cyclone report; Hurricane Katrina 23–30 August 2005.* National Hurricane Center, National Weather Service, National Oceanic and Atmospheric Administration. http://www.nhc.noaa.gov/2005atlan.shtml. Accessed 4/9/07.
4. **The White House.** 2006. *The federal response to Hurricane Katrina: Lessons learned.* Washington, DC: The White House. http://www.whitehouse.gov/reports/katrina-lessons-learned/. Accessed 3/18/07.
5. **Committee on Homeland Security and Governmental Affairs.** 2006. *Hurricane Katrina; A nation still unprepared.* Washington, DC: United States Senate.
6. **Seed, R. B., Bea, R. G., Abdelmalak, R. I., Athanasopoulos, A. G., Boutwell, G. P., Bray, J. D., Briaud, J.-L., Cheung, C., Cobos-Roa, D., Cohen-Waeber, J., Collins, B. D., Ehrensing, L., Farber, D., Hanemann, M., Harder, L. F., Inkabi, K. S., Kammerer, A. M., Karadeniz, D., Kayen, R. E., Moss, R. E. S., Nicks, J., Nimmala, S., Pestana, J. M., Porter, J., Rhee, K., Riemer, M. F., Roberts, K., Rogers, J. D., Storesund, R., Govindasamy, A. V., Vera-Grunauer, X., Wartman, J. E., Watkins, C. M., Wenk, E. Jr.**, and **Yim, S. C.** 2006. *Investigation of the performance of the New Orleans flood protection systems in Hurricane Katrina on August 29, 2005,* Volume I: *Main text and executive summary.* Berkeley, CA: University of California at Berkeley, Department of Civil Engineering, Independent Levee Investigation Team Final Report.
7. **Waugh, W. L. Jr., ed.** 2006. Shelter from the storm: Repairing the national emergency management system after Hurricane Katrina. *The Annals of the American Academy of Political and Social Science* 806:288–332.
8. **Wolshon, B., Urbina, E., Wilmot, C.**, and **Levitan, M.** 2005. Review of policies and practices for hurricane evacuation. 1: Transportation planning, preparedness, and response. *Natural Hazards Review* 6:129–42.
9. **Johns, C., ed.** 2005. *Katrina; Why it became a man-made disaster; Where it could happen next.* National Geographic Special Edition
10. **Fitzpatrick, P. J.** 1999. *Natural disasters: Hurricanes; A reference handbook.* Santa Barbara, CA: ABC-CLIO.
11. **Stewart, S. R.** 2005. *Tropical cyclone report; Hurricane Ivan 2–24 September 2004.* National Hurricane Center, National Weather Service, National Oceanic and Atmospheric Administration. http://www.nhc.noaa.gov/2004ivan.shtml? Accessed 3/18/07.

12. **Emanuel, K.** 2005. *Divine wind; The history and science of hurricanes.* New York: Oxford University Press.
13. **Pasch, R. J., Blake, E. S., Cobb, H. D. III,** and **Roberts, D. P.** 2006. *Tropical Cyclone Report; Hurricane Wilma 15–25 October 2005.* National Hurricane Center, National Weather Service, National Oceanic and Atmospheric Administration. http://www.nhc.noaa.gov/ 2005atlan.shtml/. Accessed 3/26/07.
13. A. **Longshore, D.** 2000. Encyclopedia of hurricanes, typhoons and cyclones. New York: Checkmark Books.
14. **Vescio, M., Cooper, S.,** and **Cain, D.** 2006. *Tropical cyclones: Introduction.* National Weather Service Jetstream; An online school for weather. http://www.srh.noaa.gov/jetstream/tropics/tc.htm. Accessed 4/9/07.
15. **Ahrens, C. D.** 2005. *Essentials of meteorology; An invitation to the atmosphere.* Belmont, CA: Thomson Brooks/Cole.
16. **Aguardo, E.,** and **Burt, J. E.** 2007. *Understanding weather and climate,* 4th ed. Upper Saddle River, NJ: Pearson Prentice Hall.
17. **Ackerman, S. A.,** and **Knox, J. A.** 2003. *Meteorology; Understanding the atmosphere.* Pacific Grove, CA: Thomson Brooks/Cole.
18. **Flanagan, R.** 1993. Beaches on the brink. *Earth* 2(6):24–33.
19. **Burt, C. C.** 2004. *Extreme weather: A guide & record book.* New York: W. W. Norton.
20. **Edwards, R.** 1998. Tornado production by exiting tropical cyclones. Preprints, *23rd Conference on Hurricanes and Tropical Meteorology,* Dallas, TX: American Meteorological Society, 485–88.
21. **Stewart, S. R.** 2002. *Tropical cyclone report; Tropical storm Allison 5–17 June 2001.* National Hurricane Center, National Weather Service, National Oceanic and Atmospheric Administration. http://www.nhc.noaa.gov/2001allison.html. Accessed 4/10/07.
22. **National Weather Service Southern Regional Headquarters.** 2002. *Tropical cyclones & inland flooding for the southern states.* Forth Worth, TX: National Oceanic and Atmospheric Administration.
23. **Negri, A. J., Burkardt, N., Golden, J. H., Halverson, J. B., Huffman, G. J., Larsen, M. C., McGinley, J. A., Updike, R. G., Verdin, J. P.,** and **Wieczorek, G. F.** 2005. The hurricane–flood– landslide continuum. *Bulletin American Meteorological Society* 86:1241–47.
24. **Crossett, K. M., Culliton, T. J., Wiley, P. C.,** and **Goodspeed, T. R.** 2004. *Population trends along the coastal United States: 1980–2008.* National Ocean Service, National Oceanic and Atmospheric Administration.
25. **Intergovernmental Panel on Climate Change, Working Group II.** 2007. *Climate change 2007: Climate change impacts, adaptation and vulnerability; Summary for policy makers.* World Meteorological Organization. http://www.ipcc .ch/. Accessed 4/8/07.
26. **National Hurricane Center.** 2007. *About the National Hurricane Center.* National Weather Service, National Oceanic and Atmospheric Administration. http://www.nhc.noaa.gov/aboutnhc.shtml/. Accessed 4/8/07.
27. **Environment Canada.** 2004. *Canadian Hurricane Centre.* Meteorological Service of Canada. http://www.ns.ec.gc.ca/weather/hurricane/chc1.html. Accessed 4/8/07.
28. **National Hurricane Center.** 2006. *Hurricane preparedness; Hurricane basics.* National Weather Service, National Oceanic and Atmospheric Administration. http://www.nhc.noaa.gov/HAW2/english/ basics.shtml/. Accessed 4/8/07.
29. **U.S. Air Force 53rd Weather Reconnaissance Squad-ron.** 2004. *Tropical cyclone mission.* http://www.hurricanehunters.com/cyclone.htm/. Accessed 4/8/07.
30. **Shay, L. K., Goni, G. J.,** and **Black, P. G.** 2000. Effects of a warm oceanic feature on Hurricane Opal. *Monthly Weather Review* 128:1366–83.
31. **Scharroo, R., Smith, W. H. F.,** and **Lillibridge, J. L.** 2005. Satellite altimetry and the intensification of Hurricane Katrina. *EOS, Transactions, American Geophysical Union* 86(40):366.
32. **National Aeronautics and Space Administration.** 2005. *News: Hurricane Rita roars through a warm Gulf.* Earth Observatory, National Aeronautics and Space Administration, http://earthobservatory.nasa.gov/Newsroom/NewImages/images.php3?img_id =17041. Accessed 4/13/07.
33. **Knabb, R. D., Brown, D. P.,** and **Rhome, J. R.** 2006. *Tropical cyclone report; Hurricane Rita 18–26 September 2005.* National Hurricane Center, National Weather Service, National Oceanic and Atmospheric Administration. http://www.nhc.noaa.gov/2005atlan.shtml. Accessed 4/9/07.
34. **National Hurricane Center.** 2004. *Hurricane preparedness; Disaster supply kit.* National Weather Service, National Oceanic and Atmospheric Administration. http://www.nhc.noaa.gov/HAW2/english/ prepare/supply_kit.shtml/. Accessed 4/8/07.

CHAPTER 11

1. **Lemmon, G., Neal, W. J., Bush, D. M., Pilkey, O. H., Stutz, M.,** and **Bullock, J.** 1996. *Living with the South Carolina coast.* Durham, NC: Duke University Press.
2. **Dean, C.** 1999. *Against the tide: The battle for America's beaches.* New York: Columbia University Press.
3. **Pilkey, O. H.,** and **Dixon, K. L.** 1996. *The Corps and the shore.* Washington, DC: Island Press.
4. **Thieler, E. R., Gayes, P. T., Schwab, W. C.,** and **Harris, M. S.** 1999. Tracing sediment dispersal on nourished beaches: Two case studies. In *Coastal Sediments '99,* ed. N. C. Kraus and W. G. McDougal, pp. 2118–36. Reston, VA: American Society of Civil Engineers.
5. **Morton, R. A.,** and **Miller, T. L.** 2005. *National assessment of shoreline change,* Part 2: *Historical shoreline changes and associated coastal land loss along the U.S. Southeast Atlantic Coast.* U.S. Geological Survey Open-File Report 2005–1401.
6. **Intergovernmental Panel on Climate Change, Working Group I.** 2007. *Contribution of Working Group I to the Fourth Assessment Report of the Intergovernmental Panel on Climate Change; Summary for policy makers.* World Meteorological Organization. www.ipcc.ch/ Accessed 4/25/07.
7. **Dean, C.** 2006, June 20. Next victim of warming: The beaches. *The New York Times.*
8. **Crossett, K. M., Culliton, T. J., Wiley, P. C.,** and **Goodspeed, T. R.** 2004. *Population trends along the coastal United States: 1980– 2008.* National Ocean Service, National Oceanic and Atmospheric Administration.
9. **Davis, R. E.,** and **Dolan, R.** 1993. Nor'easters. *American Scientist* 81:428–39.
10. **Minkel, J. R.** 2004. Surf's up—way up. *Scientific American* 291(4):38.
11. **Komar, P. D.** 1998. *Beach processes and sedimentation,* 2nd ed. Upper Saddle River, NJ: Prentice Hall.
12. **Johnson, J. W.** 1956. Dynamics of nearshore sediment movement. *Bulletin of the American Association of Petroleum Geologists* 40:2211–32.
13. **Bolt, B. A.** 2006. *Earthquakes,* 5th ed. New York: W.H. Freeman and Company.
14. **Ward, S.,** and **Main, C.** 1998. *Population at risk from natural hazards.* NOAA's State of the Coast Report. Silver Spring, MD: NOAA. http://oceanservice.noaa.gov/websites/retiredsites/ sotc_pdf/ PAR.PDF Accessed 4/20/07.
15. **Garrison, T.** 2005. *Oceanography; An invitation to marine science.* Belmont, CA: Thomson Brooks/Cole.
16. **Williams, S. J., Dodd, K.,** and **Gohn, K. K.** 1991. *Coasts in crisis.* U.S. Geological Survey Circular 1075.
17. **Morton, R. A., Pilkey Jr., O. H., Pilkey Sr., O. H.,** and **Neal, W. J.** 1983. *Living with the Texas shore.* Durham, NC: Duke University Press.
18. **Norris, R. M.** 1977. Erosion of sea cliffs. In *Geologic hazards in San Diego,* ed. P. L. Abbott and J. K. Victoris. San Diego, CA: San Diego Society of Natural History.
19. **Komar, P. D.** 1997. *The Pacific Northwest coast: Living with the shores of Oregon and Washington.* Durham, NC: Duke University Press.
20. **U.S. Department of Commerce.** 1978. *State of Maryland coastal management program and final environmental impact statement.* Washington, DC: U.S. Department of Commerce.
21. **Leatherman, S. P.** 1984. Shoreline evolution of North Assateague Island, Maryland. *Shore and Beach,* July:3–10.
22. **Wilkinson, B. H.,** and **McGowen, J. H.** 1977. Geologic approaches to the determination of long-term coastal recession rates, Matagorda Peninsula, Texas. *Environmental Geology* 1:359–65.
23. **Larsen, J. I.** 1973. *Geology for planning in Lake County, Illinois.* Illinois State Geological Survey Circular 481.
24. **Buckler, W. R.,** and **Winters, H. A.** 1983. Lake Michigan bluff recession. *Annals of the Association of American Geographers* 73(1):89–110.
25. **Geological Survey of Canada.** 2006. *CoastWeb: Facts about Canada's coastline.* http://gsc.nrcan.gc.ca/coast/ facts_e.php. Accessed 4/20/07.
26. **Carter, R. W. G.,** and **Oxford, J. D.** 1982. When hurricanes sweep Miami Beach. *Geographical Magazine* 54(8):442–48.
27. **Flanagan, R.** 1993. Beaches on the brink. *Earth* 2(6):24–33.
28. **Rowntree, R. A.** 1974. Coastal erosion: The meaning of a natural hazard in the cultural and ecological context. In *Natural hazards: Local, national, global,* ed. G. F. White, pp. 70–79. New York: Oxford University Press.
29. **Leatherman, S. P.** 2003. *Dr. Beach's survival guide: What you need to know about sharks, rip currents, and more before going in the water.* New Haven, CT: Yale University Press.
30. **Committee on Coastal Erosion Zone Management.** 1990. *Managing coastal erosion.* National Research Council. Washington, DC: National Academy Press.
31. **Neal, W. J., Blakeney Jr., W. C., Pilkey JR., O. H.,** and **Pilkey Sr., O. H.** 1984. *Living with the South Carolina shore.* Durham, NC: Duke University Press.
32. **Briand J.-L., Nouri, H. R.,** and **Darby, C.** 2008. *Pointe du Hoc stabilization study.* Texas A&M University.

CHAPTER 12

1. **Mann, M. E., Zhang, Z., Rutherford, S., Bradley, R. S., Hughes, M. K., Shindell, D., Ammann, C., Faluvegi, G., Ni' F.,** 2009. Global signatures and dynamical origins of the Little Ice Age and Medieval Climate Anomaly. *Science* 326: 1256–60.
2. **Fagan, B. M.** 2008. *The great warming: Climate change and the rise and fall of civilizations.* New York. Bloomsbury Press.
3. **National Research Council.** 2006. *Surface temperature reconstructions for the last 2,000 years.* Washington, DC: The National Academy Press.
4. **Cloud, P.** 1990. Personal written communication.
5. **National Aeronautics and Space Administration (NASA).** 1990. *EOS: A mission to planet Earth.* Washington, DC: NASA.
6. **Aguado, E.,** and **Burt, J. E.** 2007. *Understanding weather & climate,* 4th ed., Upper Saddle River, NJ: Prentice Hall.
7. **Houghton, J.** 2004. *Global warming: The complete briefing,* 3rd ed. New York: Cambridge University Press.
8. **Tufnell, L.** 1984. *Glacier hazards.* New York: Longman.
9. **NOAA.** 2009. Paleo proxy data. In *Introduction to paleoclimatology.* www.ncdc.noaa.gov. Accessed 3/24/10.
10. **Weart, S. R.** 2003. *The discovery of global warming.* Cambridge, MA: Harvard University Press.
11. **Oldfield, F.** 2005. *Environmental change: Key issues and alternative approaches.* New York: Cambridge University Press.
12. **Rahmstorf, S., Cazenave, A., Church, J. A., Hansen, J. E., Keeling, R. F., Parker, D. E.,** and **Somerville, R. C. J.** 2007. Recent climate observations compared to projections. *Science* 316:709.

13. **Hansen, J.** 2004. Defusing the global warming time bomb. *Scientific American* 290(3): 68–77.
14. **IPCC.** 2007. *The physical science basis:* Working Group I. *Contribution to the Fourth Assessment Report.* IPCC. New York: Cambridge University Press.
15. **Ruddiman, W. F.** 2008. *Earth's climate past and future,* 2nd ed. New York: W. H. Freeman and Company..
16. **Hansen, J.**, and **Sato, M.** 2004. Greenhouse gas growth rates. *Proc. Natl Academy Sci (PNAS)* 101:16109–14.
17. **Hansen, J., Sato, M., Ruedy, R., Lo, K., Lea, D.**, and **Medina-Elizade, M.** 2006. Global temperature change. *Proc. Natl Academy Sci (PNAS)* 103:14288–93.
18. **Kennett, J.** 1982. *Marine geology.* Englewood Cliffs, NJ: Prentice Hall.
19. **Hansen, J.**, and **44 others.** 2005. Efficacy of climate forcing. *Geophys Res.* 110, D18104. doi: 10.1029/2005 JD 005776.
20. **Hansen, J.** 2003. Can we defuse the global warming time bomb? Edited version of presentation to the Council on Environmental Quality. June 12. Washington DC; also at http://www.naturalscience.com.
21. **Broecker, W.** 1997. Will our ride into the greenhouse future be a smooth one? *GSA Today* 7(5):1–7.
22. **Seager, R.** 2006. The source of Europe's mild climate. *American Scientist* 94:334–41.
23. **Mann, M. E., Zhang, Z., Rutherford, S., Bradley, R. S., Hughes, M. K., Shindell, D., Ammann, C., Faluvegi, G., Ni' F., 2009.** Global signatures and dynamical origins of the Little Ice Age and Medieval Climate Anomaly. *Science* 326:1256–60.
24. **Crowley, T. J.** 2000. Causes of climate change over the past 1000 years. *Science* 289:270–77.
25. **Foukal, P., Frohlich, C., Spruit, H.**, and **Wigley, T.** 2006. Variations in solar luminosity and their effect on the earth's climate. *Nature* 443(14):161–66.
26. **Charlson, R. J., Schwartz, S. E., Hales, J. M., Cess, R. D., Coakley, J. A. J., Hansen, J. E.**, and **Hofmann, D. J.** 1992. Climate forcing by anthropogenic aerosols. *Science* 255:423–30.
27. **Kerr, R. A.** 1995. Study unveils climate cooling caused by pollutant haze. *Science* 268:802.
28. **McCormick, P. P., Thomason, L. W.**, and **Trepte, C. R.** 1995. Atmospheric effects of the Mt. Pinatubo eruption. *Nature* 373:399–436.
29. **Appenzeller, T.** 2007. The big thaw. *National Geographic* 211(6):56–71.
30. **Pelto, M. S.** 1996. Recent changes in glacier and alpine runoff in the North Cascades, Washington. *Hydrological Processes* 10:1173–80.
31. **U.S. Geological Survey.** USGS repeat photography project documents retreating glaciers in Glacier National Park. http://www.nrmsc.usgs.gov/repeatphoto/. Accessed 4/10/09.
32. **Stroeve, J.**, and **7 others.** 2008. Arctic sea ice extent plummets in 2007. *EOS, Transactions, American Geophysical Union* 89(2):13–14.
33. **Steig, E. J.**, and **5 others.** Warming of the Antarctic ice-sheet surface since the 1957 International Geophysical Year. *Nature* 457:459–62.
34. **Davis, C. H.**, and **4 others.** 2005. Snowfall-driven growth in East Antarctic ice sheet mitigates recent sea-level rise. *Science* 308(5739):1898–901.
35. **Monagham, A. J.**, and **15 others.** Insignificant change in Antarctic snowfall since the International Geophysics year. *Science* 313(5788):827–31.
36. **University Corporation for Atmospheric Research.** 1994. *El Niño and climate prediction.* Washington, DC: NOAA Office of Global Programs.
37. **Dennis, R. E.** 1984. A revised assessment of worldwide economic impacts: 1982–1984 El Niño/southern oscillation event. *EOS, Transactions of the American Geophysical Union* 65(45):910.
38. **Canby, T. Y.** 1984. El Niño's ill winds. *National Geographic* 165:144–81.
39. **Philander, S. G.** 1998. Who is El Niño? *EOS, Transactions of the American Geophysical Union* 79(13):170.
40. **Titus, J. G., Leatherman, S. P., Everts, C. H., Moffatt and Nichol Engineers, Kriebel, D. L.**, and **Dean, R. G.** 1985. *Potential impacts of sea level rise on the beach at Ocean City, Maryland.* Washington, DC: U.S. Environmental Protection Agency.
41. **Anderson, D. R.**, and **5 others.** 2009. *Coastal sensitivity to sea level rise.* Washington DC: U.S. Climate Change Program.
42. **Dickinson, W. R.** 2009. Pacific atoll living—how long already and until when. *GSA Today* 19(3): 4–10.
43. **Kumar, M.** 2007. Alaska melting into the sea. *Geotimes* 52(9):8–9.
44. **Pinol, J., Terradas, J.**, and **Lloret, F.** 1998. Climate warming, wildfire hazard, and wildfire occurrence in coastal eastern Spain. *Climatic Change* 38:345–57.
45. **U.S. Climate Change Program.** 2008. *Climate change and ecosystems.* Washington DC: Author.
46. **Union of Concerned Scientists.** 2003. Early warning signs: Coral reef bleaching. http://www.ucsusa.org. Accessed 4/10/09.
47. **Doney, S. C.** 2010. The growing human imprint on coastal and open-ocean biogeochemistry. *Science* 238:1512–16.
48. **Hoegh-Gulburg, O.**, and **Bruno, J. F.** 2010. The impact of climate change on the world's marine ecosystems. *Science* 238:1523–28.
49. **Hardt, M. J.**, and **Safina, C.** 2010. Threatening ocean life from the inside out. *Scientific American* 303(2):66–73.
50. **Appell, D.** 2009. Can "assisted migration" save species from Global Warming? *Scientific American* 30(3):378–30.
51. **Botkin, D. B.**, and **18 others.** 2007. Forecasting effects of global warming on biodiversity. *BioScience* 57(3):227–36.
52. **Lea, D. W.** 2004. The 100,000 year cycle in tropical SST, greenhouse forcing, and climate sensitivity. *Journal of Climate* 17(11):2170–79.
53. **Friedman, S. J.** 2003. Storing carbon in Earth. *Geotimes* 48(3):16–20.
54. **Nameroff, T.** 1997. The climate change debate is heating up. *GSA Today* 7(12):11–13.
55. **Bartlett, K.** 2003. Demonstrating carbon sequestration. *Geotimes* 48(3):22–23.
56. **Flavin, C.** 2008. *Low-carbon energy.* Worldwatch Report 178 Washington DC: Worldwatch Institute.
57. **McGeehin, J. P., Barron, J. A., Anderson, D. M.**, and **Verardo, D. J. 2008.** *Abrupt climate change.* Final Report, Synthesis and Assessment Product 3.4. Washington, DC: U.S. Climate Change Science Program.

CHAPTER 13

1. **Rossotti, H.** 1993. *Fire.* Oxford, England: Oxford University Press.
2. **Pyne, S. J., Andrews, P. L.**, and **Laven, R. D.** 1996. *Introduction to wildland fire,* 2nd ed. New York: John Wiley & Sons, Inc.
3. **Barber, C. V.**, and **Schweithelm, J.** 2000. *Trial by fire; Forest fires and forestry policy in Indonesia's era of crisis and reform.* Washington, DC: World Resources Institute.
4. **Ward, D.** 2001. Combustion chemistry and smoke. In *Forest fires: Behavior and ecological effects,* Chapter 3, ed. Johnson, E. A., and Miyanishi, K., pp. 57–77. San Diego, CA: Academic Press.
5. **Minnich, R. A.** 2002. Personal written correspondence.
6. **Arno, S. F.**, and **Allison-Bunnell, S.** 2002. *Flames in our forests.* Washington, DC: Island Press.
7. **Verdi, R. J., Tomlinson, S. A.**, and **Marella, R. L.** 2006. *The drought of 1998–2002: Impacts on Florida hydrology and landscapes.* U.S. Geological Survey Circular 1295.
8. **Florida Division of Forestry.** 2006. *Wildland fire; Fire data reports.* http://tlh-forweb1.doacs.state.fl.us/PublicReports/. Accessed 1/28/07.
9. **New York Times.** 2001, May 31. Rain clears some smoke, but fires persist in Florida. Sec. A, p. 14.
10. **U.S. Geological Survey.** 2006. *Wildfire hazards—A national threat.* U.S. Geological Survey Fact Sheet 2006–3015.
11. **Canadian Forest Service.** 2006. *Canadian Wildland Fire Information System: Historical analysis—Large fire database.* http:// cwfis.cfs.nrcan.gc.ca/en/historical/ha_lfdb_maps_e.php. Accessed 1/29/07.
12. **National Interagency Fire Center.** 2007. *Wildland fire statistics.* www.nifc.gov/stats/index.html. Accessed 2/9/07.
13. **Chandler, C., Cheney, P., Thomas, P., Trabaud, L.**, and **Williams, D.** 1983. *Fire in forestry,* Volume I: *Forest fire behavior and effects.* New York: John Wiley & Sons.
14. **DeBano, L. F.** 1981. *Water repellent soils: A state-of-the-art.* U.S. Forest Service General Technical Report INT-79.
15. **U.S. Geological Survey.** 1998. *USGS wildland fire research.* U.S. Geological Survey Fact Sheet 125–98.
16. **Beitler, J.** 2006. *Tracking nature's contribution to pollution.* NASA Earth Observatory, http://earthobservatory.nasa.gov/Study/ContributionPollution/. Accessed 1/29/07.
17. **Pfister, G., Hess, P. G., Emmons, L. K., Lamarque, J.-F., Wiedinmyer, C., Edwards, D. P., Pétron, G., Gille, J. C.**, and **Sachse, G. W.** 2005. Quantifying CO emissions from the 2004 Alaskan wildfires using MOPITT CO data. *Geophysical Research Letters* 32: L11809.
18. **Florsheim, J. L., Keller, E. A.**, and **Best, D. W.** 1991. Fluvial sediment transport in response to moderate storm flows following chaparral wild fire, Ventura County, southern California. *Geological Society of America Bulletin* 103:504–11.
19. **Keller, E. A., Valentine, D. W.**, and **Gibbs, D. R.** 1997. Hydrologic response of small watersheds following the southern California Painted Cave Fire of June 1990. *Hydrological Processes* 11: 401–4.
20. **Keeley, J. E., Fotheringham, C. J.**, and **Morais, M.** 1999. Reexamining fire suppression impacts on brushland fire regimes. *Science* 284:1829–32.
21. **Minnich, R. A.** 1983. Fire mosaics in southern California and northern Baja California. *Science* 219:1287–94.
22. **Minnich, R. A. Garbour, M. G., Burk, J. H.**, and **Sosa-Ramiriz, J.** 2000. California mixed-conifer forests under unchanged fire regimes in Sierra San Pedro Martir, Baja California, Mexico. *Journal of Biogeography* 27:105–29.
23. **Westerling, A. L., Hidalgo, H. G., Cayan, D. R.**, and **Swetnam, T. W.** 2006. Warming and earlier spring increase western U.S. forest wildfire activity. *Science* 313:940–43.
24. **Kasischke, E. S.**, and **Turetsky, M. R.** 2006. Recent changes in the fire regime across the North American boreal region—Spatial and temporal patterns of burning across Canada and Alaska. *Geophysical Research Letters* 33:L09703.
25. **Westerling, A. L., Cayan, D. R., Brown, T. J., Hall, B. L.**, and **Riddle, L. G.** 2004. Climate, Santa Ana winds and autumn wildfires in southern California. *EOS, Transactions, American Geophysical Union* 85(31):294, 296.
26. **Tolmé, P.** 2004. Will global warming cause more wildfires? *National Wildlife* 42(5):14–16.
27. **Fried, J. S., Torn, M. S.**, and **Mills, E.** 2004. The impact of climate change on wildfire severity: A regional forecast for northern California. *Climatic Change* 64:169–91.
28. **Turetsky, M. R., Harden, J. W., Friedli, H. R., Flannigan, M., Payne, N., Crock, J.**, and **Radke, L.** 2006. Wildfires threaten mercury stocks in northern soils. *Geophysical Research Letters* 33: L16403.
29. **Miller, M.** 2000. Fire autecology. In *Wildland fire in ecosystems: Effects of fire on flora,* ed. Brown, J. K., Smith, J., and Kapler, J. U.S. Forest Service General Technical Report RMRS-GTR-42-vol. 2.
30. **Fuller, M.** 1991. *Forest fires: An introduction to wildland fire behavior, management, firefighting and prevention.* New York: John Wiley.
31. **Jones, S. R.**, and **Cushman, R. C.** 2004. *A field guide to the North American prairie.* Boston: Houghton Mifflin.
32. **Arrowood, J. C.** 2003. *Living with wildfires: Prevention, preparation, and recovery.* Denver, CO: Bradford Publishing Company.

CHAPTER 14

1. **Lewis, J. S.** 1996. *Rain of iron and ice.* Redding, MA: Addison-Wesley.
2. **Rubin, A. F.** 2002. *Disturbing the solar system.* Princeton, NJ: Princeton University Press.
3. **Brown, P., Spalding, R. E., ReVelle, D. O., Tagliaferri, E.,** and **Worden, S. P.** 2002. The flux of small near-Earth objects colliding with the Earth. *Nature* 420:294–96.
4. **Cloud, P.** 1978. *Cosmos, Earth and man.* New Haven, CT: Yale University Press.
5. **Davidson, J. P., Reed, W. E.,** and **Davis, P. M.** 1997. *Exploring Earth.* Upper Saddle River, NJ: Prentice Hall.
6. **Grieve, R. A. F.** 2006. *Impact structures in Canada.* St. Johns, Newfoundland and Labrador: Geological Association of Canada GEOtext 5.
7. **Lipschutz, M. E.,** and **Schultz, L.** 2007. Meteorites. In *Encyclopedia of the solar system*, 2nd ed., ed. McFadden, L.-A., Weissman, P. R., and Johnson, T. V., pp. 251–82. New York: Elsevier.
8. **Grieve, R. A. F., Cintala, M. J.,** and **Tagle, R.** 2007. Planetary impacts. In *Encyclopedia of the solar system*, 2nd ed., ed. McFadden, L.-A., Weissman, P. R., and Johnson, T. V., pp. 813–28. New York: Elsevier.
9. **Williams, S. J., Barnes, P.,** and **Prager, E. J.** 2000. *U.S. Geological Survey coastal and marine geology research—Recent highlights and achievements.* U.S. Geological Survey Circular 1199.
10. **Dott, Jr., R. H.,** and **Prothero, D. R.** 1994. *Evolution of the Earth*, 5th ed. New York: McGraw-Hill.
11. **Stanley, S. M.** 2005. *Earth system history.* New York: W.H. Freeman.
12. **Kring, D. A.,** and **Durda, D. D.** 2003. The day the world burned. *Scientific American* 289(6):98–105.
13. **Alvarez, W.** 1997. *T. rex and the crater of doom.* New York: Vintage Books, Random House, Inc.
14. **Alvarez, L. W., Alvarez, W., Asaro, F.,** and **Michel, H. V.** 1980. Extraterrestrial cause for Cretaceous-Tertiary extinction. *Science* 208:1095–1108.
15. **Pope, K. O., Ocampo, A. C.,** and **Duller, C. E.** 1991. Mexican site for the K/T impact crater? *Nature* 351:105.
16. **Swisher, III , C. C., Grajales-Nishimura, J. N., Montanari, A., Margolis, S. V., Claeys, P., Alvarez, W., Ranne, P., Cedillo-Pardo, E., Maurrasse, F. J.-N. R., Curtis, G. H., Smit, J.,** and **McWilliams, M. O.** 1992. Coeval $^{40}Ar/^{39}Ar$ ages of 65.0 million years ago from Chicxulub crater melt rocks and Cretaceous-Tertiary boundary tektites. *Science* 257:954–58.
17. **Hildebrand, A. R., Penfield, G. T., Kring, D. A., Pilkington, N., Camargo, Z. A., Jacobsen, S. B.,** and **Boynton, W. V.** 1991. Chicxulub crater: A possible Cretaceous/Tertiary boundary impact crater on the Yucatan peninsula, Mexico. *Geology* 19:867–71.
18. **Matsui, T., Imamura, F., Tajika, E., Nakano, Y.,** and **Fujisawa, Y.** 2002. Generation and propagation of a tsunami from the Cretaceous-Tertiary impact event. In *Catastrophic events and mass extinctions*, ed. Koeberl, C., and MacLeod, K. G., pp. 69–77. Geological Society of America Special Paper 356.
19. **Firestone, R. B., and 25 others.** 2007. Evidence for an extraterrestrial impact 12,900 years ago that contributed to the megafauna extinctions and the Younger Dryas cooling. *Proceedings of the National Academy of Sciences (PNS)* 104(41):16016–21.
21. **Grayson, D. K.** 1984. Nineteenth-century explanations of Pleistocene extinctions: A review and analysis. In *Quaternary extinctions: A prehistoric revolution*, ed. P. S. Martin and R. G. Klein, pp. 5–39. Tucson: University of Arizona Press.
20. **Lange, I. M.** 2002. *Ice age mammals of North America: A guide to the big, the hairy, and the bizarre.* Missoula, MT: Mountain Press Publishing Company.
22. **Haynes, Jr., C. V.** ,2008. Younger Dryas (black mats) and the Rancholabrean termination in North America. *Proceedings of the National Academy of Sciences (PNAS)* 105(18):6520–25.
23. **Waters, M. R.,** and **Stafford, Jr., W. S.** 2007. Redefining the age of Clovis: Implications for the peopling of the Americas. *Science* 315:1122–26.
24. **Thornbury, W. D.** 1954. *Principles of geomorphology.* New York: John Wiley.
25. **Kobres, R., Howard, G. A., West, A., Firestone, R. B., Kennett, J. P., Kimbel, D.,** and **Newell, W.** 2007. Formation of the Carolina Bays: ET impact vs. wind-and-water. *EOS, Transactions, American Geophysical Union* 88(13), Joint Assembly Supplement, Abstract PP43A-10.
26. **Melton, F. A.,** and **Schriever, W.** 1933. The Carolina "Bays" . . . are they meteorite scars? *Journal of Geology* 41:52–66.
27. **Kennett, D. J., and 10 others.** 2008. Wildfire and abrupt ecosystem disruption on California's Northern Channel Islands at the Allerod–Younger Dryas boundary (13.0–12.9ka). *Quaternary Science Reviews* 27:2528–43.
28. **Kennett D. J. and** 2009. Nanodiamonds in the Younger Dryas boundary sediment layer. *Science* 323:94.
29. **Jones, A.** 2005. Meteorite impacts as triggers to large igneous provinces. *Elements* 1:277–81.
30. **Kerr, R. A.** 2007. Humongous eruptions linked to dramatic environmental changes. *Science* 316:527.
31. **McFadden, L. A.,** and **Blinzel, R. P.** 2007. Near-Earth objects. In *Encyclopedia of the solar system*, 2nd ed., ed. L.-A. McFadden, P. R. Weissman, and T. V. Johnson, pp. 283–300. New York: Elsevier.
32. **National Aeronautics and Space Administration.** 2007. Near-Earth object survey and deflection analysis of alternatives: Report to Congress. Washington, DC: Author.
33. **Morrison, D., Chapman, C. R., Steel, D.,** and **Binzel, R. P.** 2004. Impacts and the public: Communicating the nature of the impact hazard. In *Mitigation of hazardous comets and asteroids*, ed. M. J. S. Belton, T. H. Morgan, N. H. Smarasinha, and D. K. Yeomans, pp. 353–90. New York: Cambridge University Press.

Glossary

100 year flood A higher than usual stream discharge that has a 1 percent probability of occurring in any given year. The water level of this flood has been arbitrarily selected to define the regulatory floodplain for the U.S. National Flood Insurance Program.
Aa A cooler or thicker lava flow that solidifies with a blocky surface texture.
A soil horizon Uppermost soil horizon, sometimes referred to as the zone of leaching.
Absolute dating Determination of the geologic age at which time a fossil organism, earth material, or geologic feature formed; the age is usually expressed in years; commonly established by the analysis of radioactive isotopes and their decay products.
Acceptible risk Exposure to danger that an individual or society is willing to take.
Acid rain Water droplets falling from clouds that have been made artificially acidic by pollutants. In volcanoes, this acidity commonly comes from sulfate aerosols in eruption clouds, or sulfur dioxide fumes or from hydrogen chloride gas produced when lava vaporizes seawater. Rainwater normally has a slight acidity from dissolved carbon dioxide.
Active fault A fault along which future movement is thought possible. Two criteria are mappable displacement along a fracture or fractures in the past 10,000 years or several mappable displacements in the past 35,000 years.
Advance With respect to a glacier, an interval of time during which the area occupied by glacial ice increased.
Aerosol With respect to climate, a suspension of microscopic liquid and solid particles, such as mineral dust and soot, in the atmosphere.
Aftershock A subsequent, less intense earthquake that occurs a few minutes to a year or so after a main earthquake, and in the same general area. Usually one of many quakes of lesser magnitude than the main shock.
Airburst With respect to collision with extraterrestrial objects, the explosion of a meteoroid or comet in the atmosphere, generally at an altitude of 12 to 50 km (7.5 to 30 mi.).
Air mass thunderstorm An ordinary storm with lightning and thunder that develops from local convection within an unstable air mass. Commonly forms in the afternoon and is generally not severe.
Albedo A measure of the reflectivity of a material or surface to electromagnetic radiation, especially sunlight. Commonly expressed as the percentage of radiation reflected versus absorbed.
Alberta Clipper An extratropical cyclone that forms east of the Rocky Mountains in the Canadian province of Alberta and moves rapidly to the southeast across Canada and the northern United States. Commonly associated with snowstorms, especially blizzards.
Alluvial fan Fan-shaped deposit composed of coarse sediment that is dropped by a stream as it emerges from a mountain front onto flatter terrain; typically made of a variable proportion of stream and debris-flow deposits. The fan takes the shape of a segment of a cone.
Alluvium Unconsolidated sediment, such as sand, gravel, and silt, deposited by a stream.
Andesite A type of lava that has a silica content intermediate between basalt and rhyolite. Commonly, but not exclusively, associated with compostite volcanoes where it forms by the partial melting of oceanic crust in a subduction zone below the volcano.
Angle of repose Maximum angle that loose material will sustain.
Anthropogenic Produced by humans or originating in human activity.
Anticipate In respect to natural hazards and disasters, advance planning for the effects of the natural process.
Aquifer A useable accumulation of subsurface water in earth material that is saturated with water.
Artificial bypass Human action that moves sand or other sediment around an obstruction, such as dredging and pumping sand from updrift to downdrift sides of a jetty or harbor.
Artificial control of natural processes An action or structure made by humans to influence phenomena that are not caused or produced by humans.
Ash fall Fine airborne material that settles onto the land surface. The material is unconsolidated volcanic debris (ash), less than 4 mm in diameter, that has been blown out of a volcano during an eruption; it may travel a considerable distance from the volcano.
Ash flow See *Pyroclastic flow*
Asteroid Rock or metallic particle in space with a diameter between 10 m (30 ft.) and 1000 km (620 mi.).
Asteroid belt Region between orbits of Mars and Jupiter that contains most of the large rock or metallic particles in our solar system.
Asthenosphere Upper zone of Earth's mantle, located directly below the lithosphere; a hot, slowly flowing layer of relatively weak rock upon which tectonic plates move.
Atmosphere Layer of gases surrounding a planet, such as Earth.
Atmospheric pressure The force per unit area exerted by the weight of gas molecules that surround a planet; also called barometric pressure, because it is commonly measured with a barometer.
Atmospheric stability A condition of equilibrium in which a rising or sinking parcel of air tends to return to its original position. This equilibrium develops when a relatively small difference in air temperature exists between the surface and the air aloft.
Atoll A ring or horseshoe-shaped group of coral reefs, low islands, and shoals generally surrounding a large island, submerged island, bank, or underwater plateau.
Attenuation With respect to earthquakes, the loss of energy as seismic waves pass through earth materials. A process that reduces the energy of seismic waves from deep subduction-zone earthquakes before they reach Earth's surface.
Avalanche A type of landslide involving a large mass of snow, ice, or entrained rock debris that slides, flows, or falls rapidly down a mountainside.
Avulsion The process by which all or part of a river or distributary channel is abandoned in favor of a new channel.
B soil horizon Intermediate soil horizon; sometimes known as the *zone of accumulation.*
Bankfull discharge Flow condition in which water completely fills a stream channel. Many stream channels are formed and maintained by bankfull discharge.
Barometric pressure See *Atmospheric pressure*
Barrier island An elongate, offshore accumulation of sand that is above water level and is separated from the mainland by a bay or lagoon; stops large waves from reaching the shore of the mainland.
Basalt A common volcanic rock consisting of feldspar and other silicate minerals rich in iron and magnesium; has a relatively low silica content of about 50 percent; forms shield volcanoes.
Base level The theoretical lowest elevation to which a river may erode, generally at or about sea level.
Beach Associated with an ocean or lake, an accumulation of loose material, most commonly sand, gravel, or bits of shell, on a shoreline as the result of wave action. As used by river rafters, a wide sand accumulation along a low bank of the channel.
Beach budget Inventory of sediment influx and loss along a particular stretch of coastline.
Beach drift Continual movement of sand grains along the shore by the runup and backwash of waves.
Beach face The part of the shore that is normally exposed to swash action, that is, the runup and backwash of the waves; the sloping surface between the ocean or lake and the berm.
Beach nourishment Mechanical addition of sediment, generally sand, to a shore for recreational and aesthetic purposes, as well as to buffer coastal erosion.
Bedding plane A surface separating layers of sedimentary rocks. In mass wasting, this is commonly a weakness along which weathering occurs and landslides develop.
Berm The relatively flat or landward sloping part of a beach profile produced by wave-deposited sediment; the part of the beach where most people sunbathe.
Biogeochemical cycle Movement of a chemical element or compound through the major Earth systems: atmosphere, hydrosphere, biosphere, and lithosphere.
Blind fault A fracture or zone of rupture along which displacement does not extend to the Earth's surface.
Blizzard A severe winter storm producing large amounts of falling or blowing snow that create low visibilities for extended periods of time. In the United States, for a storm to be classified as a blizzard winds must exceed 56 km (35 mi.) per hour with visibilities of less than 0.4 km (0.25 mi.) for at least 3 hours in Canada, for a storm to be classified as a blizzard winds must exceed 40 km (25 mi.) per hour with visibilities of less than 1 km (0.6 mi.) for at least 4 hours.
Block With respect to a fragment ejected from a volcano, an angular particle that is longer than 64 mm (2.5 in.).
Bluff As a landform, a high steep bank or cliff along a lakeshore or river valley commonly made of weakly consolidated earth material.
Bomb With respect to a volcanic fragment, a smooth-surfaced particle that is longer than 64 mm (2.5 in.); commonly forms from a blob of magma ejected from a lava fountain; the magma cools and forms a streamlined shape as it falls to the surface.
Braided With respect to a river or stream, a channel pattern with numerous islands and sand and gravel bars that continually divide and subdivide the flow of water. This pattern is most evident during low and moderate stream flow.
Braided river See *Braided stream*
Braided stream A flowing channel of water with numerous islands and sand and gravel bars that continually divide and subdivide the flow.
Breaker zone With respect to a shoreline, that part of the beach and nearshore environment where incoming waves peak up, become unstable, and collapse toward the shore.
Breakwater A structure, such as a wall, which may be attached to a beach or located offshore, designed to protect a beach or harbor from the force of waves.
Breccia A rock or zone within a rock composed of angular fragments; may have a sedimentary, volcanic, tectonic, or impact origin.
C soil horizon Lowest soil horizon, sometimes known as the *zone of partially altered parent material.*
Caldera Giant volcanic crater produced by an extremely violent volcanic eruption or by collapse of the summit area of a volcano after an eruption.
Calving The separation of large blocks of ice from the terminus of a glacier that flows into a large body of water; calved blocks may form icebergs.
Carbon-14 Radioactive isotope of carbon with a half-life of approximately 5730 years; used in absolute dating of organic materials back to approximately 40,000 years.
Carbon dioxide (CO_2) A colorless, odorless gas that is denser than air and is produced by plant and fossil fuel combustion, animal and microbe respiration, and volcanic eruptions. In the atmosphere, it is the primary cause of global warming because, as a greenhouse gas, it heats the air by absorbing infrared radiation.
Carbon monoxide (CO) A colorless, odorless gas that forms in low-oxygen conditions with incomplete combustion or is released from volcanoes; in low concentrations it is highly toxic to humans and other animals.
Carbon sequestration The removal and isolation of carbon compounds, particularly carbon dioxide, from chemical reactions in the carbon biogeochemical cycle; a term commonly used to describe the removal of atmospheric carbon dioxide and its isolation underground or in the ocean.

Carrying capacity Maximum number of a population of a species that may be maintained within a particular environment without reducing the ability of that environment to maintain that population in the future.

Catastrophe An event or situation causing sufficient damage to people, property, or society that recovery or rehabilitation is long and complex; natural processes most likely to produce a catastrophe include floods, hurricanes, tornadoes, tsunamis, volcanoes, and large fires.

Cave A natural subterranean void commonly consisting of a series of chambers that are large enough for a person to enter; formed most often in limestone or marble.

Cave system See *Cave*

Cavern A large cave.

Cenote On the Yucatán Peninsula of Mexico, a steep-walled collapse sinkhole that extends below the water table. Some of these sinkholes are developed in fractured limestone on the perimeter of the Chicxulub impact crater.

Channel pattern The shape of a flowing stream as viewed from above ("bird's-eye view"); the three most common shapes or patterns are straight, meandering, and braided.

Channel restoration The process of returning a stream and adjacent areas to a more natural state.

Channelization Engineering technique to straighten, widen, deepen, or otherwise modify a natural stream channel.

Chute In mass wasting and landslides, a narrow path formed and maintained by the rapid downslope movement of avalanches, rock falls, or debris flows.

Cinder cone A conical volcanic hill formed by an accumulation of coarse pyroclastic material erupted from the volcano; also known as a *scoria cone*.

Clear-cutting Forestry practice of harvesting all trees from a tract of land.

Cleavage In a mineral, the smooth, easy fracture that takes place along parallel surfaces lying in one or more planes; these surfaces develop where bonds between atoms are weak or are especially long.

Climate Characteristic weather at a particular place or region during many seasons, years, or decades.

Climate forcing An imposed positive or negative change of Earth's energy balance measured in Watts per meter squared (W/m^2).

Cold front An atmospheric boundary between two air masses of contrasting temperature along which the cooler air mass advances into the warmer air mass.

Collapse sinkhole Depression in sediment or bedrock caused by a subsurface cave-in.

Colluvium Mixture of weathered rock, soil, and other, usually angular, material on a slope; produced by creep, landsliding, and other surface processes.

Color In a mineral, the name associated with the particular wavelengths of light that are reflected from the entire specimen or from its powder.

Combustion Second phase of a wildfire that follows every ignition. Includes flaming, where fine fuel and volatile gases are rapidly oxidized at high temperature, and glowing or smoldering, which take place later at lower temperature. Both flaming and glowing liberate energy as heat and light.

Comet Particle in orbit in space composed of a spongelike rocky core surrounded by ice and covered by carbon-rich dust with a tail of gas and dust that glow, as its orbit approaches the sun; ranges from a few meters to a few hundred kilometers in diameter.

Community Internet Intensity Map Online depiction of how people and structures have experienced an earthquake on the basis of e-mail reports.

Composite volcano Steep-sided volcanic cone produced by alternating layers of pyroclastic debris and lava flows; also known as a *stratovolcano*.

Conchoidal fracture In a rock or mineral, a broken surface that is smoothly curved, like the inside of a clamshell.

Conduction With respect to physics, the transfer of heat through a substance by molecular interactions.

Constructive interference In an ocean or a lake, combining two or more waves on the water surface to form a resultant higher wave.

Contact Interface or boundary between two geologic mapping units that is generally shown as a thin line on a geologic map.

Continental collision boundary The interface between two tectonic plates of similar density that are moving toward each other. Deformation and mountain formation occur along the interface because neither plate is able to slide below the other.

Continental Drift Hypothesis proposed by Alfred Wegener, in the early 20th century, that states the continental landmasses of the world "drift" about the surface of the globe and were once united to form a single super-continent, that he called Pangea. The hypothesis was supported by to geologic and paleontologic data.

Contour line On a topographic map, a series of connected points in which all points are the same elevation above sea level.

Contour interval (CI) On a topographic map, the vertical distance between each line of equal elevation; generally expressed in meters or feet.

Convection Transfer of heat by a mass of moving particles; for example, in boiling water, hot water rises to the surface and displaces cooler water that moves toward the bottom. Atmospheric heat may be transferred vertically within storm clouds. The primary way that heat is transferred from a wildfire.

Convection cell A self-contained circulation loop in a fluid in which hot, less dense fluid rises and cooler, denser fluid sinks; develops on a large scale in the atmosphere and within the Earth.

Convergence Coming together.

Convergent plate boundary Interface between two tectonic plates in which one plate generally descends below the other in a process known *as subduction*.

Core With respect to the interior of Earth, the central part of Earth below the mantle, divided into a solid inner core with a radius of approximately 1300 km and a molten outer core with a thickness of about 2000 km; the core is thought to be metallic and composed mostly of iron.

Coriolis effect An apparent deflection in the path of a moving object. This apparent deflection, to the right in the Northern Hemisphere and to the left in the Southern Hemisphere, results from Earth's rotation. The magnitude of this effect increases with increasing latitude from the equator to the poles.

Creep Slow downslope movement of soil and other weakly consolidated earth materials; characterized by slow flowing, sliding, or slipping.

Crevasse On a glacier, a large, deep fracture in the surface of the glacial ice caused by stretching of ice at and near the glacier surface.

Crown fire With respect to wildfires, flaming combustion that carries through canopies of trees; commonly driven by high wind and aided by a steep slope.

Crust The thin outermost rock layer of Earth that is differentiated from the underlying mantle by having a lower density.

Crystal form In a mineral, the characteristic geometric shape assumed by that mineral because of its orderly, internal atomic structure.

Crystallization Process by which atoms precipitate from solution such that they organize themselves into the orderly internal atomic arrangement of a mineral. The mineral may or may not possess crystal faces.

Cumulonimbus Thunderstorm cloud that grows vertically until it reaches a stable layer of air; generally with an anvil-shaped top.

Cumulus Any cloud that develops vertically; generally has a flat base with a cauliflower-like top; one of the three basic cloud types.

Cumulus stage First phase of thunderstorm development characterized by the growth of domes and towers on the top of a cumulus cloud as it becomes a cumulonimbus cloud.

Cutbank Steep or nearly vertical slope on the outside of a bend in a stream channel; typically forms by stream erosion.

Cyclone An area of low atmospheric pressure characterized by rotating winds that, in the Northern Hemisphere, circulate in a counterclockwise direction.

Cyrosphere The portion of a planet consisting of ice; on Earth this includes permafrost in soil and below the seafloor, glaciers, ice sheets, ice caps and fields, ice shelves, and icebergs.

Daughter product A stable isotope that forms by radioactive decay of an unstable parent isotope.

Debris flow Rapid downslope movement of unconsolidated, water-saturated earth material that became unstable because of torrential rain, rapid melting of snow and ice, or sudden drainage of a pond or lake; sometimes restricted to flows of this type that contain mainly coarse material.

Decompression melting The process by which solid rock melts due to a decrease in pressure that is caused by upwelling of hot rocks within the interior of the planet.

Delta Low, nearly flat area of land formed near the mouth of a stream where it enters a lake or the ocean; commonly triangular or fan-shaped and crossed by branching distributary channels of the stream that created it.

Delta plain Generally flat land surface that develops where a stream enters an ocean or lake; commonly covered with marsh or swamp and cut by distributary channels.

Dendrochronology Study of tree rings, sets of which usually form each year, to establish the absolute age of a tree or piece of wood from a tree.

Deposition In geology, the laying down or accumulation of earth material by processes that include the mechanical settling of sediment, the chemical precipitation of mineral matter, or the in-place accumulation of organic matter.

Derecho A windstorm with straight-line winds exceeding 90 km (57 mi.) per hour along a line that is at least 400 km (250 mi.) in length; damage produced by this windstorm is often equivalent to that caused by a tornado.

Desert An arid or semiarid region typified by mean annual precipitation less than 250 mm (10 in.); native animals and vegetation adapted to conserve limited water and soil that may be rich in salts.

Desertification Conversion of land from a more biotically productive state to one that more closely resembles a desert.

Desiccation crack A rupture in fine-grained sediment, typically mud, produced by complete or nearly complete drying.

Design basis ground motion One philosophy behind seismic hazard evaluation with respect to the construction of infastructure, defined as the ground motion that has a 10 percent chance of being exceeded in a 50-year period.

Differentiated With respect to meteoroids and asteroids, rock that has undergone igneous and sometimes metamorphic processes.

Digital elevation model (DEM) Three- dimensional computer depiction of a land surface as viewed from an oblique angle. Constructed with a computer program that samples land-surface elevation data on a specific spacing grid, such as every 30 m (20 ft.).

Direct effect A change that follows an event without any intervening factors. In a natural accident, disaster, or catastrophe, such effects could include people killed, injured, or dislocated and property damaged or destroyed; also referred to as a *primary effect*.

Directivity With respect to an earthquake, the increased intensity of shaking in the direction toward which a fault ruptures; observed in some moderate-to-large earthquakes.

Disappearing stream A brook, creek, or river that flows from a surface channel into the ground and continues flowing underground.

Disaster One possible effect of a hazard on society. Usually a sudden event that causes great damage or loss of life during a limited time and in a limited geographic area.

Disaster preparedness With respect to a natural hazard, actions of individuals, families, communities, states, provinces, or entire nations to minimize losses from a hazardous event before it occurs.

Discharge Quantity of water flowing past a particular point on a stream, usually measured in cubic feet per second (cfs) or cubic meters per second (cms).

Dissipative stage Final phase of a thunderstorm in which the supply of moist air is blocked by downdrafts in the lower levels of the cloud. Without a source of moisture, precipitation decreases and the storm cloud begins to disappear.

Distant tsunami A series of gravity waves in the ocean originating from a source typically thousands of kilometers away from the shoreline that is inundated. These waves are produced by the displacement of the entire water column of the ocean by an underwater earthquake, landslide, volcanic eruption, or extraterrestrial impact.

Distributary channel A waterway that branches from another waterway in a downstream direction. Typically found on stream deltas and some alluvial fans.

Divergence Spreading apart.

Divergent plate boundary Interface between tectonic plates characterized by production of new lithosphere; found along mid-oceanic ridges and

continental rift zones. Equivalent reference points on each plate will move away from each other with time.
Doppler effect Change in wave frequency that takes place when either the emitter or the detector is moving toward or away from the other. This effect can be heard in the higher-pitched sound of an approaching ambulance and the lower-pitched sound of one that is moving away.
Doppler radar With respect to weather, an electronic system that emits electromagnetic waves to detect the velocity of precipitation that is falling either toward or away from the receiver. Used to detect rotation within a thunderstorm that often precedes formation of a tornado.
Downburst Localized area of very strong winds in a downdraft; commonly beneath a severe thunderstorm.
Downdrift Located in the direction toward which a current of water, such as a longshore current, is flowing.
Downstream flood Condition in which surface runoff from a relatively wide area has caused a stream to overflow its banks—more common in the lower part of a drainage basin where tributary streams have increased the discharge of the overflowing stream.
Drainage basin Area that contributes surface water to a particular stream network.
Driving force With respect to mass wasting, an influence that tends to make earth material move downslope.
Drought Extended period of unusually low precipitation that produces a temporary shortage of water for people, other animals, and plants.
Dryline An atmospheric boundary between two air masses of contrasting moisture content where the air mass with less moisture advances into the air mass with greater moisture. Common in the south-central United States in late spring and summer, it is a source of atmospheric instability for the development of severe weather.
Dry ravel After a fire, the gravity-driven process of moving a large volume of sand, gravel, and organic debris downslope. This debris was stored upslope of brush vegetation before a fire.
Dry slot A nearly cloud-free area behind a cold front in an extratropical cyclone or a wall cloud in a severe thunderstorm that is created by sinking dry air.
Duff A layer of compacted and partially decomposed leaves, twigs, and other organic matter in a forest soil that accumulates below leaf litter and above mineral soil; generally burns by glowing and smoldering combustion.
Dust storm A weather event in which visibility at eye level drops to less than 1 km (0.6 mi.) for hours or days because of fine airborne particles of silt and clay.
Dynamic equalibrium With respect to the behavior of streams and rivers, the overall balance between the work that a river does in transporting sediment and the amount of sediment that it carries.
E-line Long, narrow band inland from the shore that marks the distance to which coastal erosion is expected to have washed away dry land in a given number of years.
E soil horizon A light-colored horizon underlying the A horizon that is leached of iron-bearing compounds.
E-zone With respect to coastal erosion, a hazard area between sea level and an e-line that is expected to erode away within a particular time period.
Earth climate system Emphasizes the interactions among the atmospheric, cryospheric, and terrestrial, and marine biospheric processes that affect the climate of Earth.
Earth fissure A large, deep crack in loose sediment at the land surface; formed by the lowering of the water table and the consequent drying out of the earth material.
Earth's energy balance Condition in which incoming solar radiant energy generally equals outgoing radiant energy from our planet; this equality exists despite changes in the energy's form as it moves through the atmosphere, oceans, land, and living things before being radiated back into outer space.
Earth system science The study of Earth as a system.
Earthquake Sudden movement of a block of Earth's crust along a geologic fault; the movement releases accumulated strain in the rocks.
Earthquake cycle A hypothesis that explains the periodic nature of earthquakes on a given fault on the basis of a drop in elastic strain after an earthquake and reaccumulation of strain before the next event.
Eddy With respect to ocean currents, the circular movement of water that develops behind an obstruction, where two currents pass in opposite directions or along the edge of a permanent current; can have a central core of either warm or cold water. Can be large enough to be detectible with satellite sensors.
Eddy vortex A peristent, circular movement of warm water that develops on the edge of the Loop Current in the Gulf of Mexico that can affect the intensity of hurricanes that cross it.
EF scale A graduated range of values from EF0 to EF5 for describing tornado intensity based on the maximum three-second wind velocity inferred from damage to buildings, towers, poles, and trees. A modification of the F-scale developed by T. Theodore Fugita in 1971.
Ejecta blanket Layer of debris surrounding an impact crater; typically forms an irregular terrain of low mounds and shallow depressions.
El Niño An event in the atmosphere and ocean during which trade winds weaken or even reverse, the eastern equatorial Pacific Ocean becomes anomalously warm, and the westward moving equatorial ocean current weakens or reverses; this change in ocean temperature can cause anomalous weather in adjacent landmasses.
Elastic rebound Instant and total recovery of a body from strain.
Elastic strain Deformation that ends when stress is released.
Electrical resistivity Degree to which earth materials are unable to conduct an electrical current. Measurements of this property are used in the exploration for energy, mineral, and water resources.
Electromagnetic energy Property of a system that enables it to do work through oscillations in an electric and magnetic field.
Electromagnetic spectrum The collection of all possible wavelengths or frequencies of electromagnetic radiation.
Electromagnetic wave An oscillation in an electric and magnetic field that transfers radiant energy.
Enhanced Fujita scale See *EF scale*
Environmental unity A principle of environmental studies that states that everything is connected to everything else.
Epicenter Point on Earth's surface directly above the focus of an earthquake; the focus is the area that first ruptures on a geologic fault in an earthquake.
Eustatic sea level The worldwide average vertical position of the ocean surface.
Evacuation With respect to a natural hazard, the movement of people away from the location of a probable dangerous or destructive natural phenomenon.
Expansive soil In-place, weathered earth materials that, upon wetting and drying, will alternately expand and contract, causing damage to foundations of buildings and other structures.
Exponential growth A type of compound growth in which a total amount or number increases at a certain percentage each year, and each year's rate of growth is added to the total from the previous year; characteristically stated in terms of a particular doubling time, that is, the time in years it will take the original number to double; commonly used in reference to population growth.
Exposure dating Determining the amount of time that sediment or rock has been at Earth's surface by measuring quantities of isotopes in the material that are produced by cosmic rays.
Extinction With reference to a wildfire, the final phase in which all combustion, including smoldering, ceases because of insufficient heat, fuel, or oxygen. With reference to organisms, the complete disappearance of a biological species.
Extratropical cyclone An area of low atmospheric pressure, characterized by rotating winds, which generally develops along a weather front between 30° and 70° latitude; commonly associated with heavy precipitation, strong winds, and other severe weather; can produce a storm surge and coastal erosion when crossing a shoreline.
Eye With respect to hurricanes and related intense tropical cyclones, the central circular area of light winds and partly cloudy skies where subsiding dry air encounters risng moist air.
Eyewall The portion of a hurricane immediately adjacent to a central circular area of nearly clear sky and light winds; usually the region having the highest winds and most intense rainfall.
F-scale See *EF scale*
Falling With respect to mass wasting and landslides, earth materials such as rocks dropping through the air from steep slopes.
Fault A fracture or fracture system in earth materials along which rocks on opposite sides of the fracture have moved relative to one another
Fault creep Slow, essentially continuous movement of blocks of Earth's crust principally along one side of a fracture; also called *tectonic creep*.
Fault gouge A clay-rich zone of pulverized rock produced by movement along a fracture or fracture system in Earth's crust; can act as an underground barrier or dam to the flow of groundwater.
Fault scarp Steep slope or small cliff that was formed during an earthquake by rupture and displacement of a block of Earth's crust.
Faulting The process of creating a rupture in Earth's crust by the movement of one crustal block in relation to another.
Fetch Distance that wind blows over a body of water; a principal factor that determines the height of wind-blown waves; in a cyclone, a contributing factor to the height of the storm surge.
Fibrous fracture A distinctive pattern for the breakage of minerals in which long slivers of the mineral break away from the specimen.
Fire management With respect to wildfires, the control of combustion to minimize loss of human life and damage to property. Accomplished through understanding the fire regime, public education, remote sensing of vegetation to determine fire potential, and prescribed burns.
Fire regime An ecosystem's potential for wildfire as inferred from fuel types, terrain, and past fire behavior.
Fissure Large fracture or crack in rock along which the opposing side have separated; may become a volcanic vent if lava or pyroclastic material is extruded.
Flash flood Overbank flow that results from a rapid increase in stream discharge; commonly occurs in the upstream part of a drainage basin and in small tributaries downstream.
Flashy discharge Stream flow characterized by a short lag time or response time between rainfall and peak stream discharge.
Flood basalt A laterally extensive, basaltic lava flow that generally is part of a thick sequence of similar flows. These flows commonly build up a plateau; also referred to as a *plateau basalt*.
Flood discharge The stream flow volume per unit of time (cms, cubic meters per second, or cfs, cubic feet per second) at which damage to personal property is likely to begin.
Flooding From a hazards perspective, high water levels in a stream, lake, or ocean that may damage human facilities. As a natural process, overbank flow that may construct a floodplain adjacent to a stream channel or a higher-than-normal water level along a coast that extends inland beyond the beach.
Floodplain Flat topography adjacent to a stream produced by overbank flow and by lateral migration of the channel and associated sand or gravel bars.
Floodplain regulation Governmental restriction of land use in an area likely to be inundated by a stream's overbank flow that could damage buildings and infrastructure.
Flood stage Water level of a stream at which damage to personal property is likely to begin.
Flow With respect to mass wasting, downslope movement of earth materials that deform as a fluid, such as the fluid movement of grains of sand, rock, snow or ice, debris, and mud.
Flowage See *Flow*
Flowstone General term for a mineral deposit, most commonly of calcium carbonate, formed by flowing water on the wall or floor of a cave.
Flux In a natural system, the rate at which a substance moves from one part of the system to another.
Focus In an earthquake, the point or location in Earth of the initial break or rupture, and from which energy is first released; during an earthquake, seismic energy radiates out from this point or location. Also known as the *hypocenter*.
Foliation plane With respect to mass wasting, a surface or zone of weakness in metamorphic rocks produced by the alignment of minerals during metamorphism.
Footwall Block of Earth'crust that is below a slanted geologic fault.
Fog A cloud that is in contact with the ground.
Force With respect to physics, an influence that tends to produce or to change the motion of a body, or one that produces stress within a body.

Forcing With respect to climate, a variable, such as solar radiation, volcanic aerosols, or greenhouse gases, that influences the direction of climate change.

Forecast With respect to a natural hazard, an announcement that states that a particular event, such as a flood or storm, is likely to occur during a particular time interval, often with some statement of the degree of its probability.

Foreshock Small to moderate earthquake that occurs before and in the same general area as the main earthquake.

Formation A body of rock or sediment composed of one or more distinctive types that can be mapped at the surface or traced in the subsurface—commonly given a geographic place name (e.g., Potsdam Sandstone) and formally defined in a scientific publication.

Fracture With respect to minerals, the breakage of a mineral in an irregular, curving or fibrous manner.

Free face A steep cliff of bare rock.

Frequency The number of waves passing a point of reference per second; commonly expressed in cycles per second or hertz (Hz); the inverse of the wave period.

Front With respect to weather, a boundary between two air masses that are distinguished by differences in temperature and, commonly, moisture content. See *Cold front*, *Warm front*, *Stationary front*, and *Occluded front*.

Frostbite Injury to body tissues caused by exposure to extreme cold.

Frost heaving The upward and lateral movement of soil particles and the land surface as the result of ice formation in the soil.

Frost-susceptible soil Surficial earth material that is likely to move and expand when frozen as the result of the accumulation of ice between particles of the material; commonly unconsolidated material with a high content of silt.

Fugita scale See *EF-scale*

Funnel cloud A narrow, rotating cloud extending downward from a thunderstorm. This cloud may become a tornado if it reaches the ground.

Gabion A metal wire basket, commonly made out of chain-link fencing that is filled with rocks. These baskets are stacked along stream banks to prevent erosion and are used as barriers to slow water flow in detention ponds.

Geochronology Science of dating earth materials, fossil organisms, and geologic features and events; and of determining the sequence of events in Earth history.

Geographic information system (GIS) Computer technology capable of storing, retrieving, transforming, and displaying spatial information about Earth and of making maps with these data.

Geologic cycle A group of interrelated sequences of Earth processes known as the hydrologic, rock, tectonic, and biogeochemical cycles.

Geologic map A two-dimensional representation of the age and composition of earth materials that are found beneath the soil within a given geographic area.

Geyser A particular type of hot spring that ejects hot water and steam above Earth's surface; perhaps the most famous is Old Faithful in Yellowstone National Park.

Glacial interval A portion of Earth history during which a cooler, wetter climate supported widespread continental glaciers.

Glacier A large mass of flowing ice that moves downslope under its own weight and persists from year to year; formed, at least in part on land by the compaction and recrystallization of snow.

Global climate model (GCM) Linked computer programs that use environmental data in mathematical equations to predict global change, such as increases in mean temperature, changes in precipitation, or some other atmospheric variable.

Global positioning system (GPS) Interacting ground control, satellites, and radio receivers that obtain extremely accurate locations and elevations on Earth's surface. Especially useful for detecting land surface changes related to a volcanic eruption, earthquake, tectonic creep, the slow movement of a landslide or land subsidence.

Global warming The increase in the mean annual temperature of the lower atmosphere and oceans in the past 150 years, primarily as a consequence of burning fossil fuels that emit greenhouse gases into the atmosphere.

Gradient Slope of the land over which a stream flows. Determined by calculating the vertical drop in elevation through some horizontal distance; commonly expressed as m/km or ft./mi.

Great earthquake A sudden rupture along a geologic fault in which the release of energy has a moment magnitude of 8 or greater.

Greenhouse effect Trapping of heat in the lower atmosphere by the absorption of infrared energy by water vapor, carbon dioxide, methane, nitrous oxides, halocarbons, and other gases.

Greenhouse gas A substance such as carbon dioxide, methane, nitrous oxides, or halocarbons (e.g., Freon) that absorb infrared radiation, and contributes to global warming of the lower atmosphere.

Groin A long, narrow structure of rock, concrete, wood, or other material generally constructed perpendicular to the shore to protect the coastline and trap sediment in the zone of littoral drift.

Groin field With respect to coastal processes, a group of long, narrow structures of rock, concrete, wood, or other material generally constructed perpendicular to the shore to protect the coastline and trap sediment in the zone of littoral drift.

Ground acceleration With respect to an earthquake, the amount that the speed or velocity of the shaking of the Earth's surface changes in a unit of time, such as a second. Ground acceleration is commonly given as a fraction or percentage of gravitational acceleration, which is 980 cm (380 in.) per second. The greatest (or peak) ground acceleration during an earthquake is a principal control on the amount of damage to structures.

Ground fire With respect to wildfires, combustion that takes place below the surface; characterized by more smoldering and little flaming; fuels typically are roots and partially decayed organic matter, such as peat.

Ground penetrating radar An electronic system that uses 10 to 1000 MHz electromagnetic waves to image Earth's shallow subsurface to locate features such as natural cavities and fissures, permafrost, contaminated groundwater, and buried pipes and tanks.

Groundwater Subsurface accumulation of water that entirely fills all voids in earth materials.

Groundwater mining The pumping of subsurface water at a faster rate than it is being replenished.

Gust front Leading edge of strong and variable surface winds from a thunderstorm downdraft.

Habitat A particular set of environmental conditions in which adapted plants and animals thrive.

Hachure With respect to topographic contours, a small mark drawn perpendicular to a contour line to indicate the downward sloping of the land into a closed depression; also referred to as a *tic mark*.

Hailstone Large rounded or irregular piece of ice that has grown while moving up and down numerous times within the clouds of a severe thunderstorm.

Half-life The amount of time necessary for one-half of the atoms of a particular radioactive element to decay.

Halocarbon A chemical compound (e.g., Freon) containing a halogen element, such as chlorine, fluorine, or bromine, that is bonded to carbon; generally synthetic in origin and in the atmosphere contributes to global warming in troposphere and ozone depletion in stratosphere.

Hanging wall A block of Earth's crust that is above a slanted geologic fault.

Hardness With respect to minerals, the resistance of a mineral to scratching; see *Relative hardness*.

Hazard See *Natural hazard*

Headland A small peninsula on an irregularly shaped shoreline that is more resistant to erosion than the surrounding earth material.

Headwaters Tributaries of a river that feed into it near its uppermost reach.

Heat energy The property of a system that enables it to do work through random molecular motion. This property may be transferred by conduction, convection, or radiation.

Heat index A value on a temperature scale that describes the human body's perception of air temperature, taking into account relative humidity.

Heatwave A prolonged period of extremely high air temperature that is both longer and hotter than normal. The criteria for establishing what constitutes longer and hotter than normal varies geographically with climate and living conditions.

High pressure center An area of elevated atmospheric pressure where the wind blows clockwise in the Northern Hemisphere. Also called an *anticyclone* or a *high*.

Hot spot A hypothesized, nearly stationary heat source located below the lithosphere that feeds overlying volcanic processes near Earth's surface.

Hot spring A natural discharge of groundwater at a temperature higher than that of the human body.

Hot tower A rain cloud that reaches the top of the troposphere in a hurricane. The release of latent heat in its formation warms the air at the high altitude.

Hummocky An irregular land surface marked by numerous small mounds and depressions.

Hurricane Tropical cyclone characterized by sustained winds of 119 km (118 km in Canada or 74 mi.) per hour; known as a typhoon in the western North Pacific and South China Sea, severe tropical cyclone in the southwest Pacific and southern Indian Oceans, and severe cyclonic storm in the northern Indian Ocean.

Hurricane warning An alert issued by weather forecasters for an area where hurricane conditions are expected in 24 hours or less. In Canada this alert is also issued for areas likely to experience high waves and coastal flooding, but not necessarily hurricane-strength winds.

Hurricane watch An alert issued by weather forecasters for an area that may experience hurricane conditions within 36 hours. In Canada this alert is not restricted to a 36-hour time frame.

Hydraulic chilling With respect to volcanoes, the use of water to cool and solidify flowing lava; a method used to divert lava flows threatening property.

Hydrogen sulfide (H_2S) A toxic, flammable gas that smells like rotten eggs and is emitted by volcanoes.

Hydrograph A graph of the discharge (flow) of a stream with time.

Hydrologic cycle Circulation of water from the oceans to the atmosphere and back to the oceans by way of precipitation, evaporation, runoff from streams, and groundwater flow.

Hydrology The study of surface and subsurface water.

Hydrophobic layer Lower layer of debris on forest floors or brushland that has accumulated water-repellant organic chemicals; especially common in drier climates after very hot fires.

Hypocenter See *Focus*

Hypothermia A medical condition characterized by rapid loss of core body heat causing intense shivering, loss of muscle coordination, mental sluggishness, and confusion. This condition can develop in an improperly dressed person in cool, wet windy weather or in one immersed water that is 25°C (77°F) or less; unless treated, the condition leads to unconsciousness and eventually death.

Ice storm A prolonged period of freezing rain during which thick layers of ice accumulate on all cold surfaces.

Iceberg Large block of floating glacial ice that has broken from the front of a glacier, dropping into the water by a process called *calving*.

Igneous rock An aggregate of minerals formed from the solidification or explosion of magma; extrusive if it crystallized on or very near the surface of Earth, and intrusive if it crystallized well beneath the surface.

Impact With respect to hazardous events, the effect or influence of an action.

Impact crater A circular-to-elliptical depression produced by collision of a large object, such as an asteroid, with a planet, moon, or other asteroid.

Impervious cover With respect to urban hydrology, surfaces covered with concrete, roofs, or other structures that impede the infiltration of water into the soil. In general, as urbanization proceeds, the percentage of the land that is under impervious cover increases.

Inactive fault A fracture in the Earth's crust along which rock has not moved in the past 2 million years.

Indirect effect A change that depends upon intervening factors. In a natural accident, disaster, or catastrophe, such effects could include emotional distress; the donation of money, goods, and services; or the payment of taxes to finance recovery; also called a *secondary effect*.

Injection well An artificial, generally cylindrical, hole excavated for pumping fluids, such as liquid hazardous waste, underground.

Inner core See *Core*

Instrumental intensity With respect to earthquakes, the direct measurement of ground motion during an earthquake with a high-quality seismograph. Used to immediately produce a map showing perceived shaking and potential damage.

Insurance The guarantee of monetary compensation for a specified loss in return for payment of a premium.

Intensity With respect to earthquakes, a measure of the effects of an earthquake at a specific place; measured as the effects on people and structures using the Modified Mercalli Scale or the direct measurement of ground motion by a sensitive seismograph; see *Instrumental intensity*. With respect to cyclones, the strength of the storm as indicated by sustained wind speeds and lowest atmospheric pressure.

Interglacial interval A portion of Earth history when continental glaciation was minimal or absent.

Intraplate earthquake Sudden movement along a fault caused by the abrupt release of accumulated strain in the interior of a lithospheric plate, far away from any plate boundary.

Ion An electrically charged atom or molecule formed by a loss or gain of electrons.

Iridium A platinum-group metal that is more abundant in meteorites than in earth materials. Its high concentration in certain clay deposits is attributed to their formation during or after impact of a large extraterrestrial body.

Isostasy The principle stating that thicker, more buoyant crust is topographically higher than crust that is thinner and denser. Also, with respect to mountains, the weight of rocks of the upper crust is compensated by buoyancy of the mass of deeper crystal rocks; that is, mountains have "roots" of lighter crustal rocks extending down into the denser mantle rocks, like icebergs in the ocean.

Isotope A variety of a chemical element distinguished by its number of neutrons; may be stable or radioactive.

Jet stream A concentrated flow of air near the top of the troposphere, characterized by strong winds.

Jetty A very long, narrow structure of rock, concrete, or other material generally constructed perpendicular to the shore; commonly constructed in pairs at the mouth of a river or inlet to a lagoon, estuary, or bay; designed to stabilize a channel, control deposition of sediment, and deflect large waves.

Joule The basic unit of energy in the metric system; equivalent to the work done by a force of one Newton when its point of application moves one meter in the direction that the force is acting.

Karst plain A generally flat land surface containing numerous sinkholes.

Karst topography A landscape characterized by sinkholes, caverns, and diversion of surface water to subterranean routes.

Kinetic energy The property of a system that enables it to do work through motion.

K soil horizon A calcium carbonate–rich horizon in which the carbonate often forms laminar layers parallel to the surface; carbonate completely fills the pore spaces between soil particles.

K-T boundary The interface in the geologic record where earth materials that formed during the Cretaceous Period are in contact with earth materials that formed during the Tertiary Period; dated as approximately 65 million years old.

Kyoto Protocol International agreement to reduce emissions of greenhouse gases in an effort to manage global warming; now legally binding in at least 175 countries, including Canada, Russia, and the European Union. The United States has not honored its original agreement to implement this protocol.

Lag time The time interval between the greatest amount of precipitation and peak discharge of a stream. Urbanization generally decreases the lag time.

Lahar A cool debris flow or mudflow on the slope of a volcano either during or between eruptions.

Landslide Specifically, rapid downslope movement of rock or soil; also a general term for all types of downslope movement.

Landslide hazard map A two-dimensional representation of the land surface that identifies areas that are likely to experience landslides and the appropriate land uses in those areas.

Landslide risk map A two-dimensional representation of the land surface that identifies areas that are likely to experience a rapid mass movement of earth material, states the probability that one of these events will occur, and assesses potential losses.

Land-use planning The preparation of an overall master plan for future development of an area; the plan may recommend zoning restrictions and infrastructure both practical and appropriate for the community and its natural environment; based on mapping and classification of existing human activities and environmental conditions, including natural hazards.

Lapilli Gravel-size pyroclastic debris ranging in size from 2 to 64 mm (0.08 to 2.5 in.).

Latent heat Energy created by random molecular motion that is either absorbed or released when a substance undergoes a phase change, such as during evaporation or condensation.

Latent heat of vaporization Energy absorbed by water molecules as they evaporate.

Lateral blast Type of volcanic eruption characterized by explosive activity that is more or less parallel to the surface of Earth. A lateral blast may take place when the side of a volcano fails.

Lava Molten material that flows onto the land surface from a volcano, or rock that solidifies from such molten material.

Lava flow Eruption of magma at Earth's surface that generally moves as a liquid mass downslope from a volcanic vent.

Lava tube A natural enclosed conduit or tunnel through which magma moves down the slope of a volcano (sometimes many kilometers); eventually the magma may again emerge at the surface; after the volcanic eruption, a tube commonly remains as an open void and forms a type of cave.

Levee A mound or embankment parallel to a stream channel; it may consist of fine sediment deposited from overbank flow during a flood or be an earthen embankment constructed by humans to protect adjacent land from flooding.

Lichenometry An absolute dating technique using measurements of the diameter of lichens on a rock surface to determine how long the surface has been exposed to the atmosphere.

Lightning A natural, high-voltage electrical discharge between a cloud and the ground, between clouds, or within clouds. The discharge takes a few tenths of a second and emits a flash of light that is followed by thunder.

Linkage With respect to natural hazards, a relationship between two phenomena.

Liquefaction Transformation of water-saturated granular material from the solid state to a liquid state.

Lithification The hardening of loose earth material into rock, commonly by cementation and compaction.

Lithosphere Outer layer of Earth, approximately 100 km thick, which comprises the tectonic plates that contain the ocean basins and continents; includes both Earth's crust and the solid, upper part of the mantle.

Littoral transport Movement of sediment in the nearshore environment as a result of return flow from waves that have washed up on the shore; sediment is moved by both the longshore current and by beach drift.

Local tsunami A series of gravity waves in the ocean originating from a source that is close to the shoreline that is inundated. These waves are produced by the displacement of the entire water column by either an underwater earthquake, landslide, volcanic eruption, or extraterrestrial impact.

Loess Deposit of windblown silt and clay.

Longitudinal profile With respect to streams, a graph that shows the decrease in elevation of a stream bed between its head and its mouth. This graph of elevation versus distance generally produces a line that is concave up.

Longshore bar and trough Elongated depression and adjacent underwater ridge of sand roughly parallel to shore; produced by wave action.

Longshore current Waterflow parallel to shore that develops in the surf zone as the result of waves that strike the land at an angle. Responsible for longshore drift.

Longshore drift The sediment transported in the nearshore environment parallel to the shore by a wave-generated current.

Longshore trough See *Longshore bar and trough*

Loose-snow avalanche A rapid downslope flow of granular snow and ice.

Love wave A type of earthquake-generated surface oscillation in which molecular vibrations within the Earth are transverse to the direction the oscillation is moving.

Low pressure center Also referred to as a *low*; see *Cyclone*.

Luster With respect to mineralogy, the way that light is reflected from a mineral; may be catagorized as metallic or nonmetallic.

Magma Naturally occurring molten rock material formed deep within the crust or in the mantle.

Magnetic reversal Involves the change of Earth's magnetic field between normal polarity and reverse polarity; also sometimes known as geomagnetic reversal.

Magnitude With respect to natural hazards, especially earthquakes, the amount of energy released.

Magnitude-frequency concept The idea that the intensity and extent of an event are inversely proportional to how often it occurs.

Mainshock The primary, highest magnitude earthquake in a series of earthquakes in a given area during a limited time interval.

Major earthquake A sudden rupture along a geologic fault in which the release of energy has a moment magnitude of 7 to 7.9.

Managed retreat With respect to coastal erosion, a strategy involving the landward movement of buildings and associated infrastructure away from a shoreline to remove the property from an area susceptible to flooding by seawater.

Mantle An internal layer of Earth approximately 3000 km thick composed of rocks that are primarily iron and magnesium-rich silicates. The lower boundary of the mantle is with the core, and the upper boundary is with the crust. The boundary is known as the Mohorovičić discontinuity (also called the Moho).

Mass extinction Sudden loss of large numbers of plant and animal species relative to new species being added.

Mass wasting A comprehensive term for any type of downslope movement of earth materials.

Material amplification Increase in the amplitude of seismic shaking caused by some earth materials. Such increase is generally associated with soft sediments, such as silt and clay deposits.

Mature stage With respect to thunderstorm development, the second phase in which downdrafts move and precipitation falls from the base of a cumulonimbus cloud. With regard to tornado development, the phase in which a visible funnel extends from the thunderstorm to the ground.

Meander With respect to streams, a bend in the channel that migrates back and forth across the floodplain, depositing sediment on the inside of the bend on a point bar, and eroding the outside of the bend into a cutbank.

Meandering A stream channel pattern that is sinuous and is characterized by gentle bends that migrate back and forth across a floodplain.

Meandering river See *Meandering stream*

Meandering stream A single, sinuous channel of flowing water that moves through gentle bends that migrate back and forth across a floodplain.

Megathrust A term used to describe large earthquakes (Magnitude 8 and greater) that occur along low angle faults at subduction zones.

Megatsunami A series of gravity waves that are about 100 times higher than the largest gravity waves produced by displacement of an entire column of ocean water from an undersea earthquake; produced by the impact of a large asteroid in the ocean.

Mesocyclone A region of rotating clouds, typically around 3 to 10 km (2 to 6 mi.) in diameter on the flank of a supercell thunderstorm.

Mesoscale convective system (MCS) A large, circular, long-lived and interacting group of thunderstorms that are commonly identifiable on satellite images; generally contain severe thunderstorms and often produce tornadoes.

Mesosphere A zone in the upper atmosphere from about 50 to 80 km (31 to 50 mi.) where the temperature decreases with altitude; directly above the stratosphere.

Metamorphic rock An aggregate of minerals that formed beneath Earth's surface from preexisting rock by the effects of heat, pressure, and chemically active fluids. In foliated metamorphic rocks, mineral grains are preferentially aligned parallel to one another or light and dark minerals are segregated into bands; nonfoliated metamorphic rocks have neither characteristic.

Meteor Extraterrestrial particle from dust to several centimeters in size that is burned up from frictional heat upon entry into Earth's atmosphere; light emitted from the burning meteor forms a "shooting star."

Meteorite Extraterrestrial particle from dust to asteroid size that hits Earth's surface.

Meteoroid Extraterrestrial particle in space that is smaller than 10 m and larger than dust size; may form from breakup of asteroids.

Meteor shower Large numbers of meteoroids entering the atmosphere and producing light from frictional heating; commonly occurs when Earth's orbit passes through the tail of a comet.

Microburst A strong, localized downburst from a thunderstorm that affects an area of 4 km (2.5 mi.) or less.

Microearthquake Sudden movement along a geologic fracture or fault in which the release of energy has a moment magnitude less than 3. Patterns of these very small quakes may be precursors to larger magnitude earthquakes; also called a *very minor earthquake*.

Microzonation With respect to earthquakes, the detailed location and land-use planning of areas having various earthquake hazards.

Mid-oceanic ridge A seafloor mountain range commonly found in the central part of oceans characterized by seafloor spreading. An example is the Mid-Atlantic Ridge.

Milankovitch cycles Natural variations in the intensity of solar radiation that reaches Earth's surface that recur at approximately 20,000-, 40,000-, and 100,000-year intervals.

Mineral A naturally occurring, inorganic crystalline substance with an ordered atomic structure and a specific chemical composition.

Mitigation The avoidance of, lessening, or compensation for anticipated harmful effects of an action, especially with respect to the natural environment.

Modified Mercalli scale A scale with 12 divisions on which the amount and severity of shaking and damage from an earthquake can be ranked.

Moho see *Mohorovičić discontinuity*.

Mohorovičić discontinuity Compositional boundary that separates the less dense crust from the underlying higher density mantle.

Moment magnitude A numerical assessment of the amount of energy released by an earthquake on the basis of its seismic moment, which is the product of the average amount of slip on the fault that produced the earthquake, the area that actually ruptured, and the shear modulus of the rocks that failed.

Mudflow A slurry of fine, unconsolidated earth materials and water that flows rapidly downslope or down a channel; also referred to as an *earth flow* or type of *debris flow*.

Multiple hypotheses Several proposed explanations for a phenomenon that are being considered concurrently; also referred to as *multiple working hypotheses*.

Natural hazard A potential danger that poses a threat to people or property that exists or is caused by nature; generally one that is not made or caused by humans.

Natural service function A benefit that arises from an event caused by nature that is also a hazard to people or the environment.

Near-Earth Asteroid Tracking Project (NEAT) NASA-supported program started in 1996 to study the size, distribution, and dynamic processes associated with near-Earth objects and to identify those objects having a diameter of about 1 km (3300 ft.) or greater.

Near-Earth object (NEO) Asteroids that reside and orbit between Earth and the sun or have orbits that intersect Earth's orbit.

Newton Measure of force in the metric system in which 1 Newton is the force necessary to accelerate a 1 kg (2.2 lb.) mass 1 m (3.3 ft.) per second each second that it is in motion.

Nor'easter An extratropical cyclone that tracks along the East Coast of the contiguous United States and Canada and which has continuously blowing northeasterly winds just ahead of the storm; commonly characterized by hurricane-strength winds, heavy snows, intense precipitation, and large waves along the coast.

Normal fault A generally steep fracture or fracture system along which the hanging wall has moved down relative to the footwall in the direction that the fracture dips.

Nowcasting With respect to weather, the near real-time prediction of the movment and behavior of weather systems, such as severe storms, that have formed or are in the process of forming.

Nuée ardente See *Pyroclastic flow*

O soil horizon Soil horizon that contains plant litter and other organic material; found above the *A* soil horizon.

Ocean acidification An increase in the acidity of seawater due primarily to dissolving additional atmospheric carbon dioxide. This condition results from higher levels of carbon dioxide in the atmosphere and is threatening the growth of corals and other marine organisms that make skeletal material from calcium carbonate.

Ocean conveyor belt A large-scale circulation pattern in the Atlantic, Indian, and southwestern Pacific Oceans that is driven by differences in water temperature and density; also referred to as a *thermohaline current*. Northward-flowing warm water in this current maintains the moderate climate of northern Europe; if this current were to slow or stop, ice-age conditions might return to portions of the Northern Hemisphere.

Occluded front An atmospheric boundary between two air masses of contrasting temperature along which the colder air mass advances into a cool air mass and elevates warmer air above the boundary.

Organic soil A surficial, in-place accumulation of partially decayed plant material, such as peat.

Organizational stage With regard to tornado development, the initial phase in which a funnel cloud forms from rotating winds within a severe thunderstorm.

Outer core See *Core*

Outflow boundary A line separating thunderstorm-cooled air from the surrounding air. This line behaves like a cold front with both a wind shift and drop in temperature. In large thunderstorm complexes this boundary may persist for more than 24 hours and provide instability that fosters development of new storms.

Overbank flow Moving floodwaters that have exceeded the capacity of a stream channel and have spread outward beyond its sides.

Overwash Flooding of the beach, dunes, and in some cases an entire barrier island by a storm surge; commonly transports sand from the beach and dunes inland; see *Washover channel*.

Ozone (O_3) A pungent, strongly reactive form of oxygen gas that is most abundant in the stratosphere, where it protects the Earth surface from high levels of ultraviolet radiation. In the lower atmosphere it is an air pollutant.

Ozone depletion Stratospheric loss of triatomic oxygen; related to release of halocarbons and other gases that destroy triatomic oxygen in the atmosphere.

P wave The fastest type of earthquake vibration within the body of Earth; also called a *primary* or *compressional wave*, this vibration oscillates in the direction that it travels.

Pahoehoe A type of lava with a smooth to ropy surface texture that forms when a flow solidifies; most common in basalt.

Paleomagnitism Refers to the study of magnetism of rocks and the intensity and direction of the magnetic field of Earth in the geologic past.

Paleoseismicity The occurrence of earthquakes in space and time in the geologic past.

Parcel With respect to weather, an imaginary balloon-like blob of air several hundred cubic meters in volume that acts independently of the surrounding air.

Parent isotope A unstable variety of a chemical element distinguished by its number of neutrons and its potential for radioactive decay. Measuring the proportions of parent and daughter isotopes in a substance is the basis for most absolute dating.

Particulate With respect to the atmosphere, a piece of dust, pollen, soot, ash, or other very fine material that is suspended in the air.

Peat Organic-rich sediment composed primarily of partially decayed plant material, such as moss, leaves, and sticks; typically forms in bogs and swamps and has a high water content.

Permafrost Earth material in the ground with a temperature that is below freezing continuously for at least two years; commonly cemented with ice; underlies about 20 percent of the world's land area.

Permanent gas A substance in the atmosphere, such as nitrogen or oxygen, that cannot be compressed with pressure alone and is present in generally constant proportions.

Permeability A measure of the ability of an earth material to transmit fluids such as water or oil.

Pineapple Express An atmospheric flow of tropical moisture and extratropical storms originating in the region of the Hawaiian Islands and directed toward the West Coast of the United States.

Plate tectonics An all-encompassing, global scientific theory that explains the characteristics and origin of most earthquakes, volcanoes, and other Earth features; based on the subdivision of the entire lithosphere into more than a dozen large, rigid blocks that move horizontally and interact to cause deformation, melting, and other changes.

Pleistocene Epoch Geologically recent interval of Earth history characterized by widespread continental glaciation and commonly referred to as the "Ice Age"; the subdivision of the Quaternary Period before the Holocene Epoch.

Plunging breaker A type of water wave in which the wave crest rises up and falls into its own trough as it strikes a relatively steep shoreline; tends to be associated with beach erosion.

Point bar Accumulation of sand and other sediment on the inside of a meander bend of a stream.

Polar desert An arid region near the North or South Pole where the annual precipitation averages less than 250 mm (10 in.) per year.

Polar jet stream A concentrated flow of air near the top of the troposphere that, in the Northern Hemisphere, typically crosses southern Canada or the northern conterminous United States and is characterized by strong winds blowing from west to east.

Pool Common bed form produced by scour in meandering and straight stream channels with relatively low channel slope; characterized at low flow by slow-moving, deep water; generally, but not exclusively, found on the outside of meander bends.

Porosity The percentage of void or empty space in earth material such as soil or rock.

Potential energy The property of a system that enables it to do work by virtue of an object's position above some reference level.

Potentially active With respect to earthquakes, a term applied to a geologic fault that shows evidence of movement in the Pleistocene Epoch, but not in the past 10,000 years.

Power With respect to physics, the rate of doing work.

Precursor With respect to natural hazards, a physical, chemical, or biological phenomenon that occurs before a hazardous event such as an earthquake, volcanic eruption, or landslide.

Precursor event See *Precursor*

Prediction With respect to a hazardous event, such as an earthquake, the advance determination of the date, time, and size of the event.

Preignition Initial phase of a wildfire during which fuel is brought to a temperature and water content that favors ignition; involves both preheating and pyrolysis.

Preparedness A state of readiness achieved by an individual or group of people, such as for a natural hazard or war; see *Disaster preparedness*.

Prescribed burn A fire purposely set and contained within a designated area to reduce the amount of fuel available for a wildfire; also called a *controlled burn*.

Primary effect See *Direct effect*

Process With respect to natural hazards, the physical, chemical, or biological ways by which events, such as volcanic eruptions, earthquakes, landslides, and floods, affect Earth's surface.

Proxy data With respect to climate, proxy data refers to data that is not strictly climate in nature, but can be used to infer, indirectly, changes in the climate record.

Punctuated uniformitarianism Concept that the geologic record of gradual mountain building, canyon erosion, and landscape construction was periodically interrupted by catastrophic events that caused rapid changes to topography or biota.

Pyroclastic activity Eruptive or explosive volcanic activity in which all types of volcanic debris, from ash to very large particles, are physically blown from a volcanic vent.

Pyroclastic deposit An accumulation of particles forcefully ejected from a volcanic vent, explosive in origin, containing volcanic ash particles and those of larger size such as blocks and bombs.

Pyroclastic flow Extremely hot and rapid flow of eruptive material down the flank of a volcano; consists of volcanic gases, ash, and other materials that move rapidly; commonly forms upon the collapse of an eruption column; may also be known as an *ash flow*, *fiery cloud*, or *nueé ardente*.

Pyroclastic rock Lithified igneous earth material composed of fragments of volcanic glass, mineral crystals, and fragments of other rocks that were ejected during a volcanic eruption.

Pyrolysis In a fire, a group of chemical processes by which heat divides or splits large fuel molecules into smaller ones. Products from these processes include volatile gases, mineral ash, tars, and carbonaceous char.

Quick clay Deposit of very fine sediment that when disturbed, such as by seismic shaking, may spontaneously liquefy and lose all shear strength.

R soil horizon Consolidated bedrock that underlies the soil.

Radiation With respect to physics, the transfer of energy as electromagnetic waves or as moving subatomic particles. A primary way that heat is transferred from the sun to Earth, the Earth to the atmosphere, and to a lesser extent from a wildfire to the atmosphere.

Radioactive isotope Unstable variety of a chemical element distinguished by its number of neutrons and its emission of ionizing radiation.

Rain band Precipitation-producing clouds that spiral around a hurricane.

Reactive With respect to natural hazards, a person's action or response to a hazard that occurs after an event, such as a natural accident, disaster, or catastrophe.

Recurrence interval The time between natural events, such as floods or earthquakes. Commonly given as the average recurrence interval, which is determined by averaging a series of intervals between events.

Red flag warning With respect to wildfires, a notice from the National Weather Service or other governmental agency that an extreme fire condition is occurring or is likely to occur within 24 hours.

Refraction Bending of a waveform. With respect to coastal processes, the bending of surface waves as they enter shallow water. This takes place where part of the wave "feels the bottom" and slows down while the remainder of the wave continues to move forward at a faster velocity.

Relative dating Placing geologic features, objects, earth materials, or events in chronological order in the geologic time scale without determining absolute ages.

Relative hardness With respect to minerals, the ability of one mineral or object to scratch another. Generally determined on the Mohs Hardness Scale in which diamond is assigned the highest value of 10 and talc the lowest value of 1. Other minerals, such as corundum, quartz, and calcite have intermediate values of relative hardness.

Relative humidity A measure of the amount of water vapor in the air compared to the amount that would saturate the air at a given temperature and pressure; commonly expressed as a percentage.

Relative sea level The local or regional position of the sea at the shore that is influenced by uplift or subsidence of the land and changes in global eustatic sea level.

Relief With respect to topography, the difference in elevation between a higher point (as on a hilltop) and a lower one (as on a valley floor).

Residence time In natural systems, the duration that a substance spends at a specific place or in a specific compartment. A measure of how long a chemical element, compound, or other substance stays in a specific part of a system.

Resisting force With respect to mass wasting, an influence that tends to oppose downslope movement.

Resonance With respect to earthquakes, the matching of the frequency of shaking with the natural vibrational frequency of an object. With respect to storm surge and tsunamis, the superposition of waves reflecting from one side of a narrow bay or lake with waves of water moving in another direction.

Resurgent caldera Uplift by rising magma of the central collapsed block of a giant volcanic crater; the giant crater resulted from an earlier explosion or collapse of the volcano summit; see *Caldera*.

Retreat With respect to glaciers, an interval of time when the area occupied by glacial ice decreased.

Retrofitting With respect to earthquakes, the renovation and physical modification of existing structures to withstand ground shaking and other effects of an earthquake.

Return interval The recurrence interval of wildfires; see *Recurrence interval*.

Return stroke In cloud-to-ground lightning, the component that involves the downward flow of electrons. Electron flow starts near the ground and is then initiated at successively higher points in a channel of ionized air. This flow produces the first bright light in a lightning flash.

Reverse fault A generally steep fracture or fracture system in which the hanging wall has moved up relative to the footwall in a direction opposite to the direction of fault dip. A low-angle reverse fault is called a *thrust fault*.

Rhyolite A fine-grained, light-colored volcanic rock consisting of feldspar, ferromagnesian minerals, and quartz with a relatively high silica content of about 70 percent; associated with volcanic events that may be very explosive.

Ridge With respect to weather, an elongated area of high atmospheric pressure.

Riffle A section of stream channel, which at low flow is shallow, fast-flowing water moving over a gravel streambed.

Rift In tectonics, a long, narrow trough that is bounded by normal faults; a surface rupture produced by extension of the lithosphere.

Ring of Fire A popular name given to the chain of volcanoes on islands and on the continents surrounding the Pacific Ocean; actual combustion is much less common than the red-hot glow of lava fountains and flows.

Rip current A seaward flow of water in a confined narrow area from a beach to beyond the breaker zone.

Riprap Layer or assemblage of large broken stones placed to protect an embankment or shoreline against erosion by running water or breaking waves.

Risk From a natural hazards viewpoint, risk may be considered as the product of the probability of an event times the consequences.

Risk analysis An evaluation that estimates the probability that an event will occur.

River A large, natural stream that carries a considerable volume of flowing surface water.

Rock In geology, a solidified natural aggregate of one or more minerals, natural glass, or solid organic remains. In engineering or design applications, a solidified natural material that requires blasting for excavation or removal.

Rock cycle Group of interrelated processes by which igneous, metamorphic, and sedimentary rocks can each be produced from the others.

Rogue wave A single crest of an oscillation of the surface of the ocean that is much higher than usual. Commonly caused by the constructive interference of smaller waves.

Rope stage With regard to tornado development, the final decaying phase in which the upward-spiraling air in the funnel contacts downdrafts from the thunderstorm, and the tornado begins to move erratically.

Rotational With respect to landslides, a type of sliding downslope movement that develops on a well-defined, upward-curving slip surface in homogeneous earth material.

Rule of Vs As applied to topographic maps, the principle that all contour lines must form a shape similar to the letter V when they cross a stream valley and that the apex of the V points upstream.

Runoff Water moving over the surface of Earth as overland flow on slopes or as stream flow; that part of the hydrologic cycle represented by precipitation or snowmelt that results in overland flow to a stream.

Runup With respect to waves, the up-rush of the wave along the shore; also, the combined vertical and horizontal distance that a tsunami moves inland from the shoreline.

S wave One type of earthquake vibration that occurs within the body of Earth; also called a *secondary* or *shear wave*, vibrations from this type of oscillation are perpendicular to the direction that it travels.

Safety factor With respect to landsliding, the ratio of resisting to driving forces; a safety factor greater than 1 suggests that a slope is stable.

Sand storm High winds, generally in the desert, that carry fine (0.05 to 2 mm) mineral particles to heights less than 2 m (7 ft.) above the ground.

Scale In maps, the relationship between the distance between features on the map and their actual distance apart on Earth's surface. Expressed either as a ratio, such as 1:24,000, or as a bar scale, a segmented line on the map.

Scarp Steep slope or cliff at the head of a landslide or along a geologic fault; produced by the downslope movement of a slump or slide or by the displacement of earth materials in an earthquake.

Scoria cone See *Cinder cone*

Sea cliff Steep, commonly nearly vertical bluff adjoining a beach or other coastal environment; produced by a combination of wave erosion and erosion from the land, such as landsliding and runoff of surface water. Groundwater seepage may also contribute to its development.

Seafloor spreading The plate tectonics concept that new crust is continuously added to the edges of lithospheric plates at divergent plate boundaries as a result of upwelling of magma along mid-oceanic ridges.

Sea wall A structure, commonly made of concrete, large stone blocks, wood or steel pilings, or cemented sand bags, built parallel to the coastline to protect buildings and infrastructure from wave and flood damage.

Secondary effect With respect to natural hazards, phenomena that are indirectly the result of a hazardous event, such as wildfires that are ignited by lava flows or water polution that occurs after flooding has damaged a sewage treatment plant; applies to those phenomena that are caused by the primary or direct effects of a hazardous event.

Secondary succession With respect to ecology, the regular pattern of colonization and replacement of plant species in a biological community that was partially destroyed in a hazardous event, such as a wildfire, flood, or lengthy volcanic ash fall.

Sediment Transported particulate material created by weathering, biological activity, or chemical precipitation.

Sedimentary rock Hard, compacted, and cemented particulate material that was created by weathering, biological activity, or chemical precipitation.

Seismic gap Area along an active fault zone that is capable of producing a large earthquake but has not produced one recently.

Seismic source The point at which vibrations from an earthquake are generated.

Seismic wave A periodic disturbance of particles of Earth by an earthquake or other vibration that is propagated without the net movement of the particles.

Seismogram A written or digital graphic record made by a seismograph.

Seismograph Instrument that records earthquakes.

Seismology The study of earthquakes as well as the structure of Earth through the evaluation of both natural and artificially generated seismic waves.

Sensible heat Energy created by random molecular motion that may be physically sensed, such as by a thermometer.

Shake map A near-real-time, two-dimensional representation of the ground motion and potential damage from an earthquake.

Shear strength With respect to mass wasting, the internal resistance in earth material to shear stress; on slopes this strength resists failure by sliding or flowing.

Shield volcano A broad, convex volcanic mountain built up by successive lava flows; the largest type of volcanic mountain.

Shrinking stage With regard to tornado development, late phase in which the tornado thins and begins to tilt as its supply of warm moist air is reduced.

Shrink-swell clay Earth material composed of very fine mineral particles that decrease in volume (shrink) when they are dry and increase in volume (swell) when they are wet.

Sinkhole Surface depression formed by the dissolution of underlying sedimentary rock or collapse over a subterranean void such as a cave.

Slab avalanche Downslope movement of snow and ice that starts out as large cohesive blocks; generally more dangerous and damaging than loose-snow avalanches.

Sliding With respect to mass wasting, the deformation or downslope movement of a nearly intact block of earth material along a slip surface.

Slip rate Long-term rate of displacement along a fault; usually measured in millimeters or centimeters per year.

Slope The slant or incline of the land surface.

Slope stability map A two-dimensional representation of the land surface that shows its susceptibility to mass wasting; commonly based on a survey of past landslides that includes a measurement of the steepness of the slopes on which they occurred.

Slow earthquake An event in which rupture and movement along a geologic fault takes days to months to complete; also referred to as a *silent earthquake*.

Slump Type of landslide characterized by downward slip of a mass of earth material, generally along a curved slip surface.

Slump block A mass of landslide material that has moved downslope along a curved slip surface.

Slumping With respect to mass wasting, the downslope movement of rock, sediment, or soil along a curved slip surface.

Smectite A group of clay minerals with extremely small crystals that can adsorb large amounts of water; common constituent of expansive soil.

Snow avalanche Rapid downslope movement of snow, ice, and rock.

Soil In geology, a generally loose accumulation of surface earth material that has been altered by chemical weathering in the place that it accumulated and can serve as a medium for plant growth. In engineering and design applications, unconsolidated surface material that can be excavated or removed without blasting.

Soil slip A type of mass wasting event in which a thin layer of unconsolidated earth material slides downslope along a narrow path; the slip surface for this type of mass movement is commonly in weathered slope deposits that are below the soil, but above unweathered bedrock; in California shallow soil slips are commonly referred to as *mudslides*.

Soil taxonomy With respect to soil science, refers to a method of classifying soils developed by the U.S. Department of Agriculture.

Solar nebula A flattened, pancakelike rotating disk of hydrogen and helium dust. Around 5 billion years ago a solar nebula condensed to form the sun by accretion of matter from its flattened disk.

Solutional sinkhole A depression that develops by the dissolution of underlying bedrock, such as limestone.

Sorting In sediment, size distribution of the particles; poorly sorted sediment has a large range in particle size, and in well-sorted sediment most particles are close to the same size.

Spacewatch A program started in 1981 to inventory the region surrounding Earth for asteroids and comets with the goal of identifying those whose orbits cross Earth's orbit.

Specific gravity The density of a mineral or rock as compared with the density of water.

Spilling breaker A type of water wave in which the wave crest tumbles down its own forward slope as it strikes a gently sloping shoreline; tends to be associated with deposition of sand on a beach.

Spit A claw-shaped, sandy to gravelly coastal landform at the end of a peninsula or island formed by accretion of beach ridges in the direction of longshore drift.

Spot fire A small fire that is commonly ignited by wind-blown burning embers ahead of the flaming front of a wildfire.

Spreading center Synonymous with mid-oceanic ridges where new crust is continuously added to the edges of lithospheric plates.

Spring With respect to groundwater, the natural outflowing of subsurface water where the groundwater system intersects Earth's surface.

Squall line A line of thunderstorms often accompanied by high wind and heavy rain; commonly forms in advance of a cold front or dryline and can produce tornadoes.

Stable isotope A variety of a chemical element distinguished by its number of neutrons and one that does not naturally undergo radioactive decay. In absolute dating, the product of radioactive decay, the daughter isotope, is stable.

Stage With respect to flowing water, the height of the water in a stream channel. Typically used in reference to the water level at which flooding begins to occur. Flood stage is commonly the water level that is likely to cause damage to personal property.

Stalactite A cylindrical or conical mineral deposit, most commonly of calcium carbonate, that has grown downward from the roof of a cave or other overhanging feature.

Stalagmite A column or ridge of mineral deposits, most commonly calcium carbonate, on the floor of a cave; forms as water drips from stalactites on the cave roof.

Stationary front Transition zone between two different air masses, commonly with contrasting temperature, that moves very little.

Stepped leader In cloud-to-ground lightning, a channel of ionized air that approaches the ground in a series of nearly invisible bursts. This channel becomes the path for the luminous return stroke.

Storm surge Wind-driven ocean waves, usually accompanying a hurricane, nor'easter, or similar storm, that pile water up on a coastline.

Strain Change in shape or size of a material as a result of applied stress.

Stratosphere Zone in Earth's atmosphere above the troposphere where the air temperature is either constant or warms with increasing altitude; contains significant quantities of ozone, which protects life from high levels of ultraviolet radiation.

Stratovolcano See *Composite volcano*

Streak The color of a mineral in powdered form; commonly produced by rubbing a mineral specimen on an unglazed ceramic plate.

Stream With respect to surface water, a body of water that flows in a channel; includes a brook, creek, and river.

Strength In soil, the ability to resist deformation; the resistance results from cohesive and frictional forces in the soil. In rock, the mechanically applied compression necessary to break or fracture a specimen.

Strike-slip fault A fracture or fracture system along which a block of Earth's crust moves horizontally relative to the block on the opposite side of the fracture.

Strike and dip Two parameters measured by geologists to describe the three-dimensional orientation of features, including layering in rocks, that intersect Earth's surface.

Strong earthquake Earthquake in which the release of energy has a moment magnitude of 6 to 6.9.

Subduction zone Convergence of tectonic plates where one plate dives beneath another and is consumed in the mantle.

Submarine landslide An underwater mass movement, often caused by an earthquake that can occur on a sloping seafloor, such as the wall of an underwater canyon, the continental slope, or the seafloor surrounding an island.

Submarine trench A relatively narrow, long (often several 1000 km), deep (often several km) depression on the ocean floor that forms as a result of convergence of two tectonic plates with subduction of one.

Subsidence Sinking, settling, or other lowering of parts of the crust of Earth.

Subtropical jet stream A concentrated flow of air near the top of the troposphere that, in the Northern Hemisphere, typically crosses Mexico and Florida and is characterized by strong winds blowing from west to east.

Suction vortices Small, intense whirls within a larger tornado; these rapidly rotating winds may cause much of the intense damage done by tornadoes.

Sulfur dioxide (SO_2) A colorless gas with a pungent odor similar to a burnt match. This gas is emited from volcanoes and burning of fossil fuels, and when dissolved in rainwater produces acid rain.

Supercell An unusually long-lived thunderstorm that has a rotating updraft that continually redevelops on the storm's flank.

Supercell storm See *Supercell*

Supershear Amplification of seismic energy that occurs when earthquake rupture propagation along the fault plane exceeds the shear wave velocity.

Surf zone The area between the breaking waves and the swash zone along the shore.

Surface fire Combustion in a wildfire that takes place primarily along the ground and consumes fuels such as grass, shrubs, dead and downed limbs, and leaf litter.

Surface wave One type of vibration produced by an earthquake in which the oscillation moves along the boundary between the air and land; generally strongest close to the epicenter where it causes much of the damage to buildings and other structures.

Swash zone Area along the shore where waves run up on the beach face and then back again into the water.

Swell With respect to coastal processes, the sorting out of storm-produced waves into wave sets having more or less uniform heights and lengths. This sorting allows groups of waves to move long distances from storms to coastal areas with relatively little loss of energy.

Talus Pieces of rock that have accumulated at the base of a cliff or steep slope; may form individual piles or a continuous slope composed of blocks of rock that have fallen from above.

Tectonic Pertaining to forces involved in large-scale crustal movement.

Tectonic creep Slow, more or less continuous, movement along a fault; also called fault creep.

Tectonic cycle A repetitive sequence of events and processes that create and destroy the Earth's crust and its ocean basins and mountain ranges.

Tectonic framework The geometry and spatial patterns of faults and seismic sources that define a specific region.

Tectonic plate A very large, fault-bounded block of Earth's crust and underlying upper mantle that moves as a coherent mass on top of the asthenosphere; also called a *lithospheric plate*. This type of block commonly forms at a mid-oceanic ridge and is destroyed in a subduction zone.

Tectonic creep Slow, more or less continuous movement along a fault.

Tephra Any material ejected and physically blown out of a volcano; mostly ash.

Texture In rock, the size, shape, and arrangement of mineral grains and other particles. In soil, the size distribution of mineral and rock particles.

Thermal expansion With respect to the ocean, the increase in volume of seawater that occurs upon heating.

Thermal spring A place where heated groundwater flows from rock, sediment, or soil onto the land surface or into a body of surface water such as a river, lake, or ocean. Water temperature must be appreciably higher than the mean annual air temperature; includes both warm springs and hot springs.

Thermokarst An irregular terrain of sinkholes and mounds formed by the melting of permafrost.

Thermosphere With respect to weather, the outermost part of the atmosphere starting at about 80 km (50 mi.) in altitude, characterized by an upward increase in temperature and very little gas; directly above the mesosphere.

Thrust fault A low-angle fracture or fractures in Earth's crust where older rocks are displaced over younger rocks.

Topographic map A two-dimensional representation of the relief of Earth's surface using contour lines of equal elevation.

Topographic profile A sideways view of the land surface with a curving line drawn to show how it changes in elevation from one point to another point.

Tornado A destructive, commonly funnel-shaped cloud of violently rotating winds that extends downward from a severe thunderstorm to reach Earth's surface.

Total load With respect to stream processes, the sum of the dissolved, suspended, and bed load that a stream or river carries.

Tower karst A landscape of steep-sided hills rising above a plain or above sinkholes. Commonly developed by extensive dissolution of limestone in a humid tropical climate.

Transform boundary Interface between two tectonic plates where one plate slides horizontally past another, such as along the San Andreas fault in California; synonymous with a *transform fault*.

Transform fault Type of fracture of Earth's crust associated with offsets of mid-oceanic ridges where one plate is displaced horizontally; may form a plate boundary, such as the San Andreas fault in California.

Translational With respect to landslides, a type of sliding downslope movement that develops on a well-defined, planar slip surface, most commonly in layered or jointed rock or along a weak clay layer.

Triangulation The location of a point using distances from three points.

Tributary A small stream that flows into a larger stream.

Triple junction Areas where three tectonic plates and their boundaries join.

Troglobite An organism, such as a blind salamander, that lives its entire life underground in a cave.

Tropical cyclone General term for large thunderstorm complexes rotating around an area of low pressure that formed over warm water in the tropics or subtropics; includes low-intensity tropical depressions and tropical storms and high-intensity storms called hurricanes, typhoons, severe tropical cyclones, or severe cyclonic storms.

Tropical depression A circular atmospheric low-pressure center in the tropics whose winds are less than 63 km (39 mi.) per hour; the threshold for a tropical storm.

Tropical disturbance An area of disorganized, but persistent thunderstorms and associated low-pressure trough that may be in the formative stages of a tropical depression.

Tropical storm An organized system of thunderstorms around a circular low-pressure center that derives its energy from warm ocean waters and has winds between 63 km (39 mi.) and 119 km (74 mi.) per hour; stronger than a tropical depression, but weaker than a hurricane.

Tropopause The boundary between the troposphere and the stratosphere.

Troposphere Lowermost layer of the atmosphere defined by a general decrease in temperature with increasing height above Earth's surface.

Trough With respect to weather, an elongated area of low atmospheric pressure.

Tsunami A series of gravity waves caused by the large-scale displacement of the entire water column of the ocean or rarely a large lake. Most commonly the result of a large earthquake deforming the seafloor, although may also be caused by a large underwater or sea-cliff landslide, collapse of part of a volcano into the ocean, a submarine volcanic explosion, or the ocean impact of an asteroid. These waves are not tidal waves and, in a few cases, are not seismic sea waves.

Tsunami ready A certification issued by the National Oceanic and Atmospheric Administration or other governmental agency that a community has taken appropriate preparedness measures for future tsunamis.

Tsunami runup map A two-dimensional representation of the level to which a tsunami has inundated or is projected to inundate a coastal zone.

Tsunami warning An emergency notice from a governmental agency that a tsunami has been detected and is heading toward the area receiving the alert.

Tsunami watch An emergency notice from a governmental agency that an event has taken place—such as an earthquake, landslide, or volcanic exposion—which could cause a tsunami to strike the alerted area.

Typhoon See *Hurricane*

Unified soil classification system Classification of soils, widely used in engineering practice, based on amount of coarse particles, fine particles, or organic material.

Uniformitarianism Concept that the present is the key to the past; that is, we can read the geologic record by studying present processes.

Updrift Located in the direction from which a water current, such as a longshore current, is flowing.

Uplift An increase in land elevation that is commonly the result of tectonic or volcanic processes.

Urban heat island effect A local climatic condition resulting from various design and land-use practices in a city, large town, or other extensively developed area. This condition can intensify heatwaves and cause temperatures in a metropolitan area to be up to 12°C (22°F) warmer than the surrounding rural area.

Valley fever A potentially fatal respiratory illness produced by fungal spores found in desert soils. Mass wasting caused by earthquakes can produce dust containing these spores.

Variable gas With respect to the atmosphere, a substance such as carbon dioxide or water vapor that cannot be compressed with pressure alone and is present in changing proportions.

Varve A very thin sedimentary layer or couplet of layers deposited in a single year; in glacial lake deposits it consists of a coarse layer deposited when ice melts in the summer and a fine layer deposited when very fine suspended sediment finally settles out in the winter.

Vein In geology, the filling of a fracture with earth material, especially minerals, that differs from the surrounding rock.

Vertical exaggeration (VE) Vertical scale of a diagram or model that is larger than the horizontal scale; used in many geologic cross sections to show relationships more clearly.

Very minor earthquake See *Microearthquake*

Viscosity The ease of flow of a fluid; controlled by the magnitude of the internal friction within a fluid. Thicker fluids have greater viscosities.

Vog A volcanic fog or smog of sulfur dioxide and other volcanic gases combined with aerosols of sulfuric acid and dust. In Hawai'i, vog is commonly an acrid blue haze produced by gases emitted from Kilauea volcano.

Volatiles Chemical compounds, such as water (H_2O) or carbon dioxide (CO_2), that evaporate easily and exist in a gaseous state at the Earth's surface.

Volcanic ash eruption An explosive volcanic event in which a tremendous quantity of fine-grained rock and volcanic glass is shattered and expelled in clouds of very fine particles into the atmosphere.

Volcanic crisis A condition in which the eruption or prospect of an eruption of a volcano produces immense difficulty, trouble, or danger in the affected society.

Volcanic dome Type of volcano characterized by very viscous magma with a high silica content; activity is generally explosive.

Volcanic vent Circular or elongated opening through which lava and pyroclastic debris are erupted.

Vortex With regard to weather, a spinning column of wind.

Wadati-Benioff Zone Inclined zone of earthquakes produced as a tectonic plate is subducted.

Wall cloud A localized, persistent, and often abrupt lowering of condensed water vapor from the rain-free base of a severe thunderstorm. These clouds may rotate and are commonly associated with tornado formation.

Warning With respect to natural hazards, the announcement of a possible hazardous event, such as a large earthquake or flood, that could occur in the near future.

Warm front An atmospheric boundary between two air masses of contrasting temperature along which the warmer air mass advances into the cooler air mass.

Washover channel A narrow conduit of water eroded through a beach or coastal dunes by a storm surge.

Watch With respect to weather, an alert issued by a forecast office that meteorological conditions are favorable for severe weather, such as a tornado, hurricane, severe thunderstorm, winter storm, or flash flood.

Watershed Land area that contributes water to a particular stream system; generally used to describe the catchment area for a smaller tributary stream; see *Drainage basin*.

Waterspout A weak, generally funnel-shaped, wind vortex that most commonly develops under fairweather conditions and has all or part of its track over the sea or a lake; some of these whirlwinds are tornadoes that form or track over the water.

Water vapor The gas phase of water.

Watt The basic unit of power; generally expressed as units of energy per unit of time.

Wave front A long, continuous crest of a single oscillation, such as one on the surface of a lake or ocean.

Wave height Vertical distance between the crest and preceding trough of an oscillation, such as one on the surface of a lake or ocean.

Wave normal An imaginary line perpendicular to the crest of an oscillation, such as one on the surface of a lake or ocean.

Wave period The time in seconds for successive crests of an oscillation, such as one on the surface of a lake or ocean, to pass a reference point; the inverse of the frequency of the oscillation.

Wavelength With respect to physics, the horizontal distance between two successive crests or troughs of an oscillation. Used to describe characteristics of seismic, electromagnetic, water, and other oscillations.

Weather Atmospheric conditions, such as air temperature, humidity, and wind speed, at any given time and place.

Weathering Changes that take place in rocks and minerals at or near Earth's surface in response to physical, chemical, and biological activity; the physical, chemical, and biological breakdown of rocks and minerals.

Wide Area Augmentation System (WAAS) A network of global positioning system (GPS) receivers at fixed locations that act as reference points for transmitting positional corrections by satellite to mobile GPS receivers. Operated by the Federal Aviation Administration, this system currently improves the accuracy of locations obtained from WAAS-enabled GPS receivers in the United States and parts of Canada.

Wildfire Uncontrolled combustion or burning of plants in a natural setting, such as a forest, grassland, brushland, or tundra.

Wildland/urban interface The area where natural undeveloped grassland, brushland, or forest is adjacent to land developed for homes, businesses, or recreation.

Wind chill The influence of wind on the perception and actual effect of cold air. Meteorological agencies in the United States and Canada have developed a wind chill index based on how the human body loses heat in the cold and wind.

Wind shear A change in the speed or direction of the wind along a given direction.

Work In physics, the exertion of a force to overcome resistance or to produce molecular change.

Younger Dryas A previously unexplained glacial interval from around 13,000 to 11,600 years ago that coincides with the extinction of many large mammals in the Northern Hemisphere and the disappearance of the Clovis culture. A recent hypothesis proposes that these events were caused by the airburst of a comet over the Candian continental ice sheet.

Index

F

G

M